Biocontrol Agents and Secondary Metabolites

Biocontrol Agents and Secondary Metabolites

Applications and Immunization for Plant Growth and Protection

Edited by

Sudisha Jogaiah
Laboratory of Plant Healthcare and Diagnostics, P.G. Department of Biotechnology and Microbiology, Karnatak University, Dharwad, Karnataka, India

ELSEVIER
WP
WOODHEAD PUBLISHING
An imprint of Elsevier

Woodhead Publishing is an imprint of Elsevier
The Officers' Mess Business Centre, Royston Road, Duxford, CB22 4QH, United Kingdom
50 Hampshire Street, 5th Floor, Cambridge, MA 02139, United States
The Boulevard, Langford Lane, Kidlington, OX5 1GB, United Kingdom

Notices

Knowledge and best practice in this field are constantly changing. As new research and experience broaden our understanding, changes in research methods, professional practices, or medical treatment may become necessary.

Practitioners and researchers must always rely on their own experience and knowledge in evaluating and using any information, methods, compounds, or experiments described herein. In using such information or methods they should be mindful of their own safety and the safety of others, including parties for whom they have a professional responsibility.

To the fullest extent of the law, neither the Publisher nor the authors, contributors, or editors, assume any liability for any injury and/or damage to persons or property as a matter of products liability, negligence or otherwise, or from any use or operation of any methods, products, instructions, or ideas contained in the material herein.

Library of Congress Cataloging-in-Publication Data
A catalog record for this book is available from the Library of Congress

British Library Cataloguing-in-Publication Data
A catalogue record for this book is available from the British Library

ISBN: 978-0-12-822919-4
ISBN: 978-0-12-823094-7

Publisher: Charlotte Cockle
Acquisitions Editor: Nancy Maragioglio
Editorial Project Manager: Alice Grant
Production Project Manager: Surya Narayanan Jayachandran
Cover Designer: Miles Hitchen

Typeset by SPi Global, India

Laboratory to Farming Community.

Contents

Contributors

K.N. Basavesha Department of Agricultural Microbiology, College of Agriculture, University of Agricultural Sciences, Dharwad, Karnataka, India

S.A. Belorkar Department of Microbiology and Bioinformatics, Atal Bihari Vajpayee University, Bilaspur, India

Piero Attilio Bianco Department of Agricultural and Environmental Sciences—Production, Landscape, Agroenergy (DISAA), University of Milan, Milano, Italy

S. Brindhadevi Department of Plant Pathology, Centre for Plant Protection Studies, Tamil Nadu Agricultural University, Coimbatore, India

Paola Casati Department of Agricultural and Environmental Sciences—Production, Landscape, Agroenergy (DISAA), University of Milan, Milano, Italy

Raju Krishna Chalannavar Department of Applied Botany, Mangalore University, Mangalagangotri, Karnataka, India

Jishuang Chen Bioresource Institute for Healthy Utilization and Key Laboratory of Basic Pharmacology and Joint International Research Laboratory of Ethnomedicine of Ministry of Education, Zunyi Medical University, Zunyi; College of Biotechnology and Pharmaceutical Engineering, Nanjing Tech University, Nanjing, China

K.S. Divya Department of Microbiology, Yuvaraja's College, University of Mysore, Mysore, Karnataka, India

Subrata Dutta Department of Plant Pathology, Bidhan Chandra Krishi Viswavidyalaya, Mohanpur, Nadia, West Bengal, India

Jinggui Fang College of Horticulture, Nanjing Agricultural University, Nanjing, P.R. China

N. Geetha Nanobiotechnology Laboratory, DOS in Biotechnology, University of Mysore, Mysore, Karnataka, India

Ankit Kumar Ghorai Department of Plant Pathology, Bidhan Chandra Krishi Viswavidyalaya, Mohanpur, Nadia, West Bengal, India

Swapan Kumar Ghosh Molecular Mycopathology Laboratory, Biocontrol Unit, PG Department of Botany, Ramakrishna Mission Vivekananda Centenary College (Autonomous), Kolkata, India

C. Gourav Centre for Rural Development and Technology, Indian Institute of Technology Delhi, New Delhi, India

H.G. Gowtham Centre for Rural Development and Technology, Indian Institute of Technology Delhi, New Delhi, India

Muhammad Salman Haider College of Horticulture, Nanjing Agricultural University, Nanjing, P.R. China

Tran Thi Minh Hang Department of Horticulture and Landscaping, Faculty of Agronomy, Vietnam National University of Agriculture, Hanoi, Vietnam

P. Hariprasad Centre for Rural Development and Technology, Indian Institute of Technology Delhi, New Delhi, India

Vu Quynh Hoa Department of Horticulture and Landscaping, Faculty of Agronomy, Vietnam National University of Agriculture, Hanoi, Vietnam

S.K. Jadhav School of Studies in Biotechnology, Pt. Ravishankar Shukla University, Raipur, Chhattisgarh, India

Archana Jain Bioresource Institute for Healthy Utilization and Key Laboratory of Basic Pharmacology and Joint International Research Laboratory of Ethnomedicine of Ministry of Education, Zunyi Medical University, Zunyi, China; Department of Microbiology, Sri Satya Sai University of Technology and Medical Sciences, Sehore, Madhya Pradesh, India

Subhendu Jash Department of Plant Pathology, Bidhan Chandra Krishi Viswavidyalaya, Mohanpur, Nadia, West Bengal, India

Muhammad Jafar Jaskani Institute of Horticultural Sciences, University of Agriculture, Faisalabad, Pakistan

Sudisha Jogaiah Laboratory of Plant Healthcare and Diagnostics, P.G. Department of Biotechnology and Microbiology, Karnatak University, Dharwad, Karnataka, India

I. Johnson Department of Plant Pathology, Centre for Plant Protection Studies, Tamil Nadu Agricultural University, Coimbatore, India

Neelu Joshi School of Biotechnology & Bioinformatics, D.Y. Patil Deemed To Be University, Navi Mumbai, Maharashtra, India

Veenu Joshi Center for Basic Sciences, Pt. Ravishankar Shukla University, Raipur, Chhattisgarh, India

G. Karthikeyan Department of Plant Pathology, Centre for Plant Protection Studies, Tamil Nadu Agricultural University, Coimbatore, Tamil Nadu, India

M. Karthikeyan Department of Plant Pathology, Centre for Plant Protection Studies, Tamil Nadu Agricultural University, Coimbatore, India

Manzoor R. Khan Section of Plant Pathology, Department of Botany, Aligarh Muslim University, Aligarh, Uttar Pradesh, India

S.M. Paul Khurana Amity Institute of Biotechnology, Amity University Haryana, Gurgaon, India

Anil Kumar Department of Plant Pathology, CCS HAU, Hisar, Haryana, India

Narendra Kumar Amity Institute of Biotechnology, Amity University Haryana, Gurgaon, India

Vipul Kumar School of Agriculture, Lovely Professional University, Phagwara, Punjab, India

Belur Satyan Kumudini Department of Biotechnology, School of Sciences (Block 1), JAIN (Deemed-to-be University), Bengaluru, Karnataka, India

Mahantesh Kurjogi Green Nanotechnology Laboratory, University of Agricultural Sciences, Dharwad, Karnataka, India

Giuliana Maddalena Department of Agricultural and Environmental Sciences—Production, Landscape, Agroenergy (DISAA), University of Milan, Milano, Italy

K. Mahendra Department of Plant Pathology, Centre for Plant Protection Studies, Tamil Nadu Agricultural University, Coimbatore, India

T. Marimuthu World Noni Research Centre, Chennai, India

S. Mahadeva Murthy Department of Microbiology, Yuvaraja's College, University of Mysore, Mysore, Karnataka, India

Muntazir Mushtaq School of Biotechnology, Sher-e-Kashmir University of Agricultural Sciences and Technology of Jammu, Jammu, Jammu and Kashmir, India

S. Nakkeeran Department of Plant Pathology, Centre for Plant Protection Studies, Tamil Nadu Agricultural University, Coimbatore, India

B. Nandini Nanobiotechnology Laboratory, DOS in Biotechnology, University of Mysore, Mysore, Karnataka, India

K. Narasimha Murthy Department of Studies in Biotechnology, University of Mysore, Mysore, Karnataka, India

Atanu Panja Molecular Mycopathology Laboratory, Biocontrol Unit, PG Department of Botany, Ramakrishna Mission Vivekananda Centenary College (Autonomous), Kolkata, India

Ghazala Parveen Section of Plant Pathology, Department of Botany, Aligarh Muslim University, Aligarh, Uttar Pradesh, India

Alessandro Passera Department of Agricultural and Environmental Sciences—Production, Landscape, Agroenergy (DISAA), University of Milan, Milano, Italy

Savita Veeranagouda Patil Department of Biotechnology, School of Sciences (Block 1), JAIN (Deemed-to-be University), Bengaluru, Karnataka, India

Rakesh Patsa Department of Plant Pathology, Bidhan Chandra Krishi Viswavidyalaya, Mohanpur, Nadia, West Bengal, India

H.V. Pavan Department of Microbiology, Yuvaraja's College, University of Mysore, Mysore, Karnataka, India

Hilal Ahmad Pir Division of Plant Breeding and Genetics, Sher-e-Kashmir University of Agricultural Sciences and Technology of Jammu, Jammu, Jammu and Kashmir, India

K. Prabakar Department of Plant Pathology, Centre for Plant Protection Studies, Tamil Nadu Agricultural University, Coimbatore, Tamil Nadu, India

Fabio Quaglino Department of Agricultural and Environmental Sciences—Production, Landscape, Agroenergy (DISAA), University of Milan, Milano, Italy

T. Raguchander Department of Plant Pathology, Centre for Plant Protection Studies, Tamil Nadu Agricultural University, Coimbatore, Tamil Nadu, India

Kushal Raj Department of Plant Pathology, CCS HAU, Hisar, Haryana, India

S. Rajamanickam Department of Plant Pathology, Centre for Plant Protection Studies, Tamil Nadu Agricultural University, Coimbatore, India

V. Rajasreelatha Department of Biochemistry, Indian Institute of Science, Bangalore, Karnataka, India

L. Rajendran Department of Plant Pathology, Centre for Plant Protection Studies, Tamil Nadu Agricultural University, Coimbatore, Tamil Nadu, India

Younes M. Rashad Plant Protection and Biomolecular Diagnosis Department, Arid Lands Cultivation Research Institute, City of Scientific Research and Technological Applications, Alexandria, Egypt

Ruby Rawal Kurukshetra University Kurukshetra (KUK), Thanesar, Haryana, India

P. Renukadevi Department of Plant Pathology, Centre for Plant Protection Studies, Tamil Nadu Agricultural University, Coimbatore, India

Richa Salwan College of Horticulture and Forestry, Dr YS Parmar University of Horticulture & Forestry, Hamirpur, Himachal Pradesh, India

Surendra Sarsaiya Bioresource Institute for Healthy Utilization and Key Laboratory of Basic Pharmacology and Joint International Research Laboratory of Ethnomedicine of Ministry of Education, Zunyi Medical University, Zunyi, China; Department of Microbiology, Sri Satya Sai University of Technology and Medical Sciences, Sehore, Madhya Pradesh, India

Santanu Sasidharan Department of Biotechnology, National Institute of Technology, Warangal, Telangana, India

Prakash Saudagar Department of Biotechnology, National Institute of Technology, Warangal, Telangana, India

V.P. Savalgi Department of Agricultural Microbiology, College of Agriculture, University of Agricultural Sciences, Dharwad, Karnataka, India

V. Sendhilvel Department of Plant Pathology, Centre for Plant Protection Studies, Tamil Nadu Agricultural University, Coimbatore, Tamil Nadu, India

Vivek Sharma University Centre for Research and Development, Chandigarh University, Mohali, Punjab, India

Jingshan Shi Bioresource Institute for Healthy Utilization and Key Laboratory of Basic Pharmacology and Joint International Research Laboratory of Ethnomedicine of Ministry of Education, Zunyi Medical University, Zunyi, China

S.D. Shruthi Microbiology and Molecular Biology Lab, BioEdge Solutions, Bangalore, Karnataka, India

Monika Sood School of Bioengineering and Biosciences, Lovely Professional University, Phagwara, Punjab, India

K. Soumya Field Marshal K M Cariappa College, Constituent College of Mangalore University, Madikeri, Karnataka, India

C. Srinivas Department of Microbiology and Biotechnology, Jnanabharathi Campus, Bangalore University, Bangalore, Karnataka, India

M. Thippeswamy Department of Botany, Davangere University, Davanagere, Karnataka, India

Silvia Laura Toffolatti Department of Agricultural and Environmental Sciences—Production, Landscape, Agroenergy (DISAA), University of Milan, Milano, Italy

Palistha Tuladhar Department of Biotechnology, National Institute of Technology, Warangal, Telangana, India

A.C. Udayashankar Department of Studies in Biotechnology, University of Mysore, Mysore, Karnataka, India

S.L. Varsha P.G. Department of Studies in Microbiology and Biotechnology, Karnatak University, Dharwad, Karnataka, India

A.B. Vedamurthy P.G. Department of Studies in Microbiology and Biotechnology, Karnatak University, Dharwad, Karnataka, India

Amber Vyas University Institute of Pharmacy, Pt. Ravishankar Shukla University, Raipur, Chhattisgarh, India

Shabir Hussain Wani Mountain Research Centre for Field Crops, Sher-e-Kashmir University of Agricultural Sciences and Technology of Kashmir, Khudwani Anantnag, Jammu and Kashmir, India

Leela Wati Department of Microbiology, CCS HAU, Hisar, Haryana, India

Vu Hai Yen Graduate School of Medical Sciences, Kyushu University, Fukuoka, Japan

Abbu Zaid Plant Physiology and Biochemistry Section, Department of Botany, Aligarh Muslim University, Aligarh, Uttar Pradesh, India

About the editor

Dr. Sudisha Jogaiah is assistant professor and program coordinator, Laboratory of Plant Healthcare and Diagnostics, PG Department of Biotechnology and Microbiology, Karnataka University, Dharwad. The technological quality and expertise of Dr. Jogaiah's lab is reflected in an extensive track record of publications with more than 81 peer-reviewed papers, one national patent, and five review articles with an H-index of 23. Also, he has contributed 18 book chapters in various book editions published by Springer, Wiley-Blackwell, and Elsevier. He had also edited three books on bioactive molecules involved in plant-microbe interactions and defense from Springer, Germany, and Elsevier Publications, UK. He is the recipient of Fellow of National Academy of Biological Sciences (FNABS) and has been rewarded with 15 and three international awards. He is editor of Scientific Reports, Plos One, BMC Plant Biology, Frontiers in Plant Science, Annals of Crop Sciences and Agriculture, Journal of Mycology and Plant Pathology, and BioMed Research International.

Foreword

Induction of host plant resistance against a variety of phytopathogens constitutes an advanced strategy of safe and effective plant disease management. Microbial agents have been extensively used quite often for enhancing the tolerance of host plants against a broad spectrum of phytopathogens, through elicitation of systemic acquired resistance or enhanced plant growth and also through active inhibition of plant pathogens. The former class of microbes are more suitably categorized as biostimulants, while the latter group of microbes are referred to as biopesticides. The two categories of microbes in future are expected to be regulated under two different acts, i.e., Fertilizer Control Order 1983 and Pesticides Management Act (presently Insecticides Act 1968). Biological agents (both biostimulants and biopesticides), however, should be integrated suitably as components of IPM with other control measures because various methods are effective at different situations and durations under varying conditions. In recent years, the contribution of biological pest control in sustainable agriculture has been increasing because of its eco-friendly nature and mechanism of action. Biocontrol agents act through mycoparasitism and/or hyperparasitism, antibiosis, competition, secondary metabolites or by inducing systemic resistance and upregulating defense responsive enzymes. Also, plants produce a large array of bioactive compounds, specialized secondary metabolites to defend themselves when facing biotic and abiotic stress conditions. Secondary metabolites play an important role in the way plants interact with their environment and are usually produced in select cell types within the plant. These compounds play an active role in defense against herbivores, fungi, bacteria, viruses, and other plants competing for resources. In spite of the fact that voluminous work has already been published, identification of new potential strains of biological control agents and plant secondary metabolites needs to be characterized and evaluated for their suitability in different climatic conditions and also their effects on various crop plants are yet to be explored.

I am pleased that the book entitled "Biocontrol Agents and Secondary Metabolites: Applications and Immunization for Plant Growth and Protection" published by Elsevier, UK, provides an up-to-date information of the newer biocontrol agents such as plant inducers and helps to understand the mechanism of biosynthesis of plant secondary metabolites for boosting immunity of plants against biotic or abiotic stresses and discusses its further application in improving crop productivity. Various aspects of biocontrol phenomenon in the book is presented through 29 chapters that have been contributed by eminent researchers who possess in-depth understanding, proficiency, and expertise on the subject at an international level. The authors have systematically organized various information about the potential microbial agents, their biocontrol and biostimulant actions displayed through competitive inhibitory or antagonistic actions as well as through induced host resistance including genomic and metabolomic approaches, in individual chapters. Overall, the book presents cutting-edge developments in the area of biocontrol agents, secondary metabolites they secrete, and their applications in immunization and plant protection, which could serve as a safer alternative to chemicals for the development of sustainable agriculture. Besides, it will also serve as a repository of ideas with many sustainable solutions that could increase the production of safe and healthy food to sustain food and nutritional security of the ever-increasing human population.

I believe that this book will be of immense interest to academia, R&D institutions, policymakers, industrialists, and unemployed youths in complementing small, medium, and even organized enterprises.

P.K. Chakrabarty
Agricultural Scientists Recruitment Board, Department of Agricultural Research & Education, Ministry of Agriculture & Farmers' Welfare, Govt. of India, Pusa, New Delhi, India

Preface

This book aims to bring together the art of findings with their cons and pros of the advances in the field of biocontrol agents and secondary metabolites in agriculture as applications and immunization for plant growth and protection. The chapters covered in this book present interesting cutting-edge developments in the area of sustainable crop improvement. It addresses numerous latest themes covering safer alternative strategies to chemicals for the development of sustainable agriculture, which are described from the latest peer-reviewed literature as reported by eminent researchers who have in-depth understanding, proficiency, vision, and have shown exceptional attributes in their scientific career at the international level. The following are the general overview of the book:

- Discusses the recent development in the practice and integration of biocontrol agents such as plant growth promoters and agrochemicals.
- Outlines the network of biocontrol agents-plants-pathogens mediated through multiple mechanisms for sustainable plant protection.
- Identifies the key challenges on the biopriming of seeds with beneficial microorganisms for enhancing growth and development of plants by regulating several biochemical and physiological processes.
- Compiles the issues that need to be addressed to conserve and to claim the IPR rights of the microbial wealth to retain soil and/or plant health.
- Includes plant secondary metabolites as the key constitutive components of plant defense against various biotic and abiotic stresses.
- Covers biosafety, regulatory requirements, marketing strategy, and exploitation of native biocontrol agents (BCAs) with the concomitant practice of their commercialization globally.

Sudisha Jogaiah

Acknowledgments

I wholeheartedly thank all the authors who have meticulously contributed to advance developments and delivered their views in consideration of both the benefits and the risks on the implementation of biocontrol agents in agriculture for safer and high yielding crops. Thanks are also due to the Karnataka University administration, Dharwad for their support. Also, thanks to Elsevier for taking up the publication of this book. I am also grateful to Dr. P. K. Chakrabarty, Member—Agricultural Scientists Recruitment Board, Government of India, New Delhi, India for his kind words in the foreword.

Introduction

Plants and fungi represent a vast pool of bioactive compounds and are more than ever a strategic source for new and successful commercial products. Recent advances made in genomics, proteomics, and combinatorial chemistry show that nature maintains compounds that have already the essence of bioactivity or function within the host and in the environment. This book intends to cover established and updated research on emerging trends in plant defense signaling in/during stress and growth at the interface of sustainable way of life, in the bifold context of human welfare and conservation of fungi as a group of organisms. Broadly the research endeavors are around the following themes, when a thrust is identified for bioactive molecules in plant signaling for defense. Finally, we provide a perspective on future directions for research in this field that could help in contributing to sustainable crop improvement with higher yields. It is believed that the book will be helpful to postgraduate and doctoral students of molecular plant Pathologists, and to botanists, microbiologists, ecologists, plant pathologists, physiologists, agronomists, molecular biologists, and entrepreneurial mycologists.

Sudisha Jogaiah

Fungi endophytes for biofactory of secondary metabolites: Genomics and metabolism

1

Surendra Sarsaiya[a,c], *Archana Jain*[a,c], *Jingshan Shi*[a], *and Jishuang Chen*[a,b]
[a]Bioresource Institute for Healthy Utilization and Key Laboratory of Basic Pharmacology and Joint International Research Laboratory of Ethnomedicine of Ministry of Education, Zunyi Medical University, Zunyi, China, [b]College of Biotechnology and Pharmaceutical Engineering, Nanjing Tech University, Nanjing, China, [c]Department of Microbiology, Sri Satya Sai University of Technology and Medical Sciences, Sehore, Madhya Pradesh, India

1.1 Introduction

Development has yielded an enormous variety of life span on Earth, trapped in a multifaceted system of dealings and determined by interspecies rivalry for limited ordinary resources. From the organic point of assessment, these associations are regularly mediated by expendable biomolecules, specifically secondary metabolites (SMs), which reflect extant and inexistent connections across progression. SMs are an important resource for the expansion of drugs that contain a heterogeneous type of low-molecular-weight compounds (Leitão and Enguita, 2014). Fungi signify a multipart set of organisms that diversify from unicellular to multicellular complex things, with a wide change of variations that permit the organisms to accept dissimilar existences successfully from saprophytes to herbs, animals to human pathogens, and from symbionts to parasites (Araujo and Maia, 2018). Several species, for example, *Aspergillus fumigatus* and *Aspergillus flavus*, are of specific concentration as a consequence of their pathogenicity as a result of both intrusiveness in immunocompromised patients and the toxic complexes they yield, for instance, gliotoxin and aflatoxin. These complexes are often the product of the SMs paths in these fungi. SMs paths offer a number of organic molecules and can also occasionally be medicinally valuable, for example, most excellently penicillin-G, cyclosporine-A, and lovastatin. Subsequently, numerous biosynthetic paths existing in the genomes of fungi form have not been connected to their artifact; this offers a prospect for encounter as additional genomes are sequenced (van Dijk and Wang, 2016).

In the environment, herbs associate with a variation of root-related fungi, which range from advantageous to detrimental. Many fungi have been stated to endorse plant development under, for example, nutrient-deficient conditions, which allows plants to deal with stressful situations (Jogaiah et al., 2013). Arbuscular mycorrhizal fungi (AMF), which are obligate biotrophs, have been considered as valuable. AMF have their place in the fungal group Glomeromycotina, which encompasses the best

Biocontrol Agents and Secondary Metabolites. https://doi.org/10.1016/B978-0-12-822919-4.00001-6

categorized mutualistic fungi indorsing plant development under nutrient-lacking conditions by transporting nutrients, for example, phosphorus, to its herb hosts. Molecular complexes fundamental symbiosis amongst AMF and 80%–90% of all terrestrial plants are supposed to be progression conserved. This symbiosis association started before 450 million years (Mya), when vegetations started to cultivate on land, as long as there was vital support for addition of plants to nutrient-lacking soil matters (Hiruma et al., 2018).

Plants are conventionally used as drugs and endure to be the basis of plant chemicals with therapeutic assets. These plants chemicals have been used and found to be advantageous to humans. Microorganisms, for example, fungi, bacteria, and to a lesser range viruses, that attack and reside inside the plant's tissues are the so-called endophytes, which accurately resource "within plants". The symbiotic association is known as mutualistic relationship when the microorganism (endosymbiont) with the host plant can share benefit to each other, while a commensalism relationship is indicated that the endosymbiont can be live inside the host without disturbing it. Fungal endophytes may exist within the host roots, leaves, or stem of the plant and incline to replicate just previously or during the senescence point of the plant as a result of which time the reproductive structure would be out. These plant endophytes often yield compounds that may be damaging or valuable to the plant, and accordingly may be taken out for their therapeutic rate or may be of agrarian value (Daley et al., 2017).

Today, we distinguish that even though most existing organisms can yield SMs, the capability to yield them is unequally dispersed (Nakkeeran et al., 2019). Amongst all recognized microbial antibiotics and comparable bioactive complexes (in total 22,500), 45% are of actinomycetes, 38% are of fungi, and 17% are of unicellular bacteria. Amongst this treasure of compounds, only about a hundred are in real use for human remedy, with the mainstream being resultant from actinomycetes. Nevertheless, it is worth revealing that in addition to penicillin, numerous other fungal SMs have positively reached the pharmacological marketplace, together through cholesterol disenchanted statins, the mycophenolic acid with immunosuppressant, and the antifungal griseofulvin (Nielsen and Nielsen, 2017).

Biosynthesis of SMs takes place from an imperfect figure of precursor metabolites from the chief metabolism. In fungi, some precursors are mostly short chain (SC) carboxylic acids (e.g., acetyl coenzyme A) or amino acids, which are allied by a backbone catalyst, for example, of nonribosomal peptide synthetase (NRPS), polyketide synthase (PKS), dimethylallyl tryptophan synthetase (DMATS), and/or terpene cyclase (TC). The subsequent oligomers are then focused on chemical alteration by tailoring enzymes which are frequently measured under a shared transcriptional parameter as the backbone enzyme. A hallmark trait of the genes complicated in an SMs pathway is that they, in the greatest cases, actual cluster in the chromosome in BGCs (biosynthetic gene clusters) (Nielsen and Nielsen, 2017).

Fungal metabolites and its action mechanisms could trigger the bio-mechanism approaches for enhancing crop development with its protection. Although the available research data (nature and function of secreted microbial metabolites) of endophytic fungi are identified predominantly, which are unbalanced in plant microbial interactions. Both fungi and bacteria yield diverse instable (or semivolatile) organic compounds (OCs), which include multiple courses of short molecular form lipophilic metabolites and their byproducts that evaporate at typical pressures and temperatures (Li et al., 2016).

Fungal SMs biosynthesis pathways are included core enzymes, which are considered as a compound backbone. Finding of novel SM genes is frequently proficient by homology explorations using identified genes of these principal enzymes. Expediently, in several cases, the supplementary genes (coding for adapting enzymes, transcription factors, resistance proteins, or transporters) in the biosynthetic route are grouped in the genome; thus, they can be recognized easily as soon as the core enzymes have been recognized. There are plentiful ways to connect the identified SM pathways with their products (van Dijk and Wang, 2016). The typical collecting genes besides the preserved motifs of mainstay genes can be oppressed for computational finding of BGCs from sequence figures. Tools like antiSMASH, SMURF, PRISM, and CASSIS/SMIPS employ these structures to dependably and with high correctness distinguish BGCs of the identified compound in endophytic fungi. Additional algorithms distinguish BGCs deprived of relying on exact motifs or the attendance of backbone genes, which allows BGCs identification beyond PKS, DMATS, NRPS, and TCs. Tools and executions of mining algorithms of BGC have been broadly reviewed (Nielsen and Nielsen, 2017).

Employing metabolic engineering to evade these limits can be greatly supported by operation of the scientific depiction of genome-scale metabolism in metabolic models, whose perceptions and applications have been reread in a different part. These models, though, often neglect SM biosynthesis, and hence their probability in studying SM has not been entirely selected. Furthermore, with the well-organized gene editing implement CRISPR Cas9 being advanced for a sum of fungal protoorganisms, a countless potential occurs for implementing the compulsory genetic alterations for the progress of improved SM producers (Nielsen and Nielsen, 2017).

1.2 Fungal endophytes frequency and transmission in plant organizations

Endophytes are frequently transmitted to the plant host vertically, specifically from generation to the next group over nonsexual propagation or spores, perhaps seen through the *Neotyphodium* spp. These endophytic are typically germ-free, though a few fungi, similar to *Epichloe* spp., are adept at acclimatizing to the environmental alteration and replicating sexually liberating spores that will certainly contaminate a neighboring host plant. Alternatively, during horizontal communication, these endophytes yield spores via asexual or sexual replica for insect and breeze dispersion. An endophyte may also be shifted straight through revelation to an adjacent infected herb. These endophytes at that time reproduce subsequently in either advantageous or deadly effects (Daley et al., 2017).

1.3 Endophytic fungus as biofactory of bioactive compounds

SMs are unique bioactive molecules formed by various organisms together with bacteria, floras, and fungi (Jogaiah et al., 2016). These compounds (as bioactive molecules) are predominantly ample in soil-residing fungi, which occur as multicellular groups competing by means of minerals, nutrients, and water (Keller et al., 2005). Contrasting primary metabolites, maximum SMs—as their term advocates—are not important for fungal development, progress, or replica under in vitro environments. They can, though, provide a shield in contradiction of various environmental pressures and throughout antagonistic communications with added soil populations or a eukaryotic mass. At a similar period, fungi are also recognized to produce plentiful mycotoxins, for example, aflatoxin, trichothecene, fumonisin, and zearalone (Khaldi et al., 2010).

1.4 Genome level secondary metabolism metabolic modeling

Several SM clusters were categorized at the molecule level together with the gliotoxin (Gardiner and Howlett, 2005), fumitremorgin, fumigaclavines (Unsold and Li, 2006), and siderophores biosynthesis clusters. The sequencing of the entire genome also discovered that the number of SMs characterized from a certain strain falls far in arrears of the clusters numbers that can be predicted based on its gene sequence figures (Chiang et al., 2008a,b). This has been accredited to the fact that not many clusters may be articulated under typical laboratory environments. Notwithstanding the agricultural and medical status of fungal SMs, maximum SM putative clusters in endophytes genomes have been foreseen by ad hoc approaches based on guide reviews of BLAST explorations created for genes backbone and their neighbors. A manual footnote of SM clusters, though, is time intense and may result in unpredictable annotation (Khaldi et al., 2010).

1.5 Gene clusters for fungal metabolism: Diversity and distribution

Gene clusters for fungal metabolism (GCFM) are loci that encompass multiple genes from diverse gene groups, which offer support to a disconnected metabolic phenotype. Currently, maximum metabolic phenotypes have been found to be programmed by GCFM contribution in nutrient acquisition, or the degradation/biosynthesis of SMs, amino acids, and cofactors. The initial GCFM was to be identified in the strain of *Saccharomyces cerevisiae* galactose consumption cluster (GAL). Ensuing nutrient acquisition GCFM discoveries contain the catabolism of quinic acid cluster in *Neurospora crassa* (QA), the catabolism of proline (PRO), and the assimilation of nitrate (HANT) clusters in *Aspergillus nidulans*, all of which revealed an extensive in-depth investigation previously to the age of comprehensive-genome sequencing. The evolutionary and functional mechanisms of the bio-synthetic genes are clustered quickly for the development and protection of the plants (Walton, 2000). More newly, nutrient operation clusters have been recognized in fractional and broad genome sequences over a manual explanation of gene purposes and by fungal analogues identification via bacterial operons (Jeffries and van Vleet, 2009). These include GCFM complex in sugars utilization (e.g., rhamnose and *N*-acetylglucosamine), catabolism of amino acid, and iron metabolism.

GCFM are also intricate in basic intracellular metabolism through contributing in the synthesis of a numeral of amino acids, vitamins, and other vital metabolites. All vitamins may be supposed to be as inimitable or complex metabolites, which cover essential metabolic ways to focus on minute numbers, but that are frequently acquired slightly than made endogenously. Likewise, rare but vital amino acids may be assimilated from other microorganisms or synthesized. Numerous pathways for amino acid/vitamin synthesis are associated by the inclination of the genes for their cluster metabolism in both prokaryotes and fungi (occasionally). Though the maximum vitamin biosynthetic routes have been adeptly clustered in bacteria, only a combine of fungal vitamin clusters have been recognized, counting the biotin cluster (Hall and Dietrich, 2007) and the functionally measured pyridoxine group in *Saccharomyces* to date *(*Li et al., *2016)*. Clustering is not entirely a functional, optional, or niche-specific pathway, however, even though that is undoubtedly the greater inclination. The AROM synthesis pathway of amino acids (aromatic) is the most commonly well-kept-up pathway amongst fungi. AROM was initially supposed to be an operon or gene cluster in fungi comparable to that in bacteria, but was later start to be a penta-property peptide that resulted from the integration of monofunctional inherited genes (Giles, 1978).

It is tough to conclude whether ecological tendencies occur in gene clustering, for the reason that fungi incline to be opportunistic and acclimatize to ecological parts rapidly. Though, the widespread gene is clustered in class of Eurotiomycetes, which is present equally in the *Aspergillus* spp. (de Vries et al., 2017) and several others microcolonial black yeasts (Teixeira et al., 2017). These are recommended GCFM in the soil saprotrophs. With the finding of novel fungal genome data and enhancements in GCFM detection algorithms, prospects to study wider outlines of fungal SM GCFM diversity are better than they have constantly been. Currently, the genomics of fungi has described the phylogenetic diversity at complex levels or within the species or genera (de Vries et al., 2017; Teixeira et al., 2017). Future exertions to sample genomes intensely within defined fungal groups will provide better power to conclude direct relations between GCFM encoded purposes and fungal biology (Slot, 2017).

Associated with other strictly connected *Aspergillus* species, *Penicillium* and *P. digitatum* produce lesser SMs with identified products, for example, phenylalanine-proline diketopiperazine and tryptoquialanines (TQA). The obtainability of the fungus's genome arrangement also enables the investigation of SM product outlines of *P. digitatum* by means of SM biosynthetic gene groups of genome mining (Sun et al., 2013). This study associated the transcriptomes of *P. digitatum* cultured in chemical medium and recovered from decomposed citrus fruit, and initiated four upcontrolled genes in enriched biosynthetic gene clusters. A wide-ranging annotation of four biosynthetic gene clusters was then made with the hybrid usage of the domain account, RNA-sequence gene expression, and genomic synteny, which recognized three SM biosynthetic paths (SMBP1–3). Amongst closely connected strains, SM biosynthesis connected gene clusters specific for biosynthesis are commonly detected between well-preserved genomic areas, for example, the clusters (gsf and vrt) in *P. aethiopicum* (Chooi et al., 2010), and disruptions in synteny typically indicate gene cluster limitations (Inglis et al., 2013), which may be owing to the function of horizontal gene transmission in gene clusters (Campbell et al., 2012). SMBP1–3 is also observed between well-maintained genomic areas and their mechanisms are controlled in a similar way. SMBP2 is the exact tqa cluster in the comparative genomics, particularly synteny analyses. It will help to the accurate annotation of gene clusters (Zhu et al., 2017).

In view of the large amount and ubiquity of polyketides (aromatic) in nature, moderately few of their biosynthetic paths have been explained (Griffiths et al., 2016) Amongst the finest known are the paths of the emodin-derivative monodictyphenone as well as the carcinogenic aflatoxin in some *Aspergillus* strains (Yu et al., 2004). The main enzymes in the fungal biosynthesis aromatic polyketides are polyketide synthases, great multidomain schemes (type I) that extend its polyketide products (Crawford and Townsend, 2010). The assemblage of modest carboxylic acid structure blocks to a polyketide (aromatic) is attained by a series of three core areas: an ACP: Acyl-carrier protein that helps as a chain of the mounting polyketide, a malonyl CoA-ACP transacylase that chooses and allocates the extender part malonyl-CoA and a ketosynthase (KS) which catalyzes recurrent decarboxylative strengthening in directive to consecutively extend the polyketide support. Typically, a thioesterase (TE) field forms the catalyst C-terminus, cyclizes, and releases the end product. A unique characteristic of the fungal PKSs aromatic is the nonappearance of any reductive area, which is why they are referred to as NR-PKSs (nonreductive PKSs). Additionally, NR-PKSs comprise two other unique areas: the starter unit: ACP transacylase that chooses the (typically nonmalonyl) appetizer unit (Crawford et al., 2006) and the product pattern (PT-) domain that arbitrates the regioselective cyclization of the extremely sensitive poly-β-keto intermediates and orders the final construction of the produce (Jahn et al., 2017).

Some of the key databases have shown a very great variety of experimentally consideration such as IMGABC, ClusterMine360, and MIBiG. Separately from the sequence data and catalytic domain association, the major usefulness of these databases is to find the chemical buildings of the secondary metabolite. A recent type of ClusterMine360 has evidence on almost 290 gene clusters complicated in biosynthesis of over 200 nonribosomal peptides and polyketides. In accumulation of a genes sequence, catalytic domain group, and chemical assembly of a secondary metabolite end product, IMG-ABC30 has also catalogued data on genomics locus for a huge number of SM gene clusters. The MIBiG31 database has been established for determine the SM biosynthetic paths information while MIBiG-submissive reannotation has been used for SM biosynthetic gene clusters. An alternative example of a valuable database for SM is NORINE, which has many chemical forms for 1168 nonribosomal peptides. On the basis of bioinformatics investigation of experimentally considered NRPS and PKS gene clusters, a numeral of computational approaches has been technologically advanced for joining "metabolites genes." Taking into consideration the notable conservation of a total biosynthetic model for nonribosomal peptides and polyketides, these computational approaches have fundamentally used a knowledge-grounded approach for deriving forecast rules founded on experimentally considered NRPS and PKS gene clusters (Khater et al., 2016).

The apparatuses like NRPS-PKS, SBSPKS, ASMPKS/MAPSI, ClustScan, NP. Searcher, NRPSpredictor, PKS/NRPS, and PKMiner permit semireflex identification and explanation of PKS, NRPS, or PKS-NRPS hybrid gene clusters. Along with annotating the areas of multidomain NRPS and PKS, many of these tackles also forecast the acyltransferase (AT) and adenylation domains substrate specificity. Apart from identification of dissimilar catalytic areas of PKS and NRPS, SBSPKS can be perfect three-dimensional constructions of comprehensive PKS modules and foresee the directive of substrate channeling in the event of PKS clusters containing numerous ORFs. Bioinformatics tools have as well been developed for examination of a specific type of SM gene clusters. SMURF43 permits identification of gene clusters in a fungal genome for biosynthesis, even though PKMiner42 supports in mining of gene clusters category II PKS. Bioinformatics tools for examination of SM biosynthetic genes have also been established for the study of metagenomic information (Khater et al., 2016).

Metagenomic models can be rapidly scanned for different natural products by PCR primers precise for SM biosynthetic gene groups. This PCR-built sequence tag method has been joined with in silico genomic tools to examine for putative SMs. eSNaPD has been specifically advanced to analyze bulky meta genomic arrangement tag datasets and support in the detection of diverse secondary metabolite clusters. Another bioinformatics which admits sequence identifiers from metagenomic data lengthwise with protein and genomic sequences is NaPDoS. It uses genomic information to explore and categorize NRPS Adenylation and/or PKS Ketosynthase domains (Khater et al., 2016).

The recently established antiSMASH47 channel can recognize the biosynthetic loci cover the entire range of identified SM compound classes (nonribosomal peptides, polyketides, terpenes, aminocoumarins, indolocarbazoles, aminoglycosides, bacteriocins, nucleosides, lantibiotics, beta-lactams, siderophores, melanins, butyrolactones, and others). antiSMASH48 is also combined with other tools corresponding to ClusterFinder49 which permits identification of putative SM gene clusters programming a novel type of SM. It is used for the PFAM domain 50 to examine the enzymes present in the biosynthesis of SM. It also permits a comparison of recognized clusters with experimentally considered clusters via cluster BLAST. The latest information of antiSMASH can recognize active site deposits of core PKS areas like KS, AT, DH, ACP, KR, TE and tailoring areas like cytochrome P450 oxygenase by means of the "Active Site Finder" segment. antiSMASH also uses the domain info of linked NRPS and PKS to foresee the linear polyketides shaped by the enquiry cluster. Though the chemical construction prediction feature comprises a

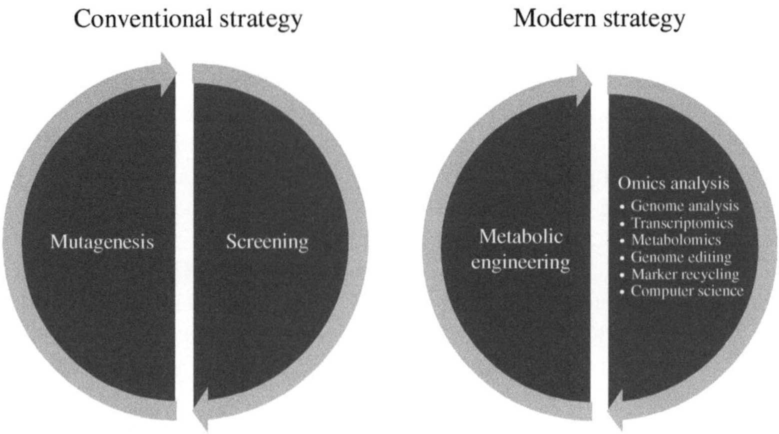

Fig. 1.1 Conventional and modern approaches for microbial metabolites.

consequence of reductive areas DH, KR, and ER on the polyketide construction, forecasts of post-PKS/NRPS variations and cyclizations are not yet obtainable in antiSMASH (Khater et al., 2016) (Fig. 1.1).

1.6 Methodological and technological advancement of genome for metabolites

1.6.1 Strategies for targeted genome editing

The SM genome editing is the alteration of a predetermined locus inside a genome with the specific region. Genome editing approaches must have the precondition of producing genetically constant organisms, be informal to use in contradiction of a series of DNA arrangements, and preferably must be also informal to perform in an extensive range of organisms (Kim and Kim, 2014). Targeted genomic alterations are built on the insertion or deletion of genetic evidence, and are strongly reliant on genetic recombination. Classical methods used for genome excision are built on homologous recombination, a procedure with an enormously low efficiency, particularly in higher eukaryotes which stalled its routine application. However, several resolutions have been established to increase the competence of genetic recombination, together with the use of tenable and directed nucleases. This enzyme family is used to breakdowns DNA double-strand at specific positions within the genome, and also activated the nonhomologous inference joining development (Boettcher and McManus, 2015). The double-strand breakdowns generation at precise DNA loci can be also utilized to simplify homologous recombination to supplement DNA fragments and produce recombinant strains. There are two main programmable nucleases family, a protein-directed nuclease, composed by Zn-finger in addition to transcription activator-comparable effector nucleases, with a nucleic acid-guided nuclease intimate, mainly signified by the CRISPR-Cas9 scheme (Leitão et al., 2017).

1.6.2 Protein-directed nucleases

The primary programmable nucleases were technologically advanced at the close of the 1990s as designed by the restriction enzyme to cut at all DNA in a preresolute arrangement. Zinc-finger domains were concocted by a plan based on a modular assemblage of adapted DNA binding areas, each of which distinguishes an exact three base pair arrangement within the DNA, to produce a set of 64 unlike domains able to identify any desired DNA arrangement when collective in the proper direction (Leitão et al., 2017; Segal et al., 2003).

1.6.3 Nucleic acid-guided nucleases

A third cluster of the genomes editing way is established by the RNA-directed nucleases characterized by the CRISPR-Cas9 scheme. The extensively recommended CRISPR-Cas9 genome editing scheme is a minimalist type derived from a primeval bacterial immune scheme that regulates the operative response in contradiction of bacteriophage contaminations isolated from the Gram-positive *Streptococcus pyogenes* bacteria (Barrangou et al., 2007). For effective binding, the ds DNA is cuts by the enzyme Cas9 necessary for the occurrence of the PAM (Protospacer adjacent motif) (Mojica et al., 2009). In an attempt to evade these target assets, a careful strategy of the gRNA arrangement and the usage of specific Cas9 modifications is recommended (Cho et al., 2014). Numerous computer procedures such as Cas-OFFinder, CHOP–CHOP, CCTop, sgRNAcas9, and COSMID were precisely advanced to support in gRNA strategy and off-target prophecy (Leitão et al., 2017).

1.6.4 Further tools for genome editing: Integrases and recombinases

The application of nucleases for the exact insertion of DNA wreckages into a genome is eventually reliant on a recombination response, which is essentially incompetent in eukaryotes. However, the exogenous DNA insertion in precise genomic loci can be improved by the operation of recombinases or integrases. A recombinase-intermediated cassette conversation is a multipurpose genome editing way which exactly substitutes a genomic target holder by a well-matched donor concept by the assistance of specific recombinases. A targeted genome operation by RMCE is stringently dependent on the reality of specific recombination positions (RT or attB) in the designated genome (Rutherford and Van Duyne, 2014). Nucleases such as CRISPR-Cas9 or TALEN can be also applied to present integrase recombination positions in any genome, provided that specific stages for heterologous gene countenance with comprehensive control ended content, course, and copy quantity of implanted genes. As an applied example, this approach has been positively employed in pluripotent stem cells of humans to supplement genes of concentration flanked through attB locations of Bxb1 and phiC31 integrases in a precise locus (Zhu et al., 2014). Further requests of RMCE procedure are predictable in the subsequent years, particularly in the pitch of gene appearance and metabolite production (Leitão et al., 2017).

1.7 Production of SMs by pathway-specific overexpression regulatory genes

Pathway-specific regulatory proteins of fungoid SM gene groups are frequently determined inside or directly together with the specific gene cluster. Characteristically, these proteins regulate the countenance of the complete gene cluster. They are not articulated under cluster noninducing situations. Thus, overexpression of their coding genes is an informal way to trigger the transcription of altogether pathway-precise genes. The benefit of this method is the supervision of only a solitary, relatively minor gene and the opportunity for both a specific and ectopic addition of the overexpression concept. Stimulatingly, the recent data has exposed over expression of a single pathway-precise regulator for the stimulation of SM gene (Bergmann et al., 2010). The fungi alcAp (alcohol dehydrogenase supporter) and gpdAp (glycerinaldehyde-3-phosphate dehydrogenase promoter) supporters were effectively applied to overexpress pathway-precise regulatory genes. As a consequence, novel mixtures were recovered and product produces of known compounds amplified (Chen et al., 2010). A number of supporters are obtainable for the protein's overexpression in filamentous fungus (Nützmann et al., 2012).

In *Trichoderma* strain, the numerous genes are responsible for the biosynthetic gene clusters, including core enzymes, for example, NRPSs (nonribosomal peptide synthetases), PKSs (polyketide synthases), or cyclases/terpene synthases, and addition enzymes (like cytochrome P450s, methyl transferases, oxidoreductases, etc.) (Bansal and Mukherjee, 2016a,b; Jogaiah et al., 2018). Six extra genomes (the mycotrophic *Trichoderma harzianum*, *Trichoderma asperellum*, *Trichoderma parareesei*, *Trichoderma gamsii*, and the devious human pathogens *Trichoderma citrinoviride* and *Trichoderma longibrachiatum*) were afterward supplementary to the public records (Zeilinger et al., 2016). The fungal SMs biosynthesis often involves exclusive and uncommon biochemical paths. These may be to some extent wide-ranging to yield a higher substances diversity from only a limited key precursor resultant from primary metabolism (Keller et al., 2005). *Trichoderma*-resultant SMs encompass nonribosomal peptides, for example, siderophores, diketopiperazines, and peptaibiotics like gliovirin, gliotoxin, polyketides, pyrones, terpenes, and isocyane metabolites (Zeilinger et al., 2016).

Comparable to other fungal strains, the appearance of SM-connected genes in *Trichoderma* spp. is recognized to be measured by connections with pH signaling, the velvet-intricate proteins, and other (micro)organisms. Atanasova et al. (Atanasova et al., 2013) deliberate the transcriptomic rejoinders of *T. atroviride*, *T. reesei*, and *T. virens* to the occurrence of *R. solani*. Two PKSs are controlled by the genes present in the *R. solani, T. atroviride* and *R. solani, T. reeseie*, while the genes are present in the synthesis group of gliotoxin are uncontrolled in the *T. virens*. *T. atroviride* is additional described for the lipoxygenase genetic factor to be complicated in the 6-PP biosynthesis system (Kubicek et al., 2011), while the development of *T. arundinaceum* in coculturing with *B. cinerea* has led to improved expression stages of the novel "Tri" biosynthetic genetic factor (Malmierca et al., 2015).

1.8 Genetic makeup of fungal secondary metabolism

In microbes, it is extensively alleged that SMs are biochemical signaling things synthesized for messaging and performing a chief part in competitors' self-consciousness (Brakhage and Schroeckh, 2011). The SM biosynthesis coding genes in fungi are also settled in clusters that can extend over 10kb, while there is a limited exception (Lo et al., 2012). The SM gene's arrangements are restricted in cluster's usage of fermentation practice as a suitable and viable alternative at large-scale production of metabolites (Wu and Chappell, 2008). These clusters typically code for enzymes complex, for example, the NRPS: nonribosomal peptide synthetases, or PKS: polyketide synthases that consist of numerous domains and components with defined purposes (Strieker et al., 2010). These multimodular enzymes display high resemblance in their mechanisms and architecture, intricate in the product assemblage, thus aiding in employment of different matters. To produce the structural support of the particular NRPS, SMs and PKS use malonyl clusters and amino acids and their byproducts as the structure blocks (Brakhage and Schroeckh, 2011). Nonribosomal peptides (NRP) and polyketides (PK) form a support for most of the SMs. Many of the NRP derivatives comprise clinically significant antibiotics, for example, cephalosporin, penicillin, and immune suppressants identical to cyclosporine, while lovastatin results from the polyketide support. Some complexes fall underneath the diverse NRP-PKS hybrid derivation like aspyridones (Bergmann et al., 2007) though some others are comparable to oxylipins (derivatives of fatty acid) with a gibberellins (terpene derivative) consequence from additional paths. In the previous study, the endophytes have received much consideration due to their maximum metabolic adaptability. Many reviews have described the attention of endophytes SMs. While endophytes have been energetically screened for their ability to elicit host-resultant phytochemicals, other information clearly designates that endophytic fungi also form a miscellaneous variety of different SMs (Deepika et al., 2016).

1.9 Identifying gene clusters of fungi

With the accessibility of a growing quantity of fungal genomes, there has been a rapid growth in foreseeing and recognizing putative SM formation genes accountable. Bioinformatic algorithms, perhaps SMURF [SM Unknown Regions Finder] (Khaldi et al., 2010), antiSMASH, and FungiFun, are accessible for gene clusters identification and responsible for SM. These tools support in the genes coding prediction for core NRPS and/or PKS enzymes, laterally with the foreseen function of the together genes to support in identification of SM clusters in the entire genome (Bergmann et al., 2007). These processes have been effectively utilized to recognize approximately 50 gene clusters in fungi, *Aspergillus* strain (28–40Mb genome size) and in *Arthroderma* sp. with 27 gene clusters (22Mb genome size) (Burmester et al., 2011). Since the efficient structural physiognomies of the metabolites determined by these groups continue to be largely unidentified, these are called "orphan" or "cryptic" clusters (Bergmann et al., 2007; Deepika et al., 2016).

1.10 Applications for secondary metabolites through genome editing and metabolic engineering

Efficient genomics is a division of the molecular science established by unit approaches that analyze the active genotype–phenotype relations. In order to regulate the genetic source accountable for an assumed phenotype, onward genetics is the most secondhand classical method in functional genomics. Primarily, forward genetics depends on the random formation of mutants by chemical or radiation mutagens and additional mapping of genomic positions. This way was applied for the explanation of genes intricate in the formation of antibiotics and many supplementary metabolites with medical uses (Lewis, 2013). Currently, advancing genetics functional transmissions can be intended more reasonably, taking the gain of the current increase in obtainable genomic statistics. Furthermore, we can also methodize the practical genomics subsequent viewpoint of the opposite genetics to determine the functional values of a genetic alarm in a specific genomic locus (Shah et al., 2015). Mutually forward with reverse inheritances can benefit from genome excision protocols (Leitão et al., 2017).

1.10.1 Forward genetics applications

Despite the presence of much genome excision information, the CRISPR-Cas9 system is used for the numerous applications. The RNA-guided directing of the Cas9 endonuclease and its alternatives to every genomic locus are found ideal to achieve high-throughput data that could be possibly explored the detection of new SM with biological action or for the regulation of the governing mechanisms leading to the formation of the previously known substances. Libraries of arrayed or pooled gRNAs can be applied together with innate Cas9 or its useful variants to achieve the functional screenings covering a comprehensive genome (Evers et al., 2016). The appearance of Cas9 alternatives in the target cells can be attained by using combined stable cells or bypassing expression arbitrated by transformation by plasmids encrypting Cas9 variants measured by specific supporters. The general approach, advantages and weaknesses, and uses for the useful CRISPR-based transmissions in eukaryotic cells, has been freshly reviewed in a different region (Housden and Perrimon, 2016; Shalem et al., 2015). Expansion of function screenings could be achieved by the usage of protein combinations comprising the inactive dCas9 mutant and an effector protein. The functional effectors, which could relate to the transcription factors family, will be directed to specific genome sections by the bonded dCas9 protein and the consistent gRNA. This approach has been effectively working in human cells with the help of the VP64 transcriptional activator merged to the nonfunctional dCas9 collected with a library of mutually synthetic gRNAs (Gilbert et al., 2014), and exhibited increased compassion and specificity when associated with other approaches, for example, recombination-built promoter production (Leitão et al., 2017; Montiel et al., 2015).

1.10.2 Reverse genetics applications

After the description of a cluster of genes in control for the generation of an SM, reverse genetics can be applied to moderate and adjust its biosynthesis. Together, CRISPRCas9 and protein-directed nucleases can be used to persuade selective and balanced modifications of regulatory or biosynthetic genes to recover or modify SM production. Genes encrypting a secondary metabolite catalyst can be target nucleases for the genome editing system (Kim et al., 2015). A certain application of this impression is the bioengineering of genes encrypting multidomain shortening enzymes that are usually present in SM gene clusters. Multidomain shortening enzymes are higher-molecular mass enzymes composed of linked domains able to trigger and abridge a limited number of metabolic forerunners, and signified by polyketide and polyene synthases. Engineering of precise domains for the making of metabolic chimeras has been formerly performed by traditional recombination approaches, demonstrating its influence on the making of new metabolites (Marsden et al., 1998). To exemplify some of the usages of genome excision in reverse genetics, recent research showed the operation of microorganisms like *Escherichia coli* as a stage for metabolic bioengineering using the CRISPR-Cas9 scheme for exact insertions and deletions of genes encrypting the catabolic catalyst. Moreover, the transcriptional interference mediated through the CRISPR-dCas9 scheme (CRISPRi) has been also functional to control the metabolic stream in an anabolic path in *E. coli*. In many investigations, the authors validate the possibility of CRISPRi by using numerous custom intended gRNAs to regulate the expression stages of a rate-restrictive enzyme in the polyhydroxyalkanoate (PHA) biosynthesis and the arrangement of the polymer (Leitão et al., 2017; Lv et al., 2015).

1.10.3 Gene deletions

A “traditional” strategy is to associate the metabolic outline of the omission mutant with that of the uninhabited type by means of LC–MS or HPLC to identify new metabolites (Brakhage and Schroeckh, 2011). Chemical outlining of single gene omission mutants in *A. nidulans* has led to the description of nearly 25 SM synthetases/synthases. Chiang et al. (Chiang et al., 2008a,b) recognized emericellamide synthesis gene cluster in *A. nidulans* and therefore explicated its biosynthesis paths involving equally NRPS as well as PKS. Deletion research has also recognized the character of LaeA and VeA (light-controlled developmental factor), core mechanisms of a nuclear composite (Bayram et al., 2008) in SM biosynthesis. A major disadvantage of this procedure is that it could be functional only to comparatively fewer metabolites that continue to be “turned on” below laboratory circumstances and not only for those gene clusters that continue to be silent under artificial situations. Genome footnote studies have led to the detection of closely 30–40 SM gene clusters in each species of aspergilli. Other methods include the over-appearance of pathway-precise gene controllers that are mostly existing in all SM gene groups. This approach was found to be possible as SM genes are typically clustered and measured by a single controller. In spite of problems involved in posttranslational alterations of cluster-precise regulators, a pathway-precise regulatory gene apdR encrypting for the Zn2Cys6 transcription feature was overexpressed in *A. nidulans*, leading to the detection of two original metabolites aspyridones A and B (Bergmann et al., 2007). Promoter conversation or substitution of the putative PKS or NRPS gene with a robust regulatable agent resulted in the biosynthesis of asperfuranone (Deepika et al., 2016).

1.10.4 Variation in epigenetic mechanisms

Chromatin-controlling mediators are also known to cause the gene clusters (cryptic) in numerous fungi. Studies elect LaeA to be a main controller complex in synthetic cluster opening through chromatin over modification (Keller et al., 2005). Add-on chemical epigenetic modifiers that avert HDAC (Histone deacetylases) and DNMT (DNA methyltransferases) have also been uncovered to activate silenced gene clusters. Williams et al. (Williams et al., 2008) described the making of novel cladochromes and calphostin B by buildup of SAHA (suberoylanilide hydroxamic acid), an HDAC fungus inhibitor in *Cladosporium cladosporioides*, and a de novo combination of numerous lunalides and oxylipins by the addition of DNMT inhibitor 5-azacytidine in Diatrype sp.. Henrikson et al. (Henrikson et al., 2009) established that SAHA can provoke the production of nygerone-A in *A. niger* cultures (Shwab et al., 2007). Lately, Vasanthakumari et al. (Vasanthakumari et al., 2015) described the successful repair of production of camptothecin in a weakened endophytic fungus through DNMT inhibitor 5-azacytidine action (Deepika et al., 2016).

1.10.5 Proteomic approach

Proteome examination has also helped in determining natural forms and their particular paths. Proteomic Investigation of Secondary Metabolism (PrISM), a mass spectrometry-built method, has been used to examine for novel metabolites in Streptomyces and Bacillus by synchronized discovery of peptides and polyketides encoded by NRPS and PKS individually (Brakhage and Schroeckh, 2011; Deepika et al., 2016).

1.10.6 Genome mining

Genome mining has turned out to be a promising way toward detection of natural compounds and consideration of cryptic synthetic clusters because of the wealth of data imminent from genome arrangement analysis information. Using bioinformatics techniques and variations in culture constraints, Scherlach and Hertweck (Scherlach and Hertweck, 2006) revealed new metabolites aspoquinolones A–D. Bioinformatics has also extended the horizons to permit an accurate estimate of the key players in alleyways involving NRPS or PKS as well as consideration of the target substrate and physicochemical possessions of the termination product (Brakhage and Schroeckh, 2011). A genomisotopic approach combining bioinformatic examination of genome arrangements and isotope-directed fractionation was used to identify possible end products of orphan gene clusters. Zerikly and Challis (Zerikly and Challis, 2009) recommended an in vitro reconstitution method for substrates estimate in cryptic synthetic paths by incubating a recombinant biosynthetic catalyst from gene clusters with foretold substrates to recognize protein products. These approaches involve less genetic handling and aids in differentiating foreseen SMs from those mixtures bearing comparable physiochemical possessions (Deepika et al., 2016).

1.10.7 Combined omics approach

The initiation of more influential metabolomics methods coupled with genomics examination has permitted the detection of new natural compounds and the examination of their position in nature, thus accelerating natural compounds detection. Metabologenomics is involved in uniting genome sequencing and automatic gene cluster forecast with mass spectrometry-founded metabolomics (Goering et al., 2016). Genomic information is used to cross-examine the chemical information and vice versa to expose families of particles with phenotypes of attention in large strain assortments. Using this method, the joint genomics-metabolics summary of the aquatic strain *Streptomyces* sp. MP131–18 led to the documentation of new biologically dynamic compounds, for example, the bisindole pyrroles spiroindimicins E and F and two novel a-pyrone lagunapyrone (Paulus et al., 2017). To recognize novel phosphonic acids, a large-scale genome mining investigation and screening for exceptional phosphoenolpyruvate mutase gene (pepM) was achieved on 10,000 actinomycetes by Metcalf and his colleagues (Ju et al., 2015). In 278 strains, 64 dissimilar groups of phosphonate synthetic gene groups were identified. By description of strains inside these clusters, a new archetypical path for phosphonate biosynthesis and 11 formerly unexplained phosphonic acid usual products were exposed. A widespread examination of lanthipeptide-associated biosynthetic gene groups in *Actinobacteria* discovered that lanthipeptide synthetases can form natural compounds other than lanthipeptides. Actually, MS data and genomics advocate cross talk amongst lanthipeptide synthetic enzymes and NRPS and PKS systems and thus natural compounds with new scaffolds. A multiomics method was also effectively applied to recognize how BGCs are exploited in microbe-animal symbiosis in a mutually marine and terrestrial ecosystem (Palazzotto and Weber, 2018).

1.11 Perspectives and conclusions

Despite the accretion of genomic statistics, the application of genetically planned endophytes for the formation of SMs with many biological actions is not very extended. Many possible explanations are connected with the intrinsic complications to manipulate these endophytes, due to their high genomic variability and resistance to be altered with exogenous DNA. The original genome editing practices grounded on programmable nucleases will develop an essential way for the utilization of the synthetic biology ethics to secondary metabolism, avoiding the difficulties detected in the genetic handling of endophytes producing biologically natural secondary metabolites. Targeted genomic editing through programmable nucleases suggest many potentials ranging from the formation of modest gene knockouts to the advancing genetics efficient screenings achieved in a high-throughput way and epigenomic editing. Numerous platforms for the usage of genome editing practices are presently under development, specifically in secondary metabolite formations like endophytic fungi, which will be an important basis for the impending application of genomic technique to the encounter of biologically active new metabolites, refining the currently prevailing ones and serving the rational strategy of original ones with amplified or novel biological actions.

Acknowledgments

The authors are grateful for the financial support under Distinguished High-Level Talents Research Grant from a Guizhou Science and Technology Corporation Platform Talents Fund (Grant No.: [2017]5733-001 and CK-1130-002) and Zunyi Medical University, Zunyi, China for their advanced research facilities. We are also thankful to our key laboratory colleagues and research staff members for their constructive advice and help.

References

Araujo, R., Maia, B.S., 2018. Fungal genomes and genotyping. In: Sariaslani, S., Gadd, G.M. (Eds.), Advances in Applied Microbiology. In: vol. 102. Academic Press, pp. 37–81. https://doi.org/10.1016/bs.aambs.2017.10.003 Chapter Two.

Atanasova, L., Le Crom, S., Gruber, S., Coulpier, F., SeidlSeiboth, V., Kubicek, C.P., Druzhinina, I.S., 2013. Comparative transcriptomics reveals different strategies of *Trichoderma* mycoparasitism. BMC Genomics 14, 121.

Bansal, R., Mukherjee, P., 2016a. Identification of novel gene clusters for secondary metabolism in *Trichoderma* genomes. Microbiology 85 (2), 185–190.

Bansal, R., Mukherjee, P., 2016b. The terpenoid biosynthesis toolkit of *Trichoderma*. Nat. Prod. Commun. 11, 431–434.

Barrangou, R., Fremaux, C., Deveau, H., Richards, M., Boyaval, P., Moineau, S., Romero, D.A., Horvath, P., 2007. CRISPR provides acquired resistance against viruses in prokaryotes. Science 315, 1709–1712.

Bayram, O., Krappmann, S., Ni, M., Bok, J.W., Helmstaedt, K., Valerius, O., et al., 2008. VelB/VeA/LaeA complex coordinates light signal with fungal development and secondary metabolism. Science 320, 1504–1506.

Bergmann, S., Schumann, J., Scherlach, K., Lange, C., Brakhage, A.A., Hertweck, C., 2007. Genomics-driven discovery of PKS-NRPS hybrid metabolites from *Aspergillus nidulans*. Nat. Chem. Biol. 3, 213–217.

Bergmann, S., Funk, A.N., Scherlach, K., Schroeckh, V., Shelest, E., Horn, U., et al., 2010. Activation of a silent fungal polyketide biosynthesis pathway through regulatory cross talk with a cryptic nonribosomal peptide synthetase gene cluster. Appl. Environ. Microbiol. 76, 8143–8149.

Boettcher, M., McManus, M.T., 2015. Choosing the right tool for the job: RNAi, TALEN, or CRISPR. Mol. Cell 58, 575–585.

Brakhage, A.A., Schroeckh, V., 2011. Fungal secondary metabolites–strategies to activate silent gene clusters. Fungal Genet. Biol. 48, 15–22.

Burmester, A., Shelest, E., Glockner, G., Heddergott, C., Schindler, S., Staib, P., et al., 2011. Comparative and functional genomics provide insights into the pathogenicity of dermatophytic fungi. Genome Biol. 12, R7.

Campbell, M.A., Rokas, A., Slot, J.C., 2012. Horizontal transfer and death of a fungal secondary metabolic gene cluster. Genome Biol. Evol. 4 (3), 289–293.

Chen, Y.P., Yuan, G.F., Hsieh, S.Y., Lin, Y.S., Wang, W.Y., Liaw, L.L., et al., 2010. Identification of the mokH gene encoding transcription factor for the upregulation of monacolin K biosynthesis in *Monascus pilosus*. J. Agric. Food Chem. 58, 287–293.

Chiang, Y.M., et al., 2008a. Molecular genetic mining of the *Aspergillus* secondary metabolome: Discovery of the emericellamide biosynthetic pathway. Chem. Biol. 15, 527–532.

Chiang, Y.M., Szewczyk, E., Nayak, T., Sanchez, J.F., Lo, H., Simityan, H., et al., 2008b. Discovery of the emericellamide gene cluster: An efficient gene targeting system for the genomic mining of natural products in *Aspergillus nidulans*. Chem. Biol. 15, 527–532.

Cho, S.W., Kim, S., Kim, Y., Kweon, J., Kim, H.S., Bae, S., Kim, J.S., 2014. Analysis of off-target effects of CRISPR/Cas-derived RNA-guided endonucleases and nickases. Genome Res. 24, 132–141.

Chooi, Y.H., Cacho, R., Tang, Y., 2010. Identification of the viridicatumtoxin and griseofulvin gene clusters from *Penicillium aethiopicum*. Chem. Biol. 17 (5), 483–494.

Crawford, J.M., Townsend, C.A., 2010. New insights into the formation of fungal aromatic polyketides. Nat. Rev. Microbiol. 8, 879–889. https://doi.org/10.1038/nrmicro2465.

Crawford, J.M., Dancy, B.C.R., Hill, E.A., Udwary, D.W., Townsend, C.A., 2006. Identification of a starter unit acyl-carrier protein transacylase domain in an iterative type I polyketide synthase. Proc. Natl. Acad. Sci. U. S. A. 103, 16728–16733. https://doi.org/10.1073/pnas.0604112103.

Daley, D.K., Brown, K.J., Badal, S., 2017. Chapter 20. Fungal metabolites. In: Badal, S., Delgoda, R. (Eds.), Pharmacognosy. Academic Press, pp. 413–421. https://doi.org/10.1016/B978-0-12-802104-0.00020-2.

de Vries, R.P., Riley, R., Wiebenga, A., Osorio, A.G., Amillis, S., Uchima, C.A., et al., 2017. Comparative genomics reveals high biological diversity and specific adaptations in the industrially and medically important fungal genus *Aspergillus*. Genome Biol. 18, 28.

Deepika, V.B., Murali, T.S., Satyamoorthy, K., 2016. Modulation of genetic clusters for synthesis of bioactive molecules in fungal endophytes: a review. Microbiol. Res. 182, 125–140. https://doi.org/10.1016/j.micres.2015.10.009.

Evers, B., Jastrzebski, K., Heijmans, J.P., Grernrum, W., Beijersbergen, R.L., Bernards, R., 2016. CRISPR knockout screening outperforms shRNA and CRISPRi in identifying essential genes. Nat. Biotechnol. 34, 631–633.

Gardiner, D.M., Howlett, B.J., 2005. Bioinformatic and expression analysis of the putative gliotoxin biosynthetic gene cluster of *Aspergillus fumigatus*. FEMS Microbiol. Lett. 248, 241–248.

Gilbert, L.A., Horlbeck, M.A., Adamson, B., Villalta, J.E., Chen, Y., Whitehead, E.H., Guimaraes, C., Panning, B., Ploegh, H.L., Bassik, M.C., Qi, L.S., Kampmann, M., Weissman, J.S., 2014. Genome-scale CRISPR-mediated control of gene repression and activation. Cell 159, 647–661.

Giles, N.H., 1978. The organization, function, and evolution of gene clusters in eukaryotes. Am. Nat. 112 (986), 641–657.

Goering, A.W., McClure, R.A., Doroghazi, J.R., Albright, J.C., Haverland, N.A., Zhang, Y., Ju, K.S., Thomson, R.J., Metcalf, W.W., Kelleher, N.L., 2016. Metabologenomics: correlation of microbial gene clusters with metabolites drives discovery of a nonribosomal peptide with an unusual amino acid monomer. ACS Cent. Sci. 2, 99–108.

Griffiths, S., Mesarich, C.H., Saccomanno, B., Vaisberg, A., De Wit, P.J.G.M., Cox, R., Collemare, J., 2016. Elucidation of cladofulvin biosynthesis reveals a cytochrome P450 monooxygenase required for anthraquinone dimerization. Proc. Natl. Acad. Sci. U. S. A. 113, 6851–6856. https://doi.org/10.1073/pnas.1603528113.

Hall, C., Dietrich, F.S., 2007. The reacquisition of biotin prototrophy in *Saccharomyces cerevisiae* involved horizontal gene transfer, gene duplication and gene clustering. Genetics 177, 2293–2307.

Henrikson, J.C., Hoover, A.R., Joyner, P.M., Cichewicz, R.H., 2009. A chemical epigenetics approach for engineering the in situ biosynthesis of a cryptic natural product from *Aspergillus niger*. Org. Biomol. Chem. 7, 435–438.

Hiruma, K., Kobae, Y., Toju, H., 2018. Beneficial associations between *Brassicaceae* plants and fungal endophytes under nutrient-limiting conditions: evolutionary origins and host–symbiont molecular mechanisms. Curr. Opin. Plant Biol. 44, 145–154. https://doi.org/10.1016/j.pbi.2018.04.009.

Housden, B.E., Perrimon, N., 2016. Comparing CRISPR and RNAi-based screening technologies. Nat. Biotechnol. 34, 621–623.

Inglis, D.O., Binkley, J., Skrzypek, M.S., Arnaud, M.B., Cerqueira, G.C., Shah, P., Wymore, F., Wortman, J.R., Sherlock, G., 2013. Comprehensive annotation of secondary metabolite biosynthetic genes and gene clusters of *Aspergillus nidulans, A. fumigatus, A. niger* and *A. oryzae*. BMC Microbiol. 13, 91.

Jahn, L., Schafhauser, T., Wibberg, D., Rückert, C., Winkler, A., Kulik, A., Weber, T., Flor, L., van Pée, K.H., Kalinowski, J., Müller, J.L., Wohlleben, W., 2017. Linking secondary metabolites to biosynthesis genes in the fungal endophyte *Cyanodermella asteris*: the anti-cancer bisanthraquinone skyrin. J. Biotechnol. 257, 233–239. https://doi.org/10.1016/j.jbiotec.2017.06.410.

Jeffries, T.W., van Vleet, J.R.H., 2009. Pichia stipitis genomics, transcriptomics, and gene clusters. FEMS Yeast Res. 9, 793–807.

Jogaiah, S., Shetty, H.S., Ito, S-i., Tran, L.-S.P., 2016. Enhancement of downy mildew disease resistance in pearl millet by the G_app7 bioactive compound produced by Ganoderma applanatum. Plant Physiol. Biochem. 105, 109–117.

Jogaiah, S., Mostafa, A., Tran, L.S.P., Shin-ichi, I., 2013. Characterization of rhizosphere fungi that mediate resistance in tomato against bacterial wilt disease. J. Exp. Bot. 64, 3829–3842.

Jogaiah, S., Abdelrahman, M., Tran, L.-S.P., Ito, S.-I., 2018. Different mechanisms of *Trichoderma virens*-mediated resistance in tomato against *Fusarium* wilt involve the jasmonic and salicylic acid pathways. Mol. Plant Pathol. 19, 870–882.

Ju, K.S., Gao, J., Doroghazi, J.R., Wang, K.K., Thibodeaux, C.J., Li, S., Metzger, E., Fudala, J., Su, J., Zhang, J.K., et al., 2015. Discovery of phosphonic acid natural products by mining the genomes of 10,000 actinomycetes. Proc. Natl. Acad. Sci. U. S. A. 112, 12175–12180.

Keller, N.P., Turner, G., Bennett, J.W., 2005. Fungal secondary metabolism from biochemistry to genomics. Nat. Rev. Microbiol. 3, 937–947.

Khaldi, N., Seifuddin, F.T., Turner, G., Haft, D., Nierman, W.C., Wolfe, K.H., Fedorova, N.D., 2010. SMURF: genomic mapping of fungal secondary metabolite clusters. Fungal Genet. Biol. 47 (9), 736–741. https://doi.org/10.1016/j.fgb.2010.06.003.

Khater, S., Anand, S., Mohanty, D., 2016. In silico methods for linking genes and secondary metabolites: the way forward. Synth. Syst. Biotechnol. 1 (2), 80–88. https://doi.org/10.1016/j.synbio.2016.03.001.

Kim, H., Kim, J.S., 2014. A guide to genome engineering with programmable nucleases. Nat. Rev. Genet. 15, 321–334.

Kim, E., Moore, B.S., Yoon, Y.J., 2015. Reinvigorating natural product combinatorial biosynthesis with synthetic biology. Nat. Chem. Biol. 11, 649–659.

Kubicek, C.P., Herrera-Estrella, A., Seidl-Seiboth, V., Martinez, D.A., Druzhinina, I.S., Thon, M., Zeilinger, S., CasasFlores, S., Horwitz, B.A., Mukherjee, P.K., Mukherjee, M., Kredics, L., Alcaraz, L.D., Aerts, A., Antal, Z., Atanasova, L., Cervantes-Badillo, M.G., Challacombe, J., Chertkov, O., McCluskey, K., Coulpier, F., Deshpande, N., von Dohren, H., Ebbole, D.J., Esquivel-Naranjo, E.U., Fekete, E., Flipphi, M., Glaser, F., Gomez-Rodriguez, E.Y., Gruber, S., Han, C., Henrissat, B., Hermosa, R., Hernandez-Onate, M., Karaffa, L., Kosti, I., Le Crom, S., Lindquist, E., Lucas, S., Lubeck, M., Lubeck, P.S., Margeot, A., Metz, B., Misra, M., Nevalainen, H., Omann, M., Packer, N., Perrone, G., Uresti-Rivera, E.E., Salamov, A., Schmoll, M., Seiboth, B., Shapiro, H., Sukno, S., Tamayo-Ramos, J.A., Tisch, D., Wiest, A., Wilkinson, H.H., Zhang, M., Coutinho, P.M., Kenerley, C.M., Monte, E., Baker, S.E., Grigoriev, I.V., 2011. Comparative genome sequence analysis underscores mycoparasitism as the ancestral life style of *Trichoderma*. Genome Biol. 12, R40.

Leitão, A.L., Enguita, F.J., 2014. Fungal extrolites as a new source for therapeutic compounds and as building blocks for applications in synthetic biology. Microbiol. Res. 169 (9–10), 652–665. https://doi.org/10.1016/j.micres.2014.02.007.

Leitão, A.L., Costa, M.C., Enguita, F.J., 2017. Applications of genome editing by programmable nucleases to the metabolic engineering of secondary metabolites. J. Biotechnol. 241, 50–60. https://doi.org/10.1016/j.jbiotec.2016.11.009.

Lewis, K., 2013. Platforms for antibiotic discovery. Nat. Rev. Drug Discov. 12, 371–387.

Li, N., Alfiky, A., Vaughan, M.M., Kang, S., 2016. Stop and smell the fungi: fungal volatile metabolites are overlooked signals involved in fungal interaction with plants. Fungal Biol. Rev. 30 (3), 134–144. https://doi.org/10.1016/j.fbr.2016.06.004.

Lo, H.C., Entwistle, R., Guo, C.J., Ahuja, M., Szewczyk, E., Hung, J.H., et al., 2012. Two separate gene clusters encode the biosynthetic pathway for the meroterpenoids austinol and dehydroaustinol in *Aspergillus nidulans*. J. Am. Chem. Soc. 134, 4709–4720.

Lv, L., Ren, Y.L., Chen, J.C., Wu, Q., Chen, G.Q., 2015. Application of CRISPRi for prokaryotic metabolic engineering involving multiple genes, a case study: controllable P(3HB-co-4HB) biosynthesis. Metab. Eng. 29, 160–168.

Malmierca, M.G., McCormick, S.P., Cardoza, R.E., Alexander, N.J., Monte, E., Gutierrez, S., 2015. Production of trichodiene by *Trichoderma harzianum* alters the perception of this biocontrol strain by plants and antagonized fungi. Environ. Microbiol. 17, 2628–2646.

Marsden, A.F., Wilkinson, B., Cortes, J., Dunster, N.J., Staunton, J., Leadlay, P.F., 1998. Engineering broader specificity into an antibiotic-producing polyketide synthase. Science 279, 199–202.

Mojica, F.J., Diez-Villasenor, C., Garcia-Martinez, J., Almendros, C., 2009. Short motif sequences determine the targets of the prokaryotic CRISPR defence system. Microbiology 155, 733–740.

Montiel, D., Kang, H.S., Chang, F.Y., Charlop-Powers, Z., Brady, S.F., 2015. Yeast homologous recombination-based promoter engineering for the activation of silent natural product biosynthetic gene clusters. Proc. Natl. Acad. Sci. U. S. A. 112, 8953–8958.

Nakkeeran, S., Vinodkumar, S., Renukadevi, P., Rajamanickam, S., Jogaiah, S., 2019. Bioactive molecules from *Bacillus* spp.: an effective tool for plant stress management. In: Jogaiah, S., Abdelrahman, M. (Eds.), Bioactive Molecules in Plant Defense. Springer, Berlin, pp. 1–23.

Nielsen, J.C., Nielsen, J., 2017. Development of fungal cell factories for the production of secondary metabolites: linking genomics and metabolism. Synth. Syst. Biotechnol. 2 (1), 5–12. https://doi.org/10.1016/j.synbio.2017.02.002.

Nützmann, H.W., Schroeckh, V., Brakhage, A.A., 2012. Chapter 16. Regulatory cross talk and microbial induction of fungal secondary metabolite gene clusters. In: Hopwood, D.A.

(Ed.), Methods in Enzymology. In: vol. 517. Academic Press, pp. 325–341. https://doi.org/10.1016/B978-0-12-404634-4.00016-4.

Palazzotto, E., Weber, T., 2018. Omics and multi-omics approaches to study the biosynthesis of secondary metabolites in microorganisms. Curr. Opin. Microbiol. 45, 109–116. https://doi.org/10.1016/j.mib.2018.03.004.

Paulus, C., Rebets, Y., Tokovenko, B., Nadmid, S., Terekhova, L.P., Myronovskyi, M., Zotchev, S.B., Rückert, C., Braig, S., Zahler, S., et al., 2017. New natural products identified by combined genomicsmetabolomics profiling of marine *Streptomyces* sp. MP131–18. Sci. Rep. 7, 42382.

Rutherford, K., Van Duyne, G.D., 2014. The ins and outs of serine integrase site-specific recombination. Curr. Opin. Struct. Biol. 24, 125–131.

Scherlach, K., Hertweck, C., 2006. Discovery of aspoquinolones A–D prenylated quinoline-2-one alkaloids from *Aspergillus nidulans* motivated by genome mining. Org. Biomol. Chem. 4, 3517–3520.

Segal, D.J., Beerli, R.R., Blancafort, P., Dreier, B., Effertz, K., Huber, A., Koksch, B., Lund, C.V., Magnenat, L., Valente, D., Barbas 3rd, C.F., 2003. Evaluation of a modular strategy for the construction of novel polydactyl zinc finger DNA-binding proteins. Biochemistry 42, 2137–2148.

Shah, A.N., Davey, C.F., Whitebirch, A.C., Miller, A.C., Moens, C.B., 2015. Rapid reverse genetic screening using CRISPR in zebrafish. Nat. Methods 12, 535–540.

Shalem, O., Sanjana, N.E., Zhang, F., 2015. High-throughput functional genomics using CRISPR-Cas9. Nat. Rev. Genet. 16, 299–311.

Shwab, E.K., Bok, J.W., Tribus, M., Galehr, J., Graessle, S., Keller, N.P., 2007. Histone deacetylase activity regulates chemical diversity in *Aspergillus*. Eukaryot. Cell 6, 1656–1664.

Slot, J.C., 2017. Fungal gene cluster diversity and evolution. Chapter Four. In: Townsend, J.P., Wang, Z. (Eds.), Advances in Genetics. In: vol. 100. Academic Press, pp. 141–178. https://doi.org/10.1016/bs.adgen.2017.09.005.

Strieker, M., Tanovic, A., Marahiel, M.A., 2010. Nonribosomal peptide synthetases: structures and dynamics. Curr. Opin. Struct. Biol. 20, 234–240.

Sun, X.P., Ruan, R.X., Lin, L.Y., Zhu, C.Y., Zhang, T.Y., Wang, M.S., Li, H.Y., Yu, D.L., 2013. Genome wide investigation into DNA elements and ABC transporters involved in imazalil resistance in *Penicillium digitatum*. FEMS Microbiol. Lett. 348 (1), 11–18.

Teixeira, M.M., Moreno, L.F., Stielow, B.J., Muszewska, A., Hainaut, M., Gonzaga, L., et al., 2017. Exploring the genomic diversity of black yeasts and relatives (*Chaetothyriales, Ascomycota*). Stud. Mycol. 86, 1–28.

Unsold, I.A., Li, S.M., 2006. Reverse prenyltransferase in the biosynthesis of fumigaclavine C in *Aspergillus fumigatus*: gene expression, purification, and characterization of fumigaclavine C synthase FGAPT1. ChemBioChem 7, 158–164.

van Dijk, J.W.A., Wang, C.C.C., 2016. Heterologous expression of fungal secondary metabolite pathways in the *Aspergillus nidulans* host system. Chapter Six. In: O'Connor, S.E. (Ed.), Methods in Enzymology. In: vol. 575. Academic Press, pp. 127–142. https://doi.org/10.1016/bs.mie.2016.02.021.

Vasanthakumari, M.M., Jadhav, S.S., Sachin, N., Vinod, G., Shweta, S., Manjunatha, B.L., et al., 2015. Restoration of camptothecine production in attenuated endophytic fungus on re-inoculation into host plant and treatment with DNA methyltransferase inhibitor. World J. Microbiol. Biotechnol. 31, 1629–1639.

Walton, J.D., 2000. Horizontal gene transfer and the evolution of secondary metabolite gene clusters in fungi: an hypothesis. Fungal Genet. Biol. 30, 167–171.

Williams, R.B., Henrikson, J.C., Hoover, A.R., Lee, A.E., Cichewicz, R.H., 2008. Epigenetic remodeling of the fungal secondary metabolome. Org. Biomol. Chem. 6, 1895–1897.

Wu, S., Chappell, J., 2008. Metabolic engineering of natural products in plants; tools of the trade and challenges for the future. Curr. Opin. Biotechnol. 19, 145–152.

Yu, J.J., Chang, P.K., Ehrlich, K.C., Cary, J.W., Bhatnagar, D., Cleveland, T.E., Payne, G.A., Linz, J.E., Woloshuk, C.P., Bennett, J.W., 2004. Clustered pathway genes in aflatoxin biosynthesis. Appl. Environ. Microbiol. 70, 1253–1262. https://doi.org/10.1128/Aem.70.3.

Zeilinger, S., Gruber, S., Bansal, R., Mukherjee, P.K., 2016. Secondary metabolism in *Trichoderma*—chemistry meets genomics. Fungal Biol. Rev. 30 (2), 74–90. https://doi.org/10.1016/j.fbr.2016.05.001.

Zerikly, M., Challis, G.L., 2009. Strategies for the discovery of new natural products by genome mining. ChemBioChem 10, 625–633.

Zhu, F., Gamboa, M., Farruggio, A.P., Hippenmeyer, S., Tasic, B., Schule, B., Chen-Tsai, Y., Calos, M.P., 2014. DICE, an efficient system for iterative genomic editing in human pluripotent stem cells. Nucleic Acids Res. 42, 34.

Zhu, C., Sheng, D., Wu, X., Wang, M., Hu, X., Li, H., Yu, D., 2017. Identification of secondary metabolite biosynthetic gene clusters associated with the infection of citrus fruit by *Penicillium digitatum.* Postharvest Biol. Technol. 134, 17–21. https://doi.org/10.1016/j.postharvbio.2017.07.011.

Impact of potassium solubilizing fungi as biopesticides and its role in crop improvement

2

Mahantesh Kurjogi[a], *K.N. Basavesha*[b], *and V.P. Savalgi*[b]
[a]Green Nanotechnology Laboratory, University of Agricultural Sciences, Dharwad, Karnataka, India, [b]Department of Agricultural Microbiology, College of Agriculture, University of Agricultural Sciences, Dharwad, Karnataka, India

2.1 Introduction

Fertilizers are defined as the important products used for enhancing the levels of available plant nutrients in soil, thereby directly influencing the quality and quantity of crop production. Knowledge of soil fertility is time immemorial, and it is mentioned that crop cultivation was introduced in the fertile lands of Indus valley civilization. Egyptians recognized the fertile soil on the bank of Nile River, Babylonians acknowledged the advantages of manure in their fields, and Romans were familiar with the importance of legumes in the maintenance of soil fertility. This antique knowledge led to the discovery of modern agriculture which significantly increased the production of agricultural products that enable us to feed the world population. However, the current problem in agriculture is compounded by the indiscriminate use of synthetic fertilizer leading to the widening of nitrogen, phosphorus, and potassium (NPK) ration which adversely affect the soil fertility and also natural microflora. Nitrogen is one of the major nutrients for plant growth but the supply of nitrogen alone is not sufficient for the prosperous growth of the plants. Therefore, the management of balanced fertilizer with the optimum ration of NPK is inevitable for sustained agricultural production. Since India is a country of diverse agroclimatological conditions, the regional disparities in fertilizer use are also limiting the overall agricultural production. It is estimated that about 17.5 million tons of plant nutrients are annually being used in India and is likely to increase up to 45 million tons by 2025. Despite the application of NPK, a considerable decline in crop productivity was observed due to the incidence of soilborne pathogens and the emerging deficiencies of other nutrients required for plant growth. The results of long-term chemical fertilizer showed a decline in organic carbon and potassium level from the native soil, thus the production efficiency gone down considerably which indicates that potassium is the limiting factor for crop yield (Murugappan et al., 2002).

Biocontrol Agents and Secondary Metabolites. https://doi.org/10.1016/B978-0-12-822919-4.00002-8

2.2 Importance of soil potassium

Potassium (K) has been identified as one of the major essential nutrients for plant growth (Verma et al., 2018). Potassium is abundant in nature; where 98% of potassium is bound in mineral form and nonexchangeable phases which is not directly available for plants and only 2% of potassium could be availed by plants in the form of solution and exchangeable phases (Schroeder, 1979; Bertsch and Thomas, 1985). However, a dynamic process takes place between the number of factors like oxalic acids, citric acids, other organic, and inorganic colloids present in the soil convert potassium from its minerals and nonexchangeable form to soluble form (Song and Huang, 1971). The reaction in the release of potassium involves the replacement of cations such as Ca from the interlattice position of micaceous clay minerals.

2.3 Role of potassium in plants

Potassium plays an important role in the conversion of ammonium ions into amino acids and proteins, which are absorbed by roots from the soil. It helps in the transportation of carbohydrates and proteins from the leaves to other parts of the plants. It also regulates the permeability of the cellular membrane; it acts as a cofactor for the number of cell metabolism. Potassium is responsible for enhancing the resistance of crops to adverse conditions during abiotic stress like hot and dry climate and biotic stress like the incidence of insect pests and diseases (Maqsood et al., 2013; Pettigrew, 2008). It increases the stiffness of straw in cereals and therefore the lodging of cereals is reduced and improves the quality of fruits and grains (Goldstein, 1994). Deficiency of potassium in plants negatively affects the photosynthesis process and it leads to stunted and bushy growth; older leaves develop chlorosis between veins and light gray to brown coloration along with the leaf margin, which leads to the poor development of roots (Sheng and Huang, 2002).

2.4 Role of microorganisms in potassium solubility and uptake

The rhizosphere is defined as the region of the soil adhering to the roots of the plant. The microbial load in the rhizosphere region is always greater than that of the non-rhizosphere soil. On the one hand, the plant roots produce the number of organic compounds that attract many microbes and on the other hand microorganisms' secret various organic acids that enables the mobilization of nutrient and also protect host plant from various pathogens and harmful pest by biological control process (Aleksandrov et al., 1967).

Inorganic colloids in the form of chemical K were supplied at the right time for the appropriate growth of the plants; but less effort was made to understand the role of organic colloidal and soil microorganisms in the potassium availability and uptake.

However, a study conducted in 2003 by Saraswathy et al. observed that potassium availability at rhizosphere soil was higher than that of unplanted soil, further potassium level enhanced with the increased population of microbes in a rice field under acidic conditions (Saraswathy et al., 2003). Similarly, Kuchenbuch reported that when potassium content was depleted there was an increase in the potassium flux toward the root mainly through diffusion (Kuchenbuch, 1987). Hence increased microbial activities in the root zone, organic acid secretion by root, and dissolution effect of organic acids are the probable reasons for the mobilization of potassium in the root zone (Prajapati et al., 2012, 2013; Sangeeth et al., 2012). Therefore, it is very clear that the microbial population plays an important role in the establishment of plant and also remarkably contributes to the potassium mobility from the soil toward plant roots. Moreover, it is evident that good microbial population found in various biofertilizers and plant growth-promoting substances secreted from microbes contribute to the potassium nutrient cycling (Rajan et al., 1996). Although not much study was made so far in this direction, it is evident that incorporating beneficial microbes in the potassium cycle will lead to much potassium availability for plants. Intensive cultivation with the mere use of high-dose fertilizers without adequate organic manures depletes the micronutrient status of the soil as well. Further, it leads to a decline in organic carbon status of soil resulting in retarded soil biological activity thus minimizing the natural nitrogen fixation and nutrient solubilization or mobilization of potassium and micronutrients (Rawlings, 2002).

In recent years, practicing organic farming seems to provide the entire crop nutrients through biofertilizer and organic manures. The fixed and immobile phosphate has been made available to plants by the inoculation of phosphate solubilizing microorganisms and phosphate mobilizers (Chen et al., 2006). Nitrogen-fixing biofertilizers are capable of fixing atmospheric nitrogen for plants. In the case of potassium, a group of soil microbes acts on soil clay minerals like illite, muscovite, microcline, leucite etc. and releases potassium which can be utilized by the plants (Anthoni, 2002). The solubilization of rock potassium by microorganisms was first reported by Muentz (Muentz, 1890). Microorganisms such as *Bacillus extorquens*, *Clostridium pasteurianum*, and *Aspergillus niger* were found to grow on biotite, muscovite, and mica (Sindhu et al., 2010). The number of potassium solubilizing fungi strains, viz., *Aspergillus* spp., *A. terreus*, *A. niger*, *Penicillium* sp. was also reported to solubilize inorganic and organic potassium and the release of structural potassium from rocks and minerals (Archana et al., 2013). *A. terreus* and *A. niger* were reported to release potassium when grown in liquid medium containing insoluble potassium; however, *A. terreus* shows the highest solubilization as well as acid production when supplemented with a different source of insoluble potassium (Verma et al., 2014). It was observed that the availability of other trace elements enhances the mobilization of potassium by fungi through the production of acids. Therefore, potassium solubilizing microorganisms present in the soil could provide an alternative source to make potassium available for the uptake by plants (Rogers et al., 1998). Thus, the identification of fungal strains capable of solubilizing potassium minerals quickly can conserve our existing resources and avoid environmental pollution hazards caused by the heavy application of chemical fertilizers. Furthermore, beneficial functions of potassium solubilizing fungi which includes the formation of essential soil humus for plant growth by decomposing organic residues, production of novel compounds, synthesis of plant growth-promoting compounds, increasing root surface area for nutrition absorption, improving soil aggregation by releasing binding agents such as glomalin, enhancing soil aeration, and water infiltration. In addition, potassium solubilizing fungi acts as a potential biological control agent against harmful insects, plant pathogens, and weeds. Furthermore, it also induces a defense mechanism in host plants during abiotic stress conditions.

2.5 Role of potassium solubilizing fungi as biofertilizer

Another growing concern in the agriculture sector is the indiscriminate application of chemical fertilizers and pesticides for crop improvement and disease management which not only leads to the emergence of resistant pests but also hazardous to the environment and human health. Thus, the control of agriculturally important plant pathogens without disturbing the ecosystem is of paramount importance. Therefore, researchers are urged to explore the feasibility of tapping the natural reserve of soil to sustain soil fertility and crop productivity by augmenting the biodissolution of nutrients in the soil. Potassium solubilizing fungi could serve as inoculants that mobilize potassium in the soil by converting insoluble potassium into a soluble form. This is a promising strategy for the improvement of plant absorption of potassium and so reducing the use of the costly and hazardous chemicals. Potassium biofertilizers in agriculture play a major role in improving soil fertility and yield attributing characters (Epstein and Bloom, 2008). In addition, their application in soil improves soil biota and minimizes the sole use of chemical fertilizers. Therefore, the inoculation of potassium solubilizing fungi in the soil is inevitable for integrated crop improvement in various crops.

2.6 Role of potassium solubilizing fungi as biopesticide/ biocontrol agent

In agriculture, the management of plant disease is economically important for ensuring high quality and quantity of food and commercial crops. Till date, several approaches have been made to mitigate the challenges in the agriculture sector, but farmers rely heavily on chemical pesticides which contributed significantly to the spectacular improvements in crop productivity and quality over the past many years. However, the excessive use of chemicals and pesticides in agricultural crops is being deplored by the scientific community because of the fact of emergence of resistant new pests and pathogens which are responsible for environmental pollution (Damalas, 2009; Carvalho, 2017; Damalas and Koutroubas, 2018). This led to considerable changes in people's attitudes toward the use of pesticides in agriculture (Pelaez and Mizukawa, 2017). Consequently, researchers have focused their efforts on developing substitute inputs to the synthetic chemical for controlling phytopathogens. The cultural practices, developing resistant varieties and biological control methods were suggested alternative techniques to get rid of the synthetic chemicals. However, breeding resistant varieties and cultural methods have several limitations which failed to cope with the emerging novel virulent pathogens. Therefore, the use of a biological control agent is the most preferred alternative method for the management of plant disease.

Biocontrol research has progressed rapidly in the last few decades, and a variety of biological control agents (BCAs) are available for use, but further development and effective adoption will require a greater understanding of the complex interactions between plants, pathogen, and environment. Biocontrol is a common term used in many allied fields of agriculture especially in entomology and plant pathology. Entomologist defines biocontrol agent as a live predatory which can suppress the population of different pest insect. In plant pathology, the term applies to the use of microbial antagonists to suppress disease as well as the use of host-specific pathogens to control weed populations. In both the fields, the organism that suppresses the pest or pathogen is referred to as the biological control agent.

Biocontrol of soilborne disease is particularly complex because the development of these diseases takes place in a very dynamic environment at the interface of pathogen, root, and rhizosphere soil. The rhizosphere is characterized by intense microbial activity and their interactions with plants. Plants release metabolically active cells from their roots and deposit much of the carbon allocated to roots in the rhizosphere, suggesting a highly evolved relationships between the plant and rhizosphere microorganisms. The use of this existing plant-microbe interaction for plant growth and as a biocontrol mechanism reduces the use of synthetic fertilizer and pesticide leading to cost-effective and ecofriendly treatment (Boddey et al., 2003; Whipps and Gerhardson, 2007; Elmer and Reglinski, 2006). The complexity of the root-soil interface must be accommodated in the study of biocontrol, which must involve whole organisms and ultimately entire communities, and thus it helps us to understand the essential interactions of soil in the field. This chapter aims at enhancing our understanding of the biology of the interactions that result in disease suppression. Thus, it is evident to use the microorganisms in the management of plant disease in agroecosystems.

2.7 Biocontrol agents

Conventional methods for crop disease management have been largely based on the use of chemical pesticides. However, due to adverse effect, reduction, or elimination of these chemical pesticides in agriculture is highly desirable. One of the most promising means to achieve this goal is by the use of new tools based on biocontrol agents for disease control (Chet and Inbar, 1994; Harman and Kubicek, 1998). To date numerous studies have demonstrated the variety of microbial biocontrol agents and also the number of BCAs have been registered, which are available as commercial products but no comprehensive investigations were done regarding the role of potassium solubilizing fungal species as a biocontrol agent. Therefore, this chapter focuses on the mechanism of potassium solubilizing fungal species as a biocontrol agent.

2.7.1 Penicillium

Penicillium is an ascomycetous fungus found in all-natural environments. *Penicillium* is an opportunistic invader, a fast-growing fungus, a strong spore producer, a source of several enzymes, and economically important fungi in the food and drug industry. Numerous strains of this genus are rhizosphere competent and are able to degrade hydrocarbons, chlorophenolic compounds, polysaccharides, and the xenobiotic pesticides used in agriculture (Singh and Walker, 2006; Hofrichter et al., 1994; Boonchan et al., 2000). Certain strains of penicillium are also used as a potential biocontrol agent in plant disease management. A report published in 2015 indicating the use of *Penicillium adametzioides*, as a potential biocontrol agent for ochratoxin-producing fungus in grapes (Ahmed et al., 2015). The greatest advantage of biological control in pre and postharvest can be achieved by the artificial introduction of a large number of natural microflora with known antagonists (Wisniewski and Wilson, 1992). Similarly, *P. griseofulvum* has been reported to be an important inhibitor of bacterial growth; the inhibition activity of penicillium is attributed with the production of griseofulvin and patulin (Jiménez et al., 1987). Likewise, the preliminary study also showed the ability of penicillium with the capacity to control harmful fungi and insects (Santamarina et al., 2002). Recent studies showed that *Penicillium citrinum* was used as a biocontrol agent to control the charcoal rot of sorghum and *Botrytis cinerea* in chickpea (Meesala and Subramaniam, 2016; Sreevidya et al., 2015). The antagonistic activity of *P. citrinum* may be associated with its capability to produce hydrolytic enzymes, siderophore, indole acetic acid (IAA), hydrocyanic acid (HCN), lipase, protease, and β-1,3-glucanase. Siderophores play significant roles in providing iron to plants and also help in disease suppression (Indiragandhi et al., 2008). IAA is known to stimulate seed germination, root formation, and elongation in plants (Ahemad and Kibret, 2014), whereas HCN was also reported to help in disease suppression (Haas et al., 1991). Similarly, lytic enzymes produced by several beneficial microbes not only help in lysis of pathogenic fungal cell walls but also play an important role in nutrient mineralization and thus promote the growth of the plants (Ekta and Archana, 2017; Compant et al., 2005). *P. citrinum* VFI-51 was also reported to induce defense mechanism to host plant during abiotic stress conditions such as high salinity, high pH, and high temperatures (Sreevidya et al., 2015). In another study, *Penicillium funiculosum* was used as a biocontrol agent against phytophthora root rot of azalea and citrus (Fang and Tsao, 1995). Bark infection on the stem of orange seedlings and lemon trees were also reported to be controlled by *Penicillium funiculosum* (Fang and Tsao, 1995a). Black-rot disease of *Allium cepa* L. (onion) caused by *Aspergillus niger* was controlled by *Penicillium* species isolated from the rhizosphere region (Khokhar et al., 2013). The results indicated that the mycoparasitism of *Alternaria*

Table 2.1 List of *Penicillium* species used as biocontrol agent.

Sl. No.	Biocontrol agent	Disease causing agent
1.	*P. adametzioides*	Ochratoxin-producing fungus
2.	*P. griseofulvum*	Bacterial pathogens
3.	*Penicillium* spp.	Harmful insects
4.	*P. citrinum*	Charcoal rot of sorghum and *Botrytis cinerea* in chickpea
5.	*P. funiculosum*	Phytophthora
6.	*Penicillium* species	*Aspergillus niger*
7.	*Penicillium* species	Bayoud disease
8.	*P. oxalicum*	Aphid *Ceratovacuna lanigera*
9.	*P. decumbens*	Fungicidal, bactericidal and insecticidal
10.	*P. oxalicum*	*Alternaria alternate*

alternata by *P. oxalicum* is a possible biological control agent for the important rice pathogens (Sempere and Santamarina, 2010). However, the molecular biology of interaction between pathogen and antagonist needs to be studied to determine the details of the underlying mechanism of their antagonistic nature and their ecosystem effects in cases of in vivo application (Table 2.1).

2.7.2 *Aspergillus*

Aspergillus includes a set of fungi that are generally considered asexual, although perfect forms have been reported. *Aspergillus* species are ubiquitous and have been observed in a broad range of habitats because they can colonize a wide variety of substrates. *Aspergillus* species are commonly found as a saprophyte growing on dead and decaying matter. The spores are widespread and are often associated with organic materials and soil. Rhizosphere associated with *Aspergillus niger* was found effective as a biocontrol agent against several diseases caused by *Fusarium oxysporum*, *Cicero*, *Macrophomina phaseolina*, *Pythium aphanidermatum*, and *Rhizoctonia solani* (Mondal et al., 2000; Mondal, 1998). Similarly, several *Aspergillus* species, isolated from indigenous soil, displayed significant antagonistic activity against *Fusarium oxysporum f.* sp. *lycopersici*, a causal agent of wilt disease in tomato crops. *A. niger*, *A. flavus*, *A. fumigatus*, and *A. tamarii* are known to inhibit the growth of *Fusarium oxysporum*. Apart from the antagonistic activity against pathogens, the biocontrol agents are also known to produce plant growth-promoting hormones (Kumar and Garampallin, 2014). Three potential *Aspergillus* species such as *A. fumigatus*, *A. repens*, and *A. niger* were used as biological control agents against *Phytophthora palmivora*, the pathogen of cocoa black pod disease (Adebola and Amadi, 1999). Likewise, several *Aspergillus* species isolated from native soil and compost are used as a biocontrol agent in controlling *Fusarium sambucinum* and *Phytophthora erythroseptica* that causes postharvest diseases in potatoes (Ben Abdallah et al., 2015). Studies also confirm that soil borne and compost-borne *Aspergillus* species significantly inhibited the pathogens such as *Pythium* species and *Fusarium* species causing potato leak and potato dry rot (Daami-Remadi et al., 2006; Aydi et al., 2013). Furthermore, several previous studies showed the potential antagonistic nature of *Aspergillus*, Barocio-Ceja et al. used *Aspergillus* for controlling several pathogens such as *F. oxysporum*, *F. subglutinans*, and Rhizoctonia (Barocio-Ceja et al., 2013). Similarly, *A. versicolor isolated from* naturally or solarized cruciferous residue-amended soils showed hyperparasitism against *F. oxysporum cumini* (Israel and Lodha, 2005). Furthermore, Adebola and Amadi isolated three *Aspergillus* species, namely *A. niger*, *A. fumigatus*, and *A. repens*, from the rhizosphere of black pod-infected cocoa trees and showed their potential ability to inhibit the growth of *Phytophthora palmivora* (Adebola and Amadi, 2010). Correspondingly *A. niger* isolated from the tomato field was successful in controlling the growth of *F. solani* and *F. oxysporum f.* sp. *lycopersici* (Dwivedi, 2013). In the same way, Tiwari et al. showed that *A. niger* was efficient in controlling various fungal species such as *Trametes*, *Stereum*, *Pycnoporus*, *Phellinus*, *Lenzites*, *Phellinus*, *Earliella Gloeophyllum*, *Flavodon*, and *Daedalea* (Tiwari et al., 2011). Further Venkatasubbaiah and Safeeulla showed that *A. niger* isolated from the rhizosphere of coffee seedlings, inhibited the growth of *Rhizoctonia solani* indicating hyperparasitism against this collar rot pathogen (Venkatasubbaih and Safeeulla, 1984). In addition, several soilborne pathogens such as *Macrophomina phaseolina*, *Pythium aphanidermatum*, and *Rhizoctonia solani* were able to inhibit the application of compost-borne *A. niger* as a biocontrol agent (Ramzan et al., 2014). Moreover, *A. niger* isolated from Chinese fermented soybean found to inhibit the spore formation of *A. flavus* and also degrade aflatoxin (Xu et al., 2013). For the first time, *Sclerotinia sclerotiorum* was successfully controlled by *A. terreus* in both laboratory and field conditions (Melo et al., 2006). Overall, the mycoparasite characteristics of *Aspergillus* spp. were attributed to its ability to produce the wide range of extracellular enzymes (Hu et al., 2013). In addition, the soil application of *A. niger* suppressed *R. solani* infection and also significantly increased eggplant yield (Khan and Anwer, 2007). The production of NH3, HCN, and siderophore may have contributed to the suppression of *R. solani*. Furthermore, Wang et al. indicated that the increased activity of defense-related enzymes in *A. flavipes* might be the part of the mechanism of controlling *P. capsici* (Wang et al., 2015). In other studies, significant increases in plant (*Brassica chinensis* Linn.) dry weight and N and P contents were observed with the addition of *A. niger* (Chuang et al., 2007). Tiwari et al. used *A. niger* as a potential biocontrol agent for inhibiting wood decay fungi (Tiwari et al., 2011). Thus, *Aspergillus* species may represent a potential source of biologically active compounds and constitute a promising alternative to manage these increasingly important diseases. Moreover, *Aspergillus* species have been reported as endophytes with antifungal activity (Zhang et al., 2008; Patil et al., 2015; Soltani and Hosseyni, 2015) and able to produce several metabolites such as phenolic and bioactive flavonoid compounds. *Aspergillus* spp. may be further

Table 2.2 List of *Aspergillus* species used as biological control agent.

Sl. No.	Biocontrol agent	Disease causing agent
1.	*A. niger*	*Fusarium oxysporum, cicero, macrophomina phaseolina, Pythium aphanidermatum and Rhizoctonia solani, Phytophthora palmivora, F. solani, Trametes, Stereum, Pycnoporus, Phellinus, Lenzites, Phellinus, Earliella Gloeophyllum, Flavodon, Daedalea, A. flavus* and wood decay fungi
2.	*A. flavus*	*Fusarium oxysporum*
3.	*A. fumigatus*	*Fusarium oxysporum, Phytophthora palmivora*
4.	*A. tamari*	*Fusarium oxysporum*
6.	*A. repens*	*Phytophthora palmivora*
7.	*A. versicolor*	*F. oxysporum cumini*
8.	*A. terreus*	*Sclerotinia sclerotiorum*
9.	*A. flavipes*	*P. capsici*

stimulated by soil amendments and fertilizers, which result in the release of volatile and nonvolatile metabolites that are toxic to several phytopathogens but can stimulate the growth and proliferation of *Aspergillus* species and other soilborne mycoparasites (Huang et al., 1997; Yang et al., 2011). Overall, the use of these bioagents is not only safe for the farmers and consumers but also ecofriendly, cost-effective, and easy to produce and apply the formulations (Dwivedi, 2013) (Table 2.2).

2.8 Mode of action

The mode of action of biological control agents against pathogens or harmful insects may be functioned in various mechanisms like competition, mycoparasitism, antibiosis, and induced resistance in host plants.

2.8.1 Competition

Nonpathogenic biocontrol agents and the pathogen compete for nutrients; space and infection site which results in control or inhibiting the growth of the pathogen. Competition for nutrients is a natural process and most important among all living organisms. It is generally believed that pathogenic and nonpathogenic microbial population competes for the colonization of rhizosphere. The siderophores produced by certain microorganisms deprive the pathogen of iron. Khan and Anwar demonstrated the role of siderophore in the biological control of *R. solani* infection by *A. niger*, which promotes the plant growth significantly by enhancing crop productivity (Khan and Anwer, 2007). In nonpathogenic *Aspergillus* spp., competition for space and nutrients is the main action involved directly between the tested antagonist and the pathogens (Ben Abdallah et al., 2015). Fang and Tsao reported the colonization of *penicillium funiculosum* on the stem of orange seedlings and lemon trees for the management of Bark infection (Fang and Tsao, 1995a) (Sempere and Santamarina, 2010).

2.8.2 Mycoparasitism

Mycoparasitism is an antagonism interaction in which one fungus attack and feed on other fungal strains. Mycoparasitism is a very common biological control process in soilborne pathogens. The first study of mycoparasites was reported with the discovery of *Trichoderma lignorum* parasitizing many soilborne fungi (Weindling, 1932). The term mycoparasitism is also generally used as hyperparasitism, direct parasitism, or interfungus parasitism. The mode of action involves several stages like the coiling of hyphae, penetration, and production of haustoria, and finally lysis of host hyphae. Different types of mycoparasitism have been identified such as hypoviruses, facultative parasites, an obligate bacterial pathogen, and predators. Antagonistic fungi which act as biological control agents penetrate the pathogens hyphae and produce lytic enzymes causing the death of the pathogen; this is an important and powerful tool for controlling the plant disease (Harman et al., 2004; Viterbo et al., 2007). A biological control agent releases the number of lytic enzymes such as chitinases, proteases, and beta-1,3-glucanases (Gajera and Vakharia, 2012). Hu et al. reported that the mycoparasite characteristics of *Aspergillus* sp. may be enhanced by its ability to produce extracellular enzymes, like chitinases and other antifungal extrolites which help the colonization of *S. sclerotiorum* sclerotia; as well as thought to be the mode of action of *A. terreus*. In fact, chitinases are commonly produced by several *Aspergillus* spp. (Hu et al., 2013) (Sempere and Santamarina, 2010).

2.8.3 Antibiotic production

It is a process in which biocontrol agent releases toxic metabolites or antibiotic which retards the growth of other microbes. Antibiotics released by biocontrol agents are highly toxic even at low concentrations they can kill the harmful insects or pathogens (Fravel, 1988). Till date thousands of antibiotic substances produced by bacteria and fungi are identified and mass culture for commercialization. Gomathi and Ambikapathy observed the zone of inhibition in the culture plate suggesting the production of antibiotic substances either by the pathogen against antagonistic fungi or vice versa (Gomathi and Ambikapathy, 2011). However, Khokhar et al. revealed that *Penicillium* species isolated in this study possessed a high antagonistic effect on the pathogen *A. niger* (Khokhar et al., 2013). *P. oxalicum* has been reported to be a pathogenic fungus of the aphid *Ceratovacuna lanigera*, an insect pest on sugarcane (Michereff et al., 1995). In addition, the mortality of insects increased when treated with *P. oxalicum* and also *P. decumbens* showed varied antagonistic activity like bactericidal, fungicidal, and insecticidal (Santamarina et al., 2002). The antibiotic activity of the biological agent is likely to be a more attractive approach for the control pathogenic fungi, bacteria, and harmful insects. Compounds such as brasiliamides isolated from penicillium species showed succesful convulsive activity against silkworm (Fujita et al., 2002).

2.8.4 Induced resistance

Biological control agents produce wide variety chemicals that stimulate the defense mechanism of plants. Such chemicals either stimulate the development of complex antioxidant defense mechanisms in host plants or induce plant defenses through biochemical changes in plant systems to cope with subsequent infections by pathogens. The induction of host defenses can be localized or systemic resistance (Jogaiah et al., 2018). Biocontrol agents added to the rhizosphere protect plants against many pathogens including viruses, bacteria, and fungi, because of the induction of resistance mechanisms similar to the hypersensitive response (HR), systemic acquired resistance (SAR), and induced systemic resistance (ISR) in plants (Jogaiah et al., 2013; Murali et al., 2013; Nagaraju et al., 2012a,b). Studies revealed that *Penicillium citrinum* not only controls pathogens but also induces defense mechanism to host plant during adverse environmental conditions like drought, high salinity, and high temperatures (Sreevidya et al., 2015). Date palm compost enriched with *Aspergillus* and *Penicillium* species are effective in controlling Bayoud disease due to the combined effect of several modes of action viz. competition, restraint of pathogen enzymes, and induced resistance (Chakroune et al., 2008; El Hassni et al., 2007).

2.9 Conclusions

Potassium is one of the major plant nutrients influencing not only plant growth, development but also enhances resistance against abiotic and biotic stresses. In this context, microorganisms with dual nature as biocontrol agents and plant growth promoters are of keen interest as an alternative strategy for agrochemicals with no risk to humans and the environment. Therefore, research on integrating biological agents into mass production would boost the commercialization and use of biopesticides.

An application of potassium solubilizing fungi may be still an improved technique for achieving better crop yield. The *Penicillium* and *Aspergillus* species may represent a promising source of biological control agent and constitute a positive alternative to manage important agricultural pest and pathogens. The use of *Penicillium* and *Aspergillus* spp. showed their potential antagonistic activity against various pathogens and may be further stimulated or enhanced by soil amendments like plant growth-promoting compounds and fertilizers. Future use of biocontrol fungi will expand if scientists can successfully enhance the virulence of potassium solubilizing fungi by genetic manipulation and explore these fungi as potential biocontrol agents for agriculture disease management. Potassium solubilizing fungi could play a vital role in agriculture by providing an indigenous source of potassium for plant uptake, maintaining potassium status in soils for sustaining crop production, and controlling pathogens and pests, and thereby improving the quality of crops.

References

Adebola, M.O., Amadi, J.E., 1999. Biological control of *Protomycopsis phaseoli*, the causal agent of leaf smut of cowpea. J. Phytopathol. 147, 371–375.

Adebola, M.O., Amadi, J.E., 2010. Screening three *Aspergillus* species for antagonistic activities against the cocoa black pod organism (*Phytophthora palmivora*). Agric. Biol. J. N. Am., 362–365.

Ahemad, M., Kibret, M., 2014. Mechanisms and applications of plant growth promoting rhizobacteria: current perspective. J King Saud. Univ. Sci. 26, 1–20.

Ahmed, H., Strub, C., Hilaire, F., Schorr-Galindo, S., 2015. First report: *Penicillium adametzioides* a potential biocontrol agent for ochratoxin-producing fungus in grapes, resulting from natural product pre-harvest treatment. Food Control 51, 23–30.

Aleksandrov, V.G., Blagodyr, R.N., Iiiev, I.P., 1967. Liberation of phosphoric acid from apatite by silicate bacteria. Mikrobiyol Zh (Kiev) 29, 111–114.

Anthoni, R., 2002. Microbial dissolution of silicates. In: Anthoni, R.A., Subramani, S., Subramanium, P. (Eds.), Biodissolution of Nutrients in Rice Ecosystem. Madurai Press, pp. 111–113.

Archana, D.S., Nandish, M.S., Savalagi, V.P., Alagawadi, A.R., 2013. Characterization of potassium solubilizing bacteria (KSB) from rhizosphere soil. Bioinfolet 10, 248–257.

Aydi, R., Hassine, M., Jabnoun-Khiareddine, H., Ben Jannet, H., Daami-Remadi, M., 2013. Valorization of *Aspergillus* spp. as biocontrol agents against *Pythium* and optimization of their antagonistic potential in vitro and in vivo. Tunis. J. Med. Plant. Nat. Prod. 9, 70–82.

Barocio-Ceja, N.B., Ceja-Torres, L.F., Morales-Garcia, J.L., Silva-Rojas, H.V., Flores-Magallon, R., et al., 2013. In vitro biocontrol of tomato pathogens using antagonists isolated from chicken-manure vermicompost. Int. J. Exp. Bot. 82, 15–22.

Ben Abdallah, R.A., Khiareddine, H.J., Trabelsi, B.M., Remadi, M.D., 2015. Soil-borne and compost-borne *Aspergillus* species for biologically controlling post-harvest diseases of potatoes incited by *Fusarium sambucinum* and *Phytophthora erythroseptica*. J. Plant Pathol. Microbiol. 6, 10.

Bertsch, P.M., Thomas, G.W., 1985. Potassium status of temperate region soils. In: Munson, R.E. (Ed.), Potassium in Agriculture. American Society of Agronomy, Crop Science Society of America and Soil Science Society of America.

Boddey, R.M., Urquiaga, S., Alves, B.J.R., Reis, V., 2003. Endophytic nitrogen fixation in sugarcane: present knowledge and future applications. Plant and Soil 252, 139–149.

Boonchan, S., Britz, M.L., Stanley, G.A., 2000. Degradation and mineralization of high-molecular-weight polycyclic aromatic hydrocarbons by defined fungal-bacterial cocultures. Appl. Environ. Microbiol. 66, 1007–1019.

Carvalho, F.P., 2017. Pesticides, environment and food safety. Food Energy Secur 6, 48–60.

Chakroune, K., Bouakka, M., Lahlali, R., Hakkou, A., 2008. Suppressive effect of mature compost of date palm by-products on *Fusarium oxysporum* f. sp. *albedinis*. Plant Pathol. J. 7, 148–154.

Chen, Y.P., Rekha, P.D., Arun, A.B., Shen, F.T., Lai, W.A., Young, C.C., 2006. Phosphate solubilizing bacteria from subtropical soil and their tricalcium phosphate solubilizing abilities. Appl. Soil Ecol. 34, 33–41.

Chet, I., Inbar, J., 1994. Biological control of fungal pathogens. Appl. Biochem. Biotechnol. 48 (1), 37–43.

Chuang, C.C., Kuo, Y.L., Chao, C.C., Chao, W.L., 2007. Solubilization of inorganic phosphates and plant growth promotion by *Aspergillus niger*. Biol. Fertil. Soils 43, 575–584.

Compant, S., Duffy, B., Nowak, J., Clement, C., Barka, E.A., 2005. Use of plant growth promoting bacteria for biocontrol of plant diseases: principles, mechanisms of action and future prospects. Appl. Environ. Microbiol. 71 (9), 4951–4959.

Daami-Remadi, M., Jabnoun-Khiareddine, H., Ayed, F., Hibar, K., Znaidi, I.E.A., 2006. In vitro and in vivo evaluation of individually compost fungi for potato *Fusarium* dry rot biocontrol. J. Biol. Sci. 6, 572–580.

Damalas, C.A., 2009. Understanding benefits and risks of pesticide use. Sci. Res. Essays 4, 945–949.

Damalas, C.A., Koutroubas, S.D., 2018. Current status and recent developments in biopesticide use. Agriculture 8, 13.

Dwivedi, S.K., Enespa, S., 2013. In vitro efficacy of some fungal antagonists against *Fusarium solani* and *Fusarium oxysporum* f. sp. *lycopersici* causing brinjal and tomato wilt. Int. J. Biol. Pharm. Res. 4, 46–52.

Ekta, K., Archana, Y., 2017. The role of microbial enzyme systems in plant growth promotion. Climate Change Environ. Sustain. 5 (2), 122–145.

El Hassni, M., El Hadrami, A., Daayf, F., Cherif, M., Ali Barka, E., El Hadrami, I., 2007. Biological control of bayoud disease in date palm: selection of microorganisms inhibiting the causal agent and inducing defense reaction. Environ. Exp. Bot. 59, 224–234.

Elmer, P.A.G., Reglinski, T., 2006. Biosuppression of *Botrytis cinerea* in grapes. Plant Pathol. 55, 155–177.

Epstein, E., Bloom, A.J., 2008. Mineral Nutrition of Plants: Principles and Perspectives, second ed. Sinauer, Sunderland MA.

Fang, J.G., Tsao, P.H., 1995. Efficacy of *Penicillium funiculosum* as a biological control agent against *Phytophthora* root rots of azalea and citrus. Phytopathology 85, 87–98.

Fang, J.G., Tsao, P.H., 1995a. Evaluation of *Pythium nunn* as a potential bio control agent against *Phytophthora* root rots of azalea and sweet orange. Phytopathology 85, 29–36.

Fravel, D.R., 1988. Role of antibiosis in the biocontrol of plant diseases. Annu. Rev. Phytopathol. 26, 75–91.

Fujita, T., Makishima, D., Akiyama, K., Hayashi, H., 2002. New convulsive compounds, brasiliamides A and B, from *Penicillium brasilianum* Batista JV-379. Biosci. Biotechnol. Biochem. 66 (8), 1697–1705.

Gajera, H.P., Vakharia, D.N., 2012. Production of lytic enzymes by *Trichoderma* isolates during in vitro antagonism with *Aspergillus niger*, the causal agent of collar rot of peanut. Braz. J. Microbiol. 43–52.

Goldstein, A.H., 1994. Involvement of the quinoprotein glucose dehydrogenase in the solubilization of exogenous phosphates by gram-negative bacteria. In: Phosphate in Microorganisms: Cellular and Molecular Biology. ASM Press, Washington DC, pp. 197–203.

Gomathi, S., Ambikapathy, V., 2011. Antagonistic activity of fungi against *Pythium debaryanum* (hesse) isolated from chilli field soil. Adv. Appl. Sci. Res. 2, 291–297.

Haas, D., Keel, C., Laville, J., Maurhofer, M., Obensli, H., Schnider, U., Voisard, C., Defago, G., 1991. Secondary metabolites of *Pseudomonas fruorescens* strain CHAO involved in the suppression of root diseases. In: Hennecke, I.H., Verma, D.P.S. (Eds.), Advances in Molecular Genetics of Plant-Microbe Interactions. In: vol. I. Kluwer Academic Publishers, Dordrecht, Boston, and London, pp. 450–456.

Harman, G.E., Kubicek, C.P., 1998. *Trichoderma* and *Gliocladium*. Enzymes, Biological Control and Commercial Applications, vol. 2. Taylor and Francis, London UK.

Harman, G.E., Howell, C.R., Viterbo, A., Chet, I., Lorito, M., 2004. *Trichoderma* species—opportunistic, avirulent plant symbionts. Nat. Rev. 2, 43–56.

Hofrichter, M., Bublitz, F., Fritsche, W., 1994. Unspecific degradation of halogenated phenols by the soil fungus *Penicillium frequentans* Bi 7/2. J. Basic Microbiol. 34, 163–172.

Hu, X., Webster, G., Xie, L., Yu, C., Li, Y., Liao, X., 2013. A new mycoparasite, *Aspergillus* sp. ASP-4, parasitizes the sclerotia of *Sclerotinia sclerotiorum*. Crop Prot. 54, 15–22.

Huang, Z., White, D.G., Payne, G.A., 1997. Corn seed proteins inhibitory to *Aspergillus flavus* and *Aflatoxin* biosynthesis. Phytopathology 87, 622–627.

Indiragandhi, P., Anandham, R., Madhaiyan, M., Sa, T.M., 2008. Characterization of plant growth-promoting traits of bacteria isolated from larval guts of diamondback moth *Plutella xylostella* (Lepidoptera; Plutellidae). Curr. Microbiol. 56, 327–333.

Israel, S., Lodha, S., 2005. Biological control of *Fusarium oxysporum* f. sp. *cumini* with *Aspergillus versicolor*. Phytopathol. Mediterr. 44, 3–11.

Jiménez, M., Santamarina, M.P., Sanchis, V., Hernández, E., 1987. Investigación de la micoflora mycotoxinas de cereales almacenados. Microbiol. Aliment Nutr. 5, 105–109.

Jogaiah, S., Mostafa, A., Tran, L.S.P., Shin-ichi, I., 2013. Characterization of rhizosphere fungi that mediate resistance in tomato against bacterial wilt disease. J. Exp. Bot. 64, 3829–3842.

Jogaiah, S., Abdelrahman, M., Tran, L.S.P., Ito, S.-I., 2018. Different mechanisms of *Trichoderma virens*-mediated resistance in tomato against *Fusarium* wilt involve the jasmonic and salicylic acid pathways. Mol. Plant Pathol. 19, 870–882.

Khan, M.R., Anwer, M.A., 2007. Molecular and biochemical characterization of soil isolates of *Aspergillus niger* aggregate and an assessment of their antagonism against *Rhizoctonia solani*. Phytopathol. Mediterr. 46, 304–315.

Khokhar, I., Haider, M.S., Mukhtar, I., Mushtaq, S., 2013. Biological control of *Aspergillus niger* the cause of black-rot disease of *Allium cepa* L. (onion) by *Penicillium* species. J. Agrobiol. 29 (1), 23–28.

Kuchenbuch, R.O., 1987. Potassium dynamics in the rhizosphere and potassium availability. In: Methodology in Soil Potassium Research, Proceedings of 20th Colloquim of the International Potash Institute. Baden bei Wien.

Kumar, A.R., Garampallin, R.H., 2014. Antagonistic effect of rhizospheric *Aspergillus* species against *Fusarium oxysporum* f. sp. *lycopersici*. Int. J. Chem. Anal. Sci. 5 (1), 39–42.

Maqsood, M., Shehzad, M.A., Wahid, A., Butt, A.A., 2013. Improving drought tolerance in maize (*Zea mays*) with potassium application in furrow irrigation systems. Int. J. Agric. Biol. 15, 1193–1198.

Meesala, S., Subramaniam, G., 2016. *Penicillium citrinum* VFI-51 as biocontrol agent to control charcoal rot of sorghum (*Sorghum bicolor* (L.) Moench). Afr. J. Microbiol. Res. 10 (19), 669–674.

Melo, I.S., Faull, J.L., Nascimento, R.S., 2006. Antagonism of *Aspergillus terreus* to *Sclerotiana sclerotiorum*. Braz. J. Microbiol. 37, 417–419.

Michereff, S.J., Silveira, N.S.S., Reis, A., Mariano, L.R., 1995. Greenhouse screening of *Trichoderma* isolates for control of *Curvularia* leaf spot of yam. Mycopathologia 130, 103–108.

Mondal, G., 1998. In vitro evaluation of *Aspergillus niger* AN27 against soil borne fungal pathogens and field testing against *Macropomina phaseolina* on potato. Ph.D. Thesis, IARI, New Delhi.

Mondal, G., Dureja, P., Sen, B., 2000. Fungal metabolites from *Aspergillus niger* AN27 related to plant growth promotion. Indian J. Exp. Biol. 38, 84–87.

Muentz, A., 1890. Surla decomposition desroches etla formation de la terre arable. C.R. Acad. Sci. 2, 110–137.

Murali, M., Jogaiah, S., Amruthesh, K.N., Shin-ichi, I., Shekar Shetty, H., 2013. Rhizosphere fungus *Penicillium chrysogenum* promotes growth and induces defense-related genes and downy mildew disease resistance in pearl millet. Plant Biol. 15, 111–118.

Murugappan, V., Latha, M.R., Jagadeeswaran, R., Sheeba, S., 2002. Nutrient sorption as the basis for identifying nutrient deficiencies in soils. J. Agric. Res. Manag. 1 (2), 70–76.

Nagaraju, A., Jogaiah, S., Mahadeva Murthy, S., Shin-Ichi, I., 2012a. Seed priming with *Trichoderma harzianum* isolates enhances plant growth and induces resistance against *Plasmopara halstedii*, an incitant of sunflower downy mildew disease. Aust. J. Plant Path. 41, 609–620.

Nagaraju, A., Murali, M., Jogaiah, S., Amruthesh, K.N., Mahadeva Murthy, S., 2012b. Beneficial microbes promote plant growth and induce systemic resistance in sunflower against downy mildew disease caused by *Plasmopara halstedii*. Curr. Bot. 3, 12–18.

Patil, M.P., Patil, R.H., Maheshwari, V.L., 2015. Biological activities and identification of bioactive metabolite from endophytic *Aspergillus flavus* L7 isolated from *Aegle marmelos*. Curr. Microbiol. 71, 39–48.

Pelaez, V., Mizukawa, G., 2017. Diversification strategies in the pesticide industry: from seeds to biopesticides. Cienc. Rural 47, e20160007.

Pettigrew, W.T., 2008. Potassium influences on yield and quality production for maize, wheat, soybean and cotton. Physiol. Plant. 133, 670–681.

Prajapati, K., Sharma, M.C., Modi, H.A., 2012. Isolation of two potassium solubilizing fungi from ceramic industry soils. Life Sci. Leaflets 5, 71–75.

Prajapati, K., Sharma, M.C., Modi, H.A., 2013. Growth promoting effect of potassium solubilizing microorganisms on *Abelmoschus esculentus*. Int. J. Agric. Sci. 3 (1), 181–188.

Rajan, S.S., Watkinson, J.H., Sinclair, A.G., 1996. Phosphate rock of for direct application to soils. Adv. Agron. 57, 77–159.

Ramzan, N., Noreen, N., Shahzad, S., 2014. Inhibition of in vitro growth of soil borne pathogens by compost-inhabiting indigenous bacteria and fungi. Pak. J. Bot. 46, 1093–1099.

Rawlings, D.E., 2002. Heavy metal mining using microbes. Annu. Rev. Microbiol. 56, 65–91.

Rogers, J.R., Bennett, P.C., Choi, W.J., 1998. Feldspars as a source of nutrients for microorganisms. Am. Mineral. 83, 1532–1540.

Sangeeth, K.P., Bhai, R.S., Srinivasan, V., 2012. *Paenibacillus glucanolyticus* a promising potassium solubilizing bacterium isolated from black pepper (*Piper nigrum* L.) rhizosphere. J Spices Aromat. Crops 21 (2), 118–124.

Santamarina, M.P., Rosello, J., Llacer, R., Sanchis, V., 2002. Antagonistic activity of *Penicillium oxalicum* Corrie and Thom, *Penicillium decumbens* Thom and *Trichoderma harzianum* Rifai isolates against fungi, bacteria and insects *in vitro*. Rev. Iberoam. Micol. 19, 99–103.

Saraswathy, R., Arunachalam, G., Jagadeeshwari, P.V., 2003. Potassium content and its flux in rhizosphere region. Crop Res. 25 (1), 66–68.

Schroeder, D., 1979. Structure and weathering of potassium containing minerals. Proc. Congr. Int. Potash inst. 11, 43–63.

Sempere, F., Santamarina, M.P., 2010. Study of the interactions between *Penicillium oxalicum currie* and *thom* and *Alternaria alternata* (fr.) Keissler. Braz. J. Microbiol. 41 (3), 700–706.

Sheng, X.F., Huang, W.Y., 2002. Study on the conditions of potassium release by strain NBT of silicate bacteria scientia. Agricul. Sinica 35 (6), 673–677.

Sindhu, S.S., Dua, S., Verma, M.K., Khandelwal, A., 2010. Growth Promotion of Legumes by Inoculation of Rhizosphere Bacteria. Springer, Wien, pp. 195–235.

Singh, B.K., Walker, A., 2006. Microbial degradation of organophosphorus compounds. FEMS Microbiol. Rev. 30 (3), 428–471.

Soltani, J., Hosseyni, M.M.S., 2015. Fungal endophyte diversity and bioactivity in the Mediterranean cypress *Cupressus sempervirens*. Curr. Microbiol. 70, 580–586.

Song, S.K., Huang, P.M., 1971. Influence of certain pedogenic factors on potassium reserves of selected Canadian prairie soils. Soil Sci. Soc. Am. Proc. 35, 500–505.

Sreevidya, M., Gopalakrishnan, S., Melo, T.M., Simic, N., Bruheim, P., Sharma, M., Srinivas, V., Alekhya, G., 2015. Biological control of *Botrytis cinerea* and plant growth-promotion potential by *Penicillium citrinum* in chickpea (*Cicer arietinum* L.). Biocontrol Sci. Technol. 25, 739–755.

Tiwari, C.K., Paribar, J., Verma, R.K., 2011. Potential of *Aspergillus niger* and *Trichoderma viride* as biocontrol agents of wood decay fungi. J. Indian Acad. Wood Sci. 8, 169–172.

Venkatasubbaih, P., Safeeulla, K.M., 1984. *Aspergillus niger* for biological control of *Rhizoctonia solani* on coffee seedling. Trop. Pest Manag. 30, 401–406.

Verma, J.P., Yadav, J., Tiwari, K.N., Jaiswal, D.K., 2014. Evaluation of plant growth promoting activities of microbial strains and their effect on growth and yield of chickpea (*Cicer arietinum* L.) in India. Soil Biol. Biochem. 70, 33–37.

Verma, P., Yadav, A.N., Khannam, K.S., Saxena, A.K., Suman, A., 2018. Potassium solubilizing microbes: diversity, distribution and role in plant growth promotion. In: Panapatte, D.G., Jhala, Y.K., Vyas, R.V., Shelat, H.N. (Eds.), Microorganisms for Green Revolution. Springer Press.

Viterbo, A., Inbar, J., Hadas, Y., Chet, I., 2007. Plant disease biocontrol and induced resistance via fungal mycoparasites: the mycota a comprehensive treatise on fungi as experimental systems for basic and applied research. In: Environmental and Microbial Relationships, second ed. Springer. 350 pp.

Wang, W.H., Zhao, X., Liu, C., Liu, L., Yu, S., 2015. Effects of the biocontrol agent *Aspergillus flavipes* on the soil microflora and soil enzymes in the rooting zone of pepper plants infected with *Phytophthora capsici*. J. Phytopathol. 163, 513–521.

Weindling, R., 1932. Trichoderma lignorum as a parasite of other soil fungi. Phytopathology 22, 837–845.

Whipps, J.M., Gerhardson, B., 2007. In: Van Elsas, J.D., Jansson, J.D., Trevors, J.T. (Eds.), Biological Pesticides for Control of Seed and Soil Borne Plant Pathogens, second ed. CRC Press, pp. 479–501.

Wisniewski, M.E., Wilson, C.L., 1992. Biological control of postharvest diseases of fruits and vegetables: recent advances. HortScience 27 (2), 94–98.

Xu, D., Wang, H., Zhang, Y., Yang, Z., Sun, X., 2013. Inhibition of non-toxigenic *Aspergillus niger* FS10 isolated from Chinese fermented soybean on growth and aflatoxin B1 production by *Aspergillus flavus*. Food Control 32, 359–365.

Yang, Y., Jin, D., Wang, G., Wang, S., Jia, X., 2011. Competitive biosorption of Acid Blue 25 and Acid Red 337 onto unmodified and CDAB-modified biomass of *Aspergillus oryzae*. Bioresour. Technol. 102, 7429–7436.

Zhang, C.L., Zheng, B.Q., Lao, J.P., Mao, L.J., Chen, S.Y., 2008. Clavatol and patulin formation as the antagonistic principle of *Aspergillus clavatonanicus* an endophytic fungus of *Taxus mairei*. Appl. Microbiol. Biotechnol. 78, 833–840.

Trichoderma-plant-pathogen interactions for benefit of agriculture and environment

3

Narendra Kumar and S.M. Paul Khurana
Amity Institute of Biotechnology, Amity University Haryana, Gurgaon, India

3.1 Introduction

The new technological (chemical) advancements in different domains of agriculture have boosted agricultural production but these practices affect the environment. We have also learned how these advanced farming practices are playing a bad role for achieving higher production. To have no harm or gain more in an eco-friendly manner, we need to explore why it is not happening. To do so, we need more experiments on a large scale to help boost wider use of biocontrol microbes. The microbial species acting as biocontrol microbes (fungi, bacteria) produce many types of enzymes which play an active role in biocontrol activity, like tolerance induced against biotic and abiotic stresses, hyphal growth, cell wall degradation, etc. There has been a continuous study on *Trichoderma* for various plant growth boosting effects since the last three decades (Nagaraju et al., 2012; Jogaiah et al., 2013) The structural and functional genomics study of *Trichoderma* revealed that it can be used as a model to study mechanisms in the plant-microbes environment. A world population of 9.1 billion people in 2050 would require the raising of overall food production by some 70% (FAO, 2009). The substantial increase in food grains production over the years has helped to meet the food security needs of the country. But the number of biotic and abiotic stresses causes yield losses to a large extent. Biotic constraints include fungi, bacteria, virus, nematodes, weeds, and pest insects causing yield losses of 31%–42% (Agrios, 2005). Therefore, pesticide consumption increased year by year from 45.39 thousand tons used in the year 2012–13 (Krishijagran, 2015). The Indian pesticides market was worth INR 197 Billion in 2018 and is now projected to reach INR 316 Billion by 2024, growing at a CAGR of 8.1% during 2019-2024 (https://www.imarcgroup.com/indian-pesticides-market). While analyzing the trend of pesticide consumption in 29 states and Union Territories (UTs) of India for the period 2000–13, a positive growth trend has been observed in 17 states/UTs. The positive growth has been observed to be highest in Jammu & Kashmir, Andaman & Nicobar Islands, and Tripura, while Uttar Pradesh, Maharashtra, Andhra Pradesh, Punjab, and Haryana are the states that accounted for 70% of total pesticide consumption (Devi et al., 2017; Satapute et al., 2019).

Biocontrol Agents and Secondary Metabolites. https://doi.org/10.1016/B978-0-12-822919-4.00003-X

Pesticides are substances or a mixture of substances intended for preventing, destroying, repelling, or mitigating any pest. Pesticides represent the last input in an agricultural operation and are applied for preventing the spoilage of crops from pests such as insects, fungi, weeds, etc., thereby increasing the agricultural productivity (https://www.imarcgroup.com/indian-pesticides-market).

The significance of pesticides has been rising over the last few decades catalyzed by the requirement to enhance the overall agricultural production and need to safeguard adequate food availability for the continuously growing population in the country. In India, pests and diseases, on an average, take a toll of 20%–25% of the total food produced (https://www.imarcgroup.com/indian-pesticides-market).

Besides affecting the environment and nontarget organisms, continuous and tremendous use of chemical pesticides creates high selection pressure on pathogens, which are forced to mutate inside the pathogens, and development of pesticide resistance races such as dodine and metalaxyl resistance in *Venturia inequalis* and *Phytophthora infestans*, respectively. Erosion of effectiveness of conventional plant protection methods has been widely studied in the past. For example, resistance to chemical pesticides in plant pathogens or pests has been extensively documented. The durability of biological control has often been assumed to be higher than that of chemical control (Bardin et al., 2015; Jogaiah et al., 2018). Pesticide resistance and environment threat due to injudicious use of chemical pesticides for disease management necessitates the introduction of new alternatives such as biological control (Murali et al., 2013). The use of biofertilizers and biopesticides is an important alternative for sustaining high production with minimum ecological damage (Govind et al., 2016). Besides the classic mycorrhizal fungi and *Rhizobium* bacteria, other plant-growth-promoting fungi such as *Trichoderma* spp. (Teleomorph: Hypocrea) can stimulate plant growth by suppressing plant diseases (Van Wees et al., 2008; Jogaiah et al., 2013). Biological control agents comprise a number of fungi and 90% of such applications have been carried out by different strains of *Trichoderma,* the antagonistic properties of which are based on the activation of multiple mechanisms (Jogaiah et al., 2018). Biological control is a key constituent of integrated pest management that has generated interest among farmers for ecological and sustainable disease management. *Trichoderma* is a free-living, common fungus in soil and root ecosystems. This interacts well in root, soil, and foliar environments (Kumar and Khurana, 2019). The genus *Trichoderma* possesses reasonable biological control attributes belonging to species *T. harzianum, T. resseyi, T. asperellum, T. viridae, T. virense*. It can be freely mentioned that *Trichoderma* spp. are found widely in organic matter, litter, soil, and the rhizospheric ecosystem of all climatic zones as saprophytes. Recent discoveries show that they are opportunistic, avirulent plant symbionts, as well as being parasites of other fungi. They fall in the category of endophytes establishing robust and long-lasting colonizations of root surfaces and penetrate into the epidermis. However, the ability of these fungi to sense, invade, and destroy other fungi has been the major driving force behind their commercial success as biopesticides (Joshi et al., 2019). *Trichoderma spp* defend the plants by their direct and indirect effects on plant-pathogen-soil environment interactions. They have not only a protective role for plants by killing fungal pathogens and nematodes. This can not only induce resistance against plant pathogens and impart tolerance to abiotic stresses but also can improve plant growth and vigor (Nagaraju et al., 2012). This increases nutrient uptake and can help in bioremediation of heavy metals and serious pollutants in the soil environment. This is because of production of secondary metabolites having clinical value and also results in production of enzymes which have industrial uses. They also result in production of a variety of compounds which induce systemic or localized resistance which brings about changes in plant metabolism. Plants are also protected from numerous pathogens by responses that are similar to systemic acquired resistance and rhizobacteria-induced systemic resistance (Jogaiah et al., 2010; Babu et al., 2015). *Trichoderma* inoculation for improving rice crop production has been documented, including growth and yield enhancement as well as alleviation of biotic and abiotic stresses (Khadka and Uphoff, 2019). Root colonization by *Trichoderma* spp. also frequently enhances root growth and development, increasing crop productivity, uptake and use of nutrients, and resistance to biotic and abiotic stresses. A detailed study as to how *Trichoderma* has evolved to interact with other fungi and/or plants will help improve and expand their applications in the biological control program. The ability to attack on soilborne plant pathogens dominated the interest in *Trichoderma* for many years (Sudisha et al., 2006). For the past few years, researchers showed interest in induced plant disease resistance, say induced systemic resistance (ISR); to some extent, systemic acquired resistance (SAR) induced by the *Trichoderma*-root symbiosis (Harman et al., 2004; Hermosa et al., 2012; Khadka and Uphoff, 2019). *Trichoderma* spp. have the ability to colonize the rhizosphere of plants from Brassicaceae, promoting growth and development as well as stimulating systemic defenses (Poveda et al., 2019).

In agriculture, various problems such as disease and pest, drought, and loss of soil fertility due to excessive use of pesticides are at a peak now. These in turn are resulting in a high level of pollution and finally global warming. So ecofriendly bioagents are needed to resolve these burning problems. Zachow et al. (2016) mentioned that *Trichoderma* are not only found in soil environments. They may develop their colony on trees, branches, shoots. The species of *Trichoderma* are well studied and explored for plant growth promotion. This reproduces asexually through forming conidia and chlamydospores, even in wild habitats, through the ascospores. These species are mycoparasitic and produce secondary metabolites which have clinical value. They have been used for biological plant protection and are known as bio-fungicides. These can also help in bioremediation. The enzymes produced by *Trichoderma* are used in producing animal feed, as well as in the wine making and brewery industries.

Trichoderma spp. come in the category of the most successful biocontrol agents (Pandey et al., 2019). Verma et al. (2007) mentioned that up to 60% of the registered bio-fungicides used in today's agriculture are *Trichoderma*-based formulations (Verma et al., 2007). There are 250 products available in India for field use but the percentage share of bio-fungicides is less. It is dominated by synthetic chemicals (Singh et al., 2009). Actually, these bio-fungicides are affected by the environment and slow in action. So their efficacy in the field is restricted. Genetic manipulations may be done for designing new strains which are more effective in comparison to the native ones.

3.2 *Trichoderma*-plant interaction

The most widely studied biocontrol agents are the *Trichoderma* spp. Their role in creation of plant defenses for fighting pathogens is well recorded (Abdelrahman et al., 2016; Jogaiah et al., 2018). But the literature is less on their activity against insects. To fill this research gap, transcriptome changes in tomato cultivar "Dwarf San Marzano" induced by the strain T22 of *Trichoderma harzianum* colonization and subsequent infestation by the aphid *Macrosiphum euphorbiae* were studied. It was observed that there was regulation of gene expression and defense responses. This was multitrophic interaction of *Trichoderma*-tomato-aphid, which was most evident with defense genes. The genes (Early and late) associated with direct defense against insects were activated (i.e., *GST, kinases, peroxidase, miraculin, polyphenol oxidase, and chitinase*) along with indirect defense genes, viz., *sesquiterpene synthase*, *geranyl phosphate synthase*. The metabolome analysis of Targeted and untargeted genes noticed through metabolomic changes in the form of an increased accumulation of isoprenoids in *Trichoderma*- inoculated plants. The transcriptomic and metabolomic alterations nicely fit with aphids' higher mortality when feeding on *Trichoderma*-treated plants. In spite of this, *Trichoderma*-treated plants projected an overexpression of transcripts which codes defense-related transcription factors (bZIP, MYB, NAC, AP2-ERF, WRKY) of several gene families. This suggests that the fungus contributes in priming responses of plants in fighting pest insects. It can be concluded that the plants treated with *Trichoderma* develop metabolomic and transcriptomic alterations, which in turn can support the plant defense responses both directly and indirectly (Abdelrahman et al., 2016; Coppola et al., 2019).

It is evident from the literature that the plant rhizosphere helps *Trichoderma* spp. growth and has the capacity to penetrate as well as colonize in plant roots internally (Harman et al., 2004; Khadka and Uphoff, 2019). This opportunistic/facultative symbiosis is driven by the ability of *Trichoderma* to derive sucrose or other nutrients from the plants and in return boost plant immunity against invading pathogens and improve photosynthetic abilities (Shoresh and Harman, 2008). The presence of *Trichoderma* in the rhizosphere evokes a coordinated transcriptomic, proteomic, and metabolomic response in the plants (Vargas et al., 2011; Brotman et al., 2012; Abdelrahman et al., 2016). The reprogramming is beneficial in developing resistance against pathogens. This helps improve the yield and growth of the plants.

3.2.1 Colonization in plant roots

Trichoderma spp. have the potential to colonize plant roots both externally and internally. This interaction is the result of a chemical signals interplay between two partners, although it is a primary step of *Trichoderma*-plant interaction, yet studied less in comparison to that normally followed, viz., attachment, penetration, and internal colonization of plant roots. *Trichoderma* spp. can produce hormonal signals for facilitating the colonization of roots. This secretes auxins that accelerate root growth that may facilitate more colonization due to the increased surface area (Contreras-Cornejo et al., 2009). The function of *accd* which directs ACC deaminase for controlling root growth of canola through *T. asperellum* was observed by gene knockout (Viterbo et al., 2010). It is very clear from studies that *Trichoderma* secretes cysteine-rich hydrophobin proteins for facilitating for attachment. Two proteins TasHyd1 (*T. asperellum*) and Qid74 (*T. harzianum*) are recorded which boosts in attaching to roots (Samolski et al., 2012). *Trichoderma* spp. secrete expansin for enhancing/boosting root penetration through cellulose-binding modules along with endopolygalacturonase (Morán-Diez et al., 2009). When these fungi are within roots, they may grow intercellularly with limits to the outer cortex and epidermal layer. It has been observed that *T. koningii* suppresses production of phytoalexins at the time of colonization of *Lotus japonicus* roots (Masunaka et al., 2011).

3.2.2 Promotion of plant growth

There has been a positive effect of *Trichoderma* on plants development. They secrete many Secondary Metabolites, viz., 6-pentyl-a-pyrone, Koninginins, trichocaranes A–D, harzianopyridone, harzianic acid, cyclonerodiol, harzianolide. They project their effects in a dose-dependent manner (Vinale et al., 2014). This is used as biological control of Shisham wilt (Kumar and Khurana, 2016), protects seed-borne fungi of mung bean *Vigna radiata* (Kumar and Khurana, 2018), and has the potential to combat fusarial wilt of banana (Kumar et al., 2018; Kumar and Khurana, 2019).

3.2.3 Induction of plant defense responses

Plants respond soon upon invasion of *Trichoderma*. This is due to the rapid ion fluxes, oxidative burst, and in turn dealing with callose deposition and synthesis of polyphenols (Shoresh et al., 2010). Thereafter, salicylate (SA) and jasmonate/ethylene (JA/ET)-signaling results in plants acquiring different degrees of tolerance to the pathogens (Shoresh et al., 2010; Jogaiah et al., 2018). This is actually JA/ET induced systemic resistance (ISR). This resemblance in response is like that triggered by plant-growth-promoting rhizobacteria (PGPR). It has been observed that higher doses of *Trichoderma* inocula can boost SA-mediated systemic acquired resistance (SAR) response. This is similar to that of necrotrophic pathogens (Yoshioka et al., 2012). The signaling events that result in induced resistance are not so well known. There is a hint that it occurs due to nitrogen-activated protein kinase from cucumber and an MAPK from *T. virens* molecular cross talk in plant and *Trichoderma*. This triggers the downstream defense response (Shoresh et al., 2006).

Trichoderma spp. also release Xylanase and peptaibols (peptaibiotics with a high content of alpha amino isobutyric acid) like alamethicin and trichovirin II. This project elicits an immune response in plants (Druzhinina et al., 2011). Mukherjee et al. (2012) identified a PKS/NRPS hybrid enzyme that plays a role in the defense response of maize. The elicitor Sm1/Epl1 is produced by *Trichoderma* spp. This is the small cysteine-rich hydrophobin-like protein of the cerato-platanin (CP) family (Djonović et al., 2006).

It has been recorded that in maize, *Trichoderma* gene deletion impairs elicitation of ISR. The glycosylated state of Sm1 is monomeric. The nonglycosylated state is sensitive for oxidative dimerization of plants which results in inactive Sm1 as an inducer of ISR (Vargas et al., 2008). The 3-D structure of the *Ceratocystis platani* cerato-platanin is now well studied as an oligomer of *N*-acetyl glucosamine (carbohydrate residue) which binds to it and has been confirmed (Oliveira et al., 2011).

3.2.4 *Soil environment vis-à-vis* Trichoderma*-plant interaction*

Trichoderma reproduces asexually and is found frequently in temperate and tropical soils that harbor 101–103 propagules per gram which can be cultured under common laboratory conditions. These can produce colonies on herbaceous and woody plants where sexual teleomorphs (genus *Hypocrea*) are observed. Many *Trichoderma* strains which have biocontrol efficacy do not have sexual stages. The *Trichoderma* has peculiar features, viz., fast growing, opportunistic invaders, and prolific producers of spores. This is powerful in antibiotic production in a highly competitive environment for light, space, and nutrients (Schuster and Schmoll, 2010; Montero-Barrientos et al., 2011). These features make *Trichoderma* ecologically fit. They grow in agricultural areas, marshy habitats, forests, salty lands, deserts of different climatic zones (covering the tundra, tropical regions, and Antarctica) (Montero-Barrientos et al., 2011; Mukherjee et al., 2013). The studies have been conducted on marine *Trichoderma* isolates for evaluating their potential in the form of halotolerant biocontrol agents. They are active against *Rhizoctonia solani* which induce systemic defense responses in plants (Gal-Hemed et al., 2011; Lorito et al., 2010; Verma et al., 2007). They have the ability to antagonize a wide spectrum of soilborne plant pathogens. They reduce the incidence of diseases by pathogens in many crops. The mechanisms which *Trichoderma* applies to antagonize phytopathogenic fungi include mycoparasitism, competition, antibiosis, and colonization. Their antagonistic power is effective for biocontrol. Use of different *Trichoderma* strains is an alternative for chemicals which control a wide variety of plant pathogens (Jogaiah et al., 2018; Nandini et al., 2020).

3.2.5 *Activity related to plant growth promotion, help in seed germination, and effect on plant morphology*

Worldwide, there are 50 products registered in the agriculture market, based on *Trichoderma*, for better yields of various crops (Woo et al., 2006). They are evaluated as new types of BCAs(Spk) (biocontrol agents). It is recorded that *Trichoderma* covers 60% of all fungal-based BCAs. It may be noted that *T. harzianum* is the most active agent coming in the category of available biopesticides and biofertilizers used recently (Lorito et al., 2010). The inherent qualities of *Trichoderma*-based BCAs are cumulative as driving factors for constant success (Verma et al., 2007).

3.2.5.1 *Effect on seed germination*

Seed biopriming or seed treatment with *Trichoderma* spp. triggers the release and/or production of enzymes and phytohormones which are involved in seed germination. It also enhances the speed of germination and seedling vigor. Enhanced percent seed germination has been found in okra, maize, beans, mustard, chili, soybean, chickpea, tomato, sunflower, etc. (Nagaraju et al., 2012; Kumar et al., 2014; Babychan and Simon, 2017). Many seed invading pathogens such as *Pythium* are unable to attack on host seed due to faster germination and seedling vigor (Matsouri et al., 2010). Bezuidenhout et al. (2012) reported that *Trichoderma harzianum* produces a metabolite such as gliotoxin that may mimic the plant growth hormone gibberellic acid, which is involved in the seed germination process. Seed and/or soil treatment with *Trichoderma* enhances seed germination percentage, directly by activating enzymes and phytohormones and indirectly by altering soil microflora and nutrient availability in soil. In this way, farmers get a better field stand.

The rhizosphere supports microbial populations capable of exerting beneficial, neutral, or detrimental effects on plant growth. *Trichoderma* spp. can enhance rice growth by boosting many parameters, viz., tiller number, leaf number, root length, root fresh weight, and plant height. In maize (*Zea mays*) plants, *Trichoderma* inoculation affects root system architecture including enhanced root biomass and increased root hair development (Harman et al., 2004). The root system adds to plant fitness by providing anchorage, which improves water use efficiency. This ups the acquisition of soil mineral nutrients. The *harzianum* and *virens* species of *Trichoderma* have growth boosting action which has been correlated with lateral roots formed prolifically (Contreras-Cornejo et al., 2009). The potential of *Trichoderma* spp. to yield phytohormones is the main factor for increasing the rice plant height (Chowdappa et al., 2013).

The evidences are strongly in view that auxins have a role in root system architecture. *T. virens* is able to produce auxins such as IEt (indole-3-ethanol), IAA (indole-3-acetic acid), IAAld (indole-3-acetaldehyde). These plays roles in the growth of plant and development (Contreras-Cornejo et al., 2009). *Trichoderma* spp. produce harzianolide, which can improve the early stage of plant development (Cai et al., 2013) by enhancing root length. When it was treated with *Trichoderma spp.*, the leaf number and tiller number got increased significantly when compared to NPK treatment in rice plants in comparison to control (Doni et al., 2014). Line of evidence is there for enhancement of number of tillers, leaves, branches, fruits, flowers, even plant height through *Trichoderma spp.* (Sajeesh, 2015). *Trichoderma spp.* produce harzianolide, which may improve early stages of plant development by enhancement of root length (Cai et al., 2013).

3.3 Effect on plant physiology, effect on yield and quality of produce

The *Trichoderma* spp. have the power to changing the plant physiology, viz., net photosynthetic rate, stomatal conductance, internal CO_2 concentration, transpiration, nutrient uptake, and water use efficiency. Thus, nutrient uptake enhances physiological processes in plants that lead to good growth (Doni et al., 2014). Micronutrients such as Nitrogen and Magnesium are the components of chlorophyll and enzymes of m-RNA synthesis and gene regulation engaged in photosynthesis. *Trichoderma harzianum* significantly increased the ability of rice plants to tolerate drought stress and increase in rice plant's water-holding capacity (Shukla et al., 2012). An approximately threefold increase in the net photosynthetic rate and stomatal conductance and twofold increase in water use efficiency in *Trichoderma*-treated rice plants as compared to NPK-treated plants has been observed (Doni et al., 2014). The increased photosynthetic rates with less transpiration in *Trichoderma* treats projects efficiency of high water use (Doni et al., 2014). The high photosynthetic potential in *Trichoderma*-treated plants with less CO_2 concentration can be correlated as an activity of carboxylation through CO_2 fixation for glucose production at the time of carbohydrate metabolism, which is very active (Thakur et al., 2010). The prolonged photosynthetic activity supporting delayed senescence in rice is due to activity of *Trichoderma* spp. (Mishra and Salokhe, 2011). *Trichoderma* releases cellulases which degrade cellulose and enhance the organic matter and nutrients in the rhizosphere. Jiang et al. (2011) reported that increase of plant physiological processes such as solubilization and minerals chelation may enhance nutrient availability that can upgrade plant metabolism (Harman et al., 2004).

3.3.1 Effect on yield and quality of produce

Trichoderma effect on seed germination, plant morphology, and physiology leads to a better field stand and it also accelerates the vegetative and reproductive growth of plants. It enhances the number of branches, spikes, flowers, and fruits per plant. In many cases, the average weight of individual fruit is also comparatively higher. Higher yield by the application of *Trichoderma* species in mustard, wheat, corn, tuberose, sugarcane, tomato, okra, etc. has been found (Haque et al., 2012; El-Katatny and Idres, 2014; Naznin et al., 2015; Idowu et al., 2016).

3.4 Induced resistance against biotic and abiotic stresses

Plants have an immune system (Singh et al., 2018) that is able to detect motifs or domains with conserved structural traits typical of entire classes of microbes, called MAMPs (microbe-associated molecular patterns) (Hermosa et al., 2012). The ability of *Trichoderma* spp. hyphae to release MAMPs (Table 3.1) for molecular recognition may contribute to signal cascade by signaling molecule within the plant such as jasmonic acid (JA), salicylic acid (SA), and ethylene (ET). Earlier, it was believed that JA and/or ET are the signaling molecule for *Trichoderma*- induced resistance but recent findings show the involvement of SA also. Colonization of the Arabidopsis root by *T. asperellum* produces a clear ISR through an SA signaling cascade, and both the SA and JA/ET signaling pathways combine in the ISR triggered by cell-free culture filtrates of *Trichoderma*. This has triggering power and can upregulate SA GENE in plants. The work of Tucci et al. (2011) revealed that SA-dependent gene expression can be modulated when plants are infected with *B. cinerea*. *Trichoderma* spp. release elicitors within the plant which can trigger expressions of defense proteins (Thakur and Sohal, 2013). Due to this, plant immunity against pathogens may be induced, which can improve the plant growth. A common feature of induced resistance responses of beneficial microbes is priming for enhanced defense. In primed plants,

Table 3.1 A glimpse of identified *Trichoderma* microbe-associated molecular patterns in different species.

***Trichoderma* species**	**Microbe-associated molecular patterns**	**Activity**	**References**
T. longibrachiatum	Cellulases	Activates ET and SA signaling route	Martınez et al. (2001)
T. viride	Alamethicin (20 mer peptaibol)	SA biosynthesis and elicitation of JA in lima bean	Engelberth et al. (2001)
T. viride	Xylanase Xyn2/Eix	In tobacco leaf tissues elicits ET biosynthesis and produces xylanase that results hypersensitive response	Rotblat et al. (2002)
T. atroviride/T. virens	Cerato-platanins Epl1/Sm1	Induces expression of defense responses in maize and cotton through Hydrophobin-like SSCP orthologues	Djonović et al. (2006) and Seidl et al. (2006)
T. virens	18 mer peptaibols	Induces against *P. syringae* for elicitation of cucumber systemic defenses	Viterbo et al. (2007)
T. asperelloides	Swollenin TasSwo	Protects in fighting *P. syringae*, and *B. cinere* in cucumber roots and leaves through releasing expansin protein and cellulose-binding domain which stimulates local defense responses	Brotman et al. (2008)
T. pseudokoningii	Trichokonin (20 mer peptaibol)	Induces virus resistance in tobacco plants through release of ROS and accumulation of phenolics by multiple defense signaling pathways	Luo et al. (2010)
Five *Trichoderma* species	Cellulase, Amylase, Lipase and Pectinase	*Trichoderma harzianum* (32.63% at 30 min) showed minimum pectinase activity than others. The extracellular hydrolytic enzyme activities were exhibited at the different rates by different Trichoderma species	Bhale and Rajkond (2012)

defense responses are not activated directly, but are accelerated upon attack by pathogens or other stress resulting in faster and stronger resistance to the attacker encountered (Van et al., 2008). *Trichoderma* acts locally and systemically, which involves signaling cascade and activation and accumulation of defense-related antimicrobial compounds which include enzymes as Phenyl ammonia lyase (PAL), Peroxidase, polyphenol oxidase, Lipoxygenase; Proteins as PR (pathogenesis-related protein) (Jogaiah et al., 2013), Terpenoid, phytoallexin as Rishitin, Lubimin, Phytotuberol, coumarin, solevetivone, Resveratol, and antioxidant as ascorbic acid, glutathione, etc. (Howell et al., 2000).

3.5 *Trichoderma*-pathogen interactions

Trichoderma-pathogen interactions: *Trichoderma* shows antagonistic interaction with pathogen by following modes of action (Fig. 3.1).

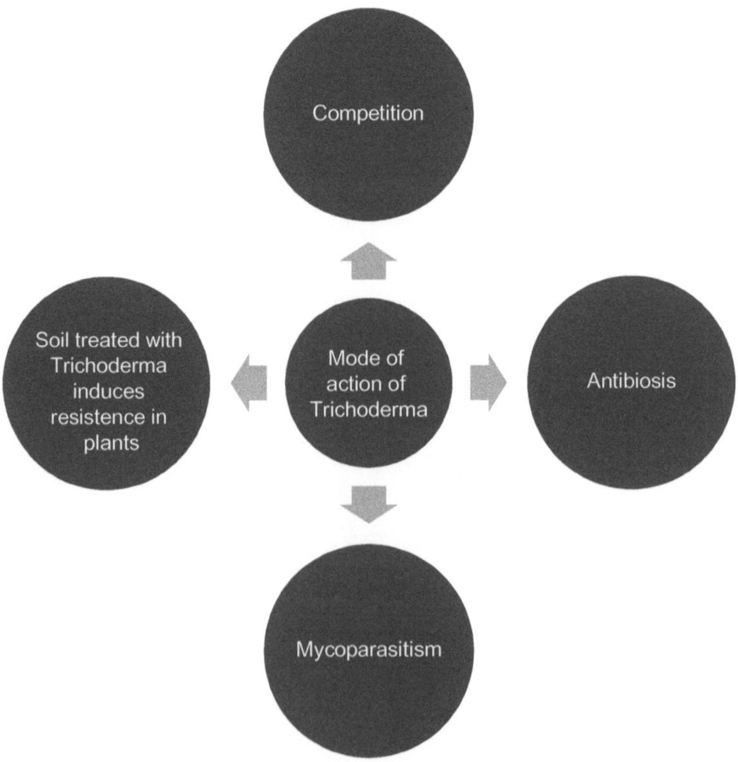

Fig. 3.1 Depicting modes of action of *Trichoderma* spp. against pathogen (suggested by Waghunde et al., 2016).

3.5.1 Mycoparasitism and lytic enzymes

3.5.1.1 Mycoparasitism

Mycoparasitism deals with the direct involvement of attack of one fungal species on another. In dealing with antagonisms of *Trichoderma*, mycoparasitism comes in the category of main mechanisms. The events leading to mycoparasitism are complex, and take place as follows: Chemotropic growth of *Trichoderma*, recognition of the host, coiling and appressoria formation, secretion of hydrolytic enzymes, penetrations of the hyphae, and lysis of the host (Fig. 3.2). This complex process includes sequential events, involving cycle of recognition by the binding of carbohydrates in the *Trichoderma* cell wall to lectins on the target fungus, hyphal coiling, and appressoria formation which contains a higher amount of osmotic solutes such as glycerol and induces penetration, attack on cellular machinery via producing several fungitoxic cell-wall-degrading enzymes, glucanases, chitinases, and proteases (Harman et al., 2004); cumulative action of these compounds results in parasitism of the target fungus and dissolution of the cell walls. At the site of appressoria formation, holes can be

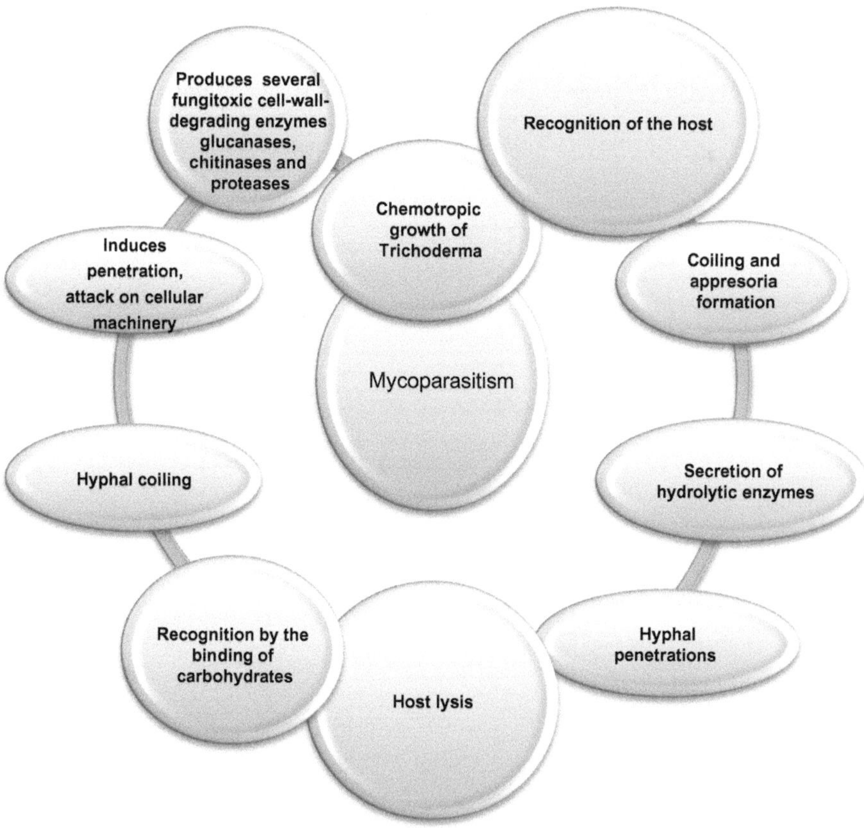

Fig. 3.2 The events leading to mycoparasitism.

produced in the target fungus leading to direct entry of *Trichoderma* hyphae into the host lumen and finally killing of the host (Kumar, 2013). Actually, this is the cumulative action of 20–30 genes, metabolites, and proteins that are involved directly in this pathway. The roles of different chitinases and glucanases in the process of mycoparasitism are well investigated using gene for gene experiments in *Trichoderma* spp. But still more studies are needed to understand this complex phenomenon (Daguerre et al., 2014).

3.5.2 Antibiosis and secondary metabolites

Antibiosis is an antagonistic interaction incorporating low molecular weight diffusible secondary metabolites. In this, the antibiotics get released by microorganisms that are detrimental to the growth of pathogens. *Trichoderma* spp. produce a variety of antibiotics (Fig. 3.3), such as gliovirin, gliotoxin, viridin, viridol, koninginins, pyrones, and peptaibols, against fungal phytopathogens (Howell, 2003; Harman et al., 2004). Trichokonins VI, a type of peptaibol from *Trichoderma pseudokoningii* SMF2, exhibited antibiotic activities by inducing extensive apoptotic programmed cell death in fungal pathogens (Shi et al., 2012). Gliovirin and Gliotoxin are the two most important *Trichoderma* secondary metabolites belonging to Q and P group strains, respectively. "P group strains of *Trichoderma* (*Gliocladium*) *virens* is active

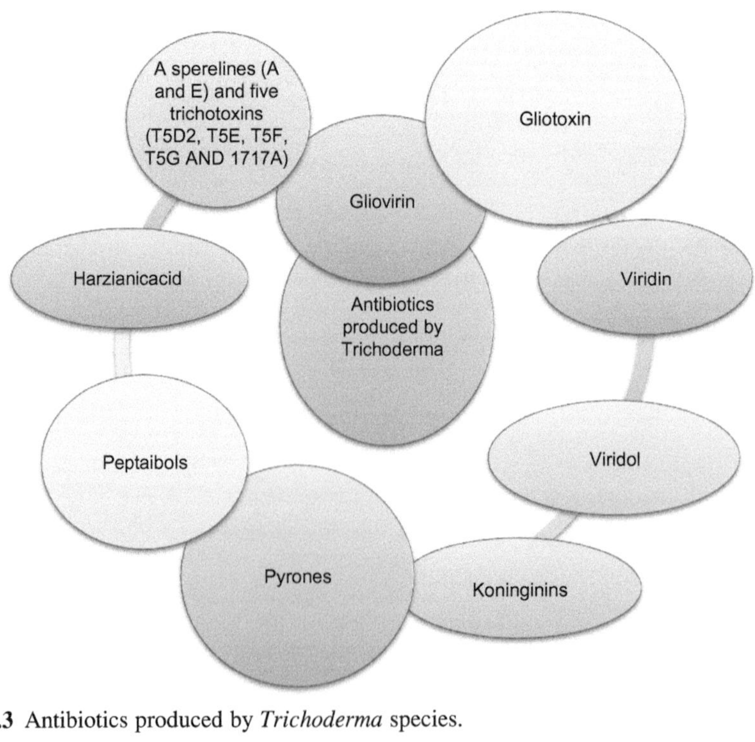

Fig. 3.3 Antibiotics produced by *Trichoderma* species.

against *P. ultimum*, but not against *R. solani* and Q group is very active against *R. solani*, but less against *P. ultimum*" (Howell, 2000). "The T. virens *ve*A ortholog *vel*1 (VELVET protein Vel1) is involved in regulation of gliotoxin biosynthesis, biocontrol activity and many other secondary metabolism-related genes" (Mukherjee et al., 2004). "Koninginin D also inhibited the growth of soil-borne plant pathogens, such as *Rhizoctonia solani, Phytophthora cinnamomi, Pythium middletonii, Fusarium oxysporum* and *Bipolaris sorokiniana*" (Dunlop et al., 1989). Viridins obtained by diverse *Trichoderma* spp. (*T. viride, T. koningii, T. virens*) inhibit spore germination of *Penicillium expansum, Botrytis allii, Fusarium caeruleum, Colletotrichum lini, Stachybotrys atra*, and *Aspergillus niger* (Singh et al., 2005). Harzianic acid obtained through the *T. harzianum* strain depicted in vitro antibiotic action against *Sclerotinia sclerotiorum, R. solani*, and *Pythium irregulare* (Vinale et al., 2009). Thc isolation of trichodiene synthase (*tri5*) and its overexpression in the *T. brevicompactum Tb41tri5* transformant have enhanced the trichodermin released and thereby increased antifungal action against *Fusarium* spp. and *Aspergillus fumigates* (Tijerino et al., 2011). This Trichoderma genus covers a group of filamentous fungi which is living freely. These species act as BCAs (biological control agents) against many pathogenic fungi. This power is the outcome of several mechanisms acting synergistically in these organisms, which in turn involves the release of secondary metabolites. They produce peptaibols, which belongs to the antibiotic peptide group. They have present nonproteinogenic amino acids such as Aib (α-aminoisobutyrate) along with N-terminal modifications and amino alcohols in the C-terminal region. The strains of *T. asperellum* resulted in production of five trichotoxins (T5D2, T5E, T5F, T5G, and 1717) and two asperelines (A and E) which are antibiosis in nature (Brito et al., 2014).

3.5.3 Competition through pathogens and soil microbial community

Substrates competition covers most important factor for fungal flora as it deals with the competition for light in comparison to evolution of plants (Garrette, 1956). This is the scarcity of limiting nutrients for rhizospheric colonization. Due to this, the death of many microorganisms occurs in the vicinity of *Trichoderma* strains. It is well recorded that the rhizosphere and root exudates are potential sources of rich nutrients such as amino acids, sugar iron, organic acids, vitamins, etc. The competition for carbon is an effective mode not only in *Trichoderma* but also in other strains of *F. oxysporum, Rhizoctonia solani* (Alabouvette et al., 2009; Sarrocco et al., 2009). The utilization dealing with immobile nutrients proficient mobilization can make it more efficient. Its competitive power may be increased in comparison with other soil microbes. This process could be related also to the production of organic acids, such as gluconic, citric, and fumaric acids, which decreases soil pH and allows the solubilization of phosphates, micronutrients, and mineral cations like iron, manganese, and magnesium (Vinale et al., 2008). In aerobic environmental conditions with oxygen and neutral pH, the iron exists mainly as Fe^{3+} and tends to form insoluble ferric oxide. This results in making it in an unavailable form for root absorption and microbial growth

(Miethke, 2013). Iron acts as a cofactor of numerous enzymes and an essential nutrient for growth of plants and other microorganisms. *Trichoderma* secretes siderophore, an iron chelating compound which binds with insoluble iron (FeIII) and converts to soluble form (FeII) for plant absorption. This inhibits growth of plant pathogens by depriving them from iron sources (Leong, 1986).

Siderophore bound Fe(III) undergoes reduction to Fe(II), which is catalyzed by free extracellular or membrane-standing ferric chelate reductases. If not already released extra cytoplasmatically, the iron has to be removed from the Fe(III)-siderophore complex in the cytosol. This is mediated either by intracellular ferric-siderophore reductases, while in others it is due to ferric siderophore hydrolases. Siderophores released through some *Trichoderma* isolates are very powerful chelators for iron. This siderophore may outcompete with *Pythium* in soil for the available iron that can control its growth.

3.6 The three-way interaction: *Trichoderma*-plant-pathogen

3.6.1 Trichoderma-*pathogen networking*

The seven transmembrane G protein coupled receptor *Gpr1* is involved in sensing the fungal prey in the nearby vicinity (Omann et al., 2012). It was observed that *Trichoderma virens* is involved in hydrophobin expression for adherence to the host and mycoparasitism (Zeilinger et al., 2005). Ligand binding with such receptors leads to downstream signaling events via activation of G-protein cascades. *Trichoderma* sp. covers three MAPK cascades comprising MAPKK, MAPKKK, and MAPK signaling pathways. This plays its action in biocontrol and mycoparasitism (Kumar et al., 2010). The production of antibiotics, CWDEs (Cell wall degrading enzymes) gives them the power to kill pathogens. Chitin and Glucan synthases are produced by the *Trichoderma* to repair own self cell wall damage through pathogen during *Trichoderma* pathogen interaction. Likely genes which encode hydrolytic enzymes like glucanases and chitinases along with secondary metabolism like NRPSs involves in inhibiting the pathogen development (Kubicek et al., 2011). Certain species like *T. atroviride* produce 6-PP (6-pentyl-2*H*-pyran-2-one), a volatile metabolite which plays a role in fungal interactions of *Trichoderma* (Hasan et al., 2008; Vinale et al., 2009). Genetic evidence has been recorded for the assembly peptaibols of 11- and 14-modules through single NRPS Tex2 of *T. virens* (Mukherjee et al., 2011). It has been recorded that peptaibiotics are strongly antimicrobial (being able to form voltage-gated membrane channels as peptaibol trichokonin VI of *T. pseudokoningii*). These are involved in induction of programmed cell death of *Fusarium oxysporum* (Shi et al., 2012).

Trichoderma spp. harbor many attributes and qualities which project many potential uses in agriculture, viz., improving germination thus helps in plant growth development, increasing nitrogen use efficiency, physiological changes by increasing nutrients uptake in plants, assisting in improving photosynthetic potential. Many procedures have been investigated and applied to identify the main genes involved in the fungal antagonist, the plant, and microbial pathogens. What are the compounds involved in a three-way cross talk? They also stimulate defense in response of biotic and abiotic stresses and antagonism including mycoparasitism, antibiosis, and competition against pathogens. *Trichoderma* elicits ISR by ET/JA-dependent pathways. This triggers priming responses in the plant. The *Trichoderma* plant interaction is expression of defense-related genes of ET/JA. The SA pathways may overlap depending on the *Trichoderma* strains and their concentration. This also deals with the developmental stage of the plant, the plant material, and even timing of the interaction. *Trichoderma* also produces the phytohormones ET and IAA, which play roles in interconnecting plant development and defense responses. The *Trichoderma* spp. Genome, which has been well investigated, has useful genes. They have the power to produce a great variety of expression patterns. This permits the fungi to adapt in many environmental conditions (dead tissues, soil, water, inside the plants, etc.). The metabolomics of *Trichoderma* spp. are incredibly complex in terms of antibiotics and secondary metabolites production (Abdelrahman et al., 2016). But due to the advancement in genomic and proteomic methods, it is now possible to explore new procedures and even new functions of compounds produced by this genus. The expression of *Trichoderma* genes in plants yields positive results mainly in management of plant diseases and in developing resistance to various adverse environmental situations. The experimental data indicates that *Trichoderma and host* plant interactions have common features as other beneficial microbe associations have, but they also project their own characteristics because of *Trichoderma*'s particular lifestyle. But still there is a need for more experiments to be conducted to gain insight in signaling transduction pathways, covering defense and development from *Trichoderma*-plant interactions in the presence of pathogens' various biotic and abiotic stresses.

3.7 Future prospects

The genetics of a three-way cross talk among Plant-*Trichoderma*-Pathogen interaction should be investigated including the process of regulation and induction of enzyme expression. The environmental clues must be considered to check how transmission of the respective signal to the promoters takes place in signaling pathways. This will improve enzyme production in the microbial cell factory, which in turn may increase the mycoparasitic potential of biocontrol strains.

The identification of effectors that may reprogram the host's genetic machinery and the plant receptors for *Trichoderma* elicitor proteins needs more emphasis. This can serve as the foundation for investigating how these symbionts trigger host defense reaction. *Trichoderma* compatibility with chemical fungicides as a part of integrated disease management should be evaluated. There is a need to enhance the popularity of *Trichoderma*-based formulations among farmers for the eco-friendly management of diseases. The ecological impact of large-scale applications of a single fungal species and their secondary metabolites for biocontrol should be evaluated to corelate a knowledge base of *Trichoderma* for the safe and sustainable use.

3.8 Conclusions

Trichoderma spp. have many uses in agriculture, viz., assists to improve photosynthetic efficiency, enhances nitrogen intake efficiency, amends abiotic stresses, improves physiological response to stresses. This is now getting popularized as general plant protectants and growth enhancers worldwide besides their application in a variety of industrial processes. The genome of *Trichoderma* spp. contains many useful genes which gives them the ability for production of a great variety of expression patterns. This gives them the capacity to adapt in many different environments (inside the plants, soil, dead tissues, water). For secondary metabolites production, *Trichoderma* spp. metabolomics are incredibly complex. Due to recent developments in adoption of advanced molecular and proteomic approaches, new pathways can be explored. The new compounds can be isolated and their functions and potential applications can be worked out. The mapping has been done for this genus growing in different environmental conditions. The information obtained has been applied for development of new products. The new products have been developed based on combinations of living fungus with its secreted metabolites. The new formulations can combine biocontrol activity with use as biofertilizer which may be considered more effective than older products. It is active against many plant pathogens. Being biotechnologically important, both basic biology and field applications are well studied. Although many studies on the genetics of *Trichoderma* with other organisms (notably fungi and plants) have been done, depth of understanding the mechanisms is still lacking. This is because of the lack of whole genome sequences. But this scenario is now expected to change due to the availability of five *Trichoderma* genomes. Some researchers have made progress in genome-wide expression studies (Seidl et al., 2009; Druzhinina et al., 2012). So a worldwide initiative may be undertaken to confirm the functions of each gene by high gene knockouts as was done in *N. crassa* (Dunlap et al., 2007). Transcriptome analyses of *Trichoderma* during mycoparasitism, root colonization with plant will help identify new candidate genes which are involved in the interactions with plants, plant pathogens, and *Trichoderma* spp. If this is achieved, it will be easy to engineer tailor-made strains for biotechnological applications and optimal biocontrol.

Acknowledgment

We are grateful to the authorities of Amity University Haryana for the facilities and encouragement.

References

Abdelrahman, M., Abdel-Motaal, F., El-Sayed, M., Jogaiah, S., Shigyo, M., Ito, S.-i., Tran, L.-S.P., 2016. Dissection of *Trichoderma longibrachiatum*-induced defense in onion (*Allium cepa* L.) against *Fusarium oxysporum* f. sp. *Cepa* by target metabolite profiling. Plant Sci. 246, 128–138.

Agrios, G.N., 2005. Plant Pathology, fifth ed. Elsevier, New Delhi.

Alabouvette, C., Olivain, C., Migheli, Q., Steinberg, C., 2009. Microbiological control of soil-borne phytopathogenic fungi with special emphasis on wilt-inducing *Fusarium oxysporum*. New Phytol. 184, 529–544.

Babu, A.N., Jogaiah, S., Ito, S.-i., Nagaraj, A.K., Tran, L-S.P., 2015. Improvement of growth, fruit weight and early blight disease protection of tomato plants by rhizosphere bacteria is correlated with their beneficial traits and induced biosynthesis of antioxidant peroxidase and polyphenol oxidase. Plant Sci. 231, 62–73.

Babychan, M., Simon, S., 2017. Efficacy of *Trichoderma* spp. against Fusarium oxysporum f. sp. lycopersici. (FOL) infecting pre-and post-seedling of tomato. J. Pharmacogn. Phytother. 6, 616–619.

Bardin, M., Ajouz, S., Comby, M., Lopez-Ferber, M., Graillot, B., Siegwart, M., Nicot, P.C., 2015. Is the efficacy of biological control against plant diseases likely to be more durable than that of chemical pesticides. Front. Plant Sci. 6, 566. https://doi.org/10.3389/fpls.2015.00566.

Bezuidenhout, C.N., Van Antwerpen, R., Berry, S.D., 2012. An application of principal component analyses and correlation graphs to assess multivariate soil health properties. Soil Sci. 177, 498–505.

Bhale, U.N., Rajkond, J.N., 2012. Enzymatic activity of Trichoderma species. Nova Nat. Sci. Res. 1 (4), 1–8.

Brito, J.P., Ramada, M.H., de Magalhães, M.T., Silva, L.P., Ulhoa, C.J., 2014. Peptaibols from Trichoderma asperellum TR356 strain isolated from Brazilian soil. Springerplus 3, 600. https://doi.org/10.1186/2193-1801-3-600.

Brotman, Y., Briff, E., Viterbo, A., Chet, I., 2008. Role of swollenin, an expansin-like protein from Trichoderma, in plant root colonization. Plant Physiol. 147, 779–789.

Brotman, Y., Lisec, J., Meret, M., Chet, I., Willmitzer, L., Viterbo, A., 2012. Transcript and metabolite analysis of the *Trichoderma* induced systemic resistance response to *Pseudomonas syringe* in *Arabidopsis thaliana*. Microbiology 158, 139–146.

Cai, F., Yu, G., Wang, P., Wei, Z., Fu, L., Shen, Q., Chen, W., 2013. Harzianolide, a novel plant growth regulator and systemic resistance elicitor from *Trichoderma harzianum*. Plant Physiol. Biochem. 73, 106–113.

Chowdappa, P., Kumar, S.P.M., Lakshmi, M.J., Upreti, K.K., 2013. Growth stimulation and induction of systemic resistance in tomato against early and late blight by *Bacillus subtilis* OTPB1 or *Trichoderma harzianum* OTPB3. Biol. Control 65, 109–117.

Contreras-Cornejo, H., Macías-Rodríguez, L., Cortés-Penagos, C., 2009. *Trichoderma virens*, a plant beneficial fungus enhances biomass production and promotes lateral root growth through an auxin dependent mechanism in *Arabidopsis*. Plant Physiol. 149, 1579–1592.

Coppola, M., Diretto, G., Digilio, M.C., Woo, S.L., Giuliano, G., Molisso, D., Pennacchio, F., Lorito, M., Rao, R., 2019. Transcriptome and Metabolome reprogramming in tomato plants by*Trichoderma harzianum strain* T22 primes and enhances Defense responses against aphids. Front. Physiol. 10, 745. https://doi.org/10.3389/fphys.2019.00745.

Daguerre, Y., Siegel, K., Edel-Hermann, V., Steinberg, C., 2014. Fungal proteins and genes associated with biocontrol mechanisms of soil-borne pathogens: a review. Fungal Biol. Rev. 28, 97–125.

Devi, P.I., Thomas, J., Raju, R.K., 2017. Pesticide consumption in India: a spatiotemporal analysis. Agric. Econ. Res. Rev. 30 (1), 163–172.

Djonović, S., Pozo, M.J., Dangott, L.J., Howell, C.R., Kenerley, C.M., 2006. Sm1, a proteinaceous elicitor secreted by the biocontrol fungus *Trichodermavirens* induces plant defense responses and systemic resistance. Mol. Plant-Microbe Interact. 19, 838–853.

Doni, F., Isahak, A., Radziah, C., Zain, C.M., Ariffin, S.M., Nurashiqin, W., Mohamad, W., Mohtar, W., Wan, Y., 2014. Formulation of *Trichoderma* sp. SL2 inoculants using different carriers for soil treatment in rice seedling growth. Springer Plus 3, 1–5.

Druzhinina, I.S., Seidl-Seiboth, V., Herrera-Estrella, A., Horwitz, B.A., Kenerley, C.M., Monte, E., Mukherjee, P.K., Zeilinger, S., Grigoriev, I.V., Kubicek, C.P., 2011. *Trichoderma*—the genomics of opportunistic success. Nat. Rev. Microbiol. 9, 749–759.

Druzhinina, I.S., Shelest, E., Kubicek, C.P., 2012. Novel traits of *Trichoderma* predicted through the analysis of its secretome. FEMS Microbiol. Lett. https://doi.org/10.1111/j.1574-6968.2012.02665. 67.

Dunlap, J.C., Borkovich, K.A., Henn, M.R., Turner, G.E., Sachs, M.S., Glass, N.L., McCluskey, K., Plamann, M., Galagan, J.E., Birren, B.W., Weiss, R.L., Townsend, J.P., Loros, J.J., Nelson, M.A., Lambreghts, R., Colot, H.V., Park, G., Collopy, P., Ringelberg, C., Crew, C., Litvinkova, L., DeCaprio, D., Hood, H.M., Curilla, S., Shi, M., Crawford, M., Koerhsen, M., Montgomery, P., Larson, L., Pearson, M., Kasuga, T., Tian, C., Baştürkmen, M., Altamirano, L., Xu, J., 2007. Enabling a community to dissect an organism: overview of the *Neurospora* functional genomics project. Adv. Genet. 57, 49–96. https://doi.org/10.1016/S0065-2660(06)57002-6.

Dunlop, R.W., Simon, A., Siwasithamparam, D., Ghisalberti, E.L., 1989. An antibiotic from *Trichoderma koningii* active against soil-borne plant pathogens. J. Nat. Prod. 52, 67–74.

El-Katatny, M.H., Idres, M.M., 2014. Effects of single and combined inoculations with *Azospirillum brasilense* and *Trichoderma harzianum* on seedling growth or yield parameters of wheat (*Triticum vulgaris L.,* Giza 168) and corn (*Zea mays L.,* hybrid 310). J. Plant Nutr. 37, 1913–1936.

Engelberth, J., Koch, T., Schü ler, G., Bachmann, N., Rechtenbach, J. & Boland, W., 2001. Ion channel-forming alamethicin is a potent elicitor of volatile biosynthesis and tendril coiling. Cross talk between jasmonate and salicylate signaling in lima bean. Plant Physiol. 125, 369–377.

FAO, 2009. http://www.fao.org/fileadmin/templates/wsfs/docs/issues_papers/hlef2050global_agrculture.pdf, retrieved on 20.01.18.

Gal-Hemed, I., Atanasova, L., Komon-Zelazowska, M., Druzhinina, I.S., Viterbo, A., Yarden, O., 2011. Marine isolates of *Trichoderma* spp. as potential halotolerant agents of biological control for arid-zone agriculture. Appl. Environ. Microbiol. 77, 5100–5109.

Garrette, S.D., 1956. Biology of Root-Infecting Fungi. Cambridge University Press, Cambridge, UK, p. 293.

Govind, S.R., Jogaiah, S., Abdelrahman, M., Shetty, H.S., Tran, L.-S.P., 2016. Exogenous trehalose treatment enhances the activities of defense-related enzymes and triggers resistance against downy mildew disease of pearl millet. Front. Plant Sci. 7, 1593. https://doi.org/10.3389/fpls.2016.01593.

Haque, M.M., Ilias, G.N.M., Molla, A.H., 2012. Impact of *Trichoderma*-enriched biofertilizer on the growth and yield of mustard (*Brassica rapa* L.) and tomato (*Solanum lycopersicum* Mill.). The Ariculturists 10, 109–119.

Harman, G.E., Howell, C.R., Viterbo, A., Chet, I., Lorito, M., 2004. *Trichoderma* species opportunistic, avirulent plant symbionts. Nat. Rev. Microbiol. 2, 43–56.

Hasan, E.A., Walker, F., Buchenauer, H., 2008. *Trichoderma harzianum* and its metabolite 6 pentyl-alpha-pyrone suppress fusaric acid produced by *Fusarium moniliforme*. J. Phytopathol. 156, 79–87.

Hermosa, R., Viterbo, A., Chet, I., Monte, E., 2012. Plant-beneficial effects of *Trichoderma* and of its genes. Microbiology 158, 17–25.

Howell, C.R., Hanson, L.E., Stipanovic, R.D., Puckhaber, L.S., 2000. Induction of terpenoid synthesis in cotton roots and control of *Rhizoctonia solani* by seed treatment with *Trichoderma virens*. Phytopathology 90, 248–252.

Howell, C.R., 2003. Mechanisms employed by *Trichoderma* species in the biological control of plant diseases: the history and evolution of current concepts. Plant Dis. 87, 4–10.

Indian Pesticides Market, 2019–2024. Indian Pesticides Market: Industry Trends, Share, Size, Growth, Opportunity and Forecast 2019–2024, https://www.imarcgroup.com/indian-pesti cides-market retrieved on 20/2/20.

Idowu, O.O., Olawole, O.I., Idumu, O.O., Salami, A.O., 2016. Bio-control effect of *Trichoderma asperellum* (Samuels) Lieckf. and *Glomus intraradices* Schenk on okra seedlings infected with *Pythium aphanidermatum* (Edson) Fitzp and *Erwinia carotovora* (Jones). Am. J. Exp. Agri. 10, 1–12.

Jiang, X., Geng, A., He, N., Li, Q., 2011. New isolate of *Trichoderma* viride strain for enhanced cellulolytic enzyme complex production. J. Biosci. Bioeng. 111, 121–127.

Jogaiah, S., Roopa, K.S., Pushpalatha, H.G., Shekar Shetty, H., 2010. Evaluation of plant growth-promoting rhizobacteria for their efficiency to promote growth and induce systemic resistance in pearl millet against downy mildew disease. Arch. Phytopath. Plant Prot. 43, 368–378.

Jogaiah, S., Mostafa, A., Tran, L.S.P., Shin-ichi, I., 2013. Characterization of rhizosphere fungi that mediate resistance in tomato against bacterial wilt disease. J. Exp. Bot. 64, 3829–3842.

Jogaiah, S., Abdelrahman, M., Tran, L.-S.P., Ito, S.-I., 2018. Different mechanisms of *Trichoderma virens*-mediated resistance in tomato against Fusarium wilt involve the jasmonic and salicylic acid pathways. Mol. Plant Pathol. 19, 870–882.

Joshi, S.M., De Britto, S., Jogaiah, S., Ito, S., 2019. Mycogenic selenium nanoparticles as potential new generation broad spectrum antifungal molecules. Biomol. Ther. 9 (9), 419.

Khadka, R.B., Uphoff, N., 2019. Effects of *Trichoderma* seedling treatment with system of Rice intensification management and with conventional management of transplanted rice. PeerJ. 7, e5877. https://doi.org/10.7717/peerj.5877.

Krishijagran, 2015. Outlook of Pesticide Consumption in India - Krishi Jagran. Available online: http://www.krishijagran.com/corporate-watch/Industry-Profile/2014/11/Outlook of Pesticide-Consumption-in-India (retrieved on 20/02/2020).

Kubicek, C.P., Herrera-Estrella, A., Seidl-Seiboth, V., Martinez, D.A., et al., 2011. Comparative genome sequence analysis underscores mycoparasitism as the ancestral life style of *Trichoderma*. Genome Biol. 12, 1–15.

Kumar, S., 2013. *Trichoderma*: a biological control weapon for managing plant diseases and promoting sustainability. Int. J. Agric. Sci. Vet. Med. 1, 106–121.

Kumar, N., Khurana, S.M.P., 2016. Biomanagement of wilting of a valuable timber and medicinal plant of shisham (*Dalbergia sissoo* Roxb.)-a review. Int. J. Curr. Microbiol. App. Sci. 5 (1), 32–54.

Kumar, N., Khurana, S.M. Paul, 2018. Biocontrol of Seed Borne Fungi of Mung Bean *Vigna radiata* (L.) Wiczek.Nat Con. on Diversity and Utilization of Tropical Plants (NCDUTP 22–23 Feb, 2018). ABSTRACT NO-88.

Kumar, N., Khurana, S.M.P., 2019. Pathogen and management of fungal wilt of banana through biocontrol agents. In: Varma, A., Tripathi, S., Prasad, R. (Eds.), Plant Microbe Interface. Springer International Publishing. ISBN 978-3-030-19831-2, pp. 177–194. https://doi.org/10.1007/978-3-030-19831-2.

Kumar, A., Scher, K., Mukherjee, M., Pardovitz-Kedmi, E., Sible, G.V., Singh, U.S., Kale, S.P., Mukherjee, P.K., Horwitz, B.A., 2010. Overlapping and distinct functions of two

Trichoderma virens MAP kinases in cell-wall integrity, antagonistic properties and repression of conidiation. Biochem. Biophys. Res. Commun. 398, 765–777.

Kumar, V., Shahid, M., Singh, A., Srivastava, M., Mishra, A., Srivastava, Y.K., Pandey, S., Shrarma, A., 2014. Effect of biopriming with biocontrol agents *Trichoderma harzianum* (Th.Azad) and *Trichoderma* viride on chickpea genotype (Radhey). J. Plant Pathol. Microbiol. 5, 1–4.

Kumar, N., Chaturvedi, S., Khurana, S.M.P., 2018. Strategies for biomanagement of *Fusariam* wilt of banana. In: International Conference on Role of Soil and Plant Health Towards Achieving Sustainable Development Goals 21–25 November 2018, Technical Session V; Eco-friendly approaches for Soil and Plant Health Management. L.04: p 37 Organizers Indian Phytopathological Society (IPS), New Delhi; Asia-Pacific Association of Agricultural Research Institutions (APAARI), Bangkok, Thailand; Department of Agriculture, Bangkok, Thailand.

Leong, J., 1986. Siderophores: their biochemistry and possible role in biocontrol of plant pathogen. Annu. Rev. Phytopathol. 24, 187–209.

Lorito, M., Woo, S.L., Harman, G.E., Monte, E., 2010. Translational research on *Trichoderma*: from omics to the field. Annu. Rev. Phytopathol. 48, 395–417.

Luo, Y., Zhang, D.D., Dong, X.W., Zhao, P.B., Chen, L.L., Song, X.Y., Wang, X.J., Chen, X.L., Shi, M., Zhang, Y.Z., 2010. Antimicrobial peptaibols induce defense responses and systemic resistance in tobacco against tobacco mosaic virus. FEMS Microbiol. Lett. 313, 120–126.

Martınez, C., Blanc, F., Le Claire, E., Besnard, O., Nicole, M., Baccou, J.C., 2001. Salicylic acid and ethylene pathways are differentially activated in melon cotyledons by active or heatdenatured cellulase from *Trichoderma longibrachiatum. Plant Physiol.* 127, 334–344.

Masunaka, A., Hyakumachi, M., Takenaka, S., 2011. Plant growth-promoting fungus *Trichoderma koningii* suppresses isoflavonoid phytoalexin vestitol production for colonization on/in the roots of *Lotus japonicus*. Microbes Environ. 26, 128–134.

Matsouri, F., Björkman, T., Harman, G.E., 2010. Seed treatment with *Trichoderma harzianum* alleviates biotic, abiotic and physiological stresses in germinating seeds and seedlings. Phytopathology 100, 1213–1221.

Miethke, M., 2013. Molecular strategies of microbial iron assimilation: from high affinity complexes to cofactor assembly systems. Metallomics 5, 15–28.

Mishra, A., Salokhe, V.M., 2011. Rice growth and physiological responses to SRI water management and implications for crop productivity. Paddy Water Environ. 9, 41–52.

Montero-Barrientos, M., Hermosa, R., Cardoza, R.E., Gutiérrez, S., Monte, E., 2011. Functional analysis of the *Trichoderma harzianum nox1* gene, encoding an NADPH oxidase, relates production of reactive oxygen species to specific biocontrol activity against *Pythium ultimum*. Appl. Environ. Microbiol. 77, 3009–3016.

Morán-Diez, E., Hermosa, R., Ambrosino, P., Cardoza, R.E., Gutiérrez, S., Lorito, M., Monte, E., 2009. The ThPG1 endopolygalacturonase is required for the *Trichodermaharzianum*–plant beneficial interaction. Mol. Plant-Microbe Interact. 22, 1021–1031.

Mukherjee, P.M., Latha, J., Hadar, R., Horwitz, B.A., 2004. Role of two Gprotein alpha subunits, TgaA and TgaB, in the antagonism of plant pathogens by *Trichoderma virens*. Appl. Environ. Microbiol. 70, 542–549.

Mukherjee, P.K., Wiest, A., Ruiz, N., Keightley, A., Moran-Diez, M.E., McCluskey, K., Pouchus, Y.F., Kenerley, C.M., 2011. Two classes of new peptaibols are synthesized by a single non-ribosomal peptide synthetase of *Trichoderma virens*. J. Biol. Chem. 286, 4544–4554.

Mukherjee, P.K., Buensanteai, N., Moran-Diez, M.E., Druzhinina, I.S., Kenerley, C.M., 2012. Functional analysis of non-ribosomal peptide synthetases (NRPSs) in *Trichoderma virens* reveals a polyketide synthase (PKS)/NRPS hybrid enzyme involved in induced systemic resistance response in maize. Microbiology 158, 155–165.

Mukherjee, P.K., Horwitz, B.A., Singh, U.S., Mukherjee, M., Schmoll, M., 2013. *Trichoderma* in agriculture, industry and medicine: an overview. In: Mukherjee, P.K., Horwitz, B.A., Singh, U.S., Mukherjee, M., Schmoll, M. (Eds.), *Trichoderma:* Biology and Applications. CABI, Nosworthy, Way, Wallingford, Oxon, UK, pp. 1–9.

Murali, M., Jogaiah, S., Amruthesh, K.N., Shin-ichi, I., Shekar Shetty, H., 2013. Rhizosphere fungus *Penicillium chrysogenum* promotes growth and induces defense-related genes and downy mildew disease resistance in pearl millet. Plant Biol. 15, 111–118.

Nagaraju, A., Jogaiah, S., Mahadeva Murthy, S., Shin-ichi, I., 2012. Seed priming with *Trichoderma harzianum* isolates enhances plant growth and induces resistance against *Plasmopara halstedii*, an incitant of sunflower downy mildew disease. Aust. J. Plant Path. 41, 609–620.

Nandini, B., Puttaswamy, H., Prakash, H.S., Adhikari, S., Jogaiah, S., Geetha, N., 2020. Elicitation of novel trichogenic-lipid nanoemulsion signaling resistance against pearl millet downy mildew disease. Biomol. Ther. 10 (1), 25.

Naznin, A., Hossain, M.M., Anju-Man Ara, K., Hoque, A., Islam, M., Hasan, T., 2015. Influence of organic amendments and bio-control agent on yield and quality of tuberose. J. Hortic. 2, 4. https://doi.org/10.4172/2376-0354.1000156.

Oliveira, A.L., Gallo, M., Pazzagli, L., Benedetti, C.E., Cappugi, G., Scala, A., Pantera, B., Spisni, A., Pertinhez, T.A., Cicero, D.O., 2011. The structure of the elicitor cerato-platanin (CP), the first member of the CP fungal protein family, reveals a double ψβ-barrel fold and carbohydrate binding. J. Biol. Chem. 286, 17560–17568.

Omann, M.R., Lehner, S., Rodriguez, E.C., Brunner, K., Zeilinger, S., 2012. The seven transmembrane receptor Gpr1 governs processes relevant for the antagonistic interaction of *Trichoderma atroviride* with its host. Microbiology 158, 107–118.

Pandey, N., Adhikhari, M., Bhantana, B., 2019. Trichoderma and its prospects in agriculture of Nepal: an overview. Int. J. Appl. Sci. Biotechnol. 7 (3), 309–316.

Poveda, J., Hermosa, R., Monte, E., et al., 2019. *Trichoderma harzianum* favours the access of arbuscular mycorrhizal fungi to non-host Brassicaceae roots and increases plant productivity. Sci. Rep. 9, 11650. https://doi.org/10.1038/s41598-019-48269-z.

Rotblat, B., Enshell-Seijffers, D., Gershoni, J.M., Schuster, S., Avni, A., 2002. Identification of an essential component of the elicitation active site of the EIX protein elicitor. Plant J. 32, 1049–1055.

Sajeesh, P.K., 2015. Cu-Chi-Tri: A triple combination for the management of late blight disease of potato (*Solanum tuberosum* L.). PhD Thesis, GBPUA&T, Pantnagar, India.

Samolski, I., Rincon, A.M., Pinzon, L.M., Viterbo, A., Monte, E., 2012. The *qid74* gene from *Trichoderma harzianum* has a role in root architecture and plant biofertilization. Microbiology 158, 129–138.

Sarrocco, S., Guidi, L., Fambrini, S., DesI"Innocenti, E. and Vannacci, G., 2009. Competition for cellulose exploitation between *Rhizoctonia solani* and two *Trichoderma* isolated in the decomposition of wheat straw. J. Plant Pathol. 91, 331–338.

Satapute, P., Milan, V.K., Shivakantkumar, S.A., Jogaiah, S., 2019. Influence of triazole pesticides on tillage soil microbial populations and metabolic changes. Sci. Total Environ. 651, 2334–2344.

Schuster, A., Schmoll, M., 2010. Biology and biotechnology of Trichoderma. Appl. Microbiol. Biotechnol. 87, 787–799.

Seidl, V., Marchetti, M., Schandl, R., Allmaier, G., Kubicek, C.P., 2006. Epl1, the major secreted protein of Hypocrea atroviridis on glucose, is a member of a strongly conserved protein family comprising plant defense response elicitors. FEBS J. 273, 4346–4359.

Seidl, V., Song, L., Lindquist, E., Gruber, S., Koptchinskly, A., Zeilinger, S., Schmoll, M., Martinez, P., Sun, J., Grigoriev, I., Herrera-Estrella, H., Baker, S.E., Kubicek, C.P., 2009. Transcriptomic response of the mycoparasitic fungus *Trichoderma atroviride* to the presence of fungal prey. BMC Genomics 10, 567.

Shi, M., Chen, L., Wang, X.W., Zhang, T., Zhao, P.B., Song, X.Y., Sun, C.Y., Chen, X.L., Zhou, B.C., Zhang, Y.Z., 2012. Antimicrobial peptaibols from *Trichoderma pseudokoningii* induce programmed cell death in plant fungal pathogens. Microbiology 158, 166–175.

Shoresh, M., Harman, G.E., 2008. The molecular basis of shoot responses of maize seedlings to *Trichoderma harzianum* T22 inoculation of the root: a proteomic approach. Plant Physiol. 147, 2147–2163.

Shoresh, M., Gal-On, A., Leibman, D., Chet, I., 2006. Characterization of a mitogen-activated protein kinase gene from cucumber required for *Trichoderma*-conferred plant resistance. Plant Physiol. 142, 1169–1179.

Shoresh, M., Harman, G.E., Mastouri, F., 2010. Induced systemic resistance and plant responses to fungal biocontrol agents. Annu. Rev. Phytopathol. 48, 21–43.

Shukla, N., Awasthi, R.P., Rawat, L., Kumar, J., 2012. Biochemical and physiological responses of rice (*Oryza sativa* L.) as influenced by *Trichoderma harzianum* under drought stress. Plant Physiol. Biochem. 54, 78–88.

Singh, S., Dureja, P., Tanwar, R.S., Singh, A., 2005. Production and antifungal activity of secondary metabolites of *Trichoderma* virens. Pestic. Res. J. 17, 26–29.

Singh, H.B., Singh, B.N., Singh, S.P., Singh, S.R., Sarma, B.K., 2009. Biological control of plant diseases: current status and future prospects. In: Johri, J.K. (Ed.), Recent Advances in Biopesticides: Biotechnological Applications. New India Pub, New Delhi, p. 32.

Singh, A., Shukla, N., Kabadwal, B.C., Tewari, A.K., Kumar, J., 2018. Review on plant-*Trichoderma*-pathogen interaction. Int. J. Curr. Microbiol. App.Sci. 7 (02), 2382–2397.

Sudisha, J., Niranjana, S.R., Umesha, S., Prakash, H.S., Shekar Shetty, H., 2006. Transmission of seed-borne infection of muskmelon by *Didymella bryoniae* and effect of seed treatments on disease incidence and fruit yield. Biol. Control 37, 196–205.

Thakur, M., Sohal, B.S., 2013. Role of elicitors in inducing resistance in plants against pathogen infection: a review. ISRN *Biochem.* 1–10. https://doi.org/10.1155/2013/762412.

Thakur, A.K., Uphoff, N., Antony, E., 2010. An assessment of physiological effects of system of rice intensification (sri) practices compared with recommended rice cultivation practices in India. Exp. Agric. 46, 77–98.

Tijerino, A., Cardoza, R.E., Moraga, J., Malmierca, M.G., Vicente, F., Aleu, J., Collado, I.G., Gutierrez, S., Monte, E., Hermosa, R., 2011. Overexpression of the trichodiene synthase gene tri5 increases trichodermin production and antimicrobial activity in *Trichoderma brevicompactum*. Fungal Genet. Biol. 48, 285–296.

Tucci, M., Ruocco, M., De Masi, L., De Palma, M., Lorito, M., 2011. The beneficial effect of *Trichoderma* spp. on tomato is modulated by the plant genotype. Mol. Plant Pathol. 12, 341–354. https://doi.org/10.1111/j.1364-3703.2010.00674.

Van Wees, S.C.M., Van der Ent, S., Pieterse, C.M.J., 2008. Plant immune responses triggered by beneficial microbes. Curr. Opin. Plant Biol. 11, 443–448.

Vargas, W.A., Djonović, S., Sukno, S.A., Kenerley, C.M., 2008. Dimerization controls the activity of fungal elicitors that trigger systemic resistance in plants. J. Biol. Chem. 283, 19804–19815.

Vargas, W.A., Crutcher, F.K., Kenerley, C.M., 2011. Functional characterization of a plant-like sucrose transporter from the beneficial fungus *Trichodermavirens*. Regulation of the symbiotic association with plants by sucrose metabolism inside the fungal cells. *New Phytol.* 189, 777–789.

Verma, M., Brar, S.K., Tyagi, R.D., Surampalli, R.Y., Valero, J.R., 2007. Antagonistic fungi, *Trichoderma* spp.: panoply of biological control. Biochem. Eng. J. 37, 1–20.

Vinale, F., Sivasithamparam, K., Ghisalberti, E.L., Marra, R., Woo, S.L., Lorio, M., 2008. *Trichoderma*-plant-pathogen interactions. Soil Biol. Biochem. 40, 1–10.

Vinale, F., Ghisalberti, E.L., Sivasithamparam, K., Marra, R., Ritieni, A., Ferracane, R., Woo, S., Lorito, M., 2009. Factors affecting the production of *Trichoderma harzianum* secondary metabolites during the interaction with different plant pathogens. Lett. Appl. Microbiol. 48, 705–711.

Vinale, F., Marra, R., Ruocco, M., 2014. *Trichoderma* secondary metabolites active on plants and fungal pathogens. Open Mycol. J. 8, 127–139.

Viterbo, A., Wiest, A., Brotman, Y., Chet, I., Kenerley, C., 2007. The 18mer peptaibols from Trichoderma virens elicit plant defence responses. Mol. Plant Pathol. 8, 737–746.

Viterbo, A., Landau, U., Kim, S., Chernin, L., Chet, I., 2010. Characterization of ACC deaminase from the biocontrol and plant growthpromoting agent *Trichodermaasperellum* T203. FEMS Microbiol. Lett. 305, 42–48.

Waghunde, R.R., Rahul, M.S., Sabalpara, A.N., 2016. *Trichoderma*: a significant fungus for agriculture and environment. Afr. J. Agric. Res. 11 (22), 1952–1965.

Woo, S.L., Scala, F., Ruocco, M., Lorito, M., 2006. The molecular biology of the interactions between *Trichoderma* spp., pathogenic fungi, and plants. Phytopathology 96, 181–185.

Yoshioka, Y., Ichikawa, H., Naznin, H.A., Kogure, A., Hyakumachi, M., 2012. Systemic resistance induced in *Arabidopsis thaliana* by *Trichoderma asperellum* SKT-1, a microbial pesticide of seedborne diseases of rice. Pest. Manag. Sci. 68, 60–66.

Zachow, C., Berg, C., Müller, H., Monk, J., Berg, G., 2016. Endemic plants harbour specific*Trichoderma* communities with an exceptional potential for biocontrol of phytopathogens. J. Biotechnol. 235, 162–170.

Zeilinger, S., Reithner, B., Scala, V., Peiss, I., Lorito, M., Mach, R.L., 2005. Signal transduction by Tga3, a novel G protein alpha subunit of *Trichoderma atroviride*. Appl. Environ. Microbiol. 71, 1591–1597.

Trichoderma: From gene to field

4

B. Nandini and N. Geetha
Nanobiotechnology Laboratory, DOS in Biotechnology, University of Mysore, Mysore, Karnataka, India

4.1 Introduction

Trichoderma spp. are one of the most widespread free-living saprophytic fungi in the rhizosphere region, which extensively inhabits the foremost share of fungal biocontrol agents in the biopesticide industry (Woo et al., 2014). The genus *Trichoderma* belongs to the kingdom Fungi, with class of Sordariomycetes, Hypocreales order and a member of Hypocreaceae family. Root colonization by *Trichoderma* spp. causes significant changes to the plant proteome and metabolome, and consequentially improves the crop productivity and tolerability to abiotic and biotic stresses (Harman et al., 2004; Shoresh et al., 2010). Additionally, *Trichoderma* beneficial effects vary with respect to plant genotypes, fungal strains, and functional concentrations (Islam et al., 2013; Segarra et al., 2009). It has been known that tropical soils have 10^1–10^3 *Trichoderma* culturable propagules per gram (Druzhinina et al., 2011; Etschmann et al., 2015; Moran-Diez et al., 2015). *Trichoderma* spp. are the most adaptable opportunistic plant symbionts that can cause substantial modification in the metabolism of host plants, thus improving plant growth and eliciting the plant defense responses to various diseases (Nagaraju et al., 2012; Abdelrahman et al., 2016).

Trichoderma spp. are eminent biocontrol agents against phytopathogens (Jogaiah et al., 2013). Competition between *Trichoderma* spp. and phytopathogens for infectivity sites and nutrients is known as niche exclusion. The aptitude of some *Trichoderma* spp. to control plant diseases and motivate the growth and development of plants reveals their extensive and long phrase use as biocontrol agents in cultivation of various crops (Hermosa et al., 2000). For example, *T. harzianum*, *T. virens*, and *T. viride* are presently economically marketed as biocontrol agents.

The biological control mechanism involves antagonism, antibiosis, mycoparasitism, and elicitation of plant defense responses, all of which have been recognized in *Trichoderma* spp. (Harman et al., 2004; Jogaiah et al., 2018). Antiphytopathogen mechanisms activities include antibiosis, mycoparasitism, induced systemic resistance, and niche exclusion (Bae, 2011). Antibiosis involves the production of different antimicrobial compounds that act as inhibitors of phytopathogen growth (Vinale et al., 2008). Cellulose is one of the main components in the *Trichoderma* cell walls and consists of homopolymer β-1,4-glucan chains produced at the plasma membrane. During the mycoparasitism course of action, cell wall degrading enzymes produced from *Trichoderma* will degrade cell walls of phytopathogen (Reithner et al., 2011).

Biocontrol Agents and Secondary Metabolites. https://doi.org/10.1016/B978-0-12-822919-4.00004-1

Trichoderma produced secondary metabolites and cell wall degrading enzymes (CWDEs) lead to the process called mycoparasitism. *Trichoderma* spp. affix to the hyphae of the target fungal pathogens through coiling and enter the cell wall by the production of CWDEs (Viterbo et al., 2002; Gajera et al., 2012). The CWDEs penetrate the target pathogen and make use of its cell wall material (β-1, 3 glucan, and chitin) as nutrients and support the growth of *Trichoderma* (Kubicek et al., 2001). Hence, *Trichoderma* spp. are extensively used as biocontrol agents that improve plant growth as well as prevent colonization of phytopathogen. Efficiency of biopriming with *T. asperellum* BHUT8 was assessed for plant growth improvement in pea (Singh et al., 2016).

Trichoderma spp. are endophytic plant symbionts that are extensively used as seed treatments to control and manage diseases and also to enhance plant growth and yield (Mastouri et al., 2010). *Trichoderma* and its direct interaction with pathogen involves extracellular CWDEs, diffusible substances including antibiotics as the most important factors for mycoparasitism and antibiosis (Sivasithamparam and Ghisalberti, 1998; Kredics et al., 2001; Benitez et al., 2004; Harman et al., 2004). The production of CWDEs such as chitinase, cellulase, protease, and β-(1–3) glucanase by *Trichoderma* spp. has a vital role in the inhibition of the fungal pathogens (Vinale et al., 2008; Gajera et al., 2012) and induced resistance of the host plant system. Additionally, enzymes can offer various benefits to the host plants by interaction and degradation of hydrocarbons and pesticides used in agriculture (Harzianic acid: a novel siderophore from *T. harzianum*) (Vinale et al., 2013). It was illustrated that, few released cell wall degraded products form host plants are a group of elicitors, which enables the host plant to activate damage-associated molecular pattern molecules (DAMPs) effectively against pathogens (Boller and Felix, 2009). *Trichoderma*-mediated plant immunity in cotton, rice, and *A. thaliana* is associated with reactive oxygen species (ROS) such as hydrogen peroxide (H_2O_2) and nitric oxide (Contreras-Cornejo et al., 2014).

Trichoderma spp. are able to form endophytic relations and interact with other microbes in the rhizosphere, thus influencing disease protection, plant growth, and yield (Nagaraju et al., 2012b). During the process of *Trichoderma* root colonization, an extensive exchange of molecular messages takes place, including the fungal elicitor's deposition in the root cell apoplast (Hermosa et al., 2012; Gupta et al., 2014; Contreras-Cornejo et al., 2014; Jogaiah et al., 2018). Some strains have the ability to decrease the severity of plant diseases by reducing plant pathogens, primarily in the soil or on plant roots, during their high antagonistic and mycoparasitic approaches (Viterbo and Horwitz, 2010). The plant-microbe relationship involves molecular recognition among the two partners through a signaling system arbitrated by the plant hormones jasmonic acid (JA), salicylic acid (SA), and ethylene (ET). JA and ET have been depicted as signal transduction molecules for induced systemic resistance (ISR) owing to the effect of valuable microbes, and the signal transduction pathway through SA accumulation is observed in the systemic acquired resistance (SAR) stimulated by the attack by pathogens.

A general characteristic feature of ISR responses to beneficial microbes is by priming to accelerate or enhance defense responses. In the bio-primed plants, defense responses are not elicited directly, but are a step up ahead assault by pathogens or insects, consequential in faster and stronger resistance to the invader (Van Wees et al., 2008). Managing a broad range of plant pathogens, including fungi, oomycetes, bacterial, and viral diseases, through elicitation of ISR by *Trichoderma* or localized resistance is well evident (Harman et al., 2004). The recent comparative genome sequence analysis of two documented biocontrol species, *T. atroviride* and *T. virens*, shows a better mycoparasitism mechanism arising from a common *Trichoderma* ancestor in their lifestyle of the genus (Kubicek et al., 2011). The existence of a fungal victim and the accessibility of root-derived nutrients might have been the main attractors for the ancestors of *Trichoderma* to set them up in the rhizosphere and to make possible the evolution of positive connections with plants (Druzhinina et al., 2011). Several *Trichoderma* rhizosphere-competent strains have been revealed to possess direct effects on plants, increasing their growth potential and nutrient uptake, fertilizer use effectiveness, percentage and rate of seed germination, and motivation of plant defenses against biotic and abiotic injuries (Shoresh et al., 2010).

4.2 *Trichoderma*-mediated genes and elicitors-induced disease resistance in plant host system

Trichoderma is well preferred as a biocontrol agent against plant pathogens, because of its capacity to control disease in plants (Dubey et al., 2007; Vinale et al., 2008). *Trichoderma* produced metabolites act as biological control agents in inhibiting *Phytophthora* pathogens (Bae et al., 2016). Singh (2016) reviewed the efficiency of seed biopriming as a comprehensive approach toward agricultural sustainability. *Trichoderma* spp. are found in widespread soils and in root inhabitants that have been broadly considered due to their ability to produce antibiotics, parasitize other fungi, and compete with harmful plant microorganisms. These fungi generate various secondary metabolites such as nonribosomal peptides, terpenoids, pyrones, and indole-derived compounds. There are 373 different secondary metabolites molecules, which have been known; however, in most of the cases, the exact activity of these molecules is mysterious (Reino et al., 2008; Mukherjee et al., 2012a,b; Crutcher et al., 2013). In the rhizosphere region, the exchange and detection of signaling molecules by *Trichoderma* and plants might change the physiological and biochemical features in both the systems. For instance, numerous *Trichoderma* strains persuade root branching and enhance shoot biomass as a result of cell division, development, and differentiation by the incidence of fungal auxin-like compounds. Moreover, *Trichoderma* association with plant roots will trigger systemic resistance and enhance plant nutrient uptake (Contreras-Cornejo et al., 2016).

Numerous kinds of transcriptional factors are also responsible for the biocontrol action by *Trichoderma* (Brunner et al., 2005; Jogaiah et al., 2018). A comparative genome analysis of the mycoparasitic species of *T. atroviride* (teleomorph *Hypocrea atroviridis*) and *T. virens* (formerly *Gliocladium virens*) and the saprophytic *T. reesei* (teleomorph *Hypocrea jecorina*) discloses significant differences in their lifestyle (Kubicek et al., 2011). Thus far, a minimum of 1100 *Hypocrea* (sexual telemorphic stage)/*Trichoderma* (asexual anamorphic stage) strains have been recognized from 75 molecularly characterized species and numerous new species are being identified (Druzhinina et al., 2011). Genome sequencing projects have also been focused on the species *T. harzianum*, *T. asperellum*, *T. longibrachiatum*, and *T. citrinoviride* (Mukherjee et al., 2013). Isolates of *Trichoderma* spp. induce systemic resistance against *Xanthomonas vesicatoria* and *Alternaria solani* pathogens (Fontenelle et al., 2011). *Trichoderma* spp. are able to interact directly with roots, escalating plant growth potential, resistance to disease, and tolerance to abiotic stresses (Hermosa et al., 2012).

Inhibitory activity of *T. harzianum* and *T. viride* culture filtrates against *Fusarium moniliforme* pathogens due to the production of volatile compounds and release of extracellular enzymes, such as those with amylolytic, pectinolytic, proteolytic, and cellulolytic activities (Calistru et al., 1997). Earlier reports reveal the efficiency of *Trichoderma* spp. as biofertilizer or biocontrol agents for crop production in the field or greenhouse agriculture farming systems (Harman et al., 2004; Brunner et al., 2005) as an alternative choice to the chemical fungicides (Harman and Kubicek, 1998). Root colonization by the advantageous antagonist *Trichoderma* has been studied using conventional microbiological techniques (Gebarowska and Pietr, 2006). *T. harzianum*-T39 reduced downy mildew disease up to 63% on susceptible grapevine cultivars caused by *Plasmopara viticola* under greenhouse conditions when applied before 48–72 h of inoculation (Perazzolli et al., 2008). Yoshioka et al. (2012) demonstrates the capability of a cell-free culture filtrate of *T. asperellum* STK-1 to induce systemic resistance in *A. thaliana* by reducing lesion development and growth of the pathogen *P. syringae* pv. tomato DC3000.

Cellulysin from the fungus *T. viride* is the first example of a fungal elicitor which induces disease resistance via the octadecanoid signaling cascade (Piel et al., 1997; Pushpalatha et al., 2011). Cellulysin isolated from efficient biocontrol *T. virens* stimulates resistance and related compounds in the roots of cotton (Hanson and Howell, 2004). Terpenoid synthesis was induced in cotton roots for the protection of plants against *Rhizoctonia solani* infection (Howell et al., 2000). Formulations of *Trichoderma* spp. seed treatment were effective in controlling preemergence damping-off caused by *Rhizopus oryzae* and *Pythium* spp. in cotton (Howell, 2002). Oomycetes *Pythium ultimum* and the necrotrophic pathogen *Colletotrichum graminicola* in maize were effectively controlled by *Trichoderma* spp. along with better growth promotion (Harman et al., 2004a,b). *T. harzianum* isolates induces systemic resistance and enhances crop growth (Nagaraju et al., 2012). *T. virens* stimulates systemic protection associated with prominent induction of JA- and GLV-biosynthetic genes. Sm1, a proteinaceous elicitor from *T. virens*, is the main requisite for inducing systemic resistance in maize (Djonović et al., 2007). Elicitors of *Trichoderma* potentially control *Phytophthora capsici* by inducing systemic resistance in red pepper (Sriram et al., 2009). *T. harzianum* formulations successfully inhibited *F. verticillioides* and Fumonisin infection and also enhanced the seed germination, vigor index, field emergence, yield, and thousand seed weight in contrast to the control (Chandranayaka et al., 2010). *T. asperellum* significantly inhibits the *Fusarium oxysporum* f. sp. *cucumerinum* growth within roots by the probable production of the CWDEs and secondary metabolites (Lorito et al., 2010; Zhao et al., 2011). Seed biopriming with *T. harzianum* supports growth parameters and drought tolerance in *Triticum aestivum* (Shukla et al., 2015).

Trichoderma isolates are proficient in producing numerous powerful plant-degrading enzymes and more than 200 types of antibiotics are produced that are highly toxic to any macro- and microorganism (Kubicek and Penttila, 1998). Antibiotic production is possibly situated in the tips of growing hyphae, with components responsible for the antagonism activity at the summit of contact with the pathogen (Michalikova and Michrina, 1997). Diverse mechanisms have been recommended for their mycoparasitic accumulation of lytic enzymes, such as cellulase (β-1, 4-glucanase), proteases, and chitinases, which damage cell walls (Witkowsha and Maj, 2002). Application of small secondary metabolites produced by different *Trichoderma* strains induces the expression of pathogenesis-related (PR) proteins in plants with the decline of disease symptoms systemically (Vinale et al., 2008). In advance, evidence for the participation of JA and ethylene in signal transducing from *Trichoderma*-inoculated roots to the leaves arrives from gene expression. In roots, real-time reverse transcription polymerase chain reaction (RT-PCR) points toward that Lox1, which encodes an LOX involvement in JA synthesis, and prevents a feed-forward loop in jasmonate synthesis that was upregulated by inoculation with *T. asperellum*. Lox1 is activated in the roots as early as 1-h post-*Trichoderma* inoculation. A second peak was noticed 24-h postinoculation probably resulting from the opening of the octadecanoic pathway and the production of JA. Another gene established to be upregulated by *Trichoderma* inoculation is *Pal1*, which initiates phenylalanine ammonia-lyase (PAL) (Shoresh et al., 2005; Shoresh and Harman, 2008).

Engelberth et al. (2001) demonstrates that the release of ET, JA, and volatile compounds associated with the octadecanoic signaling pathway is stimulated in lima bean plants treated with the peptaibol alamethicin from *T. viride*. Besides, Viterbo et al. (2007) explained that the 18-residue peptaibols induced expression of hydroperoxide lyase (HPL), phenylalanine ammonia-lyase (PAL), and peroxidase (POX), which is implicated in formation of antimicrobial compounds, associated with a systemic increase in antimicrobial compounds in the plant. The SM1-ISR was in prominent association with induction of JA and GLV-biosynthetic genes (Djonović et al., 2007). Transcription analysis of plant interaction with *T. hamatum* failed to notice the induction of ISR markers and only one marker of SAR (PR5) was upregulated (Alfano et al., 2007). So, it might be that some *Trichoderma* spp. use other mechanisms to induce plant defense (Shoresh et al., 2010). *Trichoderma* spp. have significant ability to solubilize a range of plant nutrients, such as phosphorus and micronutrients, together with iron, copper, zinc, and manganese, thus rendering them obtainable for plants (Altomare et al., 1999). In the mycoparasitism process by *Trichoderma* spp., volatile secondary metabolites play a key role (Vinale et al., 2008; Stoppacher et al., 2010). Crutcher et al. (2013) reported a comparative catalog of the volatile organic compounds released by *T. atroviride*, *T. virens*, and *T. reesei*. Certainly, the chemical profile was incredibly different among species. A number of mono- and sesquiterpenes were produced by *T. virens* or *T. reesei*. *T. atroviride* are to produce 6-pentyl-α-pyrone, wherein, *T. virens* are not able to produce this metabolite. In comparison, *T. reesei* produces a poor blend of volatile organic compounds in contrast to *T. virens* and *T. atroviride*.

The target metabolite profiling approach was elected to reveal that *T. longibrachiatum* isolated from desert soil offers beneficial agronomic behavior in onion and stimulates defense responses against *Fusarium oxysporum* f. sp. *cepa*, through activating a number of primary and secondary metabolite pathways (Abdelrahman et al., 2016). The synergistic inhibitory effect of *Trichoderma*-derived CWDEs and fungal metabolites on the fungal pathogens was also scrutinized (Saravanakumar et al., 2016). Using a qRT-PCR approach to analyze selected marker genes, biopriming with *T. asperellum* elicits defense against virulent *Pseudomonas syringae* (Pst) in *Arabidopsis thaliana* without causing major changes in gene expression. Only some genes related to ET/JA signaling were considerably affected in leaves of plants when primed by *T. asperellum* (Brotman et al., 2012). At the metabolomic level, different amino acid precursors of plant defense metabolites (Coruzzi and Last, 2000) have been exposed to secrete to an elevated level by successive priming with *T. asperellum* and also with chemicals such as pipecolic acid (Brotman et al., 2012; Vogel-Adghough et al., 2013). Molecular studies on this feature of the biocontrol are awfully restricted. Moreover, there are no reports on *Trichoderma*-derived secondary metabolites and their interactions with the growth-promoting protein of pathogens exploiting molecular docking studies (Saravanakumar et al., 2016).

4.3 *Trichoderma*-based biocontrol formulations

Many biopesticide-based formulations of *Trichoderma* spp. are available commercially (Woo et al., 2006). A number of *Trichoderma* strains have been shown to colonize roots to elicit ISR, thus priming the host for an intense defense action against succeeding pathogen assault (Hanson and Howell, 2004; Sudisha et al., 2006; Segarra et al., 2007; Tucci et al., 2011; Reglinski et al., 2012). Priming of *T. harzianum* T39, has been reported to induce resistance in grapevine against downy mildew disease (Perazzolli et al., 2011). *T. asperellum* SKT-1, elicits systemic resistance against the bacterial pathogen *Pseudomonas syringae* pv. *Tomato* DC3000 via the SA-based signaling pathway in *Arabidopsis* (Yoshioka et al., 2012). Over 100 antimicrobial compounds have been recognized from *Trichoderma* spp. (Vinale et al., 2008).

Trichoderma produces a number of secondary metabolites with pharmaceutical and biotechnological significance that comprise nonribosomal peptides (NRPs), poliketides, peptaibols, pyrones, siderophores, and volatile and nonvolatile terpenes (Vinale et al., 2008). Presently, *Trichoderma* spp. are extensively used in industrial practices and agriculture owing to their ability to produce enzymes and secondary metabolites (Mukherjee et al., 2008; Jiang et al., 2011). *Trichoderma* spp. are well

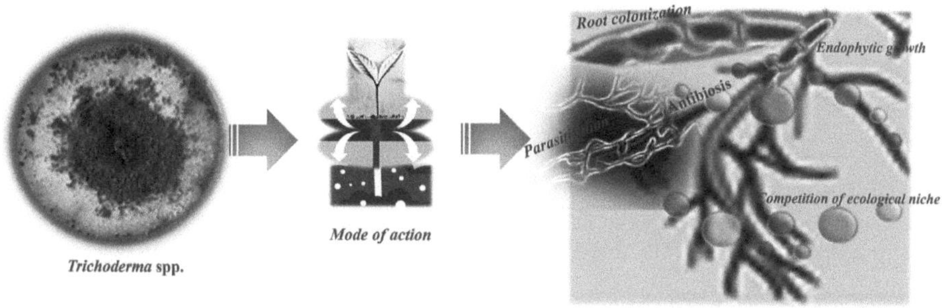

Fig. 4.1 Mode of action of *Trichoderma* spp. against phytopathogen and plant growth improvement.
From Waghunde, R.R., Shelake, R.M., Sabalpara, A.N., 2016. Trichoderma: a significant fungus for agriculture and environment. Afr. J. Agri. Res. 11, 1952–1965. https://doi.org/10.5897/AJAR2015.10584.

recognized in biofertilizer production because of their potential to establish mycorrhizal-like association with plants, prevent root and foliar pathogens, alter the microfloral symphony in roots, improve nutrient uptake, enhance root development, amplify root hair formation, help the plant in developing systemic resistance, solubilize phosphate, degrade cellulose, and generate siderophores (Fig. 4.1; Yadidia et al., 1999; Harman, 2000; Harman et al., 2004; Saba et al., 2012; Saravanakumar et al., 2013). Formulations of *T. viride* control wilt and blight diseases caused by ungal phytopathogens (Surekha et al., 2014).

A huge demand for eco-friendly based biocontrol agents has created a great potential for the incorporation of *Trichoderma*. Further, in developing countries, to ensure sustained availability to the farmers, there is a need to establish commercial production or scale-up centers at rural levels (Fig. 4.2; Table 4.1). These

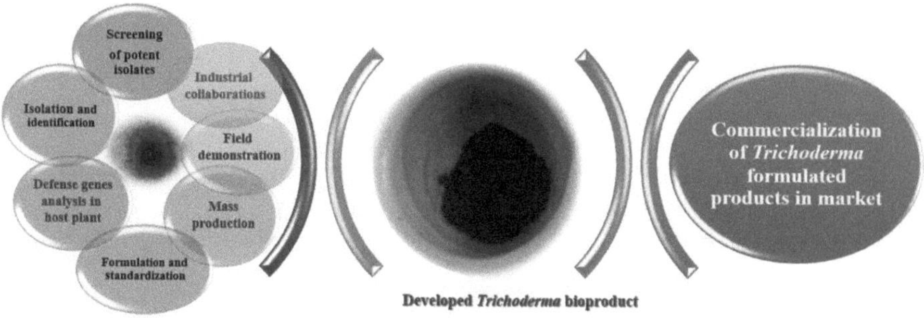

Fig. 4.2 Exploring the potential of*Trichoderma* for the Management of Seed and SoilBorne Diseases of Crops.
From Nakkeeran, S., Renukadevi, P., Aiyanathan, K.E.A. 2016. Exploring the potential of Trichoderma for the Management of Seed and Soil-Borne Diseases of Crops. In: Muniappan, R., Heinrichs, E. (eds) Integrated Pest Management of Tropical Vegetable Crops. Springer, Dordrecht.

Table 4.1 Some commercial products of *Trichoderma* spp.

Product	***Trichoderma* spp.**	**Agency/Company**
Ecofit	*T. viride*	Hoechest and Schering Agro. Evo. Ltd., Mumbai, India
Funginil	T. viride	Crop Helath Bioproduct Research Centre, Gaziabad (UP), India
Trichogourd	*T. viride*	Anu Biotech international Ltd. Bangalore, India
Defense SF	*T. viride*	Wockhardt Life Science Ltd., Mumbai India
Tricho-X	*T. viride*	Excel Industries Ltd., Mumbai, India
Biogourd	*T. viride*	Krishi Rasayan Export Pvt. Ltd., Solan (HP), India
Biocon	*T. viride*	Tocklai Experimental Station Tea Research Association, Jorhat (Assam), India
Bioderma	*T. viride + T. harzianum*	Biotech international Ltd., New Delhi, India
Bip T	*T. viride*	Poland
Pant biocontrol agent-1	*T. harzianum*	Department of Plant Pathology, GB Plant University of Agriculture and Technology, Panatnagar, Uttarakhand, India
T34 Biocontrol-product	*T. asperellum* T34	Biocontrol technologies
Trieco	*T. viride*	Ecosense Labs, India
Bioderma	*T. viride + T. harzianum*	Biotech International Ltd., New Delhi, India
Ecoderma	*T. viride + T. harzianum*	Morgo Biocontrol Pvt. Ltd., Bangalore, India
Top shield, Root shield	*T. harzianum*	Bioworks, Geneva (Switzerland) and New York (USA)
RootShieldTM, BioTrek 22GTM, SupresivitTM, T-22GTM, T-22HBTM	*T. harzianum*	BioWorks, Wilbur-Ellis, Borregaard
*Trichoderma*2000	*T. harzianum*	Mycontrol (EfA1) Ltd., Israel
PromotTM, *Trichoderma*2000, Biofungus	*Trichoderma* spp.	J.H. Biotech, Mycontrol, Ltd., De Ceuster, US, US, Belgium
Trichopel TrichopelTM, TrichojetTM, TrichodowelsTM, TrichosealTM	*T. harzianum* and *T. viride*	Agrimm Technologies Ltd., New Zealand
T-22 and T-22HB	*T. harzianum*	BioWorks (= TGT Inc.), Geneva, USA

Table 4.1 Continued

Product	*Trichoderma* spp.	Agency/Company
F-Stop	*T. harzianum*	Eastern Kodak Company TGT Inc. New York
T-35	*T. harzianum*	Makhteshim-Agan Chemical, Israel
Harzian 10 and 20	*T. harzianum*	Natural Plant Protection, Noguerres, France
Bio-Trek, Root Shield	*T. harzianum*	Borregaard and Reitzel, Denmark;
Supresivit		Fytovita, Czech Republic
Trichodowels, Trichoject, Trichoseal and others	*T. harzianum* and *T. viride*	Agrimms Biologicals, New Zealand, and others
Binab T	*T. harzianum*, *T. polysporum*	Bio-Innovation, Sweden
Trichodex	*T. harzianum T-39*	Makhteshim-Agan, Israel, several European companies, Belgium
Gilogard and Soil guard	*T. virens*	Greece-Sierre Company, Maryland
Promot PlusWP Promot PlusDD	*Trichoderma* spp. *T. koningii* *T. harzianum*	Tan Quy, Viet Nam
NLU-Tri	*T. virens*	Ho Chi Minh University of Agriculture and Forestry, Viet Nam
Biobus 1.00WP	*T. viride*	Nam Bac, Viet Nam

farmers should be encouraged to undergo skill-oriented training programs organized by scientists to implement Integrated pest management (IPM)-based innovative centers at their village levels. They have to be trained to understand the nature and pathogenesis of soil and seed-borne pathogen and role of *Trichoderma* antagonism (Nakkeeran et al., 2016).

4.4 *Trichoderma*-based effector molecules: A model system to design specific bioformulations

For a strategic implementation of plant disease management, indepth studies on the process of pathogenesis are essential to determine how the pathogens get established in plants, damage epithelial tissues, and bypass host plant defenses. One of the main events, which is a corresponding scrutiny, is a study on the early stages of the host-pathogen interactions. Pathogenesis by fungi delivers a wide range of biomolecules, which enable them to penetrate deeper plant tissues by overcoming the structural barriers, by escaping signal perception, or active defense responses (Shen et al., 2018).

These pathogen-derived molecules are termed as "effector" molecules, defined as any given molecule that can alter the physiology, structure, or function of another organism, facilitating the infection and/or triggering defense responses (Kamoun, 2006). Effector molecules have been well studied due to their involvement in harmful pathogenic interactions, which is having a negative impact on agriculture. The functional classification of the effector molecules released by *Trichoderma* is widely debated. Some recent researches on *Trichoderma* effectors in symbiotic interactions have pointed out that most effectors identified in symbiotic associations are members of the families present in phytopathogens, and involve exchanges of molecular dialogues (Ramírez-Valdespino et al., 2019).

4.5 *Trichoderma* effector proteins

The main strategy to identify *Trichoderma* effector proteins is to analyze their upregulation in the presence of host plants (Guzmán-Guzmán et al., 2017). The most copious groups of secreted proteins in *Trichoderma* spp. are the small proteins containing four or more cysteine residues, viz., the small secreted cysteine-rich proteins (SSCPs). These SSCPs are grouped into protein families with functions such as hydrophobins, cerato-platanins (Druzhinina et al., 2012). Predictions through Bioinformatic studies specified the abundant presence of SSCPs in *T. atroviridee* and *T. virens* genomes, which are present in cell surface proteins with important roles in the plant-pathogen interactions (Pérez et al., 2011; Druzhinina et al., 2012). Upregulation of hydrophobins has been reported during *Trichoderma*-plant interactions; and genomes of the *Trichoderma* species of *T. virens* and *T. atroviridee* contain 17 sequences encoding for hydrophobins (Kubicek et al., 2011). It is also reported that hydrophobins upregulate two genes related to auxin signaling, and hence this protein is involved in the promotion of growth and defense in poplar plants; also, the recombinant hydrophobin purified from *E. coli* activated both JA and SA defense signaling pathways (Huang et al., 2015). Cerato-Platanins are the noncatalytic secreted proteins, contributing to the virulence in pathogens. The major class identified that Small Proteins (Sm) Sm1/2/3 from *T. virens* and Eliciting plant-response (Epl) from *T. atroviride* are upregulated in symbiotic association with plants, having differential roles. However, the role played by Sm2/Epl2 to induce plant defense responses is still unclear, but a sturdy decrease in the protection level of maize seedlings against the pathogen, *Cochliobolus heterostrophus* was observed, when maize plants were treated with the *sm2/epl2* knockout strains in contrast with the *sm1/epl1* knockout strains (Gaderer et al., 2015).

Hunting for *Trichoderma* effector proteins has led to identifying around 20 effectors, and they have been studied during the interaction with plant systems. With the application of *Trichoderma*, transcriptome and secretome studies and analyses led to the finding that cellulases, other small proteins, and cytochrome *p*450 are highly represented (Rocha et al., 2016). Through in silico prediction, (Lamdan et al., 2015) have reported the full list of the apoplastic secretome (43 proteins), complete soluble secretome (66 proteins), and the effector candidates from the (84 proteins); also, 22 common sequences were found between the soluble secretome with the predicted effectors (Guzmán-Guzmán et al., 2017). The identification of the *Trichoderma* effector proteins will provide tools to deepen our understanding on how plant-beneficial fungi interaction is established (Ramírez-Valdespino et al., 2019).

4.6 *Trichoderma* secondary metabolites (SMs)—New effectors in plant interactions

SMs, the low molecular weight compounds, are synthesized through a great variety of pathways (Brakhage, 2013); and the synthesis of SMs is usually different between strains, greatly influenced by culture media and the growth conditions (Yu and Keller, 2005). *T. atroviridee* and *T. virens* are the well-studied *Trichoderma* sps for eliciting Sms, and it has been shown that the heterotrimeric G proteins, mitogen-activated protein kinases, and a few transcription factors are involved in the signaling pathway leading to SMs production (Mukherjee and Kenerley, 2010). The *Trichoderma* SMs synergistically act with hydrolytic enzymes and hence are associated in the efficiency of the strain producing them as a biocontrol agent (Reino et al., 2008).

4.7 Plant growth regulators (PGRs)

As *Trichoderma* is a root-associated fungus, it is able to produce PGRs that have a growth promotion effect in plants. In the model plant, *Arabidopsis thaliana*, *T. virens* produces Indole-3-acetic acid (IAA), which promotes plant growth and development (Contreras-Cornejo et al., 2009). Another PGR reported to be involved in *Trichoderma*-plant signaling is Abscisic acid (ABA); six genes related to the ABA biosynthesis pathway intermediary were identified in *T. atroviride* (Guzmán-Guzmán et al., 2019). Also, *T. virens* and *T. atroviridee* modulate ABA-regulated stress responses, like stomatal aperture functioning in *A. thaliana* (Contreras-Cornejo et al., 2015).

4.8 Nanotechnology-based *Trichoderma* formulation: Future trends for the biological control of plant diseases

There is an urgent requirement to search for efficient management of phytopathogens with smarter application approaches. *Trichoderma*-based formulation technologies are described for microencapsulation for the mycogenic synthesis of nanoparticles, with the aim of improving the biological control of pathogens, contributing immensely to sustainable agricultural practices.

4.9 Innovative technology beyond the ordinary with synthetic biology interventions: *Trichoderma* proteomics and metabolomics

Recombinant proteins and enzymes are commonly produced by cell factories for use as processing aids in agriculture, textile, and biofuel industries. *Trichoderma* spp. are referred as world champions of protein production, as they are the future trends and visions for protein cell factories. With the interventions of synthetic biology and in-depth studies on proteomics and metabolomics, these fungal strains are being strategically improved for higher production output as well as improved qualities. Newer technologies such as synthetic promoters and CRISPR/Cas9 are making it possible with precise engineering enabling higher yields with larger profits. Individual *Trichoderma* isolates differ in their resistance response to various agriculture chemicals, as they possess distinct inherent resistance mechanism process. With synthetic biology approaches, some strains have been modified to be resistant to specific agricultural chemicals. *Trichoderma reesei* is popularly used for these purposes due to its low production costs and high output yield. Cellulase from this species is used in "biostoning" of denim fabrics to provide softer, whitened fabric-stone-washed denim (Heikinheimo et al., 2000). *Trichoderma* spp. can be further employed in developing industrial technology to improve the aspects of eco-friendly biocontrol approach along with stepping forward to enhance the crop yield with an aim to expand gratified food supply.

References

Abdelrahman, M., Abdel-Motaal, F., El-Sayed, M., Jogaiah, S., Shigyo, M., Ito, S., Tran, L.S., 2016. Dissection of *Trichoderma longibrachiatum*-induced defense in onion (*Allium cepa* L.) against *Fusarium oxysporum* f. sp. cepa by target metabolite profiling. Plant Sci. 246, 128–138.

Alfano, G., Ivey, M.L.L., Cakir, C., Bos, J.I.B., Miller, S.A., Madden, L.V., Kamoun, S., Hoitink, H.A.J., 2007. Systemic modulation of gene expression in tomato by *Trichoderma hamatum* 382. Phytopathology 97, 429–437.

Altomare, C., Norvell, W.A., Bjorkman, T., Harman, G.E., 1999. Solubilization of phosphates and micronutrients by the plant growth promoting and biocontrol fungus *Trichoderma harzianum* Rifai 1295-22. Appl. Environ. Microbiol. 65, 2926–2933.

Bae, H., 2011. *Trichoderma* species as abiotic and biotic stress quenchers in plants. Res. J. Biotechnol. 6, 73–79.

Bae, S.J., Mohanta, T.K., Chung, J.Y., Ryu, M., Park, G., Shim, S., Hong, S.B., Seo, H., Bae, D.W., Bae, I., Kim, J.J., 2016. *Trichoderma* metabolites as biological control agents against *Phytophthora* pathogens. Biol. Control 92, 128–138.

Benitez, T., Rincon, A.M., Limon, M.C., Codon, A.C., 2004. Biocontrol mechanisms of *Trichoderma* strains. Int. Microbiol. 7, 249–260.

Boller, T., Felix, G., 2009. A renaissance of elicitors: perception of microbe-associated molecular patterns and danger signals by pattern-recognition receptors. Annu. Rev. Plant Biol. 60, 379–406.

Brakhage, A., 2013. Regulation of fungal secondary metabolism. Nat. Rev. Microbiol. 11, 21–32. https://doi.org/10.1038/nrmicro2916.

Brotman, Y., Lisec, J., Méret, M., Chet, I., Willmitzer, L., Viterbo, A., 2012. Transcript and metabolite analysis of the *Trichoderma*-induced systemic resistance response to *Pseudomonas syringae* in *Arabidopsis thaliana*. Microbiology 158, 139–146.

Brunner, K., Zeilinger, S., Ciliento, R., Woo, S., Lorito, M., Kubicek, C.P., Mach, R.L., 2005. Improvement of the fungal biocontrol agent *Trichoderma atroviride* to enhance both antagonism and induction of plant systemic disease resistance. Appl. Environ. Microbiol. 71, 3959–3965.

Calistru, C., McLean, M., Berjak, P., 1997. *In vitro* studies on the potential for biological control of *Aspergillus flavus*and *Fusarium moniliforme* by *Trichoderma* species: A study of the production of extracellular metabolites by *Trichoderma* species. Mycopathologia 137, 115–124.

Chandranayaka, S., Niranjana, S.R., Uday Shankar, A.C., Niranjan Raj, S., Reddy, M.S., Prakash, H.S., Mortensen, C.N., 2010. Seed biopriming with novel strain of *Trichoderma harzianum* for the control of toxigenic *Fusarium verticillioides* and *fumonisins* in maize. Arch. Phytopathol. Plant Protect. 43, 264–282.

Contreras-Cornejo, H.A., Macías-Rodríguez, L., Cortés-Penagos, C., López-Bucio, J., 2009. *Trichoderma virens*, a plant beneficial fungus, enhances biomass production and promotes lateral root growth through an auxin-dependent mechanism in*Arabidopsis*. Plant Physiol. 149, 1579–1592. https://doi.org/10.1104/pp.108.130369.

Contreras-Cornejo, H.A., Macías-Rodríguez, L., Alfaro-Cuevas, R., López-Bucio, J., 2014. *Trichoderma* spp. improve growth of *Arabidopsis* seedlings under salt stress through enhanced root development, osmolite production, and Na^+ elimination through root exudates. Mol. Plant-Microbe Interact. 27, 503–514.

Contreras-Cornejo, H.A., López-Bucio, J., Méndez-Bravo, A., Macías-Rodríguez, L., Ramos-Vega, M., Guevara-García, A.A., et al., 2015. Mitogen-Activated Protein Kinase 6 and ethylene and auxin signaling pathways are involved in*Arabidopsis* root-system architecture alteratios by*Trichoderma atroviride*. Mol. Plant-Microbe Interact. 28, 701–710. https://doi.org/10.1094/MPMI-01-15-0005-R.

Contreras-Cornejo, H.A., Macías-Rodríguez, L., Del-Val, E., Larsen, J., 2016. Ecological functions of *Trichoderma* spp. and their secondary metabolites in the rhizosphere: interactions with plants. FEMS Microbiol. Ecol. 92, 036.

Coruzzi, G., Last, R., 2000. Amino acids. In: Biochemistry and Molecular Biology of Plants, The American Society of Plant Physiologists.pp. 358–410.

Crutcher, F.K., Parich, A., Schuhmacher, R., Mukherjee, P.K., Zeilinger, S., Kenerley, C.M., 2013. A putative terpenecyclase, *vir 4*, is responsible for the biosynthesis of volatile terpene compounds in the biocontrol fungus *Trichoderma virens*. Fungal Genet. Biol. 56, 67–77.

Djonović, S., Vargas, W.A., Kolomiets, M.V., Horndeski, M., Wiest, A., Kenerley, C.M., 2007. A proteinaceous elicitor Sm1 from the beneficial fungus *Trichoderma virens* is required for induced systemic resistance in maize. Plant Physiol. 145, 875–889.

Druzhinina, I.S., Seidl-Seiboth, V., Herrera-Estrella, A., Horwitz, B.A., Kenerley, C.M., Monte, E., Mukherjee, P.K., Zeilinger, S., Grigoriev, I.V., Kubicek, C.P., 2011. *Trichoderma*: the genomics of opportunistic success. Nat. Rev. Microbiol. 9, 749–759.

Druzhinina, I.S., Shelest, E., Kubicek, C.P., 2012. Novel traits of *Trichoderma* predicted through the analysis of its secretome. FEMS Microbiol. Lett. 337, 1–9. https://doi.org/10.1111/j.1574-6968.2012.02665.x.

Dubey, S.C., Suresh, M., Birendra, S., 2007. Singh evaluation of *Trichoderma s*pecies against *Fusarium oxysporum* f. sp. *ciceris* for integrated management of chickpea wilts. Biol. Control 40, 118–127.

Engelberth, J., Koch, T., Schuler, G., Bachmann, N., Rechtenbach, J., Boland, W., 2001. Ion channel-forming alamethicin is a potent elicitor of volatile biosynthesis and tendril coiling. Cross talk between jasmonate and salicylate signaling in lima bean. Plant Physiol. 125, 369–377.

Etschmann, M.M.W., Huth, I., Walisko, R., Schuster, J., Krull, R., Holtmann, D., Wittmann, C., Schrader, J., 2015. Improving 2-phenylethanol and 6-pentyl-α-pyrone production with fungi by microparticle-enhanced cultivation (MPEC). Yeast 32, 145–157.

Fontenelle, A.D.B., Guzzo, S.D., Lucon, C.M.M., Harakaya, R., 2011. Growth promotion and induction of resistance in tomato plant against *Xanthomonas vesicatoria* and *Alternaria solani* by *Trichoderma* spp. Crop Prot. 30, 1492–1500.

Gaderer, R., Lamdan, N., Frischmann, A., Sulyok, M., Krska, R., Horwitz, B.A., et al., 2015. Sm2, a paralog of the*Trichoderma* cerato-platanin elicitor Sm1, is also highly important for plant protection conferred by the fungal-root interaction of*Trichoderma* with maize. BMC Microbiol. 15, 2. https://doi.org/10.1186/s12866-014-0333-0.

Gajera, H.P., Bambharolia, R.P., Patel, S.V., Khatrani, T.J., Goalkiya, B.A., 2012. Antagonism of *Trichoderma* spp. against *Macrophomina phaseolina*: evaluation of coiling and cell wall degrading enzymatic activities. J. Plant Pathol. Microbiol. 3, 7.

Gebarowska, E.W., Pietr, S.J., 2006. Colonization of roots and growth stimulation of cucumber by iprodione-resistant isolates of *Trichoderma* spp. applied alone and combined with fungicides. Phytopathologiapolonica 41, 51–64.

Gupta, K.J., Mur, L.A., Brotman, Y., 2014. *Trichoderma asperelloides* suppresses nitric oxide generation elicited by *Fusarium oxysporum*in *Arabidopsis* roots. Mol. Plant-Microbe Interact. 27, 307–314.

Guzmán-Guzmán, P., Alemán-Duarte, M.I., Delaye, L., Herrera-Estrella, A., Olmedo-Monfil, V., 2017. Identification of effector-like proteins in*Trichoderma* spp. and role of hydrophobin in the plant-fungus interaction and mycoparasitism. BMC Genet. 18, 16. https://doi.org/10.1186/s12863-017-0481-y.

Guzmán-Guzmán, P., Porras-Troncoso, M.D., Olmedo-Monfil, V., Herrera-Estrella, A., 2019. *Trichoderma* species: versatile plant symbionts. Phytopathology 109, 6–16. https://doi.org/10.1094/PHYTO-07-18-0218-RVW.

Hanson, L.E., Howell, C.R., 2004. Elicitors of plant defense responses from biocontrolstrains of *Trichoderma virens*. Phytopathology 94, 171–176.

Harman, G.E., 2000. Myths and dogmas of biocontrol changes in perception derived from research on *Trichoderma harzianum* T-22. Plant Dis. 84, 377–393.

Harman, G.E., Kubicek, C.P., 1998. *Trichoderma* and *Gliocladium*. Enzymes, Biological Control and Commercial Applications. vol. 2. Taylor and Francis, London, UK.

Harman, G.E., Howell, C.R., Viterbo, A., Chet, I., Lorito, M., 2004. *Trichoderma* species–opportunistic, avirulent plant symbionts. Nat. Rev. Microbiol. 2, 43–56.

Harman, G.E., Petzoldt, R., Comis, A., Chen, J., 2004a. Interactions between *Trichoderma harzianum* strain T22 and maized inbred line Mo17 and effect of this interaction on diseases caused by *Pythium ultimum*and *Colletotrichum graminicola*. Phytopathology 94, 147–153.

Heikinheimo, L., Buchert, J., Miettinen-Oinonen, A., Suominen, P., 2000. Treating denim fabrics with Trichoderma Reesei Cellulases. Text. Res. J. 70 (11), 969–973.

Hermosa, M., Grondona, I., Iturriaga, E.A., Diaz-Minguez, J.M., Castro, C.E., Monte, A., Garcia-Acha, I., 2000. Molecular characterization and identification of biocontrol isolates of *Trichoderma* spp. Appl. Environ. Microbiol. 66, 1890–1898.

Hermosa, R., Viterbo, A., Chet, I., Monte, E., 2012. Plant-beneficial effects of *Trichoderma* and of its genes. Microbiology 158, 17–25.

Howell, C.R., 2002. Cotton seedling pre-emergence damping-off incited by *Rhizopus oryzae* and *Pythium* spp. and its biological control with *Trichoderma* spp. Phytopathology 92, 177–180.

Howell, C.R., Hanson, E.L., Stipanovic, R.D., Puckhaber, L.S., 2000. Induction of terpenoid synthesis in cotton roots and control of *Rhizoctonia solani* by seed treatment with *Trichoderma virens*. Phytopathology 90, 248–252.

Huang, Y., Mijiti, G., Wang, Z., Yu, W., Fan, H., Zhang, R., et al., 2015. Functional analysis of the class II hydrophobins gene HFB2-6 from the biocontrol agent*Trichoderma asperellum* ACCC30536. Microbiol. Res. 171, 8–20. https://doi.org/10.1016/j.micres.2014.12.004.

Islam, S., Akandaa, A.M., Sultanab, F., Hossaina, M., 2013. Chilli rhizosphere fungus *Aspergillus* spp. PPA1 promotes vegetative growth of cucumber (*Cucumis sativus*) plants upon root colonization. Arch. Phytopathol. Plant Protect. 47, 1231–1238.

Jiang, X., Geng, A., He, N., Li, Q., 2011. New isolates *Trichoderma viride* strain for enhanced cellulolytic enzyme complex production. J. Biosci. Bioeng. 111, 121–127.

Jogaiah, S., Mostafa, A., Tran, L.S.P., Shin-ichi, I., 2013. Characterization of rhizosphere fungi that mediate resistance in tomato against bacterial wilt disease. J. Exp. Bot. 64, 3829–3842.

Jogaiah, S., Abdelrahman, M., Tran, L.-S.P., Ito, S.-I., 2018. Different mechanisms of *Trichoderma virens*-mediated resistance in tomato against Fusarium wilt involve the jasmonic and salicylic acid pathways. Mol. Plant Pathol. 19, 870–882.

Kamoun, S., 2006. A catalogue of the effector secretome of plant pathogenic oomycets. Annu. Rev. Phytopathol. 44, 41–60. https://doi.org/10.1146/annurev.phyto.44.070505.143436.

Kredics, L., Antal, Z., Manczinger, L., Nagy, E., 2001. Breeding of mycoparasitic *Trichoderma* strains for heavy metal resistance. Lett. Appl. Microbiol. 2, 112–116.

Kubicek, C.P., Penttila, M.E., 1998. Regulation of production of plant polysaccharide degrading enzymes by Trichoderma. In: Harman, G.E., Kubicek, C.P. (Eds.), *Trichoderma and Gliocladium*. Enzymes, Biological Control and Commercial Applications. Taylor & Francis, London, vol. 2, pp. 49–71.

Kubicek, C.P., Mach, R.L., Peterbauer, C.K., Lorito, M., 2001. *Trichoderma*: from genes to biocontrol. J. Plant Pathol. 83, 11–23.

Kubicek, C.P., Herrera-Estrella, A., Seidl-Seiboth, V., Martinez, D.A., Druzhinina, I.S., Thon, M., Zeilinger, S., Casas-Flores, S., Horwitz, B.A., Mukherjee, P.K., Mukherjee, M., 2011. Comparative genome sequence analysis underscores mycoparasitism as the ancestral life style of *Trichoderma*. Genome Biol. 12, 40.

Lamdan, N., Shalaby, S., Ziv, T., Kenerley, C.M., Horwitz, B.A., 2015. Secretome of the biocontrol fungus*Trichoderma virens* co-cultured with maize roots: tole in induced systemic resistance. Mol. Cell. Proteomics 14, 1054–1063. https://doi.org/10.1074/mcp.m114.046607.

Lorito, M., Woo, S.L., Harman, G.E., Monte, E., 2010. Translational research on *Trichoderma*: from'omics to the field. Annu. Rev. Phytopathol. 48, 395–417.

Mastouri, F., Bjorkman, T., Harman, G.E., 2010. Seed treatment with *Trichoderma harzianum* alleviates biotic, abiotic, and physiological stresses in germinating seeds and seedlings. Phytopathology 100, 1213–1221.

Michalikova, A., Michrina, J., 1997. Regulatory Potential of *T. harzianum* B1 in relation with phyto pathogenous microorganisms. In: *Proc.* XIVth Slovak and Czech Plant Protect, Conference, Nitra, pp. 291–292.

Moran-Diez, M.E., Trushina, N., Lamdan, N.L., Rosenfelder, L., Mukherjee, P.K., Kenerley, C.M., Horwitz, B.A., 2015. Host-specific transcriptomic pattern of *Trichoderma virens* during interaction with maize or tomato roots. BMC Genomics 16, 8.

Mukherjee, P.K., Kenerley, C.M., 2010. Regulation of morphogenesis and biocontrol properties in*Trichoderma virens* by a VELVET protein. Vel 1. Appl. Environ. Microbiol. 76, 2345–2352. https://doi.org/10.1128/AEM.02391-09.

Mukherjee, P.K., Nautiyal, C.S., Mukhopadhyay, A.N., 2008. Molecular mechanisms of biocontrol by *Trichoderma spp.* In: Nautiyal, C.S. (Ed.), Molecular Mechanisms of plant and microbe coexistence. In: Soil biology 15Springer-Verlag, Berlin, Heidelberg.

Mukherjee, P.K., Buensanteai, N., Moran-Diez, M.E., Druzhinina, I.S., Kenerley, C.M., 2012a. Functional analysis of non-ribosomal peptide synthetases (NRPSs) in *Trichoderma virens* reveals a polyketide synthase (PKS)/NRPS hybrid enzyme involved in induced systemic resistance response in maize. Microbiology 158, 155–165.

Mukherjee, P.K., Horwitz, B.A., Kenerley, C.M., 2012b. Secondary metabolism in *Trichoderma* – a genomic perspective. Microbiology 158 (1), 35–45.

Mukherjee, P.K., Horwitz, B.A., Herrera-Estrella, A., Schmoll, M., Kenerley, C.M., 2013. *Trichoderma* research in the genome era. Annu. Rev. Phytopathol. 51, 105–129.

Nagaraju, A., Sudisha, J., Mahadeva, M.S., Ito, S.I., 2012. Seed priming with *Trichoderma harzianum* isolates enhances plant growth and induces resistance against *Plasmopara halstedii*, an incitant of sunflower downy mildew disease. Australas. Plant Pathol. 41, 609–620.

Nagaraju, A., Murali, M., Jogaiah, S., Amruthesh, K.N., Mahadeva Murthy, S., 2012b. Beneficial microbes promote plant growth and induce systemic resistance in sunflower against downy mildew disease caused by *Plasmopara halstedii*. Current Botany 3, 12–18.

Nakkeeran, S., Renukadevi, P., Aiyanathan, K.E.A., 2016. Exploring the potential of*Trichoderma* for the Management of Seed and Soil-Borne Diseases of Crops. In: Muniappan, R., Heinrichs, E. (Eds.), Integrated Pest Management of Tropical Vegetable Crops. Springer, Dordrecht.

Perazzolli, M., Dagostin, S., Ferrari, A., Elad, Y., Ilaria, P., 2008. Induction of systemic resistance against *Plasmopara viticola* in grapevines by *T. harzianum* T39 and benzothiadiazole. Biol. Control 47, 228–234.

Perazzolli, M., Roatti, B., Bozza, E., Pertot, I., 2011. *Trichoderma harzianum* T39 induces resistance against downy mildew by priming for defense without costs for grapevine. Biol. Control 58, 74–82.

Pérez, A., Ramage, G., Blanes, R., Murgui, A., Casanova, M., Martínez, J.P., 2011. Some biological features of*Candida albicans* mutants for genes coding fungal proteins containing the CFEM domain. FEMS Yest Res. 11, 273–284. https://doi.org/10.1111/j.1567-1364.2010.00714.x.

Piel, J., Atzorn, R., Gäbler, R., Kühnemann, F., Boland, W., 1997. Cellulysin from the plant parasitic fungus *Trichoderma viride* elicits volatile biosynthesis in higher plants via the octadecanoid signalling cascade. FEBS Lett. 416, 143–148.

Pushpalatha, H.G., Sudisha, J., Geetha, N.P., Amruthesh, K.N., Shetty, H.S., 2011. Thiamine seed treatment enhances LOX expression, promotes growth and induces downy mildew disease resistance in pearl millet. Biol. Plant. 55, 522–527.

Ramírez-Valdespino, C.A., Casas-Flores, S., Olmedo-Monfil, V., 2019. *Trichoderma* as a model to study effector-like molecules. Front. Microbiol. 10, 1030. https://doi.org/10.3389/fmicb.2019.01030.

Reglinski, T., Rodenburg, N., Taylor, J.T., Northcott, G.L., Chee, A.A., Spiers, T.M., Hill, R.A., 2012. *Trichoderma atroviride* promotes growth and enhances systemic resistance to *Diplodiapinea* in radiata pine (*Pinusradiata*) seedlings. For. Pathol. 42, 75–78.

Reino, J.L., Guerrero, R.F., Hern'andez-Gal'an, R. and Collado, I. G., 2008. Secondary metabolites from species of the biocontrol agent *Trichoderma*. Phytochem. Rev. 7, 89–123.

Reithner, B., Ibarra-Laclette, E., Mach, R.L., Herrera-Estrella, A., 2011. Identification of mycoparasitism-related genes in *Trichoderma atroviride*. Appl. Environ. Microbiol. 77, 4361–4370.

Rocha, A.L., Maeda, N., Pereira, N., Kern, F., Elias, L., Simister, R., et al., 2016. Characterization of the cellulolytic secretome of*Trichoderma harzianum* during growth on sugarcane bagasse and analysis of the activity boosting effects of swwollenin. Biothechnol. Prog. 32, 327–336. https://doi.org/10.1002/btpr.2217.

Saba, H., Vibhash, D., Manisha, M., Prashant, K.S., Farham, H., Tauseff, A., 2012. *Trichoderma*-a promising plant growth stimulator and biocontrol agent. Mycosphere 8, 78–82.

Saravanakumar, K., Shanmuga Arasu, V., Kathiresan, K., 2013. Effect of *Trichoderma* on soil phosphate solubilization and growth improvement of *Avicennia marina*. Aquat. Bot. 104, 101–105.

Saravanakumar, K., Yu, C., Dou, K., Wang, M., Li, Y., Chen, J., 2016. Synergistic effect of *Trichoderma*-derived antifungal metabolites and cell wall degrading enzymes on enhanced biocontrol of *Fusarium oxysporum* f. sp. *cucumerinum*. Biol. Control. 94, 37–46.

Segarra, G., Casanova, E., Bellido, D., Odena, M.A., Oliveira, E., Trillas, I., 2007. Proteome, salicylic acid and jasmonic acid changes in cucumber plants inoculated with *Trichoderma asperellum* strain T34. Proteomics 7, 3943–3952.

Segarra, G., Van der Ent, S., Trillas, I., Pieterse, C.M., 2009. MYB72: a node of convergence in induced systemic resistance triggered by a fungal and abacterial beneficial microbe. Plant Biol. 11, 90–96.

Shen, Q., Liu, L., Wang, L., Wang, Q., 2018. Indole primes plant defense against necrotrophic fungal pathogen infection. PLoS One. 13, e0207607, https://doi.org/10.1371/journal.pone.0207607.

Shoresh, M., Harman, G.E., 2008. The molecular basis of shoot responses of maize seedlings to *Trichoderma harzianum* T22 inoculation of the root: a proteomic approach. Plant Physiol. 147, 2147–2163.

Shoresh, M., Yedidia, I., Chet, I., 2005. Involvement of the jasmonic acid/ethylene signaling pathway in the systemic resistance induced in cucumber by *Trichoderma asperellum* T203. Phytopathology 95, 76–84.

Shoresh, M., Harman, G.E., Mastouri, F., 2010. Induced systemic resistance and plantresponses to fungal biocontrol agents. Annu. Rev. Phytopathol. 48, 21–43.

Shukla, N., Awasthi, R.P., Rawat, L., Kumar, J., 2015. Seed biopriming with drought tolerant isolates of *Trichoderma harzianum* promote growth and drought tolerance in *Triticumaestivum*. Ann. Appl. Biol. 166, 171–182.

Singh, H.B., 2016. Seed biopriming: A comprehensive approach towards agricultural sustainability. Indian Phytopathol. 69, 203–209.

Singh, V., Upadhyay, R.S., Sarma, B.K., Singh, H.B., 2016. Seed bio-priming with *Trichoderma asperellum* effectively modulate plant growth promotion in pea. Int. J. Agric., Environ. Biotechnol. 9, 361.

Sivasithamparam, K., Ghisalberti, E.L., 1998. Secondary metabolism. In: *Trichoderma* and *Gliocladium*. Volume 1: Basic Biology Taxonomy and Genetics. pp. 139–191.

Sriram, S., Manasa, S.B., Savitha, M.J., 2009. Potential use of elicitors from Trichoderma in induced systemic resistance for the management of *Phytophthora capsici* in red pepper. J. Biol. Control. 23, 449–456.

Stoppacher, N., Kluger, B., Zeilinger, S., Krska, R., Schuhmacher, R., 2010. Identification and profiling of volatile metabolites of the biocontrol fungus *Trichoderma atroviride* by HS-SPME-GC-MS. J. Microbiol. Methods 81, 187–193.

Sudisha, J., Niranjana, S.R., Umesha, S., Prakash, H.S., Shekar Shetty, H., 2006. Transmission of seed-borne infection of muskmelon by *Didymella bryoniae* and effect of seed treatments on disease incidence and fruit yield. Biol. Control 37, 196–205.

Surekha, C., Neelapu, N., Prasad, B.S., Ganesh, P.S., 2014. Induction of defense enzymes and phenolic content by *Trichoderma Viride* in *Vignamungo* infested with *Fusarium oxysporum* and *Alternaria alternata*. Int. J. Agric. Sci. Res. 4, 31–40.

Tucci, M., Ruocco, M., De Masi, L., De Palma, M., Lorito, M., 2011. The beneficial effect of *Trichoderma* spp. on tomato is modulated by plant genotype. Mol. Plant Pathol. 12, 341–354.

Van Wees, S.C., Van der Ent, S., Pieterse, C.M., 2008. Plant immune responsestriggered by beneficial microbes. Curr. Opin. Plant Biol. 11, 443–448.

Vinale, F., Sivasithamparam, K., Ghisalberti, E.L., Marra, R., Barbetti, M.J., Li, H., Woo, S.L., Lorito, M., 2008. A novel role for *Trichoderma* secondary metabolites in the interactions with plants. Physiol. Mol. Plant Pathol. 72, 80–86.

Vinale, F., Nigro, M., Sivasithamparam, K., Flematti, G., Ghisalberti, E.L., Ruocco, M., Varlese, R., Marra, R., Lanzuise, S., Eid, A., Woo, S.L., Lorito, M., 2013. Harzianic acid: a novel siderophore from *Trichoderma harzianum*. FEMS Microbiol. Lett. 347, 123–129.

Viterbo, A., Horwitz, B.A., 2010. Mycoparasitism. In: Borkovich, K.A., Ebbole, D.J. (Eds.), Cellular and Molecular Biology of Filamentous Fungi, Washington.In: American Society for Microbiology, vol. 42. pp. 676–693.

Viterbo, A., Montero, M., Ramot, O., Friesem, D., Monte, E., Llobell, A., Chet, I., 2002. Expression regulation of the endochitinase chit 36 from *Trichoderma asperellum* (*T. harzianum* T-203). Curr. Genet. 42, 114–122.

Viterbo, A., Brotman, Y., Chet, I., Kenerley, C., 2007. The 18mer peptaibols from *Trichoderma virens*elicit plant defense responses. Mol. Plant Pathol. 8, 737–746.

Vogel-Adghough, D., Stahl, E., Návarová, H., Zeier, J., 2013. Pipecolic acid enhances resistance to bacterial infection and primes salicylic acid and nicotine accumulation in tobacco. Plant Signal. Behav. 11, e26366.

Witkowsha, D., Maj, A., 2002. Production of lytic enzymes by *Trichoderma* spp. and their effect on the growth of phytopathogenic fungi. Folia Microbiol. 47, 279–282.

Woo, S.L., Scala, F., Ruocco, M., Loritto, M., 2006. The molecular biology of the interactions between *Trichoderma* spp., phytopathogenic fungi, and plants. Phytopathology 96, 1061–1070.

Woo, S.L., Ruocco, M., Vinale, F., Nigro, M., Marra, R., Lombardi, N., Pascale, A., Lanzuise, S., Manganiello, G., Lorito, M., 2014. *Trichoderma*-based products and their widespread use in agriculture. Open Mycol. J. 8, 71–126.

Yadidia, I., Benhamou, N., Chet, I., 1999. Induction of defense responses in cucumber plants (*Cucumis sativus* L.) by biocontrol agent *Trichoderma harzianum*. Appl. Environ. Microbiol. 65, 1061–1070.

Yoshioka, Y., Ichikawa, H., Naznin, H.A., Kogure, A., Hyakumachi, M., 2012. Systemic resistance induced in *Arabidopsis thaliana* by *Trichoderma asperellum* SKT-1, a microbial pesticide of seed borne diseases of rice. Pest Manag. Sci. 68, 60–66.
Yu, J.H., Keller, N., 2005. Regulation of secondary metabolism in filamentous fungi. Annu. Rev. Phytopathol. 43, 437–458.
Zhao, Q.Y., Shen, Q.R., Ran, W., Xiao, T.J., Xu, D.B., Xu, Y.C., 2011. Inoculation of soil by *Bacillus subtilis* Y-IVI improves plant growth and colonization of the rhizosphere and interior tissues of muskmelon (*Cucumis melo* L.). Biol. Fertil. Soils 47, 507–514.

Potential of *Trichoderma* species in alleviating the adverse effects of biotic and abiotic stresses in plants

5

Manzoor R. Khan[a], Ghazala Parveen[a], Abbu Zaid[b], Shabir Hussain Wani[c], and Sudisha Jogaiah[d]

[a]Section of Plant Pathology, Department of Botany, Aligarh Muslim University, Aligarh, Uttar Pradesh, India, [b]Plant Physiology and Biochemistry Section, Department of Botany, Aligarh Muslim University, Aligarh, Uttar Pradesh, India, [c]Mountain Research Centre for Field Crops, Sher-e-Kashmir University of Agricultural Sciences and Technology of Kashmir, Khudwani Anantnag, Jammu and Kashmir, India, [d]Laboratory of Plant Healthcare and Diagnostics, P.G. Department of Biotechnology and Microbiology, Karnatak University, Dharwad, Karnataka, India

5.1 Introduction

Beneficial fungi with biocontrol abilities have emerged as an essential tool in sustainable agriculture. They possess the ability to reduce the adverse effects of phytopathogens and also ameliorate several abiotic stresses such as heavy metal, extreme temperature, salinity, and drought. Among the beneficial fungi, *Trichoderma* is regarded as a highly versatile biocontrol genus which has long been used for management of diseases caused by phytopathogens (Samuels, 1996). It is an opportunistic fungus which acts as an antagonistic agent against many phytopathogens to protect the plants from various diseases. The ubiquitous distribution, ease of isolation, culture, and fast growth of *Trichoderma* spp. on several substrates, and the production of secondary metabolites attracted the attention of researchers around the world to use this genus in sustainable agriculture (Harman, 2006; Contreras-Cornejo et al., 2016; Jogaiah et al., 2018). *Trichoderma* spp. have also been exploited in several biotechnological applications to ameliorate both biotic and abiotic stresses (Lorito et al., 2010; Nagaraju et al., 2012a; Hermosa et al., 2012). Numerous studies have demonstrated the biocontrol potential of *Trichoderma* spp. in the management of plant disease, and presently, commercial products of *Trichoderma* spp. as biopesticides and biofertilizers are also available (Harman et al., 2004; Sudisha et al., 2006; Vinale et al., 2008). *Trichoderma* spp. have the ability to mitigate the harmful effects of various abiotic stresses and are known to increase the uptake and efficiency of various essential nutrients. Root colonization by *Trichoderma* spp. alleviates the adverse effects of drought on the growth, stomatal conductance, green fluorescence emissions, and net photosynthesis in *Theobroma cacao* (Bae et al., 2011). *Trichoderma* spp. are known to deprive the phytopathogens of available nutrients, rendering them ineffective to cause any disease.

Biocontrol Agents and Secondary Metabolites. https://doi.org/10.1016/B978-0-12-822919-4.00005-3

5.2 Interaction, colonization, and plant growth promotion by *Trichoderma*

Trichoderma spp. colonize the roots of both monocot and dicot plants (Harman and Shoresh, 2007), and protect them from soilborne diseases (Harman, 2011). The interaction between *Trichoderma* spp. and plant roots involves recognition, attachment, and penetration of plant roots, root colonization, and nutrient transfer from the root (Mukherjee et al., 2012). Firstly, the hyphae coils around the roots, by forming appressoria-like structures (Yedidia et al., 1999) mediated by two hydrophobin-like proteins, TasHyd1 (an expansin-like protein swollen in with a cellulose-binding domain) and *qid7* (a cysteine-rich cell wall protein) which can recognize the cellulose and modify the plant cell wall architecture by producing proteolytic and cellulolytic enzymes (Viterbo and Chet, 2006; Brotman et al., 2008; Samolski et al., 2012), thereby facilitating root colonization. It has been reported that *Trichoderma harzianum* produces CWDEs such as endopolygalacturonase, which are involved in active root colonization (Morán-Diez et al., 2009). Similarly, *Trichoderma virens* produced plant growth regulators like auxins and other related compounds, which may promote root colonization, and enhanced plant growth (Contreras-Cornejo et al., 2009). Plant beneficial microbes are initially recognized as pathogenic and would trigger plant immunity through the perception of microbes-associated molecular patterns (MAMPs) by the plant-mediated receptors (Jones and Dangl, 2006; Boller and Felix, 2009; Pieterse et al., 2012). However, *Trichoderma* spp. remodel or manipulate the plant immune response by reprogramming their transcriptome and proteome (Segarra et al., 2007; Shoresh and Harman, 2008). Some *Trichoderma* spp. enter through trichomes by producing appressoria-like structures and remain in the plant as typical endophytes (Bae et al., 2009).

The interaction of plant roots with beneficial microbes in the rhizosphere improves plant growth, especially under stressful conditions (Zamioudis and Pieterse, 2012; Murali et al., 2013). In the rhizosphere, *Trichoderma* spp. proliferate and establish a beneficial relationship with plant roots, thereby improving the growth and nutrition of plants in a natural way (Shoresh et al., 2010). It has been reported that *T. harzianum* enhanced the concentration of elements, viz., Mn, N, Fe, Zn, Cu, P, K, Na, Mg, and Ca, both in the roots and shoots of cucumber and tomato seedlings (Azarmi et al., 2011) and produced various siderophores, phytohormones, and enzymes involved in nutrient solubilization (Doni et al., 2014). The antimicrobial compounds of *Trichoderma*, viz., koninginins, cyclonerodiol, harzianolide, harzianopyridone, trichocereus A–D, and harzianic acid, are known to promote plant growth (Vinale et al., 2014). The novel secondary metabolite (cerinolactone) from *T. cerinums* favorably affected the growth of tomato seedlings while isoharzianic acid produced by *T. harzianum* promoted the growth of plants, through the strong binding of iron (Vinale et al., 2014).

It has been demonstrated that *T. virens* and *T. atroviride* promoted plant growth by producing certain types of indole-3-acetic acid-related indoles, which could be one of the mechanisms employed by *Trichoderma* spp. for plant growth promotion

(Salas-Marina et al., 2011; Jogaiah et al., 2018). However, studies by Gravel et al. (2007) demonstrated that the degradation of IAA by *Trichoderma* reduced the adverse effects of IAA on the elongation of roots, which results in reduced production of ethylene, by decreasing the precursor of ethylene (1-aminocyclopropane-1-carboxylic acid, ACC), or the activity of ACC deaminase (ACCD) present in *Trichoderma.* It has also been proposed that the high activity of ACCD in *T. asperellum* results in the upregulation of the ACCD-encoding gene (Tas-acdS); however, Tas-acdS silencing reduced the potential of silenced transformants to promote the elongation of canola roots (Viterbo and Horwitz, 2010). These mechanisms have been attributed to *Trichoderma* spp. for plant growth promotion (Viterbo and Horwitz, 2010; Salas-Marina et al., 2011).

5.3 Role of *Trichoderma* spp. in alleviating biotic stress

Trichoderma has been widely used for the management of plant diseases such as seed rotting (Hadar et al., 1984), gray mold and powdery mildew (Elad et al., 1998), citrus gummosis (Bicici et al., 1992), damping-off, root rot, stem rot, crown rot, red rot, black scurf, and charcoal rot (Gnanamanickam, 2002), bacterial wilt (Yuan et al., 2016), Fusarium wilt (Jogaiah et al., 2013a), sheath blight (De França et al., 2015), mosaic virus (Luo et al., 2010), southern stem rot (Sennoi et al., 2013), downy mildew (Perazzolli et al., 2012), root-knot nematode (Al-Hazmi and Javeed, 2016), *Fusarium* wilt (Al-Ani, 2017), and stem gall (Khan and Parveen, 2018). *Trichoderma* spp. also showed antagonistic activity against Gram-positive (Shi et al., 2012) as well as Gram-negative bacteria (Li et al., 2014).

The biocontrol potential of *Trichoderma* spp. toward plant-parasitic nematodes (PPN) is exploited by different mechanisms. *Trichoderma* spp. attack eggs, juveniles, and adults of PPN. It has been reported that the unique chemical in the gelatinous matrixes of nematode eggs plays an essential role in the attachment of *Trichoderma* to nematode egg masses, eggs, and nematode juveniles (J2s) (Sharon et al., 2007, 2011). By instigating proteolytic activity, *T. harzianum* attacked and colonized the eggs and egg masses of *Meloidogyne javanica* (Sharon et al., 2001). Similarly, coiling and appressorium-like structures produced by *T. asperellum* and *T. atroviride* damaged the eggs and J2s of *M. javanica* (Sharon et al., 2007). *Trichoderma* also competes with PPN for space and feeding sites within plant roots (Hussey and Roncadori, 1978). Strains of *T. harzianum* can produce nematicidal compounds against *M. javanica*, which reduce root penetration by PPN (Sharon et al., 2001). *T. harzianum* showed antibiosis against *M. cionopagum* (Szabó et al., 2012) and stimulated the expression of patterns of the genes *PR1* against *M. incognita* (Leonetti et al., 2014). *Trichoderma* spp. stimulated resistance against PPN as a localized or systemic response (Sharon et al., 2011) and is also able to prime jasmonic acid (JA) and salicylic acid (SA) dependent defense responses (Martínez-Medina et al., 2017). The role of *Trichoderma* spp. in alleviating biotic stress has been summarized in Table 5.1.

Table 5.1 Role of *Trichoderma* spp. in ameliorating biotic stress.

***Trichoderma* spp.**	**Pathogen**	**Effect on disease**	**Reference**
T. harzianum, *T. koningii*	*F. oxysporum* f.sp. *cucumerinum*, and *Pythium* spp.	Produced toxic metabolites on seed coats pea which results in the suppression of *F. oxysporum* f.sp. *cucumerinum*, and *Pythium* spp.	Lifshitz et al. (1986)
T. harzianum	*F. oxysporum* f.sp. *lycopersici*	Enhanced the yield of tomatoes up to 26.2% over controls.	Sivan et al. (1987)
T. viride	*S. rolfsii*	Combined application of *T. viride* with PCNB provided excellent disease control with no symptoms of basal stem rot on tomato.	Wokocha (1990)
T. koningii	*R. solani* and *P. ultimum*	Reduced seedling death caused by *R. solani* and *P. ultimum*.	Harris (1999)
T. harzianum, *T. koningii*	*Macrophomina phaseolina*	Resulted in greater percentage plant stands and reduced postemergence damping off in cowpea.	Adekunle et al. (2001)
T. koningii	*Sclerotium rolfsii*	Wheat bran was the most potent inoculum form of the antagonist in reducing the viability of the sclerotia on tomato seeds.	Tsahouridou and Thanassoulopoulos (2001)
T. koningii	*S. rolfsii*	Suppressed the damping off disease on tomato.	Tsahouridou and Thanassoulopoulos (2002)
T. virens, *T. koningii*	*R. solani*	Fusants of *T. virens* and *T. koningii* showed better biocontrol activity against *R. solani* on cotton.	Hanson and Howell (2002)
T. harzianum, *T. viride*	*F. oxysporum* f.sp. *cubense*	Soil application was found most effective against the wilt disease and ass-associated vascular discoloration.	Saravanan et al. (2003)

T. asperellum (T203)	*Pseudomonas syringae* pv. *lachrymans*	*Trichoderma* preinoculation induced a potentiated state in cucumber plants to become more resistant to subsequent pathogen infection.	Shoresh et al. (2005)
T. harzianum, *T. viride*, *T. koningii*, *T. reesei* and *T. hamatum*	*Rotylenchulus reniformis*, *M. incognita*	Controls both nematode genera by a direct effect on toxic metabolites and inhibits nematode penetration and developments on eggplant.	Bokhari (2009)
T. asperellum	*F. oxysporum* f.sp. *dianthi*	Plant growth medium based on grape marc compost amended with *T. asperellum*, restores the suppressive capacity of composts against *Fusarium* wilt of carnation.	Sant et al. (2010)
T. asperellum (T-16), *T. brevicompactum*	*M. incognita*	Caused significant inhibition of nematode reproduction, suppression of root galling and an increase of tomato yield.	Affokpon et al. (2011)
T. harzianum	*F. oxysporum* f.sp. *lycopersici*	UV-C mutant Th908-5 of *T. harzianum* tolerant to fusaric acid more effectively suppressed the wilt disease on tomato than wild-type strain Th908.	Marzano et al. (2013)
T. harzianum (T-E5)	*F. oxysporum* f.sp. *cucumerinum*	Application of a bioorganic fertilizer enriched with T-E5 reduced the incidence of *Fusarium* wilt and promoted the growth of cucumber plants.	Zhang et al. (2013a,b)
T. harzianum, *T. virens*, *T. atroviride*, *T. tomentosum* and *T. rossicum*	*Xiphinema index*	Triggered the mortality of plant-parasitic nematode *X. index*, which is capable of transmitting several plant viruses.	Daragó et al. (2013)
T. pseudokoningii (SMF2)	*Pectobacterium carotovorum* subsp. *carotovorum*	Enhanced the resistance against *P. carotovorum* subsp. *carotovorum* infection in Chinese cabbage.	Li et al. (2014)

Continued

Table 5.1 Continued

Trichoderma spp.	Pathogen	Effect on disease	Reference
T. harzianum (T10)	*Meloidogyne javanica*	Caused reduction in nematode infection by increasing the accumulation of different metabolites in tomato plants.	Selim et al., (2014)
T. harzianum	*F. oxysporum* f.sp. *cubense*	Significantly decreased the incidence rate of *Fusarium* wilt on banana as compared to the control.	Zhang et al. (2014)
T. asperellum	*Phytophthora ramorum*	Eliminated *P. ramorum* propagules to nondetectable levels and showed potential to remediate *P. ramorum* infested soil.	Widmer (2014)
T. parareesei (T6)	*Botrytis cinerea*	Showed beneficial effects on tomato plants in terms of seedling lateral root development and improved defense against *Botrytis cinerea.*	Rubio et al. (2014)
T. atroviride	*Armillaria gallica*	Coniferous bark preinoculated with *T. atroviride* provided effective control of *Armillaria* root rot on strawberry plants.	Pellegrini et al. (2014)
T. harzianum (ITEM908)	*M. incognita*	Pretreatment with *T. harzianum*-induced ISR in tomato roots against *M. incognita* infestation.	Leonetti et al. (2014)
T. harzianum, (T6, T14), *T. asperellum* (T2, T12 & T20), *T. hamatum* (T3, T19), and *T. virens* (T21)	*Sclerotinia sclerotiorum*, *R. solani*, *Verticillium dahliae*, *Phytophthora nicotianae* and *P. cinnamomi*	The mycelial growth of each pathogen was affected by each *Trichoderma* isolate and its metabolites under in vitro conditions and also decreased the root rot on *Arabidopsis* caused by *R. solani* and *S. sclerotiorum.*	Aleandri et al. (2015)
T. harzianum	*M. incognita*	Improved growth and yield of soybean plants and reduced the soil nematode population and root galls.	Izuogu and Abiri (2015)

Trichoderma spp.	*M. javanica*	Caused reduction in root galling in tomatoes.	Javeed et al. (2016)
T. harzianum, *T. viride*	*M. javanica*	Suppressed the nematode reproduction and root galling and enhanced the growth of tomato plants.	Al-Hazmi and Javeed (2016)
T. atroviride	*M. javanica*	Induced systemic resistance against *M. javanica*, and significantly reduced galling and adult nematodes inside tomato roots.	De Medeiros et al. (2017)
T. viride, *T. harzianum*	*Protomyces macrosporus*	Improved the growth and yield attributes as well as reduced stem gall disease of coriander.	Khan and Parveen (2018)
T. asperellum	*Ralstonia solanacearum*	Delayed wilt development, decreased disease incidence, increased fruit yield, and improved tomato growth under field conditions.	Konappa et al. (2018)
T. harzianum, *T. viride*	*Alternaria porri*	*T. viride* showed higher efficiency in controlling purple blotch disease of onion than *T. harzianum*, and resulted in disease controlling rates that were similar to those by fungicide.	Bayoumi et al. (2019)
T. harzianum, *T. viride*, *T. virinis*, and *T. koningii*	*F. solani*, *R. solani*, and *Pythium* sp.	Reduced the growth area more than 90.6% for all tested pathogenic fungi and significantly reduced disease incidence and severity of strawberry under field conditions.	Abd-El-Kareem et al. (2019)
T. harzianum, *T. viride*, and *T. virens*	*F. solani* and *F. oxysporum*	Reduced the incidence of damping-off and root rot/wilt diseases and increased the percentage of survival plants by stimulating systemic defense responses in dry bean plants.	Abd-El-Kareem et al. (2019)
T. atroviride, *T. koningiopsis*, *T. stilbohypoxyli*, and *T. koningiopsis*	*Colletotrichum* gigasporum, *F. oxysporum*, *A. destruens* *Phoma* sp., *P. destructive* and *Pilidium concavum*	Inhibited the growth of all the pathogenic fungi under in vitro conditions and promoted the growth and improved yerba mate yield.	López et al. (2019)

Continued

Table 5.1 Continued

***Trichoderma* spp.**	**Pathogen**	**Effect on disease**	**Reference**
T. harzianum	*F. solani*, *R. solani* and *S. rolfsii*	Inhibited the mycelial growth and reduced the infection of soilborne fungi in peanut.	El-Hai and Ali (2019)
T. asperellum (T34)	*F. oxysporum* f.sp. *lycopersici*	Enhanced the growth reduced wilt disease incidence in tomato plants.	Bidellaoui et al. (2019)
T. asperellum	*S. cepivorum*	*T. asperellum* exerts efficient biocontrol against *S. cepivorum* and activates onion systemic defenses against this pathogen under greenhouse conditions, and also reduces onion white rot incidence and enhanced crop yield in field trials.	Rivera-Méndez et al. (2020)
T. pseudoharzianum, *T. koningiopsis*, *T. asperelloides*, *T. afroharzianum*, *T. citrinoviride*, *T. longibrachiatum*, *T. viridescens*	*R. solanacearum*, *Xanthomonas campestris* and *M. incognita*	The secondary metabolites from *Trichoderma* spp. inhibited the growth of both bacterial pathogens and caused significant J2s mortality and inhibition in egg hatching of *M. incognita*.	Khan et al. (2020)
T. harzianum, *T. viride* and *T. vierns*	*Orobanche crenata*	Enhanced the growth parameters and protected faba bean plants against *O. crenata* infection.	El-Dabaa and Abd-El-Khair (2020)
T. asperellum, *T. harzianum*	*M. incognita*	Both *Trichoderma* spp. induced resistance to *M. incognita* in tomato but not in cucumber.	Pocurull et al. (2020)

5.3.1 Antagonistic mechanisms employed by Trichoderma spp. in relieving biotic stress

Trichoderma spp. employs several antagonistic mechanisms such as mycoparasitism, antibiosis, and competition for nutrients and space.

5.3.1.1 Mycoparasitism

The ability of *Trichoderma* spp. to suppress, parasitize, or even kill plant pathogens has been regarded as the most essential mechanism for its biocontrol activity (Mukherjee et al., 2012). Mycoparasitism is a direct mechanism and one of the most outstanding features of *Trichoderma* spp. (Harman, 2011; Druzhinina et al., 2011) which has been widely used for the biocontrol of agricultural pests, mostly phytopathogenic fungi and PPN. *Trichoderma* spp. possess the ability to directly kill other pathogens, with a wide host range under diverse ecological conditions using many mycoparasitic strategies (Romão-Dumaresq et al., 2012; Lehner et al., 2013). The mycoparasitic abilities of *Trichoderma* spp. are very complex, and involve pathogen detection through chemotropic mechanisms followed by the pathogen's cell wall lysis (Lorito et al., 2010); hyphal penetration and secretion of the massive cell wall degrading enzymes (CWDEs), viz., β-1,3-glucanases, chitinases, and proteases (Harman et al., 2004) and peptaibols (Druzhinina et al., 2011); and finally parasitizing the cell wall contents of phytopathogens (Mukherjee et al., 2012). Genome analysis unraveled the more complex enzymatic degradation machinery of *Trichoderma* spp. (Kubicek et al., 2011).

Trichoderma spp. can recognize and respond to its hosts in diverse ecological conditions (Mendoza-Mendoza et al., 2003) by the successive expression of chitinases, proteases, and glucanases (Harman et al., 2004). Chitinases secreted by *Trichoderma* spp. degrade the cell walls of pathogens to release the oligomers, which then induce exochitinases, and finally parasitism starts (Gajera et al., 2013). The secretion of chitinases has been regarded as a critical step in mycoparasitism and biocontrol of phytopathogens (Al-Ani, 2018). *Trichoderma* spp. also produce volatile and nonvolatile secondary metabolites (Mukherjee et al., 2011). Among various species of *Trichoderma*, *T. virens* and *T. reesei* produce the highest concentration of secondary metabolites (Gruber et al., 2011), while the highest production of chitinolytic enzymes has been identified in *T. virens*, *T. atroviride*, and *T. harzianum* (Mukherjee et al., 2011). The gene clusters coding secondary metabolites have been found specific to some *Trichoderma* spp. (Mukherjee et al., 2011).

5.3.1.2 Antibiosis

Antibiosis is an antagonistic interaction between a plant, pathogen, and *Trichoderma* spp. and involves the secretion of low-molecular-weight diffusible antibiotics or secondary metabolites, which are detrimental to phytopathogens (Gajera et al., 2013). *Trichoderma* spp. produce a wide variety of antibiotics like viridol, koninginins, viridin, gliovirin, gliotoxin, pyrones, and peptaibols (Howell, 2003; Harman et al., 2004). Peptiabol (Trichokonins VI), produced by *T. pseudokoningii* SMF2, caused the programmed cell death (PCD) in fungal plant pathogens (Shi et al., 2012).

It has been reported that peptiabols work together with CWDE to inhibit the growth of phytopathogens and also elicit-induced plant resistance (Mukherjee et al., 2011). These compounds also had beneficial effects on plant growth, especially under various biotic and abiotic stresses (Keswani et al., 2014). Peptaibols showed antibacterial and antifungal activities (Szekeres et al., 2004) and stimulated the biocontrol potential of *Trichoderma* spp. through the activation of mycoparasitic gene expression (Degenkolb et al., 2012). Metabolites such as gliotoxin and gliovirin, belonging to the P and Q group strains of *T. virens*, showed antipathogenic properties in which the P group strain has been found to inhibit *P. ultimum* but not *R. solani*, while the Q group inhibits *R. solani* but not *P. ultimum* (Howell et al., 2000).

The nonvolatile metabolite (Koninginin D), secreted by *Trichoderma* spp., showed antipathogenic properties against *Phytophthora cinnamomi*, *R. solani*, *Bipolaris sorokiniana*, *Pythium middletonii*, and *Fusarium oxysporum* (Dunlop et al., 1989). Similarly, viridins produced by *T. viride*, *T. virens*, and *T. koningii* inhibited the spore germination of *Botrytis allii*, *Colletotrichum lini*, *F. caeruleum*, *Aspergillus niger*, and *Stachybotrys atra* and *Penicillium expansum* (Singh et al., 2005) and harzianic acid produced by *T. harzianum* inhibited the growth of *P. irregulare*, *Sclerotinia sclerotiorum*, and *R. solani* (Vinale et al., 2009). Further, the enhanced expression of the *tri5* (trichodiene synthase) gene in the *T. brevicompactum Tb41tri5* transformant has been found associated with the increased production of trichodermin and increased antifungal activity against fungal pathogens (Tijerino et al., 2011). The trichotoxins (T5D2, T5E, T5F, T5G, and 1717A) and asperelines (A and E) produced by *T. asperellum* also played an essential role in antibiosis (Brito et al., 2014). The elicitors released by *Trichoderma* spp. to activate plant defense include proteins or peptides (Salas-Marina et al., 2011), enzymes (Morán-Diez et al., 2009; Vos et al., 2015), and glycosphingolipids (Mukherjee et al., 2012). The production of these antipathogenic compounds correlates with the exogenous availability of nutrients such as root exudates, leakage of nutrients in the soil, and also on environmental conditions (Jacobsen, 2006).

5.3.1.3 Competition

Competition is one of the most crucial mechanisms employed for the biological control of plant pathogens (Singh et al., 2013). The death of several microbes growing in the vicinity of *Trichoderma* spp. occurs due to the scarcity of nutrients (Singh et al., 2018). The reduced nutrient concentrations in the vicinity of *Trichoderma* spp. reduced the pathogen's conidial germination and germ tube growth, decreased infection sites, and the extent of overall disease progression (Nagaraju et al., 2012b; Nassr and Barakat, 2013). *Trichoderma* spp. have a strong potential to mobilize and utilize soil nutrients, making it more efficient and competitive in comparison to other soil microbes, which could also be related to the production of organic acids that reduce the pH of soil and allow the essential nutrients (Vinale et al., 2008; Jogaiah et al., 2013a). These antagonists favorably compete with other soil microbes for carbon and iron, which results in the suppression of phytopathogens (Alabouvette et al., 2009; Segarra et al., 2010).

Under aerobic conditions, iron exists as an Fe^{3+} insoluble form which is not available for root absorption (Miethke, 2013). However, *Trichoderma* spp. produce siderophores (iron-chelating compounds), which bind the insoluble iron (Fe^{3+}) and convert it into a soluble form (Fe^{2+}) for absorption by plant roots, thereby also inhibiting the growth of phytopathogens by depriving them of iron sources (Leong, 1986). Because of the secretion of siderophores, *Trichoderma* spp. have the ability to grow and proliferate in conditions that are not rich in iron content (Miethke, 2013). Among the various species of *Trichoderma*, the highest amount of siderophores is produced by *T. harzianum*, while *T. reesei* produces a major siderophore (*cis*-fusarinine) (Al-Ani, 2018). These siderophores have many functions and effects, such as storage of iron and suppression of pathogens during competition (Miethke, 2013). There are many more examples of effective application of competition for the biocontrol of pathogens, such as *B. cinerea* (Latorre et al., 2001), which suggests that the molecular and proteomic assembly of *Trichoderma* spp. is highly efficient in mobilizing the soil nutrients as compared to many other pathogenic microorganisms.

5.4 Role of *Trichoderma* spp. in alleviating abiotic stress

Trichoderma spp. are also known to induce tolerance against abiotic stresses and improve plant growth (Yasmeen and Siddiqui, 2017). *Trichoderma* spp. are known to offer plant tolerance against drought and salinity through increased root growth, nutrient uptake, and protection against oxidative stress (Mastouri et al., 2010; Shoresh et al., 2010; Abdelrahman et al., 2016). *Trichoderma harzianum* reduced the oxidative inhibition by increasing the activities of ascorbate and glutathione-recycling enzymes and improved seedling growth of tomato under water deficit conditions (Mastouri et al., 2010). Application of *T. hamatum* DIS2196 improved plant growth by mitigating the adverse effects of drought in cacao seedling and also delayed drought-induced changes (Bae et al., 2009). Moreover, it has been reported that maize plants treated with *T. harzianum* resisted water deficit and enhanced deep rooting (Harman, 2000). Maize plants inoculated with *Trichoderma* showed high starch content in leaves under drought stress (Shoresh and Harman, 2008). *Trichoderma* application increased the root growth of rice, regardless of the water deficit, and caused a delay in drought responses by reducing proline, malondialdehyde (MDA), and hydrogen peroxide (H_2O_2) contents and increased the concentrations of phenolic compound (Shukla et al., 2012).

Application of *T. harzianum* mitigated the negative effect of NaCl stress and improved growth and physio-biochemical attributes in mustard (Ahmad et al., 2015) and tomato seedlings (Mastouri et al., 2010). *Trichoderma* treated plant restored the oil content in the leaves under NaCl stress by limiting the ABA accumulation and enhanced nutrient transportation from roots to shoots (Ahmad et al., 2015). It has been reported that under saline conditions, application of *T. harzianum* improved the chlorophyll and carotenoid content in mustard (Ahmad et al., 2015), wheat (Rawat et al., 2011), and cucumber (Zhang et al., 2013a,b). The increase in pigment concentration might be due to the phytohormone production in *Trichoderma* inoculated plants

(Martínez-Medina et al., 2014) and reduction in Na uptake (Iqbal and Ashraf, 2013). *Trichoderma* treated plants showed an increase in K^+ (Yedidia et al., 2001; Yildirim et al., 2006) and Ca^{2+} contents (Yildirim et al., 2006). The increase in the uptake and content of K^+ ameliorates the harmful effects of salinity (Shabala and Cuin, 2008) by inducing the closure of stomata (Shoresh et al., 2010). The application of *Trichoderma* enhanced the proline content in mustard (Ahmad et al., 2015) and *Arabidopsis* seedlings under salt stress (Contreras-Cornejo et al., 2014). Proline maintains the osmoregulation of cells under NaCl stress (Ahmad et al., 2010; Rasool et al., 2013) and could not only scavenge reactive oxygen species (ROS) and protect the cell from oxidative damage (Ahmad et al., 2010; Jogaiah et al., 2013) but also increase N fixation in plants. *Trichoderma* inoculated plants showed increased activities of superoxide dismutase (SOD), peroxidase (POD), glutathione-reductase (GR), and glutathione-*S*-transferase (GST) in leaves (Shoresh and Harman, 2008). *Trichoderma* T22 inoculated seedlings also restored the vigor damaged by oxidative stress (Björkman et al., 1998). Use of *T. harzianum* AK20G strain mitigates the deleterious effects of chilling stress in tomato seedlings by reducing the rate of lipid peroxidation and electrolyte leakage and increasing the accumulation of proline and the leaf water status (Ghorbanpour et al., 2018). Similarly, *T. harzianum* inoculated *Arabidopsis* seedlings showed enhanced tolerance to heat stress by producing heat shock proteins (Montero-Barrientos et al., 2010). Inoculation with *T. atroviride* F6 on soil contaminated with Cd and Ni significantly alleviates the cellular toxicity of mustard (Cao et al., 2008). Similarly, *T. longibrachiatum* (WT2) mitigates Pb^{2+}-induced oxidative stress on sunflower by enhancing the activities of antioxidant enzymes (Devi et al., 2017). *T. asperellum* inoculated plant reduced the phytotoxic effect of copper on onion (Vargas et al., 2017). The role of *Trichoderma* spp. in alleviating abiotic stress has been comprehensively summarized in Table 5.2.

5.4.1 Mechanism employed by Trichoderma spp. for abiotic stress tolerance

Plants under stressful conditions are known to produce a very high concentration of reactive oxygen species (ROS), which are generally harmful for their growth and development (Zaid and Wani, 2019). It has been demonstrated from several studies that root colonization by *T. harzianum* is associated with the enhanced activities of enzymes, such as peroxidases, chitinases, β-1,3-glucanases, and lipoxygenases, which induced the accumulation of phenolic compounds and phytoalexins (Harman, 2006; Gachomo and Kotchoni, 2008), which are also involved in the removal of toxic ROS. The increase in the concentration of toxic ROS, accompanied by enhanced accumulation of lipid peroxides, is regarded as the most common factor which adversely affects plants under stressful conditions (Zaidi et al., 2014). Seed treatment with *Trichoderma* spp. reduced the concentration of lipid peroxides in seedlings by the removal of toxic ROS (Zaidi et al., 2014). In plants, there are numerous pathways through which oxidized glutathione and ascorbate are converted into the reduced forms (Mittler, 2002) and *Trichoderma* spp. increase the activity of such pathways,

Table 5.2 Role of *Trichoderma* spp. in ameliorating abiotic stress in plants.

***Trichoderma* spp.**	**Abiotic stress**	**Effects**	**Reference**
Trichoderma harzianum	Salt (NaCl), heavy metal (Cu, Cd)	*T. harzianum* endochitinases (CHIT33 & CHIT42) generated innate defense responses and enhanced stress tolerance in tobacco transgenic plants without undesirable side effects.	Dana et al. (2006)
T. atroviride (F6)	Heavy metal (Cd, Ni)	Alleviated the cellular toxicity of cadmium and nickel and also enhanced the efficiency of phytoextraction for mustard.	Cao et al. (2008)
T. harzianum (T34)	Heat, osmotic, salt (NaCl) and oxidative stresses	*T. harzianum* T34 hsp70 confers tolerance to heat and other abiotic stresses and fungal HSP70 protein acts as a negative regulator of HSF transcriptional activity in *Arabidopsis*.	Montero-Barrientos et al. (2008, 2010)
T. hamatum (DIS2196)	Drought	Caused delayed drought-induced changes in stomatal conductance, net photosynthesis, and green fluorescence emissions.	Bae et al. (2009)
T. harzianum (T22)	Osmotic, salinity, chilling, and heat stress	Increased seedling vigor and ameliorates stresses by inducing physiological protection in tomato plants against oxidative damage.	Mastouri et al. (2010)
T. harzianum (T22)	Drought	Enhanced the redox state of colonized tomato plants under water deficit conditions by increasing the activity of ascorbate and glutathione-recycling enzymes, superoxide dismutase, catalase, and ascorbate peroxidase, in both root and shoot.	Mastouri et al. (2010)
T. harzianum	Drought	Promoted root growth in rice and delayed drought-induced changes like stomatal conductance, net photosynthesis and leaf greenness and also reduced proline, MDA and H_2O_2 contents, while increased phenolics and membrane stability index.	Shukla et al. (2012)
T. asperelloides (T203)	Salt (NaCl)	Favorably affects the expression of several genes related to osmo-protection and general oxidative stress in the roots of Arabidopsis and cucumber plants under salt stress.	Brotman et al. (2013)
T. parareesei, *T. reesei*	Salinity (NaCl)	Exerted beneficial effects in terms of lateral root development, and improved defense against *Botrytis cinerea*, promoted growth under salt stress and also primed defense responses in the tomato plants against biotic and abiotic stresses.	Rubio et al. (2014)

Continued

Table 5.2 Continued

Trichoderma spp.	Abiotic stress	Effects	Reference
T. virens, *T. atroviride*	Salt (NaCl)	Enhanced salt tolerance in Arabidopsis through auxin signaling by increasing the levels of abscisic acid, L-proline, and ascorbic acid, as well as eliminating Na^+ through root exudates.	Contreras-Cornejo et al. (2014)
T. hamatum	Salt (NaCl)	The increase in total and neutral lipids and antioxidants by *T. hamatum* treatment decreases the negative effect of NaCl on *Ochradenus baccatus*.	Hashem et al. (2014)
T. harzianum	Salinity (NaCl)	Mitigates the adverse effect of salt stress and enhanced the oil content in mustard through improved uptake of essential elements, modulation of osmolytes and antioxidants.	Ahmad et al. (2015)
T. asperellum (TaACCD)	Salt (NaCl)	Enhanced the growth performance and improved the tolerance in Arabidopsis plants under salt stress.	Zhang et al. (2015)
T. longibrachiatum (T6)	Salt (NaCl)	Ameliorated the adverse effects of salt stress by improving the antioxidative defense system and gene expression in the stressed wheat plants.	Zhang et al. (2016)
T. asperellum	Heavy metal (Cu)	Inoculation of *T. asperellum* prior to Cu exposure reduced Cu accumulation and translocation in tissues and ameliorated Cu toxicity in onion.	Vargas et al. (2017)
T. longibrachiatum (WT2)	Heavy metal (Pb^{2+})	Conferred tolerance to Pb^{2+}-induced toxicities in sunflower and enhanced the activities of antioxidant enzymes.	Devi et al. (2017)
T. harzianum (T34)	Salt (NaCl)	Application of *T. harzianum* T34 enhanced the growth and salt tolerance in unfertilized tomato plants.	Rubio et al. (2017)
T. harzianum	Drought	*T. harzianum* mitigated the negative effects of drought stress by modulation of plant secondary metabolites and improved proline and the synthesis of growth hormones in tomato.	Mona et al. (2017)
T. harzianum (Th-6)	Salinity (NaCl)	Improved the physiological responses in maize and rice under salt stress.	Yasmeen and Siddiqui (2018)

T. harzianum (AK20G)	Cold	Alleviated the adverse effects of cold stress and enhanced the photosynthetic and growth rates by reducing lipid peroxidation rate and electrolyte leakage while increasing leaf water content, proline accumulation and also improved the expression of TAS14 and P5CS.	Ghorbanpour et al. (2018)
T. atroviride (ID20G)	Drought	Colonization of tomato seedlings by *T. atroviride* mitigated the injurious effects of drought, and enhanced the drought tolerance by decreasing H_2O_2 and activating the antioxidant enzymes.	Pehlivan et al. (2018)
T. longibrachiatum (TL-6)	Salt (NaCl)	Effectively promoted the growth and enhanced tolerance to NaCl stress through the increased ACC-deaminase activity and IAA production and regulated the transcriptional levels of IAA and ethylene synthesis gene expression in wheat seedling under salt stress.	Zhang et al. (2019)
T. parareesei, *T. harzianum*	Salt (NaCl), drought	*T. parareesei* more effectively enhanced the productivity of rapeseed by increasing the expression of genes related to the hormonal pathways of abscisic acid (ABA) under drought stress, and ethylene (ET) under salt stress.	Poveda (2020)
T. longibrachiatum, *T. longibrachiatum* and *T. harzianum*	Drought	Enhanced plant dry mass, root volume, leaf water potential, and stomatal conductance as well as uptake of potassium in shoot and root of tomato under water-deficit conditions.	Khoshmanzar et al. (2020)

in part by increasing the gene expression of component enzymes (Mastouri, 2010; Mastouri et al., 2010). The increase in the activities of these pathways in chloroplasts reduces the damage caused by ROS involved in photosynthesis, thereby increasing the photosynthetic efficiency of plants (Zaidi et al., 2014). The increased photosynthetic efficiency may be attributed to the improved redox status of the plant by *Trichoderma* colonization.

It has been reported that *T. harzianum* produced cellulose, which acts as an elicitor to induce the activity of peroxidase in induced resistance (Elad, 2000). Several elicitors act as signaling molecules for plant defense mechanisms, such as lipids, proteins, glycoproteins, glycans, and synthetic molecules, which are released from the cell walls of plants or pathogens by hydrolytic enzymes secreted by either plant or pathogen (Zaidi et al., 2014). The signal transduction pathway begins with phosphorylation of proteins or activation of proteins from plasma membrane and leads to the mobilization or generation of signaling molecules like free calcium, nitric oxide, and ROS which regulate the physiological responses via changes in transcription and production of metabolites (Zaidi et al., 2014). Transcriptional regulation induces the production of phytohormones such as jasmonic acid, ethylene, abscisic acid, and salicylic acid, which leads to the transcriptional changes for the biosynthesis of phytoalexins and related compounds (Hanhong, 2011). The microbes or elicitors induce signal transduction pathways that result in cell wall reinforcement, lignification, production of antimicrobial metabolites, PR proteins, and oxidative stress protection (Lorrain et al., 2003). It has been demonstrated that seeds treated with *T. harzianum* strain T22 germinated consistently much faster and more uniformly than untreated seeds under multiple abiotic stresses (suboptimal temperatures, osmotic, and salt), physiological stress (poor seed quality induced by seed aging), and biotic stress (*P. ultimum*). Therefore, the most significant advantage of *Trichoderma* spp. to plants occurs under stressful conditions to mitigate both abiotic and biotic stresses (Zaidi and Singh, 2013). Moreover, the widespread presence and proliferation of *Trichoderma* spp. suggests their more crucial ecological role with the ability to improve agricultural sustainability and productivity, particularly under stressful conditions.

5.5 Conclusion

The information that is summarized in this chapter demonstrates that *Trichoderma* spp. are opportunistic and have the ability to proliferate, compete, and survive under complex ecological conditions. They have the potential to colonize plants and invade the superficial layers of the roots to elicit plant defense responses which protect the plants from a range of biotic and abiotic stresses. The root colonization by *Trichoderma* spp. enhances plant growth, which results in increased productivity and yield of plants. They also assist the plants to mitigate various abiotic stresses (drought, salinity, cold, heavy metal, etc.), and improve nutrient uptake. Therefore, *Trichoderma* spp. can be used as biopesticides and biofertilizers for the management of plant diseases, and also as alleviators of various biotic stresses.

References

Abd-El-Kareem, F., Elshahawy, I.E., Abd-Elgawad, M.M., 2019. Local *Trichoderma* strains as a control strategy of complex black root rot disease of strawberry in Egypt. Bull. Natl. Res. Cent. 43 (1), 1–7.

Abdelrahman, M., Abdel-Motaal, F., El-Sayed, M., Jogaiah, S., Shigyo, M., Ito, S.-I., Tran, L.-S.P., 2016. Dissection of *Trichoderma longibrachiatum*-induced defense in onion (*Allium cepa* L.) against *Fusarium oxysporum* f. sp. *Cepa* by target metabolite profiling. Plant Sci. 246, 128–138.

Adekunle, A.T., Cardwell, K.F., Florini, D.A., Ikotun, T., 2001. Seed treatment with *Trichoderma* species for control of damping-off of cowpea caused by *Macrophomina phaseolina*. Biocontrol Sci. Tech. 11 (4), 449–457.

Affokpon, A., Coyne, D.L., Htay, C.C., Agbèdè, R.D., Lawouin, L., Coosemans, J., 2011. Biocontrol potential of native *Trichoderma* isolates against root-knot nematodes in West African vegetable production systems. Soil Biol. Biochem. 43 (3), 600–608.

Ahmad, P., Jaleel, C.A., Sharma, S., 2010. Antioxidant defense system, lipid peroxidation, proline-metabolizing enzymes, and biochemical activities in two *Morus alba* genotypes subjected to NaCl stress. Russ. J. Plant Physiol. 57 (4), 509–517.

Ahmad, P., Hashem, A., Abd-Allah, E.F., Alqarawi, A.A., John, R., Egamberdieva, D., Gucel, S., 2015. Role of *Trichoderma harzianum* in mitigating NaCl stress in Indian mustard (*Brassica juncea* L.) through antioxidative defense system. Front. Plant Sci. 6, 868.

Alabouvette, C., Olivain, C., Migheli, Q., Steinberg, C., 2009. Microbiological control of soil-borne phytopathogenic fungi with special emphasis on wilt-inducing *Fusarium oxysporum*. New Phytologist Trust 184, 529–544.

Al-Ani, L.K.T., 2017. Potential of Utilizing Biological and Chemical Agents in the Control of Fusarium Wilt of Banana. PhD thesisSchool of Biology Science, Universiti Sains Malaysia, Pulau Pinang, Malaysia, p. 259.

Al-Ani, L.K.T., 2018. *Trichoderma*: beneficial role in sustainable agriculture by plant disease management. In: Plant Microbiome: Stress Response. Springer, Singapore, pp. 105–126.

Aleandri, M.P., Chilosi, G., Bruni, N., Tomassini, A., Vettraino, A.M., Vannini, A., 2015. Use of nursery potting mixes amended with local *Trichoderma* strains with multiple complementary mechanisms to control soil-borne diseases. Crop Prot. 67, 269–278.

Al-Hazmi, A.S., Javeed, T.M., 2016. Effects of different inoculum densities of *Trichoderma harzianum* and *Trichoderma viride* against *Meloidogyne javanica* on tomato. Saudi J. Biol. Sci. 23 (2), 288–292.

Azarmi, R., Hajieghrari, B., Giglou, A., 2011. Effect of *Trichoderma* isolates on tomato seedling growth response and nutrient uptake. Afr. J. Biotechnol. 10 (31), 5850–5855.

Bae, H., Sicher, R.C., Kim, M.S., Kim, S.H., Strem, M.D., Melnick, R.L., Bailey, B.A., 2009. The beneficial endophyte *Trichoderma hamatum* isolate DIS 219b promotes growth and delays the onset of the drought response in *Theobroma cacao*. J. Exp. Bot. 60 (11), 3279–3295.

Bae, H., Roberts, D.P., Lim, H.S., Strem, M.D., Park, S.C., Ryu, C.M., Bailey, B.A., 2011. Endophytic *Trichoderma* isolates from tropical environments delay disease onset and induce resistance against *Phytophthora capsici* in hot pepper using multiple mechanisms. Mol. Plant-Microbe Interact. 24 (3), 336–351.

Bayoumi, Y., Taha, N., Shalaby, T., Alshaal, T., El-Ramady, H., 2019. Sulfur promotes biocontrol of purple blotch disease via *Trichoderma* spp. and enhances the growth, yield and quality of onion. Appl. Soil Ecol. 134, 15–24.

Bicici, M., Dede, Y., Çinar, A., 1992. *Trichoderma* species against gummosis disease in lemon trees. In: Biological Control of Plant Diseases. Springer, Boston, MA, pp. 193–196.

Bidellaoui, B., Segarra, G., Hakkou, A., Trillas, M.I., 2019. Beneficial effects of *Rhizophagus irregularis* and *Trichoderma asperellum* strain T34 on growth and *Fusarium* wilt in tomato plants. J. Plant Pathol. 101 (1), 121–127.

Björkman, T., Blanchard, L.M., Harman, G.E., 1998. Growth enhancement of shrunken-2 (sh2) sweet corn by *Trichoderma harzianum* 1295-22: effect of environmental stress. J. Am. Soc. Hortic. Sci. 123 (1), 35–40.

Bokhari, F.M., 2009. Efficacy of some *Trichoderma* species in the control of *Rotylenchulus reniformis* and *Meloidogyne javanica*. Arch. Phytopathol. Plant Protect. 42 (4), 361–369.

Boller, T., Felix, G., 2009. A renaissance of elicitors: perception of microbe-associated molecular patterns and danger signals by pattern-recognition receptors. Annu. Rev. Plant Biol. 60, 379–406.

Brotman, Y., Briff, E., Viterbo, A., Chet, I., 2008. Role of swollenin, an expansin-like protein from *Trichoderma*, in plant root colonization. Plant Physiol. 147 (2), 779–789.

Brito, J.P., Ramada, M.H., de Magalhães, M.T., Silva, L.P., Ulhoa, C.J., 2014. Peptaibols from Trichoderma asperellum TR356 strain isolated from Brazilian soil. SpringerPlus 3 (1), 1–10.

Brotman, Y., Landau, U., Cuadros-Inostroza, A., Takayuki, T., Fernie, A.R., Chet, I., Willmitzer, L., 2013. *Trichoderma*-plant root colonization: escaping early plant defense responses and activation of the antioxidant machinery for saline stress tolerance. PLoS Pathog. 9(3).

Cao, L., Jiang, M., Zeng, Z., Du, A., Tan, H., Liu, Y., 2008. *Trichoderma atroviride* F6 improves phytoextraction efficiency of mustard (*Brassica juncea* (L.) Coss. var. foliosa Bailey) in Cd, Ni contaminated soils. Chemosphere 71 (9), 1769–1773.

Contreras-Cornejo, H.A., Macías-Rodríguez, L., Cortés-Penagos, C., López-Bucio, J., 2009. *Trichoderma virens*, a plant beneficial fungus, enhances biomass production and promotes lateral root growth through an auxin-dependent mechanism in *Arabidopsis*. Plant Physiol. 149 (3), 1579–1592.

Contreras-Cornejo, H.A., Macías-Rodríguez, L., Alfaro-Cuevas, R., López-Bucio, J., 2014. *Trichoderma* spp. improve growth of *Arabidopsis* seedlings under salt stress through enhanced root development, osmolite production, and Na^+ elimination through root exudates. Mol. Plant-Microbe Interact. 27 (6), 503–514.

Contreras-Cornejo, H.A., Macías-Rodríguez, L., del-Val, E., Larsen, J., 2016. Ecological functions of *Trichoderma* spp. and their secondary metabolites in the rhizosphere: interactions with plants. FEMS Microbiol. Ecol. 92 (4) fiw036.

Dana, M.l.M., Pintor-Toro, J.A., Cubero, B., 2006. Transgenic tobacco plants overexpressing chitinases of fungal origin show enhanced resistance to biotic and abiotic stress agents. Plant Physiol. 142 (2), 722–730.

Daragó, Á., Szabó, M., Hrács, K., Takács, A., Nagy, P.I., 2013. In vitro investigations on the biological control of *Xiphinema index* with *Trichoderma* species. Helminthologia 50 (2), 132–137.

De França, S.K.S., Cardoso, A.F., Lustosa, D.C., Ramos, E.M.L.S., de Filippi, M.C.C., Da Silva, G.B., 2015. Biocontrol of sheath blight by *Trichoderma asperellum* in tropical low land rice. Agron. Sustain. Dev. 35 (1), 317–324.

De Medeiros, H.A., de Araújo Filho, J.V., De Freitas, L.G., Castillo, P., Rubio, M.B., Hermosa, R., Monte, E., 2017. Tomato progeny inherit resistance to the nematode *Meloidogyne javanica* linked to plant growth induced by the biocontrol fungus *Trichoderma atroviride*. Sci. Rep. 7 (1), 1–13.

Degenkolb, T., Karimi Aghcheh, R., Dieckmann, R., Neuhof, T., Baker, S.E., Druzhinina, I.S., von Döhren, H., 2012. The production of multiple small peptaibol families by single 14-module peptide synthetases in *Trichoderma/Hypocrea*. Chem. Biodivers. 9 (3), 499–535.

Devi, S.S., Sreenivasulu, Y., Rao, K.V., 2017. Protective role of *Trichoderma logibrachiatum* (WT2) on Lead induced oxidative stress in *Helianthus annus* L. Indian J. Exp. Biol. 55, 235–241.

Doni, F., Anizan, I., Radziah, C.C., Salman, A.H., Rodzihan, M.H., Yusoff, W.M.W., 2014. Enhancement of rice seed germination and vigour by "*Trichoderma*" spp. Res. J. Appl. Sci. Eng. Technol. 7 (21), 4547–4552.

Druzhinina, I.S., Seidl-Seiboth, V., Herrera-Estrella, A., Horwitz, B.A., Kenerley, C.M., Monte, E., Kubicek, C.P., 2011. *Trichoderma*: the genomics of opportunistic success. Nat. Rev. Microbiol. 9 (10), 749–759.

Dunlop, R.W., Simon, A., Sivasithamparam, K., Ghisalberti, E.L., 1989. An antibiotic from *Trichoderma koningii* active against soilborne plant pathogens. J. Nat. Prod. 52 (1), 67–74.

Elad, Y., 2000. Biological control of foliar pathogens by means of *Trichoderma harzianum* and potential modes of action. Crop Prot. 19, 709–714.

Elad, Y., Kirshner, B., Yehuda, N., Sztejnberg, A., 1998. Management of powdery mildew and gray mold of cucumber by *Trichoderma harzianum* T39 and *Ampelomyces quisqualis* AQ10. BioControl 43 (2), 241–251.

El-Dabaa, M.A.T., Abd-El-Khair, H., 2020. Applications of plant growth promoting bacteria and *Trichoderma* spp. for controlling *Orobanche crenata* in faba bean. Bull. Nat. Res. Cent. 44 (1), 1–10.

El-Hai, K.A., Ali, A.A., 2019. Research article formulation of *Trichoderma*, *Saccharomyces* and *Rhizobium* metabolites against damping-off and root rot pathogens in peanut plant. Asian J. Biol. Sci. 12 (2), 114–121.

Gachomo, E.W., Kotchoni, S.O., 2008. The use of *Trichoderma harzianum* and *T. viride* as potential biocontrol agents against peanut microflora and their effectiveness in reducing aflatoxin contamination of infected kernels. Biotechnology 7 (3), 439–447.

Gajera, H., Domadiya, R., Patel, S., Kapopara, M., Golakiya, B., 2013. Molecular mechanism of *Trichoderma* as bio-control agents against phytopathogen system—a review. Curr. Res. Microbiol. Biotechnol. 1, 133–142.

Ghorbanpour, A., Salimi, A., Ghanbary, M.A.T., Pirdashti, H., Dehestani, A., 2018. The effect of *Trichoderma harzianum* in mitigating low temperature stress in tomato (*Solanum lycopersicum* L.) plants. Sci. Hortic. 230, 134–141.

Gnanamanickam, S.S., 2002. Biological Control of Crop Diseases. Marcel Dekker, New York, p. 449.

Gravel, V., Antoun, H., Tweddell, R.J., 2007. Growth stimulation and fruit yield improvement of greenhouse tomato plants by inoculation with *Pseudomonas putidaor Trichoderma atroviride*: possible role of indole acetic acid (IAA). Soil Biol. Biochem. 39 (8), 1968–1977.

Gruber, S., Kubicek, C.P., Seidl-Seiboth, V., 2011. Differential regulation of orthologous chitinase genes in mycoparasitic *Trichoderma* species. Appl. Environ. Microbiol. 77 (20), 7217–7226.

Hadar, Y., Harman, G.E., Taylor, A.G., 1984. Evaluation of *Trichoderma koningii* and *T. harzianum* from New York soils for biological control of seed rot caused by *Pythium* spp. Phytopathology 74 (1), 106–110.

Hanhong, B., 2011. *Trichoderma* species as abiotic and biotic stress quenchers in plants. Res. J. Biotechnol. 6 (3), 73–79.

Hanson, L.E., Howell, C.R., 2002. Biocontrol efficacy and other characteristics of protoplast fusants between *Trichoderma koningii* and *T. virens*. Mycol. Res. 106 (3), 321–328.

Harman, G.E., 2000. Myths and dogmas of biocontrol changes in perceptions derived from research on *Trichoderma harzianum* T-22. Plant Dis. 84 (4), 377–393.

Harman, G.E., 2006. Overview of mechanisms and uses of *Trichoderma* spp. Phytopathology 96 (2), 190–194.

Harman, G.E., 2011. *Trichoderma*—not just for biocontrol anymore. Phytoparasitica 39 (2), 103–108.

Harman, G.E., Shoresh, M., 2007. The mechanisms and applications of symbiotic opportunistic plant symbionts. In: Novel Biotechnologies for Biocontrol Agent Enhancement and Management. Springer, Dordrecht, pp. 131–155.

Harman, G.E., Howell, C.R., Viterbo, A., Chet, I., Lorito, M., 2004. *Trichoderma* species—opportunistic, avirulent plant symbionts. Nat. Rev. Microbiol. 2 (1), 43–56.

Harris, A.R., 1999. Biocontrol of *Rhizoctonia solani* and *Pythium ultimum* on Capsicum by *Trichoderma koningii* in potting medium. Microbiol. Res. 154 (2), 131–135.

Hashem, A., Abd-Allah, E.F., Alqarawi, A.A., Al Huqail, A.A., Egamberdieva, D., 2014. Alleviation of abiotic salt stress in *Ochradenus baccatus* (Del.) by *Trichoderma hamatum* (Bonord.) Bainier. J. Plant Interact. 9 (1), 857–868.

Hermosa, R., Viterbo, A., Chet, I., Monte, E., 2012. Plant-beneficial effects of *Trichoderma* and of its genes. Microbiology 158 (1), 17–25.

Howell, C.R., 2003. Mechanisms employed by *Trichoderma* species in the biological control of plant diseases: the history and evolution of current concepts. Plant Dis. 87 (1), 4–10.

Howell, C.R., Hanson, L.E., Stipanovic, R.D., Puckhaber, L.S., 2000. Induction of terpenoid synthesis in cotton roots and control of *Rhizoctonia solani* by seed treatment with *Trichoderma virens*. Phytopathology 90 (3), 248–252.

Hussey, R.S., Roncadori, R.W., 1978. Interaction of *Pratylenchus brachyurus* and *Gigaspora margarita* on cotton. J. Nematol. 10 (1), 16–20.

Iqbal, M., Ashraf, M., 2013. Alleviation of salinity-induced perturbations in ionic and hormonal concentrations in spring wheat through seed preconditioning in synthetic auxins. Acta Physiol. Plant 35, 1093–1112.

Izuogu, N.B., Abiri, T.O., 2015. Efficacy of *Trichoderma harzianum* T22 as a biocontrol agent against root-knot nematode (*Meloidogyne incognita*) on some soybean varieties. Croat. J. Food Sci. Technol. 7 (2), 47–51.

Jacobsen, B.J., 2006. Biological control of plant diseases by phyllosphere applied biological control agents. In: Bailey, M.J., Lilley, A.K., Timms-Wilson, T.M., Spencer-Phillips, P.T.N. (Eds.), Microbial Ecology of Aerial Plant Surfaces. Athenaeum Press, Gateshead, pp. 133–147.

Javeed, M.T., Al-Hazmi, A.S., Molan, Y.Y., 2016. Antagonistic effects of some indigenous isolates of *Trichoderma* spp. against *Meloidogyne javanica*. Pak. J. Nematol. 2, 183–191.

Jogaiah, S., Govind, S.R., Tran, L.S.P., 2013. Systems biology-based approaches toward understanding drought tolerance in food crops. Crit. Rev. Biotechnol. 33 (1), 23–39.

Jogaiah, S., Mostafa, A., Tran, L.S.P., Shin-ichi, I., 2013a. Characterization of rhizosphere fungi that mediate resistance in tomato against bacterial wilt disease. J. Exp. Bot. 64, 3829–3842.

Jogaiah, S., Abdelrahman, M., Tran, L.-S.P., Ito, S.-I., 2018. Different mechanisms of *Trichoderma virens*-mediated resistance in tomato against Fusarium wilt involve the jasmonic and salicylic acid pathways. Mol. Plant Pathol. 19, 870–882.

Jones, J.D., Dangl, J.L., 2006. The plant immune system. Nature 444 (7117), 323–329.

Keswani, C., Mishra, S., Sarma, B.K., Singh, S.P., Singh, H.B., 2014. Unraveling the efficient applications of secondary metabolites of various *Trichoderma* spp. Appl. Microbiol. Biotechnol. 98 (2), 533–544.

Khan, M.R., Parveen, G., 2018. Supplementing biocontrol agents with botanicals improved growth and yield of coriander (*Coriandrum sativum* L.) infected with *Protomyces macrosporus* Unger. Curr. Plant Biol. 15, 44–50.

Khan, R.A.A., Najeeb, S., Mao, Z., Ling, J., Yang, Y., Li, Y., Xie, B., 2020. Bioactive secondary metabolites from *Trichoderma* spp. against phytopathogenic bacteria and root-knot nematode. Microorganisms 8 (3), 401. https://doi.org/10.3390/microorganisms8030401.

Khoshmanzar, E., Aliasgharzad, N., Neyshabouri, M.R., Khoshru, B., Arzanlou, M., Lajayer, B.A., 2020. Effects of *Trichoderma* isolates on tomato growth and inducing its tolerance to water-deficit stress. Int. J. Environ. Sci. Technol. 17 (2), 869–878.

Konappa, N., Krishnamurthy, S., Siddaiah, C.N., Ramachandrappa, N.S., Chowdappa, S., 2018. Evaluation of biological efficacy of *Trichoderma asperellum* against tomato bacterial wilt caused by *Ralstonia solanacearum*. Egypt. J. Biol. Pest Co. 28 (1), 63.

Kubicek, C.P., Herrera-Estrella, A., Seidl-Seiboth, V., Martinez, D.A., Druzhinina, I.S., Thon, M., Mukherjee, M., 2011. Comparative genome sequence analysis underscores mycoparasitism as the ancestral life style of *Trichoderma*. Genome Biol. 12 (4), R40.

Latorre, B.A., Lillo, C., Rioja, M.E., 2001. Eficacia de los tratamientos fungicidas para el control de *Botrytis cinerea* de la vid en función de la época de aplicación. Ciencia e Investigacion Agraria. 28, 61–66.

Lehner, S.M., Atanasova, L., Neumann, N.K., Krska, R., Lemmens, M., Druzhinina, I.S., Schuhmacher, R., 2013. Isotope-assisted screening for iron-containing metabolites reveals a high degree of diversity among known and unknown siderophores produced by *Trichoderma* spp. Appl. Environ. Microbiol. 79 (1), 18–31.

Leonetti, P., Costanza, A., Zonno, M., Molinari, S., Altomare, C., 2014. How fungi interact with nematode to activate the plant defence response to tomato plants. Commun. Agric. Appl. Biol. Sci. 79, 357–362.

Leong, J., 1986. Siderophores: their biochemistry and possible role in biocontrol of plant pathogen. Annu. Rev. Phytopathol. 24, 187–209.

Li, H.Y., Luo, Y., Zhang, X.S., Shi, W.L., Gong, Z.T., Shi, M., Song, X.Y., 2014. Trichokonins from *Trichoderma pseudokoningii* SMF2 induce resistance against Gram-negative *Pectobacterium carotovorum* subsp. *carotovorum* in Chinese cabbage. FEMS Microbiol. Lett. 354 (1), 75–82.

Lifshitz, R., Windham, M.T., Baker, R., 1986. Mechanism of biological control of preemergence damping-off of pea by seed treatment with *Trichoderma* spp. Phytopathology 76 (7), 720–725.

López, A.C., Alvarenga, A.E., Zapata, P.D., Luna, M.F., Villalba, L.L., 2019. *Trichoderma* spp. from Misiones, Argentina: effective fungi to promote plant growth of the regional crop *Ilex paraguariensis* St. Hil. Mycology 10 (4), 210–221.

Lorito, M., Woo, S.L., Harman, G.E., Monte, E., 2010. Translational research on *Trichoderma*: from omics to the field. Annu. Rev. Phytopathol. 48, 395–417.

Lorrain, S., Vailleau, F., Balagué, C., Roby, D., 2003. Lesion mimic mutants: keys for deciphering cell death and defense pathways in plants? Trends Plant Sci. 8 (6), 263–271.

Luo, Y., Zhang, D.D., Dong, X.W., Zhao, P.B., Chen, L.L., Song, X.Y., Zhang, Y.Z., 2010. Antimicrobial peptaibols induce defense responses and systemic resistance in tobacco against tobacco mosaic virus. FEMS Microbiol. Lett. 313 (2), 120–126.

Martínez-Medina, A., Alguacil, M.D.M., Pascual, J.A., Van Wees, S.C., 2014. Phytohormone profiles induced by *Trichoderma* isolates correspond with their biocontrol and plant growth-promoting activity on melon plants. J. Chem. Ecol. 40 (7), 804–815.

Martínez-Medina, A., Fernandez, I., Lok, G.B., Pozo, M.J., Pieterse, C.M., Van Wees, S.C., 2017. Shifting from priming of salicylic acid- to jasmonic acid-regulated defences by *Trichoderma* protects tomato against the root knot nematode *Meloidogyne incognita*. New Phytol. 213 (3), 1363–1377.

Marzano, M., Gallo, A., Altomare, C., 2013. Improvement of biocontrol efficacy of *Trichoderma harzianum* vs. *Fusarium oxysporum* f. sp. *lycopersici* through UV-induced tolerance to fusaric acid. Biol. Control 67 (3), 397–408.

Mastouri, F., 2010. Use of *Trichoderma* spp. to Improve Plant Performance Under Abiotic Stress. PhD thesisCornell University, Ithaca, NY, USA.

Mastouri, F., Björkman, T., Harman, G.E., 2010. Seed treatment with *Trichoderma harzianum* alleviates biotic, abiotic, and physiological stresses in germinating seeds and seedlings. Phytopathology 100 (11), 1213–1221.

Mendoza-Mendoza, A., Pozo, M.J., Grzegorski, D., Martínez, P., García, J.M., Olmedo-Monfil, V., Herrera-Estrella, A., 2003. Enhanced biocontrol activity of *Trichoderma* through inactivation of a mitogen-activated protein kinase. Proc. Natl. Acad. Sci. 100 (26), 15965–15970.

Miethke, M., 2013. Molecular strategies of microbial iron assimilation: from high-affinity complexes to cofactor assembly systems. Metallomics 5 (1), 15–28.

Mittler, R., 2002. Oxidative stress, antioxidants and stress tolerance. Trends Plant Sci. 7 (9), 405–410.

Mona, S.A., Hashem, A., Abd_Allah, E.F., Alqarawi, A.A., Soliman, D.W.K., Wirth, S., Egamberdieva, D., 2017. Increased resistance of drought by *Trichoderma harzianum* fungal treatment correlates with increased secondary metabolites and proline content. J. Integr. Agric. 16 (8), 1751–1757.

Montero-Barrientos, M., Hermosa, R., Nicolás, C., Cardoza, R.E., Gutiérrez, S., Monte, E., 2008. Overexpression of a Trichoderma HSP70 gene increases fungal resistance to heat and other abiotic stresses. Fungal Genet. Biol. 45 (11), 1506–1513.

Montero-Barrientos, M., Hermosa, R., Cardoza, R.E., Gutierrez, S., Nicolas, C., Monte, E., 2010. Transgenic expression of the *Trichoderma harzianum* hsp70 gene increases *Arabidopsis* resistance to heat and other abiotic stresses. J. Plant Physiol. 167 (8), 659–665.

Morán-Diez, E., Hermosa, R., Ambrosino, P., Cardoza, R.E., Gutiérrez, S., Lorito, M., Monte, E., 2009. The ThPG1 endopolygalacturonase is required for the *Trichoderma harzianum*–plant beneficial interaction. Mol. Plant-Microbe Interact. 22 (8), 1021–1031.

Mukherjee, P.K., Wiest, A., Ruiz, N., Keightley, A., Moran-Diez, M.E., McCluskey, K., Kenerley, C.M., 2011. Two classes of new peptaibols are synthesized by a single non-ribosomal peptide synthetase of *Trichoderma virens*. J. Biol. Chem. 286 (6), 4544–4554.

Mukherjee, P.K., Buensanteai, N., Moran-Diez, M.E., Druzhinina, I.S., Kenerley, C.M., 2012. Functional analysis of non-ribosomal peptide synthetases (NRPSs) in *Trichoderma virens* reveals a polyketide synthase (PKS)/NRPS hybrid enzyme involved in the induced systemic resistance response in maize. Microbiology 158 (1), 155–165.

Murali, M., Jogaiah, S., Amruthesh, K.N., Shin-ichi, I., Shekar Shetty, H., 2013. Rhizosphere fungus *Penicillium chrysogenum* promotes growth and induces defense-related genes and downy mildew disease resistance in pearl millet. Plant Biol. 15, 111–118.

Nagaraju, A., Jogaiah, S., Mahadevamurthy, S., Shin-Ichi, I., 2012a. Seed priming with *Trichoderma harzianum* isolates enhances plant growth and induces resistance against

Plasmopara halstedii, an incitant of sunflower downy mildew disease. Aust. J. Plant Path. 41, 609–620.

Nagaraju, A., Murali, M., Jogaiah, S., Amruthesh, K.N., Mahadeva Murthy, S., 2012b. Beneficial microbes promote plant growth and induce systemic resistance in sunflower against downy mildew disease caused by *Plasmopara halstedii*. Curr. Bot. 3, 12–18.

Nassr, S., Barakat, R., 2013. Effect of factors on conidium germination of *Botrytis cinerea* in vitro. Methods 67, 68.

Pehlivan, N., Saruhan-Güler, N., Alpay-Karaoğlu, Ş., 2018. The effect of *Trichoderma* seed priming to drought resistance in tomato (*Solanum lycopersicum* L.) plants. Hacettepe J. Biol. Chem. 46 (2), 263–272.

Pellegrini, A., Prodorutti, D., Pertot, I., 2014. Use of bark mulch pre-inoculated with *Trichoderma atroviride* to control Armillaria root rot. Crop Prot. 64, 104–109.

Perazzolli, M., Moretto, M., Fontana, P., Ferrarini, A., Velasco, R., Moser, C., Pertot, I., 2012. Downy mildew resistance induced by *Trichoderma harzianum* T39 in susceptible grapevines partially mimics transcriptional changes of resistant genotypes. BMC Genomics 13 (1), 660.

Pieterse, C.M., Van der Does, D., Zamioudis, C., Leon-Reyes, A., Van Wees, S.C., 2012. Hormonal modulation of plant immunity. Annu. Rev. Cell Dev. Biol. 28, 489–521.

Pocurull, M., Fullana, A.M., Ferro, M., Valero, P., Escudero, N., Saus, E., Sorribas, F.J., 2020. Commercial formulates of *Trichoderma* induce systemic plant resistance to *Meloidogyne incognita* in tomato and the effect is additive to that of the Mi-1.2 resistance gene. Front. Microbiol. 10, 3042.

Poveda, J., 2020. *Trichoderma parareesei*favors the tolerance of rapeseed (*Brassica napus* L.) to salinity and drought due to a chorismate mutase. Agronomy 10 (1), 118.

Rasool, S., Ahmad, A., Siddiqi, T.O., Ahmad, P., 2013. Changes in growth, lipid peroxidation and some key antioxidant enzymes in chickpea genotypes under salt stress. Acta Physiol. Plant. 35, 1039–1050. https://doi.org/10.1007/s11738-012-1142-4.

Rawat, L., Singh, Y., Shukla, N., Kumar, J., 2011. Alleviation of the adverse effects of salinity stress in wheat (*Triticum aestivum* L.) by seed biopriming with salinity tolerant isolates of *Trichoderma harzianum*. Plant and Soil 347 (1–2), 387.

Rivera-Méndez, W., Obregón, M., Morán-Diez, M.E., Hermosa, R., Monte, E., 2020. *Trichoderma asperellum* biocontrol activity and induction of systemic defenses against *Sclerotium cepivorum* in onion plants under tropical climate conditions. Biol. Control 141, 104145.

Romão-Dumaresq, A.S., de Araujo, W.L., Talbot, N.J., Thornton, C.R., 2012. RNA interference of endochitinases in the sugarcane endophyte *Trichoderma virens* 223 reduces its fitness as a biocontrol agent of pineapple disease. PLoS One. 7(10).

Rubio, M.B., Quijada, N.M., Pérez, E., Domínguez, S., Monte, E., Hermosa, R., 2014. Identifying *Trichoderma parareesei* beneficial qualities for plants. Appl. Environ. Microbiol. 80 (6), 1864–1873.

Rubio, M.B., Hermosa, R., Vicente, R., Gómez-Acosta, F.A., Morcuende, R., Monte, E., Bettiol, W., 2017. The combination of *Trichoderma harzianum* and chemical fertilization leads to the deregulation of phytohormone networking, preventing the adaptive responses of tomato plants to salt stress. Front. Plant Sci. 8, 294.

Salas-Marina, M.A., Silva-Flores, M.A., Uresti-Rivera, E.E., Castro-Longoria, E., Herrera-Estrella, A., Casas-Flores, S., 2011. Colonization of *Arabidopsis* roots by *Trichoderma atroviride* promotes growth and enhances systemic disease resistance through jasmonic acid/ethylene and salicylic acid pathways. Eur. J. Plant Pathol. 131 (1), 15–26.

Samolski, I., Rincón, A.M., Pinzón, L.M., Viterbo, A., Monte, E., 2012. The qid74 gene from *Trichoderma harzianum* has a role in root architecture and plant biofertilization. Microbiology 158 (1), 129–138.

Samuels, G.J., 1996. *Trichoderma*: a review of biology and systematics of the genus. Mycol. Res. 100 (8), 923–935.

Sant, D., Casanova, E., Segarra, G., Avilés, M., Reis, M., Trillas, M.I., 2010. Effect of *Trichoderma asperellum* strain T34 on *Fusarium* wilt and water usage in carnation grown on compost-based growth medium. Biol. Control 53 (3), 291–296.

Saravanan, T., Muthusamy, M., Marimuthu, T., 2003. Development of integrated approach to manage the fusarial wilt of banana. Crop Prot. 22 (9), 1117–1123.

Segarra, G., Casanova, E., Bellido, D., Odena, M.A., Oliveira, E., Trillas, I., 2007. Proteome, salicylic acid, and jasmonic acid changes in cucumber plants inoculated with *Trichoderma asperellum* strain T34. Proteomics 7 (21), 3943–3952.

Segarra, G., Casanova, E., Avilés, M., Trillas, I., 2010. *Trichoderma asperellum* strain T34 controls Fusarium wilt disease in tomato plants in soilless culture through competition for iron. Microb. Ecol. 59 (1), 141–149.

Selim, M.E., Mahdy, M.E., Sorial, M.E., Dababat, A.A., Sikora, R.A., 2014. Biological and chemical dependent systemic resistance and their significance for the control of root-knot nematodes. Nematology 16 (8), 917–927.

Sennoi, R., Singkham, N., Jogloy, S., Boonlue, S., Saksirirat, W., Kesmala, T., Patanothai, A., 2013. Biological control of southern stem rot caused by *Sclerotium rolfsii* using *Trichoderma harzianum* and arbuscular mycorrhizal fungi on Jerusalem artichoke (*Helianthus tuberosus* L.). Crop Prot. 54, 148–153.

Shabala, S., Cuin, T.A., 2008. Potassium transport and plant salt tolerance. Physiol. Plant. 133 (4), 651–669.

Sharon, E., Bar-Eyal, M., Chet, I., Herrera-Estrella, A., Kleifeld, O., Spiegel, Y., 2001. Biological control of the root-knot nematode *Meloidogyne javanica* by *Trichoderma harzianum*. Phytopathology 91, 687–693.

Sharon, E., Chet, I., Viterbo, A., Bar-Eyal, M., Nagan, H., Samuels, G.J., Spiegel, Y., 2007. Parasitism of *Trichoderma* on *Meloidogyne javanica* and role of the gelatinous matrix. Eur. J. Plant Pathol. 118 (3), 247–258.

Sharon, E., Chet, I., Spiegel, Y., 2011. *Trichoderma* as a biological control agent. In: Davies, K., Spiegel, Y. (Eds.), Biological Control of Plant-Parasitic Nematodes: Building Coherence Between Microbial Ecology and Molecular Mechanisms. Springer, Berlin, pp. 183–201.

Shi, M., Chen, L., Wang, X.W., Zhang, T., Zhao, P.B., Song, X.Y., Zhang, Y.Z., 2012. Antimicrobial peptaibols from *Trichoderma pseudokoningii* induce programmed cell death in plant fungal pathogens. Microbiology 158 (1), 166–175.

Shoresh, M., Harman, G.E., 2008. The molecular basis of shoot responses of maize seedlings to *Trichoderma harzianum* T22 inoculation of the root: a proteomic approach. Plant Physiol. 147 (4), 2147–2163.

Shoresh, M., Yedidia, I., Chet, I., 2005. Involvement of jasmonic acid/ethylene signaling pathway in the systemic resistance induced in cucumber by *Trichoderma asperellum* T203. Phytopathology 95 (1), 76–84.

Shoresh, M., Harman, G.E., Mastouri, F., 2010. Induced systemic resistance and plant responses to fungal biocontrol agents. Annu. Rev. Phytopathol. 48, 21–43.

Shukla, N., Awasthi, R.P., Rawat, L., Kumar, J., 2012. Biochemical and physiological responses of rice (*Oryza sativa* L.) as influenced by *Trichoderma harzianum* under drought stress. Plant Physiol. Biochem. 54, 78–88.

Singh, S., Dureja, P., Tanwar, R.S., Singh, A., 2005. Production and antifungal activity of secondary metabolites of *Trichoderma virens*. Pestic. Res. J. 17, 26–29.

Singh, R., Ong-Abdullah, M., Low, E.T.L., Manaf, M.A.A., Rosli, R., Nookiah, R., Azizi, N., 2013. Oil palm genome sequence reveals divergence of interfertile species in Old and New worlds. Nature 500 (7462), 335–339.

Singh, A., Shukla, N., Kabadwal, B.C., Tewari, A.K., Kumar, J., 2018. Review on plant-*Trichoderma*-pathogen interaction. Int. J. Curr. Microbiol. App. Sci. 7 (2), 2382–2397.

Sivan, A., Ucko, O., Chet, I., 1987. Biological control of *Fusarium* crown rot of tomato by *Trichoderma harzianum* under field conditions. Plant Dis. 71 (7), 587–592.

Sudisha, J., Niranjana, S.R., Umesha, S., Prakash, H.S., Shekar Shetty, H., 2006. Transmission of seed-borne infection of muskmelon by *Didymella bryoniae* and effect of seed treatments on disease incidence and fruit yield. Biol. Control 37, 196–205.

Szabó, M., Csepregi, K., Gálber, M., Virányi, F., Fekete, C., 2012. Control plant-parasitic nematodes with *Trichoderma* species and nematode-trapping fungi: the role of chi18-5 and chi18-12 genes in nematode egg-parasitism. Biol. Control 63 (2), 121–128.

Szekeres, A., Kredics, L., Antal, Z., Kevei, F., Manczinger, L., 2004. Isolation and characterization of protease overproducing mutants of *Trichoderma harzianum*. FEMS Microbiol. Lett. 233 (2), 215–222.

Tijerino, A., Cardoza, R.E., Moraga, J., Malmierca, M.G., Vicente, F., Aleu, J., ⋯ Hermosa, R., 2011. Overexpression of the trichodiene synthase gene tri5 increases trichodermin production and antimicrobial activity in Trichoderma brevicompactum. Fungal Genet. Biol. 48 (3), 285–296.

Tsahouridou, P.C., Thanassoulopoulos, C.C., 2001. *Trichoderma koningii* as a potential parasite of sclerotia of *Sclerotium rolfsii*. Cryptogam. Mycol. 22 (4), 289–295.

Tsahouridou, P.C., Thanassoulopoulos, C.C., 2002. Proliferation of *Trichoderma koningii* in the tomato rhizosphere and the suppression of damping-off by *Sclerotium rolfsii*. Soil Biol. Biochem. 34 (6), 767–776.

Vargas, J.T., Rodríguez-Monroy, M., Meyer, M.L., Montes-Belmont, R., Sepúlveda-Jiménez, G., 2017. *Trichoderma asperellum* ameliorates phytotoxic effects of copper in onion (*Allium cepa* L.). Environ. Exp. Bot. 136, 85–93.

Vinale, F., Sivasithamparam, K., Ghisalberti, E.L., Marra, R., Woo, S.L., Lorito, M., 2008. *Trichoderma*–plant–pathogen interactions. Soil Biol. Biochem. 40 (1), 1–10.

Vinale, F., Ghisalberti, E.L., Sivasithamparam, K., Marra, R., Ritieni, A., Ferracane, R., Lorito, M., 2009. Factors affecting the production of *Trichoderma harzianum* secondary metabolites during the interaction with different plant pathogens. Lett. Appl. Microbiol. 48 (6), 705–711.

Vinale, F., Sivasithamparam, K., Ghisalberti, E.L., Woo, S.L., Nigro, M., Marra, R., Manganiello, G., 2014. *Trichoderma* secondary metabolites active on plants and fungal pathogens. Open Mycol. J. 8(1).

Viterbo, A.D.A., Chet, I., 2006. TasHyd1, a new hydrophobin gene from the biocontrol agent *Trichoderma asperellum*, is involved in plant root colonization. Mol. Plant Pathol. 7 (4), 249–258.

Viterbo, A., Horwitz, B.A., Borkovich, K., 2010. Mycoparasitism. In: Ebbole, D.J. (Ed.), Cellular and Molecular Biology of Filamentous Fungi. ASM Press, Washington, DC, pp. 676–693.

Vos, C.M., De Cremer, K., Cammue, B.P., De Coninck, B., 2015. The toolbox of *Trichoderma* spp. in the biocontrol of *B. otrytis cinerea* disease. Mol. Plant Pathol. 16 (4), 400–412.

Widmer, T.L., 2014. Screening *Trichoderma* species for biological control activity against *Phytophthora ramorum* in soil. Biol. Control 79, 43–48.

Wokocha, R.C., 1990. Integrated control of *Sclerotium rolfsii* infection of tomato in the Nigerian Savanna: effect of *Trichoderma viride* and some fungicides. Crop Prot. 9 (3), 231–234.

Yasmeen, R., Siddiqui, Z.S., 2017. Physiological responses of crop plants against *Trichoderma harzianum* in saline environment. Acta Bot. Croat. 76 (2), 154162. https://doi.org/10.1515/botcro-2016-0054.

Yasmeen, R., Siddiqui, Z.S., 2018. Ameliorative effects of *Trichoderma harzianum* on monocot crops under hydroponic saline environment. Acta Physiol. Plant. 40 (1), 4.

Yedidia, I., Benhamou, N., Chet, I., 1999. Induction of defense responses in cucumber plants (*Cucumis sativus* L.) by the biocontrol agent *Trichoderma harzianum*. Appl. Environ. Microbiol. 65 (3), 1061–1070.

Yedidia, I., Srivastva, A.K., Kapulnik, Y., Chet, I., 2001. Effect of *Trichoderma harzianum* on microelement concentrations and increased growth of cucumber plants. Plant Soil 235 (2), 235–242.

Yildirim, E., Taylor, A.G., Spittler, T.D., 2006. Ameliorative effects of biological treatments on growth of squash plants under salt stress. Sci. Hortic. 111 (1), 1–6.

Yuan, S., Li, M., Fang, Z., Liu, Y., Shi, W., Pan, B., Shen, Q., 2016. Biological control of tobacco bacterial wilt using *Trichoderma harzianum* amended bioorganic fertilizer and the arbuscular mycorrhizal fungi *Glomus mosseae*. Biol. Control 92, 164–171.

Zaid, A., Wani, S.H., 2019. Reactive oxygen species generation, scavenging and signaling in plant defense responses. In: Bioactive Molecules in Plant Defense. Springer, Cham, pp. 111–132.

Zaidi, N.W., Singh, U.S., 2013. *Trichoderma* in plant health management. *Trichoderma: biology and applications*. In: Mukharjee, P.K., Horwitz, B.A., Singh, U.S., Schmoll, M., Mukharjee, M. (Eds.), CABI Noswority Way. Wallingford, Oxfordshire, UK, p. 315.

Zaidi, N.W., Dar, M.H., Singh, S., Singh, U.S., 2014. *Trichoderma* species as abiotic stress relievers in plants. In: Biotechnology and Biology of Trichoderma. Elsevier, pp. 515–525.

Zamioudis, C., Pieterse, C.M., 2012. Modulation of host immunity by beneficial microbes. Mol. Plant-Microbe Interact. 25 (2), 139–150.

Zhang, F., Yuan, J., Yang, X., Cui, Y., Chen, L., Ran, W., Shen, Q., 2013a. Putative *Trichoderma harzianum* mutant promotes cucumber growth by enhanced production of indole acetic acid and plant colonization. Plant Soil 368 (1–2), 433–444.

Zhang, F., Zhu, Z., Yang, X., Ran, W., Shen, Q., 2013b. *Trichoderma harzianum* T-E5 significantly affects cucumber root exudates and fungal community in the cucumber rhizosphere. Appl. Soil Ecol. 72, 41–48.

Zhang, N., Xin, H.E., Zhang, J., Raza, W., Xing-Ming, Y.A.N.G., Yun-Ze, R.U.A.N., HUANG, Q.W., 2014. Suppression of *Fusarium wilt* of banana with application of bio-organic fertilizers. Pedosphere 24 (5), 613–624.

Zhang, F., Zhang, J., Chen, L., Shi, X., Lui, Z., Li, C., 2015. Heterologous expression of ACC deaminase from *Trichoderma asperellum* improves the growth performance of *Arabidopsis thaliana* under normal and salt stress conditions. Plant Physiol. Biochem. 94, 41–47.

Zhang, S., Gan, Y., Xu, B., 2016. Application of plant-growth-promoting fungi *Trichoderma longibrachiatum* T6 enhances tolerance of wheat to salt stress through improvement of antioxidative defense system and gene expression. Front. Plant Sci. 7, 1405.

Zhang, S., Gan, Y., Xu, B., 2019. Mechanisms of the IAA and ACC-deaminase producing strain of *Trichoderma longibrachiatum* T6 in enhancing wheat seedling tolerance to NaCl stress. BMC Plant Biol. 19 (1), 22.

Further reading

Atanasova, L., Le Crom, S., Gruber, S., Coulpier, F., Seidl-Seiboth, V., Kubicek, C.P., Druzhinina, I.S., 2013. Comparative transcriptomics reveals different strategies of *Trichoderma* mycoparasitism. BMC Genomics 14 (1), 121.

Baker, S.E., Perrone, G., Richardson, N.M., Gallo, A., Kubicek, C.P., 2012. Phylogenomic analysis of polyketide synthase-encoding genes in *Trichoderma*. Microbiology 158 (1), 147–154.

Benítez, T., Rincón, A.M., Limón, M.C., Codon, A.C., 2004. Biocontrol mechanisms of *Trichoderma* strains. Int. Microbiol. 7 (4), 249–260.

Chaverri, P., Gazis, R.O., Samuels, G.J., 2011. *Trichoderma amazonicum*, a new endophytic species on *Hevea brasiliensis* and *H. guianensis* from the Amazon basin. Mycologia 103 (1), 139–151.

Contreras-Cornejo, H.A., Macías-Rodríguez, L., Beltrán-Peña, E., Herrera-Estrella, A., López-Bucio, J., 2011. *Trichoderma*-induced plant immunity likely involves both hormonal-and camalexin-dependent mechanisms in *Arabidopsis thaliana* and confers resistance against necrotrophic fungi *Botrytis cinerea*. Plant Signal. Behav. 6 (10), 1554–1563.

Coudert, Y., Périn, C., Courtois, B., Khong, N.G., Gantet, P., 2010. Genetic control of root development in rice, the model cereal. Trends Plant Sci. 15 (4), 219–226.

Degenkolb, T., Dieckmann, R., Nielsen, K.F., Gräfenhan, T., Theis, C., Zafari, D., Thrane, U., 2008. The *Trichoderma brevicompactum* clade: a separate lineage with new species, new peptaibiotics, and mycotoxins. Mycol. Prog. 7 (3), 177–219.

Hoyos-Carvajal, L., Bissett, J., 2011. Biodiversity of *Trichoderma* in neotropics. In: The Dynamical Processes Of biodiversity—Case Studies of Evolution and Spatial Distribution. InTech, pp. 303–320.

Keswani, C., Singh, S.P., Singh, H.B., 2013. A superstar in biocontrol enterprise: *Trichoderma* spp. Biotech Today 3 (2), 27–30.

Montero-Barrientos, M., Hermosa, R., Cardoza, R.E., Gutiérrez, S., Monte, E., 2011. Functional analysis of the *Trichoderma harzianum* nox1 gene, encoding an NADPH oxidase, relates production of reactive oxygen species to specific biocontrol activity against *Pythium ultimum*. Appl. Environ. Microbiol. 77 (9), 3009–3016.

Osbourn, A., 2010. Secondary metabolic gene clusters: evolutionary toolkits for chemical innovation. Trends Genet. 26 (10), 449–457.

Reino, J.L., Guerrero, R.F., Hernández-Galán, R., Collado, I.G., 2008. Secondary metabolites from species of the biocontrol agent *Trichoderma*. Phytochem. Rev. 7 (1), 89–123.

Ripa, F.A., Cao, W.D., Tong, S., Sun, J.G., 2019. Assessment of plant growth promoting and abiotic stress tolerance properties of wheat endophytic fungi. Biomed. Res. Int. 2019.

Saravanakumar, K., Arasu, V.S., Kathiresan, K., 2013. Effect of *Trichoderma* on soil phosphate solubilization and growth improvement of *Avicennia marina*. Aquat. Bot. 104, 101–105.

Spiegel, Y., Chet, I., 1998. Evaluation of *Trichoderma* spp. as a biocontrol agent against soilborne fungi and plant-parasitic nematodes in Israel. Integr. Pest Manag. Rev. 3 (3), 169–175.

Tchameni, S.N., Ngonkeu, M.E.L., Begoude, B.A.D., Nana, L.W., Fokom, R., Owona, A.D., Kuaté, J., 2011. Effect of *Trichoderma asperellum* and arbuscular mycorrhizal fungi on cacao growth and resistance against black pod disease. Crop Prot. 30 (10), 1321–1327.

Tjamos, E.C., Tjamos, S.E., Antoniou, P.P., 2010. Biological management of plant diseases: highlights on research and application. J. Plant Pathol. S17–S21.

Tucci, M., Ruocco, M., De Masi, L., De Palma, M., Lorito, M., 2011. The beneficial effect of *Trichoderma* spp. on tomato is modulated by the plant genotype. Mol. Plant Pathol. 12 (4), 341–354.

Vinale, F., Sivasithamparam, K., Ghisalberti, E.L., Ruocco, M., Woo, S., Lorito, M., 2012. Trichoderma secondary metabolites that affect plant metabolism. Nat. Prod. Commun. 7 (11), 1545–1550.

Yang, C.A., Cheng, C.H., Lo, C.T., Liu, S.Y., Lee, J.W., Peng, K.C., 2011. A novel L-amino acid oxidase from *Trichoderma harzianum* ETS 323 associated with antagonism of *Rhizoctonia solani*. J. Agric. Food Chem. 59 (9), 4519–4526.

Beneficial plant-associated bacteria modulate host hormonal system enhancing plant resistance toward abiotic stress

6

P. Hariprasad, H.G. Gowtham, and C. Gourav
Centre for Rural Development and Technology, Indian Institute of Technology Delhi, New Delhi, India

6.1 Introduction

Plants are sessile organisms exposed to changing environment throughout their growth period that is often unfavorable or stressful. The plant has to adapt to changing environment which can be broadly categorized as biotic (infection by microbes and nematodes, herbivore attack, etc.), abiotic (drought, salt, heat, cold, nutrient deficiency, toxic metals, etc.), and stress (Haider et al., 2019). To survive, the plants evolved with multiple mechanisms to cope with the changing environment. In nature, several abiotic factors determine the geographical distribution pattern of plants. Among them, drought, salt, and temperature are primary factors which are also found to affect crop productivity adversely, threatening food security (Jogaiah et al., 2013).

On the other hand, plants exposed to various stresses may become more susceptible to the microbial pathogen and insect pest attack (Coakley et al., 1999; Duveiller et al., 2007). Further, some of the toxic metals accumulated in the plants enter the food chain affecting the health of plants, animals, and humans (Praveen et al., 2019). The adverse effects of these abiotic stresses are believed to amplify in future because of the unpredicted and increased frequency of extreme weather (Fedoroff et al., 2010). By 2050, about 50% of plants of arable lands will suffer due to drought stress (Kasim et al., 2013). Saline water and soil are a critical issue in arid as well as semiarid regions (Parvaiz and Satyawati, 2008) affecting approximately 1 billion hectares globally (Yensen, 2008).

6.2 Plant response and adaptation to the abiotic stress condition

Plant response to abiotic stresses is a very complex and dynamic process (Cramer, 2010; Skirycz and Inzé, 2010). This plant response may be localized or systemic, reversible or irreversible, which again depends on the type of abiotic stress, intensity,

Biocontrol Agents and Secondary Metabolites. https://doi.org/10.1016/B978-0-12-822919-4.00006-5

and duration of stress, age of plant, and many other factors (Tattersall et al., 2007; Dinneny et al., 2008; Pinheiro and Chaves, 2011; Vurukonda et al., 2016).

Plant species show some morphological, physiological, and biochemical changes against abiotic stress by scarifying their growth and productivity. Drought stress mainly affects plant-water potential and turgor (Hsiao and Xu, 2000; Haider et al., 2018). Moreover, transport and availability of nutrients are also influenced by drought stress (Barber, 1995; Selvakumar et al., 2012). Salinity inhibits the plant growth primarily by inhibiting cell division and enlargement of the plant's growing region, often resulting in stunted growth and reduced leaf area (Lu et al., 2009). The response is directly related to the total osmotic potential of soil water (Manchanda and Garg, 2008; Shrivastava and Kumar, 2015). Here, plants reduce water loss by reducing the size of leaves such that less water gets evaporated by transpiration, thereby compromising with photosynthesis. Cell dehydration is another initial response of plants under drought and salt stress. The water of the cell is moved to extracellular spaces leading to a decreased area of cytosols and vacuoles (Bartels and Sunkar, 2005).

Root has plasticity in morphology in response to physical conditions of soil (Tuberosa, 2012), which allows the plants to survive, especially under drought and salt stress (Yu et al., 2007). Change in the root architecture is one of the most important adaptations against drought and salt stress (Bacon et al., 2002; Huang et al., 2014) which includes changes in topology of roots, length and diameter of roots, and number and distribution of lateral roots (De Dorlodot et al., 2007; Vacheron et al., 2013: Comas et al., 2013). Reduction in Relative water content (RWC) creates a limitation for cell extension and thus causes a reduction in the growth of the plants in a drastic way (Ashraf, 2010; Castillo et al., 2013). Additionally, osmotic stress induces autophagy in cells, which involves a chain of degradation process at the intercellular level and transfers all the cytoplasmic constituents to a vacuole (Han et al., 2011).

For the metabolic adjustments, plants synthesize various compatible solutes such as proline, sugar, trehalose, betaines, amines, ammonium compounds, polyhydric alcohols, proteins, amino acids (glycine, alanine, leucine, valine, etc.), amides, imino acids, etc. (Yancey et al., 1982; Pervaiz et al., 2008; Sandhya et al., 2010; Yang et al., 2010; Mostafa et al., 2016) under drought and salt stress which play a significant role in osmotic adjustment and also may provide a storage form of nutrition that is reutilized in a later phase of plant recovery and growth. Exposure to different stress leads to the synthesis of a group of oxygen-containing chemically reactive species, collectively known as Reactive Oxygen Species (ROS), which includes peroxides, superoxide, hydroxyl radicals, singlet oxygen, hydrogen peroxide, etc. These compounds, which are highly reactive, cause deleterious effects on all types of biological molecules in the cell, thereby affecting plant growth and survival. In plants, controlled generation of ROS helps in signaling in key cellular functions and plant protection against various stresses (Dar et al., 2017; Abdelrahman et al., 2018). The ROS level in the plant system is tightly regulated by enzymatic and nonenzymatic ways (Mittler et al., 2004). Enzymes, such as catalase, superoxide dismutase, glutathione peroxidase, guaiacol peroxidase, ascorbate peroxidase, monodehydroascorbate reductase, dehydroascorbate reductase, glutathione reductase, etc. (Mittler et al., 2004;

Caverzan et al., 2016), and nonenzymes such as ascorbate, glutathione, tocopherol, carotenoids, and phenolic compounds (Mittler et al., 2004; Scandalios, 2005; Ahmad et al., 2010) are involved in the process of biomolecule protection from adverse effects of ROS in plants. But, under certain extreme stress conditions, ROX production crosses the safety limit, causing damage to biomolecules.

Though the plants are endowed with multiple sets of defense mechanisms at the morphological, physiological, and biochemical levels, beyond the certain intensity of abiotic stress, plants try to survive by compromising their growth and productivity; further, it may lead to death. Improving the capabilities of crop plants to withstand abiotic stress without compromising growth and yield is a necessary and appropriate solution to meet the growing demand for food and other agro-produce.

Most of the abiotic stress tolerance responses are mediated through plant hormones (Davies and Zhang, 1991; Zhang et al., 2006; Abdelrahman et al., 2017). Collectively, plant hormones play a pivotal role in regulating the different aspects of plant growth, development, and reproduction. The hormonal effects in the plant greatly depend on its concentration. Additionally, its location in the plant and other environmental conditions also affects the degree of response by plants. Usually, hormonal content in the plants is as low as $<1\ \mu M$ and anything above this is considered as a supraoptimal level (Naqvi, 2002). Plant growth regulators include five classical phytohormones, namely, auxin (AU), gibberellins (GAs), abscisic acid (ABA), cytokinins (CK), and ethylene (ET). Additionally, jasmonic acid (JA), salicylic acid (SA), brassinosteroids (BRs), strigolactones (SLs), nitric oxide (NO), small peptides are also identified as plant hormones. They also act as cellular signaling molecules which modulate the plant immune response toward insect pests, beneficial and pathogenic microbes, and play a critical role in the process of plant adaption and survival under biotic and abiotic stress.

In this regard, several researchers explored the possibility of generating genetically modified plants with varied levels of hormone production and studied their behavior under abiotic stress. Several genetically modified plants were successfully tested and found resistant to abiotic stress. However, their field application is awaited.

On the other hand, several plant-associated beneficial microbes are frequently reported to assist the plant to withstand various abiotic stresses. The discovery of the phenomenon of plant hormone synthesis or modulation by beneficial plant-associated bacteria provided new opportunities to explore their possible utilization to improve plant health and productivity under abiotic stress conditions. Hence, we reviewed here the capability of plant-associated beneficial bacteria to synthesize or modulate the host hormonal system, thereby imparting resistance against abiotic stress.

6.3 Abscisic acid (ABA)

Abscisic acid is one of the five "classical" plant hormones that regulates various physiological and biochemical aspects of plant growth and development (Kende and Zeevaart, 1997). Like other plant hormones, it has multiple effects at different stages of plant growth and plays important roles in the successful adaptation of plants to drought stress by closing the stomata, thereby reducing the transpiration water loss (Taiz and Zeiger, 2010). ABA is also known as a stress hormone which regulates various physiological and biochemical aspects of plant growth and development under favorable and unfavorable conditions (Munns and Cramer, 1996; Kende and Zeevaart, 1997; Cheng et al., 2002; Sharp, 2002; Finkelstein et al., 2002). Plants exposed to abiotic stress like drought, salt, cold, wounding, etc. are found to accumulate the increased level of endogenous ABA as compared to plants grown under normal conditions (Pēna-Cortés et al., 1989; Shinozaki and Yamaguchi-Shinozaki, 2000). Under drought stress, ABA plays an essential role in controlling multiple processes, such as reduction in transpiration by complete or partial closure of stomata, harvesting soil water by increasing root growth, and the production of osmoprotectants (Yamaguchi-Shinozaki and Shinozaki, 2006; Tuteja, 2007; Gómez-Cadenas et al., 2015). Additionally, ABA is an important modulator of the plant immune signaling network (Asselbergh et al., 2008; Ton et al., 2009; Cao et al., 2011).

Under abiotic stress, enhanced level of endogenous ABA is itself responsible for the suppression of adverse effects or indirectly by suppressing the production and action of ET on plant growth (Sharp, 2002; Sharp and LeNoble, 2002). This phenomenon is also well proved by using ABA deficiency mutants and its restoration by applying ABA exogenously (Sharp et al., 2000; Spollen et al., 2000; Beaudoin et al., 2000; Sharp and LeNoble, 2002; LeNoble et al., 2004). On the other hand, enhanced accumulation of ABA through overexpressing AtBG1 was found to resist drought conditions over its wild-type form (Lee et al., 2006). Thus, many research evidenced that, improved abiotic stress tolerance in the plant is associated with increased accumulation of ABA. Hence, ABA can be used as target molecules to fight against abiotic stress.

ABA is also known to interact with other plant hormones synergistically or antagonistically. ABA-deficient mutants exhibit an increased rate of ET production and independently affect the water balance in the plant (Morgan et al., 1990; Hussain et al., 1999; Spollen et al., 2000). Thus, it was concluded that accumulation of endogenous ABA had been found to limit ET production and to protect plants from abiotic stress damage (Sharp, 2002; Sharp and LeNoble, 2002). Also, ABA signaling was reported to suppress SA-mediated defense in plants, thereby increasing the susceptibility of the host plant, whereas along with JA it synergistically protects the host plant from herbivory or wounding (Abe et al., 2003; Anderson et al., 2004; Yasuda et al., 2008; De Torres-Zabala et al., 2009; Cao et al., 2011; Jiang et al., 2010). Despite this, ABA-deficient mutants often recorded decreased biomass production by exhibiting smaller leaves compared with the corresponding wild-type plants, indicating the role of endogenous ABA in keeping normal plant growth and development (Quarrie, 1987; Munns and Cramer, 1996).

Reports have also demonstrated that the application of exogenous ABA as a modulator decreases excess water loss and enhances growth and tolerance to high salinity (Amzallag et al., 1990) and drought (Wang et al., 2003) in the plant. Indeed, Khadri et al. (2006) showed that ABA application reduced the reverse effect of salinity and improved the response of plant symbiosis under salt stress conditions, including nitrogen-fixation and enzymes of ammonium assimilation and purine catabolism. There is also evidence that exogenous ABA induces accumulation of a sufficient quantity of reactive oxygen species (ROS) in plant cells that act as secondary messengers for the activation of salt-stress defensive responses (Sakamoto et al., 2008).

Endosymbiotic theory evidenced that the chloroplast is an ancient prokaryote symbiotically associated with the eukaryotic organisms. ABA synthesis starts in chloroplast and ends in releasing xanthoxin, a precursor of ABA into the cytoplasm, where it converted into ABA. In spite of these evidences, the bacterial biosynthesis of ABA is poorly understood.

Several reports are available on the bacterial species belonging to different families that are known to produce potential carotenoids. Askar et al. (2007a,b,c, 2012a,b) identified several carotenoids including astaxanthin, zeaxanthin, canthaxanthin, neoxanthin, noxtoxanthin, dihydroxy astaxanthin, etc. from several species of bacteria including the *Sphingomonas* strain TDMA-17T, *Mesoflavibacter zeaxanthinifaciens*, *Zeaxanthinibacter enoshimensis*, *Nubsella zeaxanthinifaciens*. In 2012, Askar et al. (2012a) classified carotenoid-producing bacterial isolates into 38 different species belonging to seven bacterial classes such as *Flavobacteria*, *Sphingobacteria*, *α-Proteobacteria*, *γ-Proteobacteria*, *Deinococci*, *Actinobacteria*, and *Bacilli*. Thawornwiriyanun et al. (2012) isolated zeaxanthin producing sponge-associated bacteria from sponges in the Gulf of Thailand and identified as *Sphingomonas phyllosphere* and *Sphingomonas* (*Blastomonas*) *natatoria*. Prabhu et al. (2013) reported that two novel bacterial strains (*Muricauda aquimarina* YUAB-SO-11 and *M. olearia* YUAB-SO-45), which were isolated from sandy beaches of the South West Coast of India, were found to produce zeaxanthin.

However, to date, there are no reports in the literature regarding the possible pathways for biosynthesis of ABA in bacteria. Cohen et al. (2008) feasibly hypothesized that bacteria could synthesize ABA as part of their normal metabolism. Marasco and Schmidt Dannert (2008) identified and reported only the presence of the product of carotenoid metabolism, carotenoid cleavage oxygenase (CCO) homologs in *Novosphingobium aromaticivorans* and *Bradyrhizobium* sp., which provided a possible mechanism for ABA production in bacteria.

Modulation of host plant response to abiotic stress by bacterially produced ABA is dubious and sporadic reports are available evidencing that this phytohormone promotes or regulates plant growth. Although the bacteria are known to produce ABA, there are very few reports available in the literature in connection with the possible role of ABA producing bacteria and mechanisms involved to suppress abiotic stress in plants, except the determination of the plant's ABA content after bacterial inoculation. Cohen et al. (2001) demonstrated that the effect of *A. lipoferum* inoculation partially reversed the effect of the inhibitor (such as Fluridone) in blocking the ABA synthesis of seedlings of maize and observed that the level of ABA of seedlings increased and enhanced the growth with respect to Fluridone treatment and maintained the better water status. Cohen et al. (2008) observed a twofold increase in ABA content in *Arabidopsis Thaliana* seedlings inoculated with ABA producing *Azospirillum brasilense* strain Sp 245. Similarly, Salomon et al. (2014) reported *V. vinifera* cv. Malbec treated with *B. licheniformis,* and *P. fluorescens* recorded an increased level of ABA accumulation. Further, this observation is supported by the diminished water loss rate in bacteria-treated plants.

Further, mutants of *Arabidopsis thaliana* (Col-0 and *aba*2-1) recorded increased biomass, increased lateral roots, and decreased water loss when treated with ABA producing the *Azospirillum brasilense* sp. 245 strain. Hence, there is an upsurge in the interest to understand modulation of host ABA content by plant-associated bacteria and their impact on plant growth and reproduction under abiotic stress.

6.4 Ethylene (ET)

Ethylene is a simple two-carbon gaseous hormone regulating several processes such as the ripening of fruits, abscission of leaves, flower senescence, etc. in plants (Johnson and Ecker, 1998; Glick, 2014). In plants, SAM synthase catalyzes the first step of ET synthesis in which methionine is converted to S-AdoMet. Further 1-Aminocyclopropane-1-Carboxylic acid (ACC) synthase catalyzes the conversion of S-AdoMet to ACC, which is an immediate precursor of ET. The final step involves the oxidation of ACC with the help of ACC oxidase leading to the formation of ET (Wang et al., 2002). To date, ET synthesis in plant-associated microbes is not reported.

Since a long time, ET has been considered as a stress hormone. In any plant, the synthesis of ET is dependent on several different factors such as light, temperature, climate, humidity, nutrition, gravity, plant hormones, and various other biotic stresses to which the plant is exposed (Abeles, 1992). During stress conditions, either ET concentration increases or the plant become more sensitive to ET. The increased ET concentration in plants is dangerous for the growth of the plants and responsible for several other losses such as leaf chlorosis, leaf abscission, senescence, etc. Hence, it is hypothesized that by reducing ET concentration in the plant under stress conditions, it is possible to improve the overall plant health and yield.

Evidence to use microbes to reduce ET level in plants came from the work of Honma and Shimomura (1978) where they isolated several bacteria which can degrade ACC, an immediate precursor of ET, into α-ketobutyrate and ammonia. The addition of ACC induces a bacterial cell to the expression of the ACC deaminase structural gene (acdS) gene (Jacobson et al., 1994). It was found from the sequence data of *Pseudomonas putida* UW4 that it is a very complex mechanism and is regulated by the binding of several proteins (Li et al., 2001). The upstream region of the acdS gene mainly consists of three binding sites, i.e., FNR (fumarate-nitrate reduction regulatory protein), Lrp binding site (leucine-responsive protein), and CRP (cyclic AMP receptor protein) (Li and Glick, 2001; Singh et al., 2015), which regulates the transcription of the Acds gene.

Glick et al. (1998) described the mechanism of the ACC deaminase enzyme (Fig. 6.1). According to them, the rhizobacteria bind to the roots due to the presence of exudates (Whipps, 1990) and inside the plant cells, the ACC synthase enzyme is activated by the bacterial IAA produced in the vicinity of roots, which converts adenosylmethionine to ACC (Patten and Glick, 1996). This ACC is further released

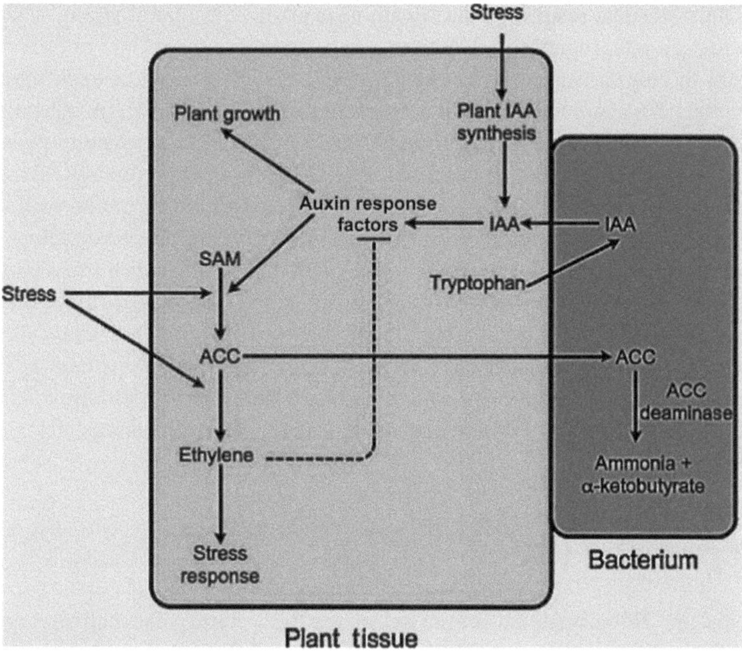

Fig. 6.1 A brief model of PGPR producing ACC deaminase (ACCd) and mechanism of lowering ethylene level in plant tissue.
Adopted from Glick, B. R. 2014. Bacteria with ACC deaminase can promote plant growth and help to feed the world. *Microbiol. Res.* 169(1), 30–39.

from plants in the form of root exudation to soil and immediately taken up by the ACC deaminase producing rhizobacteria. Finally, these bacteria containing ACC deaminase enzymes convert ACC into α-ketobutyrate and ammonia. When the ACC concentration outside of the root is decreased due to their cleavage by ACC deaminase, more ACC is released from plant roots (Glick, 2014).

Several research experiments revealed the capabilities of ACC deaminase producing rhizobacteria in reducing the adverse effect of abiotic stress on host plants. Arshad et al. (2008) found the elimination of drought stress in pea plants when subjected to *Pseudomonas* spp. (ACCd producer) inoculation. Mayak et al. (2004a) reported that *Achromobacter piechaudii* ARV8 with ACC deaminase activity was significantly found to increase the fresh and dry weights of tomato and pepper seedlings which were exposed to transient water stress. The bacterium was also found to reduce the ET production in tomato seedlings, following water stress, and continued plant growth during both water stress and after watering was resumed. The plants of pepper also recorded drought tolerance after the inoculation with *Bacillus licheniformis* K11 as compared to the uninoculated one (Lim and Kim, 2013).

Ma et al. (2004) converted an ACC deaminase negative *Sinorhizobium meliloti* to produce ACC deaminase by inserting and expressing the ACC deaminase structural gene (ACDS) and the upstream regulatory gene of *Rhizobium leguminosarum* pv.

viciae 128C53K. The resulting ACC deaminase positive bacterial strain, when inoculated into Alfa alfa, recorded a 40% increase in nodule numbers and improved shoot dry weight in comparison to its wild-type strain. A similar kind of experiment with *Pseudomonas fluorescens* was found to increase the growth of the Canola plant (Wang et al., 2000). Recently, Tiwari et al. (2018) observed improved biomass characters of *Panicum maximum* treated with ACC deaminase producing rhizospheric beneficial bacteria under drought and salt stress. Further analysis of cell wall composition of these plants revealed that the bacterial treatment improved cellulose, hemicellulose, and lignin content very close to the plant's growth under normal conditions without induced stress. Similarly, rhizobacterial strains from wheat rhizosphere with ACC-deaminase activity exhibits a significant increase in grain yield, straw yield, root length, root weight, root elongation, number of tillers, nitrogen, phosphate, and potassium uptake. Here, the researchers concluded that, the growth-promoting activity of selected rhizobacteria is by decreasing the level of endogenous ET (Ahmed et al., 2004).

6.5 Cytokinins (CK)

Cytokinins are N6-substituted amino purines which control the cell division, cell cycle, cellular proliferation, and differentiation and influence the physiological and developmental process in plants (Salisbury and Ross, 1992; Srivastava, 2002). CK plays a pivotal role in regulating cell cycle (Miller et al., 1955, 1956). AU and CK are together involved in regulating cell division and morphogenesis. CK is a negative regulator of root growth and lateral root formation (Ramireddy et al., 2018; Liao et al., 2017). A decreased level of CK or poor CK signaling in plants leads to the reduced stomatal aperture, transpiration rate, and higher water content under water deficient conditions (Nishiyama et al., 2011; Nguyen et al., 2015; Liao et al., 2017).

Enhanced levels of CK in plants are reported to have a higher efficiency of photosynthesis under drought and salt stress conditions (Ghanem et al., 2011; Reguera et al., 2013; Ma et al., 2016; Oneto et al., 2016). Similarly, reduced level of CK was also found improving the plant tolerance to drought stress. Barely plants genetically modified to express CK degrading enzyme (CKX) recorded increased assimilation of CO_2, enhanced root growth, nutrient and water uptake (Ramireddy et al., 2018), and activate putative flavonoid biosynthesis gene thereby enhancing drought tolerance (Nakabayashi et al., 2014; Vojta et al., 2016). Similarly, in plants where the ipt gene (CK biosynthesis) is ectopically expressed, increased antioxidants were observed when exposed to drought stress (Rivero et al., 2007), protecting plants from the adverse effect of ROS generated.

Plant-associated bacteria are reported to biosynthesize CK (Fig. 6.2) or its analog, which may act as a tool for invading the host plant, and influence its growth and abiotic

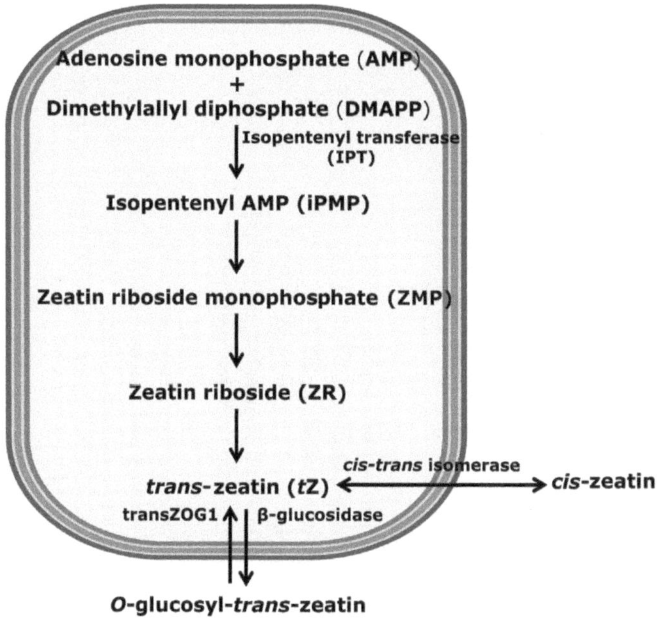

Fig. 6.2 Proposed biosynthetic and metabolic pathway for cytokinins in bacteria (e.g., *Agrobacterium tumefaciens*), including several possible modifications and the degradation of zeatin (Haberer and Kieber, 2002).

stress response (Arshad and Frankenberger, 1998). In vitro experiments evidenced the CK production ability of plant-associated bacteria (Nieto and Frankenberger, 1989; Karadeniz et al., 2006; Arkhipova et al., 2005; Ali et al., 2009). Bacteria belonging to the wide range of genus are reported to produce CK (Table 1).

rolA, *rolB*, and *rolC* genes identified in the T-DNA of *A. rhizogenes* are believed coded for glucosidase, which converts inactive stored AU and CK to an active form (Estruch et al., 1991a,b; Brzobohaty et al., 1993), thereby influencing the plant growth or initiating gall formation. *Rhodococcus fascians* strain D188, plant pathogenic actinomycetes which cause leaf galls, was well studied in detail for CK production and its impact on the host plant (Pertry, 2009). The bacterium contains a linear plasmid endowed with *fasciation* (*fas*) locus, an operon encoding genes responsible for CK bases

Table 1 List of plant-associated bacteria reported producing plant hormones.

Plant hormone	Bacteria reported synthesizing plant hormone	Reference
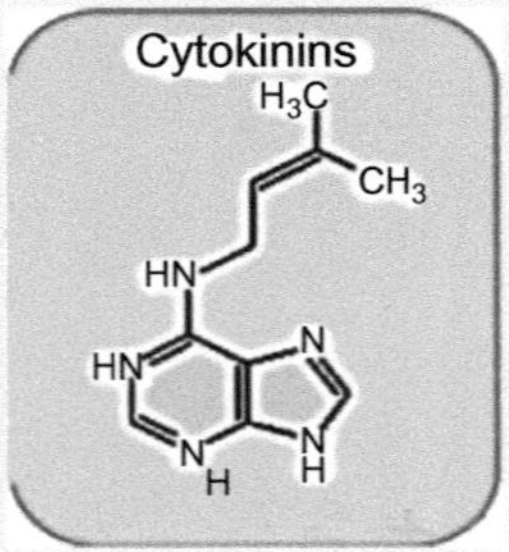	*Agrobacterium tumefaciens*	Goethals et al. (2001)
	Azospirillum sp.	Großkinsky et al. (2016)
	Azospirillum brasilense	Horemans et al. (1986), Cacciari et al. (1989), and Molina-Favero et al. (2007)
	Azotobacter chroococcum	Horemans et al. (1986)
	Arthrobacter giacomelloi	Cacciari et al. (1989) and Nieto and Frankenberger (1989)
	Azotobacter vinelandii	Horemans et al. (1986)
	Bacillus cereus	Nieto and Frankenberger (1989)
	Bacillus megaterium	Ortíz-Castro et al. (2008)
	Bacillus subtilis	Arkhipova et al. (2005)
	Escherichia coli	Nieto and Frankenberger (1989)
	Halomonas desiderata	Nieto and Frankenberger (1989)
	Klebsiella pneumonia	Nieto and Frankenberger (1989)
	Pantoea agglomerans	Goethals et al. (2001)
	Proteus mirabilis	Nieto and Frankenberger (1989)
	Paenibacillus polymyxa	Timmusk and Wagner (1999)
	Proteus vulgaris	Nieto and Frankenberger (1989)
	Pseudomonas fluorescens	García de Salamone et al. (2001) and Arkhipova et al. (2005)
	Pseudomonas savastanoi	Goethals et al. (2001)
	Rhizobium rhizogenes	Estruch et al. (1991a,b)
	Rhodococcus fascians D188	Pertry (2009)

<table>
<tr><td rowspan="20">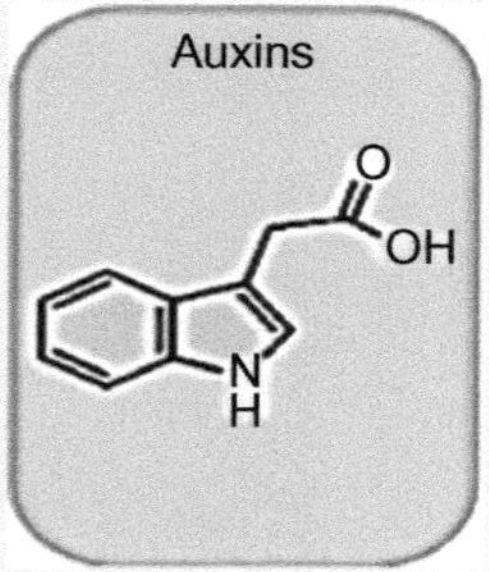

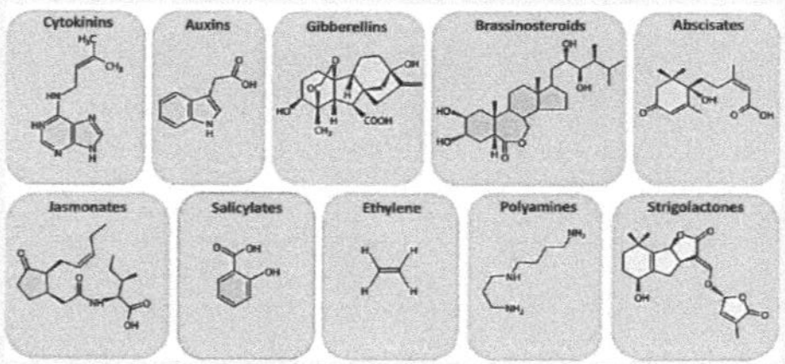
</td><td>Agrobacterium tumefaciens</td><td>Kaper and Veldstra (1958)</td></tr>
<tr><td>Azospirillum brasilense</td><td>Ali et al. (2009)</td></tr>
<tr><td>Azotobacter sp.</td><td>Hariprasad et al. (2009)</td></tr>
<tr><td>Bacillus sp.</td><td>Tien et al. (1979)</td></tr>
<tr><td>B. amyloliquefaciens</td><td>Hariprasad et al. (2009)</td></tr>
<tr><td>Escherichia sp.</td><td>Tien et al. (1979)</td></tr>
<tr><td>Micrococcus sp.</td><td>Tien et al. (1979)</td></tr>
<tr><td>Pseudomonas sp.</td><td>Tien et al. (1979)</td></tr>
<tr><td>Serratia sp.</td><td>Hariprasad et al. (2009)</td></tr>
<tr><td>Staphylococcus sp.</td><td>Tien et al. (1979)</td></tr>
<tr><td>Acetobacter diazotroph</td><td>Bastián et al. (1998)</td></tr>
<tr><td>Azospirillum lipoferum</td><td>Piccoli et al. (1999)</td></tr>
<tr><td>Azospirillum brasilense</td><td>Janzen et al. (1992)</td></tr>
<tr><td>Bacillus licheniformis</td><td>Gutiérrez Mañero et al. (2001)</td></tr>
<tr><td>Bacillus sp.</td><td>Gutiérrez Mañero et al. (2001)</td></tr>
<tr><td>Bacillus pumilus</td><td>Gutiérrez Mañero et al. (2001)</td></tr>
<tr><td>Bradyrhizobium japonicum</td><td>Nett et al. (2017)</td></tr>
<tr><td>Sinorhizobium fredii</td><td>Nett et al. (2017)</td></tr>
<tr><td>Herbaspirillum seropedicae</td><td>Bastián et al. (1998)</td></tr>
<tr><td>Rhizobium phaseoli</td><td>Atzorn et al. (1988)</td></tr>
</table>

Continued

Table 1 Continued

Plant hormone	Bacteria reported synthesizing plant hormone	Reference
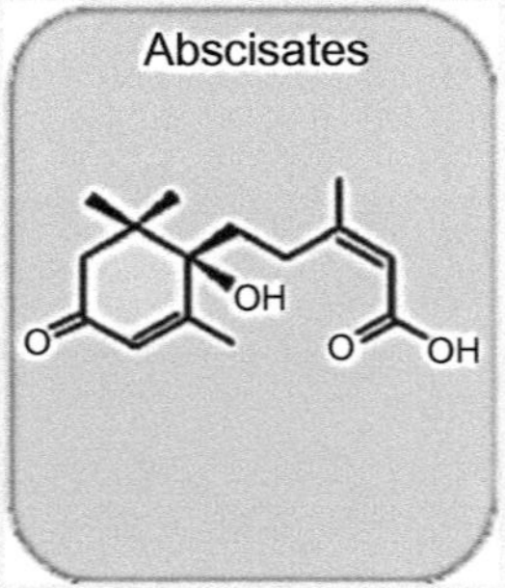	*Actinobacteria*	Askar et al. (2012a)
	Azospirillum lipoferum	Cohen et al. (2001)
	Azospirillum brasilense strain Sp 245.	Cohen et al. (2008)
	Bacillus sp.	Forchetti et al. (2007) and Askar et al. (2012a)
	Bacillus pumilus	Forchetti et al. (2007)
	Bacillus licheniformis	Salomon et al. (2014)
	Bradyrhizobium sp.	Marasco and Schmidt Dannert (2008)
	Deinococci sp.	Askar et al. (2012a)
	Flavobacteria	Askar et al. (2012a)
	Mesoflavibacter zeaxanthinifaciens	Askar et al. (2007a,b)
	Muricauda aquimarina YUAB-SO-11	Prabhu et al. (2013)
	Muricauda olearia YUAB-SO-45	Prabhu et al. (2013)
	Nubsella zeaxanthinifaciens	Askar et al. (2007a,b)
	Novosphingobium aromaticivorans	Marasco and Schmidt Dannert (2008)
	Pseudomonas fluorescens	Salomon et al. (2014)
	α-Proteobacteria	Askar et al. (2012a)
	γ-Proteobacteria	Askar et al. (2012a)
	Sphingobacteria	Askar et al. (2012a)
	Sphingomonas strain TDMA-17T	Askar et al. (2007a,b)
	Sphingomonas phyllosphere	Thawornwiriyanun et al. (2012)
	Sphingomonas (Blastomonas) natatoria	Thawornwiriyanun et al. (2012)
	Zeaxanthinibacter enoshimensis	Askar et al. (2007a,b)

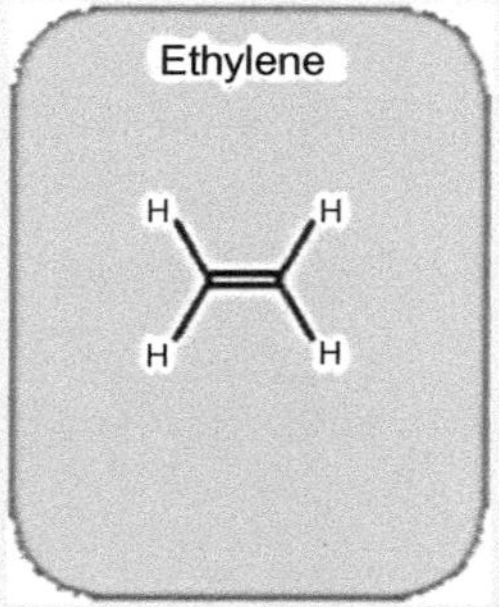 (ACC deaminase an ethylene precursor degrading enzyme)[a]	*Achromobacter piechaudii* ARV8 *Burkholderia phytofirmans* *Bacillus licheniformis* K11 *Pseudomonas fluorescens* *Pseudomonas migulae* *Pseudomonas thivervalensis* *Pseudomonas putida* UW4 *Pseudomonas* sp. *Pseudomonas fluorescens* *Pantoea agglomerans* *Ralstonia* sp. *Rhizobium leguminosarum bv. viciae 128C53K*	Mayak et al. (2004a) Sun et al. (2009) Lim and Kim (2013) Ali et al. (2014) Ali et al. (2014) Zhang et al. (2011) Li et al. (2001) Arshad et al. (2008) Wang et al. (2000) Zhang et al. (2011) Zhang et al. (2011) Ma et al. (2004)
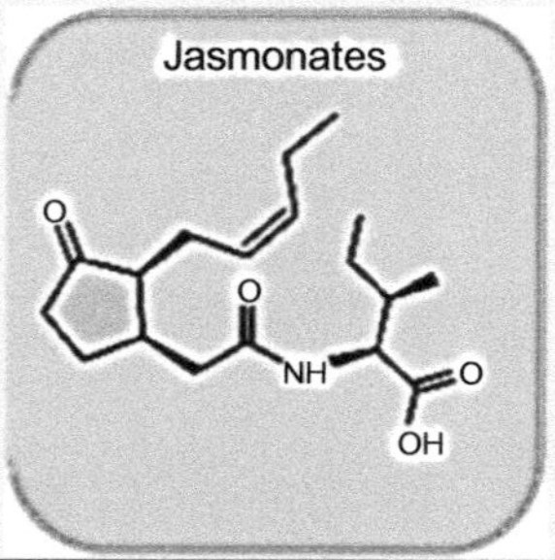	*Achromobacter* sp. *Alcaligenes* sp. *Arthrobacter* sp. *Bacillus* sp. *Bacillus pumilus* *Bacillus* sp. *E. coli*	Forchetti et al. (2007) Forchetti et al. (2007) Piccoli et al. (2011) Forchetti et al. (2007) Forchetti et al. (2007) Forchetti et al. (2007) Abdala et al. (1999)

Continued

Table 1 Continued

Plant hormone	Bacteria reported synthesizing plant hormone	Reference
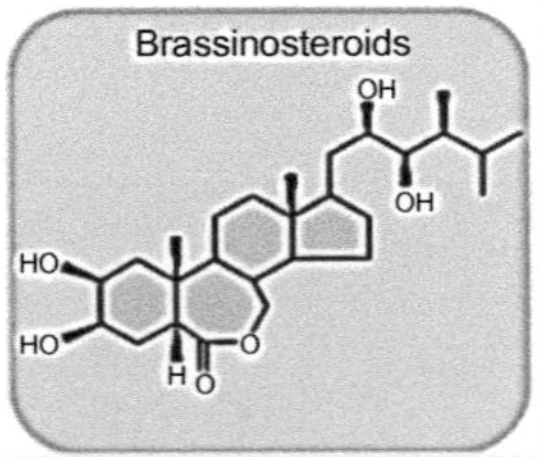	Not reported in bacteria	Not reported in bacteria
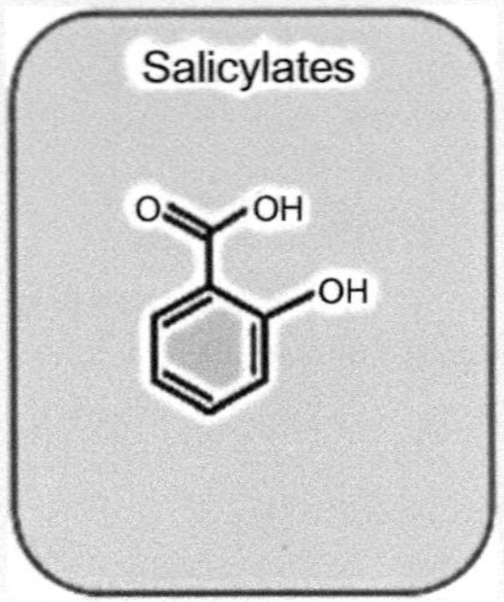	*Bacillus licheniformis* MML2501 *Pseudomonas aeruginosa* *Pseudomonas fluorescens*	Prashanth and Mathivanan (2009). Serino et al. (1995) and De Meyer and Höfte (1997) Leeman et al. (1995)

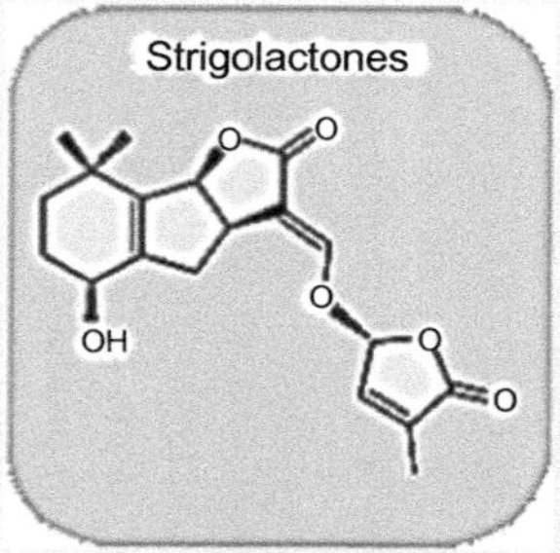	Not reported in bacteria	Not reported in bacteria
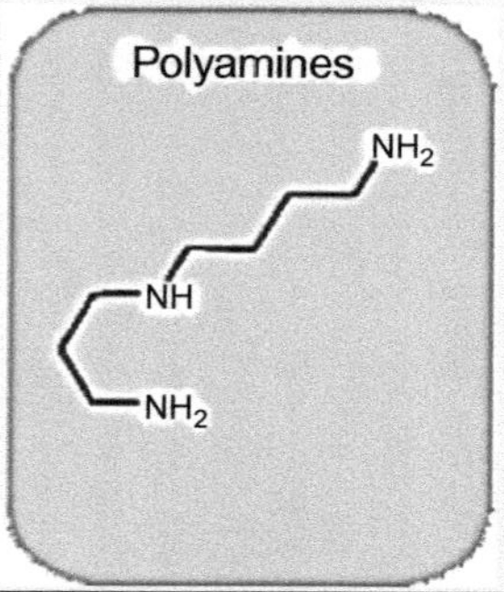	*Bacillus subtilis* OKB10	Xie et al. (2014)

[a] Ethylene is not reported to produce by bacteria but their concentration in host plant can be modified by degrading its precursor molecules with the help of ACC deaminase (discussed in detailed somewhere else in this chapter).

(2-isopentenyladenine (2-iP), cis-zeatin (cZ), trans-zeatin (tZ), and their methylthio derivatives (2MeScZ and 2MeStZ) biosynthesis and metabolism. A plasmid-free strain of *R. fascians* strain D188 was also found to produce lower concentration CK bases, indicating the presence of fas operon in the bacterial chromosome. CK has been detected and compared with other gall-causing pathogens such as *Pantoea agglomerans*, *Agrobacterium tumefaciens*, and *Pseudomonas savastanoi* (Goethals et al., 2001).

Cytokinin producing strain, *Azotobacter chroococcum*, was applied to *Zea mays* and found to increase the plant growth compared to non-inoculated control (Nieto and Frankenberger, 1989). Mixed cultures of *Azospirillum brasilense* and *Arthrobacter giacomelloi* produced a higher level of CK than single cultures (Cacciari et al., 1989). García de Salamone et al. (2001) have shown higher production of the CK, isopentenyl adenosine (IPA), zeatin riboside (ZR), and dihydroxyzeatin riboside (DHZR) by the wild-type strain *Pseudomonas fluorescens* G20-18 compared to two mutants, CNT1 and CNT2. Further, it was noted that addition of the precursor adenine increased CK production in cultures of G20-18. Previous studies by García de Salamone et al. (2001) have also shown that the ability of mutant strains CNT1 and CNT2 to promote the growth of radish plants were impaired compared to G20-18.

Timmusk and Wagner (1999) reported the production of CK by a free-living soil bacterium *Paenibacillus polymyxa* using immunoaffinity chromatography. The CK were further characterized using HPLC with online ultraviolet detection and GC-MS, which resulted in the identification of iP in the culture medium. Recently, Hussain and Hasnain (2009) analyzed the CK fractions in the bacterial extract which were separated by TLC and quantified by HPLC. The major CK detected were zeatin and zeatin riboside. *Azospirillum brasilense* has been reported to enhance cell division in the root tips of inoculated wheat (Molina-Favero et al., 2007). Bacterial extract enhanced cell division, fresh weight, and cotyledon size in the dark as well as light grown cucumber cotyledons against control. However, the impact of CK producing bacteria on plant cell division was investigated mainly in root nodules formation (Markmann and Parniske, 2009).

Ortíz-Castro et al. (2008) recently investigated the role of CK signaling in mediating the plant responses to bacterial inoculation using *Arabidopsis thaliana* mutants lacking one, two, or three of the putative CK receptors *CRE1*, *AHK2* and *AHK3*, and *RPN12*, a gene involved in CK signaling. They observed that plant growth promotion by *Bacillus megaterium* is reduced in *AHK2-2* single and double mutant combinations and in *RPN12*. Furthermore, the triple CK-receptor *CRE1-12/AHK2-2/AHK3-3* knockout was insensitive to inoculation regarding growth promotion and root developmental responses, which indicated that CK receptors play a complementary role in plant growth promotion by CK.

Several plant-associated bacteria belonging to different species can produce a lower level of CK, which was reported to modulate host immunity and induce resistance against microbial pathogens (Akhtar et al., 2020). Garcia de Salamone et al. (2001) provided the evidence for the PGPR strains found to produce the CK which is responsible for the plant growth promotion.

6.6 Gibberellins (GAs)

Gibberellins are tetracyclic diterpenoid carboxylic acids which play a crucial role in normal plant growth and reproduction, including seed germination, seedling emergence, vegetative growth, floral induction, and flower and fruit growth (Bottini et al., 2004), and are also involved in stress response (Achard et al., 2006; Sun, 2008). GAs were first isolated and characterized from a foolish rice disease-causing fungal pathogen, *Gibberella fujikuroi* (Yabuta and Sumiki, 1938), followed by several plants, fungi, and bacteria (Tudzynski, 2005; MacMillan et al., 2005; Bottini et al., 2004; Morrone et al., 2009). Involvement of GAs in normal plant growth and their interaction with other plant hormones were well studied. GAs primarily work through degrading DELLA protein which is responsible for growth suppression (Sun, 2011). Further, several researchers evidenced the involvement of GAs in abiotic stress response and adaptation in plants (Wang et al., 2018).

When proliferating plant tissues are exposed to osmotic stress, they exit mitosis (Skirycz et al., 2011), thereby limiting the number of cells leading to the growth inhibition under abiotic stress. This observation was correlated with the altered GAs metabolism in these cells (Claeys et al., 2012). This change in GAs metabolism results in the stabilization of "DELLA," a protein responsible for cell cycle inhibition, mitotic exit, and endoreduplication. Experiments with mutants (*del-1*: early onset of endoreduplication and DEL-1 overexpressing) revealed that the phenomenon of early onset of endoreduplication contributes to the induction of stress tolerance in plants (Cookson et al., 2006).

Though several bacteria are reported to produce GAs, their involvement in bacterial growth and development is poorly understood. Also, the possible participation of bacterially synthesized GAs in modulating host response to biotic and abiotic stress cannot be ruled out. Through several bioassays, production of GAs like compounds was evidenced in *Azospirillum brasialense* (Tien et al., 1979) and Rhizobium (Williams and De Mallorca, 1982). Later, Atzorn et al. (1988) identified the GA1, GA4, GA9, and GA20-like compounds in culture filtrates of mutants and wild-type strains of *Rhizobium phaseoli*. In the following year, Bottini et al. (1989) reported the capability of *Azospirillum lipoferum* to produce GA1, GA3, and iso-GA3. To date, several plant-associated bacteria have been reported for GAs production. But elucidating the biosynthesis pathway is still under progress. Genetic analysis of *Bradyrhizobium japonicum* USDA110 revealed the presence of a cluster of three complete P450 genes (CYP112, CYP114, and CYP117) plus a partial P450 gene fragment (CYP115P). Also, five open reading frames similar to a [3Fe-4S] ferredoxin, short-chain alcohol dehydrogenase family, prenyl transferases, *ent*-kaurene synthase, squalene-hopene cyclases, and the gene for synthesis of the cyclic sesquiterpene pentalenene (Tully et al., 1998). Morrone et al. (2009) reported the presence of separate ent-copalyl diphosphate and ent-kaurene synthase genes in *Bradyrhizobium japonicum* by expressing them in *E. coli*. They observed a potential homology in the gene sequence of diterpene synthase among bacteria, fungi, and plants, and concluded that the bacterial GAs biosynthesis is relative to fungi and plants. These observations were supported by the work of Méndez et al. (2014) where they demonstrated GAs oxidase activity and GA9 biosynthesis

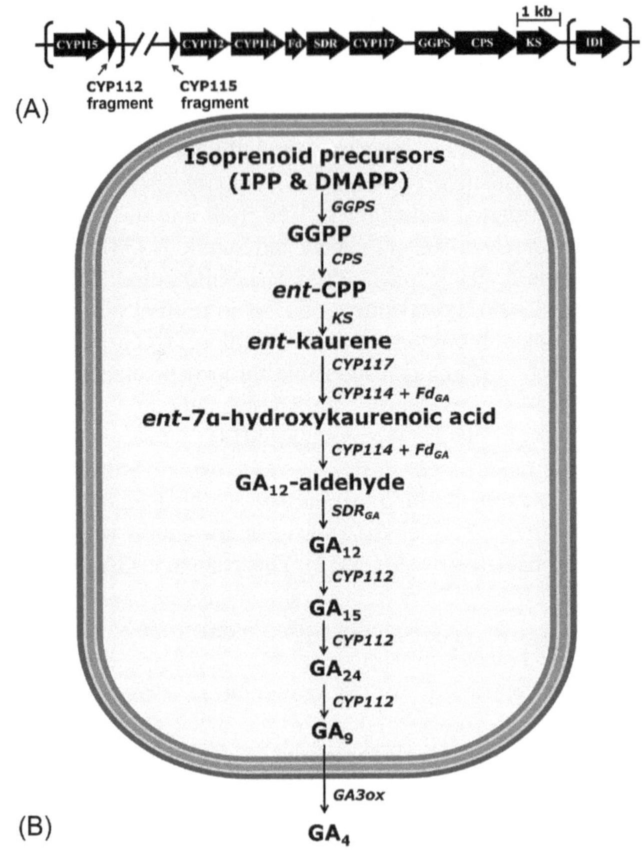

Fig. 6.3 GA biosynthesis in rhizobia. (A) GA biosynthetic operon found in *B. japonicum* and *S. fredii* (1 kb = 1 kilobase). (B) Elucidated GA biosynthetic pathway in core operon-containing rhizobia (Nett et al., 2017).

in *B. japonicum*. Through functional characterization of the putative GA biosynthetic operon/gene cluster, Nett et al. (2017) proved that the operon containing gene coded for enzymes necessary to produce GA9 present in *B. japonicum* and *S. fredii* and further elucidated GA biosynthesis in rhizobia (Fig. 6.3).

Though the GAs synthesis ability of plant-associated bacteria is well proved, their involvement in modulating host response to abiotic stress is yet to be revealed. Gutiérrez Mañero et al. (2001) reported that in a bioassay, dwarf phenotype induced in alder (*Alnus glutinosa*) by artificially treating with pactobutrazol, an inhibitor of GAs biosynthesis, was reversed when the dwarf seedlings were treated with culture filtrate of *Bacillus pumilus* and *Bacillus licheniformis*, which were originally isolated from the alder rhizosphere. Further analysis of the purified fractions of culture filtrate through GC-MS revealed the presence of GAs, with higher concentrations of GA_1

followed by GA_3. A similar observation was made in *Phaseolus lunatus* plants treated with *Bradyrhizobium* sp. which recorded a significant elongation in the internode and was correlated with increased levels of GA_1, GA_{19}, GA_{20}, and GA_{44} in nodules (Dobert et al., 1992). Cassán et al. (2001, 2003) recorded restoration of normal growth in rice dwarf mutants when treated with *Azospirillum lipoferum* and *Azospirillum brasilense* fed with deuteron GA20-glycosides in wheat. An improved drought stress response in wheat plants treated with *Azospirillum brasilene* Sp245, which is known to produce Gas, was observed by Creus et al. (2004).

6.7 Auxin (AU)

Indole acetic acid (IAA) is the main AU in plants which is virtually involved in different stages of plant development, controlling many important physiological processes including cell enlargement and division, tissue differentiation and response to light and gravity, etc. (Taiz and Zeiger, 1998; Woodward and Bartel, 2005; Teale et al., 2006; Benjamins and Scheres, 2008). It also acts synergistically or cross talks with other hormones while responding to biotic and abiotic stress (Chen et al., 2007; Wang et al., 2007; Grunewald et al., 2009; Robert-Seilaniantz et al., 2011). Additionally, AU is also known to produce by another group of organisms such as fungi and bacteria. Many plant-associated beneficial bacteria are reported to produce AU, which acts as a signaling molecule that has a significant effect on the communication between plant and bacteria and promotes plant growth (Kazan and Manners, 2009; Robert-Seilaniantz et al., 2011). On the other hand, some plant pathogenic bacteria manipulate AU signaling in plants, affecting its normal growth and development. But still, how the bacterial AU biosynthesis is regulated under the influence of plant and other abiotic stress and its impact on host plant need to be understood.

Fig. 6.4 explains the different IAA biosynthesis pathway in bacteria. Tryptophan is the main precursor in the IAA biosynthesis in bacteria (Patten and Glick, 1996). From studies on the intermediate involved in the tryptophan-dependent IAA biosynthesis, five different pathways have been identified. Although the independent tryptophan pathway is also thought to occur in bacteria (Prinsen et al., 1993), not much work has been done in this line. The same bacteria may follow the different pathway of IAA synthesis under different conditions. *Erwinia herbicola* pv. *gypsophilae* follows IAM in apoplast and IPyA on the leaf surface (Manulis et al., 1998), whereas *A. brasilense* follows IAM and IPyA (tryptophan-dependent pathways) and one tryptophan-independent pathway to synthesize IAA (Prinsen et al., 1993). Most of the beneficial plant-associated bacteria are reported to synthesize IAA via the IPyA pathway. These bacteria also contain an AU responsive element (AuxRE) in the ipdC promoter which is inducible by IAA (Lambrecht et al., 1999). Thus, IAA present in root exudates is acting as an inducer to turn on the IAA biosynthesis in bacteria at the rhizosphere (Lambrecht et al., 2000).

In bacteria, IAA is reported to be involved in signaling processes such as catabolism of amino acids under nutrient-deprived conditions (Wang et al., 2001), biofilm formation (Domka et al., 2007), and stable maintenance of multi-copy plasmids

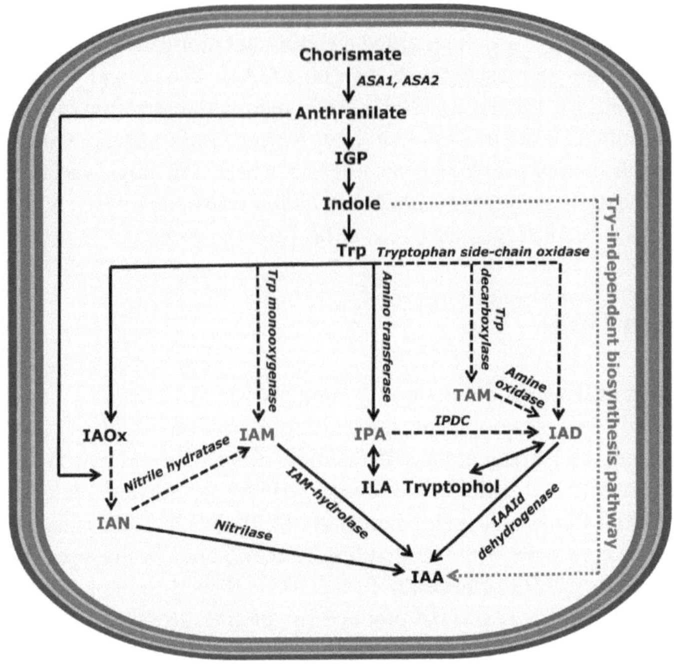

Fig. 6.4 Bacterial tryptophan-dependent and -independent IAA biosynthetic pathway. *Dashed lines* refer to the name of pathways: (a) indole-3-acetonitrile (IAN) pathway, (b) indole-3-acetamide (IAM) pathway, (c) indole-3-pyruvate (IPyA) pathway, (d) tryptamine (TAM) pathway, (e) tryptophan side chain oxidase (TSO) pathway, and (f) tryptophan-independent pathway. *IPDC*, indole-3-pyruvate decarboxylase; *IAA1d*, indole-3-acetaldehyde (Oberhänsli et al., 1991).

(Chant and Summers, 2007). According to Bianco et al. (2006), IAA acts as a signaling molecule in *E. coli* by increasing the level of alert and protection against adverse conditions. They observed an increased tolerance in IAA-treated *E. coli* against various abiotic stresses including oxidative stress, osmotic and acid shock, heat and cold shock, UV-irradiation, and toxic compounds (antibiotics, detergents, and dyes). Microarray analysis revealed the upregulation of genes involved in cell envelope component and bacterial adaptation to unfavorable environmental conditions. The authors correlated the IAA-induced protection to enhanced accumulation of trehalose and expression of Dnak gene coded for heat shock protein.

IAA produced by beneficial plant-associated bacteria at the rhizosphere modifies root architecture by increasing the number of lateral roots and its branching, and root hairs improve anchorage, water, and nutrient uptake there by helping plants to cope with water nutrient and water deficit (Mantelin and Touraine, 2004; Egamberdieva and Kucharova, 2009; Vacheron et al., 2013; Gowtham et al., 2017). Different *Bacillus* spp. capable of producing IAA isolated from the rhizosphere of Brazilian cacti showed growth promotion of maize seedlings under water stress (Kavamura et al.,

2013). Marulanda et al. (2009) observed an increase in IAA production by *P. putida* and *B. megaterium* with increased PEG-induced drought stress and at a lower level of salt concentrations. Further, these strains improved the shoot and root growth and water content of maize seedlings grown under drought stress. The adverse effect of drought and salt stress was found diminished in wheat seedlings that were grown with *A. brasiliense* Sp245, a well-studied IAA producing a free-living diazotroph (Pereyra et al., 2012). They observed an increase in the diameter of xylem vessels of coleoptile and less negative water potential under PEG-induced drought conditions. A similar observation was recorded in tomato seedlings inoculated with *A. brasilense* BNM65 under drought conditions (Romero et al., 2014).

6.8 Strigolactones (SLs)

Strigolactones and SL-like molecules are carotenoids derived terpenoid lactones which were recently identified as plant hormones which regulate the architecture of the plant by inhibiting branching (Gomez-Roldan et al., 2008; Umehara et al., 2008). Otherwise, these compounds are believed to stimulate the seed germination of the parasitic plant such as Striga and Orobanche (Matusova et al., 2005) and also act as the host detection system in plant-AM interactions including hyphal branching of AM (Akiyama et al., 2005; Besserer et al., 2006).

Several researchers observed that mutations in genes responsible for SL biosynthesis lead to enhanced branching in *Arabidopsis* (*Arabidopsis thaliana*) (Stirnberg et al., 2002), Garden pea (*Pisum sativum*) (Beveridge, 2000), Rice (Oryza sativa) (Ishikawa et al., 2005). Rice and Pea mutants (Ccd7 and Ccd8) unable to produce SL, restored their branch inhibition property when treated win GR24 (SL analog) (Johnson et al., 1981). Later, the same phenomenon was observed in mutants of Arabidopsis (Umehara et al., 2008). Earlier researchers reported that phosphorus deficiency in the growth medium induced the secretion of SL several folds in plants like clover (*Trifolium pratense*) (Yoneyama et al., 2007a) and sorghum (*Sorghum bicolor*) (Yoneyama et al., 2007b).

As the soil is the prime source of P to plant, under its deficiency, the architecture of the plant will be changed such as reduced shoot growth, branching (inhibit lateral bud growth) and primary root growth, and increased lateral root growth. These changes enable the plant to survive under P deprived conditions and harvest more P from the surroundings. As observed by McAdam et al. (2017), SL promotes infection thread formation by influencing a bacterial partner who is independent of ET action. Anca et al. (2009) reported the influence of SL on AM fungi along with endobacteria "*Candidatus Glomeribacter gigasporarum*" and found that the bacterial multiplication increases along with fungal growth stimulated by SL. Though the SL biosynthesis was not reported in bacteria, certain similar molecules were previously identified. The γ-butyrolactone moiety found in the D ring of SLs is also present in acyl-homoserine lactone which is a well-studied quorum-sensing molecule in bacteria (Ng and Bassler, 2009).

6.9 Salicylic acid (SA)

Salicylic acid is a phenolic phytohormone which plays a vital role in plant growth, development, photosynthesis, transpiration, ion uptake, transport, etc. SA plays a crucial role in the regulation of ion uptakes by the plants (Wang et al., 2011). SA is involved in endogenous signaling mediating plant defense against pathogens. Exogenously applied SA improved growth of *Zea mays* under salt stress/nonstress condition and also helped in optimization of nutrients uptakes and membrane potential (Gunes et al., 2007). Under similar conditions, improved seed germination and plant growth were observed in *Matricaria chamomilla* (Kováčik et al., 2009). Jini and Joseph (2017) and Kim et al. (2009) observed an increased shoot length, chlorophyll and biomass accumulation in rice seedlings exposed to salt stress after exogenous application of SA. Also, NaCl-induced ROS and lipid peroxidation was decreased, reducing its adverse effect. Dat et al. (1998) experimentally proved that SA-treated seedlings of *Sinapis alba* increased thermotolerance and resulting reduced the level of H_2O_2. SA increase salt tolerance in wheat plants via transcript regulation of the eight genes ASA–GSH cycle enzymes, which helps in enhancing tolerance against salt stress (Li et al., 2013). Studies of Clarke et al. (2004) revealed that SA protects the plants from heat shock, which was evidenced by the involvement of SA-dependent signaling in basal thermotolerance and through membrane thermoprotection. But, it may not be essential for acquired thermotolerance. SA and Acetyl SA induce the accumulation of glycine betaine, which works as osmoprotectants under different abiotic stress such as cold, drought, and salt stress (Jagendorf and Takabe, 2001).

Biosynthesis of SA in the plants takes place via the shikimate phenylpropanoid pathway (Métraux, 2002). Conversion of phenylalanine into SA takes place by the two passages, differing only in the hydroxylation of cinnamic acid. Firstly, cinnamic acid goes through the hydroxylation reaction converted into ortho-coumaric acid. Ortho-coumaric acid turns into SA. In the second passage, the side chain of cinnamic acid goes through oxidation, turns into benzoic acid, and is then hydrolyzed at the ortho-position converted into SA (Métraux, 2002).

Bacteria have an SA biosynthesis propensity. Biosynthesis of SA in bacteria takes place via the isochorismate pathway (Fig. 6.5) (Marshall and Ratledge, 1972; Serino et al., 1995). Some bacteria strains, such as *Bacillus licheniformis* MML250 (Prashanth and Mathivanan, 2009), *Pseudomonas aeruginosa* (De Meyer and Höfte, 1997), and *Pseudomonas fluorescens* (Leeman et al., 1995), produced SA acid, which helps in being averse to biotic stress. Though the bacterial SA biosynthesis is well studied, its regulation under the influence of root exudate/endophytic conditions, etc. is yet to be established.

6.10 Jasmonic acid (JA)

Jasmonic acid is a plants hormone which is sesquiterpene in nature and derived from lipids. Biosynthesis of JA takes places from linolenic acid via an octadecanoid pathway (Wasternack, 2007). JA has been playing vital roles insides the plant during life spans such as growth promotion, tuber formation, and ripening, although it also takes

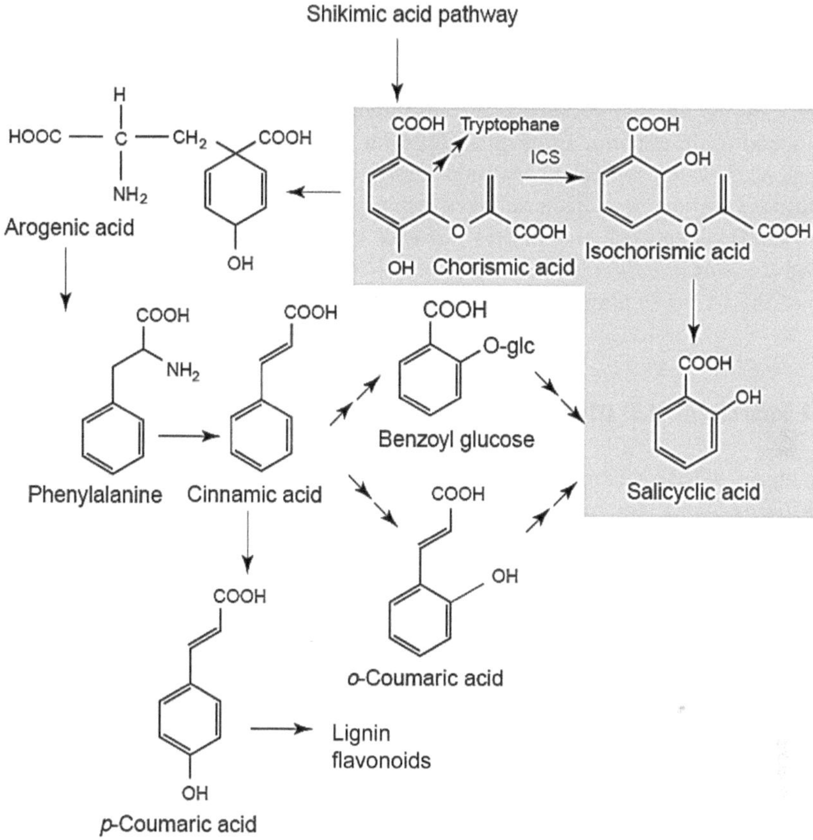

Fig. 6.5 Salicylic acid biosynthesis pathway in bacteria and plants. Plants synthesize SA through the shikimate-phenylpropanoid pathway. Whereas in bacteria, synthesis of SA occurs predominantly via the isochorismate pathway (Marshall and Ratledge, 1972; Serino et al., 1995).
Adopted from Métraux, J.P. 2002. Recent breakthroughs in the study of salicylic acid biosynthesis. *Trends Plant Sci.* 7(8), 332–334.

parts in some physiological processes such as the promotion of senescence, abscission, and inhibition of growth and seed germination (Pospíšilová, 2003). JA plays essential roles in growth and development as well as increases plant fitness during biotic and abiotic stress (Creelman and Mullet, 1995).

The level of JA begins to accumulate in wounded plants and also increases the expression level of a wound-responsive gene such as a proline-rich membrane protein, chalcone synthase, vegetative storage protein (Creelman et al., 1992). JA promotes the acquisition of ABA into the two genotypes of Maresi and wild form of *Hordeum spontaneum* and helps in combat against water deficit condition (Bandurska et al., 2003). Gao et al. (2004) reported that, exogenously applied JA significantly increased the betaine level and expression of betaine aldehyde dehydrogenase in *Pyrus bretschneideri* Redh. cv. Suli plants subjected to drought stress.

Ma et al. (2014) investigated the effect of methyl jasmonate on the wheat plant in well watered and drought condition and reported that exogenously applied methyl jasmonate influenced stomata closing, which led to balanced water status and also encouraged the activity of superoxide dismutase, peroxidase, catalase, and reduced malondialdehyde content. Exogenous application of methyl jasmonate maintained the balanced water content in leaf and improved the photosynthesis under drought conditions. Only a few bacteria have been reported to synthesize JA. Forchetti et al. (2007) identified endophytes bacteria (*Achromobacter xiloxidans, Bacillus* sp. and *Bacillus pumilus*) isolates from the *Helianthus annuus* producing higher contents of the JA while plants appeared with drought stress.

6.11 Other hormones

Zhou et al. (2016) investigated the effects of *Bacillus megaterium* strain BOFC15 on *Arabidopsis* plants. They reported that the strain was found to produce and to secrete spermidine (Spd) and polyamine. The strain was also found to induce a level of a cellular polyamine of plants that increased the free levels of Spd and spermine. It was found to abolish by addition of Spd biosynthetic inhibitor, dicyclohexylamine (DCHA). Also, inoculation with the strain was found to remarkably increase plant biomass, improve root system architecture, and augment photosynthetic capacity. The inoculated plants also were found to display a stronger ability to tolerate drought stress than non-inoculated plants. They also observed that the ABA level was higher in inoculated plants than in control plants under stress conditions. The strain BOFC15-induced ABA synthesis was markedly inhibited by addition of DCHA. They revealed that microbial Spd participated in the modulation of the ABA levels in plants and suggested that Spd-producing BOFC15 was found to improve the tolerance, which associated with the altered level of ABA and activated the plant adaptive response. Creus et al. (2005) identified the NO synthesizing ability of free-living diazotrophic bacteria, *A. brasilense* Sp245. Further, their inoculation of *A. brasilense* Sp245 into tomato plant showed enhanced lateral root formation. The bacterial effect on lateral root formation was reversed by the application of NO scavenger [2-(4-carboxyphenyl)-4,4,5,5-tetramethylimidazoline-1-oxyl-3-oxide] and the response was partially reversed by application of NO donor (sodium nitroprusside), confirming the involvement of bacterial produced NO in lateral root induction. Experiments with a mutant of *A. brasilense* Sp245 revealed that the bacteria could synthesize NO following different pathways and evidenced the involvement of NO-mediated IAA-signaling pathways in inducing lateral and adventitious root formation in tomato seedlings (Molina-Favero et al., 2008).

Yang et al. (2009) revised different mechanisms involved in plant-growth-promoting rhizobacteria (PGPR)-mediated induced systemic tolerance against abiotic stress in host plant (Fig. 6.6).

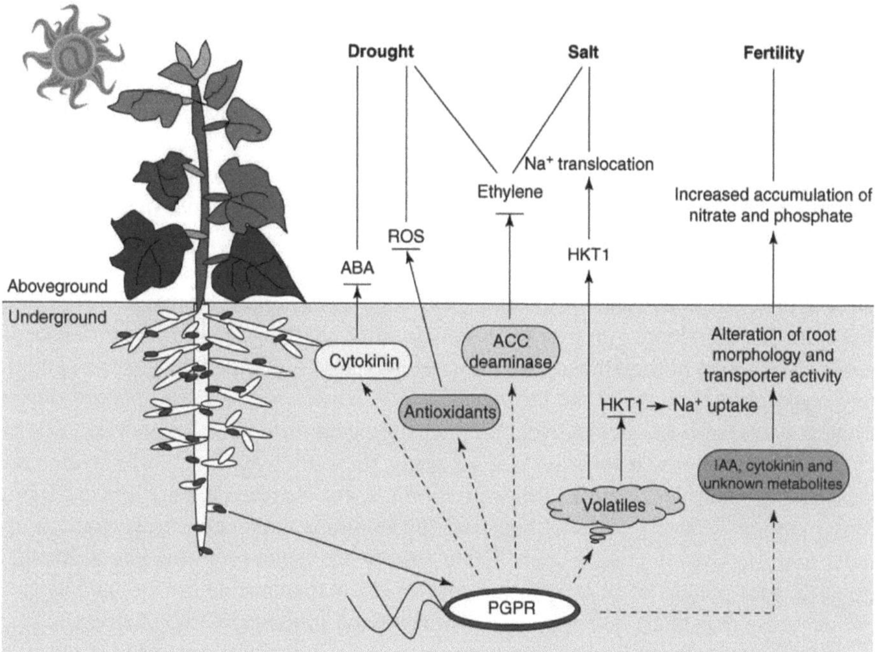

Fig. 6.6 Induced systemic tolerance (IST) elicited by plant-growth-promoting rhizobacteria (PGPR) against drought, salt, and fertility stress underground (root) and aboveground (shoot). Broken arrows indicate bioactive compounds secreted by PGPR; solid arrows indicate plant compounds affected by bacterial components. Some PGPR strains, indicated in red on the plant roots, produce cytokinins and antioxidants such as catalase, which results in ABA accumulation and ROS degradation, respectively (Figueiredo et al., 2008) and (Kohler et al., 2008). Degradation of the ethylene precursor ACC by bacterial ACC deaminase releases plant stress and rescues normal plant growth under drought and salt stresses (Kohler et al., 2008) and (Mayak et al., 2004a, 2004b). The volatiles emitted by PGPR downregulate hkt1 expression in roots but upregulate it in shoot tissues, orchestrating lower Na^+ levels and recirculation of Na^+ in the whole plant under high salt conditions (Zhang et al., 2008). Production by PGPR of IAA or unknown determinants can increase root length, root surface area, and the number of root tips, leading to enhanced uptake of nitrate and phosphorus (Gyaneshwar et al., 2002), (Mantelin and Touraine, 2004) and (Adesemoye et al., 2008). Abbreviations: *ABA*, abscisic acid; *ACC*, 1-aminocyclopropane-1-carboxylate; *HKT1*, high-affinity K^+ transporter 1; *IAA*, indole acetic acid; *IST*, induced systemic tolerance; *PGPR*, plant-growth-promoting rhizobacteria; *ROS*, reactive oxygen species.
Adopted from Yang, J., Kloepper, J.W., Ryu, C.M. 2009. Rhizosphere bacteria help plants tolerate abiotic stress. *Trends Plant Sci.* 14(1), 1–4.

6.12 Conclusion and future prospects

Overall, previous research studies evidenced that plant-associated bacteria can modify host response to a wide range of biotic and abiotic stress through modulating the host plant hormonal system. Currently, we have a wealth of information on how the plant associated beneficial bacteria modulates the hormone system to induce abiotic stress resistance. This may be achieved by directly secreting plant hormones in the proximity of plants or modifying the level of host hormone or increasing/decreasing the sensitivity of host plant to hormones. Studies on optimization of hormonal level in plants to resist abiotic stress by plant-associated bacteria opened a new avenue in developing strategies for sustainable crop production. Another challenge is to identify the regulatory mechanism of these bacterial hormone synthesis mechanisms, as these bacteria may respond to changes in the host system (e.g., varied root exudate pattern during drought stress) or bacteria may respond to changing environment. To understand such complex interactions, it is important to apply the omics approach which includes genomics, proteomics, metabolomics, lipidomics, etc. Researchers are already trying these plant-associated beneficial bacterial formulations to enhance crop productivity under marginal and degraded lands. Other studies are required in this line to identify the molecular pattern of beneficial bacteria which is responsible for the induction of abiotic stress resistance and application of the same to manage abiotic stress. At the genetic level, the possibility of expressing a specific gene from beneficial bacteria in the plant to achieve improved abiotic stress tolerance cannot be ruled out.

References

Abdala, G., Miersch, O., Correa, N., Rosas, S., 1999. Detection of jasmonic acid in cultures of *Escherichia coli* and *Saccharomyces cerevisiae*. Nat. Prod. Lett. 14, 55–63.

Abdelrahman, M., El-Sayed, M., Jogaiah, S., Burritt, D.J., Tran, L.-S.P., 2017. The "STAY-GREEN" trait and phytohormone signaling networks in plants under heat stress. Plant Cell Rep. 36, 1009–1025.

Abdelrahman, M., Jogaiah, S., Burritt, D.J., Tran, L.-S.P., 2018. Legume genetic resources and transcriptome dynamics under abiotic stress conditions. Plant Cell Environ. 41, 1972–1983.

Abe, H., Urao, T., Ito, T., Seki, M., et al., 2003. Arabidopsis AtMYC2 (bHLH) and AtMYB2 (MYB) function as transcriptional activators in abscisic acid signaling. Plant Cell 15 (1), 63–78.

Abeles, F.W., 1992. Roles and physiological effects of ethylene in plant physiology: dormancy, growth, and development. In: Ethylene in Plant Biology. Elsevier.

Achard, P., Cheng, H., De Grauwe, L., Decat, J., et al., 2006. Integration of plant responses to environmentally activated phytohormonal signals. Science 311 (5757), 91–94.

Adesemoye, A.O., Torbert, H.A., Kloepper, J.W., 2008. Enhanced plant nutrient use efficiency with PGPR and AMF in an integrated nutrient management system. Can. J. Microbiol. 54, 876–886.

Ahmad, P., Jaleel, J.A., Salem, M.A., Nabi, G., Sharma, S., 2010. Roles of enzymatic and non-enzymatic antioxidants in plants during abiotic stress. Crit. Rev. Biotechnol. 30 (3), 161–175.

Ahmed, W., Shahroona, B., Zahir, Z.A., Arshad, M., 2004. Inoculation with ACC-deaminase containing rhizobacteria for improving growth and yield of wheat. Pak. J. Agric. Sci. 41 (3/4), 119.

Akiyama, K., Matsuzaki, K.I., Hayashi, H., 2005. Plant sesquiterpenes induce hyphal branching in arbuscular mycorrhizal fungi. Nature 435 (7043), 824.

Akhtar, S.S., Mekureyaw, M.F., Pandey, C., Roitsch, T., 2020. Role of cytokinins for interactions of plants with microbial pathogens and pest insects. Front. Plant Sci. 10, 1777.

Ali, B., Sabri, A.N., Ljung, K., Hasnain, S., 2009. Auxin production by plant-associated bacteria: impact on endogenous IAA content and growth of *Triticum aestivum* L. Lett. Appl. Microbiol. 48 (5), 542–547.

Ali, S., Charles, T.C., Glick, B.R., 2014. Amelioration of high salinity stress damage by plant growth-promoting bacterial endophytes that contain ACC deaminase. Plant Physiol. Biochem. 80, 160–167.

Amzallag, G.N., Lerner, H.R., Poljakoff-Mayber, A., 1990. Exogenous ABA as a modulator of the response of sorghum to high salinity. J. Exp. Bot. 41 (12), 1529–1534.

Anca, I.A., Lumini, E., Ghignone, S., Salvioli, A., et al., 2009. The ftsZ gene of the endocellular bacterium "*Candidatus Glomeribacter gigasporarum*" is preferentially expressed during the symbiotic phases of its host mycorrhizal fungus. Mol. Plant Microbe Interact. 22 (3), 302–310.

Anderson, J.P., Badruzsaufari, E., Schenk, P.M., Manners, J.M., et al., 2004. Antagonistic interaction between abscisic acid and jasmonate-ethylene signaling pathways modulates defense gene expression and disease resistance in Arabidopsis. Plant Cell 16 (12), 3460–3479.

Arkhipova, T.N., Veselov, S.U., Melentiev, A.I., Martynenko, E.V., et al., 2005. Ability of bacterium *Bacillus subtilis* to produce cytokinins and to influence the growth and endogenous hormone content of lettuce plants. Plant and Soil 272 (1-2), 201–209.

Arshad, M., Frankenberger, W.T., 1998. Plant growth substances in the rhizosphere: microbial production and functions. Adv. Agron. 62, 46–151.

Arshad, M., Shaharoona, B., Mahmood, T., 2008. Inoculation with *Pseudomonas* spp. containing ACC-deaminase partially eliminates the effects of drought stress on growth, yield, and ripening of pea (*Pisum sativum* L.). Pedosphere 18 (5), 611–620.

Ashraf, M., 2010. Inducing drought tolerance in plants: recent advances. Biotechnol. Adv. 28 (1), 169–183.

Asker, D., Beppu, T., Ueda, K., 2007a. *Mesoflavibacter zeaxanthinifaciens* gen. nov., sp. nov., a novel zeaxanthin-producing marine bacterium of the family Flavobacteriaceae. Syst. Appl. Microbiol. 30 (4), 291–296.

Asker, D., Beppu, T., Ueda, K., 2007b. *Sphingomonas astaxanthinifaciens* sp. nov., a novel astaxanthin-producing bacterium of the family Sphingomonadaceae isolated from Misasa, Tottori, Japan. FEMS Microbiol. Lett. 273 (2), 140–148.

Asker, D., Beppu, T., Ueda, K., 2007c. *Sphingomonas jaspsi* sp. nov., a novel carotenoid-producing bacterium isolated from Misasa, Tottori, Japan. Int. J. Syst. Evol. Microbiol. 57 (7), 1435–1441.

Asker, D., Awad, T.S., Beppu, T., Ueda, K., 2012a. Novel zeaxanthin-producing bacteria isolated from a radioactive hot spring water. In: Microbial Carotenoids from Bacteria and Microalgae. Humana Press, Totowa, NJ, pp. 99–131.

Asker, D., Awad, T.S., Beppu, T., Ueda, K., 2012b. A novel radio-tolerant astaxanthin-producing bacterium reveals a new astaxanthin derivative: astaxanthin dirhamnoside. In: Microbial Carotenoids from Bacteria and Microalgae. Humana Press, Totowa, NJ, pp. 61–97.

Asselbergh, B., De Vleesschauwer, D., Höfte, M., 2008. Global switches and fine-tuning—ABA modulates plant pathogen defense. Mol. Plant Microbe Interact. 21 (6), 709–719.
Atzorn, R., Crozier, A., Wheeler, C.T., Sandberg, G., 1988. Production of gibberellins and indole-3-acetic acid by *Rhizobium phaseoli* in relation to nodulation of *Phaseolus vulgaris* roots. Planta 175 (4), 532–538.
Bacon, M.A., Davies, W.J., Mingo, D., Wilkinson, S., 2002. Root signals. In: Plant Roots. CRC Press, pp. 717–732.
Bandurska, H., Stroiński, A., Kubiś, J., 2003. The effect of JA on the accumulation of ABA, proline and spermidine and its influence on membrane injury under water deficit in two barley genotypes. Acta Physiol. Plant 25 (3), 279–285.
Barber, S.A., 1995. Soil Nutrient Bioavailability: A Mechanistic Approach. John Wiley and Sons.
Bartels, D., Sunkar, R., 2005. Drought and salt tolerance in plants. Crit. Rev. Plant Sci. 24 (1), 23–58.
Bastián, F., Cohen, A., Piccoli, P., Luna, V., et al., 1998. Production of indole-3-acetic acid and gibberellins A1 and A3 by *Acetobacter diazotrophicus* and *Herbaspirillum seropedicae* in chemically-defined culture media. Plant Growth Regul. 24 (1), 7–11.
Beaudoin, N., Serizet, C., Gosti, F., Giraudat, J., 2000. Interactions between abscisic acid and ethylene signaling cascades. Plant Cell 12 (7), 1103–1115.
Benjamins, R., Scheres, B., 2008. Auxin: the looping star in plant development. Annu. Rev. Plant Biol. 59, 443–465.
Besserer, A., Puech-Pagès, V., Kiefer, P., Gomez-Roldan, V., et al., 2006. Strigolactones stimulate arbuscular mycorrhizal fungi by activating mitochondria. PLoS Biol. 4(7), e226.
Beveridge, C.A., 2000. Long-distance signalling and mutational analysis of branching in pea. Plant Growth Regul. 32, 193–203.
Bianco, C., Imperlini, E., Calogero, R., Senatore, B., et al., 2006. Indole-3-acetic acid improves *Escherichia coli*'s defences to stress. Arch. Microbiol. 185 (5), 373–382.
Bottini, R., Fulchieri, M., Pearce, D., Pharis, R.P., 1989. Identification of gibberellins A1, A3, and iso-A3 in cultures of *Azospirillum lipoferum*. Plant Physiol. 190 (1), 45–47.
Bottini, R., Cassán, F., Piccoli, P., 2004. Gibberellin production by bacteria and its involvement in plant growth promotion and yield increase. Appl. Microbiol. Biotechnol. 65 (5), 497–503.
Brzobohaty, B., Moore, I., Kristoffersen, P., Bako, L., et al., 1993. Release of active cytokinin by a beta-glucosidase localized to the maize root meristem. Science 262 (5136), 1051–1054.
Cacciari, I., Lippi, D., Pietrosanti, T., Pietrosanti, W., 1989. Phytohormone-like substances produced by single and mixed diazotrophic cultures of *Azospirillum* and *Arthrobacter*. Plant and Soil 115 (1), 151–153.
Cao, F.Y., Yoshioka, K., Desveaux, D., 2011. The roles of ABA in plant–pathogen interactions. J. Plant Res. 124 (4), 489–499.
Cassán, F., Bottini, R., Schneider, G., Piccoli, P., 2001. *Azospirillum brasilense* and *Azospirillum lipoferum* hydrolyze conjugates of GA20 and metabolize the resultant aglycones to GA1 in seedlings of rice dwarf mutants. Plant Physiol. 125 (4), 2053–2058.
Cassán, F.D., Piccoli, P., Bottini, R., 2003. Promotion of plant growth by *Azospirillum* sp. through the production of gibberellins. An alternative model to increase agricultural production. Agricultural Microbiology A Contribution of Research in Argentina for Society, National University of Santiago del Estero, Santiago, pp. 1–16.
Castillo, P., Escalante, M., Gallardo, M., Alemano, S., et al., 2013. Effects of bacterial single inoculation and co-inoculation on growth and phytohormone production of sunflower seedlings under water stress. Acta Physiol. Plant 35 (7), 2299–2309.

Caverzan, A., Casassola, A., Brammer, S.P., 2016. Antioxidant responses of wheat plants under stress. Genet. Mol. Biol. 39 (1), 1–6.

Chant, E.L., Summers, D.K., 2007. Indole signalling contributes to the stable maintenance of *Escherichia coli* multicopy plasmids. Mol. Microbiol. 63 (1), 35–43.

Chen, Z., Agnew, J.L., Cohen, J.D., He, P., et al., 2007. *Pseudomonas syringae* type III effector AvrRpt2 alters *Arabidopsis thaliana* auxin physiology. Proc. Natl. Acad. Sci. USA 104 (50), 20131–20136.

Cheng, W.H., Endo, A., Zhou, L., Penney, J., et al., 2002. A unique short-chain dehydrogenase/reductase in Arabidopsis glucose signaling and abscisic acid biosynthesis and functions. Plant Cell 14 (11), 2723–2743.

Claeys, H., Skirycz, A., Maleux, K., Inzé, D., 2012. DELLA signaling mediates stress-induced cell differentiation in *Arabidopsis* leaves through modulation of APC/C activity. Plant Physiol. 159 (2), 739–747.

Clarke, S.M., Mur, L.A., Wood, J.E., Scott, I.M., et al., 2004. Salicylic acid dependent signaling promotes basal thermotolerance but is not essential for acquired thermotolerance in *Arabidopsis thaliana*. Plant J. 38 (3), 432–447.

Coakley, S.M., Scherm, H., Chakraborty, S., 1999. Climate change and plant disease management. Annu. Rev. Phytopathol. 37 (1), 399–426.

Cohen, A.C., Travaglia, C., Reinoso, H., Piccoli, P., et al., 2001. *Azospirillum* inoculation and inhibition of gibberellin and ABA synthesis in maize seedling under drought. In: Proceedings of Plant Growth Regulation Society of America, Annual Meetingvol. 28, pp. 88–93.

Cohen, A.C., Bottini, R., Piccoli, P.N., 2008. *Azospirillum brasilense* Sp 245 produces ABA in chemically-defined culture medium and increases ABA content in *Arabidopsis* plants. Plant Growth Regul. 54 (2), 97–103.

Comas, L.H., Becker, S.R., Cruz, V.M.V., Byrne, P.F., et al., 2013. Root traits contributing to plant productivity under drought. Front. Plant Sci. 4, 442.

Cookson, S.J., Radziejwoski, A., Granier, C., 2006. Cell and leaf size plasticity in *Arabidopsis*: what is the role of endoreduplication? Plant Cell Environ. 29 (7), 1273–1283.

Cramer, G.R., 2010. Abiotic stress and plant responses from the whole vine to the genes. Aust. J. Grape Wine R 16, 86–93.

Creelman, R.A., Mullet, J.E., 1995. JA distribution and action in plants: regulation during development and response to biotic and abiotic stress. Proc. Natl. Acad. Sci. U. S. A. 92 (10), 4114–4119.

Creelman, R.A., Tierney, M.L., Mullet, J.E., 1992. JA/methyl jasmonate accumulate in wounded soybean hypocotyls and modulate wound gene expression. Proc. Natl. Acad. Sci. USA 89 (11), 4938–4941.

Creus, C.M., Sueldo, R.J., Barassi, C.A., 2004. Water relations and yield in *Azospirillum*-inoculated wheat exposed to drought in the field. Can. J. Bot. 82 (2), 273–281.

Creus, C.M., Graziano, M., Casanovas, E.M., Pereyra, M.A., Simontacchi, M., Puntarulo, S., Barassi, C.A., Lamattina, L., 2005. Nitric oxide is involved in the *Azospirillum brasilense*-induced lateral root formation in tomato. Planta 221, 297–303.

Dar, M.I., Naitoo, M.I., Khan, F.A., Rehman, F., Green, I.D., Fauzia, N., Ansari, A.A., 2017. An introduction to reactive oxygen species metabolism under changing climate in plants. In: MIR, K., Khan, N.A. (Eds.), Reactive Oxygen Species and Antioxidant System in Plants: Role and Regulation Under Abiotic Stress. Springer, Singapore, pp. 25–52.

Dat, J.F., Lopez-Delgado, H., Foyer, C.H., Scott, I.M., 1998. Parallel changes in H_2O_2 and catalase during thermotolerance induced by salicylic acid or heat acclimation in mustard seedlings. Plant Physiol. 116 (4), 1351–1357.

Davies, W.J., Zhang, J., 1991. Root signals and the regulation of growth and development of plants in drying soil. Annu. Rev. Plant Biol. 42 (1), 55–76.

De Dorlodot, S., Forster, B., Pagès, L., Price, A., et al., 2007. Root system architecture: opportunities and constraints for genetic improvement of crops. Trends Plant Sci. 12 (10), 474–481.

De Meyer, G., Höfte, M., 1997. Salicylic acid produced by the rhizobacterium *Pseudomonas aeruginosa* 7NSK2 induces resistance to leaf infection by *Botrytis cinerea* on bean. Phytopathology 87 (6), 588–593.

De Torres-Zabala, M., Bennett, M.H., Truman, W.H., Grant, M.R., 2009. Antagonism between salicylic and abscisic acid reflects early host-pathogen conflict and molds plant defense responses. Plant J. 59 (3), 375–386.

Dinneny, J.R., Long, T.A., Wang, J.Y., Jung, J.W., et al., 2008. Cell identity mediates the response of Arabidopsis roots to abiotic stress. Science 320 (5878), 942–945.

Dobert, R.C., Rood, S.B., Blevins, D.G., 1992. Gibberellins and the legume-rhizobium symbiosis: I. Endogenous gibberellins of lima bean (*Phaseolus lunatus* L.) stems and nodules. Plant Physiol. 98 (1), 221–224.

Domka, J., Lee, J., Bansal, T., Wood, T.K., 2007. Temporal gene expression in *Escherichia coli* K12 biofilms. Environ. Microbiol. 9 (2), 332–346.

Duveiller, E., Singh, R.P., Nicol, J.M., 2007. The challenges of maintaining wheat productivity: pests, diseases, and potential epidemics. Euphytica 157 (3), 417–430.

Egamberdieva, D., Kucharova, Z., 2009. Selection for root colonizing bacteria stimulating wheat growth in saline soils. Biol. Fertil. Soils 45, 561–573.

Estruch, J.J., Chriqui, D., Grossmann, K., Schell, J., et al., 1991a. The plant oncogene rolC is responsible for the release of cytokinins from glucoside conjugates. EMBO J 10 (10), 2889–2895.

Estruch, J.J., Schell, J., Spena, A., 1991b. The protein encoded by the rolB plant oncogene hydrolyses indole glucosides. EMBO J. 10 (11), 3125–3128.

Fedoroff, N.V., Battisti, D.S., Beachy, R.N., Cooper, P.J., et al., 2010. Radically rethinking agriculture for the 21st century. Science 327 (5967), 833–834.

Figueiredo, M.V., Burity, H.A., Martínez, C.R., Chanway, C.P., 2008. Alleviation of drought stress in the common bean (*Phaseolus vulgaris* L.) by co-inoculation with *Paenibacillus polymyxa* and *Rhizobium tropici*. Appl. Soil Ecol. 40 (1), 182–188.

Finkelstein, R.R., Gampala, S.S., Rock, C.D., 2002. Abscisic acid signaling in seeds and seedlings. Plant Cell 14 (suppl 1), S15–S45.

Forchetti, G., Masciarelli, O., Alemano, S., Alvarez, D., et al., 2007. Endophytic bacteria in sunflower (*Helianthus annuus* L.): isolation, characterization, and production of jasmonates and abscisic acid in culture medium. Appl. Microbiol. Biotechnol. 76 (5), 1145–1152.

Gao, X.P., Wang, X.F., Lu, Y.F., Zhang, L.Y., et al., 2004. JA is involved in the water—stress—induced betaine accumulation in pear leaves. Plant Cell Environ. 27 (4), 497–507.

García de Salamone, I.E., Hynes, R.K., Nelson, L.M., 2001. Cytokinin production by plant growth promoting rhizobacteria and selected mutants. Can. J. Microbiol. 47 (5), 404–411.

Ghanem, M.E., Hichri, I., Smigocki, A.C., Albacete, A., et al., 2011. Root-targeted biotechnology to mediate hormonal signalling and improve crop stress tolerance. Plant Cell Rep. 30 (5), 807–823.

Glick, B.R., 2014. Bacteria with ACC deaminase can promote plant growth and help to feed the world. Microbiol. Res. 169 (1), 30–39.

Glick, B.R., Penrose, D.M., Li, J., 1998. A model for the lowering of plant ethylene concentrations by plant growth-promoting bacteria. J. Theor. Biol. 190 (1), 63–68.

Goethals, K., Vereecke, D., Jaziri, M., Van Montagu, M., et al., 2001. Leafy gall formation by *Rhodococcus fascians*. Annu. Rev. Phytopathol. 39 (1), 27–52.

Gómez-Cadenas, A., Vives, V., I Zandalinas, S., Manzi, M., et al., 2015. Abscisic acid: a versatile phytohormone in plant signaling and beyond. Curr. Protein Pept. Sci. 16 (5), 413–434.

Gomez-Roldan, V., Fermas, S., Brewer, P.B., Puech-Pagès, V., et al., 2008. Strigolactone inhibition of shoot branching. Nature 455 (7210), 189.

Gowtham, H.G., Duraivadivel, P., Hariprasad, P., Niranjana, S.R., 2017. A novel split-pot bioassay to screen indole acetic acid producing rhizobacteria for the improvement of plant growth in tomato [*Solanum lycopersicum* L.]. Sci. Hortic. 224, 351–357.

Großkinsky, D.K., Tafner, R., Moreno, M.V., Stenglein, et al., 2016. Cytokinin production by *Pseudomonas fluorescens* G20-18 determines biocontrol activity against *Pseudomonas syringae* in Arabidopsis. Sci. Rep. 6, 23310.

Grunewald, W., Vanholme, B., Pauwels, L., Plovie, E., et al., 2009. Expression of the *Arabidopsis* jasmonate signalling repressor JAZ1/TIFY10A is stimulated by auxin. EMBO Rep. 10 (8), 923–928.

Gunes, A., Inal, A., Alpaslan, M., Eraslan, F., et al., 2007. Salicylic acid induced changes on some physiological parameters symptomatic for oxidative stress and mineral nutrition in maize (*Zea mays* L.) grown under salinity. J. Plant Physiol. 164 (6), 728–736.

Gutiérrez Mañero, F.J., Ramos Solano, B., Mehouachi, J., Tadeo, F.R., et al., 2001. The plant growth promoting rhizobacteria *Bacillus pumilus* and *Bacillus licheniformis* produce high amounts of physiologically active gibberellins. Physiol. Plant. 111 (2), 206–211.

Gyaneshwar, P., James, E.K., Reddy, P.M., Ladha, J.K., 2002. *Herbaspirillum* colonization increases growth and nitrogen accumulation in aluminium tolerant rice varieties. New Phytol. 154 (1), 131–145.

Haberer, G., Kieber, J.J., 2002. Cytokinins. New insights into a classic phytohormone. Plant Physiol. 128 (2), 354–362.

Haider, M.S., Kurjogi, M., Fiaz, M., Jogaiah, S., Wang, C., Fang, J., 2018. Drought stress revealed physiological, biochemical and gene-expressional variations in 'Yoshihime' peach (*Prunus persica L*) cultivar. J. Plant Interact. 13, 83–90.

Haider, M.S., Jogaiah, S., Pervaiz, T., Yanxue, Z., Khan, N., Fang, J., 2019. Physiological and transcriptional variations inducing complex adaptive mechanisms in 2 grapevine by salt stress. Environ. Exp. Biol. 162, 455–467.

Han, S., Yu, B., Wang, Y., Liu, Y., 2011. Role of plant autophagy in stress response. Protein Cell 2 (10), 784–791.

Hariprasad, P., Navya, H.M., Chandra Nayaka, S., Niranjana, S.R., 2009. Advantage of using PSIRB over PSRB and IRB to improve plant health of tomato. Biol. Control 50 (3), 307–316.

Honma, M., Shimomura, T., 1978. Metabolism of 1-aminocyclopropane-1-carboxylic acid. Agric. Biol. Chem. 42 (10), 1825–1831.

Horemans, S., Koninck, K.D., Neuray, J., Rand, H., 1986. Production of plant growth substances by *Azospirillum* sp. and other rhizosphere bacteria [1986]. Symbiosis (USA) 2, 341–346.

Hsiao, T.C., Xu, L.K., 2000. Sensitivity of growth of roots versus leaves to water stress: biophysical analysis and relation to water transport. J. Exp. Bot. 51 (350), 1595–1616.

Huang, X.F., Chaparro, J.M., Reardon, K.F., Zhang, R., et al., 2014. Rhizosphere interactions: root exudates, microbes, and microbial communities. Botany 92 (4), 267–275.

Hussain, A., Hasnain, S., 2009. Cytokinin production by some bacteria: its impact on cell division in cucumber cotyledons. Afr. J. Microbiol. Res. 3 (11), 704–712.

Hussain, A., Black, C.R., Taylor, I.B., Roberts, J.A., 1999. Soil compaction. A role for ethylene in regulating leaf expansion and shoot growth in tomato? Plant Physiol. 121 (4), 1227–1237.

Ishikawa, S., Maekawa, M., Arita, T., Onishi, K., et al., 2005. Suppression of tiller bud activity in tillering dwarf mutants of rice. Plant Cell Physiol. 46, 79–86.

Jacobson, C.B., Pasternak, J.J., Glick, B.R., 1994. Partial purification and characterization of ACC deaminase from the plant growth-promoting rhizobacterium *Pseudomonas putida* GR12-2. Can. J. Microbiol. 40, 1019–1025.

Jagendorf, A.T., Takabe, T., 2001. Inducers of glycinebetaine synthesis in barley. Plant Physiol. 127 (4), 1827–1835.

Janzen, R.A., Rood, S.B., Dormaar, J.F., McGill, W.B., 1992. *Azospirillum brasilense* produces gibberellin in pure culture on chemically-defined medium and in co-culture on straw. Soil Biol. Biochem. 24 (10), 1061–1064.

Jiang, C.J., Shimono, M., Sugano, S., Kojima, M., et al., 2010. Abscisic acid interacts antagonistically with salicylic acid signaling pathway in rice–*Magnaporthe grisea* interaction. Mol. Plant Microbe Interact. 23 (6), 791–798.

Jini, D., Joseph, B., 2017. Physiological mechanism of salicylic acid for alleviation of salt stress in rice. Ric. Sci. 24 (2), 97–108.

Jogaiah, S., Sharathchandra, R.G., Tran, L.-S.P., 2013. Systems biology-based approaches towards understanding drought tolerance in food crops. Crit. Rev. Biotechnol. 33, 23–39.

Johnson, P.R., Ecker, J.R., 1998. The ethylene gas signal transduction pathway: a molecular perspective. Annu. Rev. Genet. 32 (1), 227–254.

Johnson, A.W., Gowda, G., Hassanali, A., Knox, J., Monaco, S., Razavi, Z., Rosebery, G., 1981. The preparation of synthetic analogs of strigol. J. Chem. Soc. Perkin Trans. 1, 1734–1743.

Kaper, J.M., Veldstra, H., 1958. On the metabolism of tryptophan by *Agrobacterium tumefaciens*. Biochim. Biophys. Acta 30 (2), 401–420.

Karadeniz, A., Topcuoğlu, Ş.F., Inan, S., 2006. Auxin, gibberellin, cytokinin and abscisic acid production in some bacteria. World J. Microbiol. Biotechnol. 22 (10), 1061–1064.

Kasim, W.A., Osman, M.E., Omar, M.N., El-Daim, I.A.A., et al., 2013. Control of drought stress in wheat using plant-growth-promoting bacteria. J. Plant Growth Regul. 32 (1), 122–130.

Kavamura, V.N., Santos, S.N., da Silvaa, J.L., Parmaa, M.M., Ávilaa, L.A., Visconti, A., Zucchi, T.D., Taketani, R.G., Andreote, F.D., de Meloa, I.S., 2013. Screening of Brazilian cacti rhizobacteria for plant growth promotion under drought. Microbiol. Res. 168, 183–191.

Kazan, K., Manners, J.M., 2009. Linking development to defense: auxin in plant–pathogen interactions. Trends Plant Sci. 14 (7), 373–382.

Kende, H., Zeevaart, J., 1997. The five "classical" plant hormones. Plant Cell 9 (7), 1197.

Khadri, M., Tejera, N.A., Lluch, C., 2006. Alleviation of salt stress in common bean (*Phaseolus vulgaris*) by exogenous abscisic acid supply. J. Plant Growth Regul. 25 (2), 110–119.

Kim, E.H., Kim, Y.S., Park, S.H., Koo, Y.J., et al., 2009. Methyl jasmonate reduces grain yield by mediating stress signals to alter spikelet development in rice. Plant Physiol. 149 (4), 1751–1760.

Kohler, J., Hernández, J.A., Caravaca, F., Roldán, A., 2008. Plant-growth-promoting rhizobacteria and arbuscular mycorrhizal fungi modify alleviation biochemical mechanisms in water-stressed plants. Funt. Plant Biol. 35 (2), 141–151.

Kováčik, J., Grúz, J., Bačkor, M., Strnad, M., Repčák, M., 2009. Salicylic acid-induced changes to growth and phenolic metabolism in *Matricaria chamomilla* plants. Plant Cell Rep. 28, 135–143.

Lambrecht, M., Vande Broek, A., Dosselaere, F., Vanderleyden, J., 1999. The *ipdC* promoter auxin-responsive element of *Azospirillum brasilense*, a prokaryotic ancestral form of the plant AuxRE? Mol. Microbiol. 32, 889–890.

Lambrecht, M., Okon, Y., Broek, A.V., Vanderleyden, J., 2000. Indole-3-acetic acid: a reciprocal signalling molecule in bacteria-plant interactions. Trends Microbiol. 8 (7), 298–300.

Lee, K.H., Piao, H.L., Kim, H.Y., Choi, S.M., et al., 2006. Activation of glucosidase via stress-induced polymerization rapidly increases active pools of abscisic acid. Cell 126 (6), 1109–1120.

Leeman, M., Van Pelt, J.A., Den Ouden, F.M., Heinsbroek, M., et al., 1995. Induction of systemic resistance against Fusarium wilt of radish by lipopolysaccharides of *Pseudomonas fluorescens*. Phytopathology 85 (9), 1021–1027.

LeNoble, M.E., Spollen, W.G., Sharp, R.E., 2004. Maintenance of shoot growth by endogenous ABA: genetic assessment of the involvement of ethylene suppression. J. Exp. Bot. 55 (395), 237–245.

Li, J., Glick, B.R., 2001. Transcriptional regulation of the *Enterobacter cloacae* UW4 1-aminocyclopropane-1-carboxylate (ACC) deaminase gene (acdS). Can. J. Microbiol. 47 (4), 359–367.

Li, J., Shah, S., Moffatt, B.A., Glick, B.R., 2001. Isolation and characterization of an unusual 1-aminocyclopropane-1-carboxylic acid (ACC) deaminase gene from *Enterobacter cloacae* UW4. Anton. Leeuw. 80 (3-4), 255–261.

Li, G., Peng, X., Wei, L., Kang, G., 2013. Salicylic acid increases the contents of glutathione and ascorbate and temporally regulates the related gene expression in salt-stressed wheat seedlings. Gene 529 (2), 321–325.

Liao, X., Guo, X., Wang, Q., Wang, Y., et al., 2017. Overexpression of Ms DREB 6.2 results in cytokinin deficient developmental phenotypes and enhances drought tolerance in transgenic apple plants. Plant J. 89 (3), 510–526.

Lim, J.H., Kim, S.D., 2013. Induction of drought stress resistance by multi-functional PGPR *Bacillus licheniformis* K11 in pepper. Plant Pathol. J. 29 (2), 201–208.

Lu, G., Gao, C., Zheng, X., Han, B., 2009. Identification of OsbZIP72 as a positive regulator of ABA response and drought tolerance in rice. Planta 229 (3), 605–615.

Ma, W., Charles, T.C., Glick, B.R., 2004. Expression of an exogenous 1-aminocyclopropane-1-carboxylate deaminase gene in *Sinorhizobium meliloti* increases its ability to nodulate alfalfa. Appl. Environ. Microbiol. 70 (10), 5891–5897.

Ma, C., Wang, Z.Q., Zhang, L.T., Sun, M.M., et al., 2014. Photosynthetic responses of wheat (*Triticum aestivum* L.) to combined effects of drought and exogenous methyl jasmonate. Photosynthetica 52 (3), 377–385.

Ma, D., Sun, D., Wang, C., Qin, H., et al., 2016. Silicon application alleviates drought stress in wheat through transcriptional regulation of multiple antioxidant defense pathways. J. Plant Growth Regul. 35 (1), 1–10.

MacMillan, C.P., Blundell, C.A., King, R.W., 2005. Flowering of the grass *Lolium perenne*. Effects of vernalization and long days on gibberellin biosynthesis and signaling. Plant Physiol. 138 (3), 1794–1806.

Manchanda, G., Garg, N., 2008. Salinity and its effects on the functional biology of legumes. Acta Physiol. Plant 30 (5), 595–618.

Mantelin, S., Touraine, B., 2004. Plant growth promoting bacteria and nitrate availability: impacts on root development and nitrate uptake. J. Exp. Bot. 55 (394), 27–34.

Manulis, S., Haviv-Chesner, A., Brandl, M.T., Lindow, S.E., Barash, I., 1998. Differential involvement of indole-3-acetic acid biosynthetic pathways in pathogenicity and epiphytic fitness of *Erwinia herbicola* pv. gypsophilae. Mol. Plant Microbe Interact. 11 (7), 634–642.

Marasco, E.K., Schmidt Dannert, C., 2008. Identification of bacterial carotenoid cleavage dioxygenase homologues that cleave the interphenyl α, β double bond of stilbene derivatives via a monooxygenase reaction. Chembiochem 9 (9), 1450–1461.

Markmann, K., Parniske, M., 2009. Evolution of root endosymbiosis with bacteria: How novel are nodules? Trends Plant Sci. 14 (2), 77–86.

Marshall, B.J., Ratledge, C., 1972. Salicylic acid biosynthesis and its control in *Mycobacterium smegmatis*. Biochim. Biophys. Acta (BBA) Gen. Subjects 264 (1), 106–116.

Marulanda, A., Barea, J.M., Azcón, R., 2009. Stimulation of plant growth and drought tolerance by native microorganisms (AM fungi and bacteria) from dry environments: mechanisms related to bacterial effectiveness. J. Plant Growth Regul. 28 (2), 115–124.

Matusova, R., Rani, K., Verstappen, F.W., Franssen, M.C., et al., 2005. The strigolactone germination stimulants of the plant-parasitic *Striga* and *Orobanche* spp. are derived from the carotenoid pathway. Plant Physiol. 139 (2), 920–934.

Mayak, S., Tirosh, T., Glick, B.R., 2004a. Plant growth-promoting bacteria confer resistance in tomato plants to salt stress. Plant Physiol. Biochem. 42 (6), 565–572.

Mayak, S., Tirosh, T., Glick, B.R., 2004b. Plant growth-promoting bacteria that confer resistance to water stress in tomatoes and peppers. Plant Sci. 166 (2), 525–530.

McAdam, E.L., Hugill, C., Fort, S., Samian, E., et al., 2017. Determining the site of action of strigolactones during no dulation. Plant Physiol. 175, 529–542.

Méndez, C., Baginsky, C., Hedden, P., Gong, F., et al., 2014. Gibberellin oxidase activities in *Bradyrhizobium japonicum* bacteroids. Phytochemistry 98, 101–109.

Métraux, J.P., 2002. Recent breakthroughs in the study of salicylic acid biosynthesis. Trends Plant Sci. 7 (8), 332–334.

Miller, C.O., Skoog, F., Von Saltza, M.H., Strong, F., 1955. Kinetin, a cell division factor from deoxyribonucleic acid. J. Am. Chem. Soc. 77, 1392–1393.

Miller, C.O., Skoog, F., Okumura, F.S., Von Saltza, M.H., et al., 1956. Isolation, structure and synthesis of kinetin, a substance promoting cell division. J. Am. Chem. Soc. 78 (7), 1375–1380.

Mittler, R., Vanderauwera, S., Gollery, M., Van Breusegem, F., 2004. Reactive oxygen gene network of plants. Trends Plant Sci. 9, 490–498.

Molina-Favero, C., Creus, C.M., Lanteri, M.L., Correa Aragunde, N., et al., 2007. Nitric oxide and plant growth promoting rhizobacteria: common features influencing root growth and development. Adv. Bot. Res. 46, 1–33.

Molina-Favero, C., Creus, C.M., Simontacchi, M., Puntarulo, S., Lamattina, L., 2008. Aerobic nitric oxide production by *Azospirillum brasilense* Sp245 and its influence on root architecture in tomato. Mol. Plant Microbe Interact 21, 1001–1009.

Morgan, P.W., He, C.J., De Greef, J.A., Maurice, P., 1990. Does water deficit stress promote ethylene synthesis by intact plants? Plant Physiol. 94 (4), 1616–1624.

Morrone, D., Chambers, J., Lowry, L., Kim, G., et al., 2009. Gibberellin biosynthesis in bacteria: Separate ent-copalyl diphosphate and ent-kaurene synthases in *Bradyrhizobium japonicum*. FEBS Lett. 583 (2), 475–480.

Mostafa, A., Abdel-Motaal, F., El-Sayed, M., Jogaiah, S., Shigyo, M., S-i, I., Tran, L.-S.P., 2016. Dissection of*Trichoderma longibrachiatum*-induced defense in onion (*Allium cepa* L.) against*Fusarium oxysporum*f. sp. Cepa by target metabolite profiling. Plant Sci. 246, 128–138.

Munns, R., Cramer, G.R., 1996. Is coordination of leaf and root growth mediated by abscisic acid? Opinion. Plant Soil 185 (1), 33–49.

Nakabayashi, R., Yonekura-Sakakibara, K., Urano, K., Urano, K., et al., 2014. Enhancement of oxidative and drought tolerance in Arabidopsis by overaccumulation of antioxidant flavonoids. Plant J. 77 (3), 367–379.

Naqvi, S.S.M., 2002. Plant growth hormones: growth promoters and inhibitors. In: Handbook of Plant and Crop Physiology. second ed. Marcel Deker, Inc., New York, pp. 501–525.

Nett, R.S., Montanares, M., Marcassa, A., Lu, X., et al., 2017. Elucidation of gibberellin biosynthesis in bacteria reveals convergent evolution. Nat. Chem. Biol. 13 (1), 69.

Ng, W.L., Bassler, B.L., 2009. Bacterial quorum-sensing network architectures. Annu. Rev. Genet. 43, 197–222.

Nguyen, H.T., Stanton, D.E., Schmitz, N., Farquhar, G.D., et al., 2015. Growth responses of the mangrove *Avicennia marina* to salinity: development and function of shoot hydraulic systems require saline conditions. Ann. Bot. 115 (3), 397–407.

Nieto, K.F., Frankenberger, W.T., 1989. Biosynthesis of cytokinins by *Azotobacter chroococcum*. Soil Biol. Biochem. 21, 967–972.

Nishiyama, R., Watanabe, Y., Fujita, Y., Le, D.T., et al., 2011. Analysis of cytokinin mutants and regulation of cytokinin metabolic genes reveals important regulatory roles of cytokinins in drought, salt and abscisic acid responses, and abscisic acid biosynthesis. Plant Cell 23 (6), 2169–2183.

Oberhänsli, T., Defago, G., Haas, D., 1991. Indole-3-acetic acid (IAA) synthesis in the biocontrol strain CHA0 of *Pseudomonas fluorescens*: role of tryptophan side chain oxidase. Curr. Microbiol. 137 (10), 2273–2279.

Oneto, C.D., Otegui, M.E., Baroli, I., Beznec, A., et al., 2016. Water deficit stress tolerance in maize conferred by expression of an isopentenyltransferase (IPT) gene driven by a stress- and maturation-induced promoter. J. Biotechnol. 220, 66–77.

Ortíz-Castro, R., Valencia-Cantero, E., López-Bucio, J., 2008. Plant growth promotion by *Bacillus megaterium* involves cytokinin signaling. Plant Signal. Behav. 3 (4), 263–265.

Parvaiz, A., Satyawati, S., 2008. Salt stress and phyto-biochemical responses of plants—a review. Plant Soil Environ. 54 (3), 89.

Patten, C.L., Glick, B.R., 1996. Bacterial biosynthesis of indole-3-acetic acid. Can. J. Microbiol. 42 (3), 207–220.

Pēna-Cortés, H., Sánchez-Serrano, J.J., Mertens, R., Willmitzer, L., et al., 1989. Abscisic acid is involved in the wound-induced expression of the proteinase inhibitor II gene in potato and tomato. Proc. Natl. Acad. Sci. USA 86 (24), 9851–9855.

Pereyra, M.A., Garcia, P., Colabelli, M.N., Barassi, C.A., Creus, C.M., 2012. A better water status in wheat seedlings induced by Azospirillum under osmotic stress is related to morphological changes in xylem vessels of the coleoptile. Appl. Soil Ecol. 53, 94–97.

Pertry, I., 2009. How the fas Locus Contributes to *Rhodococcus fascians* Cytokinin Production: An In-depth Molecular and Biochemical Analysis. Ghent University, Doctoral Dissertation.

Piccoli, P., Masciarelli, O., Bottini, R., 1999. Gibberellin production by *Azospirillum lipoferum* cultured in chemically-defined medium as affected by oxygen availability and water status. Symbiosis 27 (2), 135–145.

Piccoli, P., Travaglia, C., Cohen, A., Sosa, L., Cornejo, P., Masuelli, R., Bottini, R., 2011. An endophytic bacterium isolated from roots of the halophyte *Prosopis strombulifera* produces ABA, IAA, gibberellins A1 and A3 and jasmonic acid in chemically-defined culture medium. Plant Growth Regul. 64, 207–210.

Pinheiro, C., Chaves, M.M., 2011. Photosynthesis and drought: can we make metabolic connections from available data? J. Exp. Bot. 62 (3), 869–882.

Pospíšilová, J., 2003. Participation of phytohormones in the stomatal regulation of gas exchange during water stress. Biol. Plant. 46 (4), 491–506.

Prabhu, S., Rekha, P.D., Young, C.C., Hameed, A., et al., 2013. Zeaxanthin production by novel marine isolates from coastal sand of India and its antioxidant properties. Appl. Biochem. Biotechnol. 171 (4), 817–831.

Prashanth, S., Mathivanan, N., 2009. Optimization and production of salicylic acid by rhizobacterial strain *Bacillus licheniformis* MML2501. Int. J. Microbiol. 6, 1–15.

Prinsen, E., Costacurta, A., Michiels, K., Vanderleyden, J., et al., 1993. *Azospirillum brasilense* indole-3-acetic acid biosynthesis: evidence for a non-tryptophan dependent pathway. Mol. Plant Microbe Interact. 6, 609–615.

Praveen, S., Murali, K.P., Kurjogi, M., Jogaiah, S., 2019. Physiological adaptation and spectral annotation of Arsenic and Cadmium heavy metal-resistant and susceptible strain *Pseudomonas taiwanensis*. Environ. Pollut. 251, 555–563.

Quarrie, S.A., 1987. Use of genotypes differing in endogenous Abscisic acid levels in studies of physiology and development. In: Hormone Action in Plant Development: A Critical Appraisal. Butterworths, London, pp. 89–105.

Ramireddy, E., Hosseini, S.A., Eggert, K., Gillandt, S., et al., 2018. Root engineering in barley: Increasing cytokinin degradation produces a larger root system, mineral enrichment in the shoot and improved drought tolerance. Plant Physiol. 177 (3), 1078–1095.

Reguera, M., Peleg, Z., Abdel-Tawab, Y.M., Tumimbang, E., et al., 2013. Stress-induced CK synthesis increases drought tolerance through the coordinated regulation of carbon and nitrogen assimilation in rice. Plant Physiol. 163 (4), 1609–1622.

Rivero, R.M., Kojima, M., Gepstein, A., Sakakibara, H., et al., 2007. Delayed leaf senescence induces extreme drought tolerance in a flowering plant. Proc. Natl. Acad. Sci. USA 104 (49), 19631–19636.

Robert-Seilaniantz, A., Grant, M., Jones, J.D., 2011. Hormone crosstalk in plant disease and defense: more than just jasmonate-salicylate antagonism. Annu. Rev. Phytopathol. 49, 317–343.

Romero, A.M., Vega, D., Correa, O.S., 2014. *Azospirillum brasilense* mitigates water stress imposed by a vascular disease by increasing xylem vessel area and stem hydraulic conductivity in tomato. Appl. Soil Ecol. 82, 38–43.

Sakamoto, H., Matsuda, O., Iba, K., 2008. ITN1, a novel gene encoding an ankyrin—repeat protein that affects the ABA mediated production of reactive oxygen species and is involved in salt stress tolerance in *Arabidopsis thaliana*. Plant J. 56 (3), 411–422.

Salisbury, F.B., Ross, C.W., 1992. Plant Physiology, fourth ed. Wadsworth, Belmont, CA, pp. 23–52.

Salomon, M.V., Bottini, R., de Souza Filho, G.A., Cohen, A.C., et al., 2014. Bacteria isolated from roots and rhizosphere of *Vitis vinifera* retard water losses, induce abscisic acid accumulation and synthesis of defense-related terpenes in *in vitro* cultured grapevine. Physiol. Plant. 151 (4), 359–374.

Sandhya, V.S.K.Z., Ali, S.Z., Grover, M., Reddy, G., et al., 2010. Effect of plant growth promoting *Pseudomonas* spp. on compatible solutes, antioxidant status and plant growth of maize under drought stress. Plant Growth Regul. 62 (1), 21–30.

Selvakumar, G., Panneerselvam, P., Ganeshamurthy, A.N., 2012. Bacterial-mediated alleviation of abiotic stress in crops. In: Bacteria in Agrobiology: Stress Management. Springer, Berlin, Heidelberg, pp. 205–224.

Serino, L., Reimmann, C., Baur, H., Beyeler, M., et al., 1995. Structural genes for salicylate biosynthesis from chorismate in *Pseudomonas aeruginosa*. Mol. Gen. Genet. 249 (2), 217–228.

Sharp, R.E., 2002. Interaction with ethylene: changing views on the role of abscisic acid in root and shoot growth responses to water stress. Plant Cell Environ. 25 (2), 211–222.

Sharp, R.E., LeNoble, M.E., 2002. ABA, ethylene and the control of shoot and root growth under water stress. J. Exp. Bot. 53 (366), 33–37.
Sharp, R.E., LeNoble, M.E., Else, M.A., Thorne, E.T., 2000. Endogenous ABA maintains shoot growth in tomato independently of effects on plant water balance: evidence for an interaction with ethylene. J. Exp. Bot. 51 (350), 1575–1584.
Shinozaki, K., Yamaguchi-Shinozaki, K., 2000. Molecular responses to dehydration and low temperature: differences and cross-talk between two stress signaling pathways. Curr. Opin. Plant Biol. 3 (3), 217–223.
Shrivastava, P., Kumar, R., 2015. Soil salinity: a serious environmental issue and plant growth promoting bacteria as one of the tools for its alleviation. Saudi J. Biol. Sci. 22 (2), 123–131.
Singh, R.P., Shelke, G.M., Kumar, A., Jha, P.N., 2015. Biochemistry and genetics of ACC deaminase: a weapon to "stress ethylene" produced in plants. Front. Microbiol. 6, 937.
Skirycz, A., Inzé, D., 2010. More from less: plant growth under limited water. Curr. Opin. Biotechnol. 21 (2), 197–203.
Skirycz, A., Claeys, H., De Bodt, S., Oikawa, A., et al., 2011. Pause-and-stop: the effects of osmotic stress on cell proliferation during early leaf development in Arabidopsis and a role for ethylene signaling in cell cycle arrest. Plant Cell 23 (5), 1876–1888.
Spollen, W.G., LeNoble, M.E., Samuels, T.D., Bernstein, N., et al., 2000. Abscisic acid accumulation maintains maize primary root elongation at low water potentials by restricting ethylene production. Plant Physiol. 122 (3), 967–976.
Srivastava, L.M., 2002. Plant Growth and Development: Hormones and Environment. Elsevier.
Stirnberg, P., van De Sande, K., Leyser, H.O., 2002. MAX1 and MAX2 control shoot lateral branching in Arabidopsis. Development 129 (5), 1131–1141.
Sun, T.P., 2008. Gibberellin metabolism, perception and signaling pathways in Arabidopsis. Arabidopsis Book. 6, e0103.
Sun, T.P., 2011. The molecular mechanism and evolution of the GA–GID1–DELLA signaling module in plants. Curr. Biol. 21 (9), 338–345.
Sun, Y., Cheng, Z., Glick, B.R., 2009. The presence of a 1-aminocyclopropane-1-carboxylate (ACC) deaminase deletion mutation alters the physiology of the endophytic plant growth-promoting bacterium *Burkholderia phytofirmans* PsJN. FEMS Microbiol. Lett. 296 (1), 131–136.
Taiz, L., Zeiger, E., Taiz, L., Zeiger, E., 1998. Auxins. In: Plant Physiology. Sunderland, MA, Sinayer Associates, pp. 543–589.
Taiz, L., Zeiger, E., 2010. Plant Physiology, fifth ed. Sinauer Associates, Sunderland, MA.
Tattersall, E.A., Grimplet, J., DeLuc, L., Wheatley, M.D., et al., 2007. Transcript abundance profiles reveal larger and more complex responses of grapevine to chilling compared to osmotic and salinity stress. Funct. Integr. Genomics 7 (4), 317–333.
Teale, W.D., Paponov, I.A., Palme, K., 2006. Auxin in action: signalling, transport and the control of plant growth and development. Nat. Rev. Mol. Cell Biol. 7 (11), 847.
Thawornwiriyanun, P., Tanasupawat, S., Dechsakulwatana, C., Techkarnjanaruk, S., 2012. Identification of newly zeaxanthin-producing bacteria isolated from sponges in the gulf of Thailand and their zeaxanthin production. Appl. Biochem. Biotechnol. 167 (8), 2357–2368.
Tien, T.M., Gaskins, M.H., Hubbell, D.H., 1979. Plant growth substances produced by *Azospirillum brasilense* and their effect on the growth of pearl millet (*Pennisetum americanum* L.). Appl. Environ. Microbiol. 37 (5), 1016–1024.
Timmuskv, S., Wagner, E.G.H., 1999. The plant-growth-promoting rhizobacterium *Paenibacillus polymyxa* induces changes in Arabidopsis thaliana gene expression: a possible connection between biotic and abiotic stress responses. Mol. Plant Microbe Interact. 12 (11), 951–959.

Tiwari, G., Duraivadivel, P., Sharma, S., Hariprasad, P., 2018. 1-Aminocyclopropane-1-carboxylic acid deaminase producing beneficial rhizobacteria ameliorate the biomass characters of *Panicum maximum* Jacq. by mitigating drought and salt stress. Sci. Rep. 8 (1), 1–12.

Ton, J., Flors, V., Mauch-Mani, B., 2009. The multifaceted role of ABA in disease resistance. Trends Plant Sci. 14 (6), 310–317.

Tuberosa, R., 2012. Phenotyping for drought tolerance of crops in the genomics era. Front. Physiol. 3, 347.

Tudzynski, B., 2005. Gibberellin biosynthesis in fungi: genes, enzymes, evolution, and impact on biotechnology. Appl. Microbiol. Biotechnol. 66 (6), 597–611.

Tully, R., van Berkum, P., Lovins, K., Keister, D., 1998. Identification and sequencing of a cytochrome P450 gene cluster from *Bradyrhizobium japonicum*1. BBA Gene Struct. Exp. 1398 (3), 243–255.

Tuteja, N., 2007. Abscisic acid and abiotic stress signaling. Plant Signal. Behav. 2 (3), 135–138.

Umehara, M., Hanada, A., Yoshida, S., Akiyama, K., et al., 2008. Inhibition of shoot branching by new terpenoid plant hormones. Nature 455 (7210), 195.

Vacheron, J., Desbrosses, G., Bouffaud, M.L., Touraine, B., et al., 2013. Plant growth-promoting rhizobacteria and root system functioning. Front. Plant Sci. 4, 356.

Vojta, P., Kokáš, F., Husičková, A., Grúz, J., et al., 2016. Whole transcriptome analysis of transgenic barley with altered cytokinin homeostasis and increased tolerance to drought stress. N. Biotechnol. 33 (5), 676–691.

Vurukonda, S.S.K.P., Vardharajula, S., Shrivastava, M., SkZ, A., 2016. Enhancement of drought stress tolerance in crops by plant growth promoting rhizobacteria. Microbiol. Res. 184, 13–24.

Wang, C., Knill, E., Glick, B.R., Défago, G., 2000. Effect of transferring 1-aminocyclopropane-1-carboxylic acid (ACC) deaminase genes into *Pseudomonas fluorescens* strain CHA0 and its gac A derivative CHA96 on their growth-promoting and disease-suppressive capacities. Can. J. Microbiol. 46 (10), 898–907.

Wang, Y., Mopper, S., Hasenstein, K.H., 2001. Effects of salinity on endogenous ABA, IAA, JA, and SA in *Iris hexagona*. J. Chem. Ecol. 27 (2), 327–342.

Wang, K.L.C., Li, H., Ecker, J.R., 2002. Ethylene biosynthesis and signaling networks. Plant Cell 14 (suppl 1), 131–151.

Wang, W., Vinocur, B., Altman, A., 2003. Plant responses to drought, salinity and extreme temperatures: towards genetic engineering for stress tolerance. Planta 218 (1), 1–14.

Wang, D., Pei, K., Fu, Y., Sun, Z., et al., 2007. Genome-wide analysis of the auxin response factors (ARF) gene family in rice (*Oryza sativa*). Gene 394 (1), 13–24.

Wang, C., Zhang, S., Wang, P., Hou, J., et al., 2011. Salicylic acid involved in the regulation of nutrient elements uptake and oxidative stress in *Vallisneria natans* (Lour.) Hara under Pb stress. Chemosphere 84 (1), 136–142.

Wang, C., Jogaiah, S., Zhang, W.Y., Abdelrahman, M., Fang, J.G., 2018. Spatio-temporal expression of miRNA159 family members and their GAMYB target gene during the modulation of gibberellin-induced grapevine parthenocarpy. J. Exp. Bot. 69, 3639–3650.

Wasternack, C., 2007. Jasmonates: an update on biosynthesis, signal transduction and action in plant stress response, growth and development. Ann. Bot. 100 (4), 681–697.

Whipps, J.M., 1990. Carbon economy. In: The Rhizosphere. John Wiley & Sons, pp. 59–97.

Williams, P.M., De Mallorca, M.S., 1982. Abscisic acid and gibberellin-like substances in roots and root nodules of *Glycine max*. Plant and Soil 65 (1), 19–26.

Woodward, A.W., Bartel, B., 2005. Auxin: regulation, action, and interaction. Ann. Bot. 95 (5), 707–735.

Xie, S.S., Wu, H.J., Zang, H.Y., Wu, L.M., et al., 2014. Plant growth promotion by spermidine-producing *Bacillus subtilis* OKB105. Mol. Plant Microbe Interact. 27 (7), 655–663.

Yabuta, T., Sumiki, Y., 1938. Crystallization of gibberellins A and B. J. Agric. Chem. Soc. Japan 14, 1526.

Yamaguchi-Shinozaki, K., Shinozaki, K., 2006. Transcriptional regulatory networks in cellular responses and tolerance to dehydration and cold stresses. Annu. Rev. Plant Biol. 57, 781–803.

Yancey, P.H., Clark, M.E., Hand, S.C., Bowlus, R.D., et al., 1982. Living with water stress: evolution of osmolyte systems. Science 217 (4566), 1214–1222.

Yang, J., Kloepper, J.W., Ryu, C.M., 2009. Rhizosphere bacteria help plants tolerate abiotic stress. Trends Plant Sci. 14 (1), 1–4.

Yang, S., Vanderbeld, B., Wan, J., Huang, Y., 2010. Narrowing down the targets: towards successful genetic engineering of drought-tolerant crops. Mol. Plant 3 (3), 469–490.

Yasuda, M., Ishikawa, A., Jikumaru, Y., Seki, M., et al., 2008. Antagonistic interaction between systemic acquired resistance and the abscisic acid–mediated abiotic stress response in Arabidopsis. Plant Cell 20 (6), 1678–1692.

Yensen, N.P., 2008. Halophyte uses for the twenty-first century. In: Ecophysiology of High Salinity Tolerant Plants. Springer, Dordrecht, pp. 367–396.

Yoneyama, K., Yoneyama, K., Takeuchi, Y., Sekimoto, H., 2007a. Phosphorus deficiency in red clover promotes exudation of orobanchol, the signal for mycorrhizal symbionts and germination stimulant for root parasites. Planta 225, 1031–1038.

Yoneyama, K., Xie, X., Kusumoto, D., Sekimoto, H., et al., 2007b. Nitrogen deficiency, as well as phosphorus deficiency in sorghum, promotes the production and exudation of 5-deoxystrigol, the host recognition signal for arbuscular mycorrhizal fungi and root parasites. Planta 227, 125–132.

Yu, J.N., Huang, J., Wang, Z.N., Zhang, J.S., et al., 2007. An Na^+/H^+ antiporter gene from wheat plays an important role in stress tolerance. J. Biosci. 32 (2), 1153–1161.

Zhang, J., Jia, W., Yang, J., Ismail, A.M., 2006. Role of ABA in integrating plant responses to drought and salt stresses. Field Crop Res 97 (1), 111–119.

Zhang, H., Kim, M.S., Sun, Y., Dowd, S.E., et al., 2008. Soil bacteria confer plant salt tolerance by tissue-specific regulation of the sodium transporter HKT1. Mol. Plant Microbe Interact. 21 (6), 737–744.

Zhang, Y.F., He, L.Y., Chen, Z.J., Wang, Q.Y., et al., 2011. Characterization of ACC deaminase-producing endophytic bacteria isolated from copper-tolerant plants and their potential in promoting the growth and copper accumulation of *Brassica napus*. Chemosphere 83 (1), 57–62.

Zhou, C., Ma, Z., Zhu, L., Xiao, X., et al., 2016. Rhizobacterial strain *Bacillus megaterium* BOFC15 induces cellular polyamine changes that improve plant growth and drought resistance. Int. J. Mol. Sci. 17 (6), 976.

Biocontrol potential of plant growth-promoting rhizobacteria (PGPR) against *Ralstonia solanacearum*: Current and future prospects

7

K. Narasimha Murthy[a], *K. Soumya*[b], *A.C. Udayashankar*[a], *C. Srinivas*[c], *and Sudisha Jogaiah*[d]
[a]Department of Studies in Biotechnology, University of Mysore, Mysore, Karnataka, India, [b]Field Marshal K M Cariappa College, Constituent College of Mangalore University, Madikeri, Karnataka, India, [c]Department of Microbiology and Biotechnology, Jnanabharathi Campus, Bangalore University, Bangalore, Karnataka, India, [d]Laboratory of Plant Healthcare and Diagnostics, P.G. Department of Biotechnology and Microbiology, Karnatak University, Dharwad, Karnataka, India

7.1 Introduction

Bacterial wilt caused by *Ralstonia solanacearum* is one of the major constraints in the production of economically important crops (Hayward, 1991). It causes severe yield loss mostly in solanaceous crops as well as in other crops which are grown in tropical, subtropical, and temperate regions of the world (Devendra Kumar et al., 2017). Yield losses vary up to 91% for tomato, 10%–30% in tobacco, 33%–90% in potato, 80%–100% in banana, and up to 20% in groundnut (Yuliar et al., 2015). *R. solanacearum* is soilborne, aerobic, Gram-negative, motile with a polar flagellar tuft, non-spore, rod shaped, belongs to β-proteobacteria forming bacterial plant pathogen, and a serious threat to production of many crop plants around the world (Agrios, 2008). It has a wide host range in which approximately 450 plant species in about 50 families have been reported (Seleim et al., 2011). It causes wilt in plants by invading through xylem vessels and colonizes the xylem, prevents water movement into the upper portion of plant tissues. The blocking of pit membranes in the petioles and leaves by extracellular polysaccharide is the primary factor causing wilt (Jacobs et al., 2013). In a matter of hours, the bacteria can spread into the crown and stem through the plant's vascular system (Vasse et al., 1995).

Bacterial wilt caused by *Ralstonia solanacearum* is one of the most important bacterial diseases worldwide. The pathogen is listed in one among the "Top 10" based on scientific/economic importance plant pathogenic bacteria in molecular plant pathology worldwide (Mansfield et al., 2012). One of the reasons for this is that the

Biocontrol Agents and Secondary Metabolites. https://doi.org/10.1016/B978-0-12-822919-4.00007-7

R. solanacearum species is composed of a very large group of strains varying in their geographical origin, host range, and pathogenic behavior. This heterogeneous group is nowadays recognized as a "species complex" which has been divided into four main phylotypes (phylogenetic grouping of strains). In potato alone, it is responsible for an estimated US$1 billion in losses each year worldwide (Elphinstone, 2005). *R. solanacearum* was one of the first plant pathogen genomes to be sequenced (Salanoubat et al., 2002). The entire genome sequence necessitated the development of pathosystems with model plants, *Arabidopsis*, and *Medicago truncatula*, which has facilitated genetic and molecular studies on both the plant and bacterial partners.

To date, no effective control method has been developed for this wilt disease. Control of *R. solanacearum* is difficult once it has been established in soil due to interaction of environmental factors, aggressiveness, its broad geographical distribution, wide host range, long persistence in soil, and genetically diverse isolates (Mansfield et al., 2012; Genin and Denny, 2012). Many approaches have been developed to control bacterial wilt including breeding resistance varieties, crop rotation, chemical control, and biocontrol (Yuliar et al., 2015; Jogaiah et al., 2013). Plant breeding, field sanitation, crop rotation, and use of bactericides have met with only limited success due to high variability of the pathogen, ability of the pathogen to survive in diverse environments, and its extremely wide host range (Nguyen and Ranamukhaarachchi, 2010). One of the control methods is biocontrol, utilizing plant-associated antagonistic microorganisms (Murali et al., 2013; Safdarpour and Khodakaramian, 2018; De Britto et al., 2020).

Biological control not only suppresses the pathogen and increases the crop yield but also avoids environmental pollution due to pesticides. Since biological control is a key constituent of integrated disease management, it is important to search for plant growth-promoting rhizobacteria (PGPR) active against specific pathogens and evaluate the antagonists for wider application (Jogaiah et al., 2010; Bhat et al., 2018). Chemical pesticides may be an effective method in controlling many bacterial diseases, but these are hazardous to human health and the environment and cause the development of resistant strains (Oliveira et al., 2010). Therefore, there is a dire need to replace or at least reduce the amount of chemicals and to safeguard the environment.

Rhizobacteria have been applied to several crops to promote plant growth, seed emergence, crop yield, and control plant diseases including fungi, bacteria, and viruses (Sudisha et al., 2006; Udayashankar et al., 2009; Babu et al., 2015). PGPR are a heterogeneous group of bacteria that can be found in the rhizosphere, at root surfaces, and in association with roots, which can improve the extent or quality of plant growth directly and/or indirectly. Most of the rhizobacteria are Gram-negative rods while a very few are Gram-positive rods, cocci, or pleomorphic (Bhattacharyya and Jha, 2012). PGPR is a mixture of beneficial microorganisms which can increase the crop yield, plant growth, and also protect against soilborne pathogens (Santiago et al., 2015; Abdelwareth et al., 2012).

A number of PGPR isolated successfully against *R. solanacearum*, including *Pseudomonas fluorescens*, *Bacillus subtilis*, and *Bacillus thuringiensis*, are able to produce volatile compounds and different lytic enzymes that suppress the pathogen (Nguyen and Ranamukhaarachchi, 2010). Despite being such a varied group, only 2%–5% of rhizosphere bacteria have been found as potent PGPR (Jha et al., 2010). PGPR can protect plants against pathogens by various mechanisms such as antibiosis, siderophore production, phosphate solubilization, production of growth hormones, nitrogen fixation, secretion of antimicrobial metabolites, and niche exclusion or nutrient competition (Fig. 7.1) (Ahemad and Kibret, 2014; Babu et al., 2015). PGPR can trigger induced systemic resistance (ISR) in plants, which is phenotypically similar to the systemic acquired resistance (SAR) that occurs when plants defense mechanisms are activated in response to infection by a pathogenic agent (Babu et al., 2015; Santoyoa et al., 2016).

ISR involves jasmonate and ethylene signaling within the host and these hormones stimulate the plant's defense responses against plant pathogens (Glick, 2012; Jogaiah et al., 2018). Various reports on rhizobacteria-mediated ISR have indicated the role of common defense enzymes such as peroxidase (POX), polyphenol oxidase (PPO), phenylalanine ammonia lyase (PAL), and lipoxygenase (LOX) in induction of systemic resistance (Ramamoorthy et al., 2002; Silva et al., 2004). The beneficial effects of PGPR involve boosting key physiological processes, including water and nutrient uptake, photosynthesis, and source-sink relationships that promote growth and development (Illangumuran and Smith, 2017).

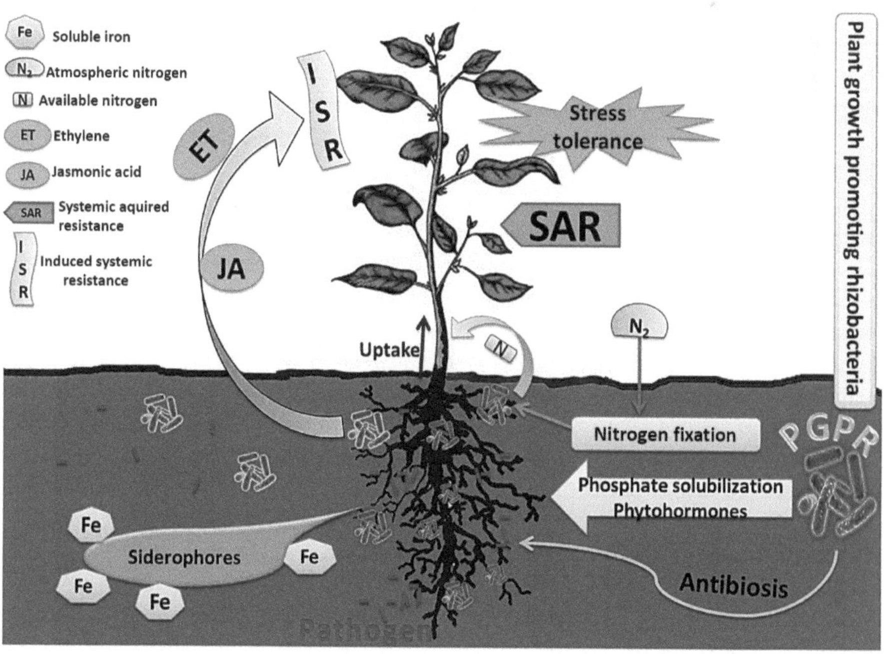

Fig. 7.1 Schematic illustration of important mechanisms of plant growth promotion by PGPR.

7.2 Mechanisms of plant growth-promoting rhizobacteria against *Ralstonia solanacearum*

7.2.1 Antibiosis

Many of the PGPR strains produce substances that are inhibitory in action against pathogenic microbial populations and suppress their growth; this phenomenon is called antibiosis (Darma et al., 2016). Antibiosis is probably the best known and perhaps the most important mechanism used by PGPR to limit the pathogen's invasion in the plant tissue (Raaijmakers and Mazzola, 2012). The production of antibiotics is the greatest effective antagonistic activity to suppress the growth of pathogens; these are organic compounds of low-molecular weight and are either derived from the ribosome or the nonribosomal peptide and/or polyketidesynthetases (NRPSs/PKS) (Bharti and Tewari, 2015). Antibiotics produced by PGPR include kanosamine, Tas A, sublancin, subtilosin, bacilysin, chlorotetain, bacillaene, Ecomycins, tensin, 2, 4-diacetylphloroglucinol (DAPG), Martínez-Viverosoligomycin A, Volatile organic compounds (VOCs), butyrolactones, Pseudomonic acid, Phenazine-1-carboxylic acid (PCA), phenazine-1-carboxamide (PCN), Pyoluteorin, Pyrrolnitrin, Oomycin A, Cepaciamide A, Viscosinamide, Butyrolactones, Zwittermycin A, Aerugine, Azomycin, Rhamnolipids, cyclic lipopeptides, Cepafungins, and viscosinamide (Bashan and de-Bashan, 2005; Goswami et al., 2016). Offensive PGPR colonization and defensive retention of rhizosphere niches are enabled by production of bacterial allele chemicals, including iron-chelating siderophores, biocidalvolatile compounds, lytic enzymes, hydrogen cyanide (HCN), and detoxification enzymes (Fig. 7.1) (Mehjabeen et al., 2017; Singh et al., 2017). Antagonistic rhizobacteria were found positive for the *phlD* gene, a key gene which is essential for synthesis DAPG, and these are able to suppress bacterial wilt disease in tomato plant under in vivo conditions in the presence of *R. solanacearum* (Ramadasappa et al., 2012).

Tomato is an important vegetable crop grown and used worldwide. Many PGPR strains have been reported worldwide to control bacterial wilt by various mechanisms. Kwon and Kim (2014) found that a major mechanism of bacterial wilt disease control in tomato by the *B. subtilis* JW-1 strain was due to the secretion of an antibacterial compound. *B. subtilis* from the rhizosphere soil of tomato as a potential biocontrol agent of bacterial wilt is caused by *R. solanacearum*. Under greenhouse conditions, seed treatment followed by soil drench application resulted in >80% reduction in bacterial wilt disease in tomato. The expression of antimicrobial *Bacillus* genes during interaction with *R. solanacearum* observed upregulation of *itu*C (coding for iturin) and *srf*AA (coding for surfactin) genes in strains Am1 and D16 (Almoneafy et al., 2014). Two tomato root colonizing strains, *B. amyloliquefaciens* CM-2 and T-5, enhanced growth of tomato seedlings and protected tomato against bacterial wilt (Tan et al., 2013).

Bacillomycin, fengycin, iturin, surfactin, bacillaene, unknown peptide, difficidin, chlorotetaine, and macrolactin have been reported as bioactive compounds produced by *B. amyloliquefaciens* (Alvarez et al., 2011). PGPR like *P. aeruginosa* T1, *Pseudomonas* sp. BH25, *Pseudomonas* sp. AM12, *Pseudomonas* sp. AM13, and *P. putida* R6 were isolated from the rhizosphere of tomato plants. Among the strains, *Pseudomonas* sp.

BH25 was promising as to the antagonistic effect on *R. solanacearum* Tom5 (Maji and Chakrabartty, 2014). The antibacterial activity exhibited by actinomycetes, which may be due to secretion of extracellular antimicrobial compounds and antibiotics diffused from the sources into the agarose medium, led to the inhibition of *R. solanacearum* (Brock, 2009). Mehjabeen et al. (2017) reported that the potential PGPR strains PR25 (*Enterobactor*) and PR9 (*Enterobactor*) strains are consistently effective in control of *R. solanacearum* wilt disease and growth promotion of tomato plant. *Pseudomonas fluorescens* PF-13 reduced incidence of bacterial wilt of chili, chickpea, and tomato and was also reported to improve plant growth (Meenakumari, 2007). Two strains of rhizobacteria, *B. amyloliquefaciens* DSBA-11 and DSBA-12, were reported to have antagonistic activity against *R. solanacearum* race 1, bv 3, phylotype I, inciting bacterial wilt of tomato under in vitro conditions, and DSBA-11 revealed maximum growth inhibition of *R. solanacearum* (Singh et al., 2016).

Xue et al. (2013) observed the different rhizobacterial antagonists against genetically diverse isolates of *R. solanacearum* and also observed that the same antagonist inhibited different *R. solanacearum* strains differentially. Even the antagonistic isolates with high homology responded differentially to the same *R. solanacearum* strain. Chakravarty and Kalita (2012) reported an indigenous strain of *P. fluorescens* in the management of bacterial wilt disease of brinjal. Carboxy methyl cellulose and substrate carrier-based formulations of *P. fluorescens* were applied by different methods in pot and field experiments to study their effectiveness in the control of bacterial wilt of brinjal. Applications performed significantly better than others, providing 83% control of bacterial wilt of brinjal in the field experiment.

Disease incidence was reduced by *Serratia* sp., fluorescent Pseudomonad, and *Bacillus* sp. compared to the control (Hua et al., 2004). Ramesh and Phadke (2012) screened rhizobacteria for their antibacterial activity against *R. solanacearum* in eggplant. They reported that *Pseudomonas* sp. and *Bacillus* sp. significantly decreased wilt incidence under greenhouse condition. Almoneafy et al. (2014) reported four *Bacillus* strains against *R. solanacearum* in liquid medium and found significant inhibition of pathogen growth. He also observed that the disruption of *R. solanacearum* cell walls resulted in severe cell lysis, and consequently cytosolic contents were also lost by cells. Tahir et al. (2016) reported that *Serratia* isolates from the potato rhizosphere showed significant antagonism against the *R. solanacearum* strain.

Five antagonistic bacterial strains, viz. *B. pseudomycoides* M3, *Brevibacillus brevis* M4, *S. maltophilia* M5, *Stenotrophomonas maltophilia* BG4, and *Streptomyces toxytricini* C5 were characterized for production of antibiotics, metabolites, and volatiles. The recovered antimicrobial compounds from the supernatant of *B. pseudomycoides* M3, *Streptomyces toxitricini* C5, and *S. maltophilia* M5 recorded a broad-spectrum antagonistic effect against *R. solanacearum* (Hassan et al., 2018). Messiha et al. (2007) reported isolates of *Stenotrophomonas maltophilia* (PD3531, PD3532, PD3533, and PD3534) from the rhizosphere of eggplant and its antagonistic effect against *R. solanacearum* race 3 biovar 2, the causal agent of potato brown rot and in vivo on potato plants. *S. maltophilia* produces various antibiotics, for example, maltophilin, a macrocyclic lactamantibiotic, which has antifungal activity, but is inactive against Gram-positive and Gram-negative bacteria (Jakobi et al., 1996).

Pseudomonas fluorescens, *P. putida*, *B. subtilis*, and *E. aerogenes* from the tomato rhizosphere were tested against *R. solanacearum* in vitro and in vivo. All PGPR isolates except *E. aerogenes* improved germination and *P. fluorescens* exhibited the highest disease reduction followed by *P. putida*, followed by *B. subtilis*. (Seleim et al., 2011). Yu et al. (2018) reported two rhizosphere-associated *B. velezensis* isolates (Y6 and F7) that have strong antagonistic activity against *R. solanacearum* under both laboratory and greenhouse conditions. They also found that lipopeptide (LP) compounds (iturin, fengycinand surfactin) production is significantly stimulated during interaction with *R. solanacearum*. *Pseudomonas fluorescens* from different potato growing areas and bacterization of potato tubers with isolates PfS2, PfWt3, and PfW1 significantly reduced the incidence of bacterial wilt and increased plant growth (Kuarabachew et al., 2007). The *B. thuringiensis* and *B. cereus* organisms are biocontrol agents capable of forming endospore and antagonistic activity against *R. solanacearum* (Raaijmakers et al., 2002). Isolates of both species were reported as effective biocontrol agents against *R. solanacearum* (Hyakumachi et al., 2013).

Tomato seed treatment with *Pseudomonas fluorescens* protected plants against soilborne infections of bacterial wilt organism. Seed treatment with the antagonistic *P. fluorescens* strain significantly improved seed germination and seedling vigor. The disease incidence was significantly reduced in plants raised from *P. fluorescens*-treated seeds followed by challenge inoculation with *R. solanacearum* (Vanitha et al., 2009). Among a total of 73 antagonists isolated from soil, eight were screened and evaluated in vitro and in vivo against *R. solanacearum*, which significantly reduced disease symptoms. Three antagonists isolated from soil, *Bacillus megaterium*, *Enterobacter cloacae*, and *Pichia guillermondii*, recorded high potential for disease suppression and also increased fruit weight, biomass, and plant height of tomato and pepper (Nguyen and Ranamukhaarachchi, 2010).

Singh et al. (2013) isolated *Bacillus* species from the rhizosphere soil of tomato plants which exhibited antagonistic capacity against *R. solanacearum* under glass house conditions with the lowest wilt incidence in Pusa Ruby and Arka Abha cultivars. *Bacillus amyloliquefaciens*, *Bacillus subtilis*, and *B. methylotrophicus* were isolated from the tomato and potato rhizosphere and examined for their antagonistic activities against *R. solanacearum* T-91, the causal agent of tomato bacterial wilt, in vitro and in vivo. In addition, they showed the ability to produce indole-3-acetic acid, siderophores and the capability of solubilizing phosphate (Almoneafy et al., 2012). Biocontrol agents from the tomato rhizosphere, viz., *P. aeruginosa*; *P. putida*, *B. subtilis*, *P. syrinagae*, *P. stutzeri*, and *B. thuringiensis*, were screened in vitro and in vivo against *R. solanacearum*. All the biocontrol agents reduced the bacterial wilt disease to variable degrees. Under greenhouse conditions, *P. aeruginosa* and *B. thuringiensis* showed the highest reduction of wilt disease followed by *P. putida*, *B. cereus*, and *P. stutzeri*, while *P. syrinagae* showed the lowest disease reduction (Makhlouf and Hamedo, 2013).

Biocontrol agent *B. amyloliquefaciens* EB13 reduced bacterial wilt in potato (Ding et al., 2013) and *B. amyloliquefaciens* SQR-7 and SQR-101 reduced bacterial wilt in tobacco (Yuan et al., 2014). Ambardar (2011) reported that the application of

Glomusmosseae (Gm1, Gm2, and Gm3), *P. fluorescens*, and *B. cereus*, separately and in combination, significantly reduced the wilt disease as compared to control in cv. Solangola and BWR-5 of tomato. Rado et al. (2015) isolated bacterial strains from rhizospheric soil samples, actinomycetes, *Pseudomonas* sp. which showed high antagonistic activity against *R. solanacearum*. Greenhouse conditions significantly reduced the percentage of wilt incidence and also improved plant growth by increase of plant height, fresh weight, and dry weight. Wydra and Semrau (2005) reported comparable *R. solanacearum* wilt disease reduction and a yield increase associated with biocontrol agents *Bacillus* sp. and *Pseudomonas fluorescens*. Elphinstone and Aley (1992) found a decline of *R. solanacearum* in the maize rhizosphere due to an increased population of *P. fluorescens*, which was antagonistic to *R. solanacearum*.

The application of *P. fluorescens* cell suspension by different treatment methods was found to be effective in controlling bacterial wilt of brinjal (Aneja, 2005). Ramesh et al. (2009) conducted the preliminary experiment in nursery, soil application of *P. fluorescens*, reduced incidence of bacterial wilt compared to control. Further, an increase of yield was observed in the *P. fluorescens*-treated plants. Chen et al. (2013a) studied the mechanisms by which strains of *B. subtilis* confer plant protection. They screened a total of 60 isolates from various locations and obtained six strains that exhibited above 50% biocontrol efficacy on tomato plants against *R. solanacearum* under greenhouse conditions.

7.2.2 Competition for root niches and nutrients

The rhizosphere is a nutrient basin which serves as an enormous range of nutrient-rich compounds. These compounds attract different microbial life forms, including phytopathogens, that compete for the available nutrients and sites or niches. For rhizospheric bacteria to claim dominance over the rest of the soil microorganisms, it must be able to compete favorably for the available nutrients and space (Lugtenberg and Kamilova, 2009). This is required to limit the incidence and severity of plant disease (Kamilova et al., 2005). Consequently, such adaptation makes the root unfit for pathogens as a result of PGPR fast colonization. Aside from the characteristic growth which PGPR obtains via competition as a result of adequate nutrient availability, other properties such as the presence of flagellium, lipopolysaccharide, chemotaxis, and the usage of secreted root exudates enhance their survival (Lugtenberg and Kamilova, 2009).

In niche competition, a physical occupation of site by PGPR is enhanced through delay tactics by preventing the colonization of pathogens until the available substrate is exhausted (Heydari and Pessarakli, 2010). Conversely, iron is one of the essential nutrients required by all microorganisms for synthesis of ATP, formation of heme, reduction of ribotide precursors of DNA, and a number of functions (Saraf et al., 2011). This feature has been an age long adaptive property exerted by beneficial soil microorganisms to occupy the root rhizosphere and make available the scarce nutrient for their upkeep (Lugtenberg et al., 2001).

7.2.3 Hydrogen cyanide (HCN) production

The production of HCN is primarily associated with *Pseudomonas* sp. (89%) (Ahmad et al., 2008). Its cyanide ion inhibits metalloenzymes, principally in copper containing cytochrome *c* oxidases. HCN produced by *Pseudomonas* strains has successfully been used to curb canker of tomato (Lanteigne et al., 2012). Cyanide occurs in solution as free cyanide, which includes the cyanide anion (CN^-) and the nondissociated HCN. Cyanide is a phytotoxic agent capable of preventing enzymes involved in major metabolic processes and is considered one of the typical features of deleterious rhizobacterial isolates (Bakker and Schippers, 1987). However, because of the aggressive colonizing strength of *P. fluorescens,* it has successfully been used in the control of soilborne plant pathogens (Lugtenberg et al., 2001). The *Pseudomonas protegens* RS-9 from the rhizosphere of a healthy tomato plant in a bacterial wilt diseased field strongly inhibited the growth of bacterial wilt pathogen *R. solanacearum.* This strain showed potential inhibition of different phytopathogenic fungi and bacteria in vitro and could produce DAPG, HCN, pyrrolnitrin, and pyoluteorin, and these main compounds were responsible for the suppression of *R. solanacearum* (Rai et al., 2017). Saber et al. (2015) reported that HCN is a broad-spectrum antimicrobial compound involved in biological control of root diseases by many plant-associated bacteria.

Pseudomonas brassicacearum J12 isolated from the rhizosphere soil of tomato, which could produce 2, 4-diacetylphloroglucinol (2, 4-DAPG), HCN, siderophore (s), and protease, strongly inhibited the growth of phytopathogenic bacteria *R. solanacearum.* Strain J12 significantly suppressed bacteria wilt by 46% in the greenhouse experiment (Zhou et al., 2012). *P. fluorescens* isolates are able to produce siderophore, HCN, and to dissolve phosphate. Higher HCN was produced by *P. fluorescens* Stj3 and *P. fluorescens* Stj8 from a tomato diseased plant grown in severely infected soil by *R. solanacearum* as well as *P. fluorescens* Stj9 and *P. fluorescens* Stj1 from the healthy tomato plant grown in infected soil by *R. solanacearum* (Nasril et al., 2018).

7.2.4 Siderophores production

Iron is essential for plant growth, development, and is required as a cofactor for proteins that are involved in a number of important metabolic processes including photosynthesis and respiration. It is an important micronutrient required by microbes and being highly insoluble is often a limiting condition in the rhizosphere (Fig. 7.1). It is present in different complexes, each having different solubilities in a natural system; therefore, the bioavailability of iron depends on the potential of siderophores to chelate the iron from its complexes (Hersman et al., 2001). Iron deficiency is a worldwide problem in crop production on calcareous soils of many crops. Under iron-limiting conditions, PGPR produce low-molecular-weight (usually below 1 kDa) compounds called siderophores to competitively acquire ferric ion. Siderophores act as determinants of ISR under iron starved conditions. It plays a key role in electron transport, oxidation–reduction reactions, detoxification of oxygen radicals, synthesis of DNA precursors, and in many other biochemical processes (Hider and Kong, 2010).

Microbes release siderophores to scavenge iron from these mineral phases by formation of soluble $Fe3^{+}$ complexes that can be taken up by active transport mechanisms (Saharan and Nehra, 2011).

Siderophore-producing rhizobacteria can prevent the proliferation of pathogenic microorganisms around the root (Vejan et al., 2016). These siderophores possess high binding affinity and specificity for iron and help to facilitate its bioavailability into the biological cell (Schalk et al., 2011). Plants take up iron produced by a large number of PGPR, including *Pseudomonas* sp. (Sulochana et al., 2014), *Aeromonas* sp. (Sayyed et al., 2013), *Bacillus* sp. (Jikare and Chavan, 2013), *Azotobacter* sp. (Ahmad et al., 2008), *Burkholderia* sp. (Jiang et al., 2008), *Rhizobium* sp. (Bano and Musarrat, 2003), and *Serratia* sp. (Selvakumar et al., 2008), as confirmed by using radiolabeled ferric siderophores as a sole source of Fc (Gupta et al., 2015). Not only iron, siderophores can also form stable complexes with other heavy metals that are of environmental concern, such as Al, Cd, Cu, Ga, In, Pb, and Zn as well as with radionuclides including U and Np (Schalk et al., 2011). Robin et al. (2008), using the iron-siderophore complex radioactive as the only source of iron, showed that plants are able to absorb the radioactive iron. Binding of the siderophore to a metal increases the soluble metal concentration (Rajkumar et al., 2010). Hence, bacterial siderophores help to alleviate the stresses imposed on plants by high soil levels of heavy metals.

The antagonistic mechanism of *P. fluorescens* to produce siderophores against *R. solanacearum* is competing with the Fe element and inhibiting the growth of pathogen by excreting pseudobactin (Budzikiewicz, 2001). Also, siderophore is responsible for dissolving insoluble-phase minerals such as phosphate to be utilized by plants (Ryder et al., 1994). The antibacterial efficiency of *P. fluorescens* depends at least partially on the production of siderophore pseudobactin, which can efficiently form complexes with iron in soils, making it unavailable to *R. solanacearum*, thus inhibiting their growth (Ran et al., 2005). Jurkevitch et al. (1992) observed one isolate of *S. maltophilia* that was able to utilize Fe^{3+} in the siderophore pseudobactin as the sole iron source, but only to a limited extent. Sayyed et al. (2013) reported in vitro phytopathogen suppression activity of siderophoregenic culture and supernatant of *Alcaligenes* sp., *Acinetobacter* sp., and *Kitazin* showed positive response by inhibiting the growth of *R. solanacearum. Alcaligenes* sp. excreted the highest amount of siderophores, while *Acinetobacter* sp. produced a lesser amount of siderophores. Biocontrol characters of the selected *B. subtilis* CIFT-MFB-4158A produced siderophore and antibiotic for growth inhibition of *R. solanacearum*. In agar well diffusion assays, two *Bacillus* strains inhibited growth of *R. solanacearum*, namely *B. cereus* C38/15 and *B. subtilis* CIFT-MFB-4158A (Yanti et al., 2017).

Pseudomonas putida WCS358r, *P. fluorescens* WCS374r, *P. fluorescens* WCS417r, and *P. aeruginosa* 7NSK2 antagonized *R. solanacearum* in vitro by siderophore-mediated competition for iron, whereas inhibition of pathogen growth by *P. fluorescens* CHA0r is antibiosis in Eucalyptus (Ran et al., 2004). The isolate *P. fluorescens*-Stj11 derived from the healthy tomato plant grown in soil infested with *R. solanacearum* produced the highest siderophore (Nasril et al., 2018). Bacillus isolates from the potato rhizosphere might have such an activity which checked the growth of *R. solanacearum* in liquid media and also seen the production of siderophores which have possibly played a role in *R. solanacearum* inhibition (Tahir et al., 2016). Three antagonistic isolates namely, *B. pseudomycoides*, *Brevibacillusbrevis,* and *S. maltophilia* were evaluated for HCN and siderophore production. Among them *B. pseudomycoides* and *S. maltophilia* revealed maximum HCN production, while maximum siderophore production was noticed in *Brevibacillusbrevis* and *S. maltophilia* (Hassan et al., 2018). Under laboratory conditions, most of the PGPR isolates produced either one or more inhibitory compounds and siderophores (Phadke, 2007). A combination of different bacterial agents with complementary biological control activities reduced the bacterial wilt incidence in potato (Shen et al., 2006) and in capsicum (Ge et al., 2004). Ramesh et al. (2009) found that most of the selected pseudomonads produced an antibiotic, DAPG, which inhibited *R. solanacearum* under in vivo conditions. They also recorded production of siderophores and IAA in culture medium by antagonists, which could be involved in biocontrol and growth promotion in crop plants. Most of the Pseudomonads which inhibited *R. solanacearum* also recorded production of siderophores and IAA in the culture medium by antagonists, which could be involved in biocontrol and growth promotion in crop plants (Ramesh et al., 2009).

7.2.5 Nitrogen fixation

Nitrogen is the most essential nutrient and its deficiency severely affects crop yields. It is necessary for all living organisms. Although 78% of the atmosphere consists of nitrogen, it cannot be used by most organisms, and consequently the availability of nitrogen in a form suitable for assimilation is often a major limiting factor for crop growth (Khan et al., 2009). Plant growth-promoting rhizobacteria that fix nitrogen in nonleguminous plants are also called diazotrophs, which are capable of forming a nonobligate interaction with the host plants (Glick et al., 1999). Symbiotic bacteria which act as PGPR are *Bradyrhizobium, Rhizobium, Sinorhizobium,* and *Mesorhizobium* with leguminous plants, *Frankia* with nonleguminous trees and shrubs (Zahran, 2001). Nonsymbiotic nitrogen-fixing rhizospheric bacteria belong to genera including *Azoarcus, Azotobacter, Acetobacter, Azospirillum, Burkholderia, Diazotrophicus, Enterobacter, Gluconacetobacter,* Pantoeaagglomerans, *Klebsiella pneumoniae, Pseudomonas, Cyanobacteria, Anabaena,* and *Nostoc* (Ahemad and Kibret, 2014). The nonsymbiotic free-living nitrogen fixers without penetrating into the host they survive around the roots and helps in contribution of nitrogen to the host plants (Goswami et al., 2016).

In symbiotic nitrogen fixation, legume crops undergo biological nitrogen fixation through symbiotic association with bacteria and meet their own needs without depending on external sources (Bhattacharyya and Jha, 2012; Gopalakrishnan et al., 2015). The process of nitrogen fixation is carried out by a complex enzyme, the nitrogenase complex (Kim and Rees, 1994). Most biological nitrogen fixation is carried out by the activity of the molybdenum nitrogenase, which is found in all diazotrophs (Bishop and Jorerger, 1990), and this enzyme complex has two component proteins encoded by the nifDK and nifH genes (Rubio and Ludden, 2008).

PGPR strains play a major role in nitrogen fixation and make it in an assimilable form for plants. Nitrogenase (*nif*) genes required for nitrogen fixation in nitrogen-fixing bacteria are more complex. The genes for nitrogen fixation, called *nif* genes, are found in both symbiotic and free-living systems (Kim and Rees, 1994). So, for improving this process, genetic strategies have been utilized to modify the genes (Souza et al., 2015). Since nitrogen fixation is a very energy demanding process, requiring at least 16 mol of ATP for each mole of reduced nitrogen, it would be beneficial if bacterial carbon resources were directed toward oxidative phosphorylation, which results in the synthesis of ATP, rather than glycogen synthesis, which results in the storage of energy in the form of glycogen (Glick, 2012). Several species of microorganisms are used in cultivation of plants of economic interest, facilitating the host plant growth without the use of nitrogenous fertilizers. *Arthrobacter* sp. can be used as a biocontrol agent, suggesting that bacterial strain SJN5 with nitrogen fixation and antagonistic properties against *R. solanacearum* can not only increase plant growth but also can inhibit plant diseases (Zhang et al., 2018).

7.2.6 Cell wall degrading enzymes

One of the main mechanisms used by biocontrol agents to control soilborne pathogens involves the production of cell wall degrading enzymes (Kobayashi et al., 2002). It is one of the important mechanisms for environment friendly control of soilborne pathogen (Sudisha et al., 2012). The growth and activities of pathogens can be suppressed by the secretion of lytic enzymes. These are cell wall degrading enzymes such as β-1,3-glucanase, chitinase, cellulase, lipases, phosphatases, proteases, hydrolases, *exo-* and *endo*-polygalacturonases, and pectinolyases secreted by several PGPR strains, which involve a direct inhibitory effect on the hyphal growth of fungal pathogens by lysis of the pathogenic bacteria cell walls (Joshi et al., 2012; Reddy, 2013).

Extracellular hydrolytic enzymes contribute directly in the parasitization of phytopathogens and rescue plants from biotic stresses (Maksimov et al., 2011). Cell wall degrading enzymes of rhizobacteria affect the structural integrity of walls of the target pathogen (Budi et al., 2000). Chitinase activity of PGPR has been well explored for suppression of different phytopathogens (Kim et al., 2008). Potential biocontrol agents with chitinolytic activities include *B. licheniformis, B. cereus, B. circulans, B. subtilis*, and *B. thuringiensis* (Sadfi et al., 2001). Among the Gram-negative bacteria, *Serratia marcescens, Enterobacter agglomerans, P. aeruginosa,* and *P. fluorescens* have been found to possess chitinolytic activities (Neiendam-Nielsen and Sørensen, 1999). The antagonistic potential of PGPR through the synthesis of β-1, 3 glucanase, chitinase, cellulase at low concentration, and proteolytic enzymes even as *Pseudomonas* sp. has been established to be a good candidate (Cattelan et al., 1999). The β-glucanases can act via two possible mechanisms; Exo-β-glucanases hydrolyze the β-glucan chain by sequentially cleaving glucose residues from the nonreducing end. Endo-β-glucanases cleave β-linkages at random sites along the polysaccharide chain, releasing smaller oligosaccharides. As multifunctional organic protein, these enzymes form protection from desiccation and against abiotic and climatic factors (Qurashi and Sabri, 2012).

The influence of PGPR in such a mechanism includes the production of hydrolytic enzymes, production of various antibiotics in response to plant pathogen or disease resistance, and so forth (Nivya, 2015; Gupta et al., 2014). Nakayama et al. (1999) reported that biocontrol agents *E. Gergoviae* EB87 and *S. maltophilia* EB35. *S. maltophilia* produce various antibiotics, such as maltophilin (Jakobi et al., 1996), xanthobaccin A, B, and C. *S. maltophilia* produces lytic enzymes that have been implicated in biological control activity against *R. solanacearum;* extracellular proteolytic enzyme may be active against *R. solanacearum* (Dunne et al., 2000). The suppression of *R. solanacearum* by *Bacillus* sp. is probably due to its capability to produce lytic enzymes like protease and synthesis of antibiotics (Mazurier et al., 2009). The *hrpB* and *PhcA* are regulatory pathways which contribute to the virulence of *R. solanacearum* and secretion of cell wall degrading enzymes via T2SS and the secretion of effector proteins via the TTSS pathway into the host cell (Hikichi et al., 2007).

7.2.7 Phosphate solubilization

Phosphorus is another important element next to nitrogen affecting growth and development of plants. Phosphorus plays a major role in almost all metabolic processes of plant such as photosynthesis, respiration, signal transduction, storage and transfer of energy, and cell division and elongation (Sagervanshi et al., 2012). Phosphorus exists in both inorganic (bound, fixed, or labile) and organic (bound) forms. It is essential for seed formation which contains the highest phosphorus content of the plant.

Approximately 95%–99% of phosphorus in soil is present in insoluble, immobilized, or precipitated form, rapid complexation with cations such as Ca, Fe, or Al; therefore, it is difficult for plants to absorb phosphorus (Jha and Saraf, 2015). Various soil microorganisms were stated to solubilize insoluble phosphorous complexes into solution and make it possible for its use by the plant (Tripura et al., 2005). Microbial solubilization of inorganic phosphate compounds is of great economic significance in plant nutrition (Sharma et al., 2013). Several PGPR involved especially in transforming insoluble phosphorus to soluble forms and quite often referred to phosphate solubilizing microorganisms (PSM) provide phosphorus to plants (Gopalakrishnan et al., 2015; Zheng et al., 2018). PGPR directly solubilize and mineralize inorganic phosphorus or facilitate the mobility of organic phosphorus through microbial turnover and/or increase the root system (Richardson and Simpson, 2011). Solubilization and mineralization of phosphorus by phosphate solubilizing bacteria is an important trait that can be attained by PGPR. Phosphate solubilizing PGPR are *Arthrobacter, Bacillus, Beijerinckia, Burkholderia, Enterobacter, Microbacterium, Pseudomonas, Erwinia, Rhizobium, Mesorhizobium, Flavobacterium, Rhodococcus,* and *Serratia* (Bhattacharyya and Jha, 2012; Oteino et al., 2015; Tensingh Baliah, 2018).

Phosphate solubilizing bacteria make the solubilization effect through the production of organic acids such as lactic acid, propionic acid, glycolic acid, 2-ketogluconic, oxalic acid, formic acid, fumaric acid, succinic acid, citric acid, gluconic acid, acetic acid, and malic acid (Zaidi et al., 2009). The primary mechanism of phosphate solubilization is based on organic acid secretion by PGPR because of sugar metabolism, which lowers the pH in the rhizosphere and thus releases the phosphorus available to plants (Kaur et al., 2016; Patel et al., 2015). PSB are also able to mineralize the insoluble organic phosphate through the secretion of extracellular enzymes such as phosphatases, phytases, and C-P lyases (Weyens et al., 2010).

Plant growth-promoting rhizobacteria phlD$^+$ isolate showed significantly fair plant growth-promoting activities like phosphate solubilization, produced indole acetic acid, and a few strains showed production of HCN and siderophore (Ramadasappa et al., 2012). Strains of PGPR *B. amyloliquefaciens* have better phosphorus solubilizing ability production under in vitro conditions (Singh et al., 2016). Potato rhizosphere isolates, *Bacillus* sp., *Pseudomonas* sp., and *Serratia* sp., recorded antagonistic activity against *R. solanacearum* isolates. These isolates recorded plant growth-promoting (PGP) traits, i.e., indole acetic acid (IAA), siderophore production, P-solubilization, root colonization, and chitinase production (Tahir et al., 2016). Application of PGPR as potential bioagents in controlling tomato bacterial wilt under field condition and *R. solanacearum* wilt disease reduction yields an increase of tomato plants after treatment by *Bacillus* sp. and *Pseudomonas fluorescens* (Guo et al., 2004). The capability of PGPR to solubilize potassium rock by generating and secreting organic acids has also been broadly studied. Potassium solubilizing PGPR, such as *Bacillus* sp., *B. edaphicus*, *B. mucilaginosus*, *Ferrooxidans* sp., *Pseudomonas* sp., *Burkholderia* sp., and *Paenibacillus* sp., have been described to release potassium in accessible form from potassium-bearing minerals in soils (Kumar et al., 2018). The highest ability to dissolve phosphate was by the isolate *P. fluorescens*-Stj4 from tomato diseased plant grown in severely infected soil by *R. solanacearum* (Nasril et al., 2018).

7.2.8 Phytohormone production

Chemicals occurring naturally inside plant tissues have a controlling, rather than a nutritional, role in growth and development. These compounds, which are usually active at very low concentrations, are known as phytohormones or plant growth substances (George et al., 2008). Phytohormones are the substances that regulate the growth, development, and physiology of plants. Generally, phytohormones affect cell enlargement, cell division, and cell extension in roots (Glick, 2014). PGPR are reported to be associated with production of phytohormones. PGPR secrete phytohormones such as Indole-3-acetic acid (IAA) (Majeed et al., 2015), cytokinins (Goswami et al., 2015), gibberellins (GBs) (Pandya and Desai, 2014), abscisic acid (ABA) (Porcel et al., 2014), and ethylene (Patten and Glick, 1996). Out of these, the production of auxin and ethylene is a very common trait among PGPR (Mishra et al., 2010).

Indole-3-acetic acid (IAA) is one of the most physiologically active auxins. It is secreted by 80% of the rhizosphere microbial populations (Patten and Glick, 1996). It has been established that enhanced root proliferation is related to bacterial IAA biosynthesis. IAA is synthesized by plant-associated microbes via L-tryptophan-dependent and independent pathways and three L-tryptophan-dependent pathways. Most of these PGPR utilize L-tryptophan, which is secreted in root exudates as a precursor for IAA production.

Upon application of plants with PGPR, a change in root architecture is observed, mainly as an increase in root hairs, lateral roots, and shortening of the root length (Glick, 2012). The plants exposed to IAA for a long time have an extremely developed root system, which in turn provides the plants greater access to nutrients. This in turn allows the plant to absorb more nutrients and hence aids in the overall growth of the plants (Aeron et al., 2011). Furthermore, downregulation of IAA as signaling is associated with the plant defense mechanisms against a number of phytopathogenic bacteria as evidenced in the enhanced susceptibility of plants to the bacterial pathogen by exogenous application of IAA or IAA produced by the pathogen (Spaepen and Vanderleyden, 2011). Different PGPR possess different routes for the synthesis of IAA. Strains of PGPR *Bacillus amyloliquefaciens* have better IAA production in in vitro conditions (Singh et al., 2016).

Cytokinins are adenine derivative phytohormones that are secreted by PGPR (Goswami et al., 2015) and are similar in role to IAA; cytokinins when applied exogenously to plants are known to control enhanced root development, enhance cell division, enhance root hair formation, shoot growth as well as branching, control of apical dominance in shoot, inhibition of root elongation, or certain other physiological responses (Jha and Saraf, 2015). Additional processes such as progressive processes, such as the formation of nutritional signaling, branching, chlorophyll production, embryo vasculature, root growth, leaf expansion and promotion of seed germination, and delay of senescence, are also heavily influenced by cytokinins (Wong et al., 2015).

Gibberellins are yet another group of essential plant hormones which influence various developmental processes such as seed germination, flowering, stem elongation, and fruit setting (Hedden and Phillips, 2000). Plant growth promotion by gibberellin producing plant growth-promoting bacteria and this progressive effect on plant biomass are regularly connected with an increased content of gibberellins in plant tissues, which was described by numerous researchers (Kang et al., 2010). The production of gibberellins by rhizobacterial strains is rare. Among *Bacillus* sp., only two strains have been documented that are capable of producing gibberellins: *B. licheniformis* and *B. pumilus* (Gutierrez-Manero et al., 2001).

Ethylene, also known as stress hormone, is a key phytohormone having a huge range of biological functions including plant growth and development. PGPR plays a key role in reduction of ethylene level which is necessary for growth of the plant at higher concentration; it induces defoliation and other cellular processes that show a negative effect on the plant's health (Bhattacharyya and Jha, 2012). It promotes root initiation, reduces wilting, inhibits root elongation, stimulates seed germination, enhances fruit ripening, and activates the production of other plant hormones (Glick et al., 2007). The enzyme-1-aminocyclopropane-1-carboxylate (ACC) deaminase is a prerequisite for ethylene production, catalyzed by ACC oxidase. A variety of PGPR produce ACC deaminase that cleaves ACC into ammonia, and α-ketobutyrate, inhibiting its transition to ethylene and which is the precursor for ethylene, is secreted into the rhizosphere and is re-adsorbed by roots, where it is converted into ethylene (Chen et al., 2013b). This accumulation of ethylene leads to a descending spiral effect, as deprived root development leads to a weakened ability to acquire water and nutrients, which in turn leads to further stress (Martinez-Viveros et al., 2010).

Production of ACC deaminase by PGPR protects plants against detrimental effects of ethylene, surviving under abiotic stresses such as water flooding, ultraviolet radiations, temperature, and heavy metals (Glick, 2014). Bacterial synthesizing ACC deaminase belongs to genera *Acinetobacter*, *Pseudomonas*, *Bacillus*, *Rhizobium*, *Azospirillum*, *Enterobacter*, *Achromobacter*, *Agrobacterium*, *Burkholderia*, *Alcaligenes*, *Serratia*, etc. (Gupta et al., 2015). However, this gaseous hormone also regulates root initiation, fruit ripening, seed germination, leaf abscission, and wilting (Kaur et al., 2016). The result of the degradation is the reduction of plant ethylene production through a range of mechanisms, while the PGPR producing ACC-deaminase regulates the ethylene level in plant and prevents the growth inhibition caused by high levels of ethylene (Noumavo et al., 2016).

Abscisic acid (ABA) plays a key role in a water stressed environment, such as is found in arid and semiarid climates, where it helps in combating the stress through stomatal closure of leaves. During water scarcity, plant cytokinin content reduces drastically with a resultant positive increase in ABA concentration. Therefore, its uptake and transport in plant and its presence in the rhizosphere could be extremely important for plant growth under water stress conditions (Frankenberger and Arshad, 1995). Assessing the production of plants hormones by different *Streptomyces* strains in broth medium shows that all strains synthesized cytokinins and gibberellins. Though this is vital for phyto development, its receptor gene in plants is often regulated by changes in osmotic conditions (Merchan et al., 2007). The role of ABA is known to be significant in biotic stress responses as increasing evidence suggests that ABA is significantly intricate in the interactions between plants and pathogens (Thaler and Bostock, 2004). ABA seems to influence biotic stress responses not only by interfering with defense signaling regulated by SA, JA, and ET, but also through shared components of stress signaling (Mauch-Mani and Mauch, 2005).

7.2.9 Induced systemic resistance (ISR)

Plant beneficial bacteria interact with plants in the rhizosphere to stimulate a defense response against a number of pathogens (Van Loon, 2007). This enhanced state of defensive ability is termed as "Induced Systemic Resistance" (ISR). The ISR utilizes organic acid and plant hormones (salicylic acid, jasmonic acid, and ethylene) in signaling and stimulation of host plant defense response against a variety of phytopathogens (Pieterse et al., 2014). PGPR response to ISR is usually felt by improved physical and mechanical strength of cell wall as well as adjustment of biochemical and physical reaction to environmental pressure (Labuschagne et al., 2010). ISR in PGPR can be in the form of salicylic acid, production of siderophores, lipopolysaccharide, flagella, *N*-acyl homoserine lactone molecules (Van Loon, 2007), and antibiotics.

Systemic acquired resistance (SAR) and ISR are two forms of induced resistance which can be differentiated on the basis of nature of elicitor and regulatory pathways involved (Choudhary et al., 2007). It is similar to pathogen induced SAR in plants as both render uninfected plant parts more resistant to a broad spectrum of pathogens. Yet, there are certain differences and similarities in SAR and ISR signaling pathways (Pieterse and Van Wees, 2015). SAR is commonly triggered by exposing the plant to virulent, avirulent, and nonpathogenic microbes and involves accumulation of pathogenesis-related proteins (chitinase and glucanase) and salicylic acid (Abramovitch et al., 2006), but it can also be induced by biological (nonmicrobial) and chemical compounds. SAR occurs over the salicylic acid (SA)-mediated signaling pathway while ISR requires jasmonic acid (JA) and ethylene (ET) for its signaling mechanism (Fu and Dong, 2013). ISR does not involve the accumulation of pathogenesis-related proteins or salicylic acid, but instead relies on pathways regulated by jasmonate and ethylene, and these hormones stimulate the host plant's defense responses against a variety of plant pathogens (Glick, 2012).

Tomato roots treated with cell-free filtrate of *B. thuringiensis* systemically suppressed bacterial wilt caused by *R. solanacearum* through systemic activation of the plant defense system (Hyakumachi et al., 2013). Bacterial wilt suppression in tomato plants was due to expression of defense-related genes such as PR-1, acidic chitinase, and β-1, 3-glucanase.

Ran et al. (2005) investigated the ability of selected strains of *Pseudomonas* sp. to induce resistance in *Eucalyptus urophylla* against *R. solanacearum.* Strains of *P. putida* WCS358r and *P. fluorescens* WCS374r activated ISR when infiltrated into two lower leaves 3–7 days before challenge with *R. solanacearum.* A mutant of strain WCS358r, defective in the biosynthesis of the fluorescent siderophore pseudobactin, did not induce resistance. The biocontrol efficiency of *B. subtilis* 4812 and *B. methylotrophicus* H8 was studied separately or in combination with two plant defense inducers, Acetyl salicylic acid (ASA) and Beta-aminobutyric acid (BABA), against tomato wilt caused by *R. solanacearum.* The combined application of *Bacillus* strains was more effective against tomato bacterial wilt (Almoneafy et al., 2013). Hyakumachi et al. (2013) recently revealed that *B. thuringiensis* induced defense-related genes, such as PR-1, acidic chitinase, and β-1, 3-glucanase, and showed resistance against *R. solanacearum.*

B. vallismortis strain EXTN-1 rhizobacteria were found to be the best in defense against bacterial wilt and disease suppression exhibited when the plants were pretreated with *B. vallismortis* EXTN-1 (Kyungseok Park et al., 2007). Jetiyanon (2007) and Kurabachew et al. (2007) observed a decrease in bacterial wilt incidence in tomato and potato plants through application of *Bacillus* strains (IN937a and IN937b) and *P. fluorescens*, respectively. Three diazotrophic rhizobacterial strains N10, SJN3, and SJN5 from the rhizosphere soil of ginger root induced resistance to *R. solanacearum* and have high inhibitory activity against *R. solanacearum* (Zhang et al., 2018). Li et al. (2008) used antagonistic bacterium *Bacillus subtilis* AR12 to control bacterial wilt of tomato in greenhouse and they found the biocontrol effectiveness to be as high as 90%. *P. fluorescens* isolated from the rhizosphere of tomato was screened against *R. solanacearum. P. fluorescens* isolates induced a significant increase in activities of peroxidase, polyphenol oxidase, phenylalanine ammonia lyase and β-1, 3-glucanase, and increase in enzyme activity challenged with

R. solanacearum (Murthy et al., 2014). Seleim et al. (2011) reported the effect of rhizobacteria to control bacterial wilt of tomato caused by *R. solanacearum* under greenhouse and field conditions and the effect of these rhizobacterial isolates in induction of POX and PPO activity. Under greenhouse condition, *P. putida* and *P. fluorescens* reduced wilt disease and increased activity of POX and PPO in tomato plants significantly (Patel et al., 2015).

Abo-Elyousr et al. (2012) showed that treatment of acibenzolar-*S*-methyl (ASM) and *P. fluorescens* (Pf2), alone or in combination, controls the tomato bacterial wilt and significant changes in the activities of PPO, glucosidase, and POX in tomato after the application of ASM and Pf2 and with *R. solanacearum*. The biocontrol agent lactic acid bacterium (LAB) from the tomato rhizosphere was used against the bacterial wilt caused by *R. solanacearum*. The LAB-treated seeds showed an increase in the germination percentage and seedling vigor index compared with control and induced the production of defense-related enzymes in tomato plants (Narasimhamurthy et al., 2016). Tomato plants treated with *P. mossellii* FS67, *P. fluorescens* FS167 significantly reduced bacteria wilt in greenhouse and increased activity of POX, PPO, and phenolics in plants treated with FS67 and FS167 strains (Safdarpour and Khodakaramian, 2018). *B. subtilis* GB03 and *B. amyloquefaciens* IN937a were able to promote plant growth indirectly through ISR. This happens through secretion of volatiles, which in turn activate an ISR pathway in potato seedlings challenged with the *R. solanacearum* by *B. subtilis* PFMRI (Ryan et al., 2001). Two PGPR strains, *Paenibacillus polymyxas* (YC0573 and YC0136), isolated from tobacco rhizospheric soil, significantly antagonized *R. solanacearum*, the tobacco bacterial wilt pathogen under greenhouse conditions. These PGPR strains can induce systemic resistance of tobacco as well as increase the activities of PAL, PPO, POX, SOD, and CAT enzymes (Wu et al., 2011).

7.3 Conclusion

PGPR in the rhizosphere soil is highly dynamic, more versatile in transforming, mobilizing, and solubilizing the nutrients. They improve plant growth, yield of different crops, and varieties. PGPR may be broadly used in plant growth promotion as it acts as a plant nourishment and enhancement source which would replenish the nutrient cycle between the soil and plant roots, exhibits detoxifying potential, controls phytopathogens, thereby exerting a positive influence on crop productivity and ecosystem functioning. PGPR prevent plant pathogens growth through the production of antagonistic compounds, induction of systemic resistance and nutrients, and space competition. They are eco-friendly, cost-effective, and nonhazardous and are the best alternatives to use of chemical fertilizers and pesticides. Also, they keep the soil health for sustainable use. Implementation of PGPR-based biopesticides to combat different plant pathogens and increase plant growth may substantially contribute to sustainable agriculture and a safe environment.

References

Abdelwareth, A., Almoneafy, G., Xie, L., Tian, W.X., Xu, L.H., Zhang, G.Q., Ibrahim, M., 2012. Characterization and evaluation of *Bacillus* isolates for their potential growth and biocontrol activities against tomato bacterial wilt. African J. Biotechnol. 11 (82), 7193–7201.

Abo-Elyousr, K.M., Ibrahim, Y.E., Balabel, N.M., 2012. Induction of disease defensive enzymes in response to treatment with acibenzolar-S-methyl (ASM) and *Pseudomonas fluorescens* Pf2 and inoculation with *Ralstonia solanacearum* race 3, biovar 2 (phylotype II). J. Phytopathol., 1–8.

Abramovitch, R.B., Anderson, J.C., Martin, G.M., 2006. Bacterial elicitation and evasion of plant innate immunity. Nat. Rev. Mol. Cell Biol. 7, 601–611.

Aeron, A., Pandey, P., Kumar, S., Maheshwari, D.K., 2011. Emerging role of plant growth promoting rhizobacteria. In: Maheshwari, D.K. (Ed.), Bacteria in Agrobiology: Crop Ecosystem. Springer Verlag, Berlin/Heidelberg, pp. 1–26.

Agrios, G.N., 2008. Plant Pathology. Elsevier Academic Press, Amsterdam, pp. 647–953.

Ahmad, F., Ahmad, I., Khan, M.S., 2008. Screening of free living rhizospheric bacteria for their multiple plant growth promoting activities. Microbiol. Res. 163, 173–181.

Almoneafy, A.A., Xie, G.L., Tian, W.X., Xu, L.H., Zhang, G.H., 2012. Characterization and evaluation of *Bacillus* isolates for their potential plant growth and biocontrol activities against tomato bacterial wilt. African J. Biotechnol. 11 (28), 7193–7201.

Almoneafy, A.A., Ojaghian, M.R., Xu, S., Ibrahim, M., Xie, G., Shi, Y., Tian, W., Li, B., 2013. Synergistic effect of acetyl salicylic acid and DL-Beta amino butyric acid on biocontrol efficacy of *Bacillus* strains against tomato bacterial wilt. Trop. Plant Pathol. 38 (2), 102–113.

Almoneafy, A.A., Kakar, K.U., Nawaz, Z., Li, B., Lan, Y.C., Xie, G.L., 2014. Tomato plant growth promotion and antibacterial related-mechanisms of four rhizobacterial *Bacillus* strains against *Ralstonia solanacearum*. Symbiosis 63, 59–70.

Alvarez, F., Castro, M., Principe, A., Borioli, G., Fisher, F., Mori, G., Jofre, E., 2011. The plant-associated *Bacillus amyloliquefaciens* strain MEP218 and ARP23 capable of producing the cyclic lipopeptides iturin or surfactin and fengycin are effective in biocontrol of sclerotinia stem rot diseases. J. Appl. Microbiol. 112, 159–174.

Ambardar, V.K., 2011. Potential of endomycorrhizae and bacterial antagonist on the growth performance of tomato cultivars against bacterial wilt. J. Res. Dev. 11, 1–12.

Ahemad, M., Kibret, M., 2014. Mechanisms and applications of plant growth promoting rhizobacteria: current perspective. J. King Saud Univ. Sci. 26, 1–20.

Aneja, K.R., 2005. Experiments in Microbiology, Plant Pathology and Biotechnology, fourth ed. New Age International Ltd. Publishers, New Delhi, pp. 278–282.

Babu, A.N., Jogaiah, S., Ito, S., Nagaraj, A.K., Tran, L.P., 2015. Improvement of growth, fruit weight and early blight disease protection of tomato plants by rhizosphere bacteria is correlated with their beneficial traits and induced biosynthesis of antioxidant peroxidase and polyphenol oxidase. Plant Sci. 231, 62–73.

Bakker, A.W., Schippers, B., 1987. Microbial cyanide production in the rhizosphere in relation to potato yield reduction and *Pseudomonas* spp. mediated plant growth stimulation. Soil Biol. Biochem. 19, 451–457.

Bano, N., Musarrat, J., 2003. Isolation and characterization of phorate degrading soil bacteria of environmental and agronomic significance. Lett. Appl. Microbiol. 36, 349–353.

Bashan, Y., de-Bashan, L.E., 2005. Bacteria: plant growth promoting. In: Encyclopedia of Soils in the Environment. vol. 1. Elsevier, Oxford, U.K., pp. 103–115. Hillel, D. Editor-in-Chief.

Bharti, P., Tewari, R., 2015. Purification and structural characterization of a phthalate antibiotic from *Burkholderia gladioli* OR1 effective against multi-drug resistant *Staphylococcus aureus*. J. Microb. Biotech. Food Sci. 5, 207.

Bhat, M.A., Rasool, R., Ramzan, S., 2018. Plant growth promoting Rhizobacteria (PGPR) for sustainable and eco-friendly agriculture. Acta Sci. Agric. 3 (1), 23–25.

Bhattacharyya, P.N., Jha, D.K., 2012. Plant growth-promoting rhizobacteria (PGPR): emergence in agriculture. World J. Microbiol. Biotechnol. 28 (4), 1327–1350.

Bishop, P.E., Jorerger, R.D., 1990. Genetics and molecular biology of an alternative nitrogen fixation system. Plant Mol. Biol. 41, 109–125.

Brock, T.D., 2009. Biology of Microorganisms, twelfth ed. Pearson Benjamin Cummings.

Budi, S.W., van Tuinen, D., Arnould, C., Dumas-Gaudot, E., Gianinazzi-Pearson, V., Gianinazzi, S., 2000. Hydrolytic enzyme activity of *Paenibacillus* sp. strain B2 and effects of the antagonistic bacterium on cell integrity of two soil borne pathogenic fungi. Appl. Soil Ecol. 15, 191–199.

Budzikiewicz, H., 2001. Siderophore-antibiotic conjugates used as trojan horses against *Pseudomonas aeruginosa*. Curr. Top. Med. Chem. 1 (1), 73–82.

Cattelan, A.J., Hartel, P.G., Fuhrmann, J.J., 1999. Screening for plant growth promoting rhizobacteria to promote early soybean growth. Soil Sci. Soc. Am. J. 63, 1670–1680.

Chakravarty, G., Kalita, M.C., 2012. Conducted experiment on biocontrol potential of *Pseudomonas fluorescens* against bacterial wilt of brinjal and its possible plant growth promoting effects. Ann. Biol. Res. 3 (11), 5083–5094.

Chen, Y., Fang, Y., Chai, Y., Hongxia, Y., Kolter, R., Losick, R., Jian, J.H., 2013a. Biocontrol of tomato wilt disease by *Bacillus subtilis* isolates from natural environments depends on conserved genes mediating biofilm formation. Environ. Micron., 15, 848–864.

Chen, L., Dodd, I.C., Theobald, J.C., Belimov, A.A., Davies, W.J., 2013b. The rhizobacterium *Variovorax paradoxus* 5C-2, containing ACC deaminase, promotes growth and development of *Arabidopsis thaliana* via an ethylene dependent pathway. Exp. Bot. 64 (6), 1565–1573.

Choudhary, D.K., Prakash, A., Johr, B.N., 2007. Induced systemic resistance (ISR) in plants: mechanism of action. Indian J. Microbiol. 47 (4), 289–297.

Darma, R., Purnamasari, I.M., Agustina, D., Pramudito, T.E., Sugiharti, M., Suwanto, A., 2016. A strong antifungal-producing bacteria from bamboo powder for biocontrol of *Sclerotium rolfsii* in melon (*Cucumis melo* var. *amanta*). J. Plant Pathol. Microbiol. 7, 334.

De Britto, S., Tanzeembanu, D.G., Praveen, S., Lalitha, S., Ramachandra, Y.L., Jogaiah, S., Ito, S., 2020. Isolation and characterization of nutrient dependent pyocyanin from *Pseudomonas aeruginosa* and its dye and agrochemical properties. Sci. Rep. 10, 1542 https://doi.org/10.1038/s41598-020-58335-6.

Devendra Kumar, C., Kumar, A., Nabi, S.U.N., 2017. *In-vitro* evaluation of *Arabidopsis thaliana* ecotypes against *Ralstonia solanacearum* Race 4. Int. J. Curr. Microbiol. App. Sci. 6 (5), 575–579.

Ding, G.C., Piceno, Y.M., Heuer, H., Weinert, N., Dohrmann, A.B., 2013. Changes of soil bacterial diversity as a consequence of agricultural land use in a semi-arid ecosystem. PLoS One 8, 59497.

Dunne, C., Moënne-Loccoz, Y., de Bruijn, F.J., O'Gara, F., 2000. Overproduction of an inducile extracellular serine protease improves biological control of *Pythium ultimum* by *Stenotrophomonas maltophilia* strain W81. Microbiology 146, 2069–2078.

Elphinstone, J.G., 2005. In: Allen, C., Prior, P., Hayward, A.C., St Paul, M.N. (Eds.), The current bacterial wilt situation: a global overview. In bacterial wilt disease and the *Ralstonia solanacearum* species complex. American Phytopathological Society, pp. 9–28.

Elphinstone, J.G., Aley, P., 1992. Integrated control of bacterial wilt of potato in the warm tropics of Peru. In: Hartman, G.L., Hayward, A.C. (Eds.), Bacterial Wilts Proceedings No 45. Kaoshuing, Korea, pp. 276–283.

Frankenberger, W.T.J., Arshad, M., 1995. Photohormones in Soil: Microbial Production and Function. Dekker, New York, p. 503.

Fu, Z.Q., Dong, X., 2013. Systemic acquired resistance: turning local infection into global defense. Annu. Rev. Plant Biol. 64, 839–863.

Ge, H.L., Guo, J.H., Qi, H.Y., Guo, Y.H., Huang, Y.X., 2004. Biological control of capsicum bacterial wilt by compound bacterial mixture AR99. Acta. Phytopathol. Sin. 34, 162–165.

Genin, S., Denny, T.P., 2012. Pathogenomics of the *Ralstonia solanacearum* species complex. Annu. Rev. Phytopathol. 50, 67–89.

George, E.F., Machakova, I., Zazimalova, E., 2008. In: George, E.F., Hall, M.A., Deklerk, G.J. (Eds.), Plant Growth Regulators. I: Auxins, Their Analogues and Inhibitors in: Plant Propagation by Tissue Culture. Springer, Dordrecht, Netherlands, pp. 175–204.

Glick, B.R., 2012. Plant growth-promoting bacteria: mechanisms and applications. Scientifica 2012, 963401. 15 pages.

Glick, B.R., 2014. Bacteria with ACC deaminase can promote plant growth and help to feed the world. Microbiol. Res. 169 (1), 30–39.

Glick, B.R., Patten, C.L., Holguin, G., Penrose, G.M., 1999. Biochemical and Genetic Mechanisms Used by Plant Growth Promoting Bacteria. Imperial College Press, London.

Glick, B.R., Cheng, Z., Czarny, J., Duan, J., 2007. Promotion of plant growth by ACC deaminase-producing soil bacteria. Eur. J. Plant Pathol. 119, 329–339.

Gopalakrishnan, S., Sathya, A., Vijayabharathi, R., Varshney, R.K., Gowda, C.L.L., Krishnamurthy, L., 2015. Plant growth promoting rhizobia: challenges and opportunities. 3. Biotech 5, 355–377.

Goswami, D., Parmar, S., Vaghela, H., Dhandhukia, P., Thakker, J., 2015. Describing *Paenibacillus mucilaginosus* strain N3 as an efficient plant growth promoting rhizobacteria (PGPR). Cogent. Food Agric. 1, 1.

Goswami, D., Thakker, J.N., Dhandhukia, P.C., 2016. Portraying mechanics of plant growth promoting rhizobacteria (PGPR): a review. Cogent. Food Agric. 2, 1–19.

Guo, J.H., Qi, H.Y., Guo, Y.H., Ge, H.L., Gong, L.Y., Zhang, L.X., Sun, P.H., 2004. Biocontrol of tomato wilt by plant growth-promoting rhizobacteria. Biol. Control 29, 66–72.

Gupta, S., Meena, M.K., Datta, S., 2014. Isolation, characterization of plant growth promoting bacteria from the plant *Chlorophytum borivilianum* and in vitro screening for activity of nitrogen fixation, phospthate solubilization and IAA production. Int. J. Curr. Microbiol. Appl. Sci. 3, 1082–1090.

Gupta, G., Parihar, S.S., Ahirwar, N.K., Snehi, S.K., Singh, V., 2015. Plant growth promoting rhizobacteria (PGPR): current and future prospects for development of sustainable agriculture. J. Microbiol. Biochem. Technol. 7, 96–102.

Gutierrez-Manero, F.J., Ramos-Solano, B., Probanza, A., Mehouachi, J.R., Tadeo, F., Talon, M., 2001. The plant-growth-promoting rhizobacteria *Bacillus pumilus* and *Bacillus licheniformis* produce high amounts of physiologically active gibberellins. Physiol. Plantarum. 111, 206–211.

Hassan, E.A., El, A.E.A., Hadidy, Balabel, N.M., Eid, N.A., Ramadan, E.S.M., 2018. Use of rhizobacteria as biocontrol agents against *Ralstonia solanacearum*: principles, mechanisms of action and characterize its bioactive compounds. Current Sci. Int. 7 (2), 242–256.

Hayward, A.C., 1991. Biology and epidemiology of bacterial wilt caused by *Pseudomonas solanacearum*. Annu. Rev. Phytopathol. 29, 65–87.

Hedden, P., Phillips, A.L., 2000. Gibberellin metabolism: new insights revealed by the genes. Trends Plant Sci. 5, 523–530.

Hersman, L.E., Folsythe, J.H., Ticknor, L.O., Maurice, P.A., 2001. Growth of *Pseudomonas mendocina* on Fe (III) (hydr) oxides. Appl. Env. Microbiol. 67, 4448–4453.

Heydari, A., Pessarakli, M., 2010. A review on biological control of fungal plant pathogens using microbial antagonists. J. Biol. Sci. 10, 273–290.

Hider, R.C., Kong, X., 2010. Chemistry and biology of siderophores. Nat. Prod. Rep. 27, 637–657.

Hikichi, Y., Yoshimochi, T., Tsujimoto, S., Shinohara, R., Nakaho, K., Kanda, A., Kiba, A., Ohnishi, K., 2007. Global regulation of pathogenicity mechanism of *Ralstonia solanacearum.* Plant Biotechnol. 24, 149–154.

Hua, J., Kersten, J., Huang, H.Y., Allen, C., 2004. *Ralstonia solanacearum* needs motility for invasive virulence on tomato and biocontrol. J. Bacteriol. 183, 3597–3605.

Hyakumachi, M., Nishimura, M., Arakawa, T., Asano, S., Yoshida, S., Tsushima, S., Takahashi, H., 2013. *Bacillus thuringiensis* suppresses bacterial wilt disease caused by *Ralstonia solanacearum* with systemic induction of defense-related gene expression in tomato. Microbiol. Environ. 28, 128–134.

Illangumuran, G., Smith, D.L., 2017. Plant growth promoting rhizobacteria in amelioration of salinity stress: a systems biology perspective. Plant Sci. 8, 1–14.

Jacobs, J.M., Milling, A., Mitra, R.M., Hogan, C.S., Ailloud, F., Prior, P., Allen, C., 2013. *Ralstonia solanacearum* requires PopS, an ancient AvrE-family effector, for virulence and to overcome salicylic acid-mediated defenses during tomato pathogenesis. MBio 4 (6). e00875–13.

Jakobi, M., Kaiser, D., Berg, G., Jung, G., Winkelmann, G., Bahl, H., 1996. Maltophilin- a new antifungal compound produced by *Stenotrophomonas maltophilia* R3089. J. Antibiot. 49, 1101–1104.

Jetiyanon, K., 2007. Defensive-related enzyme response in plants treated with a mixture of Bacillus strains (IN937a and IN937b) against different pathogens. Biol. Control 42, 178–185.

Jha, C.K., Saraf, M., 2015. Plant growth promoting rhizobacteria (PGPR): a review. E3 J. Agric. Res. Dev. 5, 108–119.

Jha, C.K., Patel, D., Rajendran, N., Saraf, M., 2010. Combinatorial assessment on dominance and informative diversity of PGPR from rhizosphere of *Jatropha curcas* L. J. Basic Microbiol. 50, 211–217.

Jiang, C., Sheng, X., Qian, M., Wang, Q., 2008. Isolation and characterization of a heavy metal resistant *Burkholderia* sp. from heavy metal-contaminated paddy field soil and its potential in promoting plant growth and heavy metal accumulation in metal polluted soil. Chemosphere 72, 157–164.

Jikare, A.M., Chavan, M.D., 2013. Siderophore produced by bacillus shackletonii GN-09 and its plant growth promoting activity. IJPBS 3, 198–202.

Jogaiah, S., Roopa, K.S., Pushpalatha, H.G., Shekar Shetty, H., 2010. Evaluation of plant growth-promoting rhizobacteria for their efficiency to promote growth and induce systemic resistance in pearl millet against downy mildew disease. Arch. Phytopath. Plant Prot. 43, 368–378.

Jogaiah, S., Mostafa, A., Tran, L.S.P., Shin-ichi, I., 2013. Characterization of rhizosphere fungi that mediate resistance in tomato against bacterial wilt disease. J. Exp. Bot. 64, 3829–3842.

Jogaiah, S., Abdelrahman, M., Tran, L.-S.P., Ito, S.-I., 2018. Different mechanisms of *Trichoderma virens*-mediated resistance in tomato against Fusarium wilt involve the jasmonic and salicylic acid pathways. Mol. Plant Pathol. 19, 870–882.

Joshi, Y.B., Chu, J., Pratico, D., 2012. Stress hormone leads to memory deficits and altered tau phosphorylation in a mouse model of Alzheimer's disease. J Alzheimer's Dis. 31, 167–176.

Jurkevitch, E., Hadar, Y., Chen, Y., 1992. Differential siderophore utilization and iron uptake by soil and rhizosphere bacteria. Appl. Environ. Microbiol. 58, 119–124.

Kamilova, F., Validov, S., Azarova, T., Mulders, I., Lugtenberg, B., 2005. Enrichment for enhanced competitive plant root tip colonizers selects for a new class of biocontrol bacteria. Environ. Microbiol. 7, 1809–1817.

Kang, B.G., Kim, W.T., Yun, H.S., Chang, S.C., 2010. Use of plant growth-promoting rhizobacteria to control stress responses of plant roots. Plant Biotechnol. Rep. 4, 179–183.

Kaur, H., Kaur, J., Gera, R., 2016. Plant growth promoting Rhizobacteria: a boon to agriculture. Int. J. Cell Sci. Biotechnol. 5, 17–22.

Khan, M.S., Zaidi, A., Wani, P.A., Oves, M., 2009. Role of plant growth promoting rhizobacteria in the remediation of metal contaminated soils. Environ. Chem. Lett. 7, 1–19.

Kim, J., Rees, D.C., 1994. Nitrogenase and biological nitrogen fixation. Biochemistry 33, 389–397.

Kim, H.S., Sang, M.K., Jeun, Y.C., 2008. Sequential selection and efficacy of antagonistic rhizaobacteria for controlling *Phytophthora* blight of pepper. Crop Prot. 27, 436–443.

Kobayashi, D.Y., Reedy, R.M., Bick, J., Oudemans, P.V., 2002. Characterization of a chitinase gene from *Stenotrophomonas maltophilia* strain 34S1 and its involvement in biological control. Appl. Environ. Microbiol. 68, 1047–1054.

Kuarabachew, H., Assefa, F., Hiskias, Y., 2007. Evaluation of Ethiopian isolates of *Pseudomonas fluorescens* as biocontrol agent against potato bacterial wilt caused by Ralstonia (pseudomonas) solanacearum. Acta Agri. Solvenica. 90 (2), 125–135.

Kumar, P., Kumar, J., Chandra, P., 2018. Advances in Microbial Biotechnology: Current Trends and Future Prospects. Apple Academic Press, p. 574.

Kurabachew, H.M., Assefa, F., Hiskias, Y., 2007. Evaluation of Ethiopian isolates of *Pseudomonas fluorescens* as biocontrol agent against potato bacterial wilt caused by *Ralstonia (Pseudomonas) solanacearum*. Acta Agric. Slov. 90, 125–135.

Kwon, J.W., Kim, S.D., 2014. Characterization of an antibiotic produced by *Bacillus subtilis* JW-1 that suppresses *Ralstonia solanacearum*. J. Microbiol. Biotechnol. 24 (1), 13–18.

Labuschagne, N., Pretorius, T., Idris, A.H., 2010. Plant growth promoting rhizobacteria as biocontrol agents against soil-borne plant diseases. In: Maheshwari, D.K. (Ed.), Plant Growth and Health Promoting Bacteria, Microbiology Monographs. Springer Verlag, Berlin/ Heldelberg, pp. 211–230.

Lanteigne, C., Gadkar, V.J., Wallon, T., Novinscak, A., Filion, M., 2012. Production of DAPG and HCN by *Pseudomonas* sp. LBUM3s00 contributes to the biological control of bacterial canker of tomato. Phytopathology 102, 967–973.

Li, S.-M., Gou, G.H., Hong-Xia, L., Jian, H.G., 2008. Analysis of defence enzymes induced by antagonistic bacterium *Bacillus subtilis* strain AR12 towards *Ralstonia solanacearum* in tomato. Ann. Microbiol. 58 (4), 573–578.

Van Loon, L.C., 2007. Plant responses to plant growth promoting rhizobacteria. Eur. J. Plant Pathol. 119, 243–254.

Lugtenberg, B., Kamilova, F., 2009. Plant-growth-promoting rhizobacteria. Ann. Rev. Microbiol. 63, 541–556.

Lugtenberg, B.J.J., Dekkers, L., Bloemberg, G.V., 2001. Molecular determinants of rhizosphere colonization by *Pseudomonas*. Annu. Rev. Phytopathol. 39, 461–470.

Majeed, A., Abbasi, M.K., Hameed, S., Imran, A., Rahim, N., 2015. Isolation and characterization of plant growth-promoting rhizobacteria from wheat rhizosphere and their effect on plant growth promotion. Front. Microbiol. 6, 198.

Maji, S., Chakrabartty, P.K., 2014. Biocontrol of bacterial wilt of tomato caused by *Ralstonia solanacearum* by isolates of plant growth promoting rhizobacteria. Aust. J. Crop. Sci. 8 (2), 208–214.

Makhlouf, A.H., Hamedo, H.A., 2013. Supression of bacterial wilt disease of tomato plants using bacterial strains. Life Sci. J. 10 (3), 1732–1739.

Maksimov, I.V., Abizgil, R.R., Pusenkova, L.I., 2011. Plant growth promoting rhizobacteria as alternative to chemical crop protectors from pathogens. Appl. Biochem. Microbiol. 47 (4), 373–385.

Mansfield, J., Genin, S., Magor, S., Citovsky, V., Sriariyanum, M., Ronald, P., Dow, M., Verdier, C., Beer, S.V., Mchado, M.A., Toth, I., Salmond, G., Foster, G.D., 2012. Top 10 plant pathogenic bacteria in molecular plant pathology. Mol. Plant Pathol. 13, 614–629.

Martinez-Viveros, O., Jorquera, M., Crowley, D.E., Gajardo, G., Mora, M.L., 2010. Mechanisms and practical considerations involved in plant growth promotion by rhizobacteria. Soil Sci. Plant Nutr. 10, 293–319.

Mauch-Mani, B., Mauch, F., 2005. The role of abscisic acid in plant–pathogen interactions. Curr. Opin. Plant Biol. 8, 409–414.

Mazurier, S., Corberand, T., Lemanceau, P., Raaijmakers, J.M., 2009. Phenazine antibiotics produced by fluorescent pseudomonads contribute to natural soil suppressiveness to *Fusarium* wilt. ISME J. 3 (8), 977–991.

Meenakumari, S.N., 2007. Conducted the experiment about suppression of bacterial wilt of chilli and tomato using native Isolates of *Pseudomonas fluorescens*. In: Paper Presented at the Sixth International Workshop on PGPR. Indian Institute of Spices Research, India.

Mehjabeen, A., Aanisia, Z., John Suchit, A., Pradeep, S.K., Ramteke, P.W., 2017. Suppression of bacterial wilt of tomato (*Lycopersicon esculentum* mill.) by plant growth promoting Rhizobacteria. Int. J. Plant Res. 30, 429.

Merchan, F., de Lorenzo, L., González-Rizzo, S., Niebel, A., Megías, M., 2007. Analysis of regulatory pathways involved in the reacquisition of root growth after salt stress in *Medicago truncatula*. Plant J. 51, 1–17.

Messiha, N.A.S., van Diepeningen, A.D., Farag, N.S., Abdallah, S.A., Janse, J.D., van Bruggen, A.H.C., 2007. *Stenotrophomonas maltophilia*: a new potential biocontrol agent of *Ralstonia solanacearum*, causal agent of potato brown rot. Eur. J. Plant Pathol. 118, 211–225.

Mishra, M., Kumar, U., Mishra, P.K., Prakash, V., 2010. Efficiency of plant growth promoting rhizobacteria for the enhancement of *Cicer arietinum* L. growth and germination under salinity. Adv. Biol. Res. 4, 92–96.

Murali, M., Jogaiah, S., Amruthesh, K.N., Shin-ichi, I., Shekar, Shetty H., 2013. Rhizosphere fungus *Penicillium chrysogenum* promotes growth and induces defense-related genes and downy mildew disease resistance in pearl millet. Plant Biol. 15, 111–118.

Murthy, K.N., Uzma, F., Chitrashree, S.C., 2014. Induction of systemic resistance in tomato against *Ralstonia solanacearum* by *Pseudomonas fluorescens*. Am. J. Plant Sci. 5, 1799–1811.

Nakayama, T., Homma, Y., Hashidoko, Y., Mizutani, J., Tahara, S., 1999. Possible role of xanthobaccins produced by *Stenotrophomonas* sp. strain SB-K88 in suppression of sugar beet damping-off disease. Appl. Environ. Microbiol. 65, 4334–4339.

Narasimhamurthy, K., Malini, M., Fazilath, U., Soumya, K.K., Chandra, N.S., Niranjana Ramachandrappa, S.R., Srinivas, C., 2016. Lactic acid bacteria mediated induction of defense enzymes to enhance the resistance in tomato against *Ralstonia solanacearum* causing bacterial wilt. Sci. Hortic. 207, 183–192.

Nasril, N., Anthoni, A., Mairawita, F.A., Armaleni, R., 2018. Isolation of *Pseudomonas fluorescens* from infected and uninfected soil by *Ralstonia solanacearum* capable of producing Siderophore, HCN and soluble phosphate. Pharm. Lett. 10 (1), 82–90.

Neiendam-Nielsen, M., Sørensen, J., 1999. Chitinolytic activity of *Pseudomonas fluorescens* isolates from barley and sugar beet rhizosphere. FEMS Microbiol. Ecol. 30, 217–227.

Nguyen, M.T., Ranamukhaarachchi, S.L., 2010. Soil-borne antagonists for biological control of bacterial wilt disease caused by *Ralstonia solanacearum* in tomato and pepper. J. Plant Pathol. 92 (2), 395–406.

Nivya, R.M.A., 2015. Study on plant growth promoting activity of the endophytic bacteria isolated from the root nodules of *Mimosa pudica* plant. Int. J. Innov. Res. Sci. Eng. Technol. 4, 6959–6968.

Noumavo, P.A., Agbodjato, N.A., Baba-Moussa, F., Adjanohoun, A., Baba-Moussa, L., 2016. Plant growth promoting rhizobacteria: beneficial effects for healthy and sustainable agriculture. Afr. J. Biotech. 15, 1452–1463.

Oliveira, M.F., da Silva, M.G., Van Der Sand, S.T., 2010. Anti-phytopathogen potential of endophytic actinobacteria isolated from tomato plants (*Lycopersicon esculentum*) in southern Brazil, and characterization of *Streptomyces* sp. R18 (6), a potential biocontrol agent. Res. Microbiol. 161, 565–572.

Oteino, N., Lally, R.D., Kiwanuka, S., Lloyd, A., Ryan, D., Germaine, K.J., Dowling, D.N., 2015. Plant growth promotion induced by phosphate solubilizing endophytic *Pseudomonas* isolates. Front. Microbiol. 6, 745.

Pandya, N.D., Desai, P.V., 2014. Screening and characterization of GA3 producing *Pseudomonas monteilii* and its impact on plant growth promotion. Int. J. Curr. Microbiol. Appl. Sci. 3, 110–115.

Park, K., Paul, D., Kim, Y.K., Nam, K.W., Lee, Y.K., Choi, H.W., Lee, S.Y., 2007. Induced systemic resistance by *Bacillus vallismortis* EXTN-1 suppressed bacterial wilt in tomato caused by *Ralstonia solanacearum*. Plant Pathol. J. 23 (1), 22–25.

Patel, K., Goswami, D., Dhandhukia, P., Thakker, J., 2015. Techniques to study microbial phytohormones. In: Maheshwari, D.K. (Ed.), Bacterial Metabolites in Sustainable Agroecosystem. Springer International, pp. 1–27.

Patten, C.L., Glick, B.R., 1996. Bacterial biosynthesis of indole-3-acetic acid. Can. J. Microbiol. 42, 207–220.

Phadke, G.S., 2007. Suppression of *Ralstonia solanacearum* (Smith) by Antagonistic Bacteria: Possible Mechanisms of Pathogen Inhibition and Plant Growth Promotion. M.Sc. Dissertation, Goa University, Goa, India, p. 79.

Pieterse, C.M.J., Van Wees, S.C.M., 2015. Induced disease resistance. Plant-Micro. Interact. 107 (1), 123–133.

Pieterse, C.M.J., Zamioudis, C., Berendsen, R.L., Weller, D.M., VanWees, S.C.M., Peter, A.H.M., 2014. Induced systemic resistance by beneficial microbes. Annu. Rev. Phytopathol. 52, 347–375.

Porcel, R., Zamarreño, Á.M., García-Mina, J.M., Aroca, R., 2014. Involvement of plant endogenous ABA in *Bacillus megaterium* PGPR activity in tomato plants. BMC Plant Biol. 14, 36.

Qurashi, A.W., Sabri, A.N., 2012. Bacterial exopolysaccharide and biofilm formation stimulate chickpea growth and soil aggregateon under salt stress. Braz. J. Microbiol. 43, 1183–1191.

Raaijmakers, J.M., Mazzola, M., 2012. Diversity and natural functions of antibiotics produced by beneficial and plant pathogenic bacteria. Annu. Rev. Phytopathol. 50, 403–424.

Raaijmakers, J.M., Vlami, M., de Souza, J.T., 2002. Antibiotic production by bacterial biocontrol agents. Antonie Van Leeuwenhoek 81, 537–547.

Rado, R., Andrianarisoa, B., Ravelomanantsoa, S., Rakotoarimanga, N., Rahetlah, V., Fienena, F.R., Andriambeloson, O., 2015. Biocontrol of potato wilt by selective rhizospheric and endophytic bacteria associated with potato plant. African Journal of Food, Agriculture, Nutrition and Development 15 (1), 9762–9776.

Rai, R., Srinivasamurthy, R., Dash, P.K., Gupta, P., 2017. Isolation characterization and evaluation of the biocontrol potential of *Pseudomonas protegens* RS-9 against *Ralstonia solanacearum* in tomato. Indian J. Exp. Biol. 55, 595–603.

Rajkumar, M., Ae, N., Prasad, M.N.V., Freitas, H., 2010. Potential of siderophore producing bacteria for improving heavy metal phytoextraction. Trends Biotechnol. 28, 142–149.

Ramadasappa, S., Ashwani, K.R., Ranjeet, S.J., Aqbal, S., Rhitu, R., 2012. Isolation and screening of phlD+ plant growth promoting rhizobacteria antagonistic to *Ralstonia solanacearum*. World J. Microbiol. Biotechnol. 28, 1681–1690.

Ramamoorthy, V., Raguchande, T., Samiyappan, R., 2002. Induction of defense-related proteins in tomato roots treated with *Pseudomonas fluorescens* Pf1 and *Fusarium oxysporum* f. sp. *lycopersici*. Plant Soil 239, 55–68.

Ramesh, R., Phadke, G.S., 2012. Rhizosphere and endophytic bacteria for the suppression of eggplant wilt caused by *Ralstonia solanacearum*. Crop Prot. 37, 35–41.

Ramesh, R., Joshi, A.A., Ghanekar, M.P., 2009. Pseudomonas: major endophytic bacteria to suppress bacterial wilt pathogen *Ralstonia solanacearum* in the eggplant (*Solanum melongena* L.). World J. Microbiol. Biotechnol. 25, 47–55.

Ran, L.X., Bakker, P.H.M., Pieterse, C.M.J., Van Loon, L.C., 2004. Understanding the involvement of rhizosphere-mediated induction of systemic resistance in biocontrol of bacterial wilt disease. Can. J. Plant Pathol. 25, 5–9.

Ran, L.X., Liu, C.Y., Wu, G.J., van Loon, L.C., Bakker, P.A.H.M., 2005. Suppression of bacterial wilt in *Eucalyptus urophylla* by fluorescent *Pseudomonas* spp. in China. Biol. Con. 32, 111–120.

Reddy, P.P., 2013. Plant growth promoting rhizobacteria (PGPR). In: Recent Advances in Crop Protection. Springer, India, pp. 131–145.

Richardson, A.E., Simpson, R.J., 2011. Soil microorganisms mediating phosphorus availability. Plant Physiol. 156, 989–996.

Robin, A., Vansuyt, G., Hinsinger, P., Meyer, J.M., Briat, J.F., Lemanceau, P., 2008. Iron dynamics in the rhizosphere. Consequences for plant health and nutrition. Adv. Agron. 99, 183–225.

Rubio, L.M., Ludden, P.W., 2008. Biosynthesis of the iron-molybdenum cofactor of nitrogenase. Annu. Rev. Microbiol. 62, 93–111.

Ryan, P.R., Delhaize, E., Jones, D.L., 2001. Function and mechanism of organic anion exudation from plant roots. Ann. Rev. Plant Physiol. Plant Mol. Biol. 52, 527–560.

Ryder, M.H., Stephens, P.M., Bowen, G.D., 1994. Improving Plant Productivity with Rhizosphere Bacteria. CSIRO, Melbourne.

Saber, F.M.A., Abdelhafez, A.A., Hassan, E.A., Ramadan, E.M., 2015. Characterization of fluorescent pseudomonads isolates and their efficiency on the growth promotion of tomato plant. Annals Agric. Sci. 60, 131–140.

Sadfi, N., Chérif, M., Fliss, I., Boudabbous, A., Antoun, H., 2001. Evaluation of bacterial isolates from salty soils and *Bacillus thuringiensis* strains for the biocontrol of fusarium dry rot of potato tubers. J. Plant Pathol. 83, 101–118.

Safdarpour, F., Khodakaramian, G., 2018. Endophytic bacteria suppress bacterial wilt of tomato caused by *Ralstonia solanacearum* and activate defense related metabolites. Biol. J. Microorg. 6 (24), 39–52.

Sagervanshi, A., Kumari, P., Nagee, A., Kumar, A., 2012. Isolation and characterization of phosphate solubilizing bacteria from an agriculture soil. Int. J. Life Sci. Pharma. Res. 23, 256–266.

Saharan, B.S., Nehra, V., 2011. Plant growth promoting Rhizobacteria: a critical review. Life Sci. Med. Res. 21, 1–30.

Salanoubat, M., Genin, S., Artiguenave, F., Gouzy, J., Mangenot, S., Arlat, M., Billault, A., Brottier, P., Camus, J.C., 2002. Genome sequence of the plant pathogen *Ralstonia solanacearum*. Nature 415, 497–502.

Santiago, T.R., Grabowski, C., Rossato, M., 2015. Biological control of eucalyptus bacterial wilt with rhizobacteria. Biol. Control 80, 14–22.

Santoyoa, G., Moreno-Hagelsiebb, G., Orozco-Mosquedac, M.D.C., Glick, B.R., 2016. Plant growth-promoting bacterial endophytes. Microbiol. Res. 183, 92–99.

Saraf, M., Jha, C.K., Patel, D., 2011. The role of ACC deaminase producing PGPR in sustainable agriculture. In: Maheshwari, D.K. (Ed.), Plant Growth and Health Promoting Bacteria Microbiology, Monographs. Springer, Germany, pp. 365–386.

Sayyed, R.Z., Chincholkar, S.B., Reddy, M.S., Gangurde, N.S., Patel, P.R., 2013. Siderophore producing PGPR for crop nutrition and phytopathogen suppression. In: Maheshwari, D.K. (Ed.), Bacteria in Agrobiology: Disease Management. Springer-Verlag, Berlin, Heidelberg.

Schalk, I.J., Hannauer, M., Braud, A., 2011. New roles for bacterial siderophores in metal transport and tolerance. Envion. Microbiol. 13, 2844–2854.

Seleim, M.A.A., Saead, F.A., Abd-El-Moneem, K.M.H., Abo-ELyousr, K.A.M., 2011. Biological control of bacterial wilt of tomato by plant growth promoting rhizobacteria. Plant Pathol. J. 10, 146–153.

Selvakumar, G., Mohan, M., Kundu, S., Gupta, A.D., Joshi, P., Nazim, S., Gupta, H.S., 2008. Cold tolerance and plant growth promotion potential of *Serratia marcescens* strain SRM (MTCC 8708) isolated from flowers of summer squash (*Cucurbita pepo*). Lett. Appl. Microbiol. 46, 171–175.

Sharma, A., Johri, B.N., Sharma, A.K., Glick, B.R., 2013. Plant growth-promoting bacterium *Pseudomonas* sp. strain GRP3 influences iron acquisition in mung bean (*Vignaradiata* L. Wilzeck). Soil Biol. Biochem. 35, 887–894.

Shen, A.R., Ji, G.H., Wei, L.F., He, Y.Q., 2006. Biological control of potato bacterial blight by combining bacterial agents. J Yunnan Agri. Univ. 21, 449–453.

Silva, H.S.A., Romeiro, R.S., Macagnan, D., Halfeld-vieira, B.A., Pereira, M.C.B., Mounteer, A., 2004. Rhizobacterial induction of systemic resistance in tomato plants: non-specific protection and increase in enzyme activities. Biol. Control 29, 288–295.

Singh, D., Yadav, D.K., Shewata, S., Mondal, K.K., Singh, G., Pandey, R.R., Singh, R., 2013. Genetic diversity of iturin producing strains of *Bacillus* species antagonistic to *Ralstonia solanacearum* causing bacterial wilt disease in tomato. Afr. J. Microbiol. Res. 7 (48), 5459–5470.

Singh, D., Yadav, D.K., Chaudhary, G., Rana, V.S., Sharma, R.K., 2016. Potential of *Bacillus amyloliquefaciens* for biocontrol of bacterial wilt of tomato incited by *Ralstonia solanacearum*. J. Plant Pathol. Microbiol. 7, 327.

Singh, H., Jaiswa, V., Singh, S., Tiwar, S.P., Singh, B., Katiyar, D., 2017. Antagonistic compounds producing plant growth promoting rhizobacteria: a tool for management of plant disease. J. Adv. Microbiol. 3 (4), 1–12.

Souza, R.D., Ambrosini, A., Passaglia, L.M.P., 2015. Plant growth-promoting bacteria as inoculants in agricultural soils. Genet. Mol. Biol. 38 (4), 401–419.

Spaepen, S., Vanderleyden, J., 2011. Auxin and plant-microbe interactions. Cold Spring Harb. Perspect. Biol. 3 (4), 1–13.

Sudisha, J., Niranjana, S.R., Umesha, S., Prakash, H.S., Shekar Shetty, H., 2006. Transmission of seed-borne infection of muskmelon by *Didymella bryoniae* and effect of seed treatments on disease incidence and fruit yield. Biol. Control 37, 196–205.

Sudisha, J., Sharathchandra, R.G., Amruthesh, K.N., Kumar, A., Shetty, H.S., 2012. Pathogenesis related proteins in plant defense response. In: Mérillon, J.M., Ramawat, K.G. (Eds.), PlantDefence: Biological Control. Progress in Biological Control, vol. 12, pp. 378–403.
Sulochana, M.B., Jayachandra, S.Y., Kumar, S.A., Parameshwar, A.B., Reddy, K.M., Dayanand, A., 2014. Siderophore as a potential plant growth-promoting agent produced by *Pseudomonas aeruginosa* JAS-25. Appl. Biochem. Biotechnol. 174, 297–308.
Tahir, I.M., Inam-ul-Haq, M., Ashfaq, M., Nadeem, A.A., Haris, B., Hira, G., 2016. Screening of effective antagonists from potato rhizosphere against bacterial wilt pathogen. Int. J. Biosci. 8 (2), 228–240.
Tan, S., Jiang, Y., Song, S., Huang, J., Ling, N., Xu, Y., Shen, Q., 2013. Two *Bacillus amyloliquefaciens* strains isolated using the competitive tomato root enrichment method and their effects in suppressing *Ralstonia solanacearum* and promoting tomato plant growth. Crop. Protect 43, 134–140.
Tensingh Baliah, N., 2018. Phosphate solubilizing bacteria (PSB) a novel PGPR for sustainable agriculture. Int. J. Inn. Res. Sci. Eng. Technol. 7 (4), 143–149.
Thaler, S.J., Bostock, R.M., 2004. Interactions between abscisic acid mediated responses and plant resistance to pathogens and insects. Ecology 85, 48–58.
Tripura, C.B., Sashidhar, B., Podile, A.R., 2005. Transgenic mineral phosphate solubilizing bacteria for improved agricultural productivity. In: Satyanarayana, T., Johri, B.N. (Eds.), Microbial Diversity Current Perspectives and Potential Applications. New Delhi, India, I.K. International Pvt. Ltd, pp. 375–392.
Udayashankar, A.C., Nayaka, C.S., Niranjan-Raj, S., Kumar, H.B., Reddy, M.S., Niranjana, S.R., Prakash, H.S., 2009. Rhizobacteria mediated resistance against *Bean common mosaic virus* strain blackeye cowpea mosaic in cowpea (*Vigna unguiculata*). Pest Manag. Sci. 65, 1059–1064.
Vanitha, S.C., Niranjana, S.R., Mortensen, C.N., Umesha, S., 2009. Bacterial wilt of tomato in Karnataka and its management by *Pseudomonas fluorescens*. BioControl 54 (5), 685–695.
Vasse, J., Frey, P., Trigalet, A., 1995. Microscopic studies of intercellular infection and protoxylem invasion of tomato roots by *Pseudomonas solanacearum*. Mol. Plant-Microbe Interact. 8, 241–251.
Vejan, P., Abdullah, R., Khadiran, T., Ismail, S., Nasrulhaq Boyce, A., 2016. Role of plant growth promoting rhizobacteria in agricultural sustainability–a review. Molecules 21, 573.
Weyens, N., Truyens, S., Dupae, J., Newman, L., van der Lelie, D., Carleer, R., Vangronsveld, J., 2010. Potential of pseudomonas putida W619-TCE to reduce TCE phytotoxicity and evapotranspiration in poplar cuttings. Environ. Pollut. 158, 2915–2919.
Wong, W.S., Tan, S.N., Ge, L., Chen, X., Yong, J.W.H., 2015. The importance of phytohormones and microbes in biofertilizers. In: Maheshwari, D.K. (Ed.), Bacterial Metabolites in Sustainable Agroecosystem. Springer International, pp. 105–158.
Wu, B., Yanqin, D., Liangtong, Y., Kai, L., Binghai, D., 2011. Disease-preventing and growth promoting effects of antifungal bacteria against *Phytophthora nicotianae* on tobacco. In: Proceedings of the 2nd Asian PGPR Conference, Beijing, PR China.
Wydra, K., Semrau, J., 2005. Phenotypic and Molecular Characterization of the Interaction of Antagonistic Bacteria with *Ralstonia solanacearum* Causing Tomato Bacterial Wilt. In: 1st International Symposium on Biological Control of Bacterial Plant Diseases, Darmstadt, Germany, pp. 112–118.
Xue, Q.Y., Ding, G.C., Li, S.M., Yang, Y., Lan, C.Z., Guo, J.H., Smalla, K., 2013. Rhizocompetence and antagonistic activity towards genetically diverse *Ralstonia solanacearum* strains an improved strategy for selecting biocontrol agents. Appl. Microbiol. Biotechnol. 97, 1361–1371.

Yanti, Y., Astuti, F.F., Habazar, T., Reflinaldon, N.C.R., 2017. Screening of rhizobacteria from rhizosphere of healthy chili to control bacterial wilt disease and to promote growth and yield of chili. Biodiversitas 18 (1), 1–9.

Yu, C., Pi, H., Chandrangsu, P., Li, Y., Wang, Y., Zhou, H., Xiong, H., Helmann, J.D., Cai, Y., 2018. Antagonism of two plant-growth promoting *Bacillus velezensis* isolates against *Ralstonia solanacearum* and *Fusarium oxysporum*. Sci. Rep. 8, 4360.

Yuan, S., Wang, L., Wu, K., Shi, J., Wang, M., Yang, X., Shen, Q., Shen, B., 2014. Evaluation of bacillus-fortified organic fertilizer for controlling tobacco bacterial wilt in greenhouse and field experiments. Appl. Soil Ecol. 75, 86–94.

Yuliar, Nion, Y.A., Toyota, K., 2015. Recent trends in control methods for bacterial wilt diseases caused by *Ralstonia solanacearum*. Microbes Environ. 30, 1–11.

Zahran, H.H., 2001. Rhizobia from wild legumes: diversity, taxonomy, ecology, nitrogen fixation and biotechnology. J. Biotechnol. 91, 143–153.

Zaidi, A., Khan, M.S., Ahemad, M., Oves, M., Wani, P.A., 2009. Recent advances in plant growth promotion by phosphate-solubilizing microbes. In: Microbial Strategies for Crop Improvement. Springer Verlag, Khan MS, Berlin Heidelberg, pp. 23–50.

Zhang, J., Guo, T., Wang, P., Tian, H., Wang, Y., Cheng, J., 2018. Characterization of diazotrophic growth promoting rhizobacteria isolated from ginger root soil as antagonists against *Ralstonia solanacearum*. Biotechnol. Equip. https://doi.org/10.1080/13102818.2018.1533431.

Zheng, B.X., Ibrahim, M., Zhang, D.P., Bi, Q.F., Li, H.Z., Zhou, G.W., Ding, K., Peñuelas, J., Zhu, Y.G., Yang, X.R., 2018. Identification and characterization of inorganic phosphate solubilizing bacteria from agricultural fields with a rapid isolation method. AMB Express 8 (1), 47.

Zhou, T., Chen, D., Li, C., Sun, Q., Li, L., 2012. Isolation and characterization of *Pseudomonas brassicacearum* J12 as an antagonist against *Ralstonia solanacearum* and identification of its antimicrobial components. Microbiol. Res. 167, 388–394.

Seed biopriming a novel method to control seed borne diseases of crops

8

Monika Sood[a], *Vipul Kumar*[b], *and Ruby Rawal*[c]
[a]School of Bioengineering and Biosciences, Lovely Professional University, Phagwara, Punjab, India, [b]School of Agriculture, Lovely Professional University, Phagwara, Punjab, India, [c]Kurukshetra University Kurukshetra (KUK), Thanesar, Haryana, India

8.1 Introduction

A seed is referred to as an embryonic plant which is shielded inside a protective covering called a seed coat. In comparison to cryptograms (lower plants) which are unable to produce seeds, phanerogams (Gymnosperms and angiosperms) show the greater survival, geographic distribution, and reproductive success. A seed possesses all the qualities which are essential for the beginning and development of a new plant, i.e., an embryo, nutritive endosperm and defensive seed coat. Seeds play significant roles in a life cycle of a plant such as nourishment of embryos, dispersal of plant population to a new location, and exhibit dormancy when the conditions are unfavorable for germination (Koornneef et al., 2002). Moreover, most of the seeds are formed as an outcome of sexual reproduction, which involves the fusion of gametes derived from both parent plants consequently, leads to genetic variation which favor the better survival and reproductive success of the future plant as a result of natural selection (Tucker and Koltunow, 2009).

Seed germination is the most critical and vulnerable stage in the life cycle of a plant. In the present scenario, the problems like climate change, population explosion, deforestation, desertification, etc. adversely affect the growth of plants at seedling stage (Meyerson, 2004). Consequently, there are several identified intrinsic and extrinsic factors which seem to be responsible for the inhibition of seed germination. Major abiotic factors accountable for germination inhibition include salinity, drought, soil pH, light, temperature, fire, flooding, etc. (Humphries et al., 2018). Similarly, biotic components of seed germination inhibition process involve pathogen infection, herbivory, etc. (Doughari, 2015). Approximately, 40%–50% of crops have been lost annually due to various kinds of stresses imposed by numerous biotic and abiotic factors (Pandey et al., 2017). It must be a serious concern to address the severe threats of food security caused by inhibition in seed germination to provide enough and quality food to a huge population of the world in present as well as in future (Misra, 2014).

Biocontrol Agents and Secondary Metabolites. https://doi.org/10.1016/B978-0-12-822919-4.00008-9

8.1.1 Seedborne diseases

Among the various kinds of plant diseases, seed/soilborne diseases (Fig. 8.1) are believed to be more constraining than others as it precisely alters the quantity as well as quality of production in several crops and consequently leads to 10%–20% annual yield losses globally (Ray et al., 2017). Because of the aggressiveness and destructive behavior of soilborne phytopathogenic fungi approximately, 50% of economically essential crops are lost annually, in India also (Pandey et al., 2018). Among the several phytopathogens responsible for causing soilborne diseases in plants fungal spp. including *Rhizoctonia solani*, *Sclerotinia sclerotium*, *Sclerotium rolfsii*, and *Fusarium oxysporum* were seemed to be worst. Infection of these spp. cause considerable damages to the plant spp. by induction of diseases such as seed rot, seedling blight, root rot, wilt, etc. with yield loss of about 60%–70% in numerous crops of economic importance. In addition to theses, seed-borne pathogens are also constantly exposing a significant hazard to crop production as they are accountable for around 10% losses in

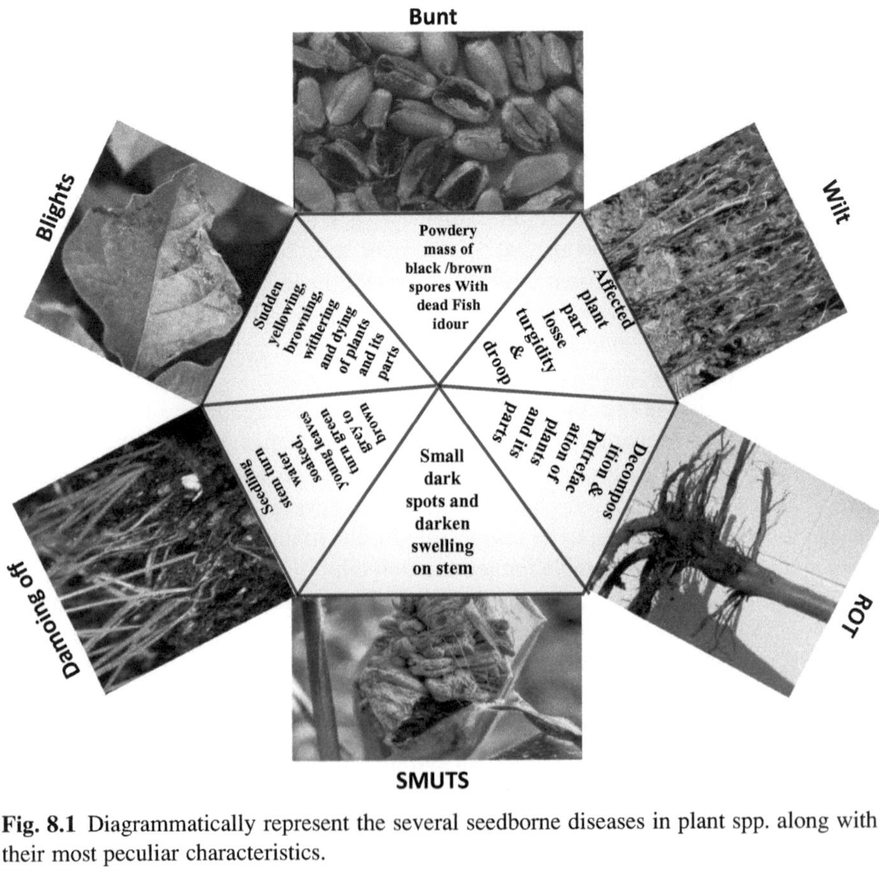

Fig. 8.1 Diagrammatically represent the several seedborne diseases in plant spp. along with their most peculiar characteristics.

main crops. Moreover, the management of these diseases is also difficult due to inadequate accessibility of efficient chemicals (Chahal, 2012).

8.1.2 Outcomes of seedborne diseases

Almost 90% of the total global crops are raised from the seeds (Schwinn, 1994). Moreover, seeds are usually spread worldwide in national as well as in international exchange. In addition to this, for breeding purpose germplasm is also spread and transacted in the form of seeds. Along with seeds several seedborne plant pathogens can also be transmitted which, consequently, leads to the invasion of these pathogens to the unaffected areas. When the environmental conditions are favorable, numerous phytopathogens including virus, bacteria, fungi, nematodes, etc. may be taken with, on, or in seeds. Presently, two feasible alternatives for improving crop productivity either involve the adoption of high-yielding varieties or avoidance of the incidence of crop diseases. However, the advantages of employing high-yielding variety may be abolished by seedborne infections. They are responsible for approximately 10% yield losses in principal crops in India (Neergaard, 1979).

Soilborne pathogens and host plants interactions cause the mortality of not only young seedlings but also damage adult trees. Pathogens display a wide range of interactions with their hosts and vary in great deal in terms of the attack on specific organs and tissues, time of attack, i.e., at different stages of growth and development, as well as the efficacy of infection toward a single species, genus, or whole family of plants (Abdulkhair and Alghuthaymi, 2016). Furthermore, several plant pathogens possessed by aboveground components of the plants persist in the soil at different phases of their life cycles. Moreover, irrespective of having a serious infection imposed by soilborne pathogens, most of the host plants do not display signs of the disease.

Modern agricultural practices undergo a transition toward the more agroecological cropping patterns like crop rotation, residue retention, intercropping, biofarming, etc. Furthermore, currently, legislation imposed strict restriction on the use of pesticides in agriculture is also a pressing issue (Özkara et al., 2016). Due to adverse economic impacts some pesticides, e.g., Aldrin, Benomyl, Diazinon, Metoxuron, and many more get banned in our country (Kathage et al., 2018). Moreover, the rising worries about their dangerous impacts on ecological sustainability and human well-being promote their lessened application in management. Hence, biological management by antagonistic microorganisms appears as a prospective, nonchemical, and eco-friendly tactic for offering safety to crops against several phytopathogens and is also beneficial for the alleviation of numerous pathosystems (Jogaiah et al., 2013). Therefore, in place of chemical pesticides, the potential of eco-friendly, cost-effective, and alternative seed treatments such as seed priming must be explored for the purpose of better germination, vigor, and growth of seedlings. It serves as a safer option against the conventional management practices which have brutally influenced the biota and agroecosystem (Abhilash et al., 2016).

8.2 Seed priming

8.2.1 History

The seed priming history dates to CE 60. There are several reports since the Ancient Greeks which describe the efforts to get better germination of seeds (Evenari, 1984). Theophrastus (371–287 BCE) (Theophrastus (371–287 BCE), n.d.) explored that the seeds imbibed with water before sowing leads to quicker germination. Gaius (1949–1954) (Gaius, 1949–1954) confirmed the significance of presoaked seeds in water to increase germination. Later, Olivier de Serres (1539–1619) (Olivier de Serres (1539–1619), 1600) uncovered the efficacy of the seed treatment on grains (*Triticum*, *Secale* and *Hordeum*spp.). Similarly, Darwin (1855) analyzed osmopriming circumstances by soaking the seeds of *Lepidium sativum* and lettuce in seawater and revealed that the treatment of seeds results in the improved germination. Later, Ells (1963) uncovered that 2 days of soaking of tomato seeds in water at 23.6°C was adequate to trigger premature germination.

8.2.2 Seed priming and its types

An adequate amount of crop production is based on the optimum range of several biotic and abiotic factors particular to that crop. Several limitations such as late sowing, low seed quality, adverse environmental conditions, diseases, etc. serve as the hurdles in the path of ample seed germination. Process of seed germination involves three primary steps, viz., (I) seed hydration, (II) activation of metabolic events, and (III) cell lengthening and the emergence of radicle from the seed. Seed priming is a simple, economical, less time-consuming technique which ensures the superior emergence of seedling, equal stand establishment, hasten flowering and, ultimately, healthier crop yields (Rehman and Farooq, 2016; Ullah et al., 2019a). Basically, seed priming involves the hydration of a seed to a level, i.e., required for the activation of metabolic activities but not adequate for the protrusion of radicle. Primarily, priming techniques can be classified into three phases,

I. In this phase, the seed is hydrated and imbibe water.
II. After adequate water intake, several biochemical processes are initiated which subsequently start seed germination.
III. In priming, the seed is carried away through Phase II and then dried, prior to the emergence of root from the seed. When situations (temperature and moisture) are suitable in the field, this phase can resume, propagation appears in a much quicker time.

8.2.3 Techniques of seed priming

Nowadays several agents can be explored for the purpose of seed priming. These can be categorized as either physical or biological methods. In this chapter, we will briefly discuss the first category and elaborate our discussion on biopriming of seeds. However, the efficacy of a priming method or agent is dependent on the type of crop plant as well as the nature of biotic or abiotic stresses (Mustafa et al., 2017).

8.2.3.1 Hydropriming

This method of seed priming is an easy, economical, and efficient approach toward the enhancement of crop production along with a better adjustment to the stress especially, drought (Nagaraju et al., 2012). The priming strategy involves the emersion of seeds in sterilized distilled water for a specific duration and under the adequate condition of temperature (Kaya et al., 2006). After appropriate imbibition of water, seeds are dried under the shade by forced air until the original weight of the seed has not been achieved (Farooq et al., 2009). In comparison to normal, hydroprimed seed exhibit quicker emergence, improved crop stand, strong and healthy plants, former flowering, greater harvest, and most importantly, superior tolerance to drought (Harris et al., 1999; Harris and Hollington, 2001). Besides these benefits the major drawback of hydropriming is that hydration of seeds can be asymmetrical, causing nonidentical propagation or crop stand (Di Girolamo and Barbanti, 2012).

Likewise, on-farm seed priming, i.e., the overnight drenching of seeds in water without air followed by dry up and then sowing has been reported to do better yield production under varied environmental situations (Rashid et al., 2002). Several reports are available which emphasize the importance of hydroprimed in comparison to non-primed seeds. Few examples are sited here, in case of onion (Caseiro et al., 2004), basil (Farahani and Maroufi, 2011), mustard (Srivastava et al., 2010a), maize (Murungu et al., 2004), etc. Furthermore, it has been also noticed that exogenous treatment of proline along with seed priming lead to the quicker seed germination at a minimal temperature as well as better fixing of stress-induced damages in *Vigna radiata* seedlings (Posmyk and Janas, 2007).

8.2.3.2 Osmopriming

As compared to hydropriming, osmopriming benefit the farmers by exhibiting advantageous characteristics like application methodology, technique, and economics (Moradi and Younesi, 2009). This method involves the socking of the seeds in aerated and low water potential solution. Osmopriming directs the activation of pregermination metabolic activities in the seed which ultimately leads to the conversion of dry as well as physiologically inactive seed to hydrated and active seed (Chen and Arora, 2011). A large range of chemicals can be utilized to obtain the minimal water potential conditions in osmopriming techniques. Due to its innocuous kind and large size Polyethylene glycol (PEG) can be used as osmopriming agent most usually. Its application in soaking solution lower the water potential without penetrating seed make it most desirable agent of osmopriming (Tavili et al., 2011; Thomas et al., 2000). In addition to PEG, other important chemicals which bring about osmopriming involve NaCl (Saha et al., 2010; Bajehbaj, 2010), mannitol (Amooaghaie, 2011), KNO_3 (Nawaz et al., 2019), $CaCl_2$, ascorbate, and KCl (Farooq et al., 2006).

8.2.3.3 Nutrient priming

It is essential for every germinating seed to possess adequate nutrient deposit in it before the root system takes up the responsibility of mineral absorption from the soil. Moreover, in a nutrient deficient rhizosphere availability of minimal nutrients limits the seed germination, growth, and vigor. Therefore, it has been observed from several studies that the accessibility of minerals ensures the seed quality capable of better germination and improved seedling establishment. This priming technique involves the seed soaking in a solution comprising macro and micronutrients (Farooq et al., 2012; Imran et al., 2013). Seed priming with nutrients like, macro- (P) and micro-(Zn, B, Mn) elements enhanced stand establishment, produce and grain biofortification in field crops such as wheat (Rehman et al., 2018; Nadeem et al., 2019; Nadeem and Farooq, 2019), rice (Johnson et al., 2005; Atique-Ur-Rehman Farooq et al., 2013), maize (Imran et al., 2013; Muhammad et al., 2015), and chickpea (Ullah et al., 2019b,c). However, the concentrations of these nutrients need to be optimized to avoid nutrient toxicity and seed losses that can impair germination (Atique-Ur-Rehman Farooq et al., 2013; Roberts, 1948; Ajouri et al., 2004).

8.2.3.4 Chemical priming

Several synthetic and natural chemical compounds have been employed in the priming process. These mainly include butanolide, Se, copper sulfate, zinc sulfate, KH_2PO_4, CH_2OH, choline, putrescine, chitosan, and paclobutrazol (Shao et al., 2005; Su et al., 2006; Foti et al., 2008; Hasanuzzaman et al., 2010; Demir et al., 2012). The butanolide is obtained from plant-originated burnt cellulose. There are various reports which signify that the application of smoke extract defends the seed as well as seedlings from the pathogenic attack in seedbeds (Demir et al., 2012). Similarly, treatment of seeds of different crops with different chemicals, e.g., *Z. mays* with $CuSO_4$ and $ZnSO_4$ (Foti et al., 2008), wheat with KH_2PO_4 (Giri and Schillinger, 2003), and Bitter gourd with Se (Chen and Sung, 2001) not only improve the seedling growth and vigor but also increase the plant tolerance to various kinds of abiotic as well as biotic stresses. Similar findings were also observed in the case of seed priming by paclobutrazol, chitosan, ethanol, putrescine, and choline (Jisha et al., 2013).

8.2.3.5 Hormopriming

Under stressful conditions, pretreatment of seeds with hormones serves as an important priming strategy (Jisha et al., 2013; Atici et al., 2003; Gratão et al., 2005; Masood et al., 2012; Hu et al., 2013). For instance, *Secale montanum* seeds primed with GA_3 exhibit enhancement in germination under water insufficiency circumstances (Ansari et al., 2013). Likewise, in *Capscum annum*, Khan et al. (2009) revealed that acetylsalicylic acid and SA pretreated seeds under elevated salinity resulted in more consistency of germination and formation of seedlings. Furthermore, ethylene was utilized to lessen the influence of excessive temperatures on *Lactuca sativa* seed germination (Nascimento et al., 2004).

8.2.3.6 Plant extract priming

Allelochemicals like phenolics, steroids, saponins, terpenoids, alkaloids, and flavonoids possessed by different tissues of several plants exhibit numerous antioxidant production of antioxidant behaviors and ultimately protect the plants from several deleterious pathogens (Satish et al., 2007). Applications of several plant-derived phytochemicals to embryos and other related structure lead to the development of improved root systems which ensure the better water absorption and ultimately result in higher vigor index (Rangaswamy et al., 1993). Some plants are reported to be the rich source of saponin and alkaloids, e.g., *Chlorophytum* leaves, while others like neem leaves are plentiful in flavonoids, terpenoids, steroids, and antiquinone (Raphael, 2012; Chakraborthy and Aeri, 2009). In a similar manner, Dawood et al. (2012) testified that combination of *Trigonellafoenum-graecum* seeds (10%) or *Psidium guajava* leaves or lantana leaves (20%) into the soil substantially increased not only carbohydrates but photosynthetic pigments of the leaf also in sunflower. Similarly, in tomato, the decline in mortality and increase in seedling vigor was described by priming tomato seeds with neem, *Chlorophytum,* and *Vinca* (Prabha et al., 2016).

8.3 Biopriming

In this priming technique, seeds are hydrated accompanied by inoculation with biocontrol agent which may be either a beneficial bacteria or fungi under warm (23° C) and moist conditions for about 20 h. After that, seeds are collected before the emergence of radicles (Callan et al., 1990). In addition, to uniformity of germination and seedling vigor as other priming techniques the major advantage of this is that it also secures the seeds against various seed born as well as soil borne pathogens. Biopriming is a new technique that integrates biological aspects along with physiological aspects for disease control was freshly used as an unconventional approach for managing many seeds and soil borne pathogens (El-Mougy and Abdel-Kader, 2008; Begum et al., 2010). Seed biopriming serve as a standard approach for the transport of a dense population of gentle microorganisms to the soil, where they can inhabit the developing roots of crop plants. This practice is immensely utilized for previous decades efficiently in the field and presents improved or similar outcomes over traditional hazardous fungicides (Callan et al., 1990; Raj et al., 2004).

8.3.1 Different agents of seed biopriming

In their rhizosphere, plant roots are exposed to the various kinds of microorganisms. Plants express a close relationship with these organisms which may be either beneficial or detrimental. There are several reports which reveal that seed biopriming with beneficial microorganisms living inside or in the vicinity of plant roots improve the seedling strength, germination ratio, rate of germination, growth as well as development (Table 8.1). Biopriming agents either directly secrete phytohormones like IAA, gibberellin, and cytokinin or accelerate their synthesis in the plant. Furthermore, they also enhance the accessibility of minerals, viz., N, P, K, Fe, etc. and improve the plant overall growth. In this chapter, we discuss mainly the different agencies of biopriming that include PGPR, mycoparasitic fungi like, Trichoderma, and mycorrhizal fungi. Furthermore, we also extend our discussion toward the impacts of biopriming on plants as well as the mechanism of action of seed biopriming. Generally applied

Table 8.1 Control of seed/soilborne diseases in plants using seed biopriming agents.

Bio-primer	Target	Crop	Seed treatment method	Effect	References
T. virens	*Fusarium wilt*	Tomato	Conidial suspension (1×10^8 spore/mL) of*T. virens* and formulation preparation at the rate of 10 g/kg	Suppress synthesis of pathogen by simultaneously improving plant growth and signaling mechanism	Chandra Nayaka et al. (2010)
Bacillus gaemokensis strain PB69	*Xanthomonas axonopodis*	Cucumber and Pepper	Seeds are submersed in autoclaved endospore producing bacterial culture	Transcriptional upregulation of signaling molecules like, SSA, JA and ethylent has been observed	Jogaiah et al. (2018)
Serratia plymuthica HRO-C48	*Verticillium longisporum* (causes Verticillium wilt) and *Phoma lingam* (causes blackleg)	Oilseed rape (OSR *Brassica napus* L.)	OSR seeds were primed with bacterial suspension by incubating them at 20°C for 5 h in rotary shaker at 150 rpm	Reduce disease incidence and improve plant growth	Fahd Shaukt (2013)
Clonostachysrosea IK726	*Alternaria dauci and A. radicina*	Carrot	*C. rosea* IK726 clay inoculum was applied at the time of seed imbibition	Reduced disease incidence and better seedling stand	Jensen et al. (2004)

Pseudomonas flurescence AB254	*Pythium ultimum*	Sweet corn	Seeds were coated with priming bacteria after that imbibed with water till the moisture content reach to the value 35%–40%	Control preemergence damping off	Callan et al. (1990)
Pseudomonas aureofaciens AB254	*Pythium ultimum*	Sweet corn	Bacterial culture was added to 1.5% methylcellulose (MC) solution followed by pouring of it upon the seeds and then constantly agitated to ensure uniform coating of the seeds	Better seedlings emergence	Mathre et al. (1994)
Trichoderma harzianum, *T. viride*, *Bacillus subtilus* and *Pseudomonas flouresences*	*Fusarium solani*, *Rhizoctonia solani*, *Sclertiumrolfsii*, and *Macrophominaphaseolina*	Soybean	During priming spore suspension of priming agents also supplemented with CMC1% individually applied to soybean seeds	Reduction in the incidence of root rot disease improved seedling defensive response against soilborne contamination	Mona et al. (2017)
Acrophialophorafusispora, *Aspergillusfumigatus*, *Aspergillus niger*, *Penicillium chrysogenum*, *Penicillium citrinum*,	*Fusarium solani*, *Rhizoctonia solani*, *Macrophominaphaseolina*, *Pythium aphanidermatum*, and *Sclerotium rolfsii*	Mungbean (*Vigna radiata* (L.)	Seeds are evenly coated with fungal and bacterial seed primers and then grown in pots	Increase biomass and yield of plants and also represent a better alternative to chemical treatments	Ramzan et al. (2016)

Continued

Table 8.1 Continued

Bio-primer	Target	Crop	Seed treatment method	Effect	References
Stachybotryscharatum, *Trichoderma harzianum*, *T. virens*, *Bacillus cereus*, *B. licheniformis*, *B. megaterium*, *B. pumilus*, *B. subtilis*, *Micrococcus varians*, and *Pseudomonas fluorescens*					
Trichoderma virens, *Trichoderma* spp. strains G-6, Fl9-4, G11-40V (*T. virens* "Q" strains), Tk-35 (*T.koningii*), T-22 (*T. harzianum*), and TKG-12 (*T. virens* × *T. koningii* hybrid)	*Pythium* spp. and *Rhizopus oryzae*	Cotton	*Trichoderma* spp. incubated at about 27°C with shaking at 150 rpm in aliquots containing 5% ground wheat bran and 1% ground peat moss in deionized water, pH was adjusted to 4.0. After 6 days, culture was centrifuged at 3500 rpm and pellet were air dried and applied to seeds	Best possible biocontrol of both pre- as well post in soil containing both pre- and postemergence damping off had been achieved	Howell (2007)
Serratia marcescens N4-5 and *Trichoderma* spp.	*Pythium ultimum*	Cucumber	Dried ethanol extract incubated with seeds for 30 s. *Trichoderma* isolates in gelatin formulation were coated on the seeds	Suppression of damping off disease incidence	Roberts et al. (2016)

T. harzianum, *T. Virenspseudomonas aeruginosa*	*Colletotrichum truncatum*	Soybean	The spore suspensions of bioagent prepared in 1.5% sodium alginate and then filtered through muslin cloth under sterile conditions	decreased pre- and postemergence damping-off, boost of seedling germination and improved seedling stand	Begum et al. (2010)
T. harzianum	*Fusarium solani*, *Rhizoctonia solani* and *Macrophominaphaseolina*	Cowpea	Bioprimed cowpea seeds were sown in contaminated soil	Considerably decrease incidence of root rot diseases and improve and germination rate	El-Mohamedy et al. (2006)
T. harzianum, *T. viride*, *T. hamatum*, *Bacillus subtilis*, *B. cereus* and *P. fluorescens*	*R. solani*, *F. solani* and *S. rolfsii*	Faba bean	Antagonistic bioagents were adhered to seeds by using carboxymethyl cellulose (CMC) and pectin as adhesives	Total decline in the incidence of root rot disease both at pre- and postemergence stages of seedlings	El-Mougy and Abdel-Kader (2008)
Pseudomonas fluorescens	*Sclerosporagraminicola*	Pearl millet	0.2% carboxymethyl cellulose (CMC) was used as adhesive agent to enable the attachment of bacterial suspension to seeds and then incubated at 27°C rotary shaker at 150 rpm for 6 h. Then treated seeds were air dried	Augmentation of: germination rate, seedling vigor, height, tillering capacity, seed weight and leaf area	Raj et al. (2004)

Continued

Table 8.1 Continued

Bio-primer	Target	Crop	Seed treatment method	Effect	References
Pseudomonas fluorescens	*Didymella bryoniae*	Muskmelon	Seeds were firstly presoaked in sterile distilled water and then coated with 0.8% powder formulation of *P. fluorescens* along with moist vermiculite in the ratio 3:1 (3 parts of vermiculite to 1 part of seed). Then seeds were dried in shade	Enhance protection of seedlings against seedborne infections	Rao et al. (2009)
Serratia plymuthica (strain HRO-C48) and *Pseudomonas chlororaphis* (strain MA 342)	*Leptosphaeria maculans*	Oilseed rape (OSR) (*Brassica napus*)	In order to obtain CFU @ $\log_{10}$ 6–7 $seed^{-1}$.Seeds were soaked in bacterial suspensions	Increased resistance against blackleg disease	Sudisha et al. (2006)
Trichoderma harzianum and *Rhizoboiummelilotii*	*Macrophominaphaseolina*, *Rhizoctonia solani* and *Fusarium* sp.	Peanut, chickpea, okra, and sunflower	Seed were immersed in Spore/cell suspension of priming agents and then air dried	Increase the growth of plants and decrease the pathogen survival	Rafi and Dawar (2015)
Bacillus spp.	*Alternaria solani*	Tomato	Bacterial formulation	Promoted seed germination along with control disease incidence	Koch et al. (2010)

PGPR, e.g., *Azotobacter*, *Azospirillum*, *Agrobacterium*, *Pseudomonas*, and *Bacillus* enhance growth and harvest, control seed or soilborne diseases, and improve abiotic stress tolerance, whereas *Trichoderma* and *Pseudomonas* develop resistance against biotic stress (Reddy, 2012; Mahmood et al., 2016).

8.3.1.1 Plant growth-promoting rhizobacteria

Kloepper (1978) for the first time make use of the word plant growth-promoting rhizobacteria (PGPR) describing them as bacteria which are intimately associated with rhizosphere. PGPR inoculation has augmented diverse crop produces in normal and stressful circumstances. Seed priming through PGPR, i.e., soaking the seeds for a precalculated period in liquid bacterial suspension, initiate the physiological processes inside the seed while radicle and plumule emergence is prohibited till the seed is propagated (Anitha et al., 2013). This spread of antagonist PGPR inside the seeds is 10-fold than infecting pathogens which enables the plant to survive those pathogens mounting the use of biopriming for the purpose of biocontrol also (Callan et al., 1990). Seed inoculation involves the use of carrier material for better transportation and application, use of adhesives to ensure the sticking of bacteria to the seeds and sometimes other materials avoiding desiccation of the inoculum (Elegba et al., 1984). Peat-based inoculants are most common and extensively used since the discovery of rhizobium for leguminous crops (Walker et al., 2004).

Most favored and commonly used method of inoculation includes application of adhesive agents on the seeds followed by inoculum spreading under shade (Vincent et al., 1962). Among the adhesive agents, most used are Arabic gum, sugar solution, methylcellulose, polyvinylpyrrolidone, caseinate salts and polyvinyl acetate (Deaker et al., 2004). Reddy (2012) explained biopriming more in biocontrol aspect as an application of beneficial bacterial inoculum to the seeds and their hydration protect these seeds against several diseases. PGPR keep on multiplying in the seed and proliferate in the spermosphere even before sowing (Taylor and Harman, 1990). Biopriming treatment is potentially able to promote quick and even germination as well as better plant growth (Moeinzadeh et al., 2010). Biopriming with rhizospheric bacteria has been reported in crops such as carrot (Jensen et al., 2001), sweet corn (Callan et al., 1990, 1991), and tomato (Warren and Bennett, 1999; Harman et al., 1989). In the case of efficacy and survival of biological agents, priming has been reported beneficial and enhance the plant growth and yield (Callan et al., 1990, 1991; Warren and Bennett, 1999; Harman et al., 1989). Germination and enhanced seedling establishment are obtained through seed priming with PGPR (Anitha et al., 2013). Bio-osmopriming can significantly enhance the uniformity of the germination and plant growth traits when associated with bacterial coating (Bennett, 1998).

8.3.1.2 Antagonistic fungi

Several reports are available which signify the protective potential of beneficial antagonistic fungi like *Trichoderma* spp. against the several soil and seedborne pathogens (Jensen et al., 2001; Harman and Taylor, 1988) (Table 8.1). *Trichoderma* spp. is one of the extremely popular free-living saprophytic fungi in rhizosphere which generally dominate the key segment of fungal biocontrol mediators in biopesticide manufacturing (Woo et al., 2014). Inoculation with different strains of *Trichoderma* spp. not only endorses plant growth but also have extensive range of antagonistic pursuits against a variety of soilborne phytopathogens through several mechanisms like competition for nutrient and space, antibiosis, parasitism, detoxification of virulence factors of pathogen, etc. (Keswani et al., 2014; Singh et al., 2015).

Due to the abovementioned reasons, Trichoderma has been widely used for seed treatment (Pill et al., 2009). In seed treatment, resting spores or conidia of *Trichoderma* are applied on the seed surface which must get geminated prior to interaction with pathogens. The presence of an active fungal agent in the spermoplane or spermosphere in enough quantity is must to protect germinating seeds (Pill et al., 2009). Pandey and Gohel (2017) revealed that treatment of *T. viride* or *T. harzianum* to soybean seeds helped in the management of charcoal rot of soybean. Similarly, inoculation of soybean seeds with BCA like, *T. harzianum*, *T. virens*, *T. atroviride*, and *Pseudomonas fluorescens* leads to enhancement of root-shoot length of magnitude 52% and 44%, dry weight by 55%, and chlorophyll content to 206% as compared to control (Entesari et al., 2013). Likewise, similar findings were also observed in chickpea (Pandey et al., 2016) and green gram (Meena et al., 2017).

8.3.1.3 Mycorrhizal fungi

Arbuscular mycorrhiza (AM) is one of the most familiar symbiotic alliance of plants with microorganisms. AM fungi occupied the ecological habitats and offer a wide range of valuable services to the plants in term of improved nutrition, resistance, and tolerance to several biotic and abiotic stresses, soil composition, and fertility. For instance, a newly detected, AM fungus *Piriformospora indica* Sav belongs to Basidiomycota serve as an important BCA. Though, this fungus has not been applied yet as a seed biopriming agent (Verma et al., 1998). In addition to the improvement in plant growth, this fungus also induces tolerance against salt stress as well as resistance against root and shoot pathogens of wheat and tomato (Fakhro et al., 2010). Hence, in seed treatments perspective its development as a priming agent may provide considerable importance in future. For example, in comparison to isolated treatments, *Pseudomonas*, *T. harzianum*, and *Glomus intraradices* (arbuscular mycorrhizal fungus) combination seems to be more effective and able to reduce Fusarium wilt incidence and severity by 74 and 67% in pots and fields respectively and also increased the yield by 20% (Srivastava et al., 2010b).

8.4 The procedure of seed biopriming

See Fig. 8.2.

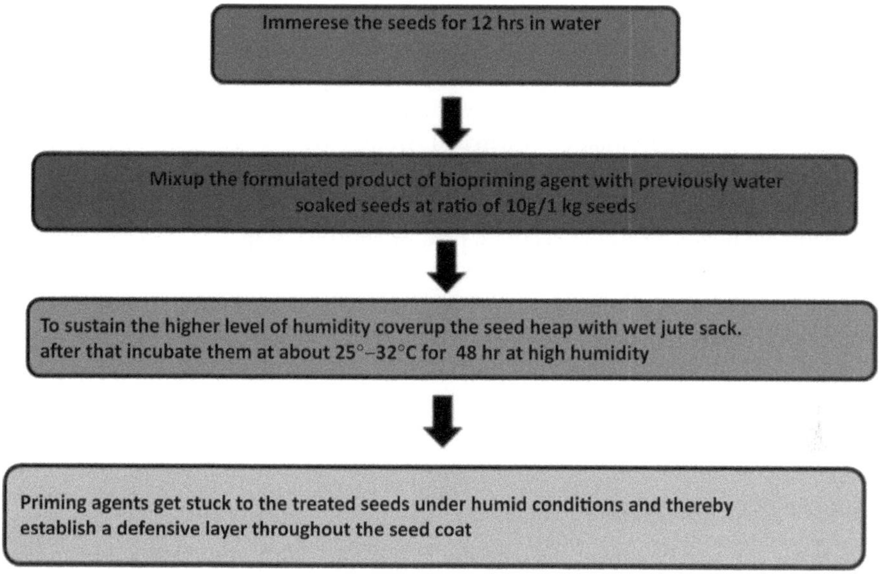

Fig. 8.2 Illustrate the steps comprising in general procedure of seed biopriming.

8.5 Mechanism of action of seed biopriming by bioagents

As previously discussed, biopriming is an evolving tendency designed to enhance seed value, seedling strength, efficiency, and challenge to biotic and abiotic stress by decreasing the usage of chemical efforts for sustainable agriculture. It is accompanying with an upsurge in hydrolytic enzyme actions, reactive oxygen species detoxifying enzymes activities, modification in internal plant hormone levels, and variable genes expression in plants that contributes to the enhanced plant growth and resistance against stressful conditions. The basic mechanism of seed biopriming involves two actions:

- Beneficial influences on seedling growth and vigor
- Antagonistic effects on disease-causing pathogens

8.5.1 Beneficial influences on seedling growth and vigor

8.5.1.1 Speed Up and synchronization of seed germination

The successful seed germination and establishment of a healthy seedling ascertain the persistence and propagation of plant species (Babu et al., 2015). Priming of seeds with bioagents not only improves the germination but also increases stress tolerance during early seedling growth (Fallahi et al., 2011; Krishna et al., 2008). Biopriming initiate the seed germination process mainly by synthesis and signaling pathway of phytohormones. Biopriming has been testified to rise GA/ABA proportion (El-Araby et al., 2006) and this may be a straightforward outcome of priming impacts on the model of gene expression (Schwember and Bradford, 2010). In primed seed, uniform endogenous GA concentration may facilitate synchronized weakening of endosperm, elongation of the embryo, and mobilization of reserves (Sung et al., 2008). Seed priming enhances the synthesis of ethylene in treated seeds which ultimately promote the production of endo-β-mannanase activity which further facilitate endosperm weakening and postpriming germination (Chen and Arora, 2013). Moreover, biopriming has been described to begin repair and reactivation of preexisting mitochondria and initiate the biogenesis of new ones (Sun et al., 2011).

Hariprasad et al. (2009) has reported that, *Bacillus* sp. BSp.3/aM (an IAA synthesizing, phosphate solubilizing, siderophore, and antibiotic-forming bacterial strain) increase seedling strength under laboratory environments and decreased prevalence of *C. capsici* under greenhouse conditions. Similarly, treatments of microbial inoculations to seed and/or transplants was observed to promote tolerance of plants against soil borne pests (Le et al., 2009; Sikora et al., 2008) and plant pathogens (Mathre et al., 1994; Pill et al., 2011; El-Bab and El-Mohamedy, 2013).

8.5.1.2 Plant growth

Biopriming directly facilitates the increase in plant growth by production or stimulation of growth-promoting compounds and mineral solubilization. Podile and Kishore (2007) accomplish several growth-promoting (PGP) mechanisms of PGPR which mainly include modified and enhanced root hair branching, advancement in seeds germination, heightened and quicker nodule performance, increased leaf area ratio, upregulation of synthesis of certain phytohormones, amplified water and nutrients absorption by plants, improved biomass production with additional growth, and healthier carbohydrate buildup, which consequently rises the growth of plant spp. Conversely, Glick (2003) classifies the bacterial strains supported plant growth in three diverse ways, comprising plant hormone synthesis (Dobbelaere et al., 2003), improved nutrient absorption by plants (Cakmakçi et al., 2006), and circumventing the diseases incidences via biological control mechanisms (Saravanakumar et al., 2008). Dey et al. (2004) propose the necessity of discovering other means of plant growth enhancement by PGPR besides previously enlisted mechanisms.

Listing of all PGPR pathways identified and investigated may include the following: (a) nutrients solubilization as well as mineralization particularly of phosphorus (Richardson, 2001; Banerjee and Yasmin, 2002); (b) symbiotic or asymbiotic N_2 fixation (Kennedy et al., 2004); (c) secretion of phytohormones like, GA_3 and cytokinins (Dey et al., 2004); IAA (Patten and Glick, 2002) and ABA (Patten and Glick, 2002); (d) synthesis of 1-aminocyclopropane-1-carboxylate (ACC)-deaminase, which assist in a decrease in the level of ethylene in roots and consequently help in increasing the root length and vigor (Li et al., 2000; Penrose and Glick, 2001); (e) synthesis of phytopathogen's antagonistic compounds like cyanides and antibiotics (Glick and Pasternak, 2003); (f) by synthesis of Fe chelating compound named siderophores enhance the obtainability of iron in the rhizosphere (Glick and Pasternak, 2003); (g) production of water-soluble vitamins comprising biotin, niacin, thiamine, and riboflavin (Revillas et al., 2000); (h) last but most important among these is biological control of pests and insects (Russo et al., 2008).

There are several reports which demonstrate that the application of biocontrol agents leads to the increase in germination, shoot length, root length, fresh weight, vigor index, and the number of nodules and importantly lessened plant mortality. For instance, priming of tomato seeds with *P. fluorescens* enhance seedling development and lowers Fusarium wilt incidence on tomato. Karthika and Vanangamudi (2013) assessed the impact of biopriming of maize hybrid seed with biofertilizers liquid for improving propagation and vigor. Biopriming with 20% *Azospirilum* for 12h conveyed greatest values of all parameters like the speed of germination, root length, shoot length, dry matter production, total dry matter production, and vigor index over the nonprime seed.

8.5.1.3 Mineral nutrition

No doubt, microorganisms in agriculture system permit a substantial decrease in the extent of minerals employed without compromising the productivity. Microbial applications play a key role in this direction by participating in the regulation of enzymatic activities pursued by nutrient subtleties in the rhizosphere. Furthermore, the defensive influence of microbes' counter to a wide array of stress has been well recognized and is the justification for their multidimensional consumption in balanced agriculture. In the modern era of sustainable agriculture, the plant-microbe collaborations in the rhizosphere perform a crucial role in conversion, deployment, solubilization, etc. of mineral nutrients from a restricted nutrient reservoir and afterward absorption of vital nutrients by plants to grasp their complete genetic budding. Numerous PGPR along with symbiotic and free-living rhizobacterial spp. are documented to produce Auxins and gibberellins in the plant rhizosphere and in this manner perform a significant part in expanding the root surface area and root tips number (Bhattacharyya and Jha, 2012).

Among the several examples of mineral nutrition by microorganisms bacterial N_2 fixation is the best. The symbiosis between legume plants and rhizobia a valuable instance for PGPR. In turn, root exudates, especially carbohydrates bacteria offer nitrogen to the host plant for the production of amino acids. Besides this, other free-living bacteria, e.g., *Stenotrophomonas*, *Azospirillum*, and *Burkholderia* are also involved in this process (Dobbelaere et al., 2003). Another important nutrient is sulfate, which can be delivered to the plant via the oxidation process (Banerjee and Yesmin, 2002). Phosphorus can be made available to plants by Bacterial mediated liberation of phosphorous from phytates like organic compounds (Unno et al., 2005). For example, application of *Azospirillum* caused an enhancement in root growth and promote acidification around the roots which in turn facilitate the uptake of P and other micro and macroelements as well (Dobbelaere and Okon, 2007). Mineral availability is also elevated via the synthesis of siderophores and siderophore uptake systems (Katiyar and Goel, 2004). Weakly soluble nutrients can be supplied through the solubilization of bacterial siderophores and the release of organic acids. In recent times, de Werra et al. (2009) demonstrated *Pseudomonas fluorescens* CHA0 competence toward glucuronic acid production as well as solubilization of mineral phosphate. Moreover, this observation focusses on the involvement of gluconic acid metabolism with antagonistic action against phytopathogens (Singh et al., 2011).

Similarly, comparison to control *Trichoderma* treated plant roots exhibit the better potential for deep penetrations as well as absorption of mineral nutrients from the soil. According to Harman et al. (2004), fungal strains produce several acids e.g., glucuronic, citric, coumaric acids which assist in the release of phosphorus ions which seems to be unavailable to plants in most of the soils (Zhao et al., 2014). The presence of *T. harzianum* strain 1295–22 in the soil increases the obtainability of phosphorus as well as other micronutrients like Fe, Zn, etc. in liquid medium (Altomare et al., 1999). Similarly, application of strain T-203 presently known as *T. asperelloides* enhance the concentrations of Fe and P to an amount of 30% and 90%, respectively, in the rhizosphere of plants. Moreover, root and shoot growth as a response to *Trichoderma* inoculation leads to an elevation in the concentration of Cu, Na, Zn as well as other micronutrients (Yedidia et al., 2001). Deficiency of Fe in alkaline soil is a major drawback for the crop production in agriculture. The potential ability of *Trichoderma* for siderophore production can be utilized to combat this problem. Application of cucumber roots with *T. asperellum* (T-6) not only increase Fe^{2+} and siderophore in soil but also increase the activity of Fe^{2+} and Fe^{3+} chelate reductase (Zhao et al., 2014). Furthermore, Colla et al. (2015) revealed that two kinds of siderophores, i.e., hydroxamate and catechol are produced by MUCL45632 strain of *T. atroviride*. These studies evidenced that *Trichoderma* application in soil assists the plant in the reduction of Fe^{3+} to Fe^{2+} which consequently boost its better solubilization and uptake.

8.5.1.4 Biopriming mediated physiological and biochemical advantages to the plants

As previously mentioned, seed biopriming with microorganisms reported heightening the root-associated characteristics and generating a robust root system. In rice, Rêgo et al. (2014) observed that biopriming induced increase in the root to leaves ratio. In addition to enhancement in root system biopriming of seeds also offer elevated root to shoot ratio, root dry weight, improved leaves number as well as area, chlorophyll concentration in crops (Priya et al., 2016; Rawat et al., 2012). For instance, Anitha and Jahagirdar (2015) testified expanded root and shoot length in soybean following biopriming. Furthermore, Rêgo et al. (2014) reported elevated contents of lignin in PGPR applied to rice. Lignin engaged in the superior association of macrofibrils which facilitate the structural strength of cell wall and the roots (Rubin, 2008). In addition to this, plant development, nutrient absorption, and nutrient use proficiency and harmonization in germination and its rate of germination, excellent plant stand under usual and stressed circumstance have been boosted by seed biopriming (Moeinzadeh et al., 2010; Yadav et al., 2013; Tanwar et al., 2013; Muruli, 2013).

In the case of bioprimed crops, boosted biomass synthesis, productivity, and yield are the consequences of growth stimulation (Karthika and Vanangamudi, 2013; Baral and Adhikari, 2013; Namvar and Khandan, 2014; Kavino et al., 2010). The micropropagated plants which have decreased photosynthetic activity, inadequately functioning of stomata and weak root and shoot system, biopriming serve as an adequate solution for improved growth. PGPR- mediated biopriming minimizes the time period requisite for lignification in micro-propagated plants and therefore, quickens production process (Ramamoorthy et al., 2002; Cyr and Bewley, 1990).

Biopriming in plants mediates several biochemical changes, which mainly include, augmented proteins production, hormones synthesis, and phenol and flavonoid accumulation which ultimately contribute to improved plant developments. In herbaceous plants, growing reactions are governed by nitrogen reserve substances such as nitrates, amino acids, and proteins (Volenec et al., 1996; Dhanya, 2014). In comparison to non-primed seeds percentage of soluble protein is higher in bioprimed seeds and seedlings (Aishwath et al., 2012). Following the PGPR biopriming, a gain in total protein content and free amino acid content was observed during various phases of plant growth (Warwate et al., 2017; Sofy et al., 2014; Gupta and Kaur, 2005). Furthermore, the content of total soluble as well as reducing sugar increased as a result of biopriming (Rolland et al., 2006). Besides, acting as compatible osmolytes to maintain cell turgor, soluble sugars also act as a primary messenger in signal transduction pathway and thereby play a crucial role in regulating the expression of a wide range of gene participate in plant growth and metabolism (Naseem and Bano, 2014; Chen and Arora, 2013). Early imbibition activity in primed seeds ensures the effective mitochondrial metabolism by boosting ATP synthesizing system (Singh et al., 2003). PGPR employed in biopriming upregulate the production of certain phenolic acid at various growth phases (Spaepen and Vanderleyden, 2011). Moreover, enhanced IAA contents in plant leads to the development of root number, root hairs, and area of the root, adventitious and lateral root formation, cell division and differentiation, development of vascular tissue, biosynthesis of phyto-pigments, and variable metabolites (Naik, 2015). Like plant growth promotion, biopriming also triggers the production of defense-related enzymes like, POD, SOD, catalase, chitinase, PAL, etc., which extend fitness assistance to plants against stressful conditions.

8.5.1.5 Yield improvement

As discussed previously, seed biopriming not only serve as a managing technique for the control of seedborne diseases but also enhance the crop yield and productivity in a sustainable style. Seed biopriming with *T. viride* (40%) for 4 h enhance the several seed quality-related parameters. In addition to this, biopriming reduce the incidence of seed rotting to a significant value. Similar findings were also observed in the case of *P. fluorescens* bioprimed seeds. Moreover, Biopriming also ensued early flowering initiation, enhanced plant height, branches/plant, pods/plant, number of seed/pods, 100 seed weight, dry pod weight, and seed yield (Naik, 2015).

8.5.1.5.1 Disease resistance

Seed priming allows plants to enable defensive responses to phytopathogens more quickly and effectively without varying plant growth and has the capability to develop as a tactical tool for modern plant safety. Biopriming is an appealing, easy, and cost-effective approach that generates systemic resistance in plants. After suitable stimulation with bioagents, defensive mechanisms in plants induced and lifted which enable it to tackle further attack by the pathogens. This reaction is called induced systemic resistance (ISR) and/or immunization and exhibit efficacy toward a wide range of pathogens (van Loon et al., 2006; Conrath et al., 2006). Microorganisms primed seed shows biochemical and physiological modifications lead to the synthesis of defense-related proteins and chemicals.

Recently, plant growth-promoting rhizobacteria (PGPR) induced ISR has collected overwhelming attention (Burdman et al., 2000; Ramamoorthy et al., 2001; Vallad and Goodman, 2004; Kuć, 2006). Further, there are several reports which signify the involvement of salicylic acid (SA), acetylsalicylic in ISR induction against a variety of pathogens under regulated environments (Saikia et al., 2003; Sarwar et al., 2005). Among the various PGPRs *Pseudomonas fluorescens* is the most valuable ISR inducer. Application of PGPR to seeds accelerate ISR in many host-pathogen interfaces and is correlated with signaling proteins that stay inactive under usual circumstances and activates after the exposure to stressful conditions (Conrath et al., 2006; Mathre et al., 1999). Various microorganisms are integrated into the seed for the period of priming activity which subsequently initiates speedy seed colonization and uniform seed surface coverage which either favor their colonization before pathogen infection or stimulate disease-resistance mechanisms (Sabalpara, 2015). No doubt, this technique of useful fungi and bacterial seed coating serves as a most recent approach for managing chief seed and soilborne pathogens (Entesari et al., 2013). This strategy has the potential to offer adequate agents in the right amount, place as well as in time and leads to mobilization, initiation, and augmentation of various cellular resistance mechanisms (Conrath et al., 2002).

The occupation of seedling roots with priming microorganisms emerges as a prospective replacement to the consumption of chemical pesticides (Manjunatha et al., 2013). Biopriming of seeds with microbes improve the physical and mechanical strength of the host cell wall. In addition to this, it also modifies the biochemical or physiological texture of plant which consequently promote the synthesis of pathogenesis-related proteins like, chitinase, β-1,3-glucanase, peroxidase, polyphenol oxidase (PPO), and phenylalanine ammonia lyase (PAL) and enhance the accumulation of secondary metabolites (Burdman et al., 2000; Ramamoorthy et al., 2001; Dutta et al., 2008). Strains of *Pseudomonas*, *Bacillus*, *Trichoderma,* and other microbes serve as outstanding bioprimers to ISR induction improvement in plant growth (Ongena et al., 2000; Gnanamanickam et al., 2002).

8.5.1.6 Mechanisms of disease resistance induction through seed priming

Initiation of basal defensive mechanism in plants after exposure to a pathogen is a normal protective approach (Reddy, 2012). After getting certain stimuli several inactivated defense-related genes get triggered (Greenberg et al., 2000). Microbial-induced production of SA is well known for the ISR in plants (Kloepper et al., 2004). PGPRs priming in seeds accelerated germination rate as well as improves seedling establishment which also accompanied by PGPRs settlement and proliferation of the spermosphere (Callan et al., 1990). Microbial-assisted seed priming triggers ISR against several diseases caused by fungi, bacteria, viruses (Liu et al., 1995), insect

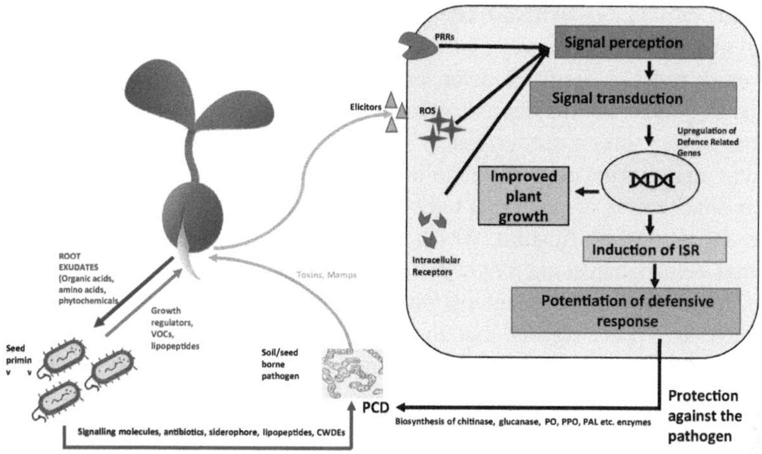

Fig. 8.3 Diagrammatically represent the mechanism of action of priming agents in the induction of defense response in plants against seedborne pathogen in rhizosphere.

(Zehnder et al., 1997), and nematode (Sikora, 1988). Seed primer microbes produced a phenolic compound, i.e., SA, which initiate a signal transduction pathway.

These events consequently lead to the production of defensive compounds in plants (Gaffney et al., 1993) (Fig. 8.3). It has been reported that lipopolysaccharides (LPSs) particularly, O-antigen side chain that exists in the outer microbial membrane serves as signaling molecule and by eliciting ISR produce several defensive compounds (Van Wees et al., 1999). In addition to SA, Jasmonic acid and ethylene biosynthesis because of application of various root colonizing bacteria are also observed as signals for induction of disease resistance in counter to microbial pathogens (Dong, 1998). Subsequently, ISR induction in plants is also correlated with the heightened activity of chitinase, β-1,3-glucanase, peroxidase (PO), polyphenol oxidase (PPO), phenylalanine ammonia-lyase (PAL), and accretion of phenolic compounds along with other PR (pathogenesis-related) proteins (Dutta et al., 2008) (Fig. 8.3).

The chitinase enzyme causes the hydrolysis of chitin, which is a major component of fungal cell wall (Jackson and Taylor, 1996). In a similar way, the function of β-1,3-glucanase enzymes is to degrade glucan present in the cell wall of pathogenic fungal spp. (Fridlender et al., 1993; Potgieter and Alexander, 1966; Velazhahan et al., 1999). Phenyl Ammonia Lyase is an important enzyme of phenylpropanoid metabolism and is associated with the biosynthesis of various defensive compounds including, phenolics, phytoalexins, and lignins, which basically serve as protective chemical barriers against pathogen invasion in plants (Kloepper et al., 2004). PO and PPO (Cu containing) enzymes bring about oxidation of phenolic compounds to quinines and also produce H_2O_2, which subsequently release free radicals which in turn accelerate the rate of conversion of phenolic compounds to lignins and ultimately attribute the tremendous role in disease resistance (Ramamoorthy et al., 2001; Meena et al., 2000).

Phenolic compounds contain one or more benzene rings along with phenolic hydroxyl groups and are widely distributed in higher plants. Broadly, phenolic compounds produced as a result of defensive metabolism classified into anthoxanthins, anthocyanins, leucoanthocyanins, glycosides, sugar esters of quinines, hydroxybenzoic acids, shikimic acid, esters or hydroxyl cinnamic acids, and coumarin derivatives make a wide class of phenolic compounds. These compounds mainly serve as donors or acceptors of H^+ during redox reactions. In addition to this, they also facilitate the oxidation lignification, which limits the proliferation of disease-inducing pathogens (Velazhahan et al., 1999; Wei et al., 1996). Studies revealed that individual strain of PGPR produces ISR against multiple diseases on one plant host (Wei et al., 1996). Several reports signify that the treatment of seeds with priming microbes results in the establishment of numerous mechanisms of biological control toward the suppression of a wide range of pathogens. For example, treatment of seeds with *P. fluorescens* safeguarded plants through ISR not only against *F. oxysporum*f. sp. *raphani* (fungal root pathogen) but also against *P. syringae*pv. Tomato (bacterial leaf pathogen) and *Alternaria brassicicola* and *F. oxysporum* (fungal leaf pathogens) (Hoffland et al., 1996).

8.5.2 Antagonistic effects on disease-causing pathogens

Seed priming with bio-inoculants helps in disease suppression by utilizing different mechanisms such as siderophore production, antimicrobial secondary metabolite, and secretion of lytic enzymes.

8.5.2.1 Destructive parasitism

Parasitism is a crucial method utilized by biopriming agents to suppress the growth of pathogens in their vicinity (Haran et al., 1996; Viterbo et al., 2002; Whipps, 2001). This type of control management necessitates nearby contact between the fungal pathogen (target) and the beneficial microbe to ensure target identification, attack initiation, infiltration, and consequent degradation as well as lyses of target spores or hyphae (Handelsman and Stabb, 1996; Benítez et al., 2004). A pathogenic fungus is recognized by rhizospheric microbes through spores, fruiting body, mycelial structure, etc. Bacterial flagella mediate their chemotactic movements either toward mycotoxins or other metabolites (Hogan et al., 2009). Furthermore, the degree of connection between host and pathogen may range between simple to the engulfment of the fungal pathogen (Hogan et al., 2009).

The extremely popular fungal colonizers bacterial strains include *Pseudomonas* and *Burkholderia* (Hogan et al., *2009*). Colonization of *Fusarium oxysporum* hyphae in tomato roots by *P. fluorescens* WCS365 and *P. chlororaphis* PCL1391 reduce the incidence of foot and root rots diseases in this plant (Bolwerk et al., 2003). The biocontrol by theses bacterial strains was mainly attributed by the cell wall hydrolytic activities of several enzymes (Chet, 1987). Furthermore, several reports indicate the hydrolytic activities exhibited by enzymes like chitinases (Wang and Chang, 1997; Radjacommare et al., 2010; Hariprasad et al., 2011) glucanases (Fridlender et al., 1993), and proteases (Dunne et al., 1997) synthesized by bacterial spp. such as, *Pseudomonas* (Nandakumar et al., 2007), *Bacillus* (Huang et al., 2005), and *Serratia* (Chet et al., 1990) (Table 8.2).

Straightforward biocontrol by *Trichoderma* spp. may be exerted by parasitizing a wide range of fungal pathogen, growing in the rhizosphere. *Trichoderma* with the help of cell wall carbohydrates adhere to the pathogens by binding with their lectins. Once this contact has been established, the *Trichoderma* attached make coils around the pathogen and form appressoria. This overall process involves the participation of CWDEs (Cell Wall Degrading Enzymes) as well as defensive metabolites (Contreras-Cornejo et al., 2015; Brotman et al., 2013). Further, investigation on the reliable signal transduction pathways of *T. atroviride* have led to the involvement of cAMP and MAP kinase, G proteins, etc. which consequently control CWDEs, antibiotic biosynthesis, and coiling around target hypha (Altomare et al., 1999).

8.5.2.2 Competition

Certainly, rhizosphere around the plant roots offers a limited supply of nutrients, space, Oxygen, and other essential factors for the growth of soil-inhabiting microorganisms. Therefore, the competition among these microbes for these resources is a natural phenomenon and serves as an ultimate mechanism for the biological control of soilborne diseases in root surface area of plants (Cawoy et al., 2011; Eisendle et al., 2004). However, competitive interaction among the pathogenic and beneficial strains occurs only under the condition of analogous requirements of these spp. toward limited resources (van Dijk and Nelson, 2000). Chemotactic substances like sugars, organic, and amino acids assist in flagellar mobility of PGPR toward root exudates in the rhizosphere (De Weger et al., 1987). In addition to this, traits like O-antigen of lipopolysaccharide, biosynthesis of amino acids (Simons et al., 1997), and vitamin B1 (Simons et al., 1996) facilitate the successful establishment and colonization of PGPRs on root surface (Tilak et al., 1999; de Weert and Bloemberg, 2007). Further, development of cellulose fibrils (*A. tumefaciens* and *R. legunisprum*) and biofilms (*Rhizobium*, *Azospirillum*, *Pseudomonas*, and nitrogen-fixing strains) provide the better advantage to adhere and survive in root surrounding as compared to pathogenic organisms (Rodríguez-Navarro et al., 2007; Meneses et al., 2011).

Despite the abundant amount of iron on earth, it is not easily accessible to microbes as well as plants. Iron is an essential nutrient for all living organisms (Neilands, 1981). Because it is easily oxidized to scantily soluble ferric ions which are very difficult to absorb as such (Neilands, 1987). BCAs produce a class of protein which exhibit a higher affinity toward Ferric ions and thereby serve as a chelating ligand for these ions in the rhizosphere (Castignetti and Smarrelli, 1986). Thus, this peculiar characteristic of beneficial microbes provides the selective advantage to them over pathogen organism and also deprives the pathogens of limited Fe available in the rhizosphere and consequently inhibits their growth (O'sullivan and O'Gara, 1992; Haas and Défago, 2005). Another mechanism by which these priming agents give serious competition to their counterparts is effectual breakdown and absorption of root exudates so that they become inaccessible for the consumption of target pathogen. For example,

Table 8.2 Compounds synthesized by seed priming agents against seed borne pathogens.

Seed priming agent	Target pathogen	Antagonistic compound	Plant	Effect	References
P. aureofaciens 30–84	*Gaeumannomycesgraminis* var. *tritici*.	Phenazine-1-carboxylic acid	Wheat	Suppression of disease incidence	Thomashow et al. (1990)
Bacillus subtilis BBG100	*Pythium aphanidermatum*	Mycosubtilin	Tomato	Increase in germination rate as well as improved fresh weight of seedlings	Leclere et al. (2005)
Pseudomonas fluorescens	*Pythium ultimum*	Pyoluteorin	Cotton	Seedling survival rate increased	Howell and Stipanovic (1980)
Trichoderma koningii, *T. harzianum*, *T. aureoviride*, *T. viride*, *T. virens*	*Pythium aphanidermatum*	Koningins, viridin, dermadin, trichoviridin, lignoren, and koningic acid	Chili	Pathogen growth inhibition as well-treated seeds exhibit maximum germination percentage, shoot length, root length, and vigor index	Muthukumar et al. (2011) and Jeleń et al. (2014)
Bacillus sp. BSp.3/aM	*Colletotrichum capsica*	Phenylalanine ammonia-lyase, peroxidase, polyphenol oxidase, lipoxygenase, chitinase	Chili	Decrease in the incidence of anthracnose due to upregulation of defensive enzymes-mediated enhanced ISR	Jayapala et al. (2019)
Bacillus spp. KFP-5, KFP-7, KFP-17	*Pyriculariaoryzae*	Superoxide dismutase, polyphenol oxidase, phenylalanine ammonia-lyase, protease, glucanase	Rice	Bioagent alleviate the oxidative damage to rice by enhancing antioxidative defense system	Rais et al. (2017)
T. *pseudokoningii* BHUR2, *T. harzianum* BHUP4, *T. viride* BHUV2, *T. longibrachiatum BHUR5*	*Sclerotium rolfsii*	Superoxide dismutase (SOD) and peroxidaes (POx), phenylalanine ammonia lyase (PAL), and total phenol content (TPC)	Tomato	Primed tomato plants were observed to be healthier and exhibit enhanced antioxidative defense mechanisms	Rajput et al. (2019)

Trichoderma viride, *T. harzianum*, *T. virens*, *and Pseudomonas fluorescens*	*Macrophominaphaseolina* and *Rhizoctonia solani*	Peroxidase, polyphenol oxidase, and phenyl alanine ammonia lyase	Mungbean	Increase in the plant over all biomass as well as defense-related proteins	Meena et al. (2016)
Trichoderma erinaceum	*Fusarium oxysporum* f. sp. *lycopersici*	Chitinases and glucanases, superoxide dismutase, Catalase, increased lignified stem tissues	Tomato	Upregulation of defense related genes	Aamir et al. (2019)
Pseudomonas aeruginosa (UPMP3), *Burkholderiacepacia* (UPMB3), and *Serratia marcescens* (UPMS3)	*Sclerotium rolfsii*	Peroxidase (PO), polyphenol oxidase (PPO), and phenylalanine ammonia-lyase (PAL) activities	Chili	Reduction in the incidence of pre- and postdamping off	Siddiqui and Meon (2009)
Bacillus subtilis	*Phytophthora capsici*	Siderophores, hydrogen cyanide, and hydrolytic enzymes	Red pepper	Reduce disease severity and improve plant growth and yield	Lee et al. (2008)
Bacillus strains	*Phytophthora capsici*, *Phytophthora citrophthora*, *Phytophthora citricola*, *Phytophthora sojae*, *Colletotrichum coccodes*, *Colletotrichum gloeosporioides*, *Colletotrichum acutatum*, *Rhizoctonia solani*, *Fusarium solani*, *Fusarium graminearum*, *Pyricularia* spp., and *Monilina* spp.	Lipopeptides like, surfactins, theiturins, and the fengycins	Red pepper	Reduce the disease incidence and increased (32.2%) the vigor index of *Capsicum annuum* L. plants	Oh et al. (2011)
Pseudomonas syringae pv. *lachrymans*	*Colletotrichum orbiculare*	Induction of ISR	Cucumber	Better pathogen control and improvement in yield	Wei et al. (1996)

priming of seeds of crops like carrot, cotton, cucumber, lettuce, radish, tomato, wheat, corn, and pea with *E. cloacae,* biocontrol the Pythium damping-off because of the fact that PGPR metabolize the long-chain fatty acids especially, linoleic acid, discharged from the seed at the time of germination and by this act decrease their readiness to sporangia of *P. ultimum* (Kageyama and Nelson, 2003). Further, the synthesis of antibiotics by priming agents also gives ecological competence to their target and stimulates the defensive response in host plants (Ongena and Jacques, 2008).

8.5.2.3 Antibiosis

The term Antibiosis refers to the mechanism by which certain metabolic substances has been produced by BCA which insert ill effects on other target organisms (Jackson, 1965). Priming microorganisms synthesize compounds such as antibiotics, toxins, VOC (volatile organic compounds), biosurfactants, etc. Berg (2009) which limit the growth of the pathogen in the rhizosphere. In general term antibiotics are secondary metabolites which are secreted by BCA in a very small amount. These are extremely precise against their target and because of their easy absorption by plant or seeds they can seriously affect even the deep-seated pathogen in the diseased tissues (Dekker, 1963). There are large number of reports which support the critical role of antibiotics in repressing the plant pathogens and their corresponding diseases. Among the several antibiotics producing PGPRs, *Pseudomonas*, *Bacillus*, *Serratia, Agrobacterium* and *Streptomyces* constitute the important bacterial strains. For instance, fluorescent *Pseudomonas* strains have been reported to produce a large number of antibiotics like, phenazines, 2, 4-diacetylphloroglucinol, pyoluteorin, pyrrolnitrin, lipopeptides, and hydrogen cyanide which exhibited antibiotic properties against numerous phytopathogenic fungi such as, *Fusarium, Macrophomina, Gaeumannomyces gamines* var. *tritici* and root-knot nematodes (Nowak-Thompson et al., 1994) (Table 8.2). Biopriming of carrot seeds *Psedomonas fluorescens* have been observed to give considerable protection to plant against *Alternaria daricia* and *A. radicina* (Koch et al., 2010).

Different strains of *Trichoderma s*ynthesize a huge variety of antibiotics, e.g., koningins, viridin, dermadin, trichoviridin, lignoren, and koningic acid were isolated from *T. koningii*, *T. harzianum*, *T. aureoviride*, *T. viride*, *T. virens*, *T. hamatum*, and *T. lignorum*cultures (Reino et al., 2008). Further, other antibacterial and fungicidal metabolites, e.g., Gliotoxin and gliovirin are among the most significant secondary metabolites of *Trichoderma.* Growth of soilborne pathogens like *R. solani, Phytophthora cinnamomi, Pythium middletonii, Fusarium oxysporum* and *Bipolarissorokiniana* is observed to be decreased by the presence of Koninginin D. (Dunlop et al., 1989). In a similar way, Viridins obtained from *Trichoderma* spp. like *T. koningii, T. viride, and T. virens* checks the spore germination of *Botrytis allii, Colletotrichum lini, Fusarium caeruleum, Penicillium expansum, Aspergillus niger* and *Stachybotrysatra* (Singh et al., *2005*). *T. harzianum*-derived Harzianic acid expressed antibiotic activity against *Pythium irregulare, Sclerotinia sclerotiorum* and *R. solani* in in vitro culture (Vinale et al., 2009). In general, antibiotic activity is combined cooperatively with lytic enzymes. Their dual action offers an advanced level of antagonism than the activity of either antibiotics or enzymes lonely (Monte, 2001). It has been reported by El-abd et al. (2013) that, biopriming of pea seed with *T. viride* combined with a supplement of mineral phosphorus caused maximum substantial expansion in vegetative growth, a yield of green pod as well as quality in sandy soil. Furthermore, Joint treatments of fluorescent *Pseudomonas*, *T. harzianum*, and *Glomus intraradis* were observed to be most effective in comparison to individual application of BCA against Fusarium wilt in tomato (Srivastava et al., 2010b).

8.6 Conclusion and future perspective

Seed priming is an old experimental tactic employed by farmers since eras to enhance germination activities in crop plants. Biopriming of seeds with microorganisms not only improve the growth and development of plants by regulating several biochemical and physiological processes but also impart stress tolerance and resistance mechanisms in plants. Biopriming initiate the specific signaling pathways during the early phases of plant phenology and leads to swifter plant defense responses. By this means, future exposure to pathogens activates a second signaling pathway which ultimately triggers and amplifies the signal transduction, which causes the rapid and more intense activation of previously acquired defense response. In addition to this, seed biopriming has several advantages such as, being cost-effective, time-saving, eco-friendly as well as provide desirable characteristics to treated seeds over chemical treatments. This technique also enables farmers to optimize yield in fewer resources, improve their socio-economic status, and solve the worldwide problem of food security. Future research on seed biopriming technology must be carried out and based on the production of microbial formulation that is more adapted to local ecological conditions. In this direction, the use of native strains of microorganisms in local condition must be explored to avoid the risk of introducing unfit nonnative microbes at the local sites.

References

Aamir, M., Kashyap, S.P., Singh, V.K., Dubey, M.K., Ansari, W.A., Upadhyay, R.S., Singh, S., 2019. Trichoderma erinaceum bio-priming modulates the WRKYs defense programming in tomato against the *Fusarium oxysporum* f. sp. lycopersici (Fol) challenged condition. Front. Plant Sci. 10, 911.

Abdulkhair, W.M., Alghuthaymi, M.A., 2016. Plant pathogens. Plant Growth. 49.

Abhilash, P.C., Dubey, R.K., Tripathi, V., Gupta, V.K., Singh, H.B., 2016. Plant growth-promoting microorganisms for environmental sustainability. Trends Biotechnol. 34 (11), 847–850.

Aishwath, O.P., Lal, G., Kant, K., Sharma, Y.K., Ali, S.F., Naimuddin, 2012. Influence of biofertilizers on growth and yield of coriander under typic haplustepts. Inter. J. Seed Spic. 2, 9–14.

Ajouri, A., Asgedom, H., Becker, M., 2004. Seed priming enhances germination and seedling growth of barley under conditions of P and Zn deficiency. J. Plant Nutr. Soil Sci. 167 (5), 630–636.

Altomare, C., Norvell, W.A., Björkman, T., Harman, G.E., 1999. Solubilization of phosphates and micronutrients by the plant-growth-promoting and biocontrol fungus *Trichoderma harzianum* Rifai 1295-22. Appl. Environ. Microbiol. 65 (7), 2926–2933.

Amooaghaie, R., 2011. The effect of hydro and osmopriming on alfalfa seed germination and antioxidant defenses under salt stress. Afr. J. Biotechnol. 10 (33), 6269–6275.

Anitha, U.M., Jahagirdar, S., 2015. Influence of seed priming agents on yield, yield parameters and purple seed stain disease in soybean. Karnataka J. Agric. Sci. 28 (1), 20–23.

Anitha, D., Vijaya, T., Reddy, N.V., Venkateswarlu, N., Pragathi, D., Mouli, K.C., 2013. Microbial endophytes and their potential for improved bioremediation and biotransformation: a review. Indo. Am. J. Pharm. Res. 3, 6408–6417.

Ansari, O., Azadi, M.S., Sharif-Zadeh, F., Younesi, E., 2013. Effect of hormone priming on germination characteristics and enzyme activity of mountain rye (*Secale montanum*) seeds under drought stress conditions. J. Stress Physiol. Biochem. 9(3).

Atici, Ö, Agar, G., Battal, P., 2003. Interaction between endogenous plant hormones and alpha-amylase in germinating chickpea seeds under cadmium exposure. Fresenius Environ. Bull. 12, 781–785.

Atique-Ur-Rehman Farooq, M., Cheema, Z.A., Wahid, A., 2013. Role of boron in leaf elongation and tillering dynamics in fine-grain aromatic rice. J. Plant Nutr. 36 (1), 42–54.

Babu, N.A., Jogaiah, S., Ito, S.-I., Nagaraj, A.K., Tran, L.-S.P., 2015. Improvement of growth, fruit weight and early blight disease protection of tomato plants by rhizosphere bacteria is correlated with their beneficial traits and induced biosynthesis of antioxidant peroxidase and polyphenol oxidase. Plant Sci. 231, 62–73.

Bajehbaj, A.A., 2010. The effects of NaCl priming on salt tolerance in sunflower germination and seedling grown under salinity conditions. Afr. J. Biotechnol. 9, 1764–1770. https://doi.org/10.5897/AJB10.1019.

Banerjee, M.R., Yasmin, L., 2002. Sulfur oxidizing rhizobacteria: an innovative environment friendly soil biotechnological tool for better canola production. In: Proceeding of AGROENVIRON. Cairo, Egypt, pp. 1–7.

Banerjee, M., Yesmin, L., 2002. Sulfur-oxidizing plant growth promoting *Rhizobacteria* for enhanced canola performance. US Patent 20080070784.

Baral, B.R., Adhikari, P., 2013. Effect of Azotobacter on growth and yield of maize. SAARC J. Agric. 11 (2), 141–147.

Begum, M.M., Sariah, M., Puteh, A.B., Abidin, M.Z., Rahman, M.A., Siddiqui, Y., 2010. Field performance of bio-primed seeds to suppress *Colletotrichum truncatum* causing damping-off and seedling stand of soybean. Biol. Control 53 (1), 18–23.

Benítez, T., Rincón, A.M., Limón, M.C., Codon, A.C., 2004. Biocontrol mechanisms of Trichoderma strains. Int. Microbiol. 7 (4), 249–260.

Bennett, M.A., 1998. The use of biologicals to enhance vegetable seed quality. Seed Technol. 198–208.

Berg, G., 2009. Plant–microbe interactions promoting plant growth and health: perspectives for controlled use of microorganisms in agriculture. Appl. Microbiol. Biotechnol. 84 (1), 11–18.

Bhattacharyya, P.N., Jha, D.K., 2012. Plant growth-promoting rhizobacteria (PGPR): emergence in agriculture. World J. Microbiol. Biotechnol. 28 (4), 1327–1350.

Bolwerk, A., Lagopodi, A.L., Wijfjes, A.H., Lamers, G.E., Chin-A-Woeng, T.F., Lugtenberg, B.J., Bloemberg, G.V., 2003. Interactions in the tomato rhizosphere of two *Pseudomonas biocontrol* strains with the phytopathogenic fungus *Fusarium oxysporum* f. sp. radicis-lycopersici. Mol. Plant-Microbe Interact. 16 (11), 983–993.

Brotman, Y., Landau, U., Cuadros-Inostroza, A., Takayuki, T., Fernie, A.R., Chet, I., et al., 2013. Trichoderma-plant root colonization: escaping early plant defense responses and activation of the antioxidant machinery for saline stress tolerance. PLoS Pathog. 9(3).

Burdman, S., Jurkevitch, E., Okon, Y., 2000. Recent advances in the use of plant growth promoting rhizobacteria (PGPR) in agriculture. In: Microbial Interactions in Agriculture and Forestry.vol. II, pp. 229–250.

Cakmakçi, R., Dönmez, F., Aydın, A., Şahin, F., 2006. Growth promotion of plants by plant growth-promoting rhizobacteria under greenhouse and two different field soil conditions. Soil Biol. Biochem. 38 (6), 1482–1487.

Callan, N.W., Mathre, D., Miller, J.B., 1990. Bio-priming seed treatment for biological control of *Pythium ultimum* preemergence damping-off in sh-2 sweet corn. Plant Dis. 74, 368–372.

Callan, N.W., Mathre, D.E., Miller, J.B., 1991. Field performance of sweet corn seed bioprimed and coated with Pseudomonas fluorescens AB254. Hort. Sci. 26 (9), 1163–1165.

Caseiro, R., Bennett, M.A., Marcos-Filho, J., 2004. Comparison of three priming techniques for onion seed lots differing in initial seed quality. Seed Sci. Technol. 32 (2), 365–375.

Castignetti, D., Smarrelli, J., 1986. Siderophores, the iron nutrition of plants, and nitrate reductase. FEBS Lett. 209 (2), 147–151.

Cawoy, H., Bettiol, W., Fickers, P., Ongena, M., 2011. Stoytcheva, M. (Ed.), Pesticides in the modern world – pesticides use and management. InTech, Rijeka, pp. 273–302.

Chahal, S., 2012. Dr Norman E Borlaug Memorial lecture: Indian agriculture: Challenges and opportunities in post Borlaug era. J. Mycol. Plant Pathol. 42 (1), 13.

Chakraborthy, G.S., Aeri, V., 2009. Phytochemical and antimicrobial studies of Chlorophytumborivilianum. Int. J. Pharm. Sci. Drug Res. 1 (2), 110–112.

Chandra Nayaka, S., Niranjana, S.R., Uday Shankar, A.C., Niranjan Raj, S., Reddy, M.S., Prakash, H.S., Mortensen, C.N., 2010. Seed biopriming with novel strain of *Trichoderma harzianum* for the control of toxigenic *Fusarium verticillioides* and fumonisins in maize. Arch. Phytopathol. Plant Protect. 43 (3), 264–282.

Chen, K., Arora, R., 2011. Dynamics of the antioxidant system during seed osmopriming, post-priming germination, and seedling establishment in spinach (*Spinacia oleracea*). Plant Sci. 180 (2), 212–220.

Chen, K., Arora, R., 2013. Priming memory invokes seed stress-tolerance. Environ. Exp. Bot. 94, 33–45.

Chen, C.C., Sung, J.M., 2001. Priming bitter gourd seeds with selenium solution enhances germinability and antioxidative responses under sub-optimal temperature. Physiol. Plant. 111 (1), 9–16.

Chet, L., 1987. Trochoderma- application, mode of action, and potential as a biocontrol agent of soil borne plant pathogenic fungi. In: Chet, I. (Ed.), Innovative Approaches to Plant Disease Control. Wiley, New York, pp. 137–160.

Chet, I., Ordentlich, A., Shapira, R., Oppenheim, A., 1990. Mechanisms of biocontrol of soil-borne plant pathogens by rhizobacteria. Plant Soil 129 (1), 85–92.

Colla, G., Nardi, S., Cardarelli, M., Ertani, A., Lucini, L., Canaguier, R., Rouphael, Y., 2015. Protein hydrolysates as biostimulants in horticulture. Sci. Hortic. 196, 28–38.

Conrath, U., Pieterse, C.M., Mauch-Mani, B., 2002. Priming in plant–pathogen interactions. Trends Plant Sci. 7, 210–216.

Conrath, U., Beckers, G.J.M., Flors, V., Garc'ıa-Agust'ın P, Jakab G, et al., 2006. Priming: getting ready for battle. Mol. Plant-Microbe Interact. 19, 1062–1071.

Contreras-Cornejo, H.A., et al., 2015. Trichoderma modulates stomatal aperture and leaf transpiration through an abscisic acid-dependent mechanism in Arabidopsis. J. Plant Growth Regul. 34 (2), 425–432.

Cyr, D.R., Bewley, J.D., 1990. Seasonal variation in nitrogen storage reserves in the roots of leafy spurge (Euphorbia esula) and responses to decapitation and defoliation. Physiol. Plant. 78 (3), 361–366.

Darwin, C., 1855. Does sea water kill seeds? In: Gardeners' Chronicle and Agricultural Gazette 15 and 21.242, pp. 356–357.

Dawood, M.G., El-Awadi, M.E., El-Rokiek, K.G., 2012. Physiological impact of fenugreek, guava and lantana on the growth and some chemical parameters of sunflower plants and associated weeds. J. Am. Sci. 8 (6), 166–174.

de Weert, S., Bloemberg, G.V., 2007. Rhizosphere competence and the role of root colonization in biocontrol. In: Plant-Associated Bacteria. Springer, Dordrecht, pp. 317–333.

De Weger, L.A., Jann, B., Jann, K., Lugtenberg, B., 1987. Lipopolysaccharides of Pseudomonas spp. that stimulate plant growth: composition and use for strain identification. J. Bacteriol. 169 (4), 1441–1446.

de Werra, P., Péchy-Tarr, M., Keel, C., Maurhofer, M., 2009. Role of gluconic acid production in the regulation of biocontrol traits of Pseudomonas fluorescens CHA0. Appl. Environ. Microbiol. 75 (12), 4162–4174.

Deaker, R., Roughley, R.J., Kennedy, I.R., 2004. Legume seed inoculation technology—a review. Soil Biol. Biochem. 36 (8), 1275–1288.

Dekker, J., 1963. Antibiotics in the control of plant diseases. Annu. Rev. Microbiol. 17, 243–262.

Demir, I., Ozuaydın, I., Yasar, F., Van Staden, J., 2012. Effect of smoke-derived butenolide priming treatment on pepper and salvia seeds in relation to transplant quality and catalase activity. S. Afr. J. Bot. 78, 83–87.

Dey, R.K.K.P., Pal, K.K., Bhatt, D.M., Chauhan, S.M., 2004. Growth promotion and yield enhancement of peanut (*Arachis hypogaea* L.) by application of plant growth-promoting rhizobacteria. Microbiol. Res. 159 (4), 371–394.

Dhanya, B.A., 2014. Evaluation of Microbial Seed Priming in Relation to Seed Germination, Plant Growth Promotion in *Morinda citrifolia* L.(Noni). Doctoral dissertation University of Agricultural Sciences, GKVK.

Di Girolamo, G., Barbanti, L., 2012. Treatment conditions and biochemical processes influencing seed priming effectiveness. Ital. J. Agron. e25.

Dobbelaere, S., Okon, Y., 2007. The plant growth-promoting effect and plant responses. In: Associative and Endophytic Nitrogen-Fixing Bacteria and Cyanobacterial Associations. Springer, Dordrecht, pp. 145–170.

Dobbelaere, S., Vanderleyden, J., Okon, Y., 2003. Plant growth-promoting effects of diazotrophs in the rhizosphere. Crit. Rev. Plant Sci. 22 (2), 107–149.

Dong, X., 1998. SA, JA, ethylene, and disease resistance in plants. Curr. Opin. Plant Biol. 1 (4), 316–323.

Doughari, J., 2015. An overview of plant immunity. J. Plant Pathol. Microbiol 6 (11), 10–4172.

Dunlop, R.W., Simon, A., Sivasithamparam, K., Ghisalberti, E.L., 1989. An antibiotic from *Trichoderma koningii* active against soilborne plant pathogens. J. Nat. Prod. 52 (1), 67–74.

Dunne, C., Crowley, J.J., Moënne-Loccoz, Y., Dowling, D.N., O'Gara, F., 1997. Biological control of Pythium ultimum by Stenotrophomonas maltophilia W81 is mediated by an extracellular proteolytic activity. Microbiology 143 (12), 3921–3931.

Dutta, S., Mishra, A.K., Kumar, B.D., 2008. Induction of systemic resistance against fusarial wilt in pigeon pea through interaction of plant growth promoting rhizobacteria and rhizobia. Soil Biol. Biochem. 40 (2), 452–461.

Eisendle, M., Oberegger, H., Buttinger, R., Illmer, P., Haas, H., 2004. Biosynthesis and uptake of siderophores is controlled by the PacC-mediated ambient-pH regulatory system in *Aspergillus nidulans*. Eukaryot. Cell 3 (2), 561–563.

El-abd, S.O., Zaki, M.F., El-Mohamedy, R.S.R., Riad, G.S., 2013. Improving growth, fresh pod yield and quality, and controlling root rot and damping off of pea grown in sandy soil by integration effect of phosphorus fertilizer with biological seed treatments. Middle East J. Agricul. Res. 2 (1), 8–15.

El-Araby, M.M., Moustafa, S.M.A., Ismail, A.I., Hegazi, A.Z.A., 2006. Hormone and phenol levels during germination and osmopriming of tomato seeds, and associated variations in protein patterns and anatomical seed features. Acta Agron. Hungarica 54 (4), 441–457.

El-Bab, T.S.F., El-Mohamedy, R.S.R., 2013. Bio-priming seed treatment for suppressive root rot soil borne pathogens and improvement growth and yield of green bean (*Phaseulas vulgaris* L.,) in new cultivated lands. J. Appl. Sci. Res. 9 (7), 4378–4387.

Elegba, M.S., Rennie, R.J., Campbell, C.A., Zentner, R.P., Amory, A.M., Cresswell, C.F., 1984. Effect of different inoculant adhesive agents on rhizobial survival, nodulation, and nitrogenase (acetylene-reducing) activity of soybeans (*Glycine max* (L.) Merrill). Can. J. Soil Sci. 64 (4), 631–636.

Ells, J.E., 1963. The influence of treating tomato seed with nutrient solution on emergence rate and seedling growth. Proc. Am. Soc. Hortic. Sci. 83, 684–687.

El-Mohamedy, R.S.R., Abd Alla, M.A., Badiaa, R.I., 2006. Soil amendment and seed bio-priming treatments as alternative fungicides for controlling root rot diseases on cowpea plants in Nobaria Province. Res. J. Agric. Biol. Sci. 2 (6), 391–398.

El-Mougy, N.S., Abdel-Kader, M.M., 2008. Long-term activity of bio-priming seed treatment for biological control of faba bean root rot pathogens. Australas. Plant Pathol. 37 (5), 464–471.

Entesari, M., Sharifzadeh, F., Ahmadzadeh, M., Farhangfar, M., 2013. Seed biopriming with *Trichoderma* species and *Pseudomonas fluorescent* on growth parameters, enzymes activity and nutritional status of soybean. Int. J. Agron. Plant Prod. 4 (4), 610–619.

Evenari, M., 1984. Seed physiology: its history from antiquity to the beginning of the 20th century. Bot. Rev. 50 (2), 119–142.

Fahd Shaukt, M., 2013. Seed bio-priming with Serratia plymuthica HRO-C48 for the control of Verticillium longisporum and Phoma lingam in Brassica napus L. spp. oleifera. Second Cycle. A2E. SLU, Dept. of Forest Mycology and Plant Pathology, Uppsala.

Fakhro, A., Andrade-Linares, D.R., von Bargen, S., Bandte, M., Büttner, C., Grosch, R., et al., 2010. Impact of Piriformosporaindica on tomato growth and on interaction with fungal and viral pathogens. Mycorrhiza 20 (3), 191–200.

Fallahi, J., Rezvani Moghaddam, P., Ghorbani, R., Amiri, M.B., Fallahpour, F., 2011. Effects of Seed Priming by biofertilizers on the growth characteristics of 412. Three wheat cultivars at the emergence period under greenhouse condition. In: 10th International Conference of the International Society for Seed Science.

Farahani, H.A., Maroufi, K., 2011. Effect of hydropriming on seedling vigour in basil (*Ocimumbasilicum* L.) under salinity conditions. Adv. Environ. Biol. 5 (5), 828–833.

Farooq, M., Basra, S.M.A., Afzal, I., Khaliq, A., 2006. Optimization of hydropriming techniques for rice seed invigoration. Seed Sci. Technol. 34 (2), 507–512.

Farooq, M., Wahid, A., Kobayashi, N., Fujita, D.B.S.M.A., Basra, S.M.A., 2009. Plant drought stress: effects, mechanisms and management. In: Sustainable Agriculture. Springer, Dordrecht, pp. 153–188.

Farooq, M., Wahid, A., Siddique, K.H., 2012. Micronutrient application through seed treatments: a review. J. Soil Sci. Plant Nutr. 12 (1), 125–142.

Foti, R., Abureni, K., Tigere, A., Gotosa, J., Gere, J., 2008. The efficacy of different seed priming osmotica on the establishment of maize (*Zea mays* L.) caryopses. J. Arid Environ. 72 (6), 1127–1130.

Fridlender, M., Inbar, J., Chet, I., 1993. Biological control of soilborne plant pathogens by a β-1, 3 glucanase-producing Pseudomonas cepacia. Soil Biol. Biochem. 25 (9), 1211–1221.

Gaffney, T., Friedrich, L., Vernooij, B., Negrotto, D., Nye, G., Uknes, S., et al., 1993. Requirement of salicylic acid for the induction of systemic acquired resistance. Science 261 (5122), 754–756.

Gaius, P.S., 1949–1954. Naturalishistoria. vol. IV–VII. Harvard University Press, Massachusetts Books 12–27 (trans: Rackham H, Jones WHS, Eichholz DE. William Heinemann, London.

Giri, G.S., Schillinger, W.F., 2003. Seed priming winter wheat for germination, emergence, and yield. Crop Sci. 43 (6), 2135–2141.

Glick, B.R., 2003. Phytoremediation: synergistic use of plants and bacteria to clean up the environment. Biotechnol. Adv. 21 (5), 383–393.

Glick, B.R., Pasternak, J.J., 2003. Plant growth promoting bacteria. In: Glick, B.R., Pasternak, J.J. (Eds.), Molecular Biotechnology—Principles and Applications of Recombinant DNA. third ed. ASM Press, Washington, DC, pp. 436–454.

Gnanamanickam, S.S., Vasudevan, P., Reddy, M.S., De, G., 2002. Principles of biological control. In: Biological Control of Crop Diseases. CRC Press, pp. 15–24.

Gratão, P.L., Polle, A., Lea, P.J., Azevedo, R.A., 2005. Making the life of heavy metal-stressed plants a little easier. Funct. Plant Biol. 32 (6), 481–494.

Greenberg, J.T., Silverman, F.P., Liang, H., 2000. Uncoupling salicylic acid-dependent cell death and defense-related responses from disease resistance in the *Arabidopsis mutant* acd5. Genetics 156 (1), 341–350.

Gupta, A.K., Kaur, N., 2005. Sugar signalling and gene expression in relation to carbohydrate metabolism under abiotic stresses in plants. J. Biosci. 30 (5), 761–776.

Haas, D., Défago, G., 2005. Biological control of soil-borne pathogens by fluorescent pseudomonads. Nat. Rev. Microbiol. 12, 1–13. https://doi.org/10.1038/nrmicro1129.

Handelsman, J., Stabb, E.V., 1996. Biocontrol of soilborne plant pathogens. Plant Cell 8 (10), 1855.

Haran, S., Schickler, H., Chet, I., 1996. Molecular mechanisms of lytic enzymes involved in the biocontrol activity of Trichoderma harzianum. Microbiology 142 (9), 2321–2331.

Hariprasad, P., Navya, H.M., Niranjana, S.R., 2009. Advantage of using PSIRB over PSRB and IRB to improve plant health of tomato. Biol. Control 50 (3), 307–316.

Hariprasad, P., Divakara, S.T., Niranjana, S.R., 2011. Isolation and characterization of chitinolytic rhizobacteria for the management of Fusarium wilt in tomato. Crop Prot. 30, 1606–1612. https://doi.org/10.1016/j.cropro.2011.02.032.

Harman, G.E., Taylor, A.G., 1988. Improved seedling performance by integration of biological control agents at favorable pH levels with solid matrix priming. Phytopathology 78 (5), 520–525.

Harman, G.E., Taylor, A.G., Stasz, T.E., Karban, R., 1989. Combining effective strains of *Trichoderma harzianum* and solid matrix priming to improve biological seed treatments. Plant Dis. 73 (8), 631–637.

Harman, G.E., Howell, C.R., Viterbo, A., Chet, I., Lorito, M., 2004. Trichoderma species—opportunistic, avirulent plant symbionts. Nat. Rev. Microbiol. 2 (1), 43–56.

Harris, D., Hollington, P.A., 2001. On-farm' seed priming–an update. Trop. Agric. Asso. (UK) Newsletter 21 (4), 7.

Harris, D., Joshi, A., Khan, P.A., Gothkar, P., Sodhi, P.S., 1999. On-farm seed priming in semi-arid agriculture: development and evaluation in maize, rice and chickpea in India using participatory methods. Exp. Agric. 35 (1), 15–29.

Hasanuzzaman, M., Hossain, M.A., Fujita, M., 2010. Selenium in higher plants: physiological role, antioxidant metabolism and abiotic stress tolerance. J. Plant Sci. 5 (4), 354–375.

Hoffland, E., Hakulinen, J., Van Pelt, J.A., 1996. Comparison of systemic resistance induced by avirulent and nonpathogenic Pseudomonas species. Phytopathology 86 (7), 757–762.

Hogan, D.A., Wargo, M.J., Beck, N., 2009. Bacterial biofilms on fungal surfaces. Lab-on-a-Chip. 10(238).

Howell, C.R., 2007. Effect of seed quality and combination fungicide-Trichoderma spp. seed treatments on pre-and postemergence damping-off in cotton. Phytopathology 97 (1), 66–71.

Howell, C.R., Stipanovic, R.D., 1980. Suppression of Pythium ultimum-induced damping-off of cotton seedlings by *Pseudomonas fluorescens* and its antibiotic, pyoluteorin. Phytopathology 70 (8), 712–715.

Hu, Y.F., Zhou, G., Na, X.F., Yang, L., Nan, W.B., Liu, X., et al., 2013. Cadmium interferes with maintenance of auxin homeostasis in Arabidopsis seedlings. J. Plant Physiol. 170 (11), 965–975.

Huang, C.J., Wang, T.K., Chung, S.C., Chen, C.Y., 2005. Identification of an antifungal chitinase from a potential biocontrol agent, Bacillus cereus 28-9. BMB Rep. 38 (1), 82–88.

Humphries, T., Chauhan, B.S., Florentine, S.K., 2018. Environmental factors effecting the germination and seedling emergence of two populations of an aggressive agricultural weed; Nassellatrichotoma. PLoS One. 13(7).

Imran, S., Afzal, I., Basra, S., Saqib, M., 2013. Integrated seed priming with growth promoting substances enhances germination and seedling vigour of spring maize at low temperature. Int. J. Agric. Biol. 15(6).

Jackson, R.M., 1965. Baker, K.F., Snyder, W.C. (Eds.), Antibiosis and fungistasis of soil microorganisms. University of California Press, Berkeley, pp. 363–369 Ecology of Soil-Borne Plant Pathogens.

Jackson, A.O., Taylor, C.B., 1996. Plant-microbe interactions: life and death at the interface. Plant Cell 8 (10), 1651.

Jayapala, N., Mallikarjunaiah, N.H., Puttaswamy, H., Gavirangappa, H., Ramachandrappa, N.S., 2019. Rhizobacteria Bacillus spp. induce resistance against anthracnose disease in chili (*Capsicum annuum* L.) through activating host defense response. Egypt. J. Biol. Pest. Control 29 (1), 45.

Jeleń, H., Błaszczyk, L., Chełkowski, J., Rogowicz, K., Strakowska, J., 2014. Formation of 6-n-pentyl-2H-pyran-2-one (6-PAP) and other volatiles by different *Trichoderma* species. Mycol. Prog. 13 (3), 589–600.

Jensen, B., Povlsen, F.V., Knudsen, I.M.B., Jensen, S.D.F., 2001. Combining microbial seed treatment with priming of carrot seeds for control of seed borne *Alternaria* spp. In: *Biocontrol agents: mode of action and interaction with other means of control: proceedings of the phytopathogens WG meeting*, pp. 197–201.

Jensen, B., Knudsen, I.M., Madsen, M., Jensen, D.F., 2004. Biopriming of infected carrot seed with an antagonist, Clonostachysrosea, selected for control of seedborne *Alternaria* spp. Phytopathology 94 (6), 551–560.

Jisha, K.C., Vijayakumari, K., Puthur, J.T., 2013. Seed priming for abiotic stress tolerance: an overview. Acta Physiol. Plant. 35 (5), 1381–1396.

Jogaiah, S., Mostafa, A., Tran, L.S.P., Shin-ichi, I., 2013. Characterization of rhizosphere fungi that mediate resistance in tomato against bacterial wilt disease. J. Exp. Bot. 64, 3829–3842.

Jogaiah, S., Abdelrahman, M., Tran, L.-S.P., Ito, S.-I., 2018. Different mechanisms of *Trichoderma virens*-mediated resistance in tomato against Fusarium wilt involve the jasmonic and salicylic acid pathways. Mol. Plant Pathol. 19, 870–882.

Johnson, S.E., Lauren, J.G., Welch, R.M., Duxbury, J.M., 2005. A comparison of the effects of micronutrient seed priming and soil fertilization on the mineral nutrition of chickpea (*Cicerarietinum*), lentil (*Lens culinaris*), rice (*Oryza sativa*) and wheat (*Triticum aestivum*) in Nepal. Exp. Agric. 41 (4), 427–448.

Kageyama, K., Nelson, E.B., 2003. Differential inactivation of seed exudate stimulation of *Pythium ultimum* sporangium germination by Enterobacter cloacae influences biological control efficacy on different plant species. Appl. Environ. Microbiol. 69 (2), 1114–1120.

Karthika, C., Vanangamudi, K., 2013. Biopriming of maize hybrid COH (M) 5 seed with liquid biofertilizers for enhanced germination and vigour. Afr. J. Agric. Res. 8 (25), 3310–3317.

Kathage, J., Castañera, P., Alonso-Prados, J.L., Gómez-Barbero, M., Rodríguez-Cerezo, E., 2018. The impact of restrictions on neonicotinoid and fipronil insecticides on pest management in maize, oilseed rape and sunflower in eight European Union regions. Pest Manag. Sci. 74 (1), 88–99.

Katiyar, V., Goel, R., 2004. Siderophore mediated plant growth promotion at low temperature by mutant of fluorescent pseudomonad⋆. Plant Growth Regul. 42 (3), 239–244.

Kavino, M., Harish, S., Kumar, N., Saravanakumar, D., Samiyappan, R., 2010. Effect of chitinolytic PGPR on growth, yield and physiological attributes of banana (Musa spp.) under field conditions. Appl. Soil Ecol. 45 (2), 71–77.

Kaya, M.D., Okçu, G., Atak, M., Cıkılı, Y., Kolsarıcı, Ö., 2006. Seed treatments to overcome salt and drought stress during germination in sunflower (*Helianthus annuus* L.). Eur. J. Agron. 24 (4), 291–295.

Kennedy, I.R., Choudhury, A.T.M.A., Kecskés, M.L., 2004. Non-symbiotic bacterial diazotrophs in crop-farming systems: can their potential for plant growth promotion be better exploited? Soil Biol. Biochem. 36 (8), 1229–1244.

Keswani, C., Mishra, S., Sarma, B.K., Singh, S.P., Singh, H.B., 2014. Unraveling the efficient applications of secondary metabolites of various *Trichoderma* spp. Appl. Microbiol. Biotechnol. 98 (2), 533–544.

Khan, H.A., Pervez, M.A., Ayub, C.M., Ziaf, K., Balal, R.M., Shahid, M.A., Akhtar, N., 2009. Hormonal priming alleviates salt stress in hot pepper (*Capsicum annuum* L.). Soil Environ. 28 (2), 130–135.

Kloepper, J.W., 1978. Plant growth-promoting rhizobacteria on radishes. In: *Proc. of the 4th Internet. Conf. on Plant Pathogenic Bacter, Station de Pathologie Vegetale et Phytobacteriologie, INRA, Angers, France*vol. 2. pp. 879–882.

Kloepper, J.W., Ryu, C.M., Zhang, S., 2004. Induced systemic resistance and promotion of plant growth by *Bacillus* spp. Phytopathology 94 (11), 1259–1266.

Koch, E., Schmitt, A., Stephan, D., Kromphardt, C., Jahn, M., Krauthausen, H.J., et al., 2010. Evaluation of non-chemical seed treatment methods for the control of *Alternaria dauci* and *A. radicina* on carrot seeds. Eur. J. Plant Pathol. 127 (1), 99–112.

Koornneef, M., Bentsink, L., Hilhorst, H., 2002. Seed dormancy and germination. Curr. Opin. Plant Biol. 5 (1), 33–36.

Krishna, A., Patil, C.R., Ragharendra, S.M., 2008. Effect of bio-fertilizers on seed germination and seedling quality of medicinal plants. Karnataka J. Agric. Sci. 21, 588–590.

Kuć, J., 2006. What's old and what's new in concepts of induced systemic resistance in plants, and its application. In: Multigenic and Induced Systemic Resistance in Plants. Springer, Boston, MA, pp. 9–20.

Le, H.T., Padgham, J.L., Sikora, R.A., 2009. Biological control of the rice root-knot nematode *Meloidogyne graminicola* on rice, using endophytic and rhizosphere fungi. Int. J. Pest. Manag. 55 (1), 31–36.

Leclere, V., Béchet, M., Adam, A., Guez, J.S., Wathelet, B., Ongena, M., et al., 2005. Mycosubtilin overproduction by Bacillus subtilis BBG100 enhances the organism's antagonistic and biocontrol activities. Appl. Environ. Microbiol. 71 (8), 4577–4584.

Lee, K.J., Kamala-Kannan, S., Sub, H.S., Seong, C.K., Lee, G.W., 2008. Biological control of Phytophthora blight in red pepper (*Capsicum annuum* L.) using Bacillussubtilis. World J. Microbiol. Biotechnol. 24 (7), 1139–1145.

Li, J., Ovakim, D.H., Charles, T.C., Glick, B.R., 2000. An ACC deaminase minus mutant of *Enterobacter cloacae* UW4No longer promotes root elongation. Curr. Microbiol. 41 (2), 101–105.

Liu, L., Kloepper, J.W., Tuzun, S., 1995. Induction of systemic resistance in cucumber against bacterial angular leaf spot by plant growth-promoting rhizobacteria. Phytopathology 85 (8), 843–847.

Mahmood, A., Turgay, O.C., Farooq, M., Hayat, R., 2016. Seed biopriming with plant growth promoting rhizobacteria: a review. FEMS Microbiol. Ecol.. 92(8).

Manjunatha, H.P., Singh, H., Chauhan, J.B., et al., 2013. Induction of resistance against sorghum downy mildew by seed treatment with Durantarepens extracts. IOSR J. Agric. Vet. Sci. 3 (6), 37–44.

Masood, A., Iqbal, N., Khan, N.A., 2012. Role of ethylene in alleviation of cadmium-induced photosynthetic capacity inhibition by sulphur in mustard. Plant Cell Environ. 35 (3), 524–533.

Mathre, D.E., Callan, N.W., Johnston, R.H., Miller, J.B., Schwend, A., 1994. Factors influencing the control of Pythium ultimum-induced seed decay by seed treatment with *Pseudomonas aureofaciens* AB254. Crop Prot. 13 (4), 301–307.

Mathre, D.E., Cook, R.J., Callan, N.W., 1999. From discovery to use: traversing the world of commercializing biocontrol agents for plant disease control. Plant Dis. 83, 972–983.

Meena, B., Ramamoorthy, V., Marimuthu, T., Velazhahan, R., 2000. Pseudomonas fluorescens mediated systemic resistance against late leaf spot of groundnut. J. Mycol. Plant Pathol. 30 (2), 151–158.

Meena, B.N., Pandey, R.N., Ram, D., 2016. Seed bio-priming for management of root rot and blight of mungbean incited by *Macrophominaphaseolina* (Tassi) goid and *Rhizoctonia solani* Kuhn. J. Pure Appl. Microbiol. 10 (2), 1223–1233.

Meena, K.K., Sorty, A.M., Bitla, U.M., Choudhary, K., Gupta, P., Pareek, A., et al., 2017. Abiotic stress responses and microbe-mediated mitigation in plants: the omics strategies. Front. Plant Sci. 8, 172.

Meneses, C.H., Rouws, L.F., Simões-Araújo, J.L., Vidal, M.S., Baldani, J.I., 2011. Exopolysaccharide production is required for biofilm formation and plant colonization by the nitrogen-fixing endophyte Gluconacetobacterdiazotrophicus. Mol. Plant-Microbe Interact. 24 (12), 1448–1458.

Meyerson, F.A., 2004. Population growth and deforestation: A critical and complex relationship. Population Reference Bureau, Washington, DC.

Misra, A.K., 2014. Climate change and challenges of water and food security. Int. J. Sustain. Built Environ. 3 (1), 153–165.

Moeinzadeh, A., Sharif-Zadeh, F., Ahmadzadeh, M., Tajabadi, F., 2010. Biopriming of sunflower (*Helianthus annuus* L.) seed with *Pseudomonas fluorescens* for improvement of seed invigoration and seedling growth. Aust. J. Crop Sci. *4* (7), 564.

Mona, M.M.R., Ashour, A.M., El-Mohamedy, R.S.R., Morsy, A.A., Hanafy, E.K., 2017. Seed bio priming as biological approach for controlling root rot soil born fungi on soybean (*Glycine max* L.) plant. Int. J. Agric. Technol. 13 (5), 771–788.

Monte, E., 2001. Understanding Trichoderma: between biotechnology and microbial ecology. Int. Microbiol. 4 (1), 1–4.

Moradi, A., Younesi, O., 2009. Effects of osmo-and hydro-priming on seed parameters of grain sorghum (*Sorghum bicolor* L.). Aust. J. Basic Appl. Sci. 3 (3), 1696–1700.

Muhammad, I., Kolla, M., Volker, R., Günter, N., 2015. Impact of nutrient seed priming on germination, seedling development, nutritional status and grain yield of maize. J. Plant Nutr. 38 (12), 1803–1821.

Muruli, C.N., 2013. Studies on seed priming and primed seed longevity in onion (*Allium cepa* L.) and capsicum (*Capsicum annuum* L.). Doctoral dissertation University of Agricultural Sciences GKVK, Bangalore.

Murungu, F.S., Chiduza, C., Nyamugafata, P., Clark, L.J., Whalley, W.R., Finch-Savage, W.E., 2004. Effects of 'on-farm seed priming'on consecutive daily sowing occasions on the emergence and growth of maize in semi-arid Zimbabwe. Field Crop Res. 89 (1), 49–57.

Mustafa, H.S.B., Mahmood, T., Ullah, A., Sharif, A., Bhatti, A.N., Muhammad Nadeem, M., Ali, R., 2017. Role of seed priming to enhance growth and development of crop plants against biotic and abiotic stresses. Bull. Biol. Allied Sci. Res. 2, 1–11.

Muthukumar, A., Eswaran, A., Sanjeevkumas, K., 2011. Exploitation of Trichoderma species on the growth of *Pythium aphanidermatum* in chilli. Braz. J. Microbiol. 42 (4), 1598–1607.

Nadeem, F., Farooq, M., 2019. Application of micronutrients in rice-wheat cropping system of South Asia. Rice Sci. 26 (6), 356–371.

Nadeem, F., Farooq, M., Nawaz, A., Ahmad, R., 2019. Boron improves productivity and profitability of bread wheat under zero and plough tillage on alkaline calcareous soil. Field Crop Res. 239, 1–9.

Nagaraju, A., Jogaiah, S., Mahadeva Murthy, S., Shin-ichi, I., 2012. Seed priming with *Trichoderma harzianum* isolates enhances plant growth and induces resistance against *Plasmopara halstedii*, an incitant of sunflower downy mildew disease. Aust. J Plant Path. 41, 609–620.

Naik, M., 2015. Doctoral dissertation. In: Effect of biopriming on seed quality and management of seed borne pathogens in Garden pea (*Pisum sativum* L.). Orissa University of Agriculture and Technology.

Namvar, A., Khandan, T., 2014. Biopriming and mineral fertilizers effects on agronomical performance of rapeseed (*Brassica napus* L.). Ekologija. 60(3).

Nandakumar, R., Babu, S., Raguchander, T., Samiyappan, R., 2007. Chitinolytic activity of native Pseudomonas fluorescens strains. J. Agric. Sci. Technol. 9, 61–68.

Nascimento, W.M., Cantliffe, D.J., Huber, D.J., 2004. Ethylene evolution and endo-beta-mannanase activity during lettuce seed germination at high temperature. Sci. Agric. 61 (2), 156–163.

Naseem, H., Bano, A., 2014. Role of plant growth-promoting rhizobacteria and their exopolysaccharide in drought tolerance of maize. J. Plant Interact. 9 (1), 689–701.

Nawaz, A., Farooq, M., Nadeem, F., Siddique, K.H., Lal, R., 2019. Rice–wheat cropping systems in South Asia: issues, options and opportunities. Crop Pasture Sci. 70 (5), 395–427.

Neergaard, P., 1979. Seed Pathology. vol. 1–2. The MacMillan Press Ltd., London, p. 1191.

Neilands, J.B., 1981. Microbial iron compounds. Annu. Rev. Biochem. 50, 715–731. https://doi.org/10.1146/annurev.bi.50.070181.003435.

Neilands, J.B., 1987. Comparative biochemistry of microbial iron assimilation. In: Iron Transport in Microbes, Plants and Animals. Elsevier Masson SAS, pp. 3–34.

Nowak-Thompson, B., Gould, S.J., Kraus, J., Loper, J.E., 1994. Production of 2, 4-diacetylphloroglucinol by the biocontrol agent *Pseudomonas fluorescens* Pf-5. Can. J. Microbiol. 40 (12), 1064–1066.

Oh, B.T., Hur, H., Lee, K.J., Shanthi, K., Soh, B.Y., Lee, W.J., et al., 2011. Suppression of Phytophthora blight on pepper (*Capsicum annuum* L.) by bacilli isolated from brackish environment. Biocontrol Sci. Tech. 21 (11), 1297–1311.

Olivier de Serres (1539–1619), 1600. Théâtre d'Agriculture. Mettayer, Paris.

Ongena, M., Jacques, P., 2008. Bacillus lipopeptides: versatile weapons for plant disease biocontrol. Trends Microbiol. 16 (3), 115–125.

Ongena, M.A.R.C., Daayf, F., Jacques, P., Thonart, P., Benhamou, N., Paulitz, T.C., Bélanger, R.R., 2000. Systemic induction of phytoalexins in cucumber in response to treatments with fluorescent pseudomonads. Plant Pathol. 49 (4), 523–530.

O'sullivan, D.J., O'Gara, F., 1992. Traits of fluorescent Pseudomonas spp. involved in suppression of plant root pathogens. Microbiol. Mol. Biol. Rev. 56 (4), 662–676.

Özkara, A., Akyıl, D., Konuk, M., 2016. Pesticides, environmental pollution, and health. In: Environmental Health Risk-Hazardous Factors to Living Species. IntechOpen.

Pandey, R., Gohel, N., 2017. Evaluation of bioagents for management of charcoal rot [*Macrophominaphaseolina* (Tassi) Goid] in Soybean [*Glycine max* (L.) Merrill] through seed treatment and soil application. Trends Biosci. 10, 2667–2670.

Pandey, V., Ansari, M.W., Tula, S., Yadav, S., Sahoo, R.K., Shukla, N., et al., 2016. Dose-dependent response of *Trichoderma harzianum* in improving drought tolerance in rice genotypes. Planta 243 (5), 1251–1264.

Pandey, P., Irulappan, V., Bagavathiannan, M.V., Senthil-Kumar, M., 2017. Impact of combined abiotic and biotic stresses on plant growth and avenues for crop improvement by exploiting physio-morphological traits. Front. Plant Sci. 8, 537.

Pandey, A.K., Burlakoti, R.R., Kenyon, L., Nair, R.M., 2018. Perspectives and challenges for sustainable management of fungal diseases of mungbean [*Vigna radiata* (L.) R. Wilczek var. radiata]: A review. Front. Environ. Sci. *6*, 53.

Patten, C.L., Glick, B.R., 2002. Role of Pseudomonas putida indoleacetic acid in development of the host plant root system. Appl. Environ. Microbiol. 68 (8), 3795–3801.

Penrose, D.M., Glick, B.R., 2001. Levels of 1-aminocyclopropane-1-carboxylic acid (ACC) in exudates and extracts of canola seeds treated with plant growth-promoting bacteria. Can. J. Microbiol. 47 (4), 368–372.

Pill, W.G., Collins, C.M., Goldberger, B., Gregory, N., 2009. Responses of non-primed or primed seeds of 'Marketmore 76' cucumber (*Cucumis sativus* L.) slurry coated with *Trichoderma* species to planting in growth media infested with *Pythium aphanidermatum*. Sci. Hortic. 121 (1), 54–62.

Pill, W.G., Collins, C.M., Gregory, N., Evans, T.A., 2011. Application method and rate of *Trichoderma* species as a biological control against Pythium aphanidermatum (Edson) Fitzp. in the production of microgreen table beets (*Beta vulgaris* L.). Sci. Hortic. 129 (4), 914–918.

Podile, A.R., Kishore, G.K., 2007. Plant growth-promoting rhizobacteria. In: Plant-Associated Bacteria. Springer, Dordrecht, pp. 195–230.

Posmyk, M.M., Janas, K.M., 2007. Effects of seed hydropriming in presence of exogenous proline on chilling injury limitation in *Vigna radiata* L. seedlings. Acta Physiol. Plant. 29 (6), 509–517.

Potgieter, H.J., Alexander, M., 1966. Susceptibility and resistance of several fungi to microbial lysis. J. Bacteriol. 91 (4), 1526–1532.

Prabha, D., Negi, S., Kumari, P., Negi, Y.K., Chauhan, J.S., 2016. Effect of seed priming with some plant leaf extract on seedling growth characteristics and root rot disease in tomato. Int. J. Agric. Syst. 4 (1), 46–51.

Priya, P., Patil, V.C., Kumar, B.A., 2016. Characterization of grain sorghum (*Sorghum bicolor* L.) for root traits associated with drought tolerance. Res. Environ. Life Sci. 9 (2), 163–165.

Radjacommare, R., Venkatesan, S., Samiyappan, R., 2010. Biological control of phytopathogenic fungi of vanilla through lytic action of Trichoderma species and *Pseudomonas fluorescens*. Arch. Phytopathol. Plant Protect. 43 (1), 1–17.

Rafi, H., Dawar, S., 2015. Role of pellets and capsules of *Acacia nilotica* and *Sapindusmukorossi* in combination of seed bio-priming with microbial antagonists in the suppression of root infecting pathogenic fungi and promotion of crop plants. J. Plant Pathol. Microbiol. 3, 2.

Rais, A., Jabeen, Z., Shair, F., Hafeez, F.Y., Hassan, M.N., 2017. Bacillus spp., a bio-control agent enhances the activity of antioxidant defense enzymes in rice against Pyriculariaoryzae. PloS one. 12(11).

Raj, S.N., Shetty, N.P., Shetty, H.S., 2004. Seed bio-priming with Pseudomonas fluorescens isolates enhances growth of pearl millet plants and induces resistance against downy mildew. Int. J. Pest Manag. 50 (1), 41–48.

Rajput, R.S., Singh, P., Singh, J., Vaishnav, A., Ray, S., Singh, H.B., 2019. Trichoderma mediated seed biopriming augments antioxidant and phenylpropanoid activities in tomato plant against *Sclerotium rolfsii*. J. Pharm. Phytochem. 8 (3), 2641–2647.

Ramamoorthy, V., Viswanathan, R., Raguchander, T., Prakasam, V., Samiyappan, R., 2001. Induction of systemic resistance by plant growth promoting rhizobacteria in crop plants against pests and diseases. Crop Prot. 20 (1), 1–11.

Ramamoorthy, V., Raguchander, T., Samiyappan, R., 2002. Induction of defense-related proteins in tomato roots treated with *Pseudomonas fluorescens* Pf1 and *Fusarium oxysporum* f. sp. lycopersici. Plant Soil 239 (1), 55–68.

Ramzan, N., Noreen, N., Perveen, Z., Shahzad, S., 2016. Effect of seed pelleting with biocontrol agents on growth and colonisation of roots of mungbean by root-infecting fungi. J. Sci. Food Agric. 96 (11), 3694–3700.

Rangaswamy, A., Purushothaman, S., Devasenapathy, P., 1993. Seed hardening in relation to seedling quality characters of crops. Madras Agric. J. 80 (9), 535–537.

Rao, M.S.L., Kulkarni, S., Lingaraju, S., Nadaf, H.L., 2009. Bio-priming of seeds: a potential tool in the integrated management of alternaria blight of sunflower/bio peleteado de semillas: una herramienta potencial en el manejo integrado de la alternaria en girasol/ bio-préparation de la graine: potentialité d'une protection intégrale du tournesol contre la fanure provoquée par le champignon alternaria. Helia *32* (50), 107–114.

Raphael, E., 2012. Phytochemical constituents of some leaves extract of Aloe vera and Azadirachtaindica plant species. Global Adv. Res. J. Environ. Sci. Toxicol. 1 (2), 014–017.

Rashid, A., Harris, D., Hollington, P.A., Khattak, R.A., 2002. On-farm seed priming: a key technology for improving the livelihoods of resource-poor farmers on saline lands. In: Prospects for Saline Agriculture. Springer, Dordrecht, pp. 423–431.

Rawat, L., Singh, Y., Shukla, N., Kumar, J., 2012. Seed biopriming with salinity tolerant isolates of trichoderma harzmnumalleviates salt stress in rice: growth, physiological and biochemical characteristics. J. Plant Pathol. 94 (2), 353–365.

Ray, K., Sen, K., Ghosh, P.P., Barman, A.R., Mandal, R., De Roy, M., Dutta, S., 2017. Dynamics of Sclerotium rolfsii as influenced by different crop rhizosphere and microbial community. J. Appl. Nat. Sci. 9 (3), 1544–1550.

Reddy, P.P., 2012. Bio-priming of seeds. In: Recent Advances in Crop Protection. Springer, New Delhi, pp. 83–90.

Rêgo, M.C.F., Ilkiu-Borges, F., Filippi, M.C.C.D., Gonçalves, L.A., Silva, G.B.D., 2014. Morphoanatomical and biochemical changes in the roots of rice plants induced by plant growth-promoting microorganisms. J. Bot. 2014.

Rehman, A., Farooq, M., 2016. Zinc seed coating improves the growth, grain yield and grain biofortification of bread wheat. Acta Physiol. Plant. 3, 238.

Rehman, A., Farooq, M., Naveed, M., Ozturk, L., Nawaz, A., 2018. Pseudomonas-aided zinc application improves the productivity and biofortification of bread wheat. Crop Pasture Sci. 69 (7), 659–672.

Reino, J.L., Guerrero, R.F., Hernández-Galán, R., Collado, I.G., 2008. Secondary metabolites from species of the biocontrol agent Trichoderma. Phytochem. Rev. 7 (1), 89–123.

Revillas, J.J., Rodelas, B., Pozo, C., Martínez-Toledo, M.V., González-López, J., 2000. Production of B-group vitamins by two Azotobacter strains with phenolic compounds as sole carbon source under diazotrophic and adiazotrophic conditions. J. Appl. Microbiol. 89 (3), 486–493.

Richardson, A.E., 2001. Prospects for using soil microorganisms to improve the acquisition of phosphorus by plants. Funct. Plant Biol. 28 (9), 897–906.

Roberts, W.O., 1948. Prevention of mineral deficiency by soaking seed in nutrient solution. J. Agric. Sci. 38 (4), 458–468.

Roberts, D.P., Lakshman, D.K., McKenna, L.F., Emche, S.E., Maul, J.E., Bauchan, G., 2016. Seed treatment with ethanol extract of *Serratia marcescens* is compatible with *Trichoderma isolates* for control of damping-off of cucumber caused by *Pythium ultimum*. Plant Dis. 100 (7), 1278–1287.

Rodríguez-Navarro, D.N., Dardanelli, M.S., Ruíz-Saínz, J.E., 2007. Attachment of bacteria to the roots of higher plants. FEMS Microbiol. Lett. 272 (2), 127–136.

Rolland, F., Baena-Gonzalez, E., Sheen, J., 2006. Sugar sensing and signaling in plants: conserved and novel mechanisms. Annu. Rev. Plant Biol. 57, 675–709.

Rubin, E.M., 2008. Genomics of cellulosic biofuels. Nature 454 (7206), 841–845.

Russo, A., Vettori, L., Felici, C., Fiaschi, G., Morini, S., Toffanin, A., 2008. Enhanced micropropagation response and biocontrol effect of Azospirillumbrasilense Sp245 on *Prunus cerasifera* L. clone Mr. S 2/5 plants. J. Biotechnol. 134 (3–4), 312–319.

Sabalpara, A.N., 2015. Biopriming- blueprint for the better crop. J. Mycol. Plant Pathol. 45, 1–3.

Saha, P., Chatterjee, P., Biswas, A.K., 2010. NaCl pretreatment alleviates salt stress by enhancement of antioxidant defense system and osmolyte accumulation in mungbean (Vigna radiata L. Wilczek). IJEB 48, 593–600.

Saikia, R., Singh, T., Kumar, R., Srivastava, J., Srivastava, A.K., Singh, K., Arora, D.K., 2003. Role of salicylic acid in systemic resistance induced by *Pseudomonas fluorescens* against *Fusarium oxysporum* f. sp. ciceri in chickpea. Microbiol. Res. 158 (3), 203–213.

Saravanakumar, D., Lavanya, N., Muthumeena, B., Raguchander, T., Suresh, S., Samiyappan, R., 2008. Pseudomonas fluorescens enhances resistance and natural enemy population in rice plants against leaffolder pest. J. Appl. Entomol. 132 (6), 469–479.

Sarwar, N.I.G.H.A.T., Haq, I., Jamil, F.F., 2005. Induction of systemic resistance in chickpea against Fusarium wilt by seed treatment with salicylic acid and Bion. Pak. J. Bot. 37 (4), 989.

Satish, S., Mohana, D.C., Ranhavendra, M.P., Raveesha, K.A., 2007. Antifungal activity of some plant extracts against important seed borne pathogens of *Aspergillus* sp. J. Agric. Technol. 3 (1), 109–119.

Schwember, A.R., Bradford, K.J., 2010. A genetic locus and gene expression patterns associated with the priming effect on lettuce seed germination at elevated temperatures. Plant Mol. Biol. 73 (1–2), 105–118.

Schwinn, F.J., 1994. Seed treatment-a panacea for plant protection? In: Seed treatment: Progress and Prospects Mono. 57, BCPC. Thornton Health, UK, pp. 3–14.

Shao, C.X., Hu, J., Song, W.J., Hu, W.M., 2005. Effects of seed priming with chitosan solutions of different acidity on seed germination and physiological characteristics of maize seedling. J. Zhejiang Univ. (Agric. Life Sci.) 31 (6), 705–708.

Siddiqui, Y., Meon, S., 2009. Effect of seed bacterization on plant growth response and induction of disease resistance in chilli. Agric. Sci. China 8 (8), 963–971.

Sikora, R.A., 1988. Interrelationship between plant health promoting rhizobacteria, plant parasitic nematodes and soil microorganisms. Mededelingen van de Faculteitlandbouw wetenschappen. Rijksuniversiteit Gent. 53 (2b), 867–878.

Sikora, R.A., Pocasangre, L., zum Felde, A., Niere, B., Vu, T.T., Dababat, A.A., 2008. Mutualistic endophytic fungi and in-planta suppressiveness to plant parasitic nematodes. Biol. Control 46 (1), 15–23.

Simons, M., Van Der Bij, A.J., Brand, I., De Weger, L.A., Wijffelman, C.A., Lugtenberg, B.J., 1996. Gnotobiotic system for studying rhizosphere colonization by plant growth-promoting Pseudomonas bacteria. Mol. Plant-Microbe Interact.: MPMI 9 (7), 600.

Simons, M., Permentier, H.P., de Weger, L.A., Wijffelman, C.A., Lugtenberg, B.J., 1997. Amino acid synthesis is necessary for tomato root colonization by Pseudomonas fluorescens strain WCS365. Mol. Plant-Microbe Interact. 10 (1), 102–106.

Singh, U.P., Sarma, B.K., Singh, D.P., 2003. Effect of plant growth-promoting rhizobacteria and culture filtrate of *Sclerotium rolfsii* on phenolic and salicylic acid contents in chickpea (*Cicer arietinum*). Curr. Microbiol. 46 (2), 0131–0140.

Singh, S., Dureja, P., Tanwar, R.S., Singh, A., 2005. Production and antifungal activity of secondary metabolites of *Trichoderma virens*. Pest. Res. J. 17, 26–29.

Singh, J.S., Pandey, V.C., Singh, D.P., 2011. Efficient soil microorganisms: a new dimension for sustainable agriculture and environmental development. Agric. Ecosyst. Environ. 140 (3–4), 339–353.

Singh, H., Jassal, R.K., Kang, J.S., Sandhu, S.S., Kang, H., Grewal, K., 2015. Seed priming techniques in field crops-A review. Agric. Rev. 36 (4), 251–264.

Sofy, A.R., Attia, M.S., Sharaf, A.M.A., El-Dougdoug, K.A., 2014. Potential impacts of seed bacterization or salix extract in FABA bean for enhancing protection against bean yellow mosaic disease. Nat. Sci. 12 (10), 67–82.

Spaepen, S., Vanderleyden, J., 2011. Auxin and plant-microbe interactions. Cold Spring Harb. Perspect. Biol. 3 (4), a001438.

Srivastava, A.K., Lokhande, V.H., Patade, V.Y., Suprasanna, P., Sjahril, R., D'Souza, S.F., 2010a. Comparative evaluation of hydro-, chemo-, and hormonal-priming methods for imparting salt and PEG stress tolerance in Indian mustard (*Brassicajuncea* L.). Acta Physiol. Plant. 32 (6), 1135–1144.

Srivastava, R., Khalid, A., Singh, U.S., Sharma, A.K., 2010b. Evaluation of arbuscular mycorrhizal fungus, fluorescent *Pseudomonas* and *Trichoderma harzianum* formulation against *Fusarium oxysporum* f. sp. lycopersici for the management of tomato wilt. Biol. Control 53 (1), 24–31.

Su, J., Hirji, R., Zhang, L., He, C., Selvaraj, G., Wu, R., 2006. Evaluation of the stress-inducible production of choline oxidase in transgenic rice as a strategy for producing the stress-protectant glycine betaine. J. Exp. Bot. 57, 1129–1135.

Sudisha, J., Niranjana, S.R., Umesha, S., Prakash, H.S., Shekar Shetty, H., 2006. Transmission of seed-borne infection of muskmelon by *Didymella bryoniae* and effect of seed treatments on disease incidence and fruit yield. Biol. Control 37, 196–205.

Sun, H., Li, L., Wang, X., Wu, S., Wang, X., 2011. Ascorbate–glutathione cycle of mitochondria in osmoprimed soybean cotyledons in response to imbibitional chilling injury. J. Plant Physiol. 168 (3), 226–232.

Sung, Y., Cantliffe, D.J., Nagata, R.T., Nascimento, W.M., 2008. Structural changes in lettuce seed during germination at high temperature altered by genotype, seed maturation temperature, and seed priming. J. Am. Soc. Hortic. Sci. 133 (2), 300–311.

Tanwar, A., Aggarwal, A., Kaushish, S., Chauhan, S., 2013. Interactive effect of AM fungi with *Trichoderma viride* and *Pseudomonas fluorescens* on growth and yield of broccoli. Plant Prot. Sci. 49 (3), 137–145.

Tavili, A., Zare, S., Moosavi, S.A., Enayati, A., 2011. Effects of seed priming on germination characteristics of Bromus species under salt and drought conditions. Am. Eurasian J. Agric. Environ. Sci. 10 (2), 163–168.

Taylor, A.G., Harman, G.E., 1990. Concepts and technologies of selected seed treatments. Annu. Rev. Phytopathol. 28 (1), 321–339.

Theophrastus (371–287 BC), n.d. Enquiry into plants, Book VII, I.6.

Thomas, U.C., Varughese, K., Thomas, A., Sadanandan, S., 2000. Seed priming—for increased vigour, viability and productivity of upland rice. Leisa India. 4(14).

Thomashow, L.S., Weller, D.M., Bonsall, R.F., Pierson, L.S., 1990. Production of the antibiotic phenazine-1-carboxylic acid by fluorescent Pseudomonas species in the rhizosphere of wheat. Appl. Environ. Microbiol. 56 (4), 908–912.

Tilak, K.V.B.R., Singh, G., Mukerji, K.G., 1999. Biocontrol—Plant growth promoting rhizobacteria: Mechanism of action. In: Biotechnological Approaches in Biocontrol of Plant Pathogens. Springer, Boston, MA, pp. 115–133.

Tucker, M.R., Koltunow, A.M., 2009. Sexual and asexual (apomictic) seed development in flowering plants: molecular, morphological and evolutionary relationships. Funct. Plant Biol. 36 (6), 490–504.

Ullah, A., Farooq, M., Hussain, M., Ahmad, R., Wakeel, A., 2019a. Zinc seed coating improves emergence and seedling growth in desi and kabuli chickpea types but shows toxicity at higher concentration. Int. J. Agric. Biol. 21, 553–559.

Ullah, A., Farooq, M., Hussain, M., Ahmad, R., Wakeel, A., 2019b. Zinc seed priming improves stand establishment, tissue zinc concentration and early seedling growth of chickpea. J. Animal Plant Sci. 29, 1046–1053.

Ullah, A., Farooq, M., Hussain, M., 2019c. Improving the productivity, profitability and grain quality of kabuli chickpea with co-application of zinc and endophyte bacteria enterobacter sp. strain MN17. Arch. Agron. Soil Sci.. .

Unno, Y., Okubo, K., Wasaki, J., Shinano, T., Osaki, M., 2005. Plant growth promotion abilities and microscale bacterial dynamics in the rhizosphere of Lupin analysed by phytate utilization ability. Environ. Microbiol. 7 (3), 396–404.

Vallad, G.E., Goodman, R.M., 2004. Systemic acquired resistance and induced systemic resistance in conventional agriculture. Crop Sci. 44 (6), 1920–1934.

van Dijk, K., Nelson, E.B., 2000. Fatty acid competition as a mechanism by which Enterobacter cloacae suppresses *Pythium ultimum* sporangium germination and damping-off. Appl. Environ. Microbiol. 66 (12), 5340–5347.

van Loon, L.C., Rep, M., Pieterse, C.M., 2006. Significance of inducible defense-related proteins in infected plants. Annu. Rev. Phytopathol. 44, 135–162.

Van Wees, S.C., Luijendijk, M., Smoorenburg, I., Van Loon, L.C., Pieterse, C.M., 1999. Rhizobacteria-mediated induced systemic resistance (ISR) in Arabidopsis is not associated with a direct effect on expression of known defense-related genes but stimulates the expression of the jasmonate-inducible gene Atvsp upon challenge. Plant Mol. Biol. 41 (4), 537–549.

Velazhahan, R., Samiyappan, R., Vidhyasekaran, P., 1999. Relationship between antagonistic activities of *Pseudomonas fluorescens* isolates against Rhizoctonia solarii and their production of lytic enzymes/Beziehungenzwischenantagonistischer Aktivität von Pseudomonas fluorescens-Isolatengegen Rhizoctonia sotaní und ihrerProduktionlytischer Enzyme. ZeitschriftfürPflanzenkrankheiten und Pflanzenschutz/J. Plant Dis. Prot. 106 (3), 244–250.

Verma, S., Varma, A., Rexer, K.H., Hassel, A., Kost, G., Sarbhoy, A., et al., 1998. Piriformosporaindica, gen. et sp. nov., a new root-colonizing fungus. Mycologia 90 (5), 896–903.

Vinale, F., Ghisalberti, E.L., Sivasithamparam, K., Marra, R., Ritieni, A., Ferracane, R., et al., 2009. Factors affecting the production of *Trichoderma harzianum* secondary metabolites during the interaction with different plant pathogens. Lett. Appl. Microbiol. 48 (6), 705–711.

Vincent, J.M., Thompson, J.A., Donovan, K.O., 1962. Death of root-nodule bacteria on drying. Aust. J. Agric. Res. 13 (2), 258–270.

Viterbo, A., Montero, M., Ramot, O., Friesem, D., Monte, E., Llobell, A., Chet, I., 2002. Expression regulation of the endochitinase chit36 from *Trichoderma asperellum* (T. harzianum T-203). Curr. Genet. 42 (2), 114–122.

Volenec, J.J., Ourry, A., Joern, B.C., 1996. A role for nitrogen reserves in forage regrowth and stress tolerance. Physiol. Plant. 97 (1), 185–193.

Walker, R., Rossall, S., Asher, M.J.C., 2004. Comparison of application methods to prolong the survival of potential biocontrol bacteria on stored sugar-beet seed. J. Appl. Microbiol. 97 (2), 293–305.

Wang, S.L., Chang, W.T., 1997. Purification and characterization of two bifunctional chitinases/lysozymes extracellularly produced by Pseudomonas aeruginosa K-187 in a shrimp and crab shell powder medium. Appl. Environ. Microbiol. 63 (2), 380–386.

Warren, J.E., Bennett, M.A., 1999. Bio-osmopriming tomato (*Lycopersiconesculentum* Mill.) seeds for improved stand establishment. Seed Sci. Technol. 27 (2), 489–499.

Warwate, S.I., Kandoliya, U.K., Bhadja, N.V., Golakiya, B.A., 2017. The effect of plant growth promoting rhizobacteria (PGPR) on biochemical parameters of coriander (*Coriandrum sativum* L.) seedling. Int. J. Curr. Microbiol. App. Sci. 6 (3), 1935–1944.

Wei, G., Kloepper, J.W., Tuzun, S., 1996. Induced systemic resistance to cucumber diseases and increased plant growth by plant growth-promoting rhizobacteria under field conditions. Phytopathology 86, 221–224.

Whipps, J.M., 2001. Microbial interactions and biocontrol in the rhizosphere. J. Exp. Bot. 52 (suppl_1), 487–511.

Woo, S.L., Ruocco, M., Vinale, F., Nigro, M., Marra, R., Lombardi, N., et al., 2014. Trichoderma-based products and their widespread use in agriculture. Open Mycol. J.. 8(1).

Yadav, S.K., Dave, A., Sarkar, A., Singh, H.B., Sarma, B.K., 2013. Co-inoculated biopriming with Trichoderma, Pseudomonas and Rhizobium improves crop growth in Cicer arietinum and Phaseolus vulgaris. Int. J. Agric., Environ. Biotechnol. 6 (2), 255–259.

Yedidia, I., Srivastva, A.K., Kapulnik, Y., Chet, I., 2001. Effect of *Trichoderma harzianum* on microelement concentrations and increased growth of cucumber plants. Plant Soil 235 (2), 235–242.

Zehnder, G., Kloepper, J., Yao, C., Wei, G., 1997. Induction of systemic resistance in cucumber against cucumber beetles (Coleoptera: Chrysomelidae) by plant growth-promoting rhizobacteria. J. Econ. Entomol. 90 (2), 391–396.

Zhao, K., Penttinen, P., Zhang, X., Ao, X., Liu, M., Yu, X., Chen, Q., 2014. Maize rhizosphere in Sichuan, China, hosts plant growth promoting Burkholderiacepacia with phosphate solubilizing and antifungal abilities. Microbiol. Res. 169 (1), 76–82.

Metabolomic profile modification and enhanced disease resistance derived from alien genes introgression in plants

9

Vu Quynh Hoa[a], *Tran Thi Minh Hang*[a], *and Vu Hai Yen*[b]
[a]Department of Horticulture and Landscaping, Faculty of Agronomy, Vietnam National University of Agriculture, Hanoi, Vietnam, [b]Graduate School of Medical Sciences, Kyushu University, Fukuoka, Japan

9.1 Introduction

Genetic improvement with the combined use of conventional breeding and genetic engineering, by which humans introduce or change DNA, RNA, or proteins in an organism to create new traits or change the expression of existing traits, developed in the 1970s, has been reported in crop varieties to express many traits, such as longer shelf life of fruit, higher nutritional content, and resistance to pests or diseases. The appearance of molecular biology in breeding programs in the 1980s enabled understanding of genetic determinants of phenotypes and marker-assisted selection (National Academies of Sciences, Engineering, and Medicine, 2016). Plant breeding is now in the genomics era, when a reference genome of nearly every major crop species has been available. In the last 15 years, various advanced technologies have been developed that permit accumulation and assessment of large-scale datasets of biological molecules, including DNA sequence (the genome), transcripts (the transcriptome; involving RNA), DNA modification (the epigenome), and, at a narrower fields, proteins and their modifications (the proteome) and metabolites (the metabolome). A review in the ongoing scientific research is undoubtedly the development of metabolomics provided a new dimension within the context of multidimensional biology enabling the in-depth study of global metabolic networks (Krishnan et al., 2005; Allwood et al., 2008; Dunn, 2008; Spratlin et al., 2009; Vinayavekhin et al., 2010). The field of metabolomics is the branch of omics technology concerned with the high-throughput identification and quantification of low-molecular-weight molecules (< 1500 Da) in a biological system (Creek and Barret, 2014; Wishart, 2008). Metabolomics is a powerful approach to gain a comprehensive perspective of how metabolic networks are regulated and has increasingly been applied by many researchers in the recent years. Therefore, metabolomics has found applications in diverse aspects of agriculture, such as the nutritional sciences (Hall et al., 2008; Saxena and Cramer, 2013; Marti et al., 2014), microbiology (Loots, 2014; Giannangelo et al., 2016;

Biocontrol Agents and Secondary Metabolites. https://doi.org/10.1016/B978-0-12-822919-4.00009-0

Yao et al., 2018), plant protection (Aliferis and Chrysayi-Tokousbalides, 2011; Brotman et al., 2012; Abdelrahman et al., 2016; Lloyd et al., 2011; Tugizimana et al., 2018; Aliferis and Jabaji, 2011; Adhikari et al., 2017), and postharvest (Gao et al., 2018; Yun et al., 2013). It can be used to elucidate the functions of genes and combine transcript profiling and proteomic in functional genomics and systems biology (Bino et al., 2004; Kromer et al., 2004). Metabolomics can be defined as "the nonbiased identification and quantification of all the metabolites in a biological system," using highly sensitive analytical procedures (Dunn et al., 2005). Thus, this approach serves as the basis for the discovery of new biomarkers for better describing the mechanisms resulting in disease resistance in plants. This chapter summarizes current metabolomic researches in elucidation of functions of plant genes related to disease resistance.

9.2 Metabolomic modification derived from genetic alteration

The advancement in genetic engineering enables the recombination of genomes of organisms belonging to distantly related species. The introduction of genes in the plant genome that encode the biosynthesis of metabolites which enhance plant tolerance to abiotic or biotic stresses has brought an evolution in crop protection (Kos et al., 2009). The close relationship with the phenotype makes metabolomics ideal for assessing the physiological state of an organism or system at a point of period (Giannangelo et al., 2016). Plants modified to express insecticidal metabolites have been introduced in order to minimize yield losses caused by insects, for example, introgression genes from *Bacillus thuringiensis* (Bt plants) as the most extensively used in agricultural practice (Betz et al., 2000). Leon et al. (2009) developed a metabolomics methodology by combining FT-ICR/MS and CE-TOF/MS for the comparison between the metabolic composition of kernels of genetically modified (GM) maize carrying the *B. thuringiensis* Cry1Ab gene and that of the corresponding isogenic lines. Analyses revealed substantial differences between GM and wild kernels, especially for metabolites implicated in amino acid biosynthesis. Application of metabolomics for the comparative study between wild and transgenic rice carrying the genes cryIAc and sck for resistance against insects revealed substantial differences in their metabolomes (Zhou et al., 2009). Le Gall et al. (2003) applied NMR metabolomics to detect metabolic differences between nontransgenic and transgenic tomatoes carrying the maize transcription factors. Similarly, application of NMR metabolomics and multivariate analysis (PCA) showed substantial metabolic differences between wild tobacco plants and plants engineered to overexpress salicylate biosynthetic genes (Choi et al., 2004). On the other hand, based on CE-TOF/MS or combining GC-TOF/MS and flow injection electrospray-MS (FIE-MS), other researchers found very few differences between conventional and transgenic varieties of soybean or no significant differences between conventional and GM potatoes (Garcia-Villalba et al., 2008; Catchpole et al., 2005). It is already clear from metabolomic analyses that significant natural variation exists within crop gene pools, accentuated by interactions with the prevailing environment. Transcript and metabolome profilings, and metabolic fingerprinting of wild-type accessions and barley transgenics show that cultivar-specific differences in transcriptome and metabolome greatly exceed effects caused by transgene expression (Kogel et al., 2010). Kusano et al. (2007) conclude that the multiplatform approach yields a wide and robust characterization of the tomato-fruit metabolome; and the differences between the transgenic lines and the control were small compared to the differences observed between ripening stages and traditional cultivars. Metabolite profiles of GM maize grown together with the non-GM near-isogenic comparators reveal that the majority of differences observed are related to natural variability and environmental factors rather than to the genetic modifications (Frank et al., 2012). Those studies suggest that use of metabolomics for assessing substantial equivalence will require testing in multiple locations and meticulous analysis to differentiate genetic from environmental effects, because there will probably be effects of gene-environment interactions. In conclusion, the characterization of natural diversity in plant metabolites using unbiased metabolite profiling approaches has provided a deeper knowledge of food composition and its variable nature both within and between species.

9.3 Genetic basis of phytochemical biosynthesis

Mapping and identifying loci or genes encoding specific metabolites are the first important steps of functional metabolomics approach (Fernie and Tohge, 2017) as a vital additional tool in existing genomics-assisted strategies for crop improvement (Fernie and Schauer, 2009). Previous works directly link expression quantitative trait loci (eQTLs) to phenotypic alterations in specific metabolic pathways, emphasizing the complexity of interactions between transcript and metabolite variation (Sonderby et al., 2007; Wentzell et al., 2007; Hansen et al., 2008). These analyses suggest significant differences between the organization of genetic regulation of transcripts and metabolites for a specific subset of *Arabidopsis thaliana* secondary metabolites (Wentzell et al., 2007). Moreover, wide spectrum analyses of metabolites allow QTL mapping of an expanded portion of the plant metabolome (Keurentjes et al., 2006; Schauer et al., 2006; Meyer et al., 2007). The lesser heritability of metabolic traits in compared with transcription levels may suggest metabolite accumulation is more sensitive with environmental conditions (Rowe et al., 2008). Clear annotations were possible for genes associated with the isoflavonoid and triterpenoid pathways of *Medicago truncatula* (Suzuki et al., 2005; Achnine et al., 2005), the methylketone pathway of tomato, pyridine alkaloid biosynthesis in tobacco (Goossens et al., 2003), and glucosinolate, flavonoid, and sterol biosynthesis in Arabidopsis (Morikawa et al., 2006; Hirai et al., 2004; Tohge et al., 2005). Germplasm collections have been utilized to explore metabolic diversity in various species and to dissect the underlying biosynthetic and regulatory pathways (Angelovici et al., 2013; Chen et al., 2014; Riedelsheimer et al., 2012; Vu et al., 2013; Ariyanti et al., 2017). Metabolic QTL (mQTL) analyses of natural variation have thus far been largely carried out in a single biparental segregating population of multiple species, including Arabidopsis (Knoch et al., 2017), tomato (Liu et al., 2016; Rambla et al., 2017), rice (Xu et al., 2016), wheat (Hill et al., 2015), and maize (Wen et al., 2015). The majority of cultivated crops carry only a small fraction of the genetic variation available in related wild species and landraces (Fernie et al., 2006). Therefore, wild genetic resources would be important to introduce specific traits into modern varieties. Metabolite profiling was applied to the identification of pericarp mQTLs (Schauer et al., 2006), using a set of well-characterized tomato introgression lines (Eshed and Zamir, 1994). Three connected biparental populations, which provides richer allelic diversity without obvious additional background noise by introducing segments from various rice subspecies, specifically *indica* MH63, *japonica* Nip, and wild accession ACC10, to the same genetic background ZS97 has been used to observed metabolic variation and thus numerous mQTL were obtained (Chen et al., 2018). Hang et al. (2004) reported the chromosomal locations of genes important for carbohydrate metabolism in *Allium cepa* species based on the set of eight *Allium fistulosum* (FF)-shallot (AA) monosomic addition lines (MALs) (FF+1A∼FF+8A) developed previously by Shigyo et al. (1996). The sugar contents in this set were evaluated monthly to determine the effect of a single alien chromosome from shallot on carbohydrate production in leaf blades of bunching onion. Their results showed that nonreducing sugars were hardly produced in leaf blades of FF+2A plant throughout the year. This finding suggested that the gene controls the production of nonreducing sugars seems to be located on the chromosome 2A of shallot. They also found the effect of shallot extra chromosome 8A on increased amounts of nonreducing sugars in the winter suggesting the gene related to nonreducing sugar production in winter season seems to be located on the chromosome 8A of shallot. Their results suggest that the complete set of MALs is vital for elucidating the mechanism of the metabolic pathway for mucilages, including fructose-oligosaccharides. The mode of inheritance for soluble carbohydrate production has been clarified by localizing candidate genes (McCallum et al., 2001) and QTLs (Havey et al., 2004) related to sucrose and soluble carbohydrate accumulations on an onion linkage map based on molecular markers (King et al., 1998). The combination of both a molecular marker map and a monosomic addition set will reveal the chromosomal locations of QTLs related to carbohydrate production in onion bulbs, as well as to an understanding of the localizations of the different QTLs on the same chromosome. Yaguchi et al. (2013) suggested that additional chromosomes (2A and 6A) would have anonymous genes related to the upregulation of polyphenol production, the antioxidative activities consequently being increased in these MALs. The recent study using the complete set of MALs reveals 50 unigenes involved in saponin biosynthesis (Abdelrahman et al., 2017a). Further studies on effect of shallot extra chromosome on the production of other metabolites, such as ascorbic acid, chlorophyll, sulfides, anthocyanins, and flavonoids, are necessary to better understanding of chromosomal locations of genes related to their productions. Recently, a complete set of *A. cepa-Allium roylei* MALs and a set of *A. fistulosum-A. roylei* chromosome addition lines have been established through our research group (Vu et al., 2012a; Ariyanti et al., 2015). These sets can be useful resources for RNA-Seq and metabolomics approaches to explore genetic and transcriptomic background of *A. roylei*, an important wild species showing bioactive metabolites toward diseases in Allium crops (Abdelrahman et al., 2015).

9.4 Active metabolites as biomarkers for disease resistance in plant breeding

Conventional plant-breeding approaches rely on the selection of plant germplasm with desirable agronomic and phenotypes from among individual plants created by using crosses and mutagenesis. Phenotype-based breeding without knowledge of the genetic composition of the plants is time intensive and resource intensive. Diverse -omics approaches with assessment of the transcriptome, proteome, and metabolome can provide information on the downstream consequences of genome changes that lead to altered phenotype. A vast number of metabolites are produced by plants, many of which are essential for plants to interact with the environment (Saito and Matsuda, 2010; Schwab, 2003). Secondary metabolites are small organic compounds (molecular masses generally less than 3000 Da), different from primary metabolites have no function in the life cycle of cells. The production of specific secondary metabolites varies between species or genera and is thus, apart from appearance and size, etc., an aspect of characterization of a species (Andersson, 2012). Today, it is widely accepted that many of them are involved in interactions between organisms, for example, in plant defense against pathogens, in toxicity of pathogens or attraction of organisms beneficial for the producer (Hartmann, 2007; Kimura et al., 2001; Bennett and Wallsgrove, 1994). The functions of the vast majority of secondary metabolites, however, are still unknown. The boundary between primary and secondary metabolites is not well defined and the areas overlap since primary metabolites such as amino acids, carbohydrates, and acetyl-coenzyme A (CoA) are used as building blocks in secondary metabolites. Some metabolic transformations in primary metabolism also have counterparts in secondary metabolism (Dewick, 2011). To date, the number of described structures exceeds 100,000 (Ribera and Zuniga, 2012; Isman and Paluch, 2011). This rich diversity results in part from an evolutionary process driven by selection for acquisition of improved defense against microbial attack or insect/animal predation (Dixon, 2001). The biosynthesis of several secondary metabolites is constitutive, whereas in many plants it can be induced and enhanced by biological stress conditions, such as wounding or infection (Jogaiah et al., 2016; Dayan et al., 2009). Kogel et al. (2010) conclude that the metabolome represents a more immediate probe of the physiological status of the plant when comparing changes in the leaf transcriptome and metabolome caused by transgenes, cultivar, or biotic interactions in the root using field-grown barley (*Hordeum vulgare*) lines. Metabolomic and transcriptomic analysis of the rice response to the bacterial blight pathogen *Xanthomonas oryzae* pv. oryzae shows enabling differentiation of metabolite and transcript responses of two different rice genotypes and bacterial strains (Sana et al., 2010). Metabolic profiles of sunflower genotypes with contrasting response to *Sclerotinia sclerotiorum* infection demonstrated a genotype-specific regulation of distinct metabolic pathways, suggesting the importance of detection of metabolic patterns rather than specific metabolite changes when looking for metabolic markers differentially responding to pathogen infection (Peluffo et al., 2010). Systemic defense response of *A. thaliana* plants to the leaf pathogen *Pseudomonas syringae* pv. tomato DC3000 (Pst) induced by the beneficial fungus *Trichoderma asperelloides* T203 includes significant changes in amino acids, polyamines, sugars, and citric acid cycle intermediates (Brotman et al., 2012). Via comparative saponin profiles of a complete *A. fistulosum-A. cepa* monosomic addition lines, Vu et al. (2012b) found the chromosome 2A of shallot might possess some of the genes related to Fusarium wilt resistance. Later, Abdelrahman et al. (2017b) were able to isolate and identify Alliospiroside A saponin compound in *A. fistulosum* with extra chromosome 2A from shallot and its role in the defense mechanism against Fusarium pathogens. A detailed primary metabolic profile of inoculated soybean roots from susceptible and resistant genotypes with contrasting resistance to *F. tucumaniae* infection showed that inoculated susceptible plants accumulated amino acids during early stages of infection, suggesting that GC-MS-based metabolomics could be a suitable approach for the rapid characterization of the cultivar response (Scandiani et al., 2015). Wojciechowska et al. (2014) reported that chlorogenic acid, a metabolite identified by untargeted metabolome analysis in resistant tomatoes, inhibits the colonization by *Alternaria alternata* by inhibiting alternariol biosynthesis. Untargeted global metabolomic analysis used to determine and compare the chemical nature of the metabolites altered during the infection of tomato plants (cv. Ailsa Craig) with *Botrytis cinerea* or *P. syringae* pv. tomato DC3000 showed significant changes in amino acids, sugars, and free fatty acids, and in primary and secondary metabolism (Camanes et al., 2015). The mapping population consisted of 210 recombinant inbred lines (RILs) derived from a cross between ZS97 and MH63 demonstrates the potential of a combined omics strategy in understanding the genetic basis of rice metabolome (Suharti et al., 2016). Li et al. (2016) indicate that numerous metabolites were activated to coordinate the banana defense response to *Fusarium oxysporum* f. sp. *cubense*. Nine primary metabolites and 16 lipids were changed in barley plants possessing durable multipathogen resistance gene *Lr34* infected with barley leaf rust (*Puccinia graminis* sp. *hordei*) with high *Lr34* expression levels grown under all three growing conditions (Bucher et al., 2017). The LC-Q/TOF-MS coupled with multivariate statistical analysis have demonstrated three metabolites from different classes of plant compounds—sugars, phenolic acids, and organic acids as potential oil palm metabolite markers for selection of partial resistance oil palm (Nurazah et al., 2017). He et al. (2018) indicated that jasmonate biosynthesis and signaling are stimulated by the fatty acid redirection of a spontaneous mutant o 'Newhall' navel orange (*Citrus sinensis* Osbeck), and participate in the tolerance of pathogenic fungi. Zhu et al. (2018) reported 90 differentially accumulated metabolites identified after *Phytothora sojae* infection; the levels of 50 metabolites differed between the resistant line and the susceptible line. Zeiss et al. (2018) indicated a total of 41 metabolites were statistically selected and annotated, consisting of amino acids, organic acids, lipids, derivatives of cinnamic acid and benzoic acids, flavonoids, and steroidal glycoalkaloids, which were especially prominent in the two highly resistant tomato cultivars to *Ralstonia solanacearum*.

9.5 Conclusion

It has been estimated that global agricultural production is annually reduced by 31%–42% due to plant pathogens and pests (Agrios, 2005). An environmentally sustainable, effective method to control disease infections in plants is the use of resistance genes. The thorough high-performing metabolite profiling technique has enabled the identification of a great number of novel molecules with putative antimicrobial, pathogen-growth inhibition, or semiotic functions. Therefore, metabolomics has provided a new aspect in the study of systems biology helping the clearer understanding of interactions in biological systems such as the interactions of plants with pests and pathogens. Nonetheless, the knowledge of plant metabolic pathways is incomplete and there is still a long way until researchers could fully understand the complexity of plant-pathogen interactions. In perspective of the importance of metabolites, it is, therefore, necessary to further study both their in planta functions and their value for humans.

References

Abdelrahman, M., Sawada, Y., Sato, S., Yamauchi, N., Shigyo, M., 2015. Integrating transcriptome and target metabolome variability in doubled haploids of *Allium cepa* for abiotic stress protection. Mol. Breed. 35, 195.

Abdelrahman, M., Abdel-Motaal, F., El-Sayed, M., Jogaiah, S., Shigyo, M., Ito, S.-i., Tran, L.-S.P., 2016. Dissection of *Trichoderma longibrachiatum*-induced defense in onion (*Allium cepa* L.) against *Fusarium oxysporum* f. sp. *Cepa* by target metabolite profiling. Plant Sci. 246, 128–138.

Abdelrahman, M., El-Sayed, M., Sato, S., Hirakawa, H., Ito, S., Tanaka, K., Mine, Y., Sugiyama, N., Suzuki, Y., Yamauchi, N., Shigyo, M., 2017a. RNA-sequencing-based transcriptome and biochemical analyses of steroidal saponin pathway in a complete set of Allium fistulosum-A. cepa monosomic addition lines. Plos One 12 (8), e0181784.

Abdelrahman, M., Mahmoud, H.Y.A.H., El-Sayed, M., Tanaka, S., Tran, L.S.P., 2017b. Isolation and characterization of Cepa2, a natural alliospiroside A, from shallot (*Allium cepa* L. Aggregatum group) with anticancer activity. Plant Physiol. Biochem. 116, 167–173.

Achnine, L., Huhman, D.V., Farag, M.A., Sumner, L.W., Blount, J.W., Dixon, R.A., 2005. Genomics-based selection and functional characterization of triterpene glycosyltransferases from the model legume *Medicago truncatula*. Plant J. 41, 875–887.

Adhikari, P., Oh, Y., Panthee, D.R., 2017. Current status of early blight resistance in tomato: an update. Int. J. Mol. Sci. 18(10), https://doi.org/10.3390/ijms18102019.

Agrios, G.N., 2005. Plant Pathology, fifth ed. Academic Press, New York.

Aliferis, K.A., Chrysayi-Tokousbalides, M., 2011. Metabolomics in pesticide research and development: review and future perspectives. Metabolomics 7, 35–53.

Aliferis, K.A., Jabaji, S., 2011. Metabolomics - A robust bioanalytical approach for the discovery of the modes-of-action of pesticides: a review. Pestic. Biochem. Physiol. 100, 105–117.

Allwood, J.W., Ellis, D.I., Goodacre, R., 2008. Metabolomic technologies and their application to the study of plants and plant-host interactions. Physiol. Plant. 132, 117–135.

Andersson, P.F., 2012. Secondary Metabolites Associated with Plant Disease, Plant Defense and Biocontrol. Doctoral thesis Swedish University of Agricultural Sciences, Uppsala.

Angelovici, R., Lipka, A.E., Deason, N., Gonzalez-Jorge, S., Lin, H.N., Cepela, J., Buell, R., Gore, M.A., DellaPenna, D., 2013. Genome-wide analysis of branched-chain amino acid levels in Arabidopsis seeds. Plant Cell 25, 4827–4843.

Ariyanti, N.A., Hoa, V.Q., Khrustaleva, L.I., Hirata, S., Abdelrahman, M., Ito, S., Yamauchi, N., Shigyo, M., 2015. Production and characterization of alien chromosome addition lines in *Allium fistulosum* carrying extra chromosomes of *Allium roylei* using molecular and cytogenetic analyses. Euphytica 206 (2), 343–355.

Ariyanti, N.A., Torikai, K., Kirana, R.P., Hirata, S., Sulistyangningsih, E., Ito, S., Yamauchi, N., Kobayashi, N., Shigyo, M., 2017. Comparative study on phytochemical variations in Japanese F1 varieties of bulb onions and south-east Asian shallot landraces. Hortic. J. 87(1), https://doi.org/10.2503/hortj.OKD-066.

Bennett, R.N., Wallsgrove, R.M., 1994. Secondary metabolites in plant defense mechanisms. New Phytol. 127 (4), 617–633.

Betz, F.S., Hammond, B.G., Fuchs, R.L., 2000. Safety and advantages of bacillus thuringiensis-protected plants to control insect pests. Regul. Toxicol. Pharmacol. 32, 156–173.

Bino, R.J., Hall, R.D., Fiehn, O., Kopka, J., Saito, K., Draper, J., Nikolau, B.J., Mendes, P., Roessner-Tunali, U., Beale, M.H., 2004. Potential of metabolomics as a functional genomics tool. Trends Plant Sci. 9, 418–425.

Brotman, Y., Lisec, J., Meret, M., Chet, I., Willmitzer, L., Viterbo, A., 2012. Transcript and metabolite analysis of the *Trichoderma*-induced systemic resistance response to *Pseudomonas syringae* in *Arabidopsis thaliana*. Microbiology 158, 139–146.

Bucher, R., Veyel, D., Willmitzer, L., Krattinger, S., Keller, B., Biglera, L., 2017. Combined GC- and UHPLC-HR-MS based metabolomics to analyze durable anti-fungal resistance processes in cereals. CHIMIA Int. J. Chem. 71 (4), 156–159.

Camanes, G., Scalschi, L., Vicedo, B., Gonzalez-Bosch, C., Garcıa-Agustın, P., 2015. Plant J. 84, 125–139.

Catchpole, G.S., Beckmann, M., Enot, D.P., Mondhe, M., Zywicki, B., Taylor, J., Hardy, N., Smith, A., King, R.D., Kell, D.B., Fiehn, O., Draper, J., 2005. Hierarchical metabolomics demonstrates substantial compositional similarity between genetically modified and conventional potato crops. Proc. Natl. Acad. Sci. U. S. A. 102, 14458–14462.

Chen, W., Gao, Y., Xie, W., Gong, L., Lu, K., Wang, W., Li, Y., Liu, X., Zhang, H., Dong, H., Zhang, W., Zhang, L., Yu, S., Wang, G., Lian, X., Luo, J., 2014. Genome-wide association analyses provide genetic and biochemical insights into natural variation in rice metabolism. Nat. Genet. 46, 714–721.

Chen, J., Wang, J., Chen, W., Sun, W., Peng, M., Yuan, Z., Shen, S., Xie, K., Jin, C., Sun, Y., Liu, X., Fernie, A.R., Yu, S., Luo, J., 2018. Metabolome analysis of multi-connected biparental chromosome segment substitution line populations. Plant Physiol. 178 (2), 612–625.

Choi, H.K., Choi, Y.H., Verberne, M., Lefeber, A.W.M., Erkelens, C., Verpoorte, R., 2004. Metabolic fingerprinting of wildtype and transgenic tobacco plants by ^{1}H NMR and multivariate analysis techniques. Phytochemistry 65, 857–864.

Creek, D.J., Barret, M.P., 2014. Determination of antiprotozoal drug mechanisms by metabolomics approaches. Parasitology 41 (Special Issue 01), 83–92.

Dayan, F.E., Cantrell, C.L., Duke, S.O., 2009. Natural products in crop protection. Bioorg. Med. Chem. 17, 4022–4034.

Dewick, P.M., 2011. Medicinal natural products: a biosynthetic approach, third. ed. John Wiley & Sons.

Dixon, R.A., 2001. Natural products and plant disease resistance. Nature 411, 843–847.

Dunn, W.B., 2008. Current trends and future requirements for the mass spectrometric investigation of microbial, mammalian and plant metabolomes. Phys. Biol. 5, 1–24.

Dunn, W.B., Bailey, N.J.C., Johnson, H.E., 2005. Measuring the metabolome: current analytical technologies. Analyst 130, 606–625.

Eshed, Y., Zamir, D., 1994. A genomic library of Lycopersicon pennellii in L. esculentum: a tool for fine mapping of genes. Euphytica 79, 175–179.

Fernie, A.R., Schauer, N., 2009. Metabolomics-assisted breeding: a viable option for crop improvement? Trends Genet. 25, 39–48.

Fernie, A.R., Tohge, T., 2017. The genetics of plant metabolism. Annu. Rev. Genet. 51, 287–310.

Fernie, A.R., Tadmor, Y., Zamir, D., 2006. Natural genetic variation for improving crop quality. Curr. Opin. Plant Biol. 9, 196–202.

Frank, T., Rohlig, R.M., Davies, H.V., Barros, E., Engel, K., 2012. Metabolite profiling of maize kernels genetic modification versus environmental influence. J. Agric. Food Chem. 60, 3005–3012.

Gao, X., Locke, S., Zhang, J., Joshi, J., Wang-Pruski, G., 2018. Metabolomics profile of potato tubers after phosphite treatment. Am. J. Plant Sci. 9, 845–864.

Garcia-Villalba, R., Leon, C., Dinelli, G., Segura-Carretero, A., Fernandez-Gutierre, A., Garcia-Canas, V., Cifuentes, A., 2008. Comparative metabolomic study of transgenic versus conventional soybean using capillary electrophoresis-time-of-flight mass spectrometry. J. Chromatogr. A 1195, 164–173.

Giannangelo, C.R., Ellis, K.M., Sexton, A.E., Stoessel, D., Creek, D.J., 2016. The role of metabolomics in antiparasitic drug discovery. In: Muller, S., Rachel Cerdan, R., Radulescu, O. (Eds.), Comprehensive Analysis of Parasite Biology: From Metabolism to Drug Discovery. Wiley-VCH Verlag GmbH & Co, KGaA.

Goossens, A., Hakkinen, S.T., Laakso, I., Seppanen-Laakso, T., Biondi, S., De Sutter, V., Lammertyn, F., Nuutila, A.M., Soderlund, H., Zabeau, M., Inze, D., Oksman-Caldentey, K.M., 2003. A functional genomics approach toward the understanding of secondary metabolism in plant cells. Proc. Natl. Acad. Sci. U. S. A. 100, 8595–8600.

Hall, R.D., Brouwer, I.D., Fitzgerald, M.A., 2008. Plant metabolomics and its potential application for human. NutritionPhysiologiaPlantarum 132, 162–175.

Hang, T.T.M., Shigyo, M., Yaguchi, S., Yamauchi, N., Tashiro, Y., 2004. Effect of single alien chromosome from shallot (*Allium cepa* L. Aggregatum group) on carbohydrate production in leaf blade of bunching onion (*A. fistulosum* L.). Genes Genet. Syst. 79, 345–350.

Hansen, B.G., Halkier, B.A., Kliebenstein, D.J., 2008. Identifying the molecular basis of QTLs: eQTLs add a new dimension. Trends Plant Sci. 13, 72–77.

Hartmann, T., 2007. From waste products to ecochemicals: fifty years research of plant secondary metabolism. Phytochemistry 68 (22–24), 2831–2846.

Havey, M.J., Galmarini, C.R., Gokce, A.F., Henson, C., 2004. QTL affecting soluble carbohydrate concentrations in stored onion bulbs and their association with flavor and health-enhancing attributes. Genome 47, 463–468.

He, Y., Han, J., Liu, R., Ding, Y., Wang, J., Li, S., Yang, X., Zeng, Y., Wen, W., Xu, J., Zhang, H., Yan, X., Chen, Z., Gu, Z., Chen, H., Tang, H., Deng, X., Cheng, Y., 2018. Horticulture Research 5, 43.

Hill, C.B., Taylor, J.D., Edwards, J., Mather, D., Langridge, P., Bacic, A., Roessner, U., 2015. Detection of QTL for metabolic and agronomic traits in wheat with adjustments for variation at genetic loci that affect plant phenology. Plant Sci. 233, 143–154.

Hirai, M.Y., Yano, M., Goodenowe, D.B., Kanaya, S., Kimura, T., Awazuhara, M., Arita, M., Fujiwara, T., Saito, K., 2004. Integration of transcriptomics and metabolomics for understanding of global responses to nutritional stresses in Arabidopsis thaliana. Proc. Natl. Acad. Sci. U. S. A. 101, 10205–10210.

Isman, M.B., Paluch, G., 2011. Needles in the haystack: exploring chemical diversity of botanical insecticides. In: Lopez, O., Fernandez-Bolanos, J.G. (Eds.), Green Trends in Insect Control. RSC, pp. 248–265.

Jogaiah, S., Shetty, H.S., Ito, S.-I., Tran, L.-S.P., 2016. Enhancement of downy mildew disease resistance in pearl millet by the G_app7 bioactive compound produced by *Ganoderma applanatum*. Plant Physiol. Biochem. 105, 109–117.

Keurentjes, J.J.B., Fu, J.Y., de Vos, C.H.R., Lommen, A., Hall, R.D., Bino, R.J., van der Plas, L.H.W., Jansen, R.C., Vreugdenhil, D., Koornneef, M., 2006. The genetics of plant metabolism. Nat. Genet. 38, 842–849.

Kimura, M., Anzai, H., Yamaguchi, I., 2001. Microbial toxins in plant-pathogen interactions: biosynthesis, resistance mechanisms, and significance. J. Gen. Appl. Microbiol. 47 (4), 149–160.

King, J.J., Bradeen, J.M., Bark, O., McCallum, J.A., Havey, M.J., 1998. A low-density genetic map of onion reveals a role for tandem duplication in the evolution of an extremely large diploid genome. Theor. Appl. Genet. 96, 52–62.

Knoch, D., Riewe, D., Meyer, R.C., Boudichevskaia, A., Schmidt, R., Altmann, T., 2017. Genetic dissection of metabolite variation in Arabidopsis seeds: evidence for mQTL hotspots and a master regulatory locus of seed metabolism. J. Exp. Bot. 68, 1655–1667.

Kogel, K., Voll, L.M., Schafer, P., Jansen, C., Wu, Y., Langen, G., Imani, J., Hofmann, J., Schmiedl, A., Sonnewald, S., Wettstein, D., Cook, R.J., Sonnewald, U., 2010. Transcriptome and metabolome profiling of field grown transgenic barley lack induced differences but show cultivar-specific variances. Proc. Natl. Acad. Sci. U. S. A. 107 (14), 6198–6203.

Kos, M., van Loon, J.J., Dicke, M., Vet, L.E., 2009. Transgenic plants as vital components of integrated pest management. Trends Biotechnol. 27, 621–627.

Krishnan, P., Kruger, N.J., Ratcliffe, R.G., 2005. Metabolite fingerprinting and profiling in plants using NMR. J. Exp. Bot. 56, 255–265.

Kromer, J.O., Sorgenfrei, O., Klopprogge, K., Heinzle, E., Wittmann, C., 2004. In-depth profiling of lysine-producing *Corynebacterium glutamicum* by combined analysis of the Transcriptome, Metabolome, and Fluxome. J. Bacteriol. 186, 1769–1784.

Kusano, M., Fukushima, A., Kobayashi, M., Hayashi, N., Jonsson, P., Moritz, T., Ebana, K., Saito, K., 2007. Application of a metabolomic method combining one-dimensional and twodimensional gas chromatography-time-of-flight/mass spectrometry to metabolic phenotyping of natural variants in rice. J Chromatogr. B Analyt. Technol. Biomed. Life Sci. 855, 71–79.

Le Gall, G., Colquhoun, I.J., Davis, A.L., Collins, G.J., Verhoeyen, M.E., 2003. Metabolite profiling of tomato (*Lycopersicon esculentum*) using ^{1}H NMR spectroscopy as a tool to detect potential unintended effects following a genetic modification. J. Agric. Food Chem. 51, 2447–2456.

Leon, C., Rodriguez-Meizoso, I., Lucio, M., Garcia-Canas, V., Ibanez, E., Schmitt-Kopplin, P., Cifuentes, A., 2009. Metabolomics of transgenic maize combining Fourier transform-ion cyclotron resonance-mass spectrometry, capillary electrophoresis-mass spectrometry and pressurized liquid extraction. J. Chromatogr. A 1216, 7314–7323.

Li, W., Li, C., Sun, J., Peng, M., 2016. Metabolomic, biochemical, and gene expression analyses reveal the underlying responses of resistant and susceptible Banana species during early infection with *Fusarium oxysporum* f. sp. *cubense*. Plant Dis. 101 (4), 534–543.

Liu, Z.Y., Alseekh, S., Brotman, Y., Zheng, Y., Fei, Z.J., Tieman, D.M., Giovannoni, J.J., Fernie, A.R., Klee, H.J., 2016. Identification of a *Solanum pennellii* chromosome 4 fruit flavor and nutritional quality-associated metabolite QTL. Front. Plant Sci. 7, 1671.

Lloyd, A.J., Allwood, W.J., Winder, C.L., Dunn, W.B., Heald, J.K., Cristescu, S.M., Sivakumaran, A., Harren, F.J., Mulema, J., Denby, K., Goodacre, R., Smith, A.R., Mur, L.A., 2011. Metabolomic approaches reveal that cell wall modifications play a major role in ethylene-mediated resistance against *Botrytis cinerea*. Plant J. 67, 852–868.

Loots, D.T., 2014. An altered *Mycobacterium* tuberculosis Metabolome induced by katG mutations resulting in isoniazid resistance. Antimicrob. Agents Chemother. 58 (4), 2144–2149.

Marti, G., Schnee, S., Andrey, Y., Simoes-Pires, C., Carrupt, P., Wolfender, J., Gindro, K., 2014. Study of leaf metabolome modifications induced by UV-C radiations in representative vitis, cissus and cannabis species by LC-MS based metabolomics and antioxidant assays. Molecules 19, 14004–14021.

McCallum, J., Leite, D., Pither-Joyce, M., Havey, M.J., 2001. Expressed sequence markers for genetic analysis of bulb onion (*Allium cepa* L.). Theor. Appl. Genet. 103, 979–991.

Meyer, R.C., Steinfath, M., Lisec, J., Becher, M., Witucka-Wall, H., Torjek, O., Fiehn, O., Eckardt, A., Willmitzer, L., Selbig, J., Altmann, T., 2007. The metabolic signature related to high plant growth rate in Arabidopsis thaliana. Proc. Natl. Acad. Sci. U. S. A. 104, 4759–4764.

Morikawa, T., Mizutani, M., Aoki, N., Watanabe, B., Saga, H., Saito, S., Oikawa, A., Suzuki, H., Sakurai, N., Shibata, D., Wadano, A., Sakata, K., Ohta, D., 2006. Cytochrome P450 CYP710A encodes the sterol C-22 desaturase in Arabidopsis and tomato. Plant Cell 18, 1008–1022.

National Academies of Sciences, Engineering, and Medicine, 2016. Genetically Engineered Crops: Experiences and Prospects. The National Academies Press, Washington, DC.

Nurazah, Z., Idris, A.S., Kushairi, A., Din, A.M., Abrizah, O., Ramli, S., 2017. Metabolomics unravel differences between Cameroon Dura and deli Dura oil palm (*Elaeis guineensis* Jacq.) genetic backgrounds against basal stem rot. J. Oil Palm. Res. 29 (2), 227–241.

Peluffo, L., Lia, V., Troglia, C., Maringolo, C., Norma, P., Escande, A., Hopp, E.H., Lytovchenko, A., Fernie, A.R., Heinz, R., Carrari, F., 2010. Metabolic profiles of sunflower genotypes with contrasting response to *Sclerotinia sclerotiorum* infection. Phytochemistry 71, 70–80.

Rambla, J.L., Medina, A., Fernandez-del-Carmen, A., Barrantes, W., Grandillo, S., Cammareri, M., Lopez-Casado, G., Rodrigo, G., Alonso, A., Garcia-Martinez, S., Primo, J., Ruiz, J.J., Fernandez-Munoz, R., Monforte, A.J., Granell, A., 2017. Identification, introgression, and validation of fruit volatile QTLs from a red-fruited wild tomato species. J. Exp. Bot. 68, 429–442.

Ribera, A.E., Zuniga, G., 2012. Induced plant secondary metabolites for phytopatogenic fungi control: a review. J. Soil Sci. Plant Nutr. 12 (4), 893–911.

Riedelsheimer, C., Czedik-Eysenberg, A., Grieder, C., Lisec, J., Technow, F., Sulpice, R., Altmann, T., Stitt, M., Willmitzer, L., Melchinger, A.E., 2012. Genomic and metabolic prediction of complex heterotic traits in hybrid maize. Nat. Genet. 44, 217–220.

Rowe, H.C., Hansen, B.G., Halkier, B.A., Kliebenstein, D.J., 2008. Biochemical networks and epistasis shape the *Arabidopsis thaliana* Metabolome. Plant Cell 20, 1199–1216.

Saito, K., Matsuda, F., 2010. Metabolomics for functional genomics, systems biology, and biotechnology. Annu. Rev. Plant Biol. 61, 463–489.

Sana, T.R., Fischer, S., Wohlgemuth, G., Katrekar, A., Jung, K., Ronald, P.C., Fiehn, O., 2010. Metabolomic and transcriptomic analysis of the rice response to the bacterial blight pathogen *Xanthomonas oryzae* pv. *oryzae*. Metabolomics 6, 451–465.

Saxena, A., Cramer, C.S., 2013. Metabolomics: a potential tool for breeding Nutraceutical vegetables. Adv Crop Sci. Tech. 1 (2), 1–4.

Scandiani, M.M., Luque, A.G., Razori, M.V., Casalini, L.C., Aoki, T., O'Donnell, K., Cervigni, G.D., Spampinato, C.P., 2015. Metabolic profiles of soybean roots during early stages of *Fusarium tucumaniae* infection. J. Exp. Bot. 66 (1), 391–402.

Schauer, N., Semel, Y., Roessner, U., Gur, A., Balbo, I., Carrari, F., Pleban, T., Perez-Melis, A., Bruedigam, C., Kopka, J., Willmitzer, L., Zamir, D., Fernie, A.R., 2006. Comprehensive metabolic profiling and phenotyping of interspecific introgression lines for tomato improvement. Nat. Biotechnol. 24, 447–454.

Schwab, W., 2003. Metabolome diversity: too few genes, too many metabolites? Phytochemistry 62, 837–849.

Shigyo, M., Tashiro, Y., Isshiki, S., Miyazaki, S., 1996. Establishment of a series of alien monosomic addition lines of Japanese bunching onion (*Allium fistulosum* L.) with extra chromosomes from shallot (*A. cepa* L. Aggregatum group). Genes Genet. Syst. 71, 363–371.

Sonderby, I.E., Hansen, B.G., Bjarnholt, N., Ticconi, C., Halkier, B.A., Kliebenstein, D.J., 2007. A systems biology approach identifies a R2R3 MYB gene subfamily with distinct and overlapping functions in regulation of aliphatic glucosinolates. PLoS One. 2, e1322.

Spratlin, J.L., Serkova, N.J., Eckhardt, S.G., 2009. Clinical applications of metabolomics in oncology: a review. Clin. Cancer Res. 15, 431–440.

Suharti, W.S., Nose, A., Zheng, S., 2016. Metabolite profiling of sheath blight disease resistance in rice: in the case of positive ion mode analysis by CE/TOF-MS. Plant Prod. Sci. 19 (2), 279–290.

Suzuki, H., Reddy, M.S., Naoumkina, M., Aziz, N., May, G.D., Huhman, D.V., Sumner, L.W., Blount, J.W., Mendes, P., Dixon, R.A., 2005. Methyl jasmonate and yeast elicitor induce differential transcriptional and metabolic re-programming in cell suspension cultures of the model legume *Medicago truncatula*. Planta 220, 696–707.

Tohge, T., Nishiyama, Y., Hirai, M.Y., Yano, M., Nakajima, J., Awazuhara, M., Inoue, E., Takahashi, H., Goodenowe, D.B., Kitayama, M., Noji, M., Yamazaki, M., Saito, K., 2005. Functional genomics by integrated analysis of metabolome and transcriptome of Arabidopsis plants over-expressing an MYB transcription factor. Plant J. 42, 218–235.

Tugizimana, F., Mhlongo, M.I., Piater, L.A., Dubery, I.A., 2018. Metabolomics in plant priming research: the way forward? Int. J. Mol. Sci. 19(6), https://doi.org/10.3390/ijms19061759.

Vinayavekhin, N., Homan, E.A., Saghatelian, A., 2010. Exploring disease through metabolomics. ACS Chem. Biol. 5, 91–103.

Vu, H.Q., Yoshimatsu, Y., Khrustaleva, L.I., Yamauchi, N., Shigyo, M., 2012a. Alien genes introgression and development of alien monosomic addition lines from a threatened species, *Allium roylei* Stearn, to *Allium cepa* L. Theor. Appl. Genet. 124 (7), 1241–1257.

Vu, H.Q., El-Sayed, M.A., Ito, S., Yamauchi, N., Shigyo, M., 2012b. Discovery of a new source of resistance to *Fusarium oxysporum*, cause of *Fusarium* wilt in *Allium fistulosum*, located on chromosome 2 of *Allium cepa* Aggregatum group. Genome 55, 797–807.

Vu, Q.H., Hang, T.T.M., Yaguchi, S., Ono, Y., Pham, T.M.P., Yamauchi, N., Shigyo, M., 2013. Assessment of biochemical and antioxidant diversities in a shallot germplasm collection from Vietnam and its surrounding countries. Genet. Resour. Crop. Evol. 60 (4), 1297–1312.

Wen, W., Li, K., Alseekh, S., Omranian, N., Zhao, L., Zhou, Y., Xiao, Y., Jin, M., Yang, N., Liu, H., Florian, A., Li, W., Pan, Q., Nikoloski, Z., Yan, J., Fernie, A.R., 2015. Genetic determinants of the network of primary metabolism and their relationships to plant performance in a maize recombinant inbred line population. Plant Cell 27, 1839–1856.

Wentzell, A.M., Rowe, H.C., Hansen, B.G., Ticconi, C., Halkier, B.A., Kliebenstein, D.J., 2007. Linking metabolic QTL with network and cis-eQTL controlling biosynthetic pathways. PLoS Genet. 3, e162.

Wishart, D.S., 2008. Applications of metabolomics in drug discovery and development. Drugs R D 9 (5), 307–322.

Wojciechowska, E., Weinert, C.H., Egert, B., Trierweiler, B., Schmidt-Heydt, M., Horneburg, B., Graeff-Honninger, S., Kulling, S.E., Geisen, R., 2014. Chlorogenic acid, a metabolite identified by untargeted metabolome analysis in resistant tomatoes, inhibits the colonization by *Alternaria alternata* by inhibiting alternariol biosynthesis. Eur. J. Plant Pathol. 139 (4), 735–747.

Xu, S.Z., Xu, Y., Gong, L., Zhang, Q.F., 2016. Metabolomic prediction of yield in hybrid rice. Plant J. 88, 219–227.

Yaguchi, S., Matsumoto, M., Date, R., Harada, T., Maeda, T., Yamauchi, N., 2013. Shigyo M (2013) biochemical analyses of the antioxidative activity and chemical ingredients in eight different allium alien monosomic addition lines. Biosci. Biotechnol. Biochem. 77 (12), 2486–2488.

Yao, Z., Guo, Z., Wang, Y., Li, W., Fu, Y., Lin, Y., Lin, W., Lin, X., 2018. Integrated Succinylome profiling and metabolomics reveal crucial role of *S*-ribosylhomocysteine lyase in quorum sensing and metabolism of *Aeromonas hydrophila*. Mol. Cell. Proteomics. https://doi.org/10.1074/mcp.RA118.001035.

Yun, Z., Gao, H., Liu, P., Liu, S., Luo, T., Jin, S., Xu, Q., Xu, J., Cheng, Y., Deng, X., 2013. Comparative proteomic and metabolomic profiling of citrus fruit with enhancement of disease resistance by postharvest heat treatment. BMC Plant Biol. 13, 44.

Zeiss, D.R., Mhlongo, M.I., Tugizimana, F., Steenkamp, P.A., Dubery, I.A., 2018. Comparative metabolic Phenotyping of tomato (*Solanum lycopersicum*) for the identification of metabolic signatures in cultivars differing in resistance to*Ralstonia solanacearum*. Int. J. Mol. Sci. 19 (9), 2558. https://doi.org/10.3390/ijms19092558.

Zhou, J., Ma, C., Xu, H., 2009. Metabolic profiling of transgenic rice with cryIAc and sck genes: an evaluation of unintended effects at metabolic level by using GC–FID and GC–MS. J. Chromatogr. B 877, 725–732.

Zhu, L., Zhou, Y., Li, X., Zhao, J., Guo, N., Xing, H., 2018. Metabolomics analysis of soybean hypocotyls in response to *Phytopthora sojae* infection. Front. Plant Sci. 9, 1530. https://doi.org/10.3389/fpls.2018.01530.

Current trend and future prospects of secondary metabolite-based products from agriculturally important microorganisms

10

Richa Salwan[a] *and Vivek Sharma*[b]
[a]College of Horticulture and Forestry, Dr YS Parmar University of Horticulture & Forestry, Hamirpur, Himachal Pradesh, India, [b]University Centre for Research and Development, Chandigarh University, Mohali, Punjab, India

10.1 Introduction

The competition from weeds, plant pathogens infection, and pests' infestation is responsible for over one-third of yield loss and are some of the major hurdles to cope the demands of agricultural produce. The applications of synthetic chemical-based fertilizers and pesticides have been extensively explored for enhanced agricultural production (Jogaiah et al., 2007). However, the side effect of these chemical-based formulations caused heavy loss due to their biomagnification in the ecosystem, followed by harmful effects of residual to the crop followed by development of resistance in pests against chemicals over the period (Shelton et al., 2002; Satapute et al., 2019a). The side effects associated with the usage of synthetic pesticides demands alternate methods for enhancing agricultural production (Mishra et al., 2015; Keswani et al., 2016; Satapute et al., 2019b). In general, several plant beneficial microorganisms are found to release a plethora of biologically active metabolites, which are classified into primary or secondary metabolites (Singh et al., 2016; Jogaiah et al., 2018).

These microbial-derived secondary metabolites have been reported for activity against harmful bacteria, fungi, insects, and plant growth promotion to act as signaling molecules (Salwan et al., 2018, 2019; Salwan and Sharma, 2020a,b,c). In addition, the basic scaffold of microbial secondary metabolites can be a foundation for chemists. The development in forward and reverse genetics tool in the form of "gene to metabolites" and vice versa offers new and unexplored areas for mining secondary metabolites genes for future applications (Salwan and Sharma, 2020a,b,c). The advent of mining genome for secondary metabolites and use of precise genome editing tools, artificial gene operons for secondary metabolites can be helpful in activating the silent/cryptic gene clusters for novel metabolites, which were otherwise poorly expressed and hence difficult to depict under in vitro conditions (Salwan and

Biocontrol Agents and Secondary Metabolites. https://doi.org/10.1016/B978-0-12-822919-4.00010-7

Sharma, 2020b). The cryptic biosynthetic pathways are potential reservoir of undiscovered small molecules.

Here in this chapter, we have discussed the role of secondary metabolites of plant beneficial microorganisms, recent development in mining the genomes for biosynthetic gene clusters of secondary metabolites among different microorganisms. The advantages of genetic engineering tools such as gene editing methods for enhanced production of secondary metabolites plays valuable role in plant disease management for enhanced crop yields in a sustainable way without the application of synthetic chemical-based approaches.

10.2 Overview of microbial metabolites

The microbial secondary metabolites are known for the plethora of biological roles and activities useful to human welfare (Manivasagan et al., 2014). The secondary metabolites of these microbes are broadly categorized into volatile compounds (VOC) and nonvolatile compounds (Nakkeeran et al., 2019). The different secondary metabolites of these microorganisms include antibiotics, VOC, siderophores, and plant growth regulators. The comprehensive list of secondary metabolites along with their origin and biological activities can be accessed at http://www.bio.nite.go.jp/pks/. Among different microorganisms, bacteria such as *Pseudomonas* (Shanmugam et al., 2008), *Bacillus* (Salwan and Sharma, 2020a,b,c), members of actinobacteria *Streptomyces* (Sharma et al., 2020) and *Frankia* as well as members of fungi *Trichoderma* (Salwan et al., 2019, 2020a,b; Salwan and Sharma, 2020a,b,d; Sharma and Salwan, 2018; Sharma et al., 2016, 2018).

The secondary metabolites are low-molecular-weight compounds and are not required for the routine growth, development, and reproductive process of an organism. However, they play a vital role in countering environmental stresses and other processes such as antimicrobial activity (Abdelrahman et al., 2016), plant growth promotion, and treating various ailments (Salwan et al., 2020a). The secondary metabolites are synthesized as nonribosomal peptides and polyketides platforms, whereas others are derived from terpenoids, shikimic acid, and amino glycosides pathways (Davies and Ryan, 2012). In Pseudomonads, antibiotics such as anthranilate, oomycin A, 2,4-diacetylphloroglucinol (DAPG), phenazine-1-carboxylic acid (PCA), pyocyanin, gluconic acid pyrrolnitrin, pyoluteorin, hydrogen cyanide (HCN), and viscosinamide have been reported for plant disease suppression (Nielsen et al., 1999, 2002; Chin A-Woeng et al., 2003; Bloemberg and Lugtenberg, 2001). Besides this, other microbes such as *Bacillus* and *Streptomyces* have been widely studied for their plant beneficial attributes (Domenech et al., 2006; Kunova et al., 2016; Thampi and Bhai, 2017; Zeng et al., 2012).

The secondary metabolites of different plant beneficial microorganisms are found to promote health and growth of the associated host plant (Babu et al., 2015). For example, the VOCs of microbial origins are active during biotic and abiotic stresses (Bitas and Kang, 2012). So far, approximately 500 microorganisms of fungal and bacterial origin have been studied for alcohols, small alkenes, ketones, esters, sesquiterpenes, lactones, thioalcohols, thioesters, and VOCs production (http://bioinformatics.charite.de/mvoc/) (Splivallo et al., 2011; Kramer and Abraham, 2012; Lemfack et al., 2013, 2014; Effmert et al., 2012). These derivatives affect the plant growth and decrease the disease severity index (Ryu et al., 2003; Vespermann et al., 2007; Zhang et al., 2008; Korpi et al., 2009; Hung et al., 2012) due to their antimicrobial nature (Strobel et al., 2001; Strobel, 2006). According to the Antibase database, the VOCs of *Trichoderma* have been explored for their roles in ecological and medical applications (Howell, 1998; Sivasithamparam and Ghisalberti, 1998; Laatsch, 2007; Reino et al., 2008). These VOCs suppress the phytopathogens (Wheatley et al., 1997; Humphris et al., 2001; Bruce et al., 2004) as well as promote plant growth (Hung et al., 2012). The field application of 2-butanone and 3-pentanol in cucumber growing fields reported to reduce *Myzus persicae* aphids infestations (Song and Ryu, 2013).

10.3 Mining platform and biochemical pathways of secondary metabolites biosynthesis

Secondary metabolites of microbial origin are commonly synthesized on polyketide platforms. Broadly, these platforms are classified into three categories:

10.3.1 Type-I PKS

Type-I polyketide synthase (type-I PKS) is modular in nature, and represents one of the most complex class amid three polyketides. Here, within each polypeptide, the module is comprised of multiple active sites. It differs from iterative class of polyketide, where a single multienzyme unit acts repeatedly. In type-I PKS multifunctional polyketide synthase complex enzymes, each module stereochemically select a specific building unit for the synthesis of metabolites. Initially, the gene clusters encoding these complex arrays were difficult to characterize. For example, the single polypeptide unit of type-I PKS biosynthesis module of erythromycin in *Streptomyces* contains multiple active sites complex of DEBS1, DEBS2, and DEBS3. Here, every single unit contains domains, which are involved in the addition of unit until a complete chain is formed. Polyketide synthase complex has been reported for the biosynthesis of various secondary metabolites including ansamycins (Berrada et al., 1987), macrolides (Murata et al., 1994; Nakkeeran et al., 2019), polyenes, and polyethers. In general, the polyketide synthase complex comprises units of ketosynthase, acyltransferase, and acyl carrier organized in modular way. These units catalyze the selection, condensation, transfer of units, and retained sequentially growing polyketide chain. The polyketide synthase complex also contains activity for ketoreductase, dehydratase, and enoylreductase, which are responsible for the diversity of metabolites. The units/domains in type-I PKS are organized in a colinear relationship with biosynthetic steps. The smallest compound 6–8 carbon compounds such as triacetic acid lactone (TAL) and 6-methylsalicylic acid (6-MSA) as well as one of the largest compounds such as maitotoxin containing 164 carbons are synthesized using such platforms (Murata et al., 1994). Several rigid and multicyclic structural compounds such as doxorubicin having multicyclic and rigid structure are produced from polyketides units via alternate folding and a series of end-to-end cyclization at intramolecular level such as erythromycin.

10.3.2 Type-II PKS

Although the exact functions of various type-II polyketides are still unclear. However, proteins unit are discrete and iterative in nature with each unit has one or more roles (Hertweck et al., 2007). The chain initiation mostly occurs through acetate and then malonate-mediated elongation through ketosynthase-like enzymes and acyl carrier protein (ACP). The thioester gradually tethers the growing chain. The MCAT complex known as malonyl-CoA: ACP transferase diverts basic units from primary metabolic pathways. The cooperation of ketoreductases, cyclases, and aromatases subunits with PKS leads to the folding and maturation of polyketide chain. The post-processing through oxygenation, glycosylation, and *C*-methyltransferase activity results remodeling of the polyketide (Rix et al., 2002).

10.3.3 Type-III PKS

The structure of type-III polyketide systems is comparatively simple. The type-III polyketide deploys single active site for repeated acetate units-mediated condensation, which ultimately results different types of aromatic products (Austin and Noel, 2003). It plays a vital role in initiation, followed by chain elongation and ultimately ring formation. The chain extension of the linear intermediate is done within active site of the same PKS cavity. The entire process is executed by the condensation at intramolecular level and then aromatization. The diversity of compounds is produced through downstream (Austin et al., 2008). The cyclodipeptide synthases (CDPSs) enzymes use different aminoacyl-tRNAs for the biosynthesis of different cyclodipeptide (Moutiez et al., 2017) and diketopiperazine, which also contribute to different biological activities (Alqahtani et al., 2015; Zazopoulos et al., 2003; James et al., 2016; Brockmeyer and Li, 2017; Salwan and Sharma, 2020a).

Although a large number of secondary metabolites have been reported from different plant beneficial microorganisms but their biosynthesis is so far least explored. These metabolites are derivatives of hydrocarbons and in general are derived from the derivatives of amino acids and proteins, carbohydrates, and lipids (Audrain et al., 2015). The VOCs biosynthesis have resulted from the condensation of the polar head groups of fatty acids (Sukovich et al., 2010), reduction of aldehyde to alcohol, which lead to release of hydroxyl groups and even through elongation-decarboxylation type of reactions (Brown and Shanks, 2012). Simultaneously, α- and β-oxidation of fatty acids is found to the release of higher carbon containing aliphatic alcohols (Hamilton-Kemp et al., 2005). On the other side, under anaerobic conditions, alcohols such as 2,3-butanediol containing less carbon chain are produced. The branch chain alcohols such as methyl-butanol are synthesized via Ehrlich pathway from amino acids (Xiao and Xu, 2007). Ketones types of VOCs such as 3-hydroxy-2-butanone and 2,3-butanedione are formed from pyruvate metabolism under anaerobic environmental conditions (Ryu et al., 2003). The sulfur-containing compounds such as dimethyl disulfide, dimethyl trisulfide, and dimethyl tetrasulfide are produced from the degradation of methionine and oxidation of methanethiol (Scholler et al., 2002). In algae and higher plants, the biosynthesis of these compounds includes removal of 3-di-methylsulfoniopropionate from L-methionine (Stefels, 2000). The secondary metabolites such as terpenoids are derived from isopentenyl diphosphate (IPP) and DMAPP dimethylallyl diphosphate units.

The volatile metabolites and terpenoids are synthesized using two pathways; mevalonic acid (MVA)-based pathway has been reported in animals and fungi, whereas methylerythritol phosphate pathway (MEP) has been reported from algae and bacteria. In plants and bacteria, both the MVA and MEP pathways have been reported (Grawert et al., 2011). In actinobacteria, the release of monoterpenes, sesquiterpenes, and their derivatives have been reported. The biosynthesis of geosmin is well documented in *Streptomyces coelicolor* (Citron et al., 2012). The sulfide and hydrogen cyanide-based inorganic VOCs are formed by the degradation of amino acids cysteine, NH_3 from peptides, L-aspartate like amino acids, deamination of aspartate to fumarate by aspartate ammonia lyase (Bernier et al., 2011), and NO from L-arginine (Mattila and Thomas, 2014). The biosynthesis of plant growth regulators such as indole-3-acetic acid (IAA) in *Streptomyces* has been proposed in our recent study (Salwan et al., 2020a). These growth regulators are involved in cell division, elongation differentiation, and increasing root hair formation and even enhancing nutrients absorption from soil. *Streptomyces* synthesize IAA from tryptophan by catabolizing indole-3-acetamide (IAM), indole-3-lactic acid (ILA), indole-3-ethanol (IEt), and indole-3-acetaldehyde (IAAid) into IAA by different pathways (Salwan and Sharma, 2020b).

10.4 Genome mining for secondary metabolites

The recent developments in next-generation sequencing enhanced the speed of mining the genomes for different candidate genes. The microbial genomes now can be explored for various gene clusters even of cryptic nature. It is now clear that majority of the BCGs of microbial origin are unexplored (Ishikawa, 2008). Therefore, genomic advancements played role in unraveling the cryptic pathways of different secondary metabolites. Initially, the identification of biosynthetic pathways and their enzymes was performed either using BLAST-based search engine or alignments of amino acid sequences followed by exploration of genes sequences. The mining of genome for BCGs was further improved using MultiGeneBlast (Weber and Kim, 2016). The further advancements in genome sequencing and computational methods such as Pan-genome-based mining and individual genome mining brought a paradigm shift in the characterization of the secondary metabolites encoding biosynthetic pathways. Pan-genome-based mining approach relies on multiple strains revealed a global scenario of gene repertoire across the strains of a species and hence is crucial in understanding the genetic diversity and molecular basis of their environmental adaptation. For example, the pan-genome analysis of 75 strains of marine actinomycetes provided comprehensive distributions and evolution of their biosynthetic pathways as well as insights of chemical diversity across these strains. On the other hand, the complete genome sequencing has shown that the abundance of gene clusters is far more than the number of metabolites identified so far (Niu, 2018).

The computational-based identification approach uses "Metabolites to Genes" based method. This retro approach uses the enzymatic reactions or biochemical transformations methods to generate a moiety from precursors and then identifies the candidate gene involved in the catalysis of the proposed transformation (Khater et al., 2016). The computational methods for the reconstruction of biosynthetic pathways have been discussed in detail (Khater et al., 2016; Weber and Kim, 2016). The databases including antiSMASH, MIBiG, ClusterMine360, https://magarveylab.ca/, IMG-ABC/Integrated Microbial Genomes: Atlas of Biosynthetic Gene Clusters (http://www.secondarymetabolites.org/databases/), and Orphan assembly line polyketide synthases have played a valuable role in mapping and linking thousands of metabolites to the corresponding BCGs. These databases permit researchers to access as well as deposit the BCGs data and develop comprehensive comparative analysis tools. A comparative overview of different actinobacteria including *Streptomyces albus* J1074, *Micromonospora* sp. ATCC 39149, *Streptomyces griseoflavus* Tu4000, *Frankia* sp., CcI3, *Streptomyces bingchenggensis* BCW-1, *Thermomonospora chromogena* strain DSM 43794, *Kribbella catacumbae* DSM 19601, *Kineococcus radiotolerans* SRS 30,216, *Salinispora pacifica* DSM 45543, and others has been described in our recent review (Salwan and Sharma, 2020a,b,c). In general, variations in the total gene pools as well as BCGs organizations, and secondary metabolites potential of actinobacteria. The tool such as ClusterMine360 has identified over 290 gene clusters involved in the biosynthesis of 200 nonribosomal types of polyketides synthesis. Other tools including Integrated Microbial Genomes-Atlas of Biosynthetic Gene Clusters like database (IMG-ABC) provide useful information about the genomic locus for BCGs. The other databases such as NP Searcher, MIBiG, NRPS predictor, ClustScan, antiSMASH integrated with tools like Cluster Finder, Prediction Informatics for Secondary Metabolomes (PRISM) permit characterization of gene clusters involved in novel metabolites. These databases also played a vital role in the identification of domains for specific adenylation, acyltransferase activity, and other tailoring reactions. The database such as Pep2Path is useful in the identification of the chemical structure of nonribosomal peptides. It relies on matching mass spectra of amino acids to corresponding gene clusters. The different BCGs provide deep understanding of operon arrangement in a cluster, e.g., eythromycin biosynthesis from L-mycarose and D-desosamine biosynthesis (Oliynyk et al., 2007), whereas BCGs for glycopeptide-based antibiotics assembly contain enzymes for nonproteinogenic amino acids 4-hydroxyphenylglycine, β-hydroxytyrosine, and 3,5-dihydroxyphenylglycine (Kahne et al., 2005).

Genome mining plays vital roles in identification of silent or orphan BCGs. Presently, over 90% of silent genes clusters responsible for the biosynthesis of diversity of metabolites have been identified from actinomycetes alone (Ishikawa, 2008). The impact of genome projects can be analyzed by the facts the microorganisms such as actinomycetes are now considered as potential powerhouse of antibiotics and other secondary metabolites (Zazopoulos et al., 2003). Different genomic tools have facilitated the mining of even rare microorganisms as a reservoir of novel antibiotics. Although, several of these BCGs are phenotypically cryptic in nature, but can be manipulated either using nonstandard fermentation or gene/genome editing approaches or amalgamation of both methods (Ishikawa, 2008; Tiwari and Gupta, 2012). Gene/genome editing using (CRISPER)/Cas (Clustered Regularly Interspaced Short Palindromic Repeats) or using synthetic biology tools of otherwise silent pathways is promising source of novel compounds discovery and enhanced production of corresponding metabolites. For successful production, the heterologous of the pathways demand expression of either single or complete operon or several genes (Lazarus et al., 2014). Significant efforts have already been made for the expression of metabolites in the heterologous host (Ongley et al., 2013; Gomez-Escribano and Bibb, 2014; He et al., 2018; Salwan and Sharma, 2020b).

10.5 Applications

10.5.1 Biocontrol potential of secondary metabolites

Nowadays, the growing concern for environment and side effects of chemical-based methods to manage plant pathogens demands alternate methods for the management of plant pathogens (Zasada et al., 2010). In this context, biocontrol agents and their metabolites have emerged as potential candidates (Jogaiah et al., 2016). In particular, the secondary metabolites and biomolecules of microbial origin are potential candidates for new generation of insecticidal, pesticidal, and antimicrobial agents. However, the cost to produce a molecule is high (USD 256 million) compared to use of biological agents (USD 20–50 million) (Olson, 2015). The other hurdles such as stability of the secondary metabolites, the target spectra of metabolites, their side effect on the environmental components including humans and other organisms need to be understood. The limitations of biocontrol agents include their survival upon introduction to new environmental conditions. A number of fungal species such as *Metarhizium, Verticillium, Penicillium, Trichoderma*, and other fungi have been explored as a potential source of secondary metabolites bioactivities against insect, fugal, and bacterial pathogens (Murali et al., 2013). Several secondary metabolites with insecticidal properties such as destruxins A and B, serinocyclin A, cytochalasins, swainsonine, and viridoxins (Gupta et al., 1993) of *Metarhizium* spp. and bassianolides origin have been found positive for insecticidal activities (Vilcinskas et al., 1997a,b; Dreyer et al., 1985; Krasnoff et al., 2006). The secondary metabolites of *Verticillium* spp. include cyclosporins, enniatins, dipicolinic acid, phomalactones, verticilides, and oosporein are reported for their insecticidal properties (Matha et al., 1988; Podsiadlowski et al., 1998; Monma et al., 2006).

10.5.2 Biological control of nematodes using microbial-derived secondary metabolites

Nematodes attacking plants, predominantly in the form of soilborne root pathogens causes over 12% worldwide loss to the agricultural yields (Askary and Martinelli, 2015). Moreover, the management of nematodes is often difficult. The worldwide predicted loss due to nematodes is around USD 157 billion per year (Singh et al., 2015). The use of chemicals for the control of plant-parasitic nematode was a preferred method over the years. However, these chemicals caused severe loss to our ecosystem. For example, earlier in the 1960s, the application of methyl bromide led to the depletion of the ozone layer. Therefore, its usage in 2005 was prohibited under Montreal Protocol (Meadows, 2013). Thereafter, efforts have been made for developing alternate methods. The biological control agents in the form of rhizobacteria in plant rhizosphere found to suppress nematode populations. For example, the nematode parasite *Pasteuria penetrans* can manage *Meloidogyne incognita* on cucumber and tomato plants, whereas *M. arenaria* can control on Snapdragon (Kokalis-Burelle, 2015). Similarly, *Bacillus nematocida* are found positive against nematodes due to the production of extracellular proteases, which can target their cuticles (Niu et al., 2006).

The nematode-pathogenic fungi comprising approximately 280 species. The majority of these fungi belong to Ascomycota and Basidiomycota (Li and Zhang, 2014). Some of them are reported for their plant pathogenic behavior (Degenkolb and Vilcinskas, 2016a). Among basidiomycetes, 77 genera representing 160 species with no side effect to the plants have been reported (Li and Zhang, 2014). The nematicidal compounds of fungi such as *Coprinus comatus* are found to produce 5-methylfuran-3-carboxylic acid and 5-hydroxy-3,5-dimethylfuran-2 (5H)-one, which are significantly effective to control the root-knot nematode pathogen *M. incognita* of crops worldwide (Degenkolb and Vilcinskas, 2016b). Beside this, nonnematophagous fungi such as wood-inhabiting basidiomycetes are also found to secrete nematicidal compounds. The secondary metabolites including sesquiterpene dichomitin B of polyporoid *Dichomitus squalens* representing poroid crust fungi are found effective phytopathogenic nematodes (Degenkolb and Vilcinskas, 2016b). Another cyclic dodecapeptides omphalotin compounds obtained from *Omphalotus olearius* mycelium and illinitone A obtained from *Limacella illinita* have been reported for nematicidal activity (Degenkolb and Vilcinskas, 2016b). Still, commercial fungal-derived nematicides are lacking compared to ivermectin obtained from actinomycete (Li and Zhang, 2014).

10.5.3 Role of secondary metabolites of fungal origin in the control of plant pathogens

The plant beneficial fungi use several mechanisms to inhibit the plant pathogens and production of secondary metabolite is one of the them (Whipps, 2004). Among fungi, the biocontrol strains of *Trichoderma* have been used against several plant pathogens worldwide. The ability to produce secondary metabolites by *Trichoderma* contributes to its success as biocontrol agent (Jogaiah et al., 2018). The ability of *Trichoderma* species to produce VOCs has been linked to their antimicrobial activities (Strobel et al., 2001; Strobel, 2006).

In addition, the fungal VOCs are used to suppress the plant pathogens growth and simultaneously promotes plant growth. Several, VOCs of *Trichoderma* origin have been found beneficial to agricultural crops (Wheatley et al., 1997; Van Loon et al., 1998;

Stoppacher et al., 2010). The role of VOCs in inducing systemic resistance through priming plants' immune response as well as in nutrients acquisition has also been studied (Van Wees et al., 2008). The application of 2-butanone and 3-pentanol in soil is found to reduce aphid *M. persicae* infestation in cucumber seedlings (Song and Ryu, 2013). The mycorrhizal association of fungi in particular of ectomycorrhiza type with majority of tree and shrub species plays key roles in the biomes (Smith and Read, 2008). A wide range of ectomycorrhizal-associated basidiomycetes species such as *Scleroderma*, *Suillus*, *Laccaria*, *Lactarius*, *Rhizopogon*, *Pisolithus*, and *Thelephora* are found positive for secondary metabolites production, which suppresses plant pathogenic fungi.

Pisolithus arhizus, one of the widespread earth-ball like fungi commercially applied an ectomycorrhizal inoculum is known for the production of antibiotics such as pisolithins A (*p*-hydroxybenzoylformic acid) and B ((*R*)-(−)-*p*-hydroxymandelic acid). *Suillus variegatus* culture as axenic produces antifungal secondary metabolites along with volatile isobutanol and isobutyric acid (Curl and Truelove, 1986; Tsantrizos et al., 1991). The culture filtrate of *Laccaria laccata* was reported to inhibit *Fusarium oxysporum* spore germination (Chakravarty and Hwang, 1991), but the responsible secondary metabolites so far has not been characterized. The antifungal bioactive compound lactarane sesquiterpene, rufuslactone obtained from *Lactarius rufus* fruit bodies was reported better than commercially available fungicide carbendazim commonly used against *Alternaria* strains (Luo et al., 2005).

10.6 Conclusion

The constant efforts on exploration of microbial secondary metabolites resulted several biologically active metabolites. Further, these metabolites are also a basis for chemists. The unraveling genome data of microorganisms for secondary metabolites encoding BCGs is a major challenge. Moreover, the abundance of repetitive DNA in the genome imposes problems in assembling the NRPS and PKS operons. The advancements in mining of microbial genomes led to the identification of a vast repository of different secondary metabolites across different microorganisms accurately across the genome. Nowadays, the identification of BCGs and linking these biosynthetic pathways of secondary metabolites is although straightforward but demands further studies for their successful commercial applications. The expression of BCGs of "cryptic" under in vitro conditions is another major challenge. Therefore, secondary metabolites production from such BCGs required development of optimized module using synthetic biology. Other tools such as plug-and-play methods are being deployed for the expression of (AGOS) artificial gene operon assembly system to express them under controlled regulated conditions. In future, synthetic biology can play major role in refactoring the cryptic gene clusters for novel metabolites biosynthesis. Artificial operon-based assembly system has been successfully assembled using AGOS for the biosynthesis of various precursors such as novobiocin biosynthesis. Alternatively, using metagenome, the genome of unculturable microorganisms can be mined for uncharacterized BCGs for novel bioactive metabolites.

Moreover, the application of targeted genome editing, codon optimization for efficient expression in heterologous systems offers new avenues. However, the expression of such BCGs is a challenging task. The application of cell-free systems can help in avoiding the complexities associated with living systems for the production of nonnatural bioactive metabolites. Despite our achievements in decoding the BCGs, the prediction of the core scaffolds of different metabolites is still at infancy stage largely due to biochemical knowledge gap. Additionally, efforts are required to develop machine learning tools for precise prediction of BCGs borders. All these studies will facilitate the discovery of novel metabolites for human welfare.

10.7 Future prospects and concerns

Environmental stresses both in the form of biotic and abiotic stresses severely affect the crop yields. The application of chemical-based method to counter these stresses resulted heavy loss to our environment. The biocontrol agents as well as their secondary metabolites have emerged as promising alternate without causing any harmful effect to our ecosystem. Use of epigenetic enzymes such as histone deacetylases (HDACs) and DNA methyl transferases (DNMTs) can play a big role in regulated biosynthesis of different clusters. However, several challenges such as availability of optimized processes for the large-scale production, development of formulation, storage and their shelf life, assessment of safety precautions, thorough assessment of these microbes to humans or the environment and their field application demands the compliance with regulatory bodies, multilocation trials followed by farmers' awareness through training must be done. Further, characterization of ectophytic or endophytic strains will help in combating plant stresses. The challenges associated with the field application of biocontrol agents necessitate the application of metabolites of these organism's origin are valuable and attractive alternative.

Acknowledgments

The authors are thankful to SEED Division, Department of Science and Technology, GOI for providing financial benefits (SP/YO/125/2017) and (SEED-TIASN-023-2018) during the completion of this work.

References

Abdelrahman, M., Abdel-Motaal, F., El-Sayed, M., Jogaiah, S., Shigyo, M., Ito, S.-i., Tran, L.-S.P., 2016. Dissection of *Trichoderma longibrachiatum*-induced defense in onion (*Allium cepa* L.) against *Fusarium oxysporum* f. sp. *Cepa* by target metabolite profiling. Plant Sci. 246, 128–138.

Alqahtani, N., Porwal, S.K., James, E.D., Bis, D.M., Karty, J.A., Lane, A.L., Viswanathan, R., 2015. Correction: synergism between genome sequencing, tandem mass spectrometry and bio-inspired synthesis reveals insights into nocardioazine B biogenesis. Org. Biomol. Chem. 13 (35), 9323. https://doi.org/10.1039/c5ob90142a.

Askary, T.H., Martinelli, P.R.P. (Eds.), 2015. Biocontrol Agents of Phytonematodes. CABI, Wallingford.

Audrain, B., Farag, M.A., Ryu, C.M., Ghigo, J.M., 2015. Role of bacterial volatile com- pounds in bacterial biology. FEMS Microbiol. Rev. 39 (2), 222–233.

Austin, M.B., Noel, J.P., 2003. The chalcone synthase superfamily of type III polyketide synthases. Nat. Prod. Rep. 20 (1), 79–110.

Austin, M.B., O'Maille, P.E., Noel, J.P., 2008. Evolving biosynthetic tangos negotiate mechanistic landscapes. Nat. Chem. Biol. 4 (4), 217–222. https://doi.org/10.1038/nchembio0408-217.

Babu, N.A., Jogaiah, S., Ito, S.-I., Nagraj, K.A., Tran, L.-S.P., 2015. Improvement of growth, fruit weight and early blight disease protection of tomato plants by rhizosphere bacteria is correlated with their beneficial traits and induced biosynthesis of antioxidant peroxidase and polyphenol oxidase. Plant Sci. 231, 62–73.

Bernier, S.P., Letoffe, S., Delepierre, M., et al., 2011. Biogenic ammonia modifies antibiotic resistance at a distance in physically separated bacteria. Mol. Microbiol. 81, 705–716.

Berrada, R., Dauphin, G., David, L., 1987. Epinigericin, a new polyether carboxylic antibiotic. Structural determination by 2D NMR methods. J. Org. Chem. 52, 2388–2391.

Bitas, V., Kang, S., 2012. Fusarium oxysporum produces volatile organic compounds that affect the growth and disease defense of *Arabidopsis thaliana*. In: APS Annual Meetingp. 588 August 4–8 Providence, USA, Poster Session: MPMI-Fungi.

Bloemberg, G.V., Lugtenberg, B.J.J., 2001. Molecular basis of plant growth promotion and biocontrol by rhizobacteria. Curr. Opin. Plant Biol. 4, 343–350.

Brockmeyer, K., Li, S.M., 2017. Mutations of residues in pocket P1 of a cyclodipeptide synthase strongly increase product formation. J. Nat. Prod. 80 (11), 2917–2922. https://doi.org/10.1021/acs.jnatprod.7b00430.

Brown, M., Shanks, J., 2012. Linear hydrocarbon producing pathways in plants, algae and microbes. In: Gopalakrishnan, K., Leeuwan, J., Brown, R. (Eds.), Sustainable Bioenergy and Bioproducts. Springer, London, pp. 1–11. https://doi.org/10.1007/978-1-4471-2324-8_1.

Bruce, A., Verrall, S., Hackett, C.A., Wheatley, R.E., 2004. Identification of volatile organic compounds (VOCs) from bacteria and yeast causing growth inhibition of sapstain fungi. Holzforschung 58, 193–198.

Chakravarty, P., Hwang, S.F., 1991. Effect of an ectomycorrhizal fungus, *Laccaria laccata*, on fusarium damping-off in *Pinus banksiana* seedlings. Eur. J. For. Path. 21, 97–106. https://doi.org/10.1111/j.1439-0329.1991.tb00949.x.

Chin A-Woeng, T.F.C., Bloemberg, G.V., Lugtenberg, B.J.J., 2003. Phenazines and their role in biocontrol by Pseudomonas bacteria. New Phytol. 157, 503–523.

Citron, C.A., Gleitzmann, J., Laurenzano, G., Pukall, R., Dickschat, J.S., 2012. Terpenoids are widespread in Actinomycetes: a correlation of secondary metabolism and genome data. ChemBioChem 13 (2), 202–214.

Curl, E.A., Truelove, B., 1986. The Rhizosphere. Advanced Series in Agricultural Sciences, vol. 15. Springer, Berlin/Heidelberg.

Davies, J., Ryan, K.S., 2012. Introducing the parvome: bioactive compounds in the microbial world. ACS Chem. Biol. 7 (2), 252–259. https://doi.org/10.1021/cb200337h.

Degenkolb, T., Vilcinskas, A., 2016a. Metabolites from nematophagous fungi and nematicidal natural products from fungi as an alternative for biological control. Part I: metabolites from nematophagous ascomycetes. Appl. Microbiol. Biotechnol. 100, 3799–3812. https://doi.org/10.1007/s00253-015-7233-6.

Degenkolb, T., Vilcinskas, A., 2016b. Metabolites from nematophagous fungi and nematicidal natural products from fungi as an alternative for biological control. Part II: metabolites from nematophagous basidiomycetes and non-nematophagous fungi. Appl. Microbiol. Biotechnol. 100, 3813–3824. https://doi.org/10.1007/s00253-015-7234-5.

Domenech, J., Reddy, M.S., Kloepper, J.W., 2006. Combined application of the biological product LS213 with *Bacillus, Pseudomonas* or *Chryseobacterium* for growth promotion and biological control of soil-borne diseases in pepper and tomato. BioControl 51, 245.

Dreyer, D.L., Jones, K.C., Molyneux, R.J., 1985. Feeding deterrency of some pyrrolizidine, indoli- zidine, and quinolizidine alkaloids towards pea aphid (*Acyrthosiphon pisum*) and evidence for phloem transport of indolizidine alkaloid swainsonine. J. Chem. Ecol. 11 (8), 1045–1051.

Effmert, U., Kalderas, J., Warnke, R., Piechulla, B., 2012. Volatile mediated interactions between bacteria and fungi in the soil. J. Chem. Ecol. 38, 665–703. https://doi.org/10.1007/s10886-012-0135-5.

Gomez-Escribano, J.P., Bibb, M.J., 2014. Heterologous expression of natural product biosynthetic gene clusters in Streptomyces coelicolor: from genome mining to manipulation of biosynthetic pathways. J. Ind. Microbiol. Biotechnol. 41 (2), 425–431.

Grawert, T., Groll, M., Rohdich, F., Bacher, A., Eisenreich, W., 2011. Biochemistry of the non-mevalonate isoprenoid pathway. Cell. Mol. Life Sci. 68 (23), 3797–3814.

Gupta, S., Krasnoff, S.B., Renwick, J.A.A., Roberts, D.W., Steiner, J.R., Clardy, J., 1993. Viridoxins A and B: novel toxins from the fungus Metarhizium flavoviride. J. Org. Chem. 58 (5), 1062–1067.

Hamilton-Kemp, T., Newman, M., Collins, R., Elgaali, H., Yu, K., Archbold, D., 2005. Production of the long-chain alcohols octanol, decanol, and dodecanol by *Escherichia coli*. Curr. Microbiol. 51, 82–86.

He, Y., Wang, B., Chen, W., Cox, R.J., He, J., Chen, F., 2018. Recent advances in reconstructing microbial secondary metabolites biosynthesis in *Aspergillus* spp. Biotechnol. Adv. 36 (3), 739–783.

Hertweck, C., Luzhetskyy, A., Rebets, Y., Bechthold, A., 2007. Type II polyketide synthases: gaining a deeper insight into enzymatic teamwork. Nat. Prod. Rep. 24 (1), 162–190.

Howell, C.R., 1998. The role of antibiosis in biocontrol. In: Harman, G.E., Kubicek, C.P. (Eds.), Trichoderma and Gliocladium. In: vol. 2. Taylor and Francis, London, pp. 173–184.

Humphris, S.N., Wheatley, R.E., Bruce, A., 2001. The effect of specific volatiles organic compounds produced by *Trichoderma* spp. on the growth of wood decay basidiomycetes. Holzforschung. 55. https://doi.org/10.1515/HF.2001.038.

Hung, R., Samantha, L., Joan, W.B., Gareth, W.G., 2012. *Arabidopsis thaliana* as a model system for testing the effect of *Trichoderma* volatile organic compounds. Fungal Ecol. 6 (1), 19–26. https://doi.org/10.1016/j.funeco.2012.09.005.

Ishikawa, J., 2008. Genome analysis system for actinomycetes: development and application. Actinomycetologica 22 (2), 46–49.

James, E.D., Knuckley, B., Alqahtani, N., Porwal, S., Ban, J., Karty, J.A., et al., 2016. Two distinct cyclodipeptide synthases from a marine actinomycete catalyze biosynthesis of the same diketopiperazine natural product. ACS Synth. Biol. 5, 547–553. https://doi.org/10.1021/acssynbio.5b00120.

Jogaiah, S., Mitani, S., Amruthesh, K.N., Shekar Shetty, H., 2007. Activity of cyazofamid against *Sclerospora graminicola*, a downy mildew disease of pearl millet. Pest Manag. Sci. 63, 722–727.

Jogaiah, S., Shetty, H.S., Ito, S.-I., Tran, L.-S.P., 2016. Enhancement of downy mildew disease resistance in pearl millet by the G_app7 bioactive compound produced by *Ganoderma applanatum*. Plant Physiol. Biochem. 105, 109–117.

Jogaiah, S., Abdelrahman, M., Tran, L.-S.P., Ito, S.-I., 2018. Different mechanisms of *Trichoderma virens*-mediated resistance in tomato against Fusarium wilt involve the jasmonic and salicylic acid pathways. Mol. Plant Pathol. 19, 870–882.

Kahne, D., Leimkuhler, C., Lu, W., Walsh, C., 2005. Glycopeptide and lipoglycopeptide antibiotics. Chem. Rev. 105, 425–448.

Keswani, C., Bisen, K., Singh, V., Sarma, B.K., Singh, H.B., 2016. Formulation technology of biocontrol agents: present status and future prospects. In: Arora, N.K., Mehnaz, S., Balestrini, R. (Eds.), Bioformulations: For Sustainable Agriculture. Springer, New Delhi, pp. 35–52.

Khater, S., Anand, S., Mohanty, D., 2016. In silico methods for linking genes and secondary metabolites: the way forward. Synth. Syst. Biotechnol. 1 (2), 80–88.

Kokalis-Burelle, N., 2015. *Pasteuria* penetrans for control of *Meloidogyne incognita* on tomato and cucumber, and *M. arenaria* on snapdragon. J. Nematol. 47 (3), 207–213.

Korpi, A., Jarnberg, J., Pasanen, A.L., 2009. Microbial volatile organic compounds. Crit. Rev. Toxicol. 39, 139–193. https://doi.org/10.1080/10408440802291497.

Kramer, R., Abraham, W.R., 2012. Volatile sesquiterpenes from fungi: what are they good for? Phytochem. Rev. 11, 15–37. https://doi.org/10.1007/s11101-011-9216-2.

Krasnoff, S.B., Sommers, C.H., Moon, Y.S., et al., 2006. Production of mutagenic metabolites by Metarhizium anisopliae [published correction appears in *J. Agric. Food Chem.* 2008 Feb 13, 1158]. J. Agric. Food Chem. 54 (19), 7083–7088. https://doi.org/10.1021/jf061405r.

Kunova, A., Bonaldi, M., Saracchi, M., 2016. Selection of *Streptomyces* against soil borne fungal pathogens by a standardized dual culture assay and evaluation of their effects on seed germination and plant growth. BMC Microbiol. 16, 272.

Laatsch, H., 2007. AntiBase 2007: The Natural Product Identifier. Wiley, VCH Verlag GmbH.

Lazarus, C.M., Williams, K., Bailey, A.M., 2014. Reconstructing fungal natural product biosynthetic pathways. Nat. Prod. Rep. 31 (10), 1339–1347.

Lemfack, M.C., Nickel, J., Dunkel, M., Preissner, R., Piechulla, B., 2013. mVOC: a database of microbial volatiles. Nucleic Acids Res. 42, D744–D748. https://doi.org/10.1093/nar/gkt1250.

Lemfack, M.C., Nickel, J., Dunkel, M., Preissner, R., Piechulla, B., 2014. VOC: a database of microbial volatiles. Nucleic Acids Res. 42, 744–748. https://doi.org/10.1093/nar/gkt1250.

Li, G.-H., Zhang, K.-Q., 2014. Nematode-toxic fungi and their nematicidal metabolites. In: Zhang, K.-Q., Hyde, K.D. (Eds.), Nematode-Trapping Fungi, Fungal Diversity Research Series. In: vol. 23. Springer, Dordrecht, pp. 313–375. https://doi.org/10.1007/978-94-017-8730-7_7.

Luo, D.Q., Shao, H.J., Zhu, H.J., Liu, J.K., 2005. Activity in vitro and in vivo against plant pathogenic fungi of grifolin isolated from the basidiomycete *Albatrellus dispansus*. Z. Naturforsch. C 60, 50–56.

Manivasagan, P., Venkatesan, J., Sivakumar, K., Kim, S.K., 2014. Pharmaceutically active secondary metabolites of marine actinobacteria. Microbiol. Res. 169 (4), 262–278.

Matha, V., Weiser, J., Olejnicek, J., 1988. The effect of tolypin in *Tolypocladium niveum* crude extract against mosquito and blackfly larvae in the laboratory. Folia Parasitol. 35 (4), 379–381.

Mattila, J.T., Thomas, A.C., 2014. Nitric oxide synthase: non-canonical expression patterns. Front Immunol. 5, 478. https://doi.org/10.3389/fimmu.2014.00478.

Meadows, R., 2013. Researchers develop alternatives to methyl bromide fumigation. Calif. Agric. 67 (3), 125–127. https://doi.org/10.3733/ca.v067n03p125.

Mishra, S., Singh, A., Keswani, C., Saxena, A., Sarma, B.K., Singh, H.B., 2015. Harnessing plant-microbe interactions for enhanced protection against phytopathogens. In: Arora, N.K. (Ed.), Plant Microbe Symbiosis—Applied Facets. Springer, New Delhi, pp. 111–125.

Monma, S., Sunazuka, T., Nagai, K., Arai, T., Shiomi, K., Matsui, R., Ōmura, S., 2006. Verticilide: elucidation of absolute configuration and total synthesis. Org. Lett. 8 (24), 5601–5604.

Moutiez, M., Belin, P., Gondry, M., 2017. Aminoacyl-tRNA-utilizing enzymes in natural product biosynthesis. Chem. Rev. 117 (8), 5578–5618. https://doi.org/10.1021/acs.chemrev.6b00523.

Murali, M., Jogaiah, S., Amruthesh, K.N., Shin-ichi, I., Shekar Shetty, H., 2013. Rhizosphere fungus *Penicillium chrysogenum* promotes growth and induces defense-related genes and downy mildew disease resistance in pearl millet. Plant Biol. 15, 111–118.

Murata, M., Naoki, H., Matsunaga, S., Satake, M., Yasumoto, T., 1994. Structure and partial Stereochemical assignments for Maitotoxin, the Most toxic and largest natural non-biopolymer. J. Am. Chem. Soc. 116, 7098–7107.

Nakkeeran, S., Vinodkumar, S., Renukadevi, P., Rajamanickam, S., Jogaiah, S., 2019. Bioactive molecules from *Bacillus* spp.: an effective tool for plant stress management. In: Jogaiah, S., Abdelrahman, M. (Eds.), Bioactive Molecules in Plant Defense. Springer, Berlin, pp. 1–23.

Nielsen, T.H., Christophersen, C., Anthoni, U., Sorensen, J., 1999. Viscosinamide, a new cyclic depsipeptide with surfactant and antifungal properties produced by Pseudomonas fluorescens DR54. J. Appl. Microbiol. 86, 80–90.

Nielsen, T.H., Sorensen, D., Tobiasen, C., Andersen, T.R., Christophersen, C., 2002. Antibiotic and biosurfactant properties of cyclic lipopeptides. Appl. Environ. Microbiol. 68, 3416–3423.

Niu, G., 2018. Genomics-driven natural product discovery in actinomycetes. Trends Biotechnol. 36 (3), 238–241.

Niu, Q., Huang, X., Zhang, L., et al., 2006. A neutral protease from *Bacillus* nematocida, another potential virulence factor in the infection against nematodes. Arch. Microbiol. 185, 439–448.

Oliynyk, M., Samborskyy, M., Lester, J.B., Mironenko, T., Scott, N., Dickens, S., et al., 2007. Complete genome sequence of the erythromycin-producing bacterium *Saccharopolyspora erythraea* NRRL 23338. Nat. Biotechnol. 25, 447–453.

Olson, S., 2015. An analysis of the biopesticide market now and where it is going. The biopesticide market. Outlooks Pest Manage. 26, 203–206. https://doi.org/10.1564/v26_oct_04.

Ongley, S.E., Bian, X., Neilan, B.A., Muller, R., 2013. Recent advances in the heterologous expression of microbial natural product biosynthetic pathways. Nat. Prod. Rep. 30 (8), 1121–1138.

Podsiadlowski, L., Matha, V., Vilcinskas, A., 1998. Detection of a P-glycoprotein related pump in Chironomus larvae and its inhibition by verapamil and cyclosporin A. Comp. Biochem. Physiol. B Biochem. Mol. Biol. 121 (4), 443–450.

Reino, J.L., Guerro, R.F., Hernandez-Galan, R., Collado, I.G., 2008. Secondary metabolites from spe- cies of the biocontrol agent *Trichoderma*. Phytochem. Rev. 7, 89–123. https://doi.org/10.1007/s11101-006-9032-2.

Rix, U., Fischer, C., Remsing, L.L., Rohr, J., 2002. Modification of post-PKS tailoring steps through combinatorial biosynthesis. Nat. Prod. Rep. 19, 542–580.

Ryu, C., Farag, M.A., Hu, C., Reddy, M.S., Wei, H., Pare, P.W., Kloepper, J.W., 2003. Bacterial volatiles promote growth in Arabidopsis. Proc. Natl. Acad. Sci. U. S. A. 100, 4927–4932. https://doi.org/10.1073/pnas.0730845100.

Salwan, R., Sharma, V., 2020a. Bioactive compounds of *Streptomyces*: biosynthesis to application. In: Bioactive Natural Products. vol. 64. Elsevier, pp. 467–491. https://doi.org/10.1016/B978-0-12-817903-1.00015-2.

Salwan, R., Sharma, V., 2020b. Genomic blueprint of *Bacillus cereus* isolate SBA12 indicates expansion of its several plant beneficial and industrial attributes. Genomics 112 (4), 2894–2902. https://doi.org/10.1016/j.ygeno.2020.03.029.

Salwan, R., Sharma, V., 2020c. Molecular and biotechnological aspects of secondary metabolites in actinobacteria. Microbiol. Res. 231, 126374. https://doi.org/10.1016/j.micres.2019.126374.

Salwan, R., Sharma, V., 2020d. Molecular mechanisms of biocontrol *Streptomyces* in plant health. In: Molecular Aspects of Plant Beneficial Microbes in Agriculture. Elsevier, pp. 89–108. https://doi.org/10.1016/B978-0-12-818469-1.00008-0.

Salwan, R., Rialch, N., Sharma, V., 2018. Bioactive volatile metabolites of *Trichoderma*: an overview. In: Secondary Metabolites of Plant Growth Promoting Rhizomicroorganisms: Discovery & Applications. Springer, pp. 87–111. https://doi.org/10.1007/978-981-13-5862-3.

Salwan, R., Sharma, A., Sharma, V., 2019. Microbes mediated plant stress tolerance in saline agricultural ecosystem. Plant and Soil 442, 1–22.

Salwan, R., Sharma, V., Sharma, A., Singh, A., 2020a. Molecular imprints of plant beneficial *Streptomyces* sp. AC30 and AC40 reveal differential capabilities and strategies to counter environmental stresses. Microbiol. Res. 235, 126449. https://doi.org/10.1016/j.micres.2020.126449.

Salwan, R., Sharma, V., Sharma, A., Ankita, S., 2020b. Genomic imprints of plant beneficial *Streptomyces* sp AC30 and *Streptomyces* sp AC40 genomes reveal differential capabilities and strategies to counter environmental stresses. Microbiol. Res. 235, 126449. https://doi.org/10.1016/j.micres.2020.126449.

Satapute, P., Milan, V.K., Shivakantkumar, S.A., Jogaiah, S., 2019a. Influence of triazole pesticides on tillage soil microbial populations and metabolic changes. Sci. Total Environ. 651, 2334–2344.

Satapute, P., Murali, K.P., Kurjogi, M., Jogaiah, S., 2019b. Physiological adaptation and spectral annotation of arsenic and cadmium heavy metal-resistant and susceptible strain *Pseudomonas taiwanensis*. Environ. Pollut. 251, 555–563.

Scholler, C.E.G., Gürtler, H., Pedersen, R., Molin, S., Wilkins, K., 2002. Volatile metabolites from actinomycetes. J. Agric. Food Chem. 50 (9), 2615–2621.

Shanmugam, V., Singh, A.N., Verma, R., Sharma, V., 2008. Diversity and differentiation among fluorescent pseudomonads in crop rhizospheres with whole-cell protein profiles. Microbiol. Res. 163 (5), 571–578. https://doi.org/10.1016/j.micres.2006.08.006.

Sharma, V., Salwan, R., 2018. Biocontrol potential and applications of actinobacteria. In: Actinobacteria: Diversity and Biotechnological Applications. Elsevier, pp. 93–108. https://doi.org/10.1016/B978-0-444-63994-3.00006-0.

Sharma, V., Salwan, R., Sharma, P.N., 2016. Differential response of extracellular proteases of *Trichoderma harzianum* against fungal phytopathogens. Curr. Microbiol. 73, 419–425. https://doi.org/10.1007/s00284-016-1072-2.

Sharma, V., Salwan, R., Shanmugam, V., 2018. Unraveling the multilevel aspects of least explored plant beneficial *Trichoderma saturnisporum* isolate GITX-Panog C. Eur. J. Plant Pathol. 152 (1), 169–183. https://doi.org/10.1007/s10658-018-1461-4.

Sharma, V., Sharma, A., Malannavar, A.B., Salwan, R., 2020. Molecular aspects of biocontrol species of *Streptomyces* in agricultural crops. Elsevier, pp. 89–109.

Shelton, A.M., Zhao, J.Z., Roush, R.T., 2002. Economic, ecological, food safety, and social consequences of the deployment of Bt transgenic plants. Annu. Rev. Entomol. 47, 845–881.

Singh, S., Singh, B., Singh, A.P., 2015. Nematodes: a threat to sustainability of agriculture. Procedia Environ. Sci. 29, 215–216.

Singh, H.B., Sarma, B.K., Keswani, C. (Eds.), 2016. Agriculturally Important Microorganisms: Commercialization and Regulatory Requirements in Asia. Springer, Singapore.

Sivasithamparam, K., Ghisalberti, 1998. Secondary metabolism in Trichoderma and Gliocladium. In: Kubicek, C., Harman, G.E. (Eds.), Trichoderma and Gliocladium Basic Biology, Taxonomy and Genetics. Taylor & Francis, London, pp. 139–191.

Smith, S.E., Read, D.J., 2008. Mycorrhizal Symbiosis, third ed. Academic, London.

Song, G.C., Ryu, C.M., 2013. Two volatile organic compounds trigger plant self-defense against a bacterial pathogen and a sucking insect in cucumber under open field conditions. Int. J. Mol. Sci. 14, 9803–9819. https://doi.org/10.3390/ijms14059803.

Splivallo, R., Ottonello, S., Mello, A., Karlovsky, P., 2011. Truffle volatiles: from chemical ecology to aroma biosynthesis. New Phytol. 189, 688–699. https://doi.org/10.1111/j.1469-8137.2010.03523.x.

Stefels, J., 2000. Physiological aspects of the production and conversion of DMSP in marine algae and higher plants. J. Sea Res. 43, 183–197.

Stoppacher, N., Kluger, B., Zeilinger, S., Krska, R., Schuhmacher, R., 2010. Identification and profiling of volatile metabolites of the biocontrol fungus Trichoderma atroviride by HS-SPME-GC-MS. J. Microbiol. Methods 81, 187–193. https://doi.org/10.1016/j.mimet.2010.03.011.

Strobel, G., 2006. *Muscodor albus* and its biological promise. J. Ind. Microbiol. Biotechnol. 33, 514–522. https://doi.org/10.1007/s10295-006-0090-7.

Strobel, G.A., Dirkse, E., Sears, J., Markworth, C., 2001. Volatile antimicrobials from *Muscodor albus* a novel endophytic fungus. Microbiology 147, 2943–2950. https://doi.org/10.1099/00221287-147-11-2943.

Sukovich, D.J., Seffernick, J.L., Richman, J.E., Gralnick, J.A., Wackett, L.P., 2010. Widespread head-to- head hydrocarbon biosynthesis in bacteria and role of OleA. Appl. Environ. Microbiol. 76, 3850–3862. https://doi.org/10.1128/AEM.00436-10.

Thampi, R.A., Bhai, S., 2017. Rhizosphere actinobacteria for combating *Phytophthora capsici* and *Sclerotium rolfsii*, the major soil borne pathogens of black pepper (Piper nigrum L.). Biol. Control 109, 1–13.

Tiwari, K., Gupta, R.K., 2012. Rare actinomycetes: a potential storehouse for novel antibiotics. Crit. Rev. Biotechnol. 32 (2), 108–132.

Tsantrizos, Y.S., Kope, H.H., Fortin, J.A., Ogilvie, K.K., 1991. Antifungal antibiotics from Pisolithus tinctorius. Phytochemistry 30, 1113–1118.

Van Loon, L.C., Bakker, P.A.H.M., Pieterse, C.M., 1998. Systemic resistance induced by rhizosphere bacteria. Annu. Rev. Phytopathol. 36, 453–483.

Van Wees, S.C., Van der Ent, S., Pieterse, C.M., 2008. Plant immune responses triggered by beneficial microbes. Curr. Opin. Plant Biol. 11 (4), 443–448. https://doi.org/10.1016/j.pbi.2008.05.005.

Vespermann, A., Kai, M., Piechulla, B., 2007. Rhizobacterial volatiles affect the growth of fungi and Arabidopsis thaliana. Appl. Environ. Microbiol. 73, 5639–5641. https://doi.org/10.1128/AEM.01078-07.

Vilcinskas, A., Matha, V., Götz, P., 1997a. Effects of the entomopathogenic fungus *Metarhizium anisopliae* and its secondary metabolites on morphology and cytoskeleton of plasmatocytes isolated from the greater wax moth, *Galleria mellonella*. J. Insect Physiol. 43 (12), 1149–1159.

Vilcinskas, A., Matha, V., Götz, P., 1997b. Inhibition of phagocytic activity of plasmatocytes isolated from *Galleria mellonella* by entomogenous fungi and their secondary metabolites. J. Insect Physiol. 43 (5), 475–483.

Weber, T., Kim, H.U., 2016. The secondary metabolite bioinformatics portal: computational tools to facilitate synthetic biology of secondary metabolite production. Synth. Syst. Biotechnol. 1 (2), 69–79.

Wheatley, R., Hackett, C., Bruce, A., 1997. Effect of substrate composition on production of volatile organic compounds from *Trichoderma* spp. inhibitory to wood decay fungi. Int. Biodeter. Biodegr. 39, 199–205. https://doi.org/10.1016/S0964-8305(97)00015-2.

Whipps, J.M., 2004. Prospects and limitations for mycorrhizas in biocontrol of root pathogens. Can. J. Bot. 82, 1198–1227. https://doi.org/10.1139/B04-082.

Xiao, Z., Xu, P., 2007. Acetoin metabolism in bacteria. Crit. Rev. Microbiol. 33, 127–140.

Zasada, I., Halbrendt, J., Kokalis-Burelle, N., LaMondia, J., McKenry, M., Noling, J., 2010. Managing nematodes without methyl bromide. Annu. Rev. Phytopathol. 48 (1), 311–328.

Zazopoulos, E., Huang, K., Staffa, A., Liu, W., Bachmann, B.O., Nonaka, K., Ahlert, J., Thorson, J.S., Shen, B., Farnet, C.M., 2003. A genomics- guided approach for discovering and expressing cryptic metabolic pathways. Nat. Biotechnol. 21, 187–190.

Zeng, W., Wang, D., Kirk, W., Hao, J., 2012. Use of *Coniothyrium minitans* and other microorganisms for reducing Sclerotinia sclerotiorum. Biol. Control 60, 225–232.

Zhang, H., Xie, X., Kim, M., Kornyeyev, D.A., Holaday, S., Pare, P., 2008. Soil bacteria augment *Arabidopsis* photosynthesis by decreasing glucose sensing and abscisic acid levels in planta. Plant J. 56, 264–273. https://doi.org/10.1111/j.1365-313X.2008.03593.x.

Antimicrobial secondary metabolites from *Trichoderma* spp. as next generation fungicides

11

S. Nakkeeran, S. Rajamanickam, M. Karthikeyan, K. Mahendra, P. Renukadevi, and I. Johnson
Department of Plant Pathology, Centre for Plant Protection Studies, Tamil Nadu Agricultural University, Coimbatore, India

11.1 Introduction

Trichoderma, the saprophytic fungi serve as the repertoire of several genes responsible for mitigating both biotic and abiotic stress, is the gift of nature given to mankind. It is omnipresent and survives at different ecological conditions. Persoon and other scientists across the globe identified that the genomic treasure hidden in *Trichoderma* has been exploited for the management of plant diseases by formulating different types of the product for the betterment of the farming community. *Trichoderma*, confront with fungal pathogens and quench their action either directly by competing for nutrients and space, producing antimicrobial substances and lytic enzymes, and enhancing mycoparasitism, or indirectly by increasing the plant vigor by aiding the mobilization of the nutrients and growth hormones, and inducing systemic resistance. Globally, owing to the potential benefits of the *Trichoderma*, several multinational stakeholders came forward and produced more than 50 different types of commercial products for the management of soilborne pathogens, nematodes, and fungal nematode complex (Nakkeeran et al., 2016; Jogaiah et al., 2018). The low-molecular-weight secondary metabolites are organic compounds which are not required for the plant growth and reproduction, but, those secondary metabolites have a diverse function in different perspectives. However, the multifaceted *Trichoderma* has not been explored for tapping the secondary metabolites for the management of fungal pathogens, bacterial pathogens, viral pathogens, and plant-parasitic nematodes in agriculture on an industrial scale and as an alternative option for the next-generation fungicides cum nematicides to improve the crop productivity.

Considering these vacuum, this chapter focuses on the scope of exploring the secondary metabolites of *Trichoderma*, as an alternative option for the next-generation fungicidal molecules to manage inimical microbes, improve crop health by triggering the immunity, and promote plant growth.

Biocontrol Agents and Secondary Metabolites. https://doi.org/10.1016/B978-0-12-822919-4.00011-9

11.2 *Trichoderma* as rhizofungi

The mycoparasitic fungi, *Trichoderma*, grow in the rhizosphere and colonize the plant roots as an opportunistic endosymbiont (Harman et al., 2004) by preventing the entry of pathogens into the root by direct confrontation and eliciting the immunity in the plant (Vargas et al., 2011; Nagaraju et al., 2012). *Trichoderma* spp. colonize the rhizosphere and induce secretome and proteome profiles resulting in growth promotion and yield improvement (Moran-Diez et al., 2012). Signaling interplay between host plant and *Trichoderma* spp. is used to colonize the rhizosphere. *Trichoderma* produces growth hormones that serve as a signal molecule for colonizing the roots and increase the colonization area of roots (Contreras et al., 2009). Cysteine-rich hydrophobin-like proteins produced by *T. asperellum* and Qid74 of *T. harzianum* aids to bind host root (Samolski et al., 2012). Swollenin, an expansin-like protein from *Trichoderma*, helps in plant root colonization (Brotman et al., 2008). After binding the roots, *Trichoderma* penetrate into the root mediated through expansin-like proteins, cellulose, and endo-polygalacturonase by *Trichoderma* spp. (Moran-Diez et al., 2009). The initial inhibition of ISR facilitate the invasion of *Trichoderma* spp. into root. Subsequently, they grow in between the cells and are further limited to the outer cortex and epidermis. *T. koningii* enter into the root of lotus by overcoming the phytoalexins during root colonization (Masunaka et al., 2011).

11.3 *Trichoderma* CWDE and MAMP molecules on improving plant health

Trichoderma produce cell-wall-degrading enzymes (CWDE) including glycosyl hydrolases (GH) and the secreted proteins which facilitate communication with root through the production of signal molecules. Carbohydrate-Active EnZymes (CAZymes) belongs to 136 GH families in eight *Trichoderma* genomes. *T. reesei* QM6a is characterized by 68 families while 75 families are observed in *T. reesei* RUTC30. *T. harzianum* and *T. guizhouense* are characterized by 130 and 136 GH families, respectively. The difference in the mixture of CAZymes among and within different species of *Trichoderma* favors the colonization of host plants and induces microbe-associated molecular pattern (MAMP)-triggered immunity (Li et al., 2015; Taniguchi et al., 2011).

Inactive cellulase and xylanase proteins were identified as the first MAMP molecule from *Trichoderma* to trigger an innate immune response on host plants (Avni et al., 1994; Martinez et al., 2001; Rotblat et al., 2002). The CWDE produce oligogalacturonides as a DAMP signal molecule, due to the enzymatic action of endopolygalacturonase ThPG1 of *Trichoderma* with Arabidopsis and tomato roots (Moran-Diez et al., 2009). The enzymatic action of *Trichoderma* chitinases-released DAMP molecule chitooligosaccharides from the chitin polymers associated with ISR (Woo et al., 2006). Though CWDE released MAMP and DAMP signal molecules, it increased the plasticity of the root cells and facilitated efficient colonization.

Root colonization in tomato by *T. harzianum* was enhanced by the production of endopolygalacturonase ThPG1. It was reflected by the eightfold increase of the *T. harzianum* DNA in the root tissues (Moran-Diez et al., 2009). Culturing of the two endophytes *T. virens* and *T. harzianum*-producing cellulases, endopolygalacturonases, and xylanases with maize or tomato upregulated polysaccharide metabolism and favored root colonization (Chacon et al., 2007; Moran-Diez et al., 2015). Thus, plant health is maintained by *Trichoderma* spp., based on the early colonization of the rhizosphere region through the production of CAZymes. Therefore, the early induction and production of CAZymes as secondary metabolites by *Trichoderma* play a pivotal role in the protection of the plant from inimical microbes and improvement of plant health.

11.4 Molecular patterns of *Trichoderma*-mediated resistance response

The molecular signals from *Trichoderma* include structural MAMPs, secreted MAMPs such as xylanases, cellulases, cerato-platanin, swollenin, avirulence proteins, lysin-motif (LysM) domains, peptaibols, 6-pentyl pyrones, trichothecenes and phytohormones, and cell-wall-degrading polygalacturonases (Hermosa et al., 2012). Fungal elicitors are generally deposited in the root cell apoplast region. The effectors may be a protein or secondary metabolite. The first recognized MAMP molecule of *Trichoderma* was xylanase, and ET-induced xylanase (Eix/Xyn2) is an elicitor produced by *T. viride*. It elicits ET biosynthesis and induces ISR in plants (Avni et al., 1994; Martinez et al., 2001). Martinez et al. (2001) reported that cellulases from *Trichoderma* induce ethylene and salicylic acid signaling pathways leading to strong peroxidase and chitinase activities. Cerato-platanin proteins are small secreted proteins that are rich in cysteines, which play an important role in root colonization as well as trigger ISR induced by JA in cotton and maize (Hermosa et al., 2012). Swollenin protein identified in *T. reesei* was similar to plant expansin and involved in root colonization but did not trigger ISR, whereas, a synthetic 36-mer carbohydrate-binding domain peptide stimulated local defense responses in cucumber roots and leaves and afforded local protection against *Botrytis cinerea* infections, proving its MAMP activity (Brotman et al., 2008). The avirulence proteins such as Avr4 and Avr9 reported from *T. atroviride* and *T. harzianum* are SSCP homologs stimulate SAR through the activation of EDS1 (enhanced disease susceptibility-1) gene against *Rhizoctonia solani* and *B. cinerea*, respectively (Marra et al., 2006; Harman et al., 2004). These avirulence genes also protect fungal cell walls from plant chitinase enzymes (Stergiopoulos and De Wit, 2009). The first DAMP molecule identified in *Trichoderma*-root interaction was oligogalacturonides released due to cell-wall-degrading enzyme polygalacturonase *ThPG1* produced by *T. harzianum* (Moran-Diez et al., 2009). Woo et al. (2006) reported the DAMP activity of chitooligosaccharides by chitinase activity of plant or *Trichoderma* on rhizospheric fungi.

11.5 Nonribosomal peptides and their antifungal activity

Non-ribosomal peptides (NRPs) are the large group of secondary metabolites formed by the fusion of amino acids in the presence of non-ribosomal peptide synthetases (NRPSs), known as a multi-modular megazyme. It is formed outside the ribosome, coupled with secondary modifications. NRPs comprise both proteinogenic and non-proteinogenic amino acids, either linearly or cyclically. The NRPs are produced by *Trichoderma* spp. including peptaibiotics, epidithiodioxopiperazines (ETPs), and siderophores.

11.5.1 Peptaibiotics

Peptaibiotics are characterized either with linear or cyclic peptides comprising of 4–21 residues with a molecular mass of 500–2100 Da. Peptaibiotics are made up of non-proteinaceous amino acid, α-aminoisobutyric acid (Aib) and isovaline α,α-dialkylated amino acids. It also comprises peptaibols, lipopeptaibols, lipoaminopeptides, and cyclic peptaibiotics (Neumann et al., 2015). Peptaibols are the peptides with α-aminoisobutyric acid and C-terminal alcohol. It is the largest group characterized by an acylated N-terminus and an amide group bound with amino alcohol at the C-terminus. Peptaibols are antimicrobials with cytotoxic activities resulting in the formation of channel and permeabilization of the membrane (Bortolus et al., 2013).

Trichoderma species are the richest source of peptaibols. Peptaibols with 11 residues are the most common secondary peptaibols produced by different *Trichoderma* spp. (Degenkolb et al., 2012). More than 80% of peptaibols in the "Comprehensive Peptaibiotics Database" pertains to *T. viride*, *T. brevicompactum*, *T. virens*, *T. parceramosum*, *T. ghanense*, and *T. harzianum* (Neumann et al., 2015). Peptaibiome of *T. atroviride* (strain P1) comprises 20 trichorzianines with 19 amino acids and 15 trichoatrokontins, a novel family with 7–9 amino acids (Stoppacher et al., 2008).

Analysis of the peptaibols of 28 different *Trichoderma* species pertaining to different clades indicated that certain peptaibol groups were produced only by certain clades and not by all. Even, the composition of peptaibol residues and amino acids varies between the strains of *T. atroviride* (Neuhof et al., 2007). Analysis of the five different commercial products of *T. harzianum* comprised of new and three groups of recurrent peptaibols with 18, 14, and 11 peptaibol residues (Degenkolb et al., 2015).

11.5.2 Epipolythiodioxopiperazines

The highly reactive secondary metabolite, epipolythiodioxopiperazines (ETPs) derived from a cyclic peptide is characterized with a diketopiperazine ring. The disulfide bridge in the diketopiperazine ring binds to the thiol group and results in the generation of ROS through redox cycling and inactivate proteins of the pathogens (Gardiner et al., 2005). Gliotoxin produced by Q strains of *T. virens* is an ETP effective against several fungal pathogens (Scharf et al., 2016). Mukherjee et al. (2012), in his review, reported gliovirin as an ETP with antifungal action produced by "P" strain of *T. virens*.

11.5.3 Siderophores

Siderophores are iron-chelating secondary metabolites playing a vital role in the suppression of plant pathogens. Fungal antagonists like *T. atroviride*, *T. asperellum*, *T. gamsii*, *T. hamatum*, *T. virens*, *T. harzianum*, *T. polysporum*, and *T. reesei* produce siderophores like dimerum acid, coprogen, fusigen, fusarinine A, and the intracellular siderophore ferricrocin (Lehner et al., 2013). Three different NRPSs responsible for the synthesis of siderophore have been identified in *Trichoderma* spp. (Mukherjee et al., 2012). Harzianic acid, the N-heterocyclic compound derived from tetramic acid is a novel siderophore isolated from *T. harzianum*. It promoted plant growth and suppresses the availability of iron to plant pathogens (Vinale et al., 2009a, 2013).

11.6 Polygalacturonase ThPG1

Cell-wall-degrading enzyme (CWDE)—endopolygalacturonase is a prerequisite for the colonization of rhizosphere by *Trichoderma* spp. In general, tritrophic interaction between, *Trichoderma* spp., host plants, and pathogens induces the expression of ThPG1 gene known to produce endopolygalacturonase. Endopolygalacturonase degrades plant pectin and produces oligogalacturonides. Oligogalacturonides behave as damage-associated molecular pattern (DAMPs)-based molecules. These molecules were recognized by pattern recognition receptors (PRRs) wall-associated kinase-1 (De Lorenzo et al., 2011). PRRs induce the synthesis and accumulation of polygalacturonase inhibitor protein (Moran-Diez et al., 2009). Tritrophic interaction between *Trichoderma harzianum*, tomato seedlings, and *Rhizoctonia solani* or *Pythium ultimum* induced the expression of endopolygalacturonase ThPG1 gene. Expression of ThPG1 in the plant roots, along with GRP19 and ATA20 genes, was involved in the downregulation of proteins that are rich in glycine. These genes suppress defense mechanisms associated with salicylic acid (SA) (Park et al., 2001). In addition, chs (chalcone synthase) and Ltp genes (lipid transfer proteins) responding to SA and the genes associated with root hair development and root strength were also downregulated. The downregulation of chs genes reduced the accumulation of flavonoid and isoflavonoid synthesized through phenylpropanoid pathway. Besides, it coordinates SA pathway involved in SAR (Jogaiah et al., 2018).

11.7 Xylanase Eix/Xyn2

The secondary metabolites from *Trichoderma* origin known as microbe-associated molecular pattern (MAMP) are actively involved in triggering the innate immunity of the plants via the recognition of MAMP molecules by their respective PRR, leading to the cascade of events in the downstream and regulate the immune response of the plant. The fungal antagonist cum plant growth regulator, *Trichoderma viride* was first recognized to produce the MAMP molecule, ethylene-induced xylanase (Eix), the potential elicitor of plant defense and hypersensitive response in tomato and tobacco (Ron and Avni, 2004). The MAMP molecule-mediated immune response is independent of xylan degradation. The induction of ethylene by xylanase activates the expression of the ACC synthase gene (Matarasso et al., 2005). The xylanase produced by other *T. reesei*-induced cell death followed by the activation of oxidative burst and expression of defense genes in tobacco (Yano et al., 1998).

11.8 Cellulases

The mixture of *Trichoderma* cellulases, on interaction with the host, triggers ISR via the induction of JA and ethylene (Piel et al., 1997). Interaction of cellulase from *T. longibrachiatum* with melon cotyledon induced the defense genes like chitinase and glucanase through rapid oxidative burst via the activation of signaling pathways involving the signal molecules like ethylene and SA (Marra et al., 2006). Similarly, treating with mixed cellulases to the cut end of the petioles of corn and lima bean activated the defense pathways involving JA and ethylene (Piel et al., 1997). In certain cases, the defense genes associated with SAR is also triggered by the cellulase mixtures of *Trichoderma*. Thus, the cellulases from *Trichoderma* origin, not only activate ISR but also SAR. Colonization of rice roots by *T. asperellum*, activates the SAR pathway in rice plants and suppress the host-pathogen relationship of *P. syringae* with rice leaves. However, foliar application of the culture filtrates of *T. asperellum* on rice leaves triggers both ISR and SAR due to the activation of several elicitors at the same time leading to the induction of defense genes through SA and JA/ET (Yoshioka et al., 2012).

11.9 Cerato-platanins in ISR and rhizosphere competence

Cerato-platanins are hydrophobin-like SSCPs. Root colonization by *T. virens* and *T. atroviride* result in the accumulation of cerato-platanins in the hyphae. Thus, the abundance of SSCPs in *Trichoderma* spp., decides rhizosphere competence (Kubicek et al., 2011). The gene Sm1 of *T. virens* and Epl1 of *T. atroviride* coding for SSCPs act as elicitors of JA and trigger ISR in cotton and maize (Hermosa et al., 2012). *Trichoderma* gene Sm1/Epl1 secrete cysteine-rich hydrophobin-like protein in abundance belonging to cerato-platanin (CP) family. It triggers ISR in maize (Mukherjee et al., 2012; Djonovic et al., 2007). The glycosylated monomeric form of Sm1 triggers ISR, while the nonglycosylated forms of Sm1 is unable to trigger ISR. Sm1 an oligomer of *N*-acetyl glucosamine coding for the cerato-platanin with carbohydrate-binding properties might be responsible for ISR (de Oliveira et al., 2011).

11.10 Swollenin-mediated root colonization and resistance

The secondary metabolite swollenin of *Trichoderma* origin resembles expansin of plant origin, which have a pivotal role in the colonization of roots by *Trichoderma*. Swollenin proteins though does not induce ISR, it acts as a MAMP molecule to confer local defense response. The N-terminal carbohydrate-binding domain of Swo1, from *T. reesei* binds to cellulose, while the C-terminal end binds to expansin-like domain. The inoculation of *T. asperellum*, into the rhizosphere region of cucumber, favors the colonization of the root portions within 6h, due to the presence of swollenin. MAMP molecule, swollenin is perceived by the PRR (36-mer carbohydrate-binding domain peptide) in cucumber and triggers localized systemic resistance in leaves and cucumber roots against *Botrytis cinerea* (Brotman et al., 2008).

11.11 Peptaibols: An inducer of signal molecules

The secondary metabolite, peptaibols with antimicrobial activity is a peptide with 5–20 amino-acids. Besides, it also has 2-aminoisobutyric acid and non-proteinogenic amino acids. Alamethicin peptaibol with 20-mer, induces the synthesis of VOCs in lima bean, using JA pathway. Besides, it upregulates SA (Hermosa et al., 2012). The increase in SA production induces defense against pathogens, controls the insect pests, and alters the behavior of parasitoids, predators, and pollinators (Pineda et al., 2010). The peptaibol, trichokonins (20 mer) produced by *T. pseudokoningii*, induces the production of ROS resulting in the accumulation of phenolic compounds in tobacco plants on the site of application. Trichokonin production, also induces the production of SA, JA, and ET and thus the suppressing the infection of the tobacco mosaic virus (Pineda et al., 2010). *T. virens* produce 18-mer peptaibols and elicit systemic resistance against *P. syringae* in cucumber. Challenging with *Trichoderma*, peptaibols elicits SA gene marker PAL and hydroperoxide lyase, the JA gene marker (Viterbo et al., 2007). The peptaibols with antimicrobial nature and with the capacity to induce MAMP triggered immunity could be explored well for the management of plant diseases.

11.12 6-Pentyl pyrones trigger ISR/SAR and plant growth

Trichoderma emits the characteristics of coconut odor and produces yellow pigmentation when cultured in potato dextrose agar medium. It is attributed to the production of volatile pentyl analog, pyrone 6-pentyl-2H pyran-2-one (6PP). Besides, it contributes to the antifungal action against *Fusarium oxysporum* (50). Pyrone derivatives induce auxin production, enhance seed germination, and increase the root and shoot architecture and plant height at low concentrations (Rubio et al., 2009). *T. atroviride*, *T. citrinoviride*, *T. hamatum*, *T. viride*, *T. asperellum*, *T. harzianum*, and *T. koningii* produces 6PP. The lateral roots formation in *A. thaliana* was enhanced due to the production of 6PP by *T. atroviride* (Nieto-jacobo et al., 2017). Linoleic acid is the precursor for the synthesis of 6PP. Linoleic acid undergoes various processes like reduction, β-oxidation, and isomerization in a sequential manner and resulted in the synthesis of 6PP. Different polyketide synthase (PKS) were involved in enhancing the catalytic activities resulting in the formation of pyrone moiety.

The application of purified 6PP at low concentrations, followed by the challenge inoculation of *B. cinerea* and *Leptosphaeria maculans* in tomato and canola seedlings suppressed the establishment of host-pathogen relationship. Similarly, the application of the increased concentrations of pyridine harzianopyridone and butenolide harzianolide suppressed the infection of fungal pathogens. The application of pentyl derivatives on to the cotyledons of canola induced the expression of defense genes (PR1), via SA-mediated SAR. IT was also responsible for the induction of PR-3 gene mediated through JA-dependent ISR (Vinale et al., 2008a). Thus, pentyl analogs can trigger both SAR and ISR-mediated defense genes and suppress the establishment of infection by plant pathogens. Therefore, *Trichoderma* can promote plant growth and suppress the establishment of plant pathogens through the simultaneous induction of ISR/SAR through the production of secondary metabolite 6PP.

11.13 Antifungal activity of trichothecenes

Trichothecenes are C15 terpene compounds belonging to sesquiterpenes. They are mycotoxins known for their toxicity in plants, animals, and humans. *T. brevicompactum* produces trichodermin, a trichothecene that causes phytotoxicity and also has antimicrobial property (Tijerino et al., 2011). A trichothecene-based compound harzianum A, produced by *T. arundinaceum* has antifungal action against *B. cinerea* and *R. solani* (Malmierca et al., 2013) through the induction of systemic resistance in tomato cultivars by the elicitation of JA/SA responsive in tomato (Malmierca et al., 2012).

11.14 Volatile organic compounds and plant defense

Trichoderma strains are known to produce diverse VOC responsible for plant growth promotion, induction of systemic resistance, SAR, and antimicrobial action. *Trichoderma* produces a group of VOCs including esters, alcohols, small alkenes, ketones, sesquiterpenes, and monoterpenes (Zhang et al., 2008; Hung et al., 2012). Soil application of 2-butanone and 3-pentanol reduced the damage by *Myzus persicae* and increased the population of predatory coccinellids in cucumber seedlings (Song and Ryu, 2013).

11.15 Antifungal activity of terpenoids

Terpenes comprise over 40,000 compounds and are the largest group of secondary metabolites (Bohlmann and Keeling, 2008). Terpenoids are classified into different types based on the number of carbon isoprene units. Some have five isoprene units (C5-hemiterpenes), 10 isoprene units (C10-monoterpenes), 15 isoprene units (C15-sesquiterpenes), 20 isoprene units (C20-diterpenes), 25 isoprene units

(C25-sesquiterpenes), 30 isoprene units (C30-triterpenes), and tetraterpenes (C40) classes. The basic building blocks for the synthesis of all these terpenes are isopentenyl diphosphate (IPP) and dimethylallyl diphosphate (DMAPP), which are synthesized by mevalonic acid (MVA) pathway. The MVA pathway starts with the combination of three units of acetyl-coenzyme A to form a six-carbon MVA which is transformed into the five-carbon IPP through a series of events such as phosphorylation, decarboxylation, and dehydration. IPP is isomerized and forms DMAPP. Sesquiterpenes were detected from *T. virens* Gv29.8 and *T. asperellum* LU1370 along with the mixture of other VOCs (Nieto-jacobo et al., 2017). Terpenoids are used as flavors, fragrances, pharmaceuticals, and food additives (Dewick, 2009). Besides, several terpenes have antifungal activity against several pathogens.

11.16 Lytic enzymes

11.16.1 Serine protease

Trichoderma species are known well for protease production. The proteolytic activity of *T. viride* is responsible for the biocontrol of *Sclerotium rolfsii*. Serine proteases of *T. harzianum* aid in mycoparasitism of plant pathogens (Geremia et al., 1993). Serine proteases are induced by heat-killed mycelium, cell wall of fungi, or chitin, but it is repressed by glucose. Proteases degrade cell wall, cell membranes, and released proteins due to the lysis of the pathogen (Goldman et al., 1994).

11.16.2 β-1,3 Glucanases

Lytic enzymes favors mycoparasitism and degrade the cell wall of pathogen. Chitinases and β-1,3 glucanases lyse the host wall and leaks the protoplasmic contents and used as food for the proliferation of antagonist (Cherif and Benhamou, 1990; Tronsmo et al., 1993). Strains of *T. harzianum* produced a mixture of chitinolytic and gluconolytic enzymes (Sivan and Chet, 1989). β-1,3 glucanase was excreted into the soil by *T. harzianum* (Elad et al., 1982). Secretion of β-1,3 glucanase was induced by *T. harzianum* in the presence of laminarin, pustulan, *R. solani* cell walls, and mycelia of fungi amended in the enzyme production medium (Jacobs et al., 1991). The purified endochitinase, chitobiosidase, 1,3-β-glucosidase, and combinations lysed *Botrytis cinerea* (Lorito et al., 1994). Synergism was observed by the combination of 1,3-β-glucosidases and chitinases toward the lysis of fungal pathogens (Tronsmo et al., 1993). *T. harzianum* suppressed *R. solani* in the soybean rhizosphere due to the production of *N*-acetylglucosaminidase, endochitinase, chitobiosidase, and endo β-1,3 glucanases (Dalsoglio et al., 1998).

11.16.3 Chitinases

Chitinase, the basic protein hydrolyze chitin, a linear polymer of β-1,4-*N*-acetyl glucosamine, the second most abundant polysaccharide in nature (Nicol, 1991). The relationship between chitinases and glucanases and their significance in mycoparasitism have been well established (Elad et al., 1983). Hence, the assay of chitinase could be used for screening biocontrol agents (Elad et al., 1982). *T. harzianum* strain Tm with six chitinolytic enzymes play a vital role in the suppression of soilborne pathogens. Among them, β-1, 4-*N*-acetyl glucosaminidase hydrolyzes chitin to monomers of *N*-acetyl glucosamine. Four were endochitinases that randomly cleaved the internal sites of the chitin microfibril. Those enzymes are CHIT 52, CHIT 42, CHIT 33, and CHIT31. *T. harzianum*—P1 released 40kDa chitobiosidase responsible for the lysis of plant pathogen (Harman et al., 1993).

Lysis of fungal pathogens by *Trichoderma* chitinases was performed with different carbon sources like chitin, purified fungal cell walls, glucose, or GlcNAc (Limon et al., 1995). Complete degradation of pathogenic fungi relies on chitinolysis and degradation of chitin by *Trichoderma* spp. It assumes a greater significance in antagonism than the other mechanisms (Cherif and Benhamou, 1990). Soil with more inoculum density of *S. rolfsii* and *R. solani* was enriched with the higher concentration of chitinases and glucanases produced by *Trichoderma* isolates (Elad et al., 1983). Interaction of *T. harzianum* with *R. solani*, enhanced endochitinase CHIT 42 (Carsolio et al., 1994). *B. cinerea*, *F. solani*, *F. graminearum*, *Ustilago avenae*, and *Uncinula necator* were inhibited by the combination of chitinolytic and glycolytic enzymes from *T. harzianum* (Harman et al., 1993). The chitinolytic enzymes from *T. harzianum* were effective against the broad spectrum of soilborne pathogens (Harman et al., 1993).

11.17 Antimicrobial genes of *Trichoderma*

Gene	Function	Organism
Tvsp1	Codes for serine protease. It is used to control *Rhizoctonia solani* infecting cotton seedlings	*Trichoderma virens*
tri5	Synthesize trichothecene and inhibit mycelial growth, protein, and DNA synthesis of the pathogens	*Trichoderma harzianum*
TgaA and *TgaB*	Antifungal activity against *R. solani* and *Sclerotium rolfsii*	*Trichoderma virens*
ThPG1	Codes for endopolygalacturonase associated with the degradation of cell wall of *R. solani* and *P. ultimum*	*Trichoderma harzianum*
Th-Chit	Codes for chitinase with antifungal activity	*Trichoderma harzianum*
tri5	Responsible for trichodermin production with antifungal action against *S. cerevisiae*, *Kluyveromyces marxianus*, *Candida albicans*, *C. glabrata*, *C. tropicalis*, and *Aspergillus fumigatus*	*Trichoderma brevicompactum*
erg1	Responsible for the production of enzyme *squalene epoxidase*, required for ergosterol synthesis	*Trichoderma harzianum*
TvGST	Confers tolerance to cadmium	*Trichoderma virens*
Thkel1	Encodes for kelch-repeat protein involved in regulating glucosidase activity. Also, provide tolerance to salt and osmotic stress	*Trichoderma harzianum*

Continued

Gene	Function	Organism
egl1	Suppress cucumber damping-off caused by *P. ultimum*	*Trichoderma longibrachiatum*
qid74	Protect its own cell from mucolytic activity and help in binding to the hydrophobic surfaces of the fungus toward the mycoparasitism against *R. solani*	*Trichoderma harzianum* CECT 2413
Taabc2	This gene has a significant role in ATP-binding cassette (ABC) transporter in cell membrane pump that helps in the mycoparasitic activity	*Trichoderma atroviride*
tac1	Helps in mycoparasitism on *R. solani* and *P. ultimum*	*Trichoderma virens*

11.18 Growth promotion by *Trichoderma*

External and internal colonization of the rhizosphere by *Trichoderma* is due to the interplay of signals between the host plant and *Trichoderma* spp. The signal molecule auxin produced by *Trichoderma* as growth hormones colonizes roots and increases the surface area of colonization (Contreras et al., 2009; Jogaiah et al., 2013). *Trichoderma* binds to host roots by producing cysteine-rich hydrophobin-like proteins like TasHyd1 from *T. asperellum* and Qid74 of *T. harzianum* (Viterbo and Chet, 2006). Besides, swollenin, an expansin-like protein from *Trichoderma* plays a vital role in plant root colonization (Brotman et al., 2008). After binding onto the root, penetration into the roots is mediated through the secretion of expansin-like proteins with cellulose-binding modules and endopolygalacturonase by *Trichoderma* spp. (Moran-Diez et al., 2009).

The different species of *Trichoderma* are effective colonizers on decaying and cellulosic materials and rhizosphere soil, which utilize the substrate by the production of their own secretion capacity for secondary metabolites viz., CWDE, growth promoting, and antibiotic nature (Abdelrahman et al., 2016). In addition, it promoted the plant growth, reproduction, and biosynthesis of secondary metabolites (Schuster and Schmoll, 2010). The application of *Trichoderma* improved crop productivity, nutritional quality as well as resistance to a plant pathogen. The colonization of rhizosphere influences plant growth through nutrient uptake and by imparting resistance to biotic/abiotic stress. The application of *T. harzianum* (T-E5) as a biostimulant decreased the incidence of *Fusarium* wilt and promoted the growth of cucumber plants (Zhang et al., 2014). Likewise, onion seeds treated with *T. harzianum* significantly promoted the seedling emergence and increased seedling length, seedling root length, number of leaves, and fresh weight of seedlings (Dabire et al., 2016). Field demonstration on wheat crop applied with the different formulation of *T. harzianum* (Th3) increased plant growth parameters like rootlets and number of tillers (Sharma et al., 2012). Similarly, the soil application of *T. viride* along with FYM and NPK increased the plant growth parameters, grain yield, and biomass yield of wheat crop (Mahato et al., 2018). Hajieghrari (2010) found that maize crop treatment with *T. hamatum* T614 isolates increased the leaf area, shoot weight, and fresh roots of seedling compared with the non-inoculated seedling. Similarly, bulb treatment with *T. harzianum* enhanced the plant growth of tuberose and resulted in increased flower production and quality (Nosir, 2016). Root dipping and soil application of *T. asperellum*-based talc formulation, effectively suppressed carnation wilt by increasing plant growth, the number of shoots, stalk length, and flower yield (Vinodkumar et al., 2017). The capacity of *Trichoderma* spp. to promote plant growth is attributed due to the synthesis of secondary metabolites, production of enzymes to convert the unavailable form of nutrients to available form, that is auxin- and gibberellins-based compounds (Parra et al., 2017; https://scielo.conicyt.cl/scielo.php?script = sci_arttext&pid = S0718–583920170004 00318 - B9).

Seedlings of *Arabidopsis* treated with *T. virens* or *T. atroviride* produced auxin-related compounds such as indole-3-acetic acid, indole-3-acetaldehyde, and indole-3-ethanol which were responsible for increased biomass production (Contreras et al., 2014). Similarly, Harman et al. (2004) found that the inoculation of *Trichoderma* in maize influenced the root system architecture, which resulted in increased crop yield, enhanced root biomass, and increased root hair development. The application of *Trichoderma* strains TH1 and T4 resulted in a greater wet shoot weight, wet root weight, and dry root weight in the presence of *Pythium* compared with the *Pythium* alone. Besides, there was an increase in the growth of lateral roots, number of nodules, and lateral roots per root system in the absence of *Pythium*. The application of *Trichoderma* strains TH1, T4, and T12 significantly increased the soil fungal population and significantly reduced the activities of ß-glucosidase, NAGase, and chitobiosidase relative to the *Pythium* control. Strains TH1, T4, and T12 significantly reduced the alkaline phosphatase activity, while the strains TH1, T4, and N47 significantly reduced the urease activity, relative to the *Pythium* control. These enzymes are directly related to carbon availability, whereas P cycle enzyme activities are inversely related to P availability. Thus, it indicated that the application of *Trichoderma* reduced the leakage of plant nutrients to the soil *Pythium*-induced damage (Naseby et al., 2000).

Similarly, the application of *Trichoderma* strains promoted the significant growth of *Capsicum annuum* seedlings at 36 days after planting. In particular, *T. atroviride* increased plant height, dry shoot biomass and expressed the biocontrol activity against root-knot nematode *Meloidogyne incognita*. Increased plant growth promotion and increased seedlings emergence favored root colonization by *Trichoderma*, as one of the critical characteristics of the spp. to utilize the organic acids as a major trait to protect cucumber roots from the infection of soil-borne pathogen (Vasumathi et al., 2016). The production of organic acids alters the soil and nutrient uptake efficiency, which was useful for plant metabolism (Hermosa et al., 2012). The longer secondary root hairs were observed in cucumber plants treated with *T. harzianum*, overexpressing *qid74*, a cysteine-rich cell-wall protein gene and induced modifications in root architecture and thus increased the total absorptive surface, facilitating nutrient uptake and translocation of nutrients in the shoots, resulting in increased plant biomass through efficient use of NPK and micronutrients (Samolski et al., 2012). Thus, the research behind the molecules associated with plant growth promotion by *Trichoderma* spp., explain the involvement of secondary metabolites, CWDE, overexpression of *qid74*, a cysteine-rich cell-wall protein gene, indole-3-acetic acid, indole-3-acetaldehyde, and indole-3-ethanol.

11.19 Antimicrobial activity of *Trichoderma* secondary metabolites

The genome of the *Trichoderma* has been well characterized and the role of different genes in the production of secondary metabolites of both volatile and non-volatile nature has been intensively studied by several groups across the globe. The biodiversity of *Trichoderma* is very rich and has been adapted to different ecological conditions. It can survive at extremes of soil moisture, temperature, pH, the organic content of the soil and etc., Different *Trichoderma* spp., that have been documented well for their diverse mode of action and spectrum of action against fungal, bacterial, viral pathogens, and nematodes are *T. harzianum*, *T. viride*, *T. koningiopsis*, *T. asperellum*, *T. lignorum*, *T. koningii*, *T. virens*, *T. pseudokoningii*, *T. hamatum*, *T. citrinoviride*, *T. viridescens*, *T. atroviride*, *T. brevicompactum*, *T. atroviride*, *T. petersenii*, *T. longibrachiatum*, and *T. albolutescens*. These, diverse species of *Trichoderma* vary in the potential to produce diverse volatile and nonvolatile antimicrobial substances with nematicidal activities, which have proved to be antifungal, antibacterial, and antiviral properties.

The antifungal compounds produced by different species of *Trichoderma* effective against different soilborne pathogens are trichorzianine A IIIc, octaketide compounds, peptaibol—trichorzianines A1 and B1, peptaibols, asperelines (A and E), and trichotoxins, harzianins HC, 6-pentyl-α-pyrone, viridiofungin A, T39-butenolide, harzianolide, T22-azaphilone, harzianic acid, harzianopyridone, 6-nonylene alcohol, massoia lactone, methyl cyclopentane, methyl cyclohexane, *N*-methyl pyrrolidine, dermadin, ketotriol, koningin-A, 3-methyl-heptadecanol, 2-methyl heptadecanol, palmitic acid, 3-(2′-hydroxypropyl)4-(hexa-2′4-dineyl)-2-(5H)-furanone, 3-(propenone)-4-(hexa-2′-4′-dineyl)-2(5H)-furanone, trichoharzianol, lignoren, viridin, acorane-type sesquiterpene, 2b-hydroxytrichoacorenol bisabolane-type sesquiterpene, trichoderic acid, cyclonerodiol, cyclonerodiol oxide, sorbicillin, trichokonin VI, cytosporone S, 6-pentyl pyrones (6PP), trichodermin, tricyclic polyketide, and 6-pentyl-α- pyrone (Table 11.1).

The antibacterial compounds produced by the diverse *Trichoderma* spp., are trichotoxin, sesquiterpenoids, cytosporone S, and trichokonins. These biomolecules are effective in the suppression of different Gram-positive and Gram-negative bacteria. The *Trichoderma* spp., with antibacterial activity, are *T. harzianum*, *T. longibrachiatum*, *T. asperellum*, *T. longibrachiatum*, *Trichoderma* sp. FKI-6626, *T. pseudokoningii* SMF2, and *T. harzianum* (Table 11.1).

Trichoderma spp., effective against different plant viruses are *T. pseudokoningii*, *T. albolutescens*, *T. koningii*, and other *Trichoderma* spp., They have been proved for their antiviral activity against *Cucumber mosaic virus*, *Pepper mottle virus*, TMV, *Watermelon mosaic virus*, *Zucchini green mottle mosaic virus*, *Melon necrotic spot carmovirus*, *Turnip mosaic virus*, TSWV, *Zucchini yellow mosaic virus*, *Pepper mild mottle virus Cymbidium mosaic virus*, *Lily symptomless virus*, *Odontoglossum ringspot virus*, *Strawberry mottle virus*, *Potato leafroll virus*, *Lily mottle virus*, *Tomato ringspot virus*, *Potato virus Y*, and *Cucumber green mottle mosaic virus* (Table 11.1). The antiviral secondary metabolites produced to suppress different plant

Table 11.1 Antifungal, antibacterial, antiviral, and nematicidal properties of *Trichoderma* secondary metabolites.

S. No.	Antimicrobial metabolite	*Trichoderma* spp. Producing antimicrobial metabolite	Pathogens controlled	References
	Antifungal activity			
1	Trichorzianine A IIIc	*T. harzianum*	*Botrytis cinerea*	Bodo et al. (1985)
2	Octaketide compounds	*T. harzianum*	*Gaeumannomyces graminis* var. *tritici*	Ghisalberti and Rowland (1993)
3	Peptaibol—trichorzianines A1 and B1	*T. harzianum*	*Botrytis cinerea*	Schirmbock et al. (1994)
4	Peptaibols, asperelines (A and E), and trichotoxins	*T. asperellum* TR356	*Sclerotinia sclerotiorum*	Brito et al. (2014)
5	Harzianins HC	*Trichoderma harzianum*	*S. aureus*	Goulard et al. (1995)
6	Crude secondary metabolites	*Trichoderma harzianum*	*Sclerotium rolfsii*, *R. solani*, and *Fusarium oxysporum*	Choudary et al. (2007)
7	6-Pentyl-α- pyrone	*T. harzianum*	*F. moniliforme*	El-Hasan et al. (2008)
8	Viridiofungin A	*T. harzianum*—T23 strain	*Verticillium dahliae*, *Phytophthora infestans*, *Sclerotinia sclerotiorum*	El-Hasan et al. (2009)
9	T39-butenolide, harzianolide, T22-azaphilone, harzianic acid	*T. harzianum*	Several fungal phytopathogens	Vinale et al. (2006, 2009a,b, 2014)
10	Harzianopyridone	*T. harzianum*	Several fungal phytopathogens	Ahluwalia et al. (2015)
11	6-Nonylene alcohol Massoia lactone Methyl cyclopentane Methyl cyclohexane *N*-methyl pyrollidine Dermadin Ketotriol Koningin-A	*T. harzianum* IARI P4	Several fungal phytopathogens	Dubey et al. (2011)

	3-Methyl-heptadecanol 2-Methyl heptadecanol Palmitic acid 3-(2′-Hydroxypropyl)4-(hexa-2′-4- dineyl)-2-(5H)-furanone 3-(Propenone)-4-(hexa-2′-4′-dineyl)-2(5H)-furanone			
12	Trichoharzianol	*T. harzianum*-FO 31	*Colletotrichum gloeosporioides*	Jeerapong et al. (2015)
13	Lignoren	*T. lignorum* HKI 0257	Broad spectrum action against several fungal pathogens	Berg et al. (2004)
14	Viridin	*T. koningii* *T. viride*, *T. virens*	*S. rolfsii*, *R. solani*, and *Pythium* sp.	Singh et al. (2005) and Mukherjee et al. (2007)
15	Acorane-type sesquiterpene 2b-Hydroxytrichoacorenol bisabolane-type sesquiterpene Trichoderic acid Cyclonerodiol Cyclonerodiol oxide Sorbicillin	*Trichoderma* spp.	Several plant pathogens	Wu et al. (2011)
16	Trichokonin VI	*T. pseudokoningii* SMF2	*F. oxysporum*	Shi et al. (2012)
17	Cytosporone S	*Trichoderma* sp. FKI-6626	Fungal pathogens	Ishii et al. (2013)
18	6-Pentyl pyrones (6PP)	*T. hamatum* *T. citrinoviride* *T. viridescens* *T. atroviride* *T. viride*	Fungal pathogens	Jelen et al. (2014)
19	Trichodermin	*T. brevicompactum*	*B. cinerea* *C. lindemuthianum* *R. solani*	Shentu et al. (2014, 2015)

Continued

Table 11.1 Continued

S. No.	Antimicrobial metabolite	*Trichoderma* spp. Producing antimicrobial metabolite	Pathogens controlled	References
20	Crude secondary metabolites	*Trichoderma* H921	*Magnaporthe oryzae*	Nguyen et al. (2016)
21	Secondary metabolites	*T. atroviride/petersenii* *T. virens*	*Phytophthora capsici* *P. drechsleri* *P. infestans* *P. cactorum* *P. melonis* *P. sojae* *P. nicotianae*	Bae et al. (2016)
22	Tricyclic polyketide	*T. koningiopsis QA-3*	Several plant pathogens	Shi et al. (2017)
23	6-pentyl-α-pyrone	*T. asperellum* T23 *T. harzianum* T16	*F. graminearum*	El-Hasan et al. (2017) and Marques et al. (2018)
	Antibacterial activity			
24	6-*n*-pentyl-α-pyrone	*T. harzianum* *T. longibrachiatum*	Gram-positive and Gram-negative bacteria	Tarus et al. (2003)
25	Trichotoxin	*T. asperellum*	Antibacterial activity	Chutrakul et al. (2008)
26	Sesquiterpenoids	*T. longibrachiatum*	*Escherichia coli Staphylococcus albus* *Shigella sonnei*	Wu et al. (2011) and Shi et al. (2012)
27	Cytosporone S	*Trichoderma* sp. FKI-6626	Antibacterial	Ishii et al. (2013)
28	Trichokonins	*T. pseudokoningii* SMF2	*Pectobacterium carotovorum*	Li et al. (2015)
29	Antibacterial compounds	*T. harzianum*	Pathogenic bacteria	Anwar and Iqbal (2017)
	Antiviral activity			
30	Antiviral peptaiboltrichokonin (induce immune response and ISR)	*T. pseudokoningii* SMF2	Tobacco mosaic virus (TMV)	Luo et al. (2010)

31	Trichodermin Trichoderminol	*T. albolutescens*	Cucumber mosaic virus Pepper mottle virus TMV Watermelon mosaic virus2 Zucchini green mottle mosaic virus Melon necrotic spot carmovirus Turnip mosaic virus TSWV Zucchini yellow mosaic virus Pepper mild mottle virus Cymbidium mosaic virus Lily symptomless virus Odontoglossum ringspot virus Strawberry mottle virus Potato leafroll virus Lily mottle virus, Tomato ringspot virus Potato virus Y, Cucumber green mottle mosaic virus	Lee et al. (2014, 2017)
32	Trichodermin Trichoderminol	*T. albolutescens*	PepMoV	Ryu et al. (2017)
	Nematicidal activity			
33	Trichodermin	15 *Trichoderma* strains	Nematicidal activity against plant parasitic nematodes	Yang et al. (2010)
34	6-Pentyl pyrone 1β-Vinylcyclopentane-1α,3α-Diol 4-(2-Hydroxyethyl)phenol	*Trichoderma* sp.,	*Panagrellus redivivus* *Caenorhabditis elegans* *Bursaphelenchus xylophilus*	Yang et al. (2012)
35	Koninginins L and M	*T. koningii* 8662	Plant parasitic nematodes	Lang et al. (2015)

viruses are trichodermin, trichoderminol, 6-pentyl pyrone, 1β-vinylcyclopentane-1α, 3α-diol, 4-(2-hydroxyethyl)phenol, and koninginins L and M (Table 11.1).

11.20 Antimicrobial activity of VOC

In agriculture, VOCs of microbial origin have been used as antimicrobial compounds for the management of plant pathogens and plant growth promotion. Similarly, *Trichoderma* spp. produce several VOCs with antimicrobial properties. Those VOCs are known for the antifungal and antibacterial properties (Strobel et al., 2001; Stoppacher et al., 2010). Postharvest pathogens are managed well through myco-fumigation with VOCs of *Trichoderma* spp. VOCs have been known for triggering ISR and has been also explored for the mobilization of nutrients (Van Wees et al., 2008).

The VOCs of *Trichoderma* are known for their antifungal activity against fungal pathogens (Vinale et al., 2008b). The GC–MS analysis of *T. viride* VOCs revealed the presence of isobutyl alcohol, isopentyl alcohol, and 3-methylbutanal. These compounds inhibited basidiomycetes and plant pathogens (Bruce et al., 2004). *T. asperellum* produced 6-pentyl-α-pyrone and reduced the severity of *Alternaria brassicicola* and *B. cinerea* infection in Arabidopsis. Besides, these metabolites reduced spore germination, mycelial growth, and pigmentation of fungal pathogens. The volatiles of *T. atroviride* also induced the expression of biocontrol genes of *Pseudomonas fluorescens* (Lutz et al., 2004).

Considering, the broad-spectrum action of different biomolecules produced by *Trichoderma* could be formulated and combined for synergistic action as next-generation alternatives for fungicides to suppress the plant pathogens.

11.21 Conclusion

In the last two decades, enormous advancement has been made for understanding the molecular mechanisms of *Trichoderma* on the antagonistic properties to control plant pathogens and enhance plant growth promotion and development. A combination of biochemical, genetic, and molecular approaches has led to know the interaction between *Trichoderma* in plants and other organisms. It helped to understand signal perception and transduction processes of secondary metabolites. The integration of *Trichoderma* signaling molecules and defense responses like mycoparasitism, production of antibiotics, competence by space and nutrients, colonization of plant roots, and induction of systemic resistance are emerging as a new regulatory mode by which these fungi affect plant growth and induce immune responses.

Besides, *Trichoderma* species are known well for their diverse biological activity to produce several secondary metabolites. Many metabolites produced by *Trichoderma* spp. have been isolated and characterized functionally. Understanding the function of underlying genes, biosynthesis pathways, and their regulation is still limited. However, little information is available about the genes involved in their biosynthesis, mainly because each strain of *Trichoderma* can produce simultaneously large quantities of these kinds of compounds, which make it difficult to relate one gene with one intermediate or final product. Detailed knowledge of the biosynthetic machinery and the biotic and abiotic factors triggering secondary metabolite production in *Trichoderma* will allow a tailor-made application of these fungi in biocontrol. Currently, the chemical structure of many secondary metabolites produced by *Trichoderma* during the plant-microbe interactions remains to be determined. Molecular-level understanding of the interaction between *Trichoderma* and plants is essential for utilizing these biocontrol agents in the field. In the meantime, a new detailed exercise should be conducted to understand the mechanisms of action of *Trichoderma* secondary metabolites and their possible synergisms with other compounds used in agriculture. The application of selected *Trichoderma* secondary metabolites to suppress plant pathogens, insects, nematodes, and weeds can be explored well as an alternative to chemicals. Hence, future attention on the characterization of biomolecules of *Trichoderma* can provide new views for the management of pests and diseases.

References

Abdelrahman, M., Abdel-Motaal, F., El-Sayed, M., Jogaiah, S., Shigyo, M., Ito, S.-I., Tran, L.-S.P., 2016. Dissection of *Trichoderma longibrachiatum*-induced defense in onion (*Allium cepa* L.) against *Fusarium oxysporum* f. sp. *Cepa* by target metabolite profiling. Plant Sci 246, 128–138.

Ahluwalia, V., Kumar, J., Rana, V.S., Sati, O.P., Walia, S., 2015. Comparative evaluation of two *Trichoderma harzianum* strains for major secondary metabolite production and antifungal activity. Nat. Prod. Res. 29 (10), 914–920. https://doi.org/10.1080/14786419.2014.958739.

Anwar, J., Iqbal, Z., 2017. Effect of growth conditions on antibacterial activity of *Trichoderma harzianum* against selected pathogenic bacteria. Sarhad J. Agric. 33 (4), 501–510. https://doi.org/10.17582/journal.sja/2017/33.4.501.510.

Avni, A., Bailey, B.A., Mattoo, A.K., Anderson, J.D., 1994. Induction of ethylene biosynthesis in *Nicotiana tabacum* by a *Trichoderma viride* xylanase is correlated to the accumulation of 1-aminocyclopropane-1-carboxylic acid (ACC) synthase and ACC oxidase transcripts. Plant Physiol. 106, 1049–1055.

Bae, S., Mohant, T.K., Chung, J.Y., Ryu, M., Gweekyo, P., Shim, S., Hong, S.B., Seo, H., Bae, D.W., Bae, I., Kim, J.J., Bae, H., 2016. *Trichoderma* metabolites as biological control agents against *Phytophthora* pathogens. Biol. Control 92, 128–138. https://doi.org/10.1111/j.1472-765X.2009.02599.x.

Berg, A., Wangu, H.V.K., Nkengfack, A.E., Schlegel, B., 2004. Lignoren, a new sesquiterpenoid metabolite from *Trichoderma lignorum* HKI 0257. J. Basic Microbiol. 44, 317–319. https://doi.org/10.1002/jobm.200410383.

Bodo, B., Rebuffat, S., El Hajji, M., Davoust, D., 1985. Structure of trichorzianine a IIIc, an antifungal peptide from *Trichoderma harzianum*. J. Am. Chem. Soc. 107 (21), 6011–6017.

Bohlmann, J., Keeling, C.I., 2008. Terpenoid biomaterials. Plant J. 54 (4), 656–669.

Bortolus, M., De Zotti, M., Formaggio, F., Maniero, A.L., 2013. Alamethicin in bicelles: orientation, aggregation, and bilayer modification as a function of peptide concentration. Biochim. Biophys. Acta 1828, 2620–2627.

Brito, J.P.C., Ramada, M.H.S., de Magalhaes, M.T.Q., Silva, L.P., Ulhoa, C., 2014. Peptaibols from *Trichoderma asperellum* TR356 strain isolated from Brazilian soil. SpringerPlus 3, 600. https://doi.org/10.1186/2193-1801-3-600.

Brotman, Y., Briff, E., Viterbo, A., Chet, I., 2008. Role of swollenin, an expansin-like protein from *Trichoderma*, in plant root colonization. Plant Physiol. 147, 779–789.

Bruce, A., Verrall, S., Hackett, C.A., Wheatley, R.E., 2004. Identification of volatile organic compounds (VOCs) from bacteria and yeast causing growth inhibition of sapstain fungi. Holzforschung 58, 193–198.

Carsolio, C., Gutierrez, A., Jimenez, B., Van Montagu, M., Herrera-Estrella, A., 1994. Characterization of ech-42, a *Trichoderma harzianum* endochitinase gene expressed during mycroparasitism. Proc. Natl. Acad. Sci. U. S. A. 91, 10903–10907.

Chacon, M., Rodriuez-Galan, O., Benitez, T., Sousa, S., Rey, M., Llobell, A., Delgado-Jarana, J., 2007. Microscopic and transcriptome analysis of early colonization of tomato roots by *Trichoderma harzianum*. Int. Microbiol. 10, 19–27.

Cherif, M., Benhamou, N., 1990. Cytochemical aspects of chitin breakdown during the parasitic action of a *Trichoderma* sp. on *Fusarium oxysporum f. sp. radicis lycopersici*. Phytopathology 80, 1406–1414.

Choudary, K.A., Reddy, K.R.N., Reddy, M.S., 2007. Antifungal activity and genetic variability of *Trichoderma harzianum* isolates. J. Mycol. Plant Pathol. 37 (2), 1–6.

Chutrakul, C., Alcocer, M., Bailey, K., Peberdy, J.F., 2008. The production and characterisation of trichotoxin peptaibols, by *Trichoderma asperellum*. Chem. Biodivers. 5, 1694–1705. https://doi.org/10.1002/cbdv.200890158.

Contreras, C.H.A., Macias-Rodriguez, L., Cortes-Penagos, C., Lopez-Bucio, J., 2009. *Trichoderma virens*, a plant beneficial fungus, enhances biomass production and promotes lateral root growth through an auxin-dependent mechanism in *Arabidopsis*. Plant Physiol. 149, 1579–1592.

Contreras, C.H.A., Macias-Rodriguez, L.I., Alfaro-Cuevas, R., 2014. *Trichoderma* improves growth of *Arabidopsis* seedlings under salt stress through enhanced root development, osmolite production and Na^+ elimination through root exudates. Mol. Plant Microbe Interact. 27, 503–514.

Dabire, T.G., Bonzi, S., Somda, I., Legreve, A., 2016. Evaluation of the potential of *Trichoderma harzianum* as a plant growth promoter and biocontrol agent against *Fusarium* damping-off in onion in Burkina Faso. Asian J. Plant Pathol. 10 (4), 49–60.

Dalsoglio, F.K., Bertagnolli, B.L., Sinclair, J.B., Eastbrn, D.M., 1998. Production of chitinolytic enzymes and endoglucanases in the soybean rhizosphere in the presence of *T. harzianum and R. Solani*. Biol. Control 12, 111–117.

De Lorenzo, G., Brutus, A., Savatin, D.V., Sicilia, F., Cervone, F., 2011. Engineering plant resistance by constructing chimeric receptors that recognize damage-associated molecular patterns (DAMPs). FEBS Lett. 585, 1521–1528.

de Oliveira, A.L., Gallo, M., Pazzagli, L., Benedetti, C.E., Cappugi, G., Scala, A., Pantera, B., Spisni, A., Pertinhez, T.A., Cicero, D.O., 2011. The structure of the elicitor Cerato-platanin (CP), the first member of the CP fungal protein family, reveals a double ψβ-barrel fold and carbohydrate binding. J. Biol. Chem. 286, 17560–17568.

Degenkolb, T., Karimi Aghcheh, R., Dieckmann, R., Neuhof, T., Baker, S.E., Druzhinina, I.S., Kubicek, C.P., Bruckner, H., VonDohren, H., 2012. The production of multiple small peptaibol*derma/Hypocrea*. Chem. Biodivers. 9, 499–535.

Degenkolb, T., Fog Nielsen, K., Dieckmann, R., Branco-Rocha, F., Chaverri, P., Samuels, G.J., Thrane, U., Von Dohren, H., Vilcinskas, A., Bruckner, H., 2015. Peptaibol, secondary-metabolite, and hydrophobin pattern of commercialbiocontrol agents formulated with species of the *Trichoderma harzianum* complex. Chem. Biodivers. 12, 662–684.

Dewick, P.M. (Ed.), 2009. Medicinal Natural Products: A Biosynthetic Approach. In: vol. 3. Wiley, Chichester.

Djonovic, S., Vargas, W.A., Kolomiets, M.V., Horndeski, M., Wiest, A., Kenerley, C.M., 2007. A proteinaceus elicitor Sm1 from the benefi cial fungus *Trichoderma virens* is required for induced systemic resistance in maize. Plant Physiol. 145, 875–889.

Dubey, S.C., Tripathi, A., Dureja, P., Grover, A., 2011. Characterization of secondary metabolites and enzymes produced by *Trichoderma* species and their efficacy against plant pathogenic fungi. Indian J. Agric. Sci. 81 (5), 455–461.

Elad, Y., Chet, I., Henis, Y., 1982. Degradation of plant pathogenic fungi by *Trichoderma harzianum*. Can. J. Microbiol. 28, 719–725.

Elad, Y., Chet, I., Boyle, P., Henis, Y., 1983. Parasitism of *Trichodermia* spp. on *Rhizoctonia solani* and *Sclerotium rolfsii*. Scanning electron microscopy and fluorescent microscopy. Phytopathology 73, 85–88.

El-Hasan, A., Walker, F., Buchenauer, H., 2008. *Trichoderma harzianum* and its metabolite 6-pentyl-alpha-pyrone suppress fusaric acid produced by *Fusarium moniliforme*. J. Phytopathol. 156, 79–87. https://doi.org/10.1111/j.1439-0434.2007.01330.x.

El-Hasan, A., Walker, F., Schone, J., Buchenauer, H., 2009. Detection of viridiofungin a and other antifungal metabolites excreted by *Trichoderma harzianum* active against different plant pathogens. Eur. J. Plant Pathol. 124, 457–470.

El-Hasan, A., Schone, J., Hoglinger, B., Walker, F., Voegele, R.T., 2017. Assessment of the antifungal activity of selected biocontrol agents and their secondary metabolites against *Fusarium graminearum*. Eur. J. Plant Pathol. 150 (1), 91–103. https://doi.org/10.1007/s10658-017-1255-0.

Gardiner, D.M., Waring, P., Howlett, B.J., 2005. The epipolythio-dioxopiperazine (ETP) class of fungal toxins: distribution, mode of action, functions and biosynthesis. Microbiology 151, 1021–1032.

Geremia, R., Goldman, G.H., Jacobs, D., Ardiles, W., Vila, S.B., Van Montagu, M., Herera-Estrella, A., 1993. Molecular characterization of the proteinase-encoding gene, Prb1, related to mycoparasitism by *Trichoderma harzianum*. Mol. Microbiol. 8, 603–613.

Ghisalberti, E.I., Rowland, C.Y., 1993. Antifungal metabolites from *Trichoderma harzianum*. J. Nat. Prod. 56 (10), 1799–1804.

Goldman, H.G., Hayes, C., Harman, G.E., 1994. Molecular and cellular biology of biocontrol by *Trichoderma* spp. Trends Biotechnol. 12, 478–482.

Goulard, C., Hlimi, S., Rebuffat, S., Bodo, B., 1995. Trichorzins HA and MA, antibiotic peptides from *Trichoderma harzianum*. I. Fermentation, isolation and biological properties. J. Antibiot. 48 (11), 1248–1253. https://doi.org/10.7164/antibiotics.48.1248.

Hajieghrari, B., 2010. Effects of some Iranian *Trichoderma* isolates on maize seed germination and seedling vigor. Afr. J. Biotechnol. 9 (28), 4342–4347.

Harman, G.E., Hayes, C.K., Lorito, M., Broadway, R.M., Di-Pietro, A., Petebauer, C., Tronsmo, A., 1993. Chitinolytic enzymes of *Trichoderma harzianum*. Purification of chitobiosidase and endochitinase. Phytopathology 83, 313–318.

Harman, G.E., Howell, C.R., Viterbo, A., Chet, I., Lorito, M., 2004. *Trichoderma* species-opportunistic, avirulent plant symbionts. Nat. Rev. Microbiol. 2, 43–56.

Hermosa, R., Viterbo, A., Chet, I., Monte, E., 2012. Plant-beneficial effects of *Trichoderma* and of its genes. Microbiology 158, 17–25.

Hung, R., Samantha, L., Joan, W.B., Gareth, W.G., 2012. Arabidopsis Thaliana as a model system for testing the effect of *Trichoderma* volatile organic compounds. Fungal Ecol. 6 (1), 19–26.

Ishii, T., Nonaka, K., Suga, T., Masuma, R., Omura, S., Shiomi, K., 2013. Cytosporone S with antimicrobial activity, isolated from the fungus *Trichoderma* sp. FKI-6626. Bioorg. Med. Chem. Lett. 23, 679–681. https://doi.org/10.1016/j.bmcl.2012.11.113.

Jacobs, D., Geremia, R.A., Goldman, G.H., Kamoen, O., Van Montagu, M., Herrera-Estrella, A., 1991. Study of the expression of β-1,3-glucanase of *Trichoderma harzianum*. Petria 1, 125.

Jeerapong, C., Phupong, W., Bangrak, P., Intana, W., Tuchinda, P., 2015. Trichoharzianol, a new antifungal from *Trichoderma harzianum* F031. J. Agric. Food Chem. 63 (14), 3704–3708.

Jelen, H., Błaszczyk, L., Chełkowski, J., Rogowicz, K., Strakowska, J., 2014. Formation of 6-n-pentyl-2H-pyran-2-one (6-PAP) and other volatiles by different *Trichoderma* species. Mycol. Prog. 13, 589–600.

Jogaiah, S., Mostafa, A., Tran, L.S.P., Shin-ichi, I., 2013. Characterization of rhizosphere fungi that mediate resistance in tomato against bacterial wilt disease. J. Exp. Bot. 64, 3829–3842.

Jogaiah, S., Abdelrahman, M., Tran, L.-S.P., Ito, S.-I., 2018. Different mechanisms of *Trichoderma virens*-mediated resistance in tomato against Fusarium wilt involve the jasmonic and salicylic acid pathways. Mol. Plant Pathol. 19, 870–882.

Kubicek, C.P., Herrera-Estrella, A., Seidl-Seiboth, V., 2011. Comparative genome sequence analysis underscores mycoparasitism as the ancestral life style of *Trichoderma*. Genome Biol. 12, R40.

Lang, B.Y., Li, J., Zhou, X.X., Chen, Y.H., Yang, Y.H., Li, X.N., Zeng, Y., Zhao, P.J., 2015. Koninginins L and M, two polyketides from *Trichoderma koningii* 8662. Phytochem. Lett. 11, 1–4.

Lee, D.H., Kim, J.J., Ryu, K.H., Kim, B.S., Ryu, S.M. 2014. Novel antiviral composition, and method for controlling plant viruses by using same. WO Patent WO2016089166.

Lee, D.H., Kim, J.J., Ryu, K.H., Kim, B.S., Ryu, S.M. 2017. Antiviral composition and method for controlling plant viruses using the same. US Patent No. US20170265473A.

Lehner, S.M., Atanasova, L., Neumann, N.K., Krska, R., Lemmens, M., Druzhinina, I.S., Schuhmacher, R., 2013. Isotope-assisted screening for iron-containing metabolitesreveals a high degree of diversity among known and unknown siderophores produced by *Trichoderma* spp. Appl. Environ. Microbiol. 79, 18–31.

Li, R.X., Cai, F., Pang, G., Shen, Q.R., Li, R., Chen, W., 2015. Solubilisation of phosphate and micronutrients by *Trichoderma harzianum* and its relationship with the promotion of tomato plant growth. PLoS One. 10(6), e0130081.

Limon, M.C., Lora, J.M., Garcia, I., De LaCruz, J., Lobell, A., Benitez, T., Pintor Toro, J.A., 1995. Primary structure and expression pattern of the 33 kDa chitinase gene from the mycoparasitic fungus *Trichoderma harzianum*. Curr. Genet. 28, 478–483.

Lorito, M., Peterbauer, C., Hayes, C.K., Harman, G.E., 1994. Synergistic interaction between fungal cell wall degrading enzymes and different antifungal compounds enhances inhibition of spore germination. Microbiology 140, 623–629.

Luo, Y., Zhang, D.D., Dong, X.W., Zhao, P.B., Chen, L.L., Song, X.Y., Wang, X.J., Chen, X.L., Shi, M., Zhang, Y.Z., 2010. Antimicrobial peptaibols induce defense responses and systemic resistance in tobacco against *Tobacco mosaic virus*. *FEMS Microbiol.* Lett. 313 (2), 120–126. https://doi.org/10.1111/j.1574-6968.2010.02135.x.

Lutz, M.P., Wenger, S., Maurhofer, M., Defago, G., Duffy, B., 2004. Signaling between bacterial and fungal biocontrol agents in a strain mixture. FEMS Microbiol. Ecol. 48, 447–455.

Mahato, S., Bhuju, S., Shrestha, J., 2018. Effect of *Trichoderma viride* as biofertilizer on growth and yield of wheat. Malays. J. Sustain. Agric. 2 (2), 01–05.

Malmierca, M.G., Cardoza, R.E., Alexander, N.J., McCormick, S.P., Hermosa, R., Monte, E., Gutierrez, S., 2012. Involvement of *Trichoderma* trichothecenes in the biocontrol activity and induction of plant defense-related genes. Appl. Environ. Microbiol. 78, 4856–4868.

Malmierca, M.G., Cardoza, R.E., Alexander, N.J., McCormick, S.P., Collado, I.G., Hermosa, R., Monte, E., Gutierrez, S., 2013. Relevance of trichothecenes in fungal physiology: disruption of tri5 in *Trichoderma arundinaceum*. Fungal Genet. Biol. 53, 22–33.

Marques, E., Martins, I., de Scm, M., 2018. Antifungal potential of crude extracts of *Trichoderma* spp. Biota Neotrop.. 18(1), e20170418, https://doi.org/10.1590/1676-0611-BN-2017-0418.

Marra, R., Ambrosino, P., Carbone, V., Vinale, F., Woo, S.L., Ruocco, M., Ciliento, R., Lanzuise, S., Ferraioli, S., Soriente, I., Gigante, S., Turra, D., Fogliano, V., Scala, F., Lorito, M., 2006. Study of the three-way interaction between *Trichoderma atroviride*, plant and fungal pathogens by using a proteomic approach. Curr. Genet. 50, 307–321.

Martinez, C., Blanc, F., Le Claire, E., Besnard, O., Nicole, M., Baccou, C.J., 2001. Salicylic acid and ethylene pathways are differentially activated in melon cotyledons by active or heat-denatured cellulase from *Trichoderma longibrachiatum*. Plant Physiol. 127, 334–344.

Masunaka, A., Hyakumachi, M., Takenaka, S., 2011. Plant growth promoting fungus *Trichoderma koningii* suppresses isoflavonoid phytoalexin vestitol production for colonization on/in the roots of *Lotus japonicus*. Microbes Environ. 26, 128–134.

Matarasso, N., Schuster, S., Avni, A., 2005. A novel plant cysteine protease has a dual function as a regulator of 1-aminocyclopropane-1-carboxylic acid synthase gene expression. Plant Cell 17, 1205–1216.

Moran-Diez, E., Hermosa, R., Ambrosino, P., Cardoza, R.E., Gutierrez, S., Lorito, M., Monte, E., 2009. The ThPG1 endopolygalacturonase is required for the *Trichoderma harzianum*-plant beneficial interaction. Mol. Plant Microbe Interact. 22, 1021–1031.

Moran-Diez, E., Rubio, B., Domınguez, S., Hermosa, R., Monte, E., Nicolas, C., 2012. Transcriptomic response of *Arabidopsis thaliana* after 24 h incubation with the biocontrol fungus *Trichoderma harzianum*. J. Plant Physiol. 169, 614–620.

Moran-Diez, M.E., Trushina, N., Lamdan, N.L., Rosenfelder, L., Mukherjee, P.K., Kenerley, C.M., Horwitz, B.A., 2015. Host-specific transcriptomic pattern of *Trichoderma virens* during interaction with maize or tomato roots. BMC Genomics 16 (1), 8.

Mukherjee, M., Mukherjee, P.K., Kale, S.P., 2007. cAMP signalling is involved in growth, germination, mycoparasitism and secondary metabolism in *Trichoderma virens*. Microbiology 153, 1734–1742.

Mukherjee, P.K., Horwitz, B.A., Kenerley, C.M., 2012. Secondary metabolism in *Trichoderma*ea genomic perspective. Microbiology 158, 35–45.

Nagaraju, A., Jogaiah, S., Mahadeva Murthy, S., Ito, S.-I., 2012. Seed priming with *Trichoderma harzianum* isolates enhances plant growth and induces resistance against *Plasmopara halstedii*, an incitant of sunflower downy mildew disease. Aust. Plant Pathol. 41, 609–620.

Nakkeeran, S., Renukadevi, P., Eraivan, K.A., 2016. Exploring the potential of Trichoderma for the management of seed and soil borne diseases of crops. In: Muniyappan, K. et al., (Ed.), The Book Integrated Pest Management of Tropical Vegetable Crops. Springer Publications, Dordrecht, pp. 77–130.

Naseby, D.C., Pascual, J.A., Lynch, J.M., 2000. Effect of biocontrol strains of *Trichoderma* on plant growth, *Pythium ultimum* populations, soil microbial communities and soil enzyme activities. J. Appl. Microbiol. 88, 161–169.

Neuhof, T., Dieckmann, R., Druzhinina, I.S., Kubicek, C.P., VonDohren, H., 2007. Intact-cell MALDI-TOF mass spectrometry analysis of peptaibol formation by the genus *Trichoderma/-Hypocrea*: can molecular phylogeny of species predict peptaibol structures? Microbiology 153, 3417–3437.

Neumann, N.K., Stoppacher, N., Zeilinger, S., Degenkolb, T., Bruckner, H., Schuhmacher, R., 2015. The peptaibiotics data-base-a comprehensive online resource. Chem. Biodivers. 12, 743–751.

Nguyen, Q.T., Ueda, K., Kihara, J., Ueno, M., 2016. Culture filtrates of *Trichoderma* isolate H921 inhibit *Magnaporthe oryzae* spore germination and blast lesion formation in rice. Adv. Microbiol. 6, 521–527. https://doi.org/10.4236/aim.2016.67052.

Nicol, S., 1991. Life after death for empty shells. New Sci. 129, 46–48.
Nieto-jacobo, M.F., Steyaert, J.M., Salazar-badillo, F.B., 2017. Environmental growth conditions of Trichoderma spp. affects indole acetic acid derivatives, volatile organic compounds, and plant growth promotion. Front. Plant Sci. 8, 1–18.
Nosir, W.S., 2016. *Trichoderma harzianum* as a growth promoter and bio-control agent against *Fusarium oxysporum* f. sp. *tuberosi*. Adv. Crop Sci. Technol. 4 (2), 217.
Park, A.R., Cho, S.K., Yun, U.J., Jin, M.Y., Lee, S.H., Sachetto-Martins, G., Park, O.K., 2001. Interaction of the Arabidopsis receptor protein kinase Wak1 with a glycine-rich protein, AtGRP-3. J. Biol. Chem. 276, 26688–26693.
Parra, E.H., Alejo, J.C., Zapata, J.A.R., 2017. *Trichoderma* strains as growth promoters in *Capsicum annuum* and as biocontrol agents in *Meloidogyne incognita*. Chil. J. Agric. Res. 77 (4), 318–324.
Piel, J., Atzorn, R., Gabler, R., Kuhnemann, F., Boland, W., 1997. Cellulysin from the plant parasitic fungus *Trichoderma viride* elicits volatile biosynthesis in higher plants via the octadecanoid signalling cascade. FEBS Lett. 416, 143–148.
Pineda, A., Zheng, S.J., Van Loon, J.J.A., Pieterse, C.M.J., Dicke, M., 2010. Helping plants to deal with insects: the benefi cial soil-borne microbes. Trends Plant Sci. 15, 507–514.
Ron, M., Avni, A., 2004. The receptor for the fungal elicitor ethyleneinducing xylanase is a member of a resistance-like gene family in tomato. Plant Cell 16, 1604–1615.
Rotblat, B., Enshell-Seijffers, D., Gershoni, J.M., Schuster, S., Avni, A., 2002. Identification of an essential component of the elicitation active site of the EIX protein elicitor. Plant J. 32, 1049–1055.
Rubio, M.B., Hermosa, R., Reino, J.L., Collado, I.G., Monte, E., 2009. Thctf1 transcription factor of *Trichoderma harzianum* is involved in 6-pentyl 2H-pyran-2-one production and antifungal activity. Fungal Genet. Biol. 46, 17–27.
Ryu, S.M., Lee, H.M., Song, E.G., Seo, Y.H., Lee, J., Guo, Y., Kim, B.S., Kim, J.J., Hong, J.S., Ryu, K.H., Lee, D., 2017. Antiviral activities of trichothecenes isolated from *Trichoderma albolutescens* against *Pepper mottle virus*. J. Agric. Food Chem. 65 (21), 4273–4279. https://doi.org/10.1021/acs.jafc.7b01028.
Samolski, I., Rincon, A., Pinzon, L.M., Viterbo, A., Monte, E., 2012. The *qid74* gene from *Trichoderma harzianum* has a role in root architecture and plant biofertilization. Microbiology 158, 129–138.
Scharf, D.H., Brakhage, A.A., Mukherjee, P.K., 2016. Gliotoxin—bane or boon? Environ. Microbiol. 18, 1096–1109.
Schirmbock, M., Lorito, M., Wang, Y.L., Hayes, C.K., Arisan-Atac, I., Scala, F., Harman, G.E., Kubicek, C.P., 1994. Parallel formation and synergism of hydrolytic enzymes and peptaibol antibiotics, molecular mechanisms involved in the antagonistic action of *Trichoderma harzianum* against phytopathogenic fungi. Appl. Environ. Microbiol. 60 (12), 4364–4370.
Schuster, A., Schmoll, M., 2010. Biology and biotechnology of *Trichoderma*. Appl. Microbiol. Biotechnol. 87, 787–799.
Sharma, P., Patel, A.N., Saini, M.K., Deep, S., 2012. Field demonstration of *Trichoderma harzianum* as a plant growth promoter in wheat (*Triticum aestivum* L.). J. Agric. Sci. 4, 8.
Shentu, X., Zhan, X., Ma, Z., Yu, X., Zhang, C., 2014. Antifungal activity of metabolites of the endophytic fungus *Trichoderma brevicompactum* from garlic. Braz. J. Microbiol. 45 (1), 248–254. https://doi.org/10.1590/S1517-83822014005000036.
Shentu, X.P., Yuan, X.F., Liu, W.P., Xum, J.F., Yu, X.P., 2015. Cloning and functional analysis of tri14in *Trichoderma brevicompactum*. Am. J. Mol. Biol. 11 (3), 169–175.
Shi, M., Chen, L., Wang, X.W., Zhang, T., Zhao, P.B., Song, X.Y., Sun, C.Y., Chen, X.L., Zhou, B.C., Zhang, Y.Z., 2012. Antimicrobial peptaibols from *Trichoderma*

pseudokoningii induce programmed cell death in plant fungal pathogens. Microbiology 158, 166–175.

Shi, X.S., Wang, D.J., Li, X.M., Li, H.L., Meng, L.H., Li, X., Pi, Y., Zhou, X.W., Wang, B.G., 2017. Antimicrobial polyketides from *Trichoderma koningiopsis* QA-3, an endophytic fungus obtained from the medicinal plant *Artemisia argyi*. RSC Adv. 7, 51335–51342.

Singh, S., Dureja, P., Tanwar, R.S., Singh, A., 2005. Production and antifungal activity of secondary metabolites of *Trichoderma virens*. Pestic. Res. J. 17, 26–29.

Sivan, A., Chet, I., 1989. The possible role of competition between *Trichoderma harzianum* and *Fusarium oxysporium* on rhizosphere colonization. Phytopathology 79, 198–203.

Song, G.C., Ryu, C.M., 2013. Two volatile organic compounds trigger plant self-defense against a bacterial pathogen and a sucking insect in cucumber under open field conditions. Int. J. Mol. Sci. 14, 9803–9819.

Stergiopoulos, I., De Wit, P.J.G.M., 2009. Fungal effector proteins. Annu. Rev. Phytopathol. 47, 233–263.

Stoppacher, N., Zeilinger, S., Omann, M., Lassahn, P.G., Roitinger, A., Krska, R., Schuhmacher, R., 2008. Characterisation of the peptaibiome of the biocontrol fungus *Trichodermaatroviride* by liquid chromatography/tandem massspectrometry. Rapid Commun. Mass Spectrom. 22, 1889–1898.

Stoppacher, N., Kluger, B., Zeilinger, S., Krska, R., Schuhmacher, R., 2010. Identification and profiling of volatile metabolites of the biocontrol fungus *Trichoderma atroviride* by HS-SPME-GC-MS. J. Microbiol. Methods 81, 187–193.

Strobel, G.A., Dirkse, E., Sears, J., Markworth, C., 2001. Volatile antimicrobials from *Muscodor albus* a novel endophytic fungus. Microbiology 147, 2943–2950.

Taniguchi, N., Honke, K., Fukuda, M. (Eds.), 2011. Handbook of Glycosyltransferases and Related Genes. Springer Science & Business Media.

Tarus, P.K., Langat-Thoruwa, C.C., Wanyonyi, A.W., Chhabra, S.C., 2003. Bioactive metabolites from *Trichoderma harzianum* and *Trichoderma longibrachiatum*. Bull. Chem. Soc. Ethiop. 17 (2), 185–190.

Tijerino, A., Cardoza, R.E., Moraga, J., et al., 2011. Overexpression of the trichodiene synthase gene tri5 increases trichodermin production and antimicrobial activity in *Trichoderma brevicompactum*. Fungal Genet. Biol. 48, 285–296.

Tronsmo, A., Klemsdal, S.S., Hayes, C.K., Lorito, M., Harman, G.E., 1993. The role of hydrolytic enzymes produced by *Trichoderma harzianum* in biological control of plant diseases, *Trichoderma reesei* cellulases and other hydrolases, enzyme structure, biochemistry, genetic and applications. In: Suominen, P., Reinikainen, T. (Eds.), Foundation for Biotechnical and Industrial Fermentation Research. In: vol. 8. Helsinki, Finland, pp. 159–168.

Van Wees, S.C.M., Van der Ent, S., Pieterse, C.M., 2008. Plant immune responses triggered by beneficial microbes. Curr. Opin. Plant Biol. 11, 443–448.

Vargas, W.A., Crutcher, F.K., Kenerley, C.M., 2011. Functional characterization of a plant-like sucrose transporter from the beneficial fungus *Trichoderma virens*. Regulation of the symbiotic association with plants by sucrose metabolism inside the fungal cells. New Phytol. 189, 777–789.

Vasumathi, S., Eraivan, A.A.K., Nakkeeran, S., 2016. Exploring the diversity of *Trichoderma* spp. and the synergistic activity of delivery systems for the management of *Fusarium* wilt of cucumber under protected cultivation. J. Mycol. Plant Pathol. 46 (4), 311.

Vinale, F., Marra, R., Scala, F., Ghisalberti, E.L., Lorito, M., Sivasithamparam, K., 2006. Major secondary metabolites produced by two commercial *Trichoderma* strains active against different phytopathogens. Lett. Appl. Microbiol. 43, 143–148. https://doi.org/10.1111/j.1472-765X.2006.01939.x.

Vinale, F., Sivasithamparam, K., Ghisalberti, E.L., Marra, R., Barbetti, M.J., Li, H., Woo, S.L., Lorito, M., 2008a. A novel role for Trichoderma secondary metabolites in the interactions with plants. Physiol. Mol. Plant Pathol. 72, 80–86.

Vinale, F., Sivasithamparam, K., Ghisalberti, E.L., Marra, R., Woo, S.L., Lorito, M., 2008b. Trichoderma- plant- pathogen interactions. Soil Biol. Biochem. 40, 1–10.

Vinale, F., Ghisalberti, E.L., Sivasithamparam, K., Marra, R., Ritieni, A., Ferracane, R., Woo, S., Lorito, M., 2009a. Factors affecting the production of *Trichoderma harzianum* secondary metabolites during the interaction with different plant pathogens. Lett. Appl. Microbiol. 48, 705–711. https://doi.org/10.1111/j.1472-765X.2009.02599.x.

Vinale, F., Flematti, G., Sivasithamparam, K., Lorito, M., Marra, R., Skelton, B.W., Ghisalberti, E.L., 2009b. Harzianic acid, an antifungal and plant growth promoting metabolite from *Trichoderma harzianum*. J. Nat. Prod. 72 (11), 2032–2035. https://doi.org/10.1021/np900548p.

Vinale, F., Nigro, M., Sivasithamparam, K., Flematti, G., Ghisalberti, E.L., Ruocco, M., Varlese, R., Marra, R., Lanzuise, S., Eid, A., Woo, S.L., Lorito, M., 2013. Harzianic acid: a novel*siderophorefrom Trichodermaharzianum*. FEMS Microbiol. Lett. 347, 123–129.

Vinale, F., Manganiello, G., Nigro, M., Mazzei, P., Piccolo, A., Pascale, A., Ruocco, M., Marra, R., Lombardi, N., Lanzuise, S., Varlese, R., Cavallo, P., Lorito, M., Woo, S.L., 2014. A novel fungal metabolite with beneficial properties for agricultural applications. Molecules 19, 9760–9772. https://doi.org/10.3390/molecules19079760.

Vinodkumar, S., Indumathi, T., Nakkeeran, S., 2017. *Trichoderma asperellum* (NVTA2) as a potential antagonist for the management of stem rot in carnation under protected cultivation. Biol. Control 113, 58–64.

Viterbo, A., Chet, I., 2006. *TasHyd1*, a new hydrophobin gene from the biocontrol agent *Trichoderma asperellum*, is involved in plant root colonization. Mol. Plant Pathol. 7, 249–258.

Viterbo, A., Wiest, A., Brotman, Y., Chet, I., Kenerley, C., 2007. The 18mer peptaibols from *Trichoderma virens* elicit plant defence responses. Mol. Plant Pathol. 8, 737–746.

Woo, S.L., Scala, F., Ruocco, M., Lorito, M., 2006. The molecular biology of the interactions between *Trichoderma* spp., phytopathogenic fungi, and plants. Phytopathology 96, 181–185.

Wu, S.H., Zhao, L.X., Chen, Y.W., Huang, R., Miao, C.P., Wang, J., 2011. Sesquiterpenoids from the endophytic fungus *Trichoderma* sp. PR-35 of Paeonia delavayi. Chem. Biodivers. 8, 1717–1722.

Yang, Z.S., Li, G.H., Zhao, P.J., Zheng, X., Luo, S.L., Li, L., Niu, X.M., Zhang, K.Q., 2010. Nematicidal activity of *Trichoderma* spp. and isolation of an active compound. World J. Microbiol. Biotechnol. 26, 2297–2302. https://doi.org/10.1007/s11274-010-0410-y.

Yang, Z., Yu, Z., Lei, L., Xia, Z., Shao, L., Zhang, K., Li, G., 2012. Nematicidal effect of volatiles produced by *Trichoderma* sp. J. Asia Pac. Entomol. 15, 647–650.

Yano, S., Tokumitsu, H., Soderling, T.R., 1998. Calcium promotes cell survival through CaMK kinase activation of the protein-kinase-B pathway. Nature 396, 584–587.

Yoshioka, Y., Ichikawa, H., Naznin, H.A., Kogure, A., Hyakumachi, M., 2012. Systemic resistance induced in Arabidopsis thaliana by Trichoderma asperellum SKT-1, a microbial pesticide of seed borne diseases of rice. Pest Manag. Sci. 68, 60–66.

Zhang, H., Xie, X., Kim, M., Kornyeyev, D.A., Holaday, S., Pare, P., 2008. Soil bacteria augment Arabidopsis photosynthesis by decreasing glucose sensing and abscisic acid levels in planta. Plant J. 56, 264–273.

Zhang, F., Yang, X., Ran, W., Shen, Q., 2014. *Fusarium oxysporum* induces the production of proteins and volatile organic compounds by *Trichoderma harzianum* T-E5. FEMS Microbiol. Lett. 359, 116–123.

Microbial secondary metabolites and their role in stress management of plants

12

Ankit Kumar Ghorai, Rakesh Patsa, Subhendu Jash, and Subrata Dutta
Department of Plant Pathology, Bidhan Chandra Krishi Viswavidyalaya, Mohanpur, Nadia, West Bengal, India

12.1 Introduction

Microbes are highly diverse and natural source of indefinite chemical substances. Scientific literature says for the existence of over one trillion species of microbes on earth and yet 99.99% of them are to be discovered. Because of their diversified mechanism of adaptations, they are ubiquitous, that is, they are present even in extreme temperatures of arctic ice to boiling volcanoes and the bottom surface of oceans to the stratosphere. Soil is the thriving place for a diverse group of microorganisms which include bacteria, archaebacteria, fungi, algae, and protozoa. The soil microorganisms are primarily responsible for supporting plant lives as (i) they synthesize and mobilize various nutrients, (ii) produce metabolites that protect and promote plant growth, and (iii) are the chief recycler of various nutrients on earth.

Presently, the 6.7 billion world population is expected to shoot up to 10 billion by 2050. This 45% increase in current world population will impose greater demand for increased food supply. But with the progressive climate change, agriculture is being threatened with a growing number of plant pathogens, insects, drought, salinity, and extreme temperature conditions (Satapute et al., 2019). According to the study done by Zhang et al. (2011), approximately 9000 species of insects and mites, 50,000 species of plant pathogens and 8000 species of weeds are reported to impose biotic stress over crops. Among abiotic stresses, yield loss up to 17% due to drought, 20% due to salinity, 40% due to high temperature, 15% due to low temperature and 8% due to other stress are observed (Ashraf and Harris, 2004). Among the biotic stresses weed infestation accounts for highest yield loss (45%) followed by insect pests (30%), diseases (20%), and other pests (5%) (Subba Rao and Madhulety, 2005; Sudisha et al., 2006). Among the global farming land, 45% is affected by continuous or frequent drought (Long and Ort, 2010). FAO estimated that abiotic stress would result in 30% land degradation in the coming 25 years and failing to devise strategies to combat this will increase the risk of land degradation up to 50% by 2050 (Manus et al., 2002).

The indiscriminate application of fertilizers, chemicals, and pesticides have proved to be hazardous to man and his environment (Sudisha et al., 2005). In 2014 the share of miscellaneous pesticides peaked to 53.84%, followed by herbicides (25.10%), fungicides and bacteriocides (12.06%), insecticides (7.50%), and plant growth regulators

Biocontrol Agents and Secondary Metabolites. https://doi.org/10.1016/B978-0-12-822919-4.00012-0

(1.24%). Organophosphorus pesticides have accounted for 86.02% of fatal human cases (Zhang et al., 2011). Such indiscriminate agrochemical use has also led to the extinction of many species (Zhang et al., 2011). Insecticidal residues have been found in nearly 75% of the honey in the world of which neonicotinoids are of significant concern. In India, the estimated annual crop production losses due to pests are as high as US$ 42.66 million. Pesticide consumption rate has continuously increased with 0.29 kg/ha (GCA) during the year 2014–15, which was 50% higher as compared to 2009–10 (Sushil, 2016).

In such a threatening context, harnessing the immense potential of ecofriendly microbial metabolites is the only way to restore ecological balance, biological diversity, and promote sustainable agriculture (Jogaiah et al., 2013; Murali et al., 2013). As the future human civilization is entirely dependent on sustainable agriculture, recent scientific endeavors over identification of novel microbial metabolites and their commercialization as biopesticides have been given utmost priority. The share of biopesticides is currently $3 billion in crop protection market globally although it accounts only to 5% of the crop protection market (Damalas and Spyridon, 2018). The total global demand for microbes and their metabolites was $143.5 billion in 2014 and has been targeted to reach $ 306 billion in 2020 (Microbial Products, 2015).

The plant growth-promoting rhizobacteria (PGPR) residing in the below ground rhizosphere ecology supports the growth of plants under various environmental conditions chiefly by two broad mechanisms; (i) direct mechanism and (ii) indirect mechanism. The direct mechanism includes biological nitrogen fixation, phosphate solubilization, phytohormone production, and iron sequestration using siderophores. While indirect mechanism includes induced systemic resistance in plants and promotion of plant growth using a wide range of secondary metabolites. In this chapter, we have elaborately discussed the various secondary metabolites produced by microbes in the rhizosphere and how they stimulate plant growth promotion under different biotic and abiotic stress conditions.

12.2 Microbial metabolites

Microorganisms are the natural manufacturer of diverse metabolites. The primary metabolites are synthesized during growth phase due to energy metabolism and are responsible for carrying out primary physiological functions that support the growth and development of the producers. They include alcohol, amino acids, nucleotides, antioxidants, organic acids, polyols, and vitamins. In contrast secondary metabolites are low-molecular mass compounds formed near the stationary phase of growth when the nutrient source gets depleted. The secondary metabolites determine the ecological parameters of the microorganisms. The production of secondary metabolites depends on various regulatory mechanisms which are triggered by the nutrient, pH, and temperature. They include metabolites include antibiotics, siderophores, hormones, exopolysaccharides, pheromones, toxins, etc. The microbial secondary metabolites produced by different microrganisms and their role in plant stress management have been illustrated in Tables 12.1–12.8.

Table 12.1 Different antibiotics and their role in plant disease management.

Antibiotic	Crop	Disease	Pathogen	References
Streptomycin	Apple	Fire blight	*Erwinia amylovora*	Rangaswami and Mahadevan (1998)
	Bean	Halo bight	*P. syringae* pv. *phaseolicola*	
	Celery	Bacterial blight	*Pseudomonas cichorii*	
	Potato	Bacterial soft rot	*E. carotovora subspecies carotovora*	
	Tomato	Bacterial spot	*X. campestris* pv. *vesicatoria*	
	Tobacco	Wildfire	*P. syringae* pv. *tabaci*	
	Roses	Crown gall	*Agrobacterium tumefaciens*	
	Chrysanthemum	Bacterial wilt	*E. chrysanthemi*	
	Tobacco	Blue mold	*Peronospora tabacina*	
Tetracycline	Apple	Fire blight	*Erwinia amylovora*	Rangaswami and Mahadevan (1998)
	Nectarine	Bacterial leaf and fruit spot	*X. campestris* pv. *pruni*	
	Peach	Bacterial leaf and fruit spot	*X. campestris* pv. *pruni*	
	Pear	Fire blight	*Erwinia amylovora*	
	Sugar beet	Bacterial rot	*Erwinia sp.*	
	Palm	Lethal yellow	*Phytoplasma*	
	Chrysanthemum	Bacterial wilt	*E. chrysanthemi*	
	Citrus	HLB/citrus greening	*Candidatus liberibacter asiaticus*	
Aureofungin	Citrus	Citrus gummosis	*Phytophthora citrophthora*	Rangaswami and Mahadevan (1998)
	Mango	Rot	*Diplodia sp.*	
	Grapes	Downey mildew	*Plasmopara viticola*	
	Tomato	Alternaria rot	*Alternaria solani*	
	Rice	Blast	*Pyricularia oryzae*	
	Apple	Powdery mildew	*Podosphaera leucotricha*	
	Citurs	Citrus greening	*Candidatus liberinbacter asiaticus*	
	Peach	Rot	*Sclerotinia sclerotiorum*	
	Cucurbits	Root rot	*Pythium*	

Continued

Table 12.1 Continued

Antibiotic	Crop	Disease	Pathogen	References
Validamycin	Rice	Sheath blight	*Rhizoctonia solani*	Vurukonda et al. (2018)
	Cucumber	Damping off	*Rhizoctonia solani*	
	Potato	Black scurf	*Rhizoctonia solani*	
Cyclohexamidine	Oat	Covered smut	*Ustilago hordei*	Rangaswami and Mahadevan (1998)
	Peach	Brown rot		
Blasticidin S	Rice	Blast	*Pyricularia oryzae*	Misato et al. (1959)
Kasugamycin	Rice	Blast	*Pyricularia oryzae*	Vidhyasekaran (2004)
	Apple	Fire blight	*Erwinia amylovora*	McGhee and Sundin (2011)
	Apple	Scab	*Venturia inaequalis*	Vidhyasekaran (2004)
	Potato	Soft rot	*Pectobacterium carotovorum subspecies carotovorum*	Vidhyasekaran (2004)
Antimycin	Tomato	Early blight	*Alternaria solani*	Anderson and Gottlieb (1952)
	Rice	Blast	*Pyricularia oryzae*	Nakayama et al. (1956)
Thiolutin	Potato	Late blight	*Phytophthora infestans*	Bonde (1953)
	Broccoli	Downey mildew	*Peronospora parasitica*	Natti et al. (1956)
Bulbiformin	Pea	Fusarium wilt	*Fusarium udum*	Vasudeva et al. (1963)
Milbemycin	Citrus, brinjal, tea	Mite infestation	Citrus red mite, spider mites, and citrus leaf miner of citrus, tea, and brinjal	Vurukonda et al. (2018)
Abamectin (Avermectins)	Cotton, citrus, and vegetables	Insects	Leaf miners, suckers, and beetles	Vurukonda et al. (2018)

Table 12.2 Microbial secondary metabolites of different species of *Azospirillum* and their role in plant stress management.

Microbe	Metabolites	Stress	Mode of action	References
Azospirillum brasilense	Phenylacetic acid	Bacterial speck disease of tomato (*Pseudomonas syringae* pv. *tomato*), Crwon gall (*Agrobacterium tumifaciens*), and bacterialleaf blight of mulberry	Antibacterial activity	Bashan and De-Bashan (2002) and Somers et al. (2005)
Azospirillum sp.	Indole-3-acetic acid/indole-3-butyric acid	Limitation of nutrients like nitrogen, carbon, and phosphorous	Promote plant growth by root development	Malhotra and Srivastava (2009)
A. brasilense Ab-V5 and Ab-V6	Metabolites	Drought and plant pathogens	Upregulation of oxidative stress and PR genes	Fukami et al. (2017)
Azospirillum sp.	Exopolysaccharide	Resistance to *Pyricularia oryzae* in rice	By promoting ISR	Sankari et al. (2011)
Azospirillum sp.	IAA	Drought stress in canola	Increase in shoot length and shoot weight	Fukami et al. (2018)
Azospirillum lipoferum and *A. brsilense*	GA1 and GA3	Dwarfism in wheat, rice, and maize	3-β-hydroxylation of GA20, primary root elongation, latent root development and improve nitrogen uptake	Bottini et al. (2004) and Fulchieri et al. (1993)
A. brasilense	Gibberellic acid	Dormancy of soybean and wheat	By reducing hydrolytic enzymes, amylases, and protease	Mehnaz (2015)
Azospirillum brasilense sp245	ABA	*Arabidopsis thaliana* grown in salt and water stress	Induces twofold ABA in plants to combat stress	Cohen et al. (2008)
A. brasilense	Polyamines (cadaverine)	Environmental stress in rice	Promote root growth and minimizes osmotic stress	Cassán et al. (2009)
A. brasilense REC 2 and REC 3	Siderophores (catechol type)	Anthracnose of strawberry (*Colletotrichum acutatum*)	Antifungal activity	Tortora et al. (2011)

Table 12.3 Microbial secondary metabolites of different species of fluorescent Pseudomonads and their role in plant stress management.

Microbes	Metabolites	Stress	Mode of action	References
P. putida	Pyoverdin (siderophore)	Fusarium wilt and take all disease of wheat	By restricting iron availability to pathogens	Haas and Defago (2005)
P. fluorescens WCS358	Pseudobactin 358 (siderophore)	*Fusarium oxysporum* f. sp. *raphini* (Fusarium wilt of radish)	By scavenging iron	Leeman et al. (1996)
P. fluorescens WCS 374	Pseudobactin	Fusarium wilt in radish	By scavenging iron	Leeman et al. (1996)
P. fluorescens strain CHA0	2,4-Diacetyl fluroglucinol, HCN, pyoleuteorin, pyoveridine, pyrrolnitrin, and salicylate	Root rot of tobacco and tomato (*Thielaviopsis basicola*) and damping off of cucumber (*Pythium* spp.)	ISR	Weller (2007)
Pseudomonas fluorescens CHA0	2,4-Diacetyl fluroglucinol (Ph1)	*Meloidogyne javanica*: Rootknot nematode of tomato	ISR	Siddiqui and Shaukat (2003)
P. fluorescens Pf-5	Pyoleuteorin	Damping off of cotton seedlings *Pythium ultimum*	ISR	Howell and Stipanovic (1979)
P. putida (NH-50)	Pyoleuteorin	Red rot of sugarcane (*Colletotrichum falcatum*)	ISR	Hassan et al. (2011)
Pseudomonas aeruginosa PM389, ZNP1	IAA, cytokinin, and Exopolysaccharide	*Arabidopsis thaliana* grown under osmotic stress	Amelioration of osmotic stress	Ghosh et al. (2019)

P. cepacia B37W	Pyrrolnitrin	*Fusarium sambucinum* (causing potato dry rot), *F. nivale*, *F. graminearum*, *F. moniliforme*	Antibiosis	Burkhead et al. (1994)
P. fluorescens 2-79 and 30-84	Phenazine-1-carboxylic acid and anthranilic acid	*Gaeumannomyces graminis var. tritici* (take all of wheat)	Biofilm formation, ISN	Weller (2007)
P. fluorescens 30-84	Phenazines [phenazine-1-carboxylic acid, 2-hydroxyphenazine-1-carboxylic acid (2-OH-PCA) and 2-hydroxyphenazine (2-OH-PZ) as well] and HCN	*Gaeumannomyces graminis* var. *tritici* (take all of the wheat)	Biofilm formation, ISN	Weller (2007)
P. chlororaphis PA23	Pyrrolnitrin and HCN	*Caenorhabditis elegans*	Impairs egg laying rate and due to lethal paralysis (repellent)	Nandi et al. (2015)
P. chlororaphis PA23	Pyrrolnitrin	*Sclerotinia sclerotiorum* stem rot of canola	Antifungal	Nandi et al. (2015)
Pseudomonas fluorescens KD	Type III secretion system	*Pythium ultimum* damping-off of cucumber and sugar beet	Reduces the pectinase secreting ability of *P. ultimum*	Rezzonico et al. (2005)
P. aeruginosa	Glycine betaine and proline	Alleviation of drought stress in *Vigna radiata*	Stimulate antioxidant enzyme, osmolytes production and upregulating stress-responsive genes in plants.	Sarma et al. (2014)
P. fluorescens	Proline	Faba bean (*Vicia faba*) under salinity stress	Reducing the ROS level under salinity stress	Metwali et al. (2015)

Continued

Table 12.3 Continued

Microbes	Metabolites	Stress	Mode of action	References
P. mendocina	Proline	Lettuce grown under salt stress	The increase of antioxidant enzymes	Kohler et al. (2009)
P. aeruginosa PF23	Exopolysaccharides	Drought stress in sunflower seedlings (*Helianthus annuus*)	Activate antioxidant machinery of host plant	Sandhya et al. (2009)
P. aeruginosa PF23	Exopolysaccharides	Charcoal rot of sunflower (*Macrophomina phaseolina*)	Antifungal action	Tewari and Arora (2014a,b)
P. putida GAP-P45	Exopolysaccharides	Sunflower grown under drought conditions	Biofilm formation resulted in better soil aggregate formation and higher water content in leaves	Sandhya et al. (2009)
P. putida UW4	ACC deaminase	Enhanced growth of canola under low temperature and stress conditions	Inhibit accumulation of ethylene	Saleem et al. (2007)
P. putida H-2-3	Superoxide dismutase	Soyabean grown under salt stress	Reduce ROS and maintain cell turgidity	Kang et al. (2014)
P. stutzeri MBE05	Ascorbate peroxidise	Peanut seedlings grown under saline conditions	Reduced ROS	Sharma et al. (2016)
P. putida AKMP7	Catalase	Wheat grown under heat stress	Scavenges ROS	Shaik et al. (2011)

P. flurescens Mk 25	Auxin	Enhance salt tolerance Index in *Vigna radiate*	Enhance root growth and salt tolerant index	Ahmad et al. (2013)
P. putida Rs 198	Auxin	Cotton grown under salt stress	Increase germination and growth rate of cotton under salt stress	Yao et al. (2010)
P. aeruginosa T15	Auxin	Tomato grown under salinity conditions	Increased root and shoot length of tomato under high salinity	Tank and Saraf (2010)
P. fluorescens	Gibberellic acid	Raddish grown under salt stress	Increased fresh weight of root and leaves	Mohamed and Gomaa (2012)
P. putida H-2-3	Gibberellic acid	Soybean grown under salt stress	Increase shoot length, plant fresh weight and chlorophyll content of *Glycine* max	Kang et al. (2014)
Pseudomonas sp.	ACC deaminase	Drought tolerance to pea	Enhancing root length and water absorption efficiency by roots	Arshad et al. (2008)
P. fluorescens CHA0	HCN	Root rot of tobacco (*Thielaviopsis basicola*)	Blocking cytochrome oxidase pathway	Voisard et al. (1989)
P. fragi CS11RH1	IAA and HCN	Nutrient stress in wheat	Increase germination percentage of wheat seed and nutrient uptake	Selvakumar et al. (2009)

Table 12.4 Microbial secondary metabolites of different species of *Bacillus* and their role in plant stress management.

Microbes	Metabolites	Stress	Mode of action	References
Bacillus	Nitrogenase	Improved yield of maize (*Zea mays*) low Nitrogen stress	Nitogen fixation and supply to host plant	Kuan et al. (2016)
B. subtilis	Cytokinin	*Lactuca sativa*	Promote shoot development	Arkhipova et al. (2015)
B. subtilis CAS15	Trilactone (DHB-Gly-Thr)3 (siderophore)	*Fusarium* wilt of pepper	ISR and hindrance of chlamydospore germination of *Fusarium oxysporum* Schl. f.sp. capsici	Yu et al. (2011)
Bacillus sp. SC2b	Siderophores, ACC deaminase and IAA	Adaptation of *Sedum plumbizincicola* to metal toxicity in soil	Promoting accumulation of Cd and Zn in roots and shoots	Ma et al. (2015)
Bacillus sp. PZ-1	Hydroxamate-type siderophore	Phytoremediation of high lead containing soils by growing *Brassica juncea*	Enhances assimilation of lead by *B. juncea*	Yu et al. (2017)
B. megaterium Bm4C	Auxin and siderophore production	Protect Indian mustard under Ni toxicity in soil	Increase in shoot length and dry weight by enhanced uptake of minerals and iron	Mani and Helena (2008)

B. subtilis FZB24 and *B. subtilis* GB03	Surfactin	Black scurf of potato (*Rhizoctonia solani*) and Fusarium wilt	Cytoplasmic membrane disruption	Kong et al. (2010)
B. subtilis GB03	Surfactin	Damping off fungi, root rots caused by *Fusarium, Pythium, Phytophthora, Rhizoctonia* spp. of cole crops	Antifungal activity and ISR	Kong et al. (2010)
Bacillus amyloliquefaciens SYBC H47	Bacillomycin L, fengycin and surfactin (lipopeptides)	Protection against peach gummosis (*Botryosphaeria dothidea*)	Inhibit conidial germination	Li et al. (2016)
B. subtilis V26	Chitosanase	Against tomato fruit rot (*Botrytis cinerea*)	Antifungal activity	Kilani-Feki et al. (2016)
Bacillus amyloliquefaciens FZB42	Bacillomycin, fengimycin	*Fusarium oxysporum* wilt	ISR	Chowdhury et al. (2015)
Bacillus amyloliquefaciens FZB42	Bacillomycin D (lipopeptide)	Head blight of wheat (*Fusarium graminearum*)	Deformation of plasma membrane of hyphae and conidia by increasing reactive oxygen species	Gu et al. (2017)
Bacillus subtilis AU195	Bacillomycin D (lipopeptide)	*Aspergillus flavus*	Cytoplasmic membrane disruption	Moyne et al. (2001)

Continued

Table 12.4 Continued

Microbes	Metabolites	Stress	Mode of action	References
Bacillus endophyticus J13, *Bacillus tequilensis* J12,	IAA, cytokinin, and Exopolysaccharide	*Arabidopsis thaliana* grown under osmotic stress	Amelioration of osmotic stress	Ghosh et al. (2019)
Bacillus subtilis AH18 and *B. licheniformis* K11	Indole-3-acetic acid, indole-3-butyric acid, indole-3-propionic acid, abscisic acid, cytokinin and jasmonic acid, siderophore, β-glucanase and cellulose	Phytophthora blight of red pepper *Phytophthora capsici*	Promoting plant growth and cell wall hydrolysis of *Phytophthora capsici* and iron sequestration	Lim and Kim (2010)
Bacillus drentensis P16	Exopolysaccharides	Saline stress in Mung bean	Maintain nutrient and water availability to plants by forming biofilm	Mahmood et al. (2016)

Table 12.5 Microbial secondary metabolites of different microorganisms and their role in plant stress management.

Microbes	Metabolites	Stress	Mode of action	References
Rhizopus arrhizus	Raphorin (siderophore)	Combats iron deficiency in tomato	Increase in iron uptake by plants	Shenker et al. (1992)
Azotobacter chroococcum	Exopolysaccharide	Faba bean (*Vicia faba*)	Plant growth promotion	El-Ghany and Attia (2020)
Klebsiella oxytoca Rs-5	IAA	Salt stress in cotton	Plant growth promotion	Liu et al. (2013)
Enterobacter cloacae P6	Exopolysaccharides	Salt stress in Mung bean (*Vigna radiata*)	Maintain water and nutrient supply to plant roots by promoting biofilm formation	Mahmood et al. (2016)
Herbaspirillum seropedicae	SmR1 exopolysaccharide	Salt stress in maize, rice and sorghum	Biofilm formation	Balsanelli et al. (2014)
Citricoccus zhcaiensis B-4	ACC deaminase	Onion grown under drought stress	Stimulate IAA and GA_3 production in plants and inhibit lethal ethylene levels accumulation	Selvakumar et al. (2015)
Achromobacter piechaudii ARV8	ACC deaminase	Tomato and pepper to transient water stress	Inhibit lethal ethylene accumulation and promote plant growth	Mayak et al. (2004)
Burkholderia cepacia SE4	Gibberellic acid	*Cucumis sativus* grown under salinity stress conditions	Promote root and shoot growth parameters	Kang et al. (2014)
Sphingomonas sp. LK 11	Gibberellic acid	*Solanum lycopersicon* grown under oxidative stress in saline conditions	Increase in plant growth	Halo et al. (2015)
Arthrobacter protophormiae	ACC deaminase	Pea grown under salt stress	Reduces ACC oxidase activity and ethylene accumulation	Barnawal et al. (2014)
Mesorhizobium loti MP6	HCN	India mustard (*Brassica campestris*)	Blocks cytochrome oxidase pathway	Chandra et al. (2007)

Table 12.6 Microbial secondary metabolites of different species of *Streptomyces* and their role in plant stress management.

Microbes	Metabolites	Stress	Mode of action	References
Streptomyces strain AzR-051	IAA and siderophores	Early blight of tomato (*Alternaria alternate*)	Plant growth promotion and antifungal action	Verma et al. (2011)
Streptomyces CMU-PA101 and *Streptomyces* CMU-SK126	Antibiotics, IAA and siderophores	Phytopathogenic fungi: *Alternaria brassicicola, Colletotrichum gloeosporioides, Fusarium oxysporum, Penicillium digitatum*, and *Sclerotium rolfsii*	Antifungal activity	Khamna et al. (2009)
Streptomyces sp. SF5	Phytohormones	Salt stress in wheat (*Triticum durum*)	Improve germination	Hanane and Ghoul (2014)
Streptomyces sp.	IAA	Salt stress in wheat	Plant growth promotion	Sadeghi et al. (2012)
Streptomyces platensis	Volatile compounds phenylethyl alcohol and (+)-epi-bicyclesesquiphellandrene	Leaf blight/seedling blight of rice, leaf blight of oilseed rape and fruit rot of strawberry	Antifungal activity	Wan et al. (2008)
Streptomyces sp. PGPA 39		Stem rot of chickpea (*Sclerotinia sclerotiorum*) and salt stress	Triggered ET-mediated defense pathway in chickpea and stimulates phenylpropanoid pathway	Sathya et al. (2017)

Streptomyces acidiscabies E13	Desferridoxine E Desferridoxine B Coelichelin	Nickel stress in cowpea	Sequesters Ni and increases height and biomass	Sathya et al. (2017)
Streptomyces hygroscopicus TP-A045	Pteridic acids A and B		Induce adventitious roots formation in hypocotyls of kidney beans	Igarashi et al. (2002)

Table 12.7 Microbial secondary metabolites of different species of *Rhizobium* and *Bradyrhizobium* and their role in plant stress management.

Microbe	Metabolites	Stress	Mode of action	Reference
Rhizobium sp.	NodBC and NodBCL metabolites	Low nitrogen condition	NodB, NodC and NodA is involved in root nodulation	Spaink and Lugtenberg (1994)
Rhizobium sp. *Strain* YAS34	*Exopolysaccharides*	Sunflower grown under drought condition	Soil aggregate formation	Alami et al. (2000)
Rhizobium meliloti EFB 1	*Exopolysaccharides*	Alfa alfa (*Medicago sativa*) grown under salt stress	By inducing symbiosis and nodulation	Lloret et al. (1998)
Rhizobium leguminosarum	Cysteine rich peptide	*Brassica juncea* grown Zn contaminated medium	Reduction of Zn toxicity in root endodermis	Adediran et al. (2015)
Rhizobium sp. RP5	Phytohormones	Pea Grown in zinc and nickel Metal-Amended Soil	Nitrogen fixation, Growth promotion and reduction of metal toxicity	Wani et al. (2008)
Rhizobium tropici CIAT 899 with *Paenibacillus polymyxa* DSM36	IAA and cytokinin	*Phaseolus vulgaris* grown under stress condition in green house	Plant growth promotion	Figueiredo et al. (2008)
Rhizobium leguminosarum	IAA	canola and lettuce grown on stress condition	Plant growth condition	Noel et al. (1996)
Rhizobium sp.	phytohormones	Maize grown under salinity stress	Plant growth promotion	Bano and Fatima (2009)
Bradyrhizobium sp.	Exopolysaccharides	Cowpea	Desiccation tolerance	Hartel and Alexander (1986)
Bradyrhizobium sp.	Nitrogenase	Water stress on Cowpea	Plant growth promotion through nodulation	Fugyeuredi et al. (1999)

Table 12.8 Microbial secondary metabolites of different species of *Trichoderma* and their role in plant stress management.

Microbes	Metabolites	Stress	Mode of action	References
Trichoderma harzianum M 10	Harzianic acid (siderophore)	Foot rot of tomato (*Rhizoctonia solani*)	Induce expression of protease inhibitors, R protein like (CC-NBS-LRR) in tomato	Manganiello et al. (2018)
T. harzianum strain SQR-T037	Harzianolide	*Sclerotinia sclerotiorum*	Promote plant growth and antifungal activity and upregulate genes of SA pathway and JA/ET pathway	Cai et al. (2013)
T. koningii	Koninginin C	Take all disease of wheat (*Gaeumannomyces* var. *tritici*)	Promote plant growth and upregulate plant defense genes	Vinale et al. (2014)
T. koningii, *T. harzianum*, and *T. aureoviridae*	Koninginin D	*R. solini*, *Phytophthora cinnamomi*, *Fusarium oxysporum*, and *bipolaris sorokiniana*	Promote plant growth and upregulate plant defense genes	Vinale et al. (2014)
T. harzianum strain T22	Azaphilone	*Rhizoctonia solani*, *Pythium debaryana*, and *Gaeumannomyces* var. *tritici*	Promote plant growth and possess antibiotic activity	Keswani et al. (2014)
T. atroviride MUCL 45632	Hydroxamate and catechol-type siderophores	Stress in lettuce, melon, pepper, tomato, and zucchini	Plant growth promotion through improved plant nutritional status	Colla et al. (2015)

Among the microbial metabolites, the secondary metabolites have been proved to be of wide use in combating biotic and abiotic stress of plants. Various microbial secondary metabolites that promote the growth of plant under stress conditions have been discussed below.

12.2.1 Antibiotics

With the discovery of Penicillin by Alexander Fleming in 1929, and its mass production during World War II, the potential of microorganisms as mini and economic factories for natural manufacture of noble chemicals were identified. Since then, research has been done to explore diverse microorganisms for their potential to synthesize various noble antibiotics.

1. *Streptomycin*: Streptomycin is produced by *Streptomyces griseus*, was first discovered by Selman Waksman and coworkers and registered in the United States in 1959 (Rezzonico et al., 2005). It belongs to the aminoglycoside group of antibiotics. Formulations of Streptomycin sulfate or Streptomycin nitrate with the trade names of Agrimycin, Agristrep, Fructocin, orthoStreptomycin, and Plantomycin, are mainly applied against fire blight of pome fruits during pre-blossom, blossom, and post-blossom stages up to 30 days before harvest. But due to its broad application in the United States during the 1950s and 1960s led to the development of Streptomycin resistant strains of *Erwinia amylovora, Pseudomonas* sp. and *Xanthomonas* sp. (Manus et al., 2002). Streptomycin binds to the protein 30s subunit of the ribosome and causes the structural deformity, which in turn terminates the translation process in the target cell. Recommended concentrations for streptomycin application range from 50 to 200 ppm (50–200 μg/mL) (Vidaver, 2002).
2. *Oxytetracycline*: Oxytetracycline (Terramycin) was discovered by Alexander Finlay, is naturally produced by the soil bacterium *Streptomyces rimosus* and is highly thermostable in nature. It is formulated as oxytetracycline calcium complex or oxytetracycline hydrochloride and is marketed under the trade names Biostat, Mycoshield, and Terramycin. Oxytetracycline has bacteriostatic activity as it inhibits the multiplication of the bacterial cell by inhibiting the binding of aminoacyl-tRNA to the mRNA-ribosome complex. Oxytetracycline is also injected into the trunks of palm and elm disease to get rid of lethal yellow disease caused by a phytoplasma.
3. *Aureofungin*: Aureofungin was discovered by M. J. Thirumulachar, is naturally produced from *Streptoverticillium cinnamoneum var, terricola*. It belongs to the chemical group of polyene antibiotics and is translocated systemically in plants through xylem and phloem. Compared to other polyene antibiotics it is stable under UV radiations. It is effective against many classes of phytopathogens. It binds to the sterol containing fungal membranes and forms sterol-polyene complex, which in turn damages the cell membrane (Kataria and Thakkar, 1981). They are not effective against bacteria as the later mostly lacks sterol in cell wall and membrane.
4. *Validamycin*: It is produced from fermentation products of Streptomyces hygroscopicus var. *limoneous*, is a nonsystemic antibiotic highly effective against *Rhizoctonia solani* causing sheath blight of rice. It inhibits the activity of trehalase enzyme. The tips of the target fungal hyphae branch abnormally and restrict further growth. It is sold under the trade names of Falak, Valida-X, Paramycin, Solid, Dabang G and Sheathguard with the formulation of 3% L. (dose)
5. *Cycloheximide*: Cycloheximide is produced by *Streptomyces griseus,* have a systemic effect against fungal diseases of plants. It inhibits translation elongation through binding

to the *E*-site of the 60S ribosomal unit and interfering with deacetylated tRNA. It is sold under the trade name of Actidione, Actidione R2, and Actispray.

6. *Blasticidin S*: Blasticidin S is a highly selective fungicide which is produced by *Streptomyces griseochromogenes* (Fukunaga et al., 1996). It is highly effective against rice blast pathogen. It hinders the translation by inhibiting the peptide bond formation and inhibiting the termination step.
7. *Kasugamycin*: It is aminoglycosidic antibiotics obtained from *Streptomyces kasugaensis* effective against some fungal and bacterial pathogens. This antibiotic inhibits protein synthesis both in eukaryotic and prokaryotic cells at an initial step (Okuyama et al., 1971) without affecting translational elongation. It is effective against rice blast, bacterial leaf spot of chilli, tomato, angular leaf spot of cucumber, black rot of crucifers, citrus canker, etc.
8. *Antimycin*: They are the secondary metabolites produced mostly by *S. kitasawensis* and *S. griseus*. They inhibit the cellular respiration by arresting the oxidative phosphorylation.
9. *Thiolutin*: It is formed from the fermentation product of several strains of *Streptomyces lutesporeus* and inhibits bacterial and yeast RNA polymerase.
10. *Griseofulvin*: It is an antifungal antibiotics, named after the organism from which it was first isolated, *Penicillium griseofulvum*. It inhibits fungal mitotic cell division and nucleic acid synthesis.
11. *Avermectin*: They possess anti insecticidal and antihelminthic property. They are produced as fermentation products of *Streptomyces avermitilis*. William C Campbell and Santoshi Omura were awarded The Noble Prize in Medicine in the year 2015 for purifying avermectin proved to be effective against river blindness and elephantiasis. Avermectins block the neural transmission in nerve and muscle cells of insects by enhancing the effects of glutamate at the glutamate gated chloride channel. The hyperpolarization thus occurring as a consequence causes paralysis of neural muscle systems (Bloomquist, 1996).
12. *Milbemycin*: They are macrolides and were first isolated from *Streptomyces hygroscopicus*. Unlike avermectins, they increase the membrane permeability by binding to both glutamate and GABA activated chloride channels (Löscher et al., 2006). Use of various antibiotics in agriculture has been summarised in Table 12.1

12.2.2 1-Aminocyclopropane-1-carboxylate (ACC) deaminase

1-Aminocyclopropane-1-carboxylate (ACC) deaminase is a pyridoxal 5′ phosphate (PLP)-dependent hydrolase enzyme. ACC deaminase reduces the formation of ethylene in plant cells by cleavage of ACC (precursor of ethylene) to ammonia and α-ketobutyrate.

$$1-\text{Aminocyclopropane}-1-\text{carboxylate}+H_2O \underset{\text{ACC Deaminase}}{\rightleftharpoons} \alpha-\text{ketobutyrate}+NH_3$$

Ethylene production in plants can be induced under various stress conditions like metal toxicity, salinity, drought, extreme temperature conditions, UV radiation and damages by nematode, and plant pathogens. Ethylene is produced in two peaks and the second peak is much higher which is produced in responses to the stress conditions mentioned above. But the hike in ethylene concentration in plants under stress is responsible for early senescence, chlorosis and leaf abscission, lower yield, and plant death.

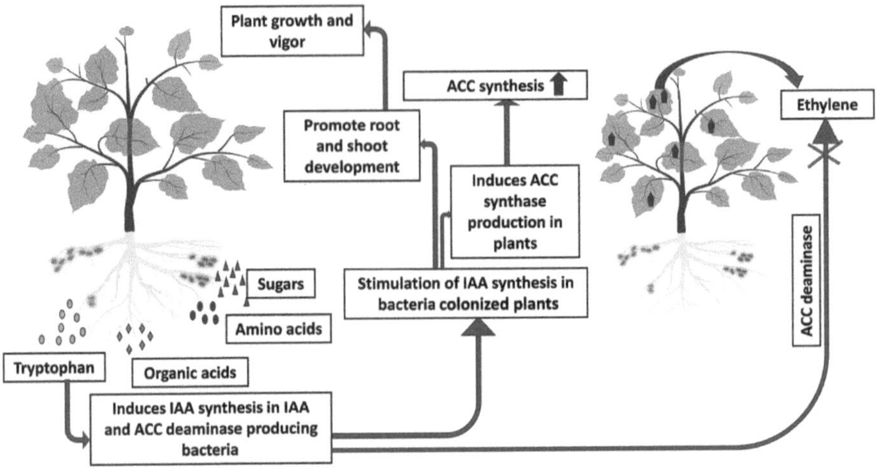

Fig. 12.1 A diagrammatic model representing the influence of rhizobacteria mediated plant growth promotion and stress mitigation through the production of 1-aminocyclopropane-1-carboxylate (ACC) deaminase and indole acetic acid (IAA) synthesis.

Among the PGPR, those can produce IAA together with ACC deaminase are of importance. These IAA and ACC deaminase producing beneficial bacteria may either colonize the roots of the plants or may choose to be endophyte, residing in the internal tissues of plants (Babu and Jogaiah, 2015). These bacteria uptake nutrients from the exudates of plant roots which are fairly composed of sugar, organic acids, amino acids, and small molecules like tryptophan (Fig. 12.1). Tryptophan helps in the induction of IAA synthesis in the bacteria. These IAA, in turn, stimulates the synthesis of IAA inside the host plant. The IAA produced by both bacteria and plant acts in two significant ways. Firstly they promote plant growth like cell elongation, root and shoot development. Secondly, the IAA induces ACC synthase production in plants. The ACC synthase induced inside the plants, in turn, synthesizes ACC (Fig. 12.1). Finally, a major portion of this ACC diffuses out through the seed, leaves, and roots and is cleaved by the ACC deaminase of bacteria (Glick, 2014). The microbial ACC deaminase enzyme is responsible for the cleavage of ACC, the plant ethylene precursor, into ammonia and α-ketobutyrate. ACC deaminase must play its role prior to induction of ACC oxidase as the later converts the ACC to ethylene. Hence, plant ethylene levels are reduced, which otherwise in high concentrations leads to plant growth inhibition or early senescence or even death (Gamalero and Glick, 2015).

12.2.3 Osmolytes

Drought, salinity, and phytopathogenic stress lead to the creation of reactive oxygen species (ROS) in plant cells (Haider et al., 2018, 2019). This ROS like H_2O_2 (hydrogen peroxide), superoxide (O_2^-), singlet oxygen ($^1O_2$), and hydroxyls ions (OH^-) causes membrane leakage of water, electrolytes, and molecular oxygen. The osmolytes are highly soluble, nontoxic small organic molecules that increase the water potential in cell and protect the cell from dehydration. They act as scavengers of ROS and as chemical chaperons by stabilizing the membrane and protein structure (Ashraf and Foolad, 2007). Osmolytes consists of three major groups: (i) amino acids (proline); (ii) quaternary amines (glycine betaine, alanine betaine, proline betaine, hydroxyprolinebetaine, choline-*O*-sulfate, and pipecolatebetaine); and (iii) sugars (trehalose and mannitol). Plants have been shown to uptake proline and sugars synthesized by the colonized rhizobacteria without any or minor chemical conversion (Schobert et al., 1998).

The ProJ-ProA-ProH biosynthetic pathway synthesizes proline under osmotic stress conditions in bacteria like *Bacillus subtilis* (Hoffmann et al., 2017). Inoculation with such rhizobacteria capable of increasing proline synthesis under salinity stress in turn stimulates the proline biosynthesis in plants by glutamate or ornithine pathway. Proline is actively uptaken by mitochondria using amino acid permease transporters which are expressed only under stress. Proline functions as an osmoprotectant by alleviating stress through (i) normalizing the ROS levels in plans by its antioxidant activity; (ii) hinders programmed cell death mediated by ROS; (iii) reduces ROS mediated efflux of K^+ under slat stress; (iv) maintains $NADP^+$-NADPH values compatible with metabolism (Hare and Cress, 1997); (v) increases the water potential in leaf cells under drought conditions; (vi) stabilizing the mitochondrial electron transport complex II under salt stress; (vii) prevents lipid peroxidation under metal stress; (viii) upregulates activities of enzymatic antioxidants like ascorbate peroxides, monohydrate ascorbate reductase, dihydro ascorbate reductase; and (ix) as chemical chaperons to stabilize protein structures and promote enzymatic activities against stress (Hayat et al., 2012) (Fig. 12.2). Pyrroline-5-Carboxylate (P_5C) is produced as an intermediate in proline anabolism and catabolism pathway. P_5C confer resistance to plant against bacterial phytopathogens like *Pseudomonas syringae* pv. *tomato*, by inducing SAR or hypersensitive response (Qamar et al., 2015).

Trehalose is xeroprotectant produced by some bacteria to fight against desiccation under drought conditions. Biosynthesis of trehalose includes the formation of trehalose-6-phospahte from UDP glucose and glucose-6-phosphate, which is finally converted to trehalose by the enzyme trehalose phosphate phosphatase. Trehalose protects the plants against stress like drought, cold, and high salinity as well as oomycete phytopathogens (Govind et al., 2016). They function by forming a gel phase upon dehydration of cells, prevent protein degradation, stabilizes membrane and protein structure. Timmusk et al. (2014) have named them as rhizobacterial drought tolerance enhancers (RDTE).

Glycine betaine (GB) is either synthesized or uptake of 13 biosynthetic precursors of glycine betaine choline can be by the bacteria like *Bacillus subtilis* under high osmolarity conditions chiefly by five major ABC transport uptake systems—OpuA, Opu B, Opu C, Opu D, and Opu E. GB can be synthesized by bacteria from the precursor choline by use of GbsB and GbsA GBsB enzymes (Holtmann and Bremer, 2004). GB has a role in protecting the thylakoid membrane of chloroplast and thereby maintaining the photosynthetic efficiency of plants. The ROS generated under drought, salinity, and other abiotic stress conditions inhibit the repair of the D1 protein, a constituent of PSII reaction center (Fig. 12.3).

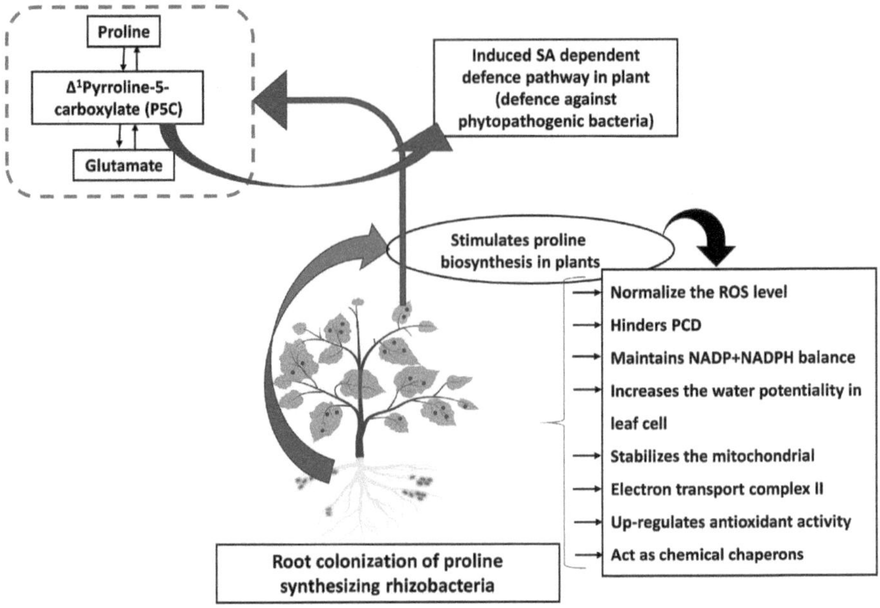

Fig. 12.2 Model representing influence of proline in bacteria colonized plant for stress mitigation.

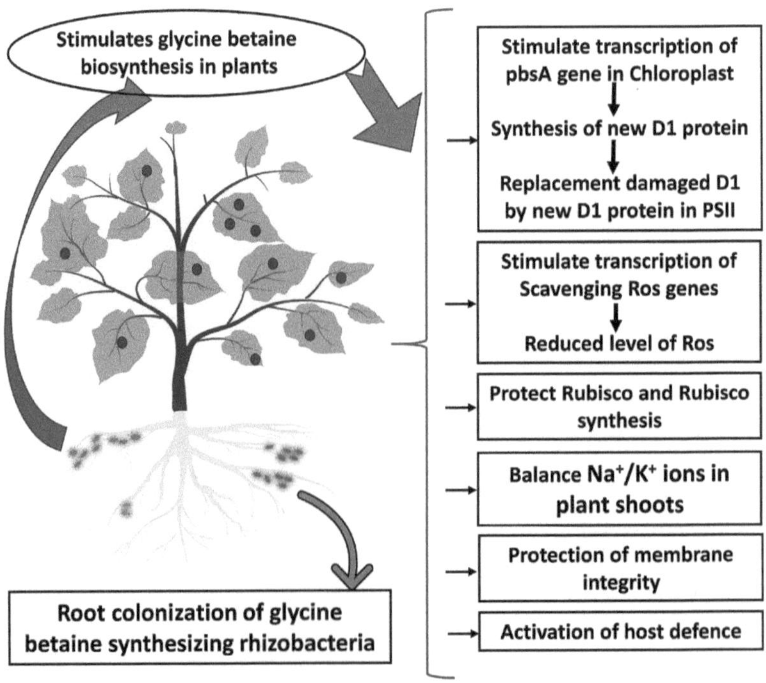

Fig. 12.3 Model representing influence of glycine betaine in bacteria colonized plant for stress mitigation.

GB induces the transcription of *psbA* gene in chloroplast which encodes for a new D1 protein and restores the function of PSII by replacing the damaged D1 protein (Aro et al., 1993). GB induces the transcription of ROS scavenging genes and reduces the level of ROS in a cell under stress. GB also maintains the photosynthetic efficiency under stress by protecting the enzymes Rubisco and Rubisco oxidase (Fariduddin et al., 2013). Further Glycine betaine increases the accumulation of K^+ ions and reduces Na^+ ions in shoots of crop plants under salt stress (Rahman et al., 2002; Abdelrahman et al., 2018). GB also helps in activation of host defence responses and thereby managing the biotic stresses (Lavanya and Amruthesh, 2017).

Cadaverine, a lysine catabolite has been shown to promote plant root growth by promoting cell division and combating against osmotic stress (Jancewicz et al., 2016).

12.2.4 Siderophores

Iron is vital for various metabolic processes like photosynthesis, DNA synthesis, respiration, nitrogen fixation, and is the component of many enzymes like cytochrome oxidase of the electron transport system. Iron is essential for the synthesis of chlorophyll and to maintain the integrity of chloroplast structure and function. In the photosynthetic apparatus, 2–3 iron atoms are associated with molecules of photosystem-II (PS-II), 12 atoms with photosystem-I (PS-I), five atoms with cytochrome complex, and two in ferredoxin molecule (Varatto et al., 2002). Although iron is the fourth most abundant element in lithosphere its bioavailability is limited as iron exists in the form of highly insoluble Fe^{3+} ions in neutral to alkaline soils (Zuo and Zhang, 2011). This makes iron the third most limiting nutrient essential for plant growth. Nutritional deficiency of iron in plants leads to interveinal chlorosis and stunted root growth which ultimately accounts for the reduction in yield parameters.

Plants can cope up with the iron stress deficiency by taking them from siderophore producing bacteria. Many fungi and bacteria have the innate property to produce siderophores under iron limited soil environmental conditions (Nagaraju et al., 2012; Babu and Jogaiah, 2015). The siderophore form complexes with the free iron and transports them into a cell with the help of membrane receptor molecules. Siderophores besides scavenging iron also have an affinity for binding with metal ions of Mo, Mn, Co, and Ni (Ahmed and Holmstrom, 2014). A diverse group of structurally and chemically distinguished siderophores are produced which differ in their capability of chelating iron. They can be broadly classified into three major families based on the ligand used to chelate the ferric ion. They are (i) hydroxamate, (ii) catechols, (iii) carboxylate type, and (iv) mixed type.

(i) *Hydroxamate siderophores*: These are the siderophores that possess the C(=O)N-(OH)R, hydroxamate functional group and are produced by both bacteria and fungi. They show very strong absorption of 425–500 nm by forming a hexadentate octahedral complex with Fe^{3+}. The hydroxamate siderophore-Fe^{3+} complex have 1:1 stability constant which is close to that found in Fe(III)-EDTA complex (Ahmed and Holmstrom, 2014). Hydroxamates include ferrichrome (*Ustilago ferrigona*), ferribactin produced by *Pseudomonas fluorescens* and vicibactin (*Rizobium leguminosarium bv. viciae*) which is a trihydroxamate and Desferridoxamine B (*Streptomyces* griseus).

(ii) *Catechol siderophores*: These types of siderophores are produced only by bacteria. They form the strongest hexadentate octahedral siderophore-Fe^{3+} complex. They possess mono or dihydroxybenzoic acid group for chelating iron. They include enterobactin (*Escherichia coli*), bacillibactin (*Bacillus subtilis*), amonabactin (*Aeromonas hydrophila*), vibriobactin (*Vibrio cholera*).

(iii) *Carboxylate siderophores*: These siderophores possesses hydroxyl or carboxyl as donor groups to iron. It includes rhizobactin containing three carboxyl and one hydroxyl group, produced by *Rhizobium melioliti* strain DM4 and staphyloferrin A, containing two citryl residues linked by amide bonds to D-ornithine, produced by *Staphylococcus hyicus* DSM20459 (Konetschny-Rapp et al., 1990). A carboxylate siderophore of fungal origin, rhizoferrin consisting of two molecules of citric acid linked to 1,4-diaminobutane through two amide bonds produced by *Rhizopus microsporus* var. *rhizopodiformis* (Drechsel et al., 1991).

(iv) *Mixed ligands*: These siderophores possesses multiple functional groups for chelating iron. They may be salicylate derivatives like pyoverdine and pyochelin (Fluorescent pseudomonas). Pyoverdine is a water-soluble siderophore with a dihydroxyquinoline fluorescent chromophores and a small dicarboxylic acid. Lamont and Martin (2003) grouped pyoverdins into three classes; Type I containing formyl-hydroxyornithines, Type II containing one formyl-hydroxyornithine and a terminal cyclized hydroxyornithine, and Type III containing two formyl-hydroxyornithines. Mycobactin is lipid-soluble siderophore produced by *Mycobacterium tuberculosis* include hydroxamate and phenol catechol functional groups.

Siderophore is synthesized via two biochemical pathways, viz., (i) nonribosomal protein (NRPs)-dependent and (ii) NRPs-independent pathway. In NRPs-dependent pathway the peptide bond is formed without the presence of an mRNA chain. Hydroxamate and catechol-type siderophores are synthesized by the NRPs-dependent pathway. While the NRPs-independent pathway linking dicarboxylic acid and diamine or amino alcohol building blocks with amide or ester bonds. Siderophores like enterobactin, yersiniabactin, pyochelin, pyoverdin, vibriobactin, and mycobactin are synthesized by NRPs-independent pathway. The siderophore synthesized is exported out from the bacterial cell by utilizing three main proteins, viz., (i) the major facilitator superfamily (MFS), (ii) the resistance, nodulation, and cell division (RND) superfamily, and (iii) the ABC superfamily. The MFS protein transports siderophores from cytoplasm to the periplasmic space. The siderophores are captured by RND proteins and guide them to export out through the outer membrane by ToIC protein. As the Fe^{3+} ions are hard acids while the siderophore donor oxygen is hard bases, which helps in the formation of Fe^{3+}-siderophore complex. The uptake of this Fe^{3+}-siderophore complex into the bacterial cell is mediated by outer membrane receptor proteins (OMP). These OMP are very specific to particular type of siderophores *viz.* FepA for enterobactin, FhuA for ferrichrome and FecA for citrate siderophores. The OMP possess β barrel structure with plugs facing towards outer and inner side of the outer membrane of Gram negative bacteria. These TonB protein complex (TonB-ExbB-ExbD) helps both the OMP proteins to undergo conformational change to facilitate the Fe^{3+}-siderophore complex transport and transduces the proton motive force for internalization of the Fe^{3+}-siderophore complex into the periplasmic space. Finally the proteins of periplasmic space (FepB for enterobactin, FhuD for ferrichrome and

FecB for citrate) and ATP dependent ABC transporter system (Fep D,G,C for enterobactin; FhuB and Fhc for ferrichrome; Fec C,D,E for citrate) present in the cytoplasmic membrane intakes the Fe^{3+}-siderophore complex into the bacterial cytoplasm (Shanmugaiah et al., 2015). The Fe^{3+} siderophore complex reduces to Fe^{2+} to release iron in the cell cytosol in case of hydroxamates and carboxylates. This scavenging of iron by siderophore benefits the plants both directly and indirectly. Iron also froms an integral structural part of virulence determining factors of this soil borne phytopathogens. Hence, scavenging of iron by siderophore producing root colonizing rhizobacteria make the phytopathogenic bacteria and fungi starve for iron and in turn protects the host plant from infection. For instance many fluroscent pseudomonads are reported to competitively suppress *Fusarium oxysourum, Gaeumannomyces tritici*, and *Pythium* sp. (Haas and Defago, 2005; Leeman et al., 1996; Weller, 2007). While competitive colonization of rhizobacteria producing siderophores promotes the plant directly by various approaches like nirogen fixation, phytohormone secretions and wide range of nutrient uptakes.

12.2.5 Exopolysaccharides

Bacteria secrets natural carbohydrate polymers called exopolysaccharides (EPS). Bacterial exopolysaccharides belong to three major classes: (i) internal storage polysaccharides (glycogen), (ii) capsular polysaccharides, and (iii) extracellular bacterial polysaccharides (xanthan, alginate, cellulose, etc.). The EPS are biosynthesized following four pathways: (i) WZX/Wzy-dependent pathway, (ii) ATP binding ABC transporter-dependent pathway, (iii) synthase dependent pathway, and (iv) extracellular synthesis using single sucrose protein (Schmid et al., 2015).

These EPS helps in the formation and maintenance of the structural and functional integrity of biofilm. Generally, bacterial cells initiate biofilm formation by coming close to each other and attaching to root surfaces or above plant part surface (leaves). The bacteria then form microcolonies by the cell to cell attachment with the exopolysaccharides secreted. The biofilm prevents desiccation of the bacterial cell under drought and salinity stress. EPS promotes association of plant roots with the rhizospheric bacteria, which is termed as rhizosheath. The rhizosheath is the site of nutrient recycling, balancing of different ions, making water and nutrient available to plant, increasing nodulation in legumes, and maintaining a symbiotic association of rhizospheric bacteria and roots. EPS helps in improving soil aggregate formation which helps to reduce evaporation losses of water under high temperature and drought conditions. Under saline conditions, EPS protects the root nodules from oxidative stress and enhances the efficiency of nitrogenase enzyme to fix N_2 (Fig. 12.4). EPS reduces the accumulation of Na^+ ions in root cells and also prevent the efflux of K^+ ions under salt stress (Mishra et al., 2018).

12.2.6 Phytohormones

Phytohormones like auxin, gibberellic acid, and ethylene promote various plant development stages right from seed germination to flowering and fruit setting. They help the plants to combat against various biotic and abiotic stresses by up and

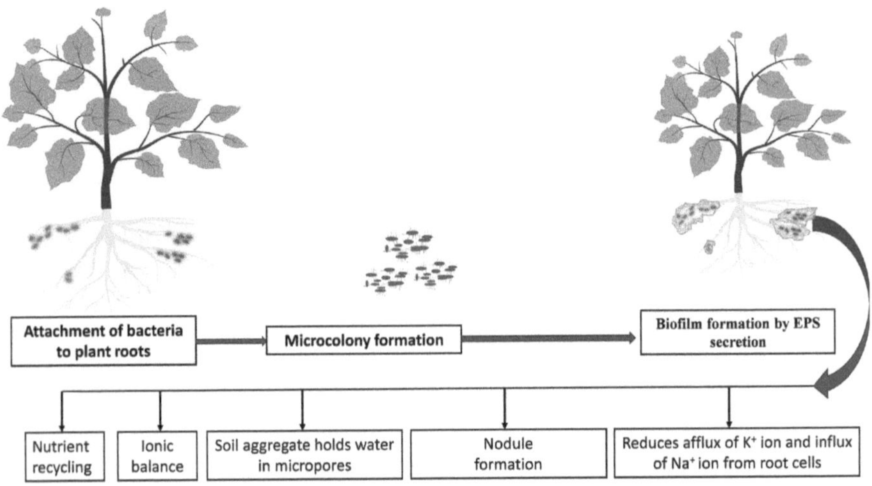

Fig. 12.4 A model representing the role of bacterial exopolysaccharides in plant microbe interaction and stress tolerance in plants.

downregulation of various genes in plants. The PGPR secretes phytohormones as secondary metabolites are either taken up by plants which in turn modulate the hormonal levels in plants and also a directly provide defense to plants against various phytopathogens (Jogaiah et al., 2010).

12.2.6.1 Auxin

Auxin has been known to promote plant growth by (i) lateral and adventitious root development, (ii) increasing absorption of nutrients by increasing root surface area and length, (iii) root cells elongation which permits more root exudation which in turn increase the beneficial microorganism in the rhizosphere, (iv) assist plants response to light and fluorescent, (v) affects photosynthesis, and (vi) helps in crucial metabolite synthesis and upregulate defense genes against phytopathogens.

Auxin is produced as secondary metabolites by over 80% of soil microbes (Patten and Glick, 1996). The production efficiency has been shown to vary among different genus of soil microorganisms such as *Pseudomonas* sp. (94%), *Azospirillum* sp. (80%), *Azotobacter* sp. (65%), and *Bacillus* sp. (40%). Different studies have proved the increase in endogenous IAA of the plant after uptake of IAA secreted by rhizosphere residing microorganisms. So far six auxin biosynthesis pathways have been identified in bacteria. In the most common indole-3-pyruvate pathway the compound chorismate generated from phosphoenolpyruvate and erythrose-4-phosphate gets converted into anthranilate. Tryptophan generated from anthranilate is acted upon by various enzymes to finally yield indole-3-acetic acid. Some of the bacteria following indole-3-pyruvate pathway are *Pseudomonas syringae, P. putida, Azospirillum brasilense, Enterobacter cloaceae, Pantoea agglomerans, Rhizobium* sp., *Bradyrhizobum* sp., *Paenibacillus polymyxa, Rhodococcus fascians* and *Arthrobacter pascens*. In case of *Bacillus cereus* and *A. brasilense* the trypamine pathway is followed where auxin is formed with intermediates like tryptamine and indole-3-acetaldehyde. Majority of phytopathogenic bacteria like *Agrobacterium tumefaciens*, *P. syringae*, *P. agglomerans*, *Rhizobium*, and *Bradyrhizobium* synthesizes auxin by formation of indole-3-acetamide pathway. Again, in Indole-3-acetonitrile pathway, *Agrobacterium tumefaciens* and *Rhizobium* sp. utilize anthranilate to synthesize IAA by formation of acetonitrile as an intermediate. While the tryptophan sidechain oxidase pathway have been reported in *P. syringae* CHAO and *P. fluorescens* HP72 (Ahmad et al., 2013; Malhotra and Srivastava, 2009; Patten and Glick, 1996).So far six auxin biosynthesis pathways have been identified in bacteria. In the most common indole-3-pyruvate pathway the compound chorismate generated from phosphoenolpyruvate and erythrose-4-phosphate gets converted into anthranilate. Tryptophan generated from anthranilate is acted upon by various enzymes to finally yield indole-3-acetic acid. Some of the bacteria following indole-3-pyruvate pathway are *Pseudomonas syringae*, *P. putida*, *Azospirillum brasilense*, *Enterobacter cloaceae*, *Pantoea agglomerans*, *Rhizobium* sp., *Bradyrhizobum* sp., *Paenibacillus polymyxa*, *Rhodococcus fascians* and *Arthrobacter pascens*. In case of *Bacillus cereus* and *A. brasilense* the trypamine pathway is followed where auxin is formed with intermediates like tryptamine and indole-3-acetaldehyde. Majority of phytopathogenic bacteria like *Agrobacterium tumefaciens*, *P. syringae*, *P. agglomerans*, *Rhizobium*, and *Bradyrhizobium* synthesizes auxin by formation of indole-3-acetamide pathway. Again, in Indole-3-acetonitrile pathway, *Agrobacterium tumefaciens* and *Rhizobium* sp. utilize anthranilate to synthesize IAA by formation of acetonitrile as an intermediate. While the tryptophan sidechain oxidase pathway have been reported in *P. syringae* CHAO and *P. fluorescens* HP72 (Ahmad et al., 2013; Malhotra and Srivastava, 2009; Patten and Glick, 1996).

12.2.6.2 Gibberellic acid

The discovery of gibberellic acid is well known to be associated with the foolish seedling disease of rice caused by the fungi *Gibberella fujikuroi*. The symptom is primarily observed in the seedling stage where some of the seedlings become unnecessarily tall because of intermodal elongation of the stem. They promote stem elongation, germination of seeds by breaking dormancy and fruit senescence (Elezar and Escamilla, 2000). Many of the soil bacteria are known to produce Gibberellic acid as secondary metabolite which modulates the gibberellin level in the plant after its uptake.

12.2.6.3 Cytokinin

Cytokinin produced as secondary metabolites by PGPR plays an essential role in cell division in plant roots and shoots, growth, and differentiation of tissues. Cytokinins delay the senescence, aging and counter the apical dominance activity of auxin.

12.2.7 HCN

Hydrogen cyanide is produced by 88.89% of species of *Pseudomonas* and 50% of *Bacillus* in the rhizosphere. HCN produced by the rhizobacteria acts as a antimicrobial metabolite for weed, insects, and many soilborne phytopathogens. HCN effectively inhibits the root growth of weeds (Heydari et al., 2008). HCN functions by blocking cytochrome oxidase pathway.

12.3 Conclusion

The microbial metabolites help plants to withstand various biotic and abiotic stress by antibiosis of soilborne phytopathogens, iron sequestration that promotes plant growth and suppress phytopathogens activity, scavenges the reactive oxygen species that otherwise causes membrane permeability under oxidative stress, inhibits the accumulation of lethal ethylene levels and prevents abscission, upregulate genes related to salicylic acid dependent pathway, promote various growth parameters of the plant by modulating plant hormone levels, increases the stability of soil aggregates in retaining water under drought stress and increases nutrient alongwith water uptake by plants.

References

Abdelrahman, M., Jogaiah, S., Burritt, D.J., Tran, L.P., 2018. Legume genetic resources and transcriptome dynamics under abiotic stress conditions. Plant Cell Environ. 41, 1949–2225.

Adediran, G.A., Ngwenya, B.T., Mosselmans, J.F.W., Heal, K.V., 2015. Bacteria–zinc co-localization implicates enhanced synthesis of cysteine-rich peptides in zinc detoxification when *Brassica juncea* is inoculated with *Rhizobium leguminosarum*. New. Phytol. 209, 280–293.

Ahmad, M., Zahir, Z.A., Nazli, F., Akram, F., Arshad, M., Khalid, M., 2013. Effectiveness of halo-tolerant, auxin producing Pseudomonas and Rhizobium strains to improve osmotic stress tolerance in mung bean (Vigna radiata L.). Braz. J. Microbiol. 44, 1341–1348.

Ahmed, E., Holmstrom, S.J., 2014. Siderophores in environmental research: roles and application applied microbiology. J. Microbial. Biotechnol. 7, 196–208.

Alami, Y., Achouak, W., Marol, C., Heulin, T., 2000. Rhizosphere soil aggregation and plant growth-promotion of sunflowers by an exopolysaccharide-producing *Rhizobium* sp. strain isolated from sunflower roots. Appl. Environ. Microbiol. 66, 3393–3398.

Anderson, H.W., Gottlieb, D., 1952. Plant disease control with antibioitics. Econ. Bot. 6, 294.

Arkhipova, T.N., Vysotskaya, L.B., Martinenko, E.V., Ivanov, I.I., Kudoyarova, G.R., 2015. Participation of cytokinins in plant response to competitors. Russ. J. Plant Physiol. 62 (4), 524–533.

Aro, E.M., Virgin, I., Andersson, B., 1993. Photoinhibition of photosystem II: inactivation, protein damage and turnover. Biochim. Biophys. Acta 1143 (2), 113–134.

Arshad, M., Shaharoona, B., Mahmood, T., 2008. Inoculation with *Pseudomonas* spp. containing ACC-deaminase partially eliminates the effects of drought stress on growth, yield, and ripening of pea (*Pisum sativum* L.). Pedosphere 18 (5), 611–620.

Ashraf, M., Foolad, M., 2007. Roles of glycine betaine and proline in improving plant abiotic stress resistance. Environ. Exp. Bot. 59 (2), 206–216.

Ashraf, M., Harris, J.C., 2004. Potential biochemical indicators of salinity tolerance in plants. Plant. Sci.. 166, 33-16.

Babu, N.A., Jogaiah, S., Ito, S.-I., Nagraj, K.A., Tran, L.-S.P., 2015. Improvement of growth, fruit weight and early blight disease protection of tomato plants by rhizosphere bacteria is correlated with their beneficial traits and induced biosynthesis of antioxidant peroxidase and polyphenol oxidase. Plant Sci. 231, 62–73.

Balsanelli, E., de Baura, V.A., FDO, P., de Souza, E.M., Monteiro, R.A., 2014. Exopolysaccharide biosynthesis enables mature biofilm formation on abiotic surfaces by*Herbaspirillum seropedicae*. PLoS One 9 (10), e110392.

Bano, A., Fatima, M., 2009. Salt tolerance in *Zea mays* (L). following inoculation with Rhizobium and Pseudomonas. Biol. Fert. Soils 45 (4), 405–413.

Barnawal, D., Bhartia, N., Majia, D., Chanotiyab, S.C, Kalraa, A., 2014. ACC deaminase-containing *Arthrobacter protophormiae* induces NaCl stress tolerance 2 through reduced ACC oxidase activity and ethylene production resulting in improved 3 nodulation and mycorrhization in *Pisum sativum*. J. Plant. Physiol. 171(11).

Bashan, Y., De-Bashan, L.E., 2002. Protection of tomato seedlings against infection by Pseudomonas syringae pv. tomato by using the plant growth-promoting bacterium *Azospirillum brasilense*. Appl. Environ. Microbiol. 68 (6), 2637–2643.

Bloomquist, J.R., 1996. Ion channels as targets for insecticides. Annu. Rev. Entomol. 41, 163–190.

Bonde, R., 1953. Studies on the control of Late blight of potato (*Phytophthoira infestans*) with antibiotics. Phytopathology 43, 463–464.

Bottini, R., Cassan, A.F., Piccoli, P., 2004. Gibberellin production bacteria and its involvement in plant growth promotion and yield increase. Appl. Microbiol. Biotech. 65 (5), 497–503.

Burkhead, K.D., Schisler, A.D., Slininger, J.P., 1994. Pyrrolnitrin production by biological control agent *Pseudomonas cepacia* B37w in culture and in colonized wounds of potatoes. Appl. Environ. Microbiol. 60 (6), 2031–2039.

Cai, F., Yu, G., Wang, P., Wei, Z., Fu, L., Shen, Q., Chen, W., 2013. Harzianolide, a novel plant growth regulator and systemic resistance elicitor from *Trichoderma harzianum*. Plant Physiol. Biochem. 73, 106–113.

Cassán, F., Maiale, S., Masciarelli, O., Vidal, A., Luna, V., Ruiz, O., 2009. Cadaverine production by *Azospirillum brasilense* and its possible role in plant growth promotion and osmotic stress mitigation. Eur. J. Soil Biol. 45, 12–19.

Chandra, S., Kamlesh, C., Ramesh, C.D., Dinesh, K.M., 2007. Rhizosphere competent *Mesorhizobiumloti* MP6 induces root hair curling, inhibits *Sclerotinia sclerotiorum* and enhances growth of Indian mustard (*Brassica campestris*). Braz. J. Microbiol. 38 (1), 124–130.

Chowdhury, S.P., Hartmann, A., Gao, X.W., Borriss, R., 2015. Biocontrol mechanism by root-associated*Bacillus amyloliquefaciens* FZB42—a review. Front. Microbiol. 6, 780.

Cohen, A.C., Bottini, R., Piccoli, P.N., 2008. *Azospirillum brasilense* Sp. 245 produces ABA in chemically-defned culture medium and increases ABA content in Arabidopsis plants. Plant Growth Regul. 54, 97–103.

Colla, G., Rouphael, Y., Di Mattia, E., El-Nakhel, C., Cardarelli, M., 2015. Co-inoculation of Glomus intraradices and *Trichoderma atroviride* acts as a biostimulant to promote growth, yield and nutrient uptake of vegetable crops. J. Sci. Food. Agric. 95 (8), 1706–1715.

Damalas, C.A., Spyridon, D.K., 2018. Current status and recent developments in biopesticide use. Agriculture. 8(13).

Drechsel, H., Metzger, J., Freund, S., Jung, G., Boelaert, J.R., Winkelmann, G., 1991. Rhizoferrin - a novel siderophore from the fungus *Rhizopus microsporus* var. *rhizopodiformis*. BioMetals 4, 238–243.

Elezar, M., Escamilla, S., 2000. Optimization of GA3 production by immobilized *G. fujikuroi* mycelium in fluidized bioreactors. J. Biotechnol. 76, 147.

El-Ghany, M.F.A., Attia, M., 2020. Effect of exopolysaccharide-producing bacteria and melatonin on faba bean production in saline and non-saline soil. Agronomy 10, 316.

Fariduddin, Q., Varshney, P., Yusuf, M., Ali, A., Ahmad, A., 2013. Dissecting the role of glycine betaine in plants under abiotic stress. Plant Sress 7 (1), 8–18.

Figueiredo, M.V.B., Burity, H.A., Martinez, C.R., Chanway, C.P., 2008. Alleviation of drought stress in common bean (*Phaseolus vulgaris* L.) by coinoculation with *Paenibacillus polymyxa* and *Rhizobium tropici*. Appl. Soil Ecol. 40, 182–188.

Fugyeuredi, M.V.B., Vilhar, J.J., Burity, H.A., de Franca, F.P., 1999. Alleviation of water stress effects in cowpea by Bradyrhizobium spp. inoculation. Plant Soil 207, 67–75.

Fukami, J., Ollero, F.J., Megías, M., Hungria, M., 2017. Phytohormones and induction of plant-stress tolerance and defense genes by seed and foliar inoculation with *Azospirillum brasilense* cells and metabolites promote maize growth. AMB Exp. 7, 153.

Fukami, J., Cerezini, P., Hungria, M., 2018. Azospirillum: benefits that go far beyond biological nitrogen fixation. AMB Exp. 8, 73.

Fukunaga, K., Misato, T., Ishii, I., Asakawa, M., 1996. Blasticidin, a new anti-phytopathogenic fungal substance. Part I. Bull. Chem. Soc. Jpn. 19 (3), 181–188.

Fulchieri, M., Lucangeli, C., Bottini, R., 1993. Inoculation withAzospirillum lipoferum affects growth and gibberellin status ofcorn seedling roots. Plant Cell Physiol 34, 1305–1309.

Gamalero, E., Glick, B.R., 2015. Bacterial modulation of plant ethylene levels. Plant Physiol 169, 13–22.

Ghosh, D., Gupta, A., Mohapatra, S., 2019. A comparative analysis of exopolysaccharide and phytohormone secretions by four drought-tolerant rhizobacterial strains and their impact on osmotic-stress mitigation in *Arabidopsis thaliana*. World. J. Microbiol. Biotechnol. 35, 90.

Glick, B.R., 2014. Bacteria with ACC deaminase can promote plant growth and help to feed the world. Microbiol. Res. 169, 30–39.

Govind, S.R., Jogaiah, S., Abdelrahman, M., Shetty, H.S., Tran, L.-S.P., 2016. Exogenous trehalose treatment enhances the activities of defense-related enzymes and triggers resistance against downy mildew disease of pearl millet. Front. Plant Sci. 7, 1593. https://doi.org/10.3389/fpls.2016.01593.

Gu, Q., Yang, Y., Yuan, Q., Shi, G., Wu, L., Lou, Z., Huo, R., Wu, H., Borriss, R., Gao, X., 2017. Bacillomycin D produced by *Bacillus amyloliquefaciens* is involved in the antagonistic interaction with the plant-pathogenic fungus *Fusarium graminearum*. Appl. Environ. Microbiol.. 83(19), e01075-17.

Haas, D., Defago, G., 2005. Biological control of soilborne pathogens by fluorescent pseudomonads. Nat. Rev. Microbiol. 3 (4), 307.

Haider, M.S., Kurjogi, M., Fiaz, M., Jogaiah, S., Wang, C., Fang, J., 2018. Drought stress revealed physiological, biochemical and gene-expressional variations in 'Yoshihime' peach (*Prunus persica* L) cultivar. J. Plant Interact. 13, 83–90.

Haider, M.S., Jogaiah, S., Pervaiz, T., Yanxue, Z., Khan, N., Fang, J., 2019. Physiological and transcriptional variations inducing complex adaptive mechanisms in 2 grapevine by salt stress. Environ. Exp. Biol. 162, 455–467.

Halo, A.B., Khan, L.A., Waqas, M., AlHarrasi, A., Hussain, J., Ali, L., Adnan, M., Lee, J.I., 2015. Endophytic bacteria (*Sphingomonas* sp. LK11) and gibberellin can improve *Solanum lycopersicum* growth and oxidative stress under salinity. J. Plant Interact. 10 (1), 117–125.

Hanane, A., Ghoul, M., 2014. Effect of salinity stress, *Streptomyces* sp. SF5 and *Salsola vermiculata* on germination of *Triticum durum* L. Sky J. Agric. Res. 3, 7–16.

Hare, P.D., Cress, W.A., 1997. Metabolic implications of stress-induced proline accumulation in plants. Plant Growth Regul. 21 (2), 79–102.

Hartel, P.G., Alexander, M., 1986. Role of extracellular polysaccharide production and clays in the desiccation tolerance of cowpea Bradyrhizobia. Soil Sci. Soc. Am. J. 50, 1193–1198.

Hassan, M.N., Afghan, S., Hafeeza, F.Y., 2011. Biological control of red rot in sugar cane by native pyoluteorin-producing *Pseudomonas putida* strain NH-50 under field conditions and its potential modes of action. Pest Manag. Sci. 67, 1147–1154.

Hayat, S., Hayat, Q., Alyemeni, M.N., Wani, A.S., Pichtel, J., Ahmad, A., 2012. Role of proline under changing environments. Plant Signal. Behav. 7 (11), 1456–1466.

Heydari, S., Moghaddam, P.R., Arab, S.M., 2008. Hydrogen Cyanide Production Ability by *Pseudomonas fluorescence* Bacteria and their Inhibition Potential on Weed Germination. Ferdowsi University, Mashhad, Iran, Agronomy, Iran, Faculty of Agriculture.

Hoffmann, T., Bleisteiner, M., Sappa, P.K., Steil, L., Mäder, U., Völker, U., Bremer, E., 2017. Synthesis of the compatible solute proline by *Bacillus subtilis*: point mutations rendering the osmotically controlled proHJ promoter hyperactive. Environ. Microbiol. 19 (9), 3700–3720.

Holtmann, G., Bremer, E., 2004. Thermoprotection of Bacillus subtilis by exogenously provided Glycine Betaine and structurally related compatible solutes: involvement of Opu transporters. J. Bacteriol. 186 (6), 1683–1693.

Howell, C.R., Stipanovic, R.D., 1979. Control of *Rhizoctonia solani* on cotton seedlings with *Pseudomonas fluorescens* and with an antibiotic produced by the bacterium. Phytopathology 69, 480–482.

Igarashi, Y., Iida, T., Yoshida, R., Furuma, T., 2002. Pteridic acids a and B, novel plant growth promoters with auxin-like activity from *Streptomyces hygroscopicus* TP-A0451. J. Antibiot. 55 (8), 764–767.

Jancewicz, A.L., Gibbs, N.M., Masson, P.H., 2016. Cadaverine's functional role in plant development and environmental response. Front. Plant Sci. 7, 870.

Jogaiah, S., Roopa, K.S., Pushpalatha, H.G., Shekar Shetty, H., 2010. Evaluation of plant growth-promoting rhizobacteria for their efficiency to promote growth and induce systemic resistance in pearl millet against downy mildew disease. Arch. Phytopathol. Plant Prot. 43, 368–378.

Jogaiah, S., Mostafa, A., Tran, L.S.P., Shin-ichi, I., 2013. Characterization of rhizosphere fungi that mediate resistance in tomato against bacterial wilt disease. J. Exp. Bot. 64, 3829–3842.

Kang, M.S., Radhakrishnan, R., Khan, L.A., Kim, J.M., Kim, J.M., Park, J.M., Kim, R.B., Shin, H.D., Lee, J.I., 2014. Gibberellin secreting rhizobacterium, Pseudomonas putida H-2-3 modulates the hormonal and stress physiology of soybean to improve the plant growth under saline and drought conditions. Plant Physiol. Biochem. 84, 115–124.

Kataria, H.R., Thakkar, R.K., 1981. Mechanism of antifungal action of Aureofungin against *Saccharomyces cerevisiae*. Biochem. Physiol. Pflanzen 176, 784–788.

Keswani, C., Mishra, S., Sarma, B.K., Singh, S.P., Singh, H.B., 2014. Unraveling the efficient applications of secondary metabolites of various *Trichoderma spp*. Appl. Microbiol. Biotechnol. 98, 533–544.

Khamna, S., Yokota, A., Lumyong, S., 2009. Actinomycetes isolated from medicinal plant rhizosphere soils: diversity and screening of antifungal compounds, indole-3-acetic acid and siderophore production. World J. Microbiol. Biotechnol. 25, 649.

Kilani-Feki, O., Ben, Khedher S., Dammak, M., Kamoun, A., Jabnoun-Khiareddine, H., Daami-Remadi, M., Touns, S., 2016. Improvement of antifungal metabolites production by *Bacillus subtilis* V26 for biocontrol of tomato postharvest disease. Biol. Control. 95, 73–82.

Kohler, J., Hernández, J.A., Caravaca, F., Roldána, A., 2009. Induction of antioxidant enzymes is involved in the greater effectiveness of a PGPR versus AM fungi with respect to increasing the tolerance of lettuce to severe salt stress. Environ. Exp. Bot. 65, 245–252.

Konetschny-Rapp, S., Jung, G., Meiwes, J., Zähner, H., 1990. Staphyloferrin A: a structurally new siderophore from staphylococci. Eur. J. Biochem. 191 (1), 65–74.

Kong, A.H.G., Kim, J.C., Choi, G.J., Lee, K.Y., Kim, H.J., Hwang, E.C., Moon, B.J., Lee, S.W., 2010. Production of surfactin and iturin by *Bacillus licheniformis* N1 responsible for plant disease control. Plant Pathol. J. 26 (2), 170–177.

Kuan, K.B., Othman, R., Rahim, K.A., Shamsuddin, Z.H., 2016. Plant growth-promoting Rhizobacteria inoculation to enhance vegetative growth, nitrogen fixation and nitrogen remobilisation of Maizeunder greenhouse conditions. PLos One 10, 1371.

Lamont, I.L., Martin, L.W., 2003. Identification and characterization of novel pyoverdine synthesis genes in *Pseudomonas aeruginosa*. Microbiology 149, 833–842.

Lavanya, N.S., Amruthesh, N.K., 2017. Glycine betaine mediated disease resistance against *Sclerospora graminicola* in Pearl Millet. J. Appl. Biol. Biotechnol. 5 (3), 45–51.

Leeman, M., den Ouden, F.M., vanPelt, J.A., Dirkx, F.P.M., Steijl, H., Bakker, P.A.H.M., Schippers, B., 1996. Iron availability affects induction of systemic resistance to Fusarium wilt of radish by Pseudomonas fluorescens. Phytopathology 86, 149–155.

Li, X., Zhang, Y., Wei, Z., Guan, Z., Cai, Y., Liao, X., 2016. Antifungal activity of isolated *Bacillus amyloliquefaciens* SYBC H47 for the biocontrol of peach gummosis. PLoS One. 11(9).

Lim, J.H., Kim, S.D., 2010. Biocontrol of Phytophthora blight of red pepper caused by Phytophthora capsici using Bacillus subtilis AH18 and *B. licheniformis* K11 formulations. J. Korean Soc. Appl. Biol. Chem. 53 (6), 766–773.

Liu, Y., Shi, Z., Yao, L., Yue, H., Li, H., Li, C., 2013. Effect of IAA produced by *Klebsiella oxytoca* Rs-5 on cotton growth under salt stress. J. Gen. Appl. Microbiol. 59 (1), 59–65.

Lloret, J., Wulff, H.B.B., Rubio, M.J., Downie, A.J., Bonilla, I., Rivilla, R., 1998. Exopolysaccharide II Production Is Regulated by Salt in the Halotolerant Strain*Rhizobium meliloti* EFB1. Appl. Environ. Microbiol. 64 (3), 1024–1028.

Long, S.P., Ort, D.R., 2010. More than taking the heat: crops and global change. Curr. Opin. Plant Biol. 13 (3), 241–248.

Löscher, W., Schirmer, M., Freichel, C., Gernert, M., 2006. Distribution of GABAergic neurons in the striatum of amygdala-kindled rats: an immunohistochemical and in situ hybridization study. Brain Res. 1083, 50–60.

Ma, Y., Oliveira, S.R., Wu, L., Luo, Y., Rajkumar, M., Rocha, I., Freitas, H., 2015. Inoculation with metal-mobilizing plant-growth-promoting rhizobacterium *Bacillus* sp. SC2b and its role in rhizoremediation. J. Toxicol. Environ. Health A 78, 931–944.

Mahmood, S., Daur, I., Al-Solaimani, G.S., Ahmad, S., Madkour, H.M., Yasir, M., Hirt, H., Ali, S., Ali, Z., 2016. Plant growth promoting rhizobacteria and silicon synergistically enhance salinity tolerance of mung bean. Front. Plant Sci. 7, 876.

Malhotra, M., Srivastava, S., 2009. Stress-responsive indole-3-acetic acid biosynthesis by *Azospirillum brasilense* SM and its ability to modulate plant growth. Eur. J. Soil Biol. 45 (1), 73–80.

Manganiello, G., Sacco, A., Ercolano, M.R., Vinale, F., Lanzuise, S., Pascale, A., Napolitano, M., Lombardi, N., Lorito, M., Woo, S.L., 2018. Modulation of tomato response to *Rhizoctonia solani* by *Trichoderma harzianum* and its secondary metabolite harzianic acid. Front. Microbiol. 9.

Mani, R., Helena, F., 2008. Effects of inoculation of plant growth promoting bacteria on Ni uptake by Indian mustard. Bioresour. Technol. 99, 3491–3498.

Manus, P.S.M., Stockwell, V.O., Sundin, G.W., Jones, A.L., 2002. Antibiotic use in plant agriculture. Annu. Rev. Phytopathol. 40, 443–465.

Mayak, S., Tirosh, T., Glick, B.R., 2004. Plant growth-promoting bacteria confer resistance in tomato plants to salt stress. Plant Physiol. Biochem. 42, 565–572.

McGhee, G.C., Sundin, G.W., 2011. Evaluation of kasugamycin for fire blight management, effect on nontarget bacteria, and assessment of kasugamycin resistance potential in *Erwinia amylovora*. Phytopathology 101, 192–204.

Mehnaz, S., 2015. Azospirillum: a biofertilizer for every crop. In: Arora, N.K. (Ed.), Plant Microbes Symbiosis: Applied Facets.pp. 297–314.

Metwali, E., Abdelmoneim, T.S., Bakheit, M.A., Kadasa, N.M.S., 2015. Alleviation of salinity stress in faba bean (*Vicia faba* L.) plants by inoculation with plant growth promoting rhizobacteria (PGPR). Plant Omics 8 (5), 449–460.

Microbial Products, Microbial products: technologies, applicationsand global markets, 2015, BIO086C, BCCResearch.

Misato, T., Ishii, I., Asakawa, M., Okimoto, Y., Fukunaga, K., 1959. Antibiotics as protectant fungicides against rice blast. Japanese J. Phytopath. 24 (5), 302–306.

Mishra, J., Fatima, T., Arora, N., 2018. Role of secondary metabolites from plant growth-promoting Rhizobacteria in combating salinity stress. In: Plant Microbiome: Stress Response.pp. 127–163.

Mohamed, I.H., Gomaa, E.Z., 2012. Effect of plant growth promoting *Bacillus subtilis* and *Pseudomonas fluorescens* on growth and pigment composition of radish plants (*Raphanus sativus*) under NaCl stress. Photosynthetica 50 (2), 263–272.

Moyne, A.L., Shelby, R., Cleveland, T.E., Tuzun, S., 2001. Bacillomycin D: an iturin with anti-fungal activity against *Aspergillus avus*. J. Appl. Microbiol. 90, 622–629.

Murali, M., Jogaiah, S., Amruthesh, K.N., Shin-ichi, I., Shekar Shetty, H., 2013. Rhizosphere fungus *Penicillium chrysogenum* promotes growth and induces defense-related genes and downy mildew disease resistance in pearl millet. Plant Biol. 15, 111–118.

Nagaraju, A., Jogaiah, S., Mahadeva Murthy, S., Shin-Ichi, I., 2012. Seed priming with *Trichoderma harzianum* isolates enhances plant growth and induces resistance against *Plasmopara halstedii*, an incitant of sunflower downy mildew disease. Aust. J. Plant Pathol. 41, 609–620.

Nandi, M., Selin, C., Brassinga, A.K.C., Belmonte, M.F., Fernando, W.G.D., Loewen, P.C., et al., 2015. Pyrrolnitrin and hydrogen cyanide production by*Pseudomonas chlororaphis* strain PA23 exhibits nematicidal and repellent activity against*Caenorhabditis elegans*. PLoS One. 10(4).

Nakayama, K., Okamoto, F., Harada, Y., 1956. Antimycin A: isolation from a new Streptomyces and activity against rice plant blast fungi. J. Antibiot. 9, 63–66.

Natti, J.J., Hervey, G.E.R., Sayre, C.B., 1956. Factors contributing to the increase of downy mildew of broccoli in New York State and its control with fungicides and Agrimycin. Plant. Dis. Rep. 40 (2), 118–124.

Noel, T.C., Sheng, C., Yost, C.K., Pharis, R.P., Hynes, M.F., 1996. *Rhizobium leguminosarum* as plant growth-promoting rhizobacterium: direct growth promotion of canola and lettuce. Can. J. Microbiol 42, 279–283.

Okuyama, A., Machiyama, N., Kinoshita, T., Tanaka, N., 1971. Inhibition by kasugamycin of initiation complex formation on 30s ribosomes. Biochem. Biophys. Res. Commun. 43, 196–199.

Patten, C.L., Glick, B.R., 1996. Bacterial biosynthesis of indole-3-acetic acid. Can. J. Microbiol. 42, 207–220.

Qamar, A., Mysore, K.S., Kumar, M.S., 2015. Role of proline and pyrroline-5-carboxylate metabolism in plant defense against invading pathogens. Front. Plant Sci. 6, 503.

Rahman, M.S., Miyake, H., Takeoka, Y., 2002. Effects of exogenous glycine-betaine on growth and ultra-structure of salt-stressed rice seedlings (*Oryza sativa* L.). Plant Prod. Sci. 5 (1), 33–44.

Rangaswami, G., Mahadevan, A., 1998. Diseases of crop plants in India. PHI Learning Pvt Ltd..

Rezzonico, F., Binder, C., Défago, G., Loccoz, Y.M., 2005. The type III secretion system of biocontrol *Pseudomonas fluorescens* KD targets the Phytopathogenic Chromista *Pythium ultimum* and promotes cucumber protection. MPMI 18 (9), 991–1001.

Sadeghi, A., Karimi, E., Dahaji, P.A., Javid, M.G., Dalvand, Y., Askari, H., 2012. Plant growth promoting activity of an auxin and siderophore producing isolate of *Streptomyces* under saline soil conditions. World J. Microbiol. Biotechnol. 28 (4), 1503–1509.

Sandhya, V., Ali, S.K.Z., Minakshi, G., Reddy, G., Venkateswarlu, B., 2009. Alleviation of drought stress effects in sunflower seedlings by the exopolysaccharides producing Pseudomonas putida strain GAP-P45. Biol. Fertil. Soils 46, 17–26.

Sankari, U., Dinakar, S., Sekar, C., 2011. Dual effect of Azospirillum exopolysaccharides (EPS) on the enhancement of plant growth and biocontrol of blast (*Pyricularia oryzae*) disease in upland rice (var. ASD-19). J. Phytology 3 (10), 16–19.

Sarma, R.K., Gogoi, A., Dehury, B., Debnath, R., Bora, T.C., Saikia, R., 2014. Community profiling of culturable fluorescent Pseudomonads in the rhizosphere of green gram (*Vigna radiata* L.). PLoS One. 9(10).

Satapute, P., Murali, K.P., Kurjogi, M., Jogaiah, S., 2019. Physiological adaptation and spectral annotation of arsenic and cadmium heavy metal-resistant and susceptible strain *Pseudomonas taiwanensis*. Environ. Pollut. 251, 555–563.

Saleem, M., Arshad, M.G., Hussain, S., Bhatti, A.S., 2007. Perspective of plant growth promoting rhizobacteria (PGPR) containing ACC deaminase in stress agriculture, Perspective of plant growth promoting rhizobacteria (PGPR) containing ACC deaminase in stress agriculture. J. Ind. Microbiol. Biot. 34 (10), 635–648.

Sathya, A., Vijayabharathi, R., Gopalakrishnan, S., 2017. Plant growth-promoting actinobacteria: a new strategy for enhancing sustainable production and protection of grain legumes. 3Biotech 7 (2), 102.

Schmid, J., Sieber, V., Rehm, B., 2015. Bacterial exopolysaccharides: biosynthesis pathways and engineering strategies. Front. Microbiol. 6, 496.

Schobert, C., Kockenberger, W., Komor, E., 1998. Uptake of amino acids by plants from the soil: A comparative study with castor bean seedlings grown under natural and axenic soil conditions. Plant Soil 109, 181.

Selvakumar, G., Joshi, P., Nazim, S., Mishra, P.K., Bisht, J.K., Gupta, H.S., 2009. Phosphate solubilization and growth promotion by *Pseudomonas fragi* CS11RH1 (MTCC 8984), a psychrotolerant bacterium isolated from a high altitude Himalayan rhizosphere. Biologia 64 (2), 239–245.

Selvakumar, G., Bhatt, M.R., Upreti, K.K., Bindu, H.G., Shweta, K., 2015. *Citricoccus zhacaiensis* B-4 (MTCC 12119) a novel osmotolerant plant growth promoting actinobacterium enhances onion (*Allium cepa* L.) seed germination under osmotic stress conditions. World J. Microbiol. Biotechnol. 31, 833–839.

Shaik, Z.A., Sandhya, V., Grover, M., Linga, V.R., Bandi, V., 2011. Effect of inoculation with a thermotolerant plant growth promoting *Pseudomonas putida* strain AKMP7 on growth of wheat (*Triticum* spp.) under heat stress. J. Plant Interact. 6 (4), 239–246.

Shanmugaiah, V., Nithya, K., Harikrishnan, H., Jayaprakashvel, M., Balasubramanian, N., 2015. Biocontrol mechanisms of siderophores against bacterial plant pathogens. In: Sustainable Approaches to Controlling Plant Pathogenic Bacteria. CRC Press, pp. 168–181.

Sharma, S., Kulkarni, J., Jha, B., 2016. Halotolerant rhizobacteria promote growth and enhance salinity tolerance in peanut. Front. Microbiol. 7, 1600.

Shenker, M., Oliver, I., Helmann, M., Hadar, Y., Chen, Y., 1992. Utilization by tomatoes of iron mediated by a siderophore produced by *Rhizopus arrhizus*. J. Plant Nutr. 15 (10), 2173–2182.

Siddiqui, I.A., Shaukat, S.S., 2003. Suppression of root-knot disease by *Pseudomonas fluorescens* CHA0 in tomato: importance of bacterial secondary metabolite, 2,4-diacetylpholoroglucinol. Soil Biol. Biochem. 35 (12), 1615–1623.

Somers, E., Ptacek, D., Gysegom, P., Srinivasan, M., Vanderleyden, J., 2005. *Azospirillum brasilense* produces the auxin-like phenylacetic acid by using the key enzyme for Indole-3-acetic acid biosynthesis. Appl. Environ. Microbiol 71 (4), 1803–1810.

Spaink, H.P., Lugtenberg, B.J.J., 1994. Role of rhizobial lipo-chitin oligosaccharide signal molecules in root nodule organogenesis. Plant Mol. Biol. 26, 1413–1422.

Subba Rao, I.V., Madhulety, T.Y., 2005. Role of herbicides in improving crop yields. In: Development in physiology, biochemistry and molecular biology of plants. New India Publishing Agency, New Delhi, pp. 203–287.

Sudisha, J., Amruthesh, K.N., Deepak, S.A., Shetty, N.P., Sarosh, B.R., Shekar Shetty, H., 2005. Comparative efficacy of strobilurin fungicides against downy mildew disease of pearl millet. Pest. Biochem. Physiol. 81, 188–197.

Sudisha, J., Niranjana, S.R., Umesha, S., Prakash, H.S., Shekar Shetty, H., 2006. Transmission of seed-borne infection of muskmelon by *Didymella bryoniae* and effect of seed treatments on disease incidence and fruit yield. Biol. Control 37, 196–205.

Sushil, S.N., , 2016. Emerging issues of plant protection in India. In: Natural Resource Management: Ecological Perspectives. International Conference, SKUAST Jammu.

Tank, N., Saraf, M., 2010. Salinity-resistant plant growth promoting rhizobacteria ameliorates sodium chloride stress on tomato plants. J. Plant Interact. 5 (1), 51–58.

Tewari, S., Arora, N.K., 2014a. Multifunctional exopolysacccharides from *Pseudomonas aeruginosa* PF23 involved in plant growth stimulation, biocontrol and stress amelioration in sunflower under stress conditions. Curr. Microbiol. 69, 484–494.

Tewari, S., Arora, N.K., 2014b. Talc based exopolysaccharides formulation enhancing growth and production of *Helcthus annuus* under saline conditions. Cell. Mol. Biol. 60 (5), 73–81.

Timmusk, S., Islam, A., El-Daim, A., Copolovici, L., Tanilas, T., Kannaste, A., Behers, L., Nevo, E., Seisenbaeva, G., Stenstrom, S., Niinemets, U., 2014. Drought-tolerance of wheat improved by rhizosphere bacteria from harsh environments: enhanced biomass production and reduced emissions of. PLoS One. 9(5), e96086.

Tortora, M.L., Díaz-Ricci, J.C., Pedraza, R.O., 2011. *Azospirillum brasilense* siderophores with antifungal activity against *Colletotrichum acutatum*. Arch. Microbiol. 193, 275–286.

Varatto, C., Maiwaid, D., Pesaresi, P., Jahns, P., Salamini, F., Leister, D., 2002. The metal ion transporter IRT1 is necessary for iron homeostasis and efficient photosynthesis in *Arabidopsis thaliana*. Plant. J. 31, 589–599.

Vasudeva, R.S., Singh, P., Sengupta, P.K., Mahmood, M., Bajaj, B.S., 1963. Further studies on the biological activity of bulbiformin. Ann. Appl. Biol. 51 (3), 415–423.

Verma, V.C., Singh, S.K., Prakash, S., 2011. Research paper bio-control and plant growth promotion potential of siderophore producing endophytic Streptomyces from *Azadirachta indica* A. Juss. J. Basic Microbiol. 51 (5), 550–556.

Vidaver, A.K., 2002. Uses of antimicrobials in plant agriculture. Clin. Infect. Dis. 34 (3), 107–110.

Vidhyasekaran, P., 2004. Concise Encyclopedia of Plant Pathology. Taylor & Francis 619p.

Vinale, F., Sivasithamparam, K., Ghisalberti, E.L., Woo, S.L., Nigro, M., Marra, R., Lombardi, N., Pascale, A., Ruocco, M., Lanzuise, S., Manganiello, G., Lorito, M., 2014. Trichoderma secondary metabolites active on plants and fungal pathogens. Open Mycol. J. 8, 127–139.

Voisard, C., Keel, C., Haas, D., Dèfago, G., 1989. Cyanide production by *Pseudomonas fluorescens* helps suppress black root rot of tobacco under gnotobiotic conditions. EMBO J. 8 (2), 351–358.

Vurukonda, S.S.K.P., Giovanardi, D., Stefani, E., 2018. Plant growth promoting and biocontrol activity of *Streptomyces spp.* as endophytes. Int. J. Mol. Sci. 19, 952.

Wan, W., Li, C., Zhang, J., Jiang, D., Huan, H.C., 2008. Effect of volatile substances of *Streptomyces platensis* F-1 on control of plant fungal diseases. Biol. Control. 46, 552–559.

Wani, P.A., Khan, M.S., Zaidi, A., 2008. Effect of metal-tolerant plant growth-promoting Rhizobium on the performance of pea grown in metal-amended soil. Arch. Environ. Contam. Toxicol. 55 (1), 33–42.

Weller, D.M., 2007. Pseudomonas biocontrol agents of soilborne pathogens: looking back over 30 years. Phytopathology 97, 250–256.

Yao, L.X., Wu, Z.S., Zheng, Y.Y., Kaleem, I., Li, C., 2010. Growth promotion and protection against salt stress by *Pseudomonas putida* Rs-198 on cotton. Eur. J. Soil Biol. 46, 49–54.

Yu, X., Ai, C., Xin, L., Zhou, G., 2011. The siderophore-producing bacterium, *Bacillus subtilis* CAS15, has a biocontrol effect on Fusarium wilt and promotes the growth of pepper. Eur. J. Soil Biol. 47 (2), 138–145.

Yu, S., Teng, C., Bai, X., Liang, J., Song, T., Dong, L., Jin, Y., Qu, J., 2017. Optimization of siderophore production by *Bacillus sp.* PZ-1 and its potential enhancement of Phytoextration of Pb from soil. J. Microbiol. Biotechnol. 27 (8), 1500–1512.

Zhang, L., Xiao, S., Li, W., Feng, W., Li, J., Wu, Z., Gao, X., Liu, F., Shao, M., 2011. Overexpression of a harpin-encoding gene *hrf1* in rice enhances drought tolerance. J. Exp. Biol. 62, 4229–4238.

Zuo, Y., Zhang, F., 2011. Soil and crop management strategies to prevent iron deficiency in crops. Plant and Soil 339, 83–95.

Further reading

Ahemad, M., Kibret, M., 2014. Mechanisms and applications of plant growth promoting rhizobacteria: current perspective. J. King Saud. Univ. Sci. 26 (1), 1–20.

Cohen, A.C., Travaglia, C.N., Bottini, R., Piccoli, P.N., 2009. Participation of abscisic acid and gibberellins produced by endophytic Azospirillum in the alleviation of drought effects in maize. Botany 87, 455–462.

Couillerot, O., Combaret, P.C., Mellado, C.J., Loccoz, M.Y., 2009. *Pseudomonas fluorescens* and closely-related fluorescent pseudomonads as biocontrol agents of soil-borne phytopathogens. Lett. Appl. Microbiol. 48 (5), 505–512.

Feki, O.K., Khedher, B.S., Dammak, M., Kamoun, A., Khiareddine, H.J., Remadi, D.M., Tounsi, S., 2016. Improvement of antifungal metabolites production by *Bacillus subtilis* V26 for biocontrol of tomato postharvest disease. Biol. Control 95, 73–82.

Jogaiah, S., Abdelrahman, M., Tran, L.-S.P., Ito, S.-I., 2018. Different mechanisms of *Trichoderma virens*-mediated resistance in tomato against Fusarium wilt involve the jasmonic and salicylic acid pathways. Mol. Plant Pathol. 19, 870–882.

Kasotia, A., Varma, A., Tuteja, N., Choudhary, D.K., 2016. Amelioration of soybean plant from saline induced condition by exopolysaccharide producing *Pseudomonas*-mediated expression of high affinity K^+ transporter (HKT1) gene. Curr. Sci. 111 (12), 25.

Kets, E.P.W., de Bont, J.A.M., Heipieper, H.J., 1996. Physiological response of *Pseudomonas putida* S12 subjected to reduced water activity. FEMS Microbiol. Lett. 139, 133–137.

Kim, K.K., Kim, Y.C., Choi, Y.W., Sin, T.S., Park, K.D., Kang, U.G., Choi, Y.L., Park, H.C., 2008. Biological control of plant pathogens by *Bacillus* sp. AB02. J. Life Sci. 18 (6), 858–864.

Kohler, J., Hernandez, J.A., Caravaca, R.A., 2008. Plant-growth-promoting rhizobacteria and arbuscular mycorrhizal fungi modify alleviation biochemical mechanisms in water-stressed plants. Funct. Plant Biol. 35, 141–151.

Kurz, M., Burch, A.Y., Seip, B., Lindow, S.E., Gross, H., 2010. Genome-driven investigation of compatible solute biosynthesis pathways of *Pseudomonas syringae* pv. syringae and their contribution to water stress tolerance. Appl. Environ. Microbiol. 76 (16), 5452–5462.

Rajkumar, M., Freitas, H., 2008. Effects of inoculation of plant-growth promoting bacteria on Ni uptake by Indian mustard. Bioresour. Technol. 99, 3491–3498.

Rezzonico, F., Stockwell, V.O., Duffy, B., 2009. Plant agricultural streptomycin formulations do not carry antibiotic resistance genes. Antimicrob. Agents Chemother. 53, 3173–3177.

Sandhya, V.S., Ali, S.Z., Grover, M., Reddy, G., Venkateswarlu, B., 2010. Effect of plant growth promoting *Pseudomonas* spp. on compatible solutes, antioxidant status and plant growth of maize under drought stress. Plant Growth Regul. 62 (1), 21–30.

Saravanakumar, D., Samiyappan, R., 2007. ACC deaminase from Pseudomonas fluorescens mediated saline resistance in groundnut (*Arachis hypogea*) plants. J. Appl. Microbiol. 102, 1283–1292.

Shaharoona, B., Arshad, M., Zahir, Z.A., 2006. Effect of plant growth promoting rhizobacteria containing ACC-deaminase on maize (*Zea mays* L.) growth under axenic conditions and on nodulation in mung bean (*Vigna radiata* L.). Lett. Appl. Microbiol. 42, 155–159.

Spaink, H.P., Wijfjes, M.H.A., van der Drift, M.K., Lugtenberg, B., 1994. Structural identification of metabolites produced by the NodB and NodC proteins of *Rhizobium leguminosarum*. Mol. Microbiol. 13 (5), 821–831.

Signatures of signaling pathways underlying plant-growth promotion by fungi

13

Swapan Kumar Ghosh and Atanu Panja
Molecular Mycopathology Laboratory, Biocontrol Unit, PG Department of Botany, Ramakrishna Mission Vivekananda Centenary College (Autonomous), Kolkata, India

13.1 Introduction

Many fertile lands or agrifields are converted to lands for urbanizations and industrialization to meet the human civilization pressure, so the limitation of fertile agricultural fields is increasing day by day (Mishra et al., 2017; Fletcher et al., 2006). At one time agriculturists openly accepted that inorganic fertilizers are the backbone of the agriculture for crop production and without them, it is very difficult to feed ever-growing population of human in the World, which is projected to be around 9.38 billion by 2050 (USCB, 2015). But later they admitted that application of this component of agriculture is not the final solution as its indiscriminate use is polluting the environment in havoc and causing disturbance of soil health and human health hazards (Ghosh and Pal, 2017; Hermosa et al., 2012; Smith et al., 2008; PAN, 2007; Powler, 2006; Ward et al., 2005). Unfortunately, to gain a target crop production with chemical fertilizers and pesticides, over 100 species of nontarget organisms are adversely affected (Alabouvette and Couteadier, 1992). According to Pelosi et al. (2013), even earthworms which are known as "Natural farmer," are at huge risk due to the use of chemical fertilizers and pesticides. This realization promptly directed scientists for an alternative to chemical fertilizers and pesticides. After recent findings, it has been shown that fungi can be used as the substitute of different chemical fertilizers. Fungi play a major role in the growth promotion in the plant. The fungi which are used for plant-growth promotion are known as plant-growth-promoting fungi (PGPF), and those which have a contribution for plant protection against diseases are called fungal biocontrol agents (FBCA). The same fungus may have a dual role. There are about 75,000 species of soil fungi which are recorded in the world (Finley, 2007). Plant-growth promotion effect is induced by different ways such as auxin production, solubilization of phosphate, iron, potassium and other minerals, siderophore formation, etc. PGPF (plant-growth-promoting factor) helps in the growth promotion by physiological and different intercellular molecular mechanisms. Several genera of fungi, e.g., *Trichoderma*, *Penicillium*, *Aspergillus*, *Fusarium*, mycorhizal fungi, etc., have been widely studied and reported as highly tolerant groups to different pollutants such as polyethylene, PCBs, chlorobenzoic acids (CBA), endosulfan, etc.

Biocontrol Agents and Secondary Metabolites. https://doi.org/10.1016/B978-0-12-822919-4.00013-2

(Ghosh et al., 2013; Garon et al., 2000; Tigini et al., 2009; Pinedo-Rivilla et al., 2009) and they are also heavy metal tolerant (Gaddie, 1993; Morley and Gadd, 1995). These fungi secrete plant-growth-promoting hormones. Auxin generally functions with the ABP1 (auxin binding protein1) receptor pathway and altering of the ABP1 gene in *Arabidopsis* has been reported for alter development (Contreras-Cornejo et al., 2009). Auxin also resumes cell cycle by Cdc2 (cell division cycle2) and CDK (cyclin-dependent kinase) synthesis. Some major auxin-induced growth regulation genes, have been identified like Aux/IAA, SAUR (small auxin-up RNA), GH3 (Growth Hormone3), etc. Plant-growth promotion pathways are mainly controlled by a network of genetic interaction, also are involved in feed-back loops where two or more genes like KNAT1 (homeobox protein knotted-1-like 1), AS1 (Asymetric Leaves 1), STM (suppressor of_to_mL), etc. regulate each other's expression, whereas cytokinin involves in CRE1 (Cytokinin receptor1), AHK2 (Arabidopsis Histidine Kinase 2), AHK3 (Arabidopsis Histidine Kinase 3) domains, etc. Abscisic acid, inducing Ca^{2+} activity within the cell, activates histidine kinase pathway. The metabolites like 6-pentyl-a-pyrone, harzianolide and harzianopyridone activate plant defense mechanisms and also trigger plant growth in pea, tomato, and canola (Vinale et al., 2008), advocating that this case is a good example as the same component induces both plant-growth promotion and defense.

In this chapter, two-way molecular interactions between plant and PGPF for plant-growth enhancement and some of the signaling pathways that are influenced by fungi for this purpose will be discussed.

13.2 Plant-growth promotion (PGP) by fungi (PGPF)

It is two-way interactions of plant and PGPF for their sustainable growth. Several workers have admitted that PGP happens with rhizospheric fungi (Jogaiah et al., 2018; Jogaiah et al., 2013; Murali et al., 2013; Nagaraju and Jogaiah, 2012). Among the PGPF, species of *Trichoderma, Penicillium, Phoma, Aspergillus, Fusarium*, and mycorrhizal fungi have gained much attention due to their effective role in plant-growth activities and disease management in agrifields (Contreras-Cornejo et al., 2009; Ghosh, 2017a; Naznin et al., 2013; Murali et al., 2012; Elsharkawy et al., 2012; Saldajeno and Hyakumachi, 2011). Windham (1986) noted a mechanism for enhanced plant-growth triggered by *Trichoderma* spp. Many spp. of *Trichoderma* are widely known for their growth promotion effect in different crops in agrisystem such as *Trichoderma harzianum* (Contreras-Cornejo et al., 2009; Jogaiah et al., 2013), *Trichoderma virens* (Jogaiah et al., 2018), *Trichoderma longibrachiatum* (Mostafa et al., 2016), *T. asperellum* (Viterbo et al., 2010), etc. *Trichoderma* induces plant growth after colonizing on the root (Brotman et al., 2008; Chang, 1986). Crop productivity in fields can be increased up to 300% after the addition of *Trichoderma hamatum* or *Trichoderma koningii* (Chet et al., 1997; Benítez et al., 1998). In our (Ghosh and Pal, 2017) experiment in minipot trial, when we treated chickpea with *Trichoderma asperellum*, there was 25% increase of seed germination and shoot length (Fig. 13.1), but 27% increase of root length and 45% increase of vigor index

Fig. 13.1 Growth promotion of chick pea by BCA. (A) Control plant (untreated). (B) Seedling growth promotion by *T. asperellum* on chick pea. (C) Measurement of plant height.

Fig. 13.2 Growth promotion of rice by *T. viride*: (A) control seed germination; (B) treated seed germination showing enhanced germination and growth vigor; (C) control seedling in pot; (D) *Trichoderma* treated pot showing better seedling growth (Ghosh, 2017a).

compared to control (untreated). In the case of Rice, when some fungi were applied in our (Ghosh, 2017a) experimental field, all were able to enhance the seed germination, shoot and root growth of seedlings (Fig. 13.2) and induction of catalase and peroxidase enzymes in rice seedlings but *T. viride* was best of all (Ghosh, 2017a). Different spp. of *Trichoderma* are reported to induce IAA production, siderophore production, phosphate solubilization, and chlorophyll production on maize plants (Kumar et al., 2016) and *Arabidopsis* (Jogaiah et al., 2018) in comparison with control plants correlated with a significant increase in root and shoot length. Enhanced seed germination and root and shoot growth response of several plants, such as tomato (Vinale et al., 2008), pea (Inbar et al., 1994), cucumber (Kleifeld and Chet, 1992), pepper, lettuce (Vinale et al., 2008), and brinjal (Ghosh, 2017b) were recorded by application of *Trichoderma*. Different sugars, such as those derived from polymers (cellulose, glucan, chitin, etc.), render glucose (Chet et al., 1997) to *Trichoderma* as a support by the isolation of a high-affinity glucose transporter, Gtt 1, in *T. harzianum* CECT 2413 (Delgado-Jarana et al., 2003). This strain is present in the environment where the nutrient is very poor and it also relies on extracellular hydrolases for survival. The Gtt1 is only expressed at very low glucose concentration, i.e., when sugar transport is expected to limit in nutrient competition (Delgado-Jarana et al., 2003). Sucrose helps *Trichoderma* to colonize in the plant root system, and solute transporter channel such as a di/tripeptide transporter and a permease/intracellular invertase system are involved in root transude acquisition (Vargas et al., 2009; Vizcaıno et al., 2006). Sneh et al. (1986) also reported that *Rhizoctonia solani* secreted the growth-promoting agents which induced growth promotion in the field experiment. Other fungi also increased plant growth by hormone production like abscisic acid (Hyakumachi and Kubota, 2003), IAA (Indole Acetic Acid), Gibberellin, etc. (Windham, 1986; Ram, 1959). A recent example of biotechnological solutions from *Trichoderma* is the *T. harzianum* Thkel1 gene, encoding a kelch-repeat protein involved in the modulation of glucosidase activity that enhanced seed germination and plant tolerance to salt and osmotic stresses when it was expressed in *Arabidopsis* (Hermosa et al., 2011).

13.3 Molecular mechanisms or cell signaling of plant-growth promotion

13.3.1 Mechanism for attachment and colonization of PGPF on plant root and protection from plant defense

It is very interesting to note how fungi attach with a root system and protect themselves from the plant defense for colonization. For *Trichoderma,* hydrophobin existing in the outermost layer of the fungal cell wall mediates the adherence with plant root surface. Viterbo and Chet (2006) gave one important finding that *T. asperellum* produces class I type hydrophobin, TasHydro 1, which promotes colonization of *T. asperellum* with plant roots and also protects the hyphal tip from plant defense system. Brotman et al. (2008) reported that another protein TasSwo (swollenin, an expansin-like protein) has a great role for attachment with root by modifying the structure of the plant cell wall. In *Trichoderma harzianum,* a protein ThPG1, which is an endopolygalacto-uranose (Moran-Diez et al., 2009), helps the fungi for attachment to the plant root and protects from plant defense. Plant responses against fungal invasion by producing phytoalexins, flavonoids, and terpenoids, phenolic derivatives, aglycones and other antimicrobial compounds. *Trichoderma* strains are generally more tolerant of these compounds than most fungi and, their capacity to colonize in rhizospheric zones and plant roots is correlated with the capacity of each strain to tolerate them. Protection from plant defense which is considered an essential requirement for plant colonization has been associated with the presence of ABC (ATP-binding cassette) transport systems in *Trichoderma* strains (Harman et al., 2004).

13.3.2 Plant-growth promotion through hormone production

Only a few reports are available for the *Trichoderma* interaction with *Arabidopsis* (Chacón et al., 2007; Shoresh et al., 2010; Korolev et al., 2008). *Trichoderma atroviride* promotes the growth of *Arabidopsis* when it colonizes on the root which is a key factor for providing nutrients and phytohormone (Contreras-Cornejo et al., 2009; Chang, 1986). It has been already reported that several beneficial microorganisms help their host plant growth by the synthesis of growth hormone which is used by plants (Chang, 1986; van Loon et al., 2006). Similarly, *T. atroviride* produces indole which is utilized by plant for its growth promotion. Indole is the precursor of auxin and may be absorbed by plant cells and later it enhances auxin synthesis. Other filamentous fungi produce phytohormones, such as indole acetic acid (IAA) and ethylene, whose metabolic pathways have been identified (Arora et al., 1992; Osiewacz, 2002). Cytokinin-like molecules, e.g., zeatyn and gibberellin GA3 or GA3-related, have been recently detected from *Trichoderma* strains (Mach and Zeilinger, 2003). Due to the wide study on *Trichoderma* as PGPF, the necessary mechanism also has been described by different works. A number of mechanisms for plant-growth promotion by *Trichoderma* have been proposed (Harman et al., 2004). Among these, fungal interaction with auxin signaling is very important, and Contreras-Cornejo et al. (2009) found that *Trichoderma virens* and *T. atroviridae* increased biomass in *Arabidopsis* sp. by altering auxin production by mutating genes such as Aux 1, BIG, EIR 1, and AXR 1. They highlighted the important role of auxin signaling for plant-growth promotion by *Trichoderma*. It is found that *Trichoderma virens* stimulates plant growth by auxin accumulation by the production of IAA or IAAid-like precursor. IAAid later converts into IAA by own enzyme of fungi (Contreras-Cornejo et al., 2009). *Trichoderma* has been reported to produce volatile substance such as 6-pentyl-2H-pyron-2-one (6-PP) which enhances plant growth and changes root architecture of *A. thaliana* via auxin transport and signaling and the ethylene response modulator EIN2 (Garnica-Vergara et al., 2016).

13.3.3 Auxin-mediated cell signaling pathway in plants

Exogenous auxin supply induces the expression of several genes (Paponov et al., 2008) and changes the cellular response via short transcription pathway (Harman et al., 2004; Segarra et al., 2009; Chapman and Estelle, 2009). Auxin-mediated cell signaling in plant (Fig. 13.3) (Leyser, 2018) starts as soon as auxin enters the nucleus. The main components of the auxin signaling pathway are three protein families such as (i) the F-box Transport Inhibitor Response 1/Auxin Signaling F-Box Protein (TIR1/AFB) which acts as auxin co-receptors, (ii) the Auxin/Indole-3-Acetic Acid (Aux/IAA) acts as transcriptional repressors, and (iii) the Auxin Response Factor (ARF) which acts as transcription factors (Leyser, 2018; Kepinski and Leyser, 2005; Salehin et al., 2015). Auxin, in the appropriate amount, mediates the binding of

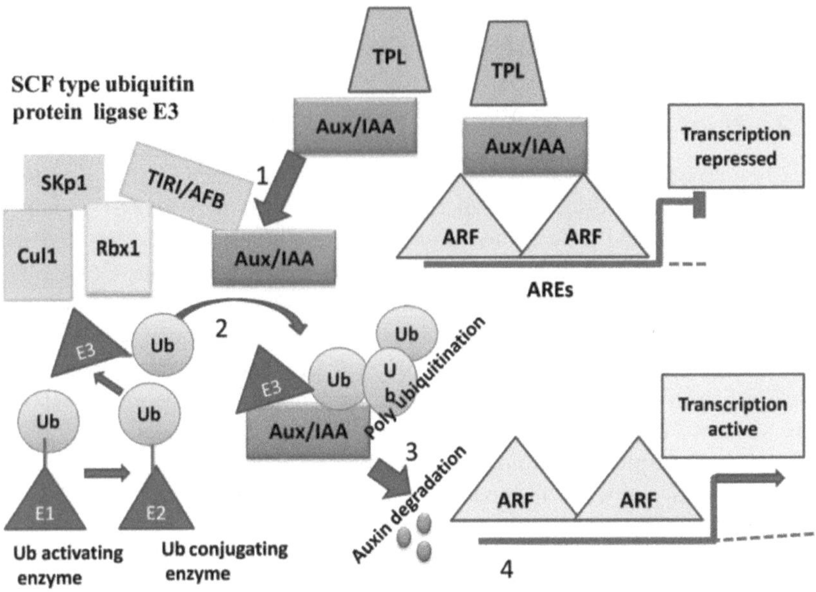

Fig. 13.3 Transcription regulation by auxin. Auxin inducible genes have auxin response element (ARE) in their promoter region, which are bound with ARF family. Aux/IAA transcription repressor interaction with the ARF prevents gene expression. TPL (TOPLESS) family co-repressor recruited by Aux/IAA family stabilizes the repressed state. 1, Molecular glue bringing together Aux/IAA and TIR/AFB family; 2, F-box protein transfers activated ubiquitin (Ub) from an E1/E2 enzyme system to E3 and E3 to substrate, Auxin/IAA; 3, degradation of Aux/IAA by polyubiquitination; 4, repression at ARE-containing promoters (Leyser, 2018). *TIR1/AFB*, transport Inhibitor response 1/auxin signaling F-Box protein; *Cul1*, CULLIN1; *Rbx1*, ring-box protein 1; *E3*, substrate recognition protein.

TIR1/AFB with Aux/IAA proteins through polyubiquitylation of Aux/IAAs. Polyubiquitylation of the Aux/IAA transcriptional repressors needs an E3 ubiquitin ligase $SCF^{TIR/AFB}$ complex. SCF complexes have an F-box protein that provides substrate recognition, an Arabidopsis Skp1 Homolog1 (ASK1) adaptor (SKP1 in animals and fungi), the scaffold protein CULLIN1 (CUL1), and Ring-Box Protein1 (Rbx1) that promotes the transfer of ubiquitin molecules to the substrate (Paponov et al., 2008; Smalle and Vierstra, 2004). The $SCF^{TIR/AFB}$ complex binds the Aux/IAA substrate in an auxin-dependent manner through the TIR1 or AFB F-box protein (Kepinski and Leyser, 2005; Salehin et al., 2015; Dharmasiri et al., 2005). So, a transcriptional repressor, Aux/IAAs become degraded and the ARF become released from repressor and start to transcribe (Salehin et al., 2015; Wang and Estelle, 2014; Lavy and Estelle, 2016) (Fig. 13.3). Gene expression associated with ARF activation is also associated with diverse processes (cellular growth, including cell elongation, cell division, and differentiation) (Perrot-Rechenmann, 2010; Takatsuka and Umeda, 2014) in land plants.

Further, auxin helps in gene expression of S-phase of the cell cycle. Here auxin influences to form CYCD—CDKA complex which phosphorylates and triggers the expression of S-phase genes in cell division pathway. Moreover, auxin upregulates some genes which are involved in the production of some enzymes or proteins which loosen the cell wall components (Perrot-Rechenmann, 2010).

13.3.4 Plant-growth promotion through soil phosphate solubilization

Out of the total phosphate (P) (0.5%) in soil, only 0.1% is available for plants (Scheffer and Schachtschabel, 1988). So, the majority of P exists in the insoluble form and plants cannot absorb (Rodriguez and Fraga, 1999) and as a result growth and yields of plants are hampered. Phosphate exists in the soil in two forms such as organic (e.g., phytate, sugar phosphates, phospholipids, nucleic acids, polyphosphates, and phosphonates) and inorganic phosphate (P_i) (such as tricalcium phosphate, dicalcium phosphate, hydroxyapatite, and rock phosphate). Organic phosphate solubilization is also called mineralization of organic phosphorus, and it comes from remnants of plant and animal and remains in the soil, which contains a large amount of organic phosphorus compounds. The decomposition of organic matter in the soil is carried out by the action of numerous saprophytes, which produce or release enzymes. Phosphorus can be released from organic compounds in soil by three categories of enzymes such as nonspecific phosphatases, which release phosphorous by dephosphorylation of phospho-ester or phosphoanhydride bonds in organic remains of plants and animals, e.g., phytases, which specifically unlocks P from phytic acid, phosphonatases, and C-P lyases enzymes that cause C-P cleavage in organophosphonates.

The mineral phosphate (Pi) solubilization process is mediated by the secretion of different kinds of organic acids like acetic acid, succinic acid, oxalic acid, etc. (Vazquez et al., 2000; Patel et al., 2008) and these acids can form a chelate compound by using their hydroxyl and carboxyl groups to bind the soil phosphate into a soluble form. The efficiency of soluble phosphate conversion depends on the type, strength, and amount of organic acid production (PAN, 2007). Aliphatic acids are found to be more effective in P solubilization than phenolic acids and citric acids. Fumaric acid has the highest P solubilizing ability. Tribasic and dibasic acids are also more superior than monobasic acids in this purpose.

13.3.4.1 Soil phosphate solubilization by organic acids produced by fungi

The solubilization of phosphate is an important factor for growth promotion. The theories for explaining P-solubilization mechanisms by microbes including fungi are grouped into three categories: (i) the organic acid theory (Cunningham and Kuiack, 1992), (ii) the sink theory (Halvorson et al., 1990), and (iii) the acidification by H^+ excretion theory (Illmer and Schinner, 1992). Out of these, the organic acid theory is wildly accepted. This theory states that insoluble P becomes solubilized by lowering the pH or by enhancing chelation of cations bound the P by fungus releasing organic acids in the environment (Whitelaw et al., 1999; Maliha et al., 2004). Omar (1998) recorded that organic acids can either directly dissolve the mineral P as a result of anion exchange of PO_4^{2-} by acid anion or can chelate both Fe and Al ions accompanied with P. Thus, the production and release of organic acid by fungi into the surrounding environment acidify the fungal cells and their surrounding environment that ultimately leads to the release of P ions from the P mineral by H^+ substitution for Ca^{2+} (Goldstein, 1994). Of the different organic acids involved in the solubilization of insoluble P, succinic citric, gluconic, a-ketogluconic, and oxalic acids are the most prominent acids which are released by fungal strains.

The mechanism of P solubilizing by *Trichoderma* has been reported (Viterbo et al., 2010) which revealed that organic acid production is the main solubilization mechanism for phosphate and its compound (Shoresh et al., 2010). In *Aspergillus niger,* citric acid (Fig. 13.4) and oxalic acid (Fig. 13.5) (Wolschek and Kubicek, 1997; Khan et al., 2007) are produced and they are majorly helpful in P-solubilization. P solubilizing property of fungi can be easily detected in culture plate containing specific medium (Pikovaskaya's agar medium) where the fungus is grown after solubilizing insoluble phosphate (e.g., Tricalcium phosphate/Rock phosphate) and as a result, a clear halo zone around fungal colony is seen (Fig. 13.6). In *Aspergillus niger* cellular mechanism has been studied for the production of the citric acid biosynthesis pathway through Gycolysis (Papagianni, 2007) which is very complex and operated under medium nutritional conditions (Wolschek and Kubicek, 1997; Kristiansen and Sinclair, 1979) and by the help of different enzymes (Soccol et al., 2006). Oxalic acid biosynthesis occurs by oxaloacetate hydrolysis and acetate catalyzed by cytosolic oxaloacetate. Glyoxylate oxidation pathway of oxalic biosynthesis was also noted in fungi

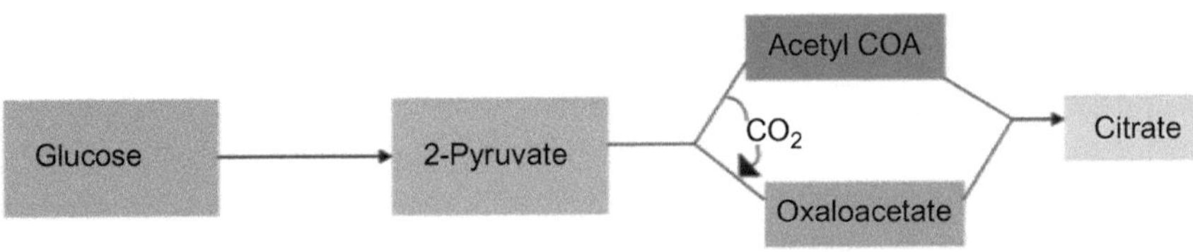

Fig. 13.4 Citric acid formation from glucose via anaplerotic carbon dioxide fixation (Khan et al., 2007).

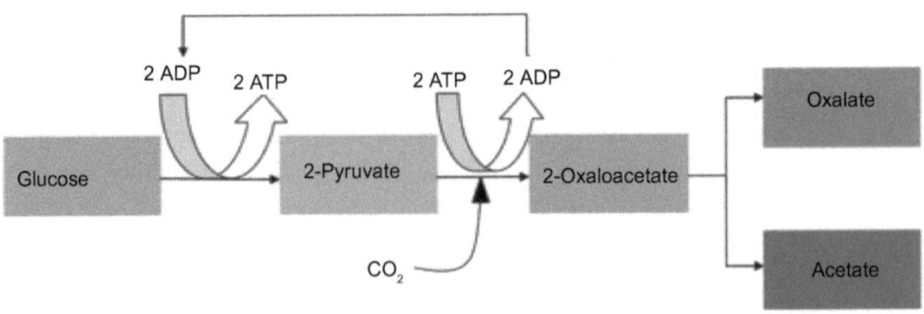

Fig. 13.5 Oxalate biosynthesis by *Aspergillus niger* (Khan et al., 2007).

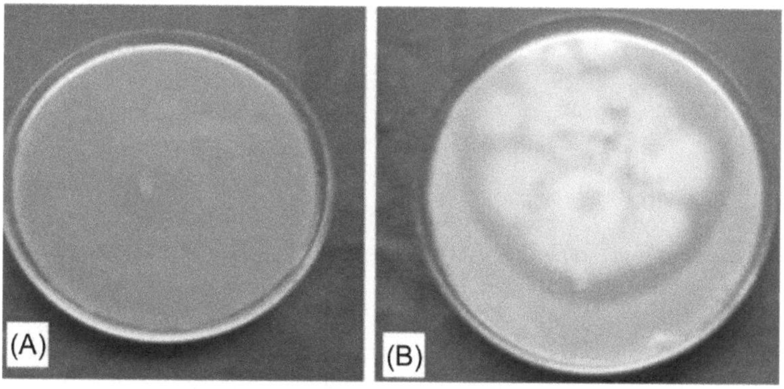

Fig. 13.6 Phosphate solubilization by fungus (A), control. (B) Phosphate solubilization by *Aspergillus* sp. showing halo zone surrounding culture in Petri plate.

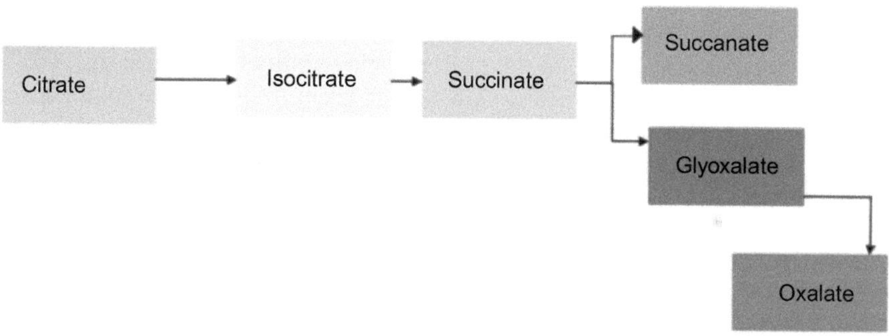

Fig. 13.7 Glyoxylate oxidation pathway of oxalic biosysthesis (Mendes et al., 2013; Wolschek and Kubicek, 1997).

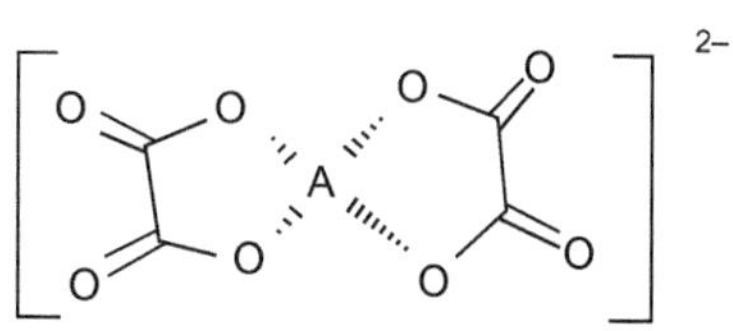

Fig. 13.8 Oxalic acid chelate formation with (A) any metal (Khan et al., 2007).

and bacteria (Wolschek and Kubicek, 1997; Mendes et al., 2013) (Fig. 13.7). Citric acid, generally after producing from TCA cycle, forms complex with metal, so citric acid can solubilize the phosphate and other minerals. Oxalic acid and carboxylic acid separately can bind with metal to form chelate (Fig. 13.8). Phosphatase enzyme is released outside the cell and participate in dissolution and mineralization of organic phosphate (Dutton et al., 1993; Shimada et al., 1997). Acid phosphatase also exists in vacuoles of some fungi and takes part in this purpose (Rodriguez and Fraga, 1999; To-O et al., 2000). Illmer and Schinner (1995) investigated the capability of

Trichoderma harzianum (T-22) to solubilize *in vitro* some insoluble minerals (P, K, Fe, etc.) via three possible mechanisms: acidification of the medium, production of chelating metabolites, and redox activity. So, the organic acids such as citric acids, oxalic acids, and carbolic acid are secreted by phosphate solubilizing fungi (PSFs) which convert insoluble rock phosphate to soluble phosphate in the soil which is available for plant absorption (Fig. 13.9). K-(potassium) solubilizing fungi (KSF) like *Aspergillus niger* and *A. terreus* were isolated from different K-rich soil (Prajapati et al., 2012). In a similar study, Lopes-Assad et al. (2010) exhibited the potential of *A. niger* as a potassic biofertilizer.

The findings gathered from different studies reveal that plant-growth-promoting fungi (e.g., *T. harzianum* T-22) have the ability to solubilize other minerals from their solid-phase compounds (e.g., MnO_2, Fe_2O_3, metallic zinc, potassium, etc.). A broad range of mechanisms including chemical entities may influence in the solubilization of different minerals. As for instances, the fungal chemical that chelated Fe is not able to chelate Mn. Furthermore, Fe^{3+} and Cu^{2+} were reduced to Fe^{2+} and Cu^{+} but these reductions were due to different fungal chemicals. In addition, the chemicals that chelate Fe are invariably not the same chemicals that reduce Fe^{3+} to Fe^{2+}; the fungal chemicals, which chelate iron, do not have reducing ability (Mostafa et al., 2016).

Penicillium bilaiae and *P. radicum* are very efficient for phosphate solubilization as tested in agrifield and so, they have been commercially cultivated and their preparations have been marketed in Canada and Australia as JumpStart (Philom Bios, Saskatoon, Canada) and PR-70 RELEASE (Bio-Care Technology, Somersby, Australia), respectively.

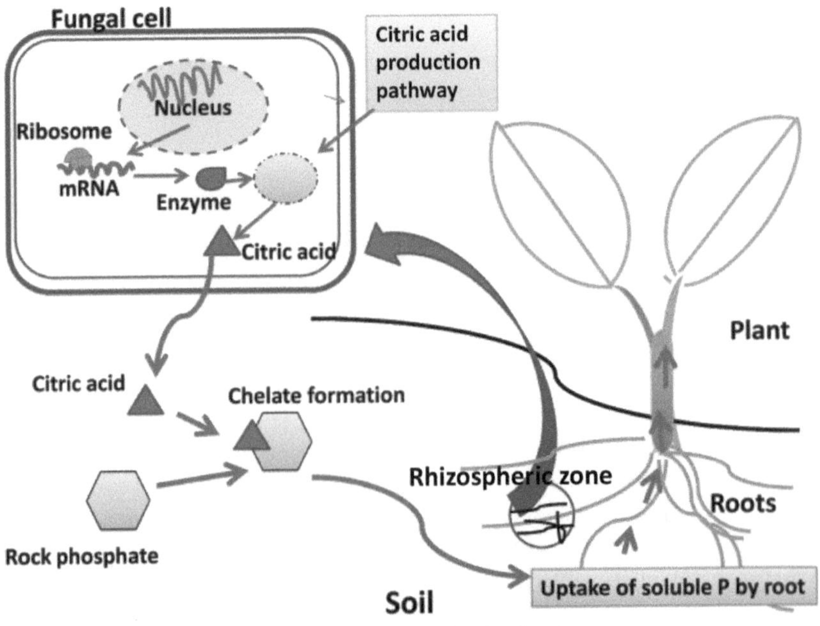

Fig. 13.9 Inorganic rock phosphate (P) solubilization by fungi.

In the plant system, Pi (inorganic phosphate) acquisition at the root periphery is coupled to proton entry (Pi: H+ symporter) and monitored by members of the *PHT1* (Phosphate Hydrogen Transporter 1) gene family (Schachtman et al., 1998). The *Arabidopsis* contains nine while rice genome has 13 PHT1 family members (Paszkowski et al., 2002; Poirier and Bucher, 2002) and these genes after expressing in roots encode proteins which act as high-affinity Pi uptake transporters (Nussaume et al., 2011; Remy et al., 2012). Regulation of this gene family takes place at the transcriptional as well as translational levels (Bouain et al., 2014; Rubio et al., 2009; Hammond and White, 2011). Now the question is whether fungi or fungal components can influence the upregulation of PHT1 gene family, but it is not reported and future research will give the answer.

13.3.5 Fungal siderophore-mediated pathway for iron solubilization, uptake, and plant-growth promotion

Although Fe is available in the earth crust (4.2%) and many soils (2%–60%), Fe availability is very often limiting for the growth of living organisms at all levels (Thomine and Lanquar, 2011). Iron deficiency is thought to limit agricultural yields on as much as 30% of arable lands. It is not accessible to plants as it exists in soil as ferric ions (Fe^{3+}) in the soil which is not very soluble (Lehner et al., 2013). Actually, iron acquisition is an object of a tough ecological vying. While plants require around 10^{-8} M Fe, the solubility of Fe^{3+} ranges from 10^{-17} M at pH 7 to 10^{-6} M at pH 3.3 (Marschner, 1997; Fox and Guerinot, 1998; Hell and Stephan, 2003). If we go through the mechanism of Fe solubilization, acquisition, and uptake of Fe in the plant root system, we found two strategies to efficiently take up Fe from the soil such as Strategy I (reductase based) and Strategy II (chelate-based).

In Strategy I (Fig. 13.10) at least three steps are involved such as the first step—acidification of the rhizospheric zone through the extrusion of protons to solubilize Fe chelates. This step is probably facilitated by fungi as fungi secrete organic acids (citric acid, oxalic acid, etc.), as a result, acidification is enhanced. The second step is the activation of ferric reductase activity while third step is the induction of a high-affinity ferrous Fe transport system. Whether second and third steps are facilitated by fungi, are not clear. *Arabidopsis* has 12 genes for encoding 12 H+-ATPases where AHA1, present at the plasma membrane, is not affected by Fe deficiency, suggesting a role in basal function independent of Fe deficiency. However, among the 12 H+-ATPase isoforms, AHA2 and AHA7 are upregulated by Fe deficiency. The solubilization of the FeIII chelates is followed by the reduction of ferric Fe to ferrous Fe performed by a ferric-chelate reductase. Among the eight members detected in *Arabidopsis*, AtFRO6 has also been localized at the plasma membrane. In the third step, the reduction of ferric chelates to ferrous Fe is followed by Fe^{2+} uptake where IRT1 (iron-regulated transporter 1) acts as transporter (Thomine and Lanquar, 2011). This strategy is found in nongraminaceous plants including dicotyledonous plants and also in rice.

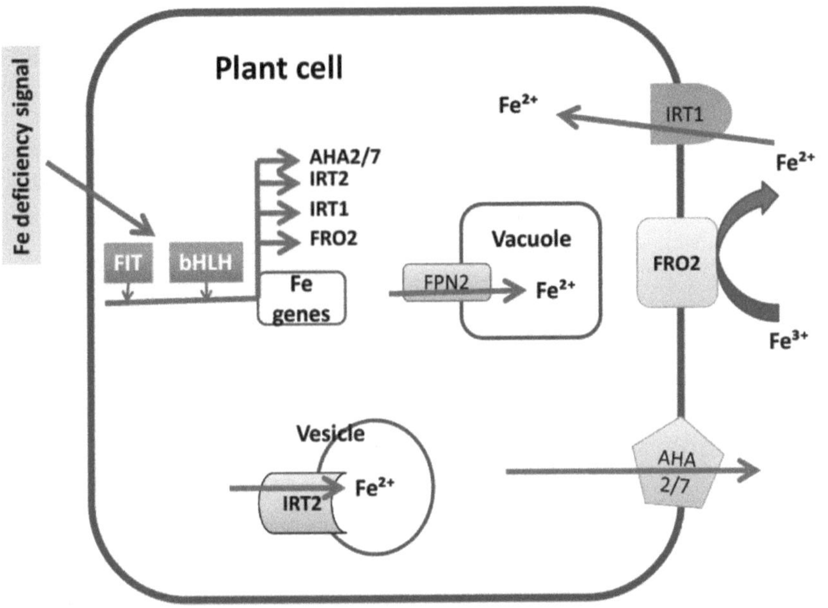

Fig. 13.10 Transport of Fe by chelation based strategy (Thomine and Lanquar, 2011). IRT1, iron regulated transporter 1 and AHA2/7, *Arabidopsis* 12 H+-ATPase 2/7 (plasma membrane H –ATPase) are upregulated by iron deficient condition; *FRO2*, ferric reductase-oxydase 2; *FPN2*, ferroportin 2; *bHLH (basic helix-loop-helix)*, a transcription factor/protein; *FIT*, *Fe-deficiency induced transcription factor 1*.

Strategy II (Fig. 13.11) occurs in graminaceous species, and mediated by phytosiderophores (PS), which bind Fe^{3+}. The expression of the genes encoding the enzymes involved for the biosynthesis of phytosiderophores is upregulated by Fe deficiency (Negishi et al., 2002). The mechanism of PS release from plant cells in the soil is not reported. However, different pathways involving transporters, channels, etc. are most probably operated here (Negishi et al., 2002; Nishizawa and Mori, 1987; Sakaguchi et al., 1999). As soon as PS are released in the rhizophere, they bind to Fe^{3+} and the complex is reimported into root cells through transporters. ZmYS1 gene encodes a plasma membrane transporter in maize while in rice 18 genes have been identified and out of them, OsYSL15 (Koike et al., 2004; Lee et al., 2009; Curie et al., 2001) and HvYS1 (Roberts et al., 2004; Schaaf et al., 2004) are primary Fe-PS uptake transporter. This is the PS-based plant's own mechanism of iron uptake. At the same way, many fungi by their own siderophores (FS) capture Fe and as a result, Fe^{2+} availability is found in the soil for plants (Vinale et al., 2013; Mendoza-Mendoza et al., 2017). Now we will discuss how FS is involved in Fe solubilization and how FS supplies Fe to plant. Eight different species of *Trichoderma*, such as *T. atroviride*, *T. asperellum*, *T. gamsii*, *T. hamatum*, *T. virens*, *T. harzianum*, *T. polysporum*, and *T. reesei* showed 12–14 siderophores, with six common to all species as recorded by Lehner et al. (2013) and are synthesized by three nonribosomal protein synthases (NRPs), which are present as a cluster in the genome (Vinale

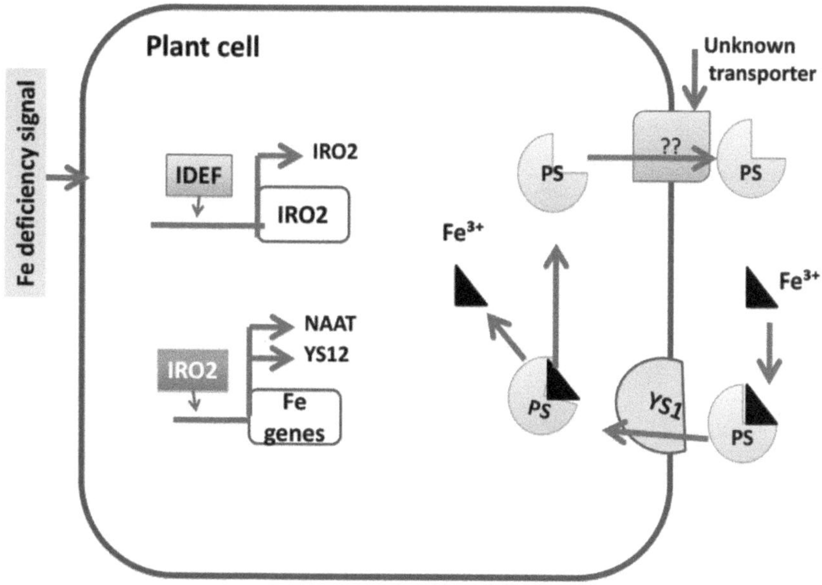

Fig. 13.11 Transport of Fe by strategy II (Thomine and Lanquar, 2011). Yellow-stripe 1 (YS1) transporter of PS (phytosiderophore); IDEF1 (iron deficiency-responsive *cis*-acting element/factor 1), transcription factor; NAAT, nicotianamine aminotransferase.

et al., 2013; Mukherjee et al., 2012; Zeilinger et al., 2016). *Although Aspergillus* has been reported to produce 55 types of siderophore, generally fungi have been reported to produce two types of siderophore like hydroxamate and carboxylate (von Wirén et al., 1994; Ghosh et al., 2015). *T. harzianum* produced a maximum percentage of siderophore (85.00%), followed by *T. viride* (65.50%), *T. asperellum* (60.27%) and *T. longibrachiatum* (45.50%). While studying the typification of siderophores, it was found that *T. harzianum* recorded maximum hydroxymate and carboxylate production whereas *T. viride*, *T. asperellum* and *T. longibrachiatum* recorded lesser production of hydroxymate and carboxylate as confirmed by color intensity. But none of the isolates was recorded with catecholate production (Ghosh et al., 2017). Hydroxamate siderophore fusarinine C and triacetyl fusarinine are utilized by *Aspergillus* for capturing extracellular iron, while ferricrocin siderophore reported intracellular iron distribution and conidial iron siderophore hydroxy-ferricrocin (Curie et al., 2001) for conidial iron storage.

Masalha et al. (2000) stated that microbial siderophores can chelate Fe from soils and then allow a ligand exchange with phytosiderophores. The chelation-based strategy has been represented in Fig. 13.12. There are different mechanisms which are suggested for siderophore-promoted Fe dissolution (Holmén and Casey, 1996). The general mechanism is that the Fe(III)-siderophore complex is formed at the mineral surface and is then transferred into the surrounding soil solution and becomes available for uptake by plants (Kalinowski et al., 2000; Kraemer, 2004). Fungi serve the plants for iron by four different mechanisms for siderophore-mediated Fe transport

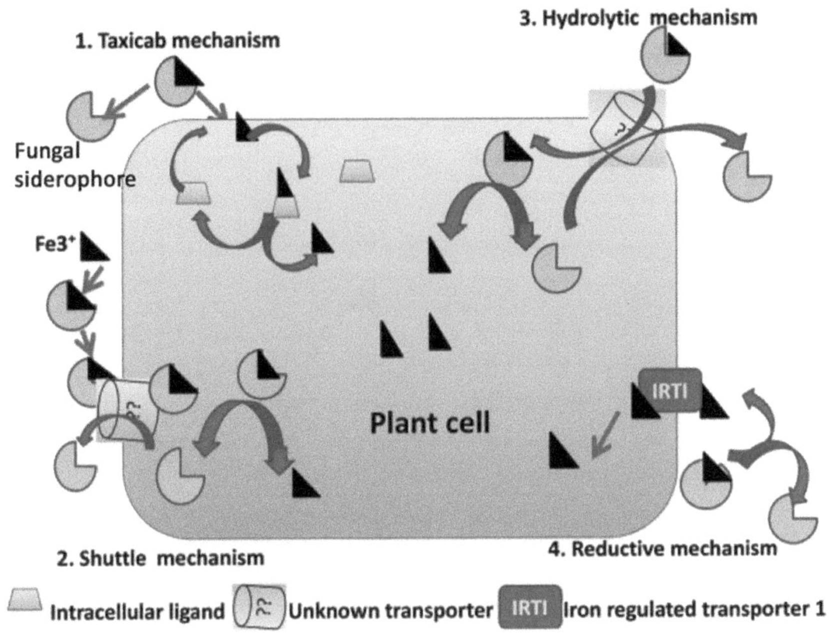

Fig. 13.12 Four different mechanisms of Fe transport in cell by siderophore (Van der Helm and Winkelmann, 1994).

systems (Fig. 13.12) (Van der Helm and Winkelmann, 1994): (i) in the shuttle mechanism, the Fe(III)-siderophore complex after attaching to the membrane, is transported across the cell membrane where the Fe(III) is dislodged from the ligand and reduced by enzyme, and the free siderophore is then back just outside the membrane of the plant and recycled, e.g., transporting ferrichrome by *Ustilago maydis* (Ardon et al., 1998), (ii) the taxicab mechanism where the Fe(III), after releasing just outside the membrane of the plant from the extracellular siderophore, is transported across the cell membrane to intracellular ligands, e.g., *Rhodotorula* species (Winkelmann and Huschka, 1987), (iii) in the hydrolytic mechanism, the whole Fe(III)-siderophore complex after attaching with plant membrane, is passed into the cell through membrane transporter and Fe(III) released by several reductive and degradative processes. The Fe(III) is reduced to Fe(II) inside the cell and the siderophore is excreted outside the cells for reuse again, e.g., Fe(III)-triacetylfusarinine complexes by *Mycelia sterilia* (Adjimani and Emery, 1988) and (iv) the reductive mechanism where the Fe(III)-siderophore complex is not transported across the cell membrane but comes near the membrane after releasing from complex, Fe(III) is reduced to Fe(II) at the cell membrane and then the reduced Fe is taken up by the cell through membrane transporter, e.g., the uptake of Fe(II) from Fe(III)-ferrichrome complex by *Ustilago sphaerogena* (Ecker and Emery, 1983). Although mineral iron dissolution or solubilization is performed by different organic acids secreted by microbes including fungi, the impact of siderophores on soil mineral weathering is more functional than that of LMMOAs (low molecular mass organic acids) as siderophores give more stable complexes with Fe. According to Matzanke (1991), 1:1 complex is formed between siderophore and Fe(III), with constants ranging between $K = 10^{30}$ and $K = 10^{52}$ but the constants of oxalic and citric acids with Fe(III) are $K = 10^{12}$ and 10^{8}, respectively (Perrin, 1979). Reichard et al. (2007) found that both LMMOAs and siderophores operate synergistically.

13.3.6 ACCD [1-aminocyclopropane-1-carboxylate (ACC) deaminase] mediated plant-growth promotion

It is mainly based on reduction of ethylene production by cleaving ACC which is the immediate precursor of ethylene with the enzyme 1-aminocyclopropane-1-carboxylate (ACC) deaminase produced by some fungi (Todorovic and Glick, 2008). This enzyme cleaves the ACC into α ketobutarate and ammonia (Fig. 13.13) (Van de Poel and Van Der Straeten, 2014). Ethylene is an important signaling molecule in plants which inhibits plant growth (Abels et al., 1992). PGPR/PGPF produces ACC deaminase (ACCD) which reduces ethylene levels and decreases inhibition of plant growth following environmental or pathogen-induced stress (Glick, 2014). ACCD activity has been observed in *Trichoderma asperellum* (Viterbo et al., 2010), *T. atroviride* (Gravel et al., 2007), and *Penicillium citrinum* (Jia et al., 2000). Two putative acdS genes were isolated from the genome of

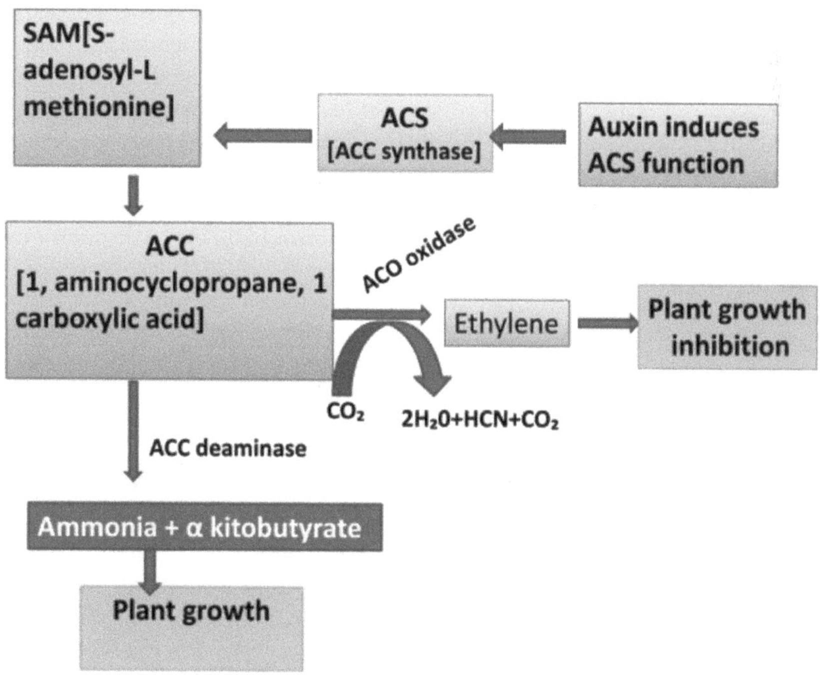

Fig. 13.13 ACC mediated pathway of Auxin induction (Van de Poel and Van Der Straeten, 2014).

Arabidopsis thaliana and data were placed validating the hypothesis that these genes can act as regulators of ACC levels in *A. thaliana* and also in tomato fruit development (McDonnell et al., 2009). Some *Trichoderma* spp. that have beneficial effects on plant growth (Harman et al., 2004), also possess ACCD putative sequences (http://genome.jgi-psf.org/Trive1/Trive1.home.html).

13.4 Mycorrhizal fungi (MF) as growth promoter

Among the different mycorrhizae, arbuscular mycorrhizae (AM) previously known as Vesicular arbuscular mycorrhiza (VAM), has an association with maximum crops in agriculture and its role in crop growth promotion has taken much-paid attention of modern scientists. On the other hand, ectomycorrhizal association is very frequent on the roots of forest plants and their contribution for the growth of forest tree has been highlighted and applied in afforestation particularly in barren and fallow lands. Harley and Smith (1983) mentioned that ectomycorrhizal seedlings are taller and the root system is larger, and phosphate absorption in plant root is more stimulated in mycorrhizal root than nonmycorrhizal root. Due to limited nutrient reserves in orchid seeds, it is recorded that, after immediate seed germination, orchideous mycorrhizal association is compulsory for continuous development of embryos (Harley and Smith, 1983). Recently in *Laccaria laccata* and *L. bicolor* (two ectomycorrhizal basidiomycetes), the principal siderophore has been reported which has the ester-containing siderophore linear fusigen in addition to coprogen, ferricrocin and triacetylfusarinine C (Haselwandter et al., 2013; Khan et al., 2018). According to Jones et al. (1998), the efficiency with which mycorrhizal plants uptake Pi, is 3.1–4.7 times higher than that of nonmycorrhizal plants. Mycorrhizal associations increase the growth and yield of many crop plants. The complex cellular relationship between host roots and AM fungi (AMF) requires a continuous exchange of signals, for proper development of mycorrhiza in the roots of a host plant (Gianinazzi-Pearson, 1996). Plant hormones such as IAA and cytokinins which are released by mycorrhizal fungi may also contribute to the enhancement of plant growth (Frankenberger Jr and Arshad, 1995). Now AMF are recognized as multifunctional (Sikes, 2010). The nutritional and nonnutritional advantages of AMF symbiosis was analyzed by Smith and Smith (2012). AMF in the family *Gigasporaceae* are very potential in enhancing plant P (Maherali and Klironomos, 2007) and by several mechanisms, influences C, P, and N dynamics (Correa et al., 2015). The mycelia of AMF penetrate inside the soil, do recycle of C and also do better soil texture (Olsson et al., 1999). Millner and Wright (2002) recorded that C-cost to the plant is balanced by access to a greater volume of soil through fungal hyphae having a larger surface area to volume ratio than to root hairs and it goes up to 8 cm beyond nutrient depletion zones around roots. AMF by "AM pathway" scavenge P from large volumes of soil and rapidly serve to cortical cells of the root (Smith et al., 2011). Other mechanisms may be operated here such as rhizosphere acidification, and they increase in root phosphatase activity or excretion of chelating agents. The rhizosphere acidification by AMF mobilizes P from sources that are not available to the control plant (Bago and Azcon-Aguilar, 1997). Fries et al. (1998) exhibited that APase and alkaline phosphatase activities in maize roots were closely correlated to levels of AMF colonization of roots. Tarafdar and Marschner (1994) found that 48%–59% of the P absorbed by wheat was due to mobilization of Pi from P_o sources by phosphatase produced by AMF. Rillig and Mummey (2006) noted that AMF produces glomalin (acts as glue), hydrophobin protein (provides mycelium attachment to surfaces, alteration of biotic or abiotic surface properties, and lowering water tension) and mucilage (provides attachment, nutrient capture, and desiccation resistance) that make better health of the soil. Further works also state that AMF colonization helps in N uptake and increases amino acids and organic acids in shoots and roots (Bucking and Kafl, 2015).

13.5 Conclusion

It is very clear from above narration that there are many rhizospheric and mycorrhizal fungi which have a definite role for plant-growth promotion. The mechanisms for plant-growth promotion relies on the capacity of hormone (auxin, GA, cytokinin) secretion and induction of auxin-mediated growth, and P-solubilization from rock phosphate by different organic acid secretion and chelate formation for easy plant absorption, capturing iron element from the soil by siderophore and serving plant roots and also changing soil texture and other quality in favor of plant root formation. Auxin-dependent cell signaling pathway is activated mainly. Auxin changes the cellular response via short transcription pathway and in response to auxin, many genes change their expression. FS-mediated pathway is operated for Fe solubilization and uptake. Mainly three mechanisms such as shuttle mechanism, taxicab mechanism, and reductase mechanism are involved here. ACCD-mediated pathway induced plant growth by few fungi has been also recorded in few plants. More research needs to decipher whether other PGPF-dependent mechanisms or cell signals are operated or not. The mechanism-based research will help for proper biofertilizer formulation, preparation, and ecofriendly application in agrifields for enhancing crop productivity and this technique will reduce the application of chemical fertilizers.

Acknowledgment

Authors are grateful to the Department of Science and Biotechnology, GoWB for financial assistance and Principal, RKMVC College, for providing Lab. facilities.

References

Abels, F.B., Morgan, P.W., Saltveit, M.E., 1992. Ethylene in Plant Biology. Academic Press, New York.

Adjimani, J.P., Emery, T., 1988. Stereochemical aspects of iron transport in *Mycelia sterilia* EP-76. J. Bacteriol. 170, 1377–1379.

Alabouvette, C., Couteadier, Y., 1992. Biological control of plant diseases: progress and challenges for the future. In: Tjamos, E.C., Papavizas, G.C., Cook, R.J. (Eds.), Biological Control of Plant Diseases. Plenum Press, New York, pp. 415–426.

Ardon, O., Nudelman, R., Caris, C., Libman, J., Shanzer, A., Chen, Y., Hadar, Y., 1998. Iron uptake in *Ustilago maydis*: tracking the iron path. J. Bacteriol. 180, 2021–2026.

Arora, D.K., Elander, R.P., Mukerji, K.G. (Eds.), 1992. Handbook of Applied Mycology: Fungal Biotechnology. In: vol 4. Marcel Dekker, New York.

Bago, B., Azcon-Aguilar, C., 1997. Changes in the rhizosphere pH induced by arbuscular mycorrhiza formation in onion (*Allium cepaa*). Z. Pflanzen. Bodenk. 160, 333–339.

Benítez, T., Delgado-Jarana, J., Rincón, A.M., Rey, M., Limón, M.C., 1998. Biofungicides: *Trichoderma* as a biocontrol agent against phytopathogenic fungi. In: Pandalai, S.G. (Ed.), Recent Research Developments in Microbiology. In: 2, Research Signpost, Trivandrum, pp. 129–150.

Bouain, N., Shahzad, Z., Rouached, A., Khan, G.A., Berthomieu, P., Abdelly, C., Poirier, Y., Rouached, H., 2014. Phosphate and zinc transport and signaling in plants: toward a better understanding of their homeostasis interaction. J. Exp. Bot. 65 (20), 5725–5741.

Brotman, Y., Briff, E., Viterbo, A., Chet, I., 2008. Role of swollenin, an expansin-like protein from *Trichoderma*, in plant root colonization. Plant Physiol. 147 (2), 779–789.

Bucking, H., Kafl, A., 2015. Role of arbuscular mycorrhizal fungi in the nitrogen uptake of plants: current knowledge and research gaps. Agronomy 5, 587–612.

Chacón, M.R., Rodríguez Galán, O., Benítez Fernández, C.T., Sousa, S., Rey, M., Llobell González, A., Delgado Jarana, J., 2007. Microscopic and transcriptome analyses of early colonization of tomato roots by "*Trichoderma harzianum*" Int. Microbiol. 10 (1), 19–27.

Chang, Y., 1986. Increased growth of plants in the presence of the biological control agent *Trichoderma harzianum*. Plant Dis. 70, 145–148.

Chapman, E.J., Estelle, M., 2009. Mechanism of auxin-regulated gene expression in plants. Annu. Rev. Genet. 43, 265–285.

Chet, I., Inbar, J., Hadar, I., Wicklow, D.T., Söderström, B., 1997. Fungal antagonists and mycoparasites. In: The Mycota IV: Environmental and Microbial Relationships. Springer-Verlag, Berlin, pp. 165–184.

Contreras-Cornejo, H.A., Macías-Rodríguez, L., Cortés-Penagos, C., López-Bucio, J., 2009. *Trichoderma virens*, a plant beneficial fungus, enhances biomass production and promotes lateral root growth through an auxin-dependent mechanism in *Arabidopsis*. Plant Physiol. 149 (3), 1579–1592.

Correa, A., Cruz, C., Ferrol, N., 2015. Nitrogen and carbon/nitrogen dynamics in arbuscular mycorrhiza: the great unknown. Mycorrhiza 25, 499–515.

Cunningham, J.E., Kuiack, C., 1992. Production of citric and oxalic acids and solubilisation of calcium phosphate by *Penicillium bilaji*. Appl. Environ. Microbiol. 58, 1451–1458.

Curie, C., Panaviene, Z., Loulergue, C., Dellaporta, S.L., Briat, J.F., Walker, E.L., 2001. Maize yellow stripe 1 encodes a membrane protein directly involved in Fe (III) uptake. Nature 409 (6818), 346.

Delgado-Jarana, J., Moreno-Mateos, M.A., Benítez, T., 2003. Glucose uptake in *Trichoderma harzianum*: role of gtt 1. Eukaryot. Cell 2 (4), 708–717.

Dharmasiri, N., Dharmasiri, S., Estelle, M., 2005. The F-box protein TIR1 is an auxin receptor. Nature 435, 441–445.

Dutton, M.V., Evans, C.S., Atkey, P.T., Wood, D.A., 1993. Oxalate production by basidiomycetes, including the white-rot species *Coriolus versicolor* and *Phanerochaete chrysosporium*. Appl. Microbiol. Biotechnol. 39 (1), 5–10.

Ecker, D.J., Emery, T., 1983. Iron uptake from ferrichrome A and iron citrate in *Ustilago sphaerogena*. J. Bacteriol. 155, 616–622.

Elsharkawy, M.M., Shimizu, M., Takahashi, H., Hyakumachi, M., 2012. The plant growth-promoting fungus *Fusarium equiseti* and the arbuscular mycorrhizal fungus *Glomus mosseae* induce systemic resistance against Cucumber mosaic virus in cucumber plants. Plant and Soil 361, 397–409.

Finley, R.D., 2007. The fungi in soil. In: van Elsas, J.D., Jansson, J.K., Travors, J.T. (Eds.), Modern Soil Microbiology. CRC Press, New York, pp. 107–146.

Fletcher, J., Bender, C., Budowle, B., Cobb, W.T., Gold, S.E., Ishimaru, C.A., Luster, D., Melcher, U., Murch, R., Scherm, H., Seem, R.C., Sherwood, J.L., Sobral, B.W., Tolin, S.A., 2006. Plant pathogen forensics: capabilities, needs, and recommendations. Microbiol. Mol. Biol. Rev. 70 (2), 450–471.

Fox, T.C., Guerinot, M.L., 1998. Molecular biology of cation transport in plants. Annu. Rev. Plant. Physiol. Plant. Mol. Biol. 49, 669–696.

Frankenberger Jr., W.T., Arshad, M., 1995. Microbial synthesis of auxins. In: Frankberger Jr., W.T., Arshad, M. (Eds.), Phytohormones in Soils. Marcel Deeker Inc., New York, pp. 35–71.

Fries, L.L.M., Pacovsky, R.S., Safir, G.R., Kaminski, J., 1998. Phosphorus effect on phosphatase activity in endomycorrhizal maize. Physiol. Plant. 103, 152–171.

Gaddie, G.M., 1993. Interaction of fungi with toxic metals. New Phytol. 124, 25–60.

Garnica-Vergara, A., Barrera-Ortiz, S., Munoz-Parra, E., Raya-Gonzalez, J., Mendez-Bravo, A., et al., 2016. The volatile 6-pentyl-2H-pyran-2-one from *Trichoderma atroviride* regulates *Arabidopsis thaliana* root morphogenesis via auxin signaling and ethylene insensitive 2 functioning. New Phytol. 209, 1496–1512.

Garon, D., Krivobok, S., Seigle-Murandi, F., 2000. Fungal degradation of fluorene. Chemosphere 40, 91–97.

Ghosh, S.K., 2017a. Study of effects of antagonistic soil microbes on rice early stand establishment, prevention of seedling mortality and induction of peroxidase and catalase enzymes. Plant Cell Biotechnol. Mol. Biol. 18 (5&6), 306–315.

Ghosh, S.K., 2017b. Study of some antagonistic soil fungi for protection of fruit rot (*Phomopsis Vexans*) and growth promotion of brinjal. Int. J. Adv. Res. 5 (7), 485–494.

Ghosh, S.K., Pal, S., 2017. Growth promotion and *Fusarium* wilt disease management eco-friendly in chickpea by *Trichoderma asperellum*. Int. J. Curr. Res. Aca Rev. 5 (1), 14–26.

Ghosh, S.K., Pal, S., Ray, S., 2013. Study of microbes having potentiality for biodegradation of plastics. Environ. Sci. Pollut. Res. 20, 4339–4355.

Ghosh, S.K., Pal, S., Chakraborty, N., 2015. The qualitative and quantitative assay of siderophore production by some microorganisms and effect of different media on its production. Int. J. Chem. Sci. 13 (4), 1621–1629.

Ghosh, S.K., Banerjee, S., Sengupta, C., 2017. Bioassay, characterization and estimation of siderophores from some important antagonistic fungi. J. Biopest. 10 (2), 105–112.

Gianinazzi-Pearson, V., 1996. Plant cell responses to arbuscular mycorrhizal fungi getting to the roots of the symbiosis. Plant Cell 8, 1871–1883.

Glick, B.R., 2014. Bacteria with ACC deaminase can promote plant growth and help to feed the world. Microbiol. Res. 169, 30–39.

Goldstein, A.H., 1994. Involvement of the quinoprotein glucose dehydrohenase in the solubilization of exogenous phosphates by Gram-negative bacteria. In: Torriani-Gorini, A., Yagil, E., Silver, S. (Eds.), Phosphate in Microorganisms: Cellular and Molecular Biology. ASM Press, Washington (DC), pp. 197–203.

Gravel, V., Antoun, H., Tweddel, R.H., 2007. Growth stimulation and fruit yield improvement of greenhouse tomato plants by inoculation with *Pseudomonas putida* or *Trichoderma atroviride*: possible role of indole acetic acid (IAA). Soil Biol. Biochem. 39, 1968–1977.

Halvorson, H.O., Keynan, A., Kornberg, H.L., 1990. Utilization of calcium phosphates for microbial growth at alkaline pH. Soil Biol. Biochem. 22, 887–890.

Hammond, J.P., White, P.J., 2011. Sugar signaling in root responses to low phosphorus availability. Plant Physiol. 156, 1033–1040.

Harley, J.L., Smith, S.E., 1983. Mycorrhizal Symbiosis. Academic Press, London and New York.

Harman, G.E., Howell, C.R., Viterbo, A., Chet, I., Lorito, M., 2004. *Trichoderma* species: opportunistic, avirulent plant symbionts. Nat. Rev. Microbiol. 2, 43–56.

Haselwandter, K., Häninger, G., Ganzera, M., Haas, H., Nicholson, G., Winkelmann, G., 2013. Linear fusigen as the major hydroxamate siderophore of the ectomycorrhizal Basidiomycota *Laccaria laccata* and *Laccaria bicolor*? Biometals 26 (6), 969–979.

Hell, R., Stephan, U.W., 2003. Iron uptake, trafficking and homeostasis in plants. Planta 216, 541–551.

Hermosa, R., Botella, L., Keck, E., Jimenez, J.A., Montero Barrientos, M., Arbona, V., Gomez-Cadenas, A., Monte, E., Nicolas, C., 2011. The over expression in *Arabidopsis thaliana* of a *Trichoderma harzianum* gene that modulates glucosidase activity, and enhances tolerance to salt and osmotic stresses. J. Plant Physiol. 168, 1295–1302.

Hermosa, R., Viterbo, A., Chet, I., Monte, E., 2012. Plant-beneficial effects of *Trichoderma* and of its genes. Microbiology 158 (1), 17–25.

Holmén, B.A., Casey, W.H., 1996. Hydroxamate ligands, surface chemistry, and the mechanism of ligandpromoted dissolution of goethite [α-FeOOH(s)]. Geochim Cosmochim Acta 60, 4403–4416.

Hyakumachi, M., Kubota, M., 2003. Fungi as plant growth promoter and disease suppressor. In: Arora, D. (Ed.), Fungal Biotechnology in Agricultural, Food and Environmental Application. CRC Press, Boca Raton.

Illmer, P.A., Schinner, F., 1992. Solubilization of inorganic phosphates by microorganisms isolated from forest soil. Soil Biol. Biochem. 24, 389–395.

Illmer, P., Schinner, F., 1995. Solubilization of inorganic calcium phosphates solubilization mechanisms. Soil Biol. Biochem. 27, 257–263.

Inbar, J., Abramsky, M., Cohen, D., Chet, I., 1994. Plant growth enhancement and disease control by *Trichoderma harzianum* in vegetable seedlings grown under commercial conditions. Eur. J. Plant Pathol. 100, 337–346.

Jia, Y.J., Ito, H., Matsui, H., Honma, M., 2000. 1-Aminocyclopropane-1-carboxylate (ACC) deaminase induced by ACC synthesized and accumulated in *Penicillium citrinum* intracellular spaces. Biosci. Biotech. Bioch. 64, 299–305.

Jogaiah, S., Mostafa, A., Tran, L.-S.P., Shin-ichi, I., 2013. Characterization of rhizosphere fungi that mediate resistance in tomato against bacterial wilt disease. J. Exp. Bot. 64, 3829–3842.

Jogaiah, S., Abdelrahman, M., Tran, L.S.P., Ito, S.I., 2018. Different mechanisms of *Trichoderma virens*-mediated resistance in tomato against *Fusarium* wilt involve the jasmonic and salicylic acid pathways. Mol. Plant Pathol. 19, 870–882.

Jones, M.D., Durall, D.M., Tinker, P.B., 1998. A comparison of arbuscular and ectomycorrhizal *Eucalyptus coccifera*: growth response, phosphorus uptake efficiency, and external hyphal production. New Phytol. 140, 125–134.

Kalinowski, B.E., Liermann, L.J., Brantley, S.L., Barnes, A.S., Pantano, C.G., 2000. X-ray photoelectron evidence for bacteria-enhanced dissolution of hornblende. Geochim. Cosmo. Chim. Acta 64, 1331–1343.

Kepinski, S., Leyser, O., 2005. The Arabidopsis F-box protein TIR1 is an auxin receptor. Nature 435, 446–451.

Khan, M.S., Zaidi, A., Wani, P.A., 2007. Role of phosphate-solubilizing microorganisms in sustainable agriculture—a review. Agron. Sustain. Dev. 27 (1), 29–43.

Khan, A., Singh, P., Srivastava, A., 2018. Synthesis, nature and utility of universal iron chelator—siderophore: a review. Microbiol. Res. 11-12, 103–111.

Kleifeld, O., Chet, I., 1992. *Trichoderma harzianum* interaction with plants and effect on growth response. Plant and Soil 144, 267–272.

Koike, S., Inoue, H., Mizuno, D., Takahashi, M., Nakanishi, H., Mori, S., Nishizawa, N.K., 2004. OsYSL2 is a rice metal-nicotianamine transporter that is regulated by iron and expressed in the phloem. Plant J. 39 (3), 415–424.

Korolev, N., David, D.R., Elad, Y., 2008. The role of phytohormones in basal resistance and Trichoderma-induced systemic resistance to *Botrytis cinerea* in *Arabidopsis thaliana*. Bio Control 53 (4), 667–683.

Kraemer, S.M., 2004. Iron oxide dissolution and solubility in the presence of siderophores. Aquat Sci. 66, 3–18.

Kristiansen, B., Sinclair, C.G., 1979. Production of citric acid in continuous culture. Biotechnol. Bioeng. 21 (2), 297–315.

Kumar, K., Manigundan, K., Natarajan, A., 2016. Influence of salt tolerant *Trichoderma* spp. on growth of maize (*Zea mays*) under different salinity conditions. J. Basic Microbiol. 56, 1–10.

Lavy, M., Estelle, M., 2016. Mechanisms of auxin signaling. Development 143, 3226–3229.

Lee, S., Chiecko, J.C., Kim, S.A., Walker, E.L., Lee, Y., Guerinot, M.L., An, G., 2009. Disruption of OsYSL15 leads to iron inefficiency in rice plants. Plant Physiol. 150 (2), 786–800.

Lehner, S.M., Atanasova, L., Neumann, N.K., Krska, R., Lemmens, M., et al., 2013. Isotope-assisted screening for iron-containing metabolites reveals a high degree of diversity among known and unknown siderophores produced by *Trichoderma* spp. Appl. Environ. Microbiol. 79, 18–31.

Leyser, O., 2018. Auxin signaling. Plant Physiol. 176, 465–479.

van Loon, L.C., Rep, M., Pieterse, C.M., 2006. Significance of inducible defense-related proteins in infected plants. Annu. Rev. Phytopathol. 44, 135–162.

Lopes-Assad, M.L., Avansini, S.H., Rosa, M.M., Carvalho, J.R.P., Ceccato-Antonini, S.R., 2010. The solubilization of potassium-bearing rock powder by *Aspergillus niger* in small-scale batch fermentations. Can. J. Microbiol. 56, 598–605.

Mach, R.L., Zeilinger, S., 2003. Regulation of gene expression in industrial fungi: *Trichoderma*. Appl. Microbiol. Biotechnol. 60, 515–522.

Maherali, H., Klironomos, J., 2007. Influence of phylogeny on fungal community assembly and ecosystem functioning. Science 316, 1746–1748.

Maliha, R., Samina, K., Najma, A., Sadia, A., Farooq, L., 2004. Organic acids production and phosphate solubilization by phosphate solubilizing microorganisms under in vitro conditions. Pak. J. Biol. Sci. 7, 187–196.

Marschner, H., 1997. Mineral Nutrition of Higher Plants, 2nd edn Academic Press, London.

Masalha, K.H., Elmaci, Ö., Mengel, K., 2000. The central role of microbial activity for iron acquisition in maize and sunflower. Biol. Fertil. Soils 30, 433–439.

Matzanke, B.F., 1991. Structures, coordination chemistry and functions of microbial iron chelates. In: Winkelmann, G. (Ed.), CRC Handbook of Microbial Iron Chelates. CRC Press, Boca Raton, FL, USA, pp. 15–64.

McDonnell, L., Plett, J.M., Andersson-Gunneras, S., Kozela, C., Dugardeyn, J., Van Der Straeten, D., Glick, B.R., Sundberg, B., Regan, S., 2009. Ethylene levels are regulated by a plant encoded 1-aminocyclopropane-1-carboxylic acid deaminase. Physiol. Plant. 136, 94–109.

Mendes, G.O., Dias, C.S., Silva, I.R., Júnior, J.I.R., Pereira, O.L., Costa, M.D., 2013. Fungal rock phosphate solubilization using sugarcane bagasse. World J. Microbiol. Biotechnol. 29 (1), 43–50.

Mendoza-Mendoza, A., Nogueira-López, G., Arizmendi, F.P., Cripps-Guazzone, N., Nieto-Jacobo, M.F., Lawry, R., Kandula, D., Salazar-Badillo, F.B., Salas-Muñoz, S.S., Jorge Armando Mauricio-Castillo, J.A., Robert Hill, R., Stewart, A., Steyaert, J., 2017. Mechanisms of growth promotion by members of the rhizosphere fungal genus *Trichoderma*. In: Singh, H.B., Sarma, B.K., Keswani, C. (Eds.), Advances in PGPR Research. CAB International.

Millner, P.D., Wright, S.F., 2002. Tools for support of ecological research on arbuscular mycorrhizal fungi. Symbiosis 33, 101–123.

Mishra, J., Singh, R., Arora, N.K., 2017. Plant growth-promoting microbes: diverse roles in agriculture and environmental sustainability. In: Kumar, V. et al., (Ed.), Probiotics and Plant Health. Springer Nature Singapore Pte Ltd, p. 71. https://doi.org/10.1007/978-981-10-3473-2.

Moran-Diez, E., Hermosa, R., Ambrosino, P., Cardoza, R.E., Gutierrez, S., Lorito, M., Monte, E., 2009. The ThPG1 endopolygalacturonase is required for the trichoderma harzianum-plant beneficial interaction. Mol. Plant Microbe Interact. 22 (8), 1021–1031.

Morley, G.F., Gadd, G.M., 1995. Sorption of toxic metals by fungi and clay minerals. Mycol. Res. 99, 1429–1438.

Mostafa, A., Abdel-Motaal, F., El-Sayed, M., Jogaiah, S., Shigyo, M., Ito, S.-I., Tran, L.S.P., 2016. Dissection of *Trichoderma longibrachiatum*-induced defense in onion (*Allium cepa* L.) against *Fusarium oxysporum* f. sp. *Cepa* by target metabolite profiling. Plant Sci. 246, 128–138.

Mukherjee, P.K., Buensanteai, N., Moran-Diez, M.E., Druzhinina, I.S., Kenerley, C.M., 2012. Functional analysis of non-ribosomal peptide synthetases (NRPSs) in *Trichoderma virens* reveals a polyketide synthase (PKS)/NRPS hybrid enzyme involved in the induced systemic resistance response in maize. Microbiology 158, 155–165.

Murali, M., Amruthesh, K.N., Sudisha, J., Niranjana, S.R., Shetty, H.S., 2012. Screening for plant growth promoting fungi and their ability for growth promotion and induction of resistance in pearl millet against downy mildew disease. J. Phytology 4 (5), 30–36.

Murali, M., Jogaiah, S., Amruthesh, K.N., Shin-ichi, I., Shekar Shetty, H., 2013. Rhizosphere fungus *Penicillium chrysogenum* promotes growth and induces defence-related genes and downy mildew disease resistance in pearl millet. Plant Biol. 15, 111–118.

Nagaraju, A., Jogaiah, S., Mahadevamurthy, S., Shin-ichi, I., 2012. Seed priming with *Trichoderma harzianum* isolates enhances plant growth and induces resistance against *Plasmopara halstedii*, an incitant of sunflower downy mildew disease. Aust. J. Plant Path. 41, 609–620.

Naznin, H.A., Kimura, M., Miyazawa, M., Hyakumachi, M., 2013. Analysis of volatile organic compounds emitted by plant growth-promoting fungus *Phoma* sp. GS8–3 for growth promotion effects on tobacco. Microbes Environ. 28, 42–49.

Negishi, T., Nakanishi, H., Yazaki, J., Kishimoto, N., Fujii, F., Shimbo, K., Yamamoto, K., Sakata, K., Sasaki, T., Kikuchi, S., Mori, S., Nishizawa, N.K., 2002. cDNA microarray analysis of gene expression during Fe-deficiency stress in barley suggests that polar transport of vesicles is implicated in phytosiderophore secretion in Fe-deficient barley roots. Plant J. 30, 83–94.

Nishizawa, N., Mori, S., 1987. The particular vesicle appearing in barley root cells and its relation to mugineic acid secretion. J. Plant Nutr. 10, 1013–1020.

Nussaume, L., Kanno, S., Javot, H., Marin, E., Pochon, N., Ayadi, A., Nakanishi, T.M., Thibaud, M.C., 2011. Phosphate import in plants: focus on the PHT1 transporters. Front. Plant Sci. 2, 83.

Olsson, P.A., Thingstrup, I., Jakobsen, I., Baath, E., 1999. Estimation of the biomass of arbuscular mycorrhizal fungi in a linseed field. Soil Biol. Biochem. 31, 1879–1887.

Omar, S.A., 1998. The role of rock phosphate solubilizing fungi and vesicular arbuscular mycorrhiza (VAM) in growth of wheat plants fertilized with rock phosphate. World J. Microbiol. Biotechnol. 14, 211–219.

Osiewacz, H.D., 2002. Genes, mitochondria and aging in filamentous fungi. Ageing Res. Rev. 1, 425–442.

PAN. 2007. Available from: http://www.paninternational.org/panit/file/WG1/Eliminatingtheworstpesticide.pdf

Papagianni, M., 2007. Advances in citric acid fermentation by *Aspergillus niger*: biochemical aspects, membrane transport and modeling. Biotechnol. Adv. 25, 244–263.

Paponov, I.A., Paponov, M., Teale, W., Menges, M., Chakrabortee, S., Murray, J.A., Palme, K., 2008. Comprehensive transcriptome analysis of auxin responses in Arabidopsis. Mol. Plant 1 (2), 321–337.

Paszkowski, U., Kroken, S., Roux, C., Briggs, S.P., 2002. Rice phosphate transporters include an evolutionarily divergent gene specifically activated in arbuscular mycorrhizal symbiosis. Proc. Natl. Acad. Sci. USA 99, 13324–13329.

Patel, D.K., Archana, G., Kumar, G.N., 2008. Variation in the nature of organic acid secretion and mineral phosphate solubilization by *Citrobacter* sp. DHRSS in the presence of different sugars. Curr. Microbiol. 56 (2), 168–174.

Pelosi, C., Barot, S., Capowiez, Y., Hedde, M., Vandenbulcke, F., 2013. Pesticides and earthworms: a review. Agron. Sustain. Dev. 34, 199–228.

Perrin, D.D., 1979. Stability Constants: Part B. IUPAC, Pergamon, Turkey.

Perrot-Rechenmann, C., 2010. Cellular responses to auxin: division versus expansion. Cold Spring Harb. Perspect. Biol. 2, a001446.

Pinedo-Rivilla, C., Aleu, J., Collado, I.G., 2009. Pollutants biodegradation by fungi. Curr. Org. Chem. 13, 1194–1214.

Poirier, Y., Bucher, M., 2002. Somerville, C.R., Meycrowitz, E.M. (Eds.), The Arabidopsis Book. American Society of Plant Biologists, Rockville, MD e0024.

Powler D (2006) Blue baby links—11 February 2006—New Scientist. Retrieved 2008-03-24.

Prajapati, K., Sharma, M.C., Modi, H.A., 2012. Isolation of two potassium solubilizing fungi from ceramic industry soils. Life Sci. Leaflets 5, 71–75.

Ram, C.V., 1959. Production of growth-promoting substances by fusaria and their action on root elongation in *Oryza sativa* L. In: Proceedings of the Indian Academy of Sciences-Section B. Vol. 49. Springer India, pp. 167–182.

Reichard, P.U., Kretzschmar, R., Kraemer, S.M., 2007. Dissolution mechanisms of goethite in the presence of siderophores and organic acids. Geochim. Cosmochim. Acta 71, 5635–5650.

Remy, E., Cabrito, T.R., Batista, R.A., Teixeira, M.C., Sa-Correia, I., Duque, P., 2012. The Pht 1; 9 and Pht 1; 8 transporters mediate inorganic phosphate acquisition by the Arabidopsis thaliana root during phosphorus starvation. New Phytol. 195, 356–371.

Rillig, M., Mummey, D.L., 2006. Mycorrhizas and soil structure. New Phytol. 171, 41–53.

Roberts, L.A., Pierson, A.J., Panaviene, Z., Walker, E.L., 2004. Yellow stripe 1. Expanded roles for the maize iron-phytosiderophore transporter. Plant Physiol. 135 (1), 112–120.

Rodriguez, H., Fraga, R., 1999. Phosphate solubilizing bacteria and their role in plant growth promotion. Biotechnol. Adv. 17 (4-5), 319–339.

Rubio, V., Bustos, R., Irigoyen, M.L., Cardona-Lopez, X., Rojas-Triana, M., Paz-Ares, J., 2009. Plant hormones and nutrient signaling. Plant Mol. Biol. 69, 361–373.

Sakaguchi, T., Nishizawa, N., Nakanishi, H., Yoshimura, E., Mori, S., 1999. The role of potassium in the secretion of mugineic acids family phytosiderophores from iron-deficient barley roots. Plant and Soil 215, 221–227.

Saldajeno, M.G.B., Hyakumachi, M., 2011. The plant growth-promoting fungus *Fusarium equiseti* and the arbuscular mycorrhizal fungus *Glomus mosseae* stimulate plant growth andreduce severity of anthracnose and damping-off diseases in cucumber (*Cucumis sativus*)seedlings. Ann. Appl. Biol. 159, 28–40.

Salehin, M., Bagchi, R., Estelle, M., 2015. SCFTIR1/AFB-based auxin perception: mechanism and role in plant growth and development. Plant Cell. pp. tpc-114.

Schaaf, G., Ludewig, U., Erenoglu, B.E., Mori, S., Kitahara, T., von Wirén, N., 2004. ZmYS1 functions as a proton-coupled symporter for phytosiderophore-and nicotianamine-chelated metals. J. Biol. Chem. 279 (10), 9091–9096.

Schachtman, D.P., Reid, R.J., Ayling, S.M., 1998. Phosphorus uptake by plants: from soil to cell. Plant Physiol. 116, 447–453.

Scheffer, F., Schachtschabel, P., 1988. Lehrbuch der Bodenkunde. Enke, Stuttgart.

Segarra, G., Van der Ent, S., Trillas, I., Pieterse, C.M.J., 2009. MYB72, a node of convergence in induced systemic resistance triggered by a fungal and a bacterial beneficial microbe. Plant Biol. 11 (1), 90–96.

Shimada, M., Akamtsu, Y., Tokimatsu, T., Mii, K., Hattori, T., 1997. Possible biochemical roles of oxalic acid as a low molecular weight compound involved in brown-rot and white-rot wood decays. J. Biotechnol. 53 (2-3), 103–113.

Shoresh, M., Harman, G.E., Mastouri, F., 2010. Induced systemic resistance and plant responses to fungal biocontrol agents. Annu. Rev. Phytopathol. 48, 21–43.

Sikes, B.A., 2010. When do arbuscular mycorrhizal fungi protect plant roots from pathogens? Plant Signal. Behav. 5 (6), 763–765.

Smalle, J., Vierstra, R.D., 2004. The ubiquitin 26S proteasome proteolytic pathway. Annu. Rev. Plant Biol. 55, 555–590.

Smith, S.E., Smith, F.A., 2012. Fresh perspectives on the roles of arbuscular mycorrhizal fungi in plant nutrition and growth. Mycologia 104, 1–13.

Smith, P., Martino, D., Cai, Z., Gwary, D., Janzen, H., Kumar, P., Mc-Carl, B., Ogle, S., O'Mara, F., Rice, C., Scholes, B., Sirotenko, O., Howden, M., McAllister, T., Pan, G., Romanenkov, V., Schbeider, U., Towprayoon, S., Wattenbach, M., Smith, J., 2008. Greenhouse gas mitigationin agriculture. Philos. Trans. Royal Soc. 363, 789–813.

Smith, S.E., Jakobsen, I., Gronlund, M., Smith, F.A., 2011. Roles of mycorrhizas in plant phosphorus nutrition: interactions between pathways of phosphorus uptake in arbuscular mycorrhizal roots have important implications for understanding and manipulating plant phosphorus acquisition. Plant Physiol. 156, 1050–1057.

Sneh, B., Zeidan, M., Ichielevich-Auster, M., Barash, I., Koltin, Y., 1986. Increased growth responses induced by a nonpathogenic *Rhizoctonia solani*. Can. J. Bot. 64 (10), 2372–2378.

Soccol, C.R., Vandenberghe, L.P., Rodrigues, C., Pandey, A., 2006. New perspectives for citric acid production and application. Food Technol. Biotech. 44 (2), 141–149.

Takatsuka, H., Umeda, M., 2014. Hormonal control of cell division and elongation along differentiation trajectories in roots. J. Exp. Bot. 65, 2633–2643.

Tarafdar, J.C., Marschner, H., 1994. Phosphatase activity in the rhizosphere and hydrosphere of VA mycorrhizal wheat supplied with inorganic and organic phosphorus. Soil Biol. Biochem. 26, 377–395.

Thomine, S., Lanquar, V., 2011. Iron transport and signaling in plants. In: Geisler, M., Venema, K. (Eds.), Transporters and Pumps in Plant Signaling, Signaling and Communication in Plants. Springer-Verlag, Berlin Heidelberg.

Tigini, V., Prigione, V., Di Toro, S., Fava, F., Varese, G.C., 2009. Isolation and characterization of polychlorinated biphenyl (PCB) degrading fungi from a historically contaminated soil. Microb. Cell Fact. 8, 5.

Todorovic, B., Glick, B.R., 2008. The interconversion of ACC deaminase and D-cysteine desulfydrase by directed mutagenesis. Planta 229, 193–205.

To-O, K., Kamasaka, H., Kuriki, T., Okada, S., 2000. Substrate selectivity in *Aspergillus niger* KU-8 acid phosphatase II using phosphoryl oligosaccharides. Biosci. Biotechnol. Biochem. 64 (7), 1534–1537.

USCB (2015) United State Census Bureau. International Database. Available from: http://www.census.gov/population/international/data/idb/informationGateway.php

Van de Poel, B., Van Der Straeten, D., 2014. 1-Aminocyclopropane-1-carboxylic acid (ACC) in plants: more than just the precursor of ethylene!. Front. Plant Sci. 5, 640.

Van der Helm, D., Winkelmann, G., 1994. Hydroxamates and polycarboxylates as iron transport agents (siderophores) in fungi. In: Winkelmann, G., Winge, D. (Eds.), Metal Ions in Fungi. Marcel Dekker, New York, USA, pp. 39–98.

Vargas, W.A., Mandawe, J.C., Kenerley, C.M., 2009. Plant-derived sucrose is a key element in the symbiotic association between *Trichoderma virens* and maize plants. Plant Physiol. 151, 792–808.

Vazquez, P., Holguin, G., Puente, M.E., Lopez-Cortes, A., Bashan, Y., 2000. Phosphate-solubilizing microorganisms associated with the rhizosphere of mangroves in a semiarid coastal lagoon. Biol. Fertil. Soils 30 (5-6), 460–468.

Vinale, F., Sivasithamparam, K., Ghisalberti, E.L., Marra, R., Barbetti, M.J., Li, H., Woo, S.L., Lorito, M., 2008. A novel role for *Trichoderma* secondary metabolites in the interactions with plants. Physiol. Mol. Plant Pathol. 72, 80–86.

Vinale, F., Nigro, M., Sivasithamparam, K., Flematti, G., Ghisalberti, E.L., et al., 2013. Harzianic acid: a novel siderophore from *Trichoderma harzianum*. FEMS Microbiol. Lett. 347, 123–129.

Viterbo, A., Chet, I., 2006. Tas Hyd1, a new hydrophobin gene from the biocontrol agent *Trichoderma asperellum*, is involved in plant root colonization. Mol. Plant Pathol. 7, 249–258.

Viterbo, A., Landau, U., Kim, S., Chernin, L., Chet, I., 2010. Characterization of ACC deaminase from the biocontrol and plant growth-promoting agent *Trichoderma asperellum* T203. FEMS Microbiol. Lett. 305 (1), 42–48.

Vizcaıno, J.A., Cardoza, R.E., Hauser, M., Hermosa, R., Rey, M., Llobell, A., Becker, J.M., Gutierrez, S., Monte, E., 2006. ThPTR2, adi/tri-peptide transporter gene from *Trichoderma harzianum*. Fungal Genet. Biol. 43, 234–246.

Wang, R., Estelle, M., 2014. Diversity and specificity: auxin perception and signaling through the TIR1/AFB pathway. Curr. Opin. Plant Biol. 21, 51–58.

Ward, M.H., De-Kok, T.M., Levallois, P., Brender, J., Gulis, G., Nolan, B.T., Van-Derslice, J., 2005. Work group report: drinking-water nitrate and health—recent findings and research needs. Environ. Health Perspect. 113 (11), 1607–1614.

Whitelaw, M.A., Harden, T.J., Helyar, K.R., 1999. Phosphate solubilization in solution culture by the soil fungus *Penicillium radicum*. Soil Biol. Biochem. 32, 655–665.

Windham, M.T., 1986. A mechanism for increased plant growth induced by *Trichoderma* spp. Phytopathology 76, 518–521.

Winkelmann, G., Huschka, H.G., 1987. Molecular recognition and transport of siderophores in fungi. In: Winkelmann, G., van der Helm, D., Neilands, J.B. (Eds.), Iron Transport in Microbes, Plants and Animals. VCH, Weinheim, Germany, pp. 317–336.

von Wirén, N., Mori, S., Marschner, H., Romheld, V., 1994. Iron inefficiency in maize mutant ys1 (*Zea mays* L. cv Yellow-Stripe) is caused by a defect in uptake of iron phytosiderophores. Plant Physiol. 106 (1), 71–77.

Wolschek, M.F., Kubicek, C.P., 1997. The filamentous fungus *Aspergillus niger* contains two "differentially regulated" trehalose-6-phosphate synthase-encoding genes, tps A and tpsB. J. Biol. Chem. 272 (5), 2729.

Zeilinger, S., Gruber, S., Bansal, R., Mukherjee, P.K., 2016. Secondary metabolism in *Trichoderma*—chemistry meets genomics. Fungal Biol. Rev. 30, 74–90.

Overproduction of ROS: underlying molecular mechanism of scavenging and redox signaling

14

Muhammad Salman Haider[a], *Muhammad Jafar Jaskani*[b], *and Jinggui Fang*[a]
[a]College of Horticulture, Nanjing Agricultural University, Nanjing, P.R. China
[b]Institute of Horticultural Sciences, University of Agriculture, Faisalabad, Pakistan

14.1 Introduction

Normally in the biosphere, molecular oxygen (O_2) is not reactive but oxygen derivatives are likely to contribute in chemical reactions to form reactive oxygen species (ROS), which are the by-products of aerobic metabolic processes and are localized in mitochondria, chloroplast, and peroxisomes (Halliwell, 2006; del Rio et al., 2006). ROS synthesis is initiated during normal cell metabolism and its production is hastened under unfavorable conditions, which affects the equipoise between oxidants/antioxidants toward oxidants, consequently leading to intracellular oxidative stress (Apel and Hirt, 2004). Under standard growth conditions, ROS are unable to pose any damage due to the presence of antioxidant scavenging system (Gill and Tuteja, 2010), but this balance between ROS synthesis and scavenging is perturbed by various biotic and environmental stress conditions, such as drought, high/low temperature, salinity, UV radiations, heavy metals, nutrients deficiency, the overdose of pesticides/herbicides, and pathogen infections (Wu et al., 2014; Leng et al., 2015; Haider et al., 2017a,b). ROS cause oxidative damage to DNA, lipids, and proteins, that reveal the evolution of an intricate pattern of enzymatic and nonenzymatic rehabilitation mechanism in plants. Despite their cytotoxic properties, ROS also act as signaling molecules in regulating the development, intracellular signal transduction, and the defense responses in plants (Bolwell et al., 2002; Foyer and Noctor, 2005, 2011).

In the living organism, under aerobic conditions, 90% of consumed O_2 is demoted undeviatingly into H_2O by cytochrome oxidase in electron chain transport (ETC) via four-electron mechanisms, without the release of ROS. However, the one-electron reduction from the remaining 10% of depleted O_2 results in the conversion of O_2 into superoxide radical (anion; $O_2^{\bullet-}$), while further dismutation of $O_2^{\bullet-}$ by superoxide dismutases (SODs) generates hydrogen peroxide (H_2O_2) (Zhang et al., 2014; Suh et al., 2000). Subsequently, H_2O_2 by accepting one electron is cleaved into hydroxyl anion (OH^-) and highly reactive hydroxyl radical ($OH^{\bullet}$). Then, $OH^{\bullet}$ is reduced into H_2O by accepting one more electron (Fig. 14.1). Collectively, $O_2^{\bullet-}$, H_2O_2, and $OH^{\bullet}$ are known as ROS, of which $O_2^{\bullet-}$ and $OH^{\bullet}$ are only the free radicals, excluding H_2O_2. Plants evolve a complex mechanism to counter the stress-induced ROS production by the

Biocontrol Agents and Secondary Metabolites. https://doi.org/10.1016/B978-0-12-822919-4.00014-4

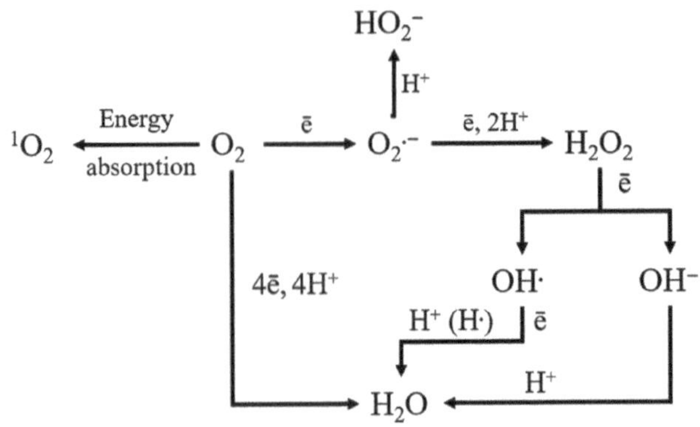

Fig. 14.1 Schematic diagram of ROS production by energy transfer.

accumulation of multiple enzymatic antioxidant scavengers, such as SOD, CAT, GPX, APX, GST, and nonenzymatic antioxidants, including GSH, ASH, flavonoids, α-tocopherols, and carotenoids (Haider et al., 2017b, 2018). Proline has recently been added in nonenzymatic antioxidants that plant, animals, and microbes require to neutralize the affectivity of ROS. This chapter summarizes the significance of ROS, its production, functional categorization, scavenging properties, and redox signaling.

14.2 ROS biochemistry

O_2 is vital for aerobic organisms to perform several biochemical reactions, which in its ground state has no harmful effects. However, the partial reduction of O_2 could lead to the production of ROS, which includes radical and nonradical oxygen species (Tripathy and Oelmuller, 2012). ROS oxidize the biomolecules and can react with DNA, lipids, and proteins (Gill and Tuteja, 2010; Apel and Hirt, 2004). The main ROS includes $O_2^{\cdot-}$, $OH^{\cdot}$, $RO^{\cdot}$, and $ROO^{\cdot}$ as free-radical molecules and $^1O_2$, H_2O_2, O_3, HCLO (hypochlorous acid) as nonradicals (del Río, 2015; Singh et al., 2016). The first step in O_2 reduction leads to the production of relatively short-lived ROS ($O_2^{\cdot-}$ and H_2O_2), where $O_2^{\cdot-}$ is enzymatically catalyzed by SOD to produce H_2O_2. H_2O_2 is then disproportionate to H_2O and O_2 by APX and AsA, while GSH oxidizes and regenerates AsA by accepting NAPDH as donor agent (Haider et al., 2017a; Quinlan et al., 2013). The electron structure of ROS has shown in the Fig. 14.2.

14.2.1 Singlet oxygen ($^1O_2$)

Singlet oxygen ($^1O_2$) is a first excited electronic reduction of O_2, which is produced by the light-harvesting complex (LHC) and photosynthetic electron transport chain (ETC) center II (PSII) in chloroplasts of the plant (Bechtold et al., 2005):

$$Chl \xrightarrow{Light} {}^3Chl^* \tag{14.1}$$

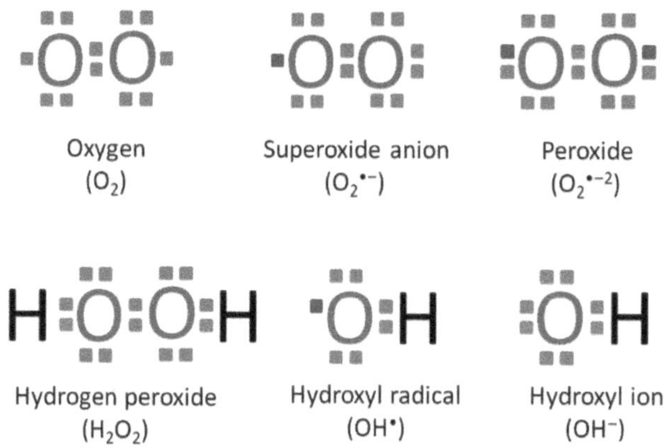

Fig. 14.2 Electron structure of most prevailing ROS with its name and chemical formula. *Red dot* (*dark gray* in print version) indicates unpaired electron.

$$^{3}\mathrm{Chl} + {}^{3}\mathrm{O}_2 \longrightarrow \mathrm{Chl} + {}^{1}\mathrm{O}_2 \tag{14.2}$$

During photosynthesis, insufficient transfer of energy results in the formation of triplet chlorophyll state ($^3Chl^*$) that further reacts with $^3O_2$ to generate highly reactive $^1O_2$ (Carmody et al., 2016), demonstrating that $^1O_2$ has malicious effects on PSI, PSII, and whole photosynthetic apparatus. The plastoquinone electron acceptors (QA and QB) in PSII ETC are over-reduced by the excess light energy that leads to tentative charge separation among Pheophytin and P680 molecules and form the triplet form of Chl (3P680), hence resulting in the formation of $^1O_2$. Under oxidative stress, stomatal closure leading to low intercellular CO_2 assimilation in the chloroplast assists $^1O_2$ production (Zhang et al., 2014). Moreover, the reaction of $O_2^{\cdot -}$ with $OH^{\cdot}$ at lower electron transfer site lead to the production of $^1O_2$ (Suh et al., 2000; Fryer et al., 2002). However, $^1O_2$ has a very short life span (about 3 μs) with the diffusion distance of approximately 100 nm that can cause cellular damage (Hatz et al., 2007). Plants can activate the antioxidant system to quench the $^1O_2$, while mostly involve nonenzymatic antioxidants, including β-carotene, plastoquinone, and tocopherols (Krieger-Liszkay et al., 2008). Besides its destructive effect, some plants generate secondary metabolites with photosensitizing attributes that utilize $^1O_2$ to hasten the antimicrobial activity in response to pathogen defense (Flors and Nonell, 2010).

14.2.2 Superoxide radical ($O_2^{\cdot -}$)

The superoxide anion ($O_2^{\cdot -}$) is generated by the partial reduction of the O_2 in different compartments of plant cells. The main site of $O_2^{\cdot -}$ production is the thylakoid membrane-bound noncyclic ETC of the chloroplast at Complex I and Complex III (Saxena et al., 2016). Light reduces O_2 to produce acetaldehyde catalyzed by CAT and ethanol. This results in the production of a photoreduced product (H_2O_2). Typically, the control of electron flow between PSII and PSI regulates the electron acceptor site of PSI (Gill and Tuteja, 2010). This regulatory mechanism of PSI is an essential factor in determining the ferredoxin pool because of ferredoxins and electron carriers can donate an electron to O_2, which results in the formation of $O_2^{\bullet-}$ (Bielski et al., 1983; Elstner, 1987). In mitochondrial ETC, cytochrome c-oxidase reacts with O_2 and produces H_2O by transferring four electrons. However, O_2 can also react with other ETC components to generate $O_2^{\bullet-}$ by transferring only one electron. The $O_2^{\bullet-}$ is the first ROS produced in plants and it has a half-life of about 2–4 μs, hence causing no harm to plant cells (Halliwell, 2006; Gill and Tuteja, 2010; del Río, 2015). Though $O_2^{\bullet-}$ may get converted into the highly ROS ($OH^{\bullet}$), which causes lipid peroxidation (LPO) and membrane damage (Gill and Tuteja, 2010).

$$O_2^{\bullet-} + Fe^{3+} \rightarrow {}^1O_2 + Fe^{2+} \tag{14.3}$$

$$2O_2^{\bullet-} + 2H^+ \xrightarrow{SOD} O_2 + H_2O_2Fe^{3+} \tag{14.4}$$

$$H_2O_2Fe^{3+} + Fe^{2+} \rightarrow OH^- + OH^{\bullet} + Fe^{3+} \tag{14.5}$$

14.2.3 Hydrogen peroxide (H_2O_2)

Hydrogen peroxide (H_2O_2) is a moderately reactive ROS, produced by the protonation of univalent reduction of $O_2^{\bullet-}$. H_2O_2 is also generated nonenzymatically by the dismutation of $O_2^{\bullet-}$ in a reaction catalyzed by SOD at lower pH (Das and Roychoudhury, 2014; Tripathi et al., 2016). The photooxidation reaction by xanthine oxidase (XOD) and NADPH also results in H_2O_2 production. Compared to other ROS, H_2O_2 has a relatively higher half-life (1 μs) and can diffuse to some distance (Sharma et al., 2012). The main sites of H_2O_2 production include the cell membrane, chloroplastic ETC, endoplasmic reticulum (ER), mitochondria, photorespiration, and β-oxidation of fatty acids. Under environmental stress, increased stomatal conductance reduces the CO_2 availability that favors the oxygenation of ribulose 1,5-biphosphate (RuBP) and results in intensified photorespiration (Das and Roychoudhury, 2014; Jogaiah et al., 2013):

$$2O_2^{\bullet-} + 2H^+ \rightarrow H_2O_2 + O_2$$

$$2O_2^{\bullet-} + 2H^+ \xrightarrow{SOD} H_2O_2 + O_2$$

In plant cells, H_2O_2 can play advantageous (low concentration) and detrimental role (high concentration). At the low level, it plays a signaling-related role in complex physiological processes, for instance, growth and development (Foreman et al., 2003), stomatal conductance (Bright et al., 2010), photosynthesis and photorespiration (Das and Roychoudhury, 2014), senescence (Peng et al., 2005), and cell cycle (Mittler, 2002) and triggers tolerance to various biotic and abiotic stresses (Quan et al., 2008). At a relatively high level, H_2O_2 acts as a second messenger for ROS-generated signals as it can diffuse across the membrane through aquaporins and can cause programmed cell death (PCD) (Quan et al., 2008; Bienert et al., 2006). H_2O_2 possesses the potential to oxidize both methionine and cysteine deposits and can deactivate the enzymes of the Calvin cycle by oxidizing their thiol group (Gill and Tuteja, 2010).

14.2.4 Hydroxyl radicals ($OH^•$)

Hydroxyl radical ($OH^•$) is the most reactive and deadliest among other ROS members. It is produced from Fenton reaction among $O_2^{•-}$ and H_2O_2 at standard temperature and neutral pH in the presence of transition metals (Fe^{2+} and Fe^{3+}) (Das and Roychoudhury, 2014). Hydroxyl radical ($OH^•$) has a single unpaired electron, which quickly makes a bond with $^3O_2$. Enzymatic antioxidant scavenging system is absent in plant cells to eliminate $OH^•$, consequently its excessive accumulation ultimately induces cell death (Gill and Tuteja, 2010). The $OH^•$ potentially reacts and damages the biological molecules like DNA, proteins, and lipids (Tripathi et al., 2016).

$$2O_2^{\bullet-} + H_2O_2Chl \overset{Fe^{2+},Fe^{3+}}{\rightarrow} OH^- + O_2 + OH^{\bullet 3}Chl \quad (14.6)$$

14.3 ROS Production in plant cell

ROS generation is a primary response of living organisms under aerobic metabolism due to the partial modification in the state of O_2. There are plenty of experimental evidence demonstrating the potential of cellular components to produce and accumulate ROS. Nevertheless, in green plants, chloroplast and peroxisomes are potential sources of ROS under high light intensity (Bhattacharjee, 2005; Tripathy and Oelmüller, 2012). Likewise, mitochondria are the primary source of ROS under darkness in nongreen plants (Sweetlove and Foyer, 2004). Generally, chloroplast produces $O_2^{•-}$ and singlet oxygen ($^1O_2$) in photosynthesis compartments, while mitochondria generate $O_2^{•-}$ at sites resides in complex I and complex III of ETC (Table 14.1). Bhattacharjee (Bhattacharjee, 2005) demonstrated that about 2%–3% of O_2 consumed by mitochondria results in the synthesis of ROS. The general mechanism of ROS production in different organelles is represented in Fig. 14.3.

14.3.1 Chloroplast and peroxisome-mediated ROS production

In plant cells, most of the cellular compartments can generate ROS via different channels. The primary light-harvesting photosynthesis process is carried out in the thylakoid membrane of the chloroplast. The photosynthetic compartments (PSI and PSII) are the vital source of ROS, including $O_2^{•-}$, $^1O_2$, and H_2O_2 (Haider et al., 2017a; Tripathy and Oelmüller, 2012). The formation of ROS initiated via the dismutation of $O_2^{•-}$ when the dearth of $NADP^+$ in PSI effects overloading in ETC, resulting in spilling of electrons from ferredoxin to O_2, thus reducing it to $O_2^{•-}$ through Mehler reaction in the chloroplast (Das and Roychoudhury, 2014). Electron spilling or excessive electron transfer to O_2 has also been witnessed in QA-QB complex of PSII, which leads to

Table 14.1 Intracellular ROS production sites and ROS-mediated response(s).

Serial no.	Sites of ROS	Type of ROS	Source of ROS	Function of ROS	Reference
1	Chloroplast	$O_2^{\bullet-}$, H_2O_2, $^1O_2$	Photosynthetic electron transport chain, PSI, and PSII	Induce chloroplast avoidance movement	Apel and Hirt (2004), Sweetlove and Foyer (2004)
2	Peroxisomes	$O_2^{\bullet-}$, H_2O_2, $OH^{\bullet}$	Flavin oxidase, fatty acid β-oxidation, glycolate oxidase	Abiotic stress response, leaf senescence	Mittler (2002); Del Rio et al. (2006)
3	Mitochondria	$O_2^{\bullet-}$, H_2O_2, $OH^{\bullet}$	Mitochnodrial respiratory electron transport chain, Complex I and Complex III	Signal transduction within organelles	Sweetlove and Foyer (2004)
4	Endoplasmic reticulum	$O_2^{\bullet-}$, H_2O_2	NADPH-dependent electron transport	Ca^{2+} signaling	Mittler (2002)
5	Apoplast	$O_2^{\bullet-}$, H_2O_2	Polyamine oxidase, peroxidase (class III), germin-like oxalate oxidase	Modulation of cell wall elasticity, ABA-induced stomatal closure	Mittler (2002)
6	Plasma membrane	$O_2^{\bullet-}$, H_2O_2	NADPH-dependent oxidases	Hypersensitive response to biotic and abiotic stress	Apel and Hirt (2004)

$^1O_2$; singlet oxygen, $O_2^{\bullet-}$; superoxide radical; H_2O_2; hydrogen peroxide; $OH^{\bullet}$, hydroxyl radical.

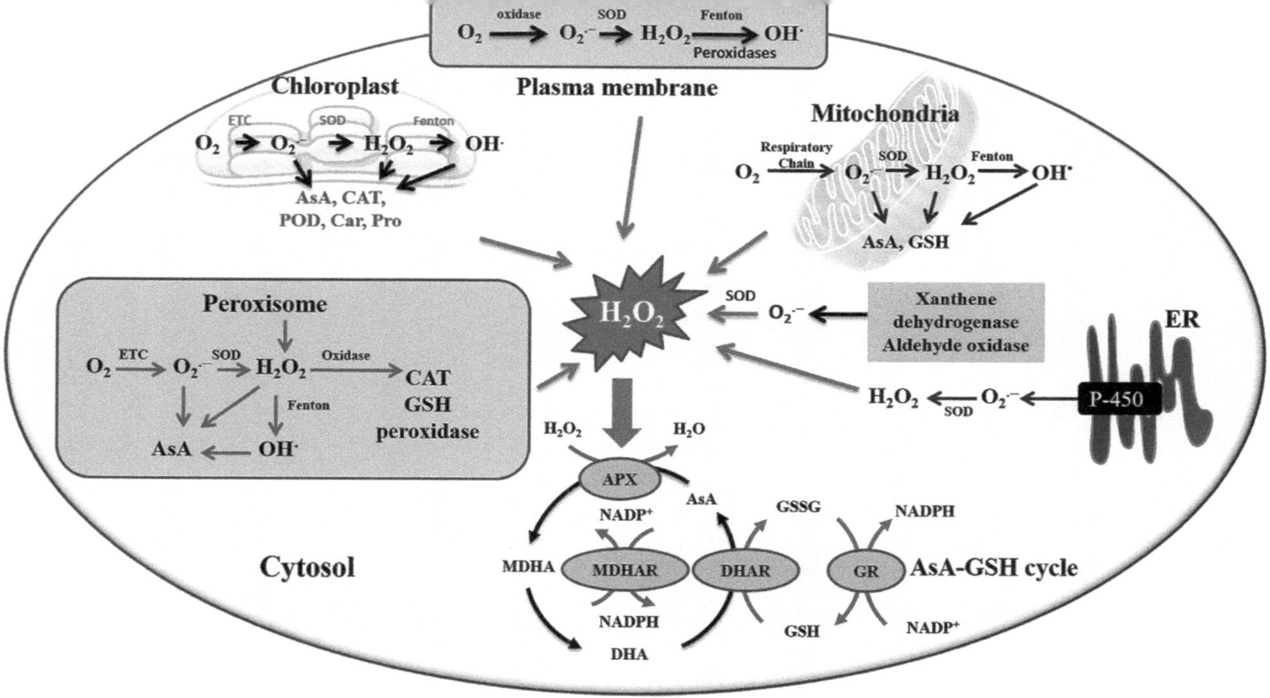

Fig. 14.3 Schematic diagram of involvement of different cellular compartments involved in production of ROS. Biotic and abiotic factors give rise to different toxic elements, such as superoxide radical ($O_2^{\cdot -}$), hydrogen peroxide (H_2O_2), and hydroxyl radicals ($OH^{\cdot}$), which cause cellular damage by interacting with different macromolecules and metabolites. The process of ROS formation commenced with the spattering of electron ($\bar{e}$) from chloroplast (ETC), mitochondria (respiratory chain), peroxisomes (ETC), and plasma membrane (NOX) resulting in conversion of O_2 to $O_2^{\cdot -}$. The $O_2^{\cdot -}$ produced is then detoxified by superoxide dismutase (SOD), thereby converted into H_2O_2. Plants possess enzymatic and nonenzymatic detoxification mechanism to diminish the ROS. Among which, acorbate-glutathione (AsA-GSH) cycle play a crucial role to disintegrate H_2O_2. Ascorbate peroxidase (APX), which is enduringly present in all cellular organelles converts H_2O_2 to water (H_2O) by using AsA as an electron donor, suggesting its protective role against oxidative damage in plants. *AsA*, ascorbic acid; *APX*, ascorbate peroxidase; *CAT*, catalase; *Car*, Carotenoids; *DHA*, dehydroascorbate; *DHAR*, dehydroascorbate reductase; *ETC*, electron transport chain; *ER*, endoplasmic reticulum; *GSH*, reduced glutathione; *GSSG*; oxidized glutathione; *GR*, glutathione reductase; *MDHA*, monodihydroascorbate; *MDHAR*, monodihydroascorbate reductase; *NADP*, nicotinamide adenine dinucleotide phosphate; *NADPH*, reduced NADP; *Pro*, proline.

the generation of $O_2^{\bullet-}$ (Sharma et al., 2012). The $O_2^{\bullet-}$ then underwent spontaneous or enzymatic (Cu-Zn SOD) dismutation to generate H_2O_2 or may be protonated to $HO_2^{\bullet-}$ (Asada, 2006; Miller et al., 2008). Moreover, singlet oxygen ($^1O_2$) is naturally produced during photosynthesis, particularly in PSII (P680) even under low light intensity. Under photochemical stress, electrons from chlorophyll are excited to a higher energy level, which inhibits the redox state of plastoquinone and QA-QB complex, resulting in the rejoining of oxidized P680 with a reduced prosthetic group (pheophytin). This favors the production of $^1O_2$, which is rapidly quenched by water, indicating that $^1O_2$ has a very short life span while, that is, a critical factor to evaluate the biological effects of ROS (Asada, 2006; Hideg et al., 2011; Wagner et al., 2004).

Peroxisomes are omnipresent multipurpose subcellular organelles defined by single lipid membrane involved in the oxidation reactions of fatty acid (FA α- and β-oxidation) of VLCFA (very long-chain FA), biosynthesis of bile acids and glycerolipids, and catabolism of purine and are critical sites of intracellular H_2O_2 generation (Baker and Graham, 2002). The primary metabolic chemistries involved in the production of H_2O_2 in peroxisomes includes the oxidation of glycolate via photosynthetic carbon oxidation cycle (PCOC), FA β-oxidation by acyl-CoA oxidase, enzymatic reaction of flavin oxidases, and dismutation of $O_2^{\bullet-}$ radicals (Foyer and Noctor, 2003; Mittler et al., 2004; Palma et al., 2009).

Previous studies have demonstrated that $O_2^{\bullet-}$ production accompany two channels in peroxisomes, including (i) peroxisomal matrix possesses xanthine oxidase (XOD) and (ii) peroxisomal membranes contain NADPH (del Rio et al., 2002). The XOD, which is a reduced form of xanthine dehydrogenase (XDH), catalyzes the oxidation of both xanthine and hypoxanthine into uric acid and gives rise to $O_2^{\bullet-}$ (Palma et al., 2002), whereas, in the peroxisomal membrane where small ETC composed of flavoprotein NADH with 32kDa MW and Cyt (cytochrome) b are found to be involved in the $O_2^{\bullet-}$ production. It is also observed that polypeptides (PMPs) with 18- and 32-kDa use NADH as an electron donor to generate $O_2^{\bullet-}$. The PMP18 is leading to the producer of $O_2^{\bullet-}$ and belongs to Cyt b group, PMP29 is linked with NADPH Cyt P450 reductase, and PMP32 is related to MDHAR. The major ROS-scavenging enzymes (SOD and CAT) are also located in the peroxisomes and play a defensive role against heavy metal stress (Palma et al., 2009; Hayashi and Nishimura, 2006; Luis and Lopez-Huertas, 2006). However, the overproduction of $O_2^{\bullet-}$ and H_2O_2 in peroxisomes leads to oxidative stress and ultimately cell death, though their role as signaling molecules mediating pathogen-triggered PCD in plants is vital when present in balance quantity (del Rio et al., 2002; McDowell and Dangl, 2000).

14.3.2 Mitochondrial ROS production

The mitochondria are also the potential source of primitive ROS ($O_2^{\bullet-}$), which is a byproduct of monoelectronic reduction of O_2 (Navrot et al., 2007). The mitochondrial ETC harbors four-electron reduction of O_2 to produce a diminutive amount of $O_2^{\bullet-}$ (reactive ROS), which is modified into H_2O_2 by Cu-Zn SOD and Mn-SOD or by APX in the intermembrane spaces (Sharma et al., 2012; Okado-Matsumoto and Fridovich, 2001; Sturtz et al., 2001). In mitochondrial ETC, several complexes, such as Complex I (NADH dehydrogenase), Complex II (succinate dehydrogenase), and Complex III (ubiquinone-Cyt region) are involved in the production of ROS (Sweetlove and Foyer, 2004; Noctor et al., 2007). The NADH dehydrogenase region of the respiratory chain instinctively reduces O_2 to $O_2^{\bullet-}$ anions, while this ROS production hastens when electrons flow in the reverse direction from succinate to ubiquinone (via Complex II) to Complex 1 due to lack of NAD^+-linked substrates, which are regulated by ATP hydrolysis (Turrens, 2003). The $O_2^{\bullet-}$ during mitochondrial ETC enhanced significantly in the presence of antimycin, which inhibits the electron transfer to Cyt c1, resulting in the accumulation of semiquinone radicals while its autooxidation produces $O_2^{\bullet-}$ (Maxwell et al., 1999; Murphy, 2009).

Several other enzymes in the mitochondrial matrix are documented as a source of ROS production, intimating their vital role within the mitochondrial microenvironment. Some enzymes produce ROS directly, such as aconitase, while some enzymes produce indirectly, such as 1-galactono-γ-lactone dehydrogenase (GAL) by passing electrons to the mitochondrial transport system (Rasmusson and Wallström, 2010). Moreover, glycerol phosphate dehydrogenase and flavin oxidase acyl-CoA dehydrogenase produce ROS by oxidizing lipid-derived substrates and may responsible for the release and elevation of ROS in some tissues (Tretter et al., 2007a,b). The dismutation of H_2O_2 generates highly reactive $OH^{\bullet}$ via the Fenton reaction and is involved in the peroxidation of mitochondrial membrane-bounded polyunsaturated fatty acids (PUFAs), which contributes in the synthesis of cytotoxic alkanes, hydroxyl alkanes, aldehydes and malondialdehyde (Gill and Tuteja, 2010; Abdelrahman et al., 2018). Under oxidative stress, the linkage of ETC and ATP synthesis gets affected, which stimulates the reduction in ubiquinone (electron carriers) pool, thus generating a higher amount of ROS (Wahid et al., 2007; Atkin and Macherel, 2009).

14.3.3 Endoplasmic reticulum-mediated ROS production

The endoplasmic reticulum (ER) is intracellular microtubules and primary organelle for the secretory pathway in which proteins are modified. The Cyt-P450 catalyzes the NADPH-dependent ETC and generates ROS ($O_2^{\bullet-}$) in the ER (Das and Roychoudhury, 2014). The flavoproteins dismutase the interaction between Cyt-P450 and an organic substrate RH to generate free-radicals intermediates (Cyt-P450R-), which further reacts with triplet oxygen ($^3O_2$) to generate an oxygenated complex (Cyt-P450-ROO-). Then Cyt-b reduces this complex to Cyt-P450-Rh to produce $O_2^{\bullet-}$ (Sharma et al., 2012).

14.3.4 Apoplastic ROS production

The apoplast is also a vital source of ROS ($O_2^{\bullet-}$ and H_2O_2) in its extracellular spaces during the apoplastic oxidative burst, while cell wall-linked peroxidases, pH-dependent peroxidases, plasma membrane NADPH oxidases, and germin-like oxalate oxidases are essential enzymes to participate in this process (Das and Roychoudhury, 2014; Bindschedler et al., 2006; Suzuki et al., 2011). Another enzyme Rboh (respiratory-burst oxidase), which is a homolog of multimeric flavocytochrome that encodes plasma membrane-localized proteins, oxidizes NADPH (cytoplasmic) and dismutases extracellular $^3O_2$ to generate $O_2^{\cdot-}$ in the cell wall. This ROS is short lived and ineffectual to cross the lipid bilayer due to its charge and outlive in the apoplast. The $O_2^{\cdot-}$ underwent spontaneous or enzymatic (SOD) dismutation and converted into H_2O_2 (Apel and Hirt, 2004). Besides the toxicity of H_2O_2, it can diffuse into the cell and contribute to the plant defenses, such as PCD (Dangl and Jones, 2001). For instance, the genetic engineering of oxalate oxidase in plants lacking H_2O_2-producing capability has been reported to provide a resilience mechanism against pathogens secreting oxalate (Donaldson et al., 2001), indicating that apoplastic H_2O_2 plays a physiological role against pathogens in plants.

14.3.5 Other sources of ROS production

ROS are also produced in the plasma membrane by NADPH-dependent oxidases, which catalyze the electron transfer from NADPH (cytoplasmic) to molecular oxygen to generate $O_2^{\cdot-}$ (Sharma et al., 2012), also known as Rboh in plants. H_2O_2 is generated by enzymatic reactions by oxidases, such as the degradation of mono, di, or polyamines by polyamine oxidases, purine by xanthine oxidases, and rarely polyphenols by polyphenol oxidases (Halliwell, 2006). In the cell wall, H_2O_2 is produced and stimulated by the cell wall-associated peroxidases during hypersensitive response (Kim et al., 2010) and cell wall-localized LOX during stress conditions (Das and Roychoudhury, 2014). Likewise, oxalate oxidase is responsible for ROS production in drought-responsive root cells, whereas, glycolate oxidase enhances nonhost pathogen defense in plants (Rojas et al., 2012). Moreover, fungi are the source of ROS as a result of oxidation of enzymatic sugar (Apel and Hirt, 2004).

14.4 ROS scavenging by the antioxidant defense system

Under normal growth period, there is a balance between production and quenching of ROS metabolites. This balance between ROS production and scavenging is interrupted by various biotic and abiotic factors, leading to the overproduction of extracellular ROS levels (Haider et al., 2018). To counter the damage severity caused by ROS toxicity, plants possess a panel of enzymatic and nonenzymatic antioxidants, which provide a defensive mechanism against ROS (Haider et al., 2017a,b). However, the molecular mechanism underlying the ROS production is universal in all types of stresses, depending on plant species, but different antioxidative strategies are involved for ROS detoxification. For instance, enzymatic antioxidants act in coordination to quench $O_2^{\cdot-}$ and H_2O_2, whereas $OH^{\cdot}$ and $^1O_2$ are detoxified by the low-molecular-weight (LMW) nonenzymatic antioxidants (Sharma et al., 2012). The enzymatic antioxidants include superoxide dismutase (SOD), catalase (CAT), ascorbate peroxidase (APX), dehydroascorbate reductase (DHAR), monodehydroascorbate reductase (MDHAR), glutathione peroxidase (GPX), and glutathione reductase (GR), while ascorbic acid (AA), glutathione (GSH), phenolics, carotenoids, tocopherols, and proline are included in LMW nonenzymatic antioxidants (Haider et al., 2018; Quinlan et al., 2013).

14.4.1 Superoxide dismutase (SOD)

The SOD (E.C.1.15.1.1) is a member of the metalloenzyme family, plays an essential role in cellular defense, and can alter $O_2^{\cdot -}$ by dismutating it in O_2 and H_2O_2 (Gill and Tuteja, 2010). Under oxidative stress, SOD provides the first line of defense against ROS ($O_2^{\cdot -}$) produced during photosynthesis and respiration and also reduces the $OH^{\cdot}$ formation due to the elimination of $O_2^{\cdot -}$ via metal-catalyzed Haber-Weiss reaction (Das and Roychoudhury, 2014; Abouzari and Fakheri, 2015). SODs have several isozymes, which are grouped according to their subcellular localization by the presence of metal cofactors. In higher plants, SODs have three isozymes: Cu/Zn-SOD (localized in the cytosol, plastids, and peroxisomes), Fe-SOD (often nondetectable, but conserved in cytosol and chloroplast), Mn-SOD (localized in the matrix of mitochondria and peroxisomes) (Alscher et al., 2002; Perry et al., 2010). Isomers of SOD are encoded in the nucleus of the plant cell and targeted to their precise subcellular locations by an amino-terminal targeting sequence (Racchi et al., 2001). These three isozymes vary in their structure, such as the eukaryotic Cu/Zn-SOD and prokaryotic Fe-SOD. Mn-SOD enzymes are dimers, while mitochondrial Mn-SOD is a tetramer. In eukaryotes, 90% of SOD activity is mainly comprised of Cu/Zn-SOD (Liu et al., 2007). The intensified SOD activity under various types of stress conditions plays a significant role in enhancing the oxidative stress to rescue the plants, intimating a positive correlation between SOD activity and ROS production (Singh et al., 2016; Boguszewska et al., 2010). Several reports have demonstrated that SOD activity was increased in plants under various environmental stresses, such as drought stress (Haider et al., 2017a, 2018), salt stress (Kukreja et al., 2005), heavy metal toxicity (Leng et al., 2015), and pesticidal stress (Wu and Liu, 2002). Genetically transformed plants overexpressing SOD have shown tolerance under various abiotic stress environments (Xu et al., 2014).

14.4.2 Catalase (CAT)

Catalase (CAT; E.C.1.11.1.6) was the pioneer antioxidant enzyme to be identified and characterized in 1900 by Loew (1900). CAT can catalyze the conversion of H_2O_2 into O_2 and H_2O (Das and Roychoudhury, 2014; Weydert and Cullen, 2010). CAT are heme-containing tetrameric enzymes located in the subcellular organelles (peroxisomes), which is the main site for producing H_2O_2 by photorespiratory oxidation, β-oxidation of FAs, and purine catabolism during oxidative stress conditions (Vellosillo et al., 2010). Among all characterized antioxidant enzymes, CAT with a high turnover rate can function without consuming cellular reducing equivalent and one molecule of CAT can catalyze six million H_2O_2 molecules into O_2 and H_2O per minute (Sharma et al., 2012; Hojati et al., 2010). Under oxidative stress, cells start to produce energy through a catabolic process, which produces H_2O_2 and CAT that can eliminate H_2O_2 in an energy-efficient manner (Mallick and Mohn, 2000). Various isozymes of CAT have been reported in angiosperms, three isozymes are found in *Zea mays*: CAT1 (localized in cytosol and peroxisomes) expressed in pollen and seeds; CAT2 (localized in cytosol and peroxisomes) expressed in leaves, roots, and seed; and CAT3 (Localized in mitochondria) expressed in seeds and seedlings

(Sharma et al., 2010). Yang and Poovaiah (2002) demonstrated that calmodulin (CaM) connects with CAT3 during calcium (Ca^{2+})- signaling and regulates the CAT activity. Also, CAT has an influential role against pathogen defense, which has been verified by either overexpressing or silencing the CAT genes in transgenic plants (Noctor and Foyer, 1998). In general, an increase in the CAT activity is deliberated as an adaptive trait in plants, helps to reduce tissue damage by H_2O_2 toxicity.

$$H_2O_2 \overset{CAT}{\rightarrow} H_2O + {}^1O_2$$

14.4.3 Ascorbate peroxidase (APX)

APX (E.C.1.1.11.1) is a vital Class 1 peroxidase member of the ascorbate-glutathione (AsA-GSH) cycle that uses AsA as an electron donor to eliminate H_2O_2 generated in cytosol and chloroplast (Haider et al., 2017a). During this reaction, APX dismutates H_2O_2 in H_2O and dehydroascorbate (DHA) using AA as a reducing agent (Gill and Tuteja, 2010). In higher plants, APX has five distinct isoforms localized in different subcellular organelles, such as cytosol (cAPX), chloroplast stromal soluble form (sAPX), thylakoid (tAPX), and glyoxysome membrane forms (gmAPX) (Noctor and Foyer, 1998; Sharma and Dubey, 2005). APX has a prominent magnetism toward the detoxification of H_2O_2 compared with POD and CAT, thus plays a significant role in the H_2O_2 exclusion and maintaining ROS level inside the cell. Many researchers have demonstrated the alleviated activity of APX in response to various abiotic stresses, including salt, drought, high/low temperature, heavy metal toxicity, and UV irradiation and to an overdose of pesticides (Leng et al., 2015; Haider et al., 2018; Miller et al., 2007; Munns and Tester, 2008).

$$H_2O_2 + AA \overset{APX}{\rightarrow} 2H_2O + DHA$$

14.4.4 Dehydroascorbate reductase (DHAR)

DHAR (E.C.1.8.5.1) is extensively present in leaves and shoots and catalyzes the recovery of oxidized DHA back to AsA by utilizing the reduced GSH as a source of an electron donor or reducing power (Haider et al., 2019). The oxidized GSH (GSSG) is reverted into GSH by GR using NADPH as an electron donor, which is crucial to tolerate various abiotic stresses. Hence, DHAR plays an important role in the regeneration of AsA pool in the cell and maintains the cellular redox level by sustaining the AsA pool in both apoplast and symplast of the plant cell (Das and Roychoudhury, 2014; Chen and Gallie, 2006)

$$DHA + 2GSH \rightarrow AsA + GSSG$$

14.4.5 Monodehydroascorbate reductase (MDHAR)

MDHAR (E.C.1.6.5.4) has two isozymes localized in cytosol and chloroplast, possesses flavin adenine dinucleotide (FAD) as a cofactor. MDHAR catalyzes the regeneration of AsA and manifests great specificity for MDHA (monodehydroascorbate) as an electron acceptor and NADH as an electron donor (del Rio et al., 2002). Similar to APX, MDHAR is also found in mitochondria and peroxisomes that scavenges H_2O_2 to balance the cellular redox level (Sharma et al., 2012; Sultana et al., 2012).

$$\text{MDHA} + \text{NADH} \rightarrow \text{AsA} + \text{NADP}^+$$

14.4.6 Glutathione peroxidase (GPX)

GPX (E.C.1.11.1.9) is a heme-containing family of enzymes with peroxidase activity that eliminates the H_2O_2, organic peroxides, and lipid peroxides by using GSH as an electron donor. Also, it employs aromatic compounds as a reducing agent to protect the plant from oxidative damage (Das and Roychoudhury, 2014). However, peroxidases play a crucial role in many developmental and physiological processes, such as biosynthesis of ethylene and secondary metabolites, auxin catabolism, pathogen infection, and abiotic stresses (Cosio and Dunand, 2009). In plants, the GPX family contains various isoforms with unique subcellular locales and have distinct spatiotemporal expression pattern in response to abiotic stresses (Passia et al., 2014). There exist eight isoforms of GPX in *Arabidopsis*, including *GPX1*, *GPX6*, and *GPX7* (localized in the chloroplast), *GPX3* (mitochondria), *GPX5* (endoplasmic reticulum), and *GPX4* and *GPX8* are localized in the cytoplasm (Mittler et al., 2004). Among these isoforms, *AtGPX1* is only expressed in vegetative and reproductive organs, except in roots, indicating its defensive role against photosynthesis-mediated ROS production (Millar et al., 2003). *AtGPX2*, *AtGPX3*, and *AtGPX8* showed higher expression levels during root differentiation, while *AtGPX4* and *AtGPX5* were only expressed in stamens and pollens (Zhai et al., 2013). Also, *AtGPX1*, *2*, *5*, and *6* were induced in response to salt stress, and *AtGPX1* was the only gene induced under high-temperature stress, while the rest of the AtGPX genes were responsive to osmotic stress (Millar et al., 2003). GPX also participates in the protein complex that works against oxidative stress. In general, the difference in subcellular locations and expression patterns of plant GPX intimate that each isoform has a multitude of functions.

$$H_2O_2 + \text{GSH} \rightarrow H_2O + \text{GSSG}$$

14.4.7 Glutathione reductase (GR)

The GR (E.C.1.6.4.2) is a flavoenzyme oxidoreductase that converts GSSG to two reduced GSH molecules by using NADPH as cofactor. GSH is further employed to revive AsA from DHA and MDHA then converted into its oxidized state (GSSG). GSH plays an antioxidative role in eliminating ROS from the cell. The GR is mostly located in the chloroplast; however, a minor quantity of GR is also reported in the peroxisomes, cytosol, glyoxysomes, and mitochondria. Like other antioxidant enzymes, GR is also the main component of AsA-GSH cycle, where it protects chloroplast against oxidative stress by balancing a high GSH/GSSG ration in the plant cell (Foyer and Noctor, 2011; Reiter et al., 2014).

$$GSSG + NADPH^+ \rightarrow 2GSH + NADP^+$$

14.5 Nonenzymatic antioxidants

The panel of nonenzymatic antioxidants includes AA, GSH, α-tocopherols, carotenoids, phenolics, flavonoids, and proline. The nonenzymatic antioxidant machinery has a higher potential to protect cellular components from the toxic effect of ROS (Haider et al., 2018). Besides their vital role in defense mechanism and as enzyme cofactor, they also play a crucial role in plant growth and development by regulating mitosis, cell division and elongation, senescence, and ultimately PDC (de Pinto et al., 2013; Semchuk et al., 2009). Among them, AsA and GSH are well-documented compounds that mediate the cellular redox state under oxidative stress.

14.5.1 Ascorbic acid (AA)

Ascorbic acid is plentiful and LMW water-soluble antioxidant that directly scavenges the enhanced production of ROS (H_2O_2), by its ability to donate the electrons in wide range of enzymatic and nonenzymatic reactions (Gill and Tuteja, 2010; Wang et al., 2005). It exists ubiquitously in photosynthetic cells, cell organelles, and apoplast, whereas, it is present in the reduced ascorbate form in the chloroplast (Smirnoff and Wheeler, 2000). Like SOD, the apoplastic AA is also pondered as the first line of defense against ROS toxicity in an aqueous phase to protect critical macromolecules (Barnes et al., 2002). The AA protects the membrane from oxidative damage by scavenging the $O_2^{\cdot -}$, $^1O_2$, and $OH^{\cdot}$, degrading H_2O_2 into H_2O by donating an electron in APX reaction while regenerating tocopheroxyl radicals and α-tocopherol (Horemans et al., 2000; Chaves et al., 2002; Zaefyzadeh et al., 2009). Plant mitochondrion not only generates AsA through the Smirnoff-Wheeler pathway catalyzed by L-galacto-r-lactone dehydrogenase but also regenerates AsA from its oxidized form (Szarka et al., 2007). The AA is oxidized into MDHA and DHA, of which MDHA is relatively unstable in an aqueous state, whereas DHA can be enzymatically reduced to AsA via GSH (Foyer and Noctor, 2002). In addition to the metabolic process, the elevated level of endogenous AA is compulsory to combat oxidative stress in plants.

14.5.2 Reduced glutathione (GSH)

Glutathione is the major low-molecular-weight thiol and essential metabolite that participates in thiol-disulfide interactions, in which GSH is oxidized to GSSG, which further converted to GSH by de novo synthesis or by enzymatic reactions using NADPH as an electron donor to maintain the cellular GSH pool (Noctor et al., 2012).

Glutathione is widely distributed in cellular organelles, such as chloroplast, peroxisomes, cytosol, mitochondria, apoplast, vacuole, and endoplasmic reticulum that eliminate intracellular ROS accumulation during oxidative stress. In plants, the synthesis of glutathione follows two ATP-dependent steps; the first step commenced with the synthesis of γ-glutamylcysteine (γ-GC) from Glu and Cys in the presence of Glu-Cys ligase (GCL) and then glutathione synthetase (GSH-S) assists Gly in γ-GC to produce GSH (Foyer and Noctor, 2005; Jogaiah et al., 2013; Noctor et al., 2012). The GSH can scavenge $O_2^{\bullet-}$, $^1O_2$, $OH^{\bullet}$, and H_2O_2 by forming adducts through glutathionylation and protects different macromolecules (Mullineaux and Rausch, 2005).

14.5.3 Tocopherols

Tocopherols (TOCs) are lipophilic antioxidants, protect the membrane from lipid peroxidation by lipid peroxyl radicals, scavenging, and quenching ROS, thus pondered as an indispensable element of the biomembrane (Munne-Bosch et al., 2013; Kiffin et al., 2006). In plants, the TOCs have four isomers (α, β, γ, and δ) that are localized in the thylakoid membrane of the chloroplast, while their capability to detoxify ROS is mainly based on the methyl group attached with the phenolic ring structure. Among these isomers, α-TOCs occupying the three methyl substituents have the highest antioxidant activity. As synthesized by photosynthetic organisms, TOCs are found only in green parts of the plants and protect PSII by quenching excess energy and chemically reacting with O_2 in order to protect lipids and other membrane components of the chloroplast (Igamberdiev et al., 2004; Hollander-Czytko et al., 2005). TOCs inhibit the chain propagation step during the lipid autooxidation process by repairing the oxidizing radicals and setting an effective free-radical trap. α-TOC reacts with lipid radicals ($RO^{\bullet}$, $ROO^{\bullet}$, and RO^*), produced from the oxidation of polyunsaturated fatty acids (PUFA) at the membrane-water interface, where α-TOC acts as hydrogen donor and form $TOH^{\bullet}$, which then recycled to its reduced form by interacting with AA and GSH (Igamberdiev et al., 2004). In addition, 5-nitro-γ-TOC (5-NgT) was reported in leaves and germinating seeds of many plant species, which prolongs early development by dismutating NOx level (Desel et al., 2007). Relative accumulation of α-TOC in plants has been shown to induce tolerance against oxidative stress, such as drought, salt, and chilling (Guo et al., 2004).

14.5.4 Carotenoids

Carotenoids are another LMW lipophilic antioxidants located in the plastids, possess the ability to absorb light (450–570 nm), dissipate excess energy as heat, maintain the structural stability of PSI, react with triplet chlorophyll (3Chl*) molecule to inhibit the synthesis of $^1O_2$, scavenge ROS ($^1O_2$), and suppress lipid peroxidation, thus play a photoprotective role in photosynthetic organisms (Collins, 2001; Niyogi et al., 2001). In photosynthetic tissues, β-carotene plays a protective role by immediate quenching of 3Chl*, which stifles $^1O_2$ synthesis and limits oxidative damage (Collins, 2001). The oxidized form of β-carotene (β-cyclocitral) is accumulated under oxidative stress and induces the expression of $^1O_2$-responsive genes, intimating that β-cyclocitral that acts as a signaling molecule in $^1O_2$ signaling in *Arabidopsis* (Ramel et al., 2012).

14.5.5 Phenolics and flavonoids

Phenolics are widespread and plentiful secondary metabolites (flavonoids, phenolic acid, tannins, and lignins) in the plant kingdom; however, very few of them possess antioxidant properties (Haider et al., 2013). In plants, phenolic compounds are mainly the by-product of benzoic acid and cinnamic acid that contains an aromatic ring with OCH_3 or OH^- substituent. Similar to AsA and α-TOC, phenolics can scavenge ROS production; inhibit LPO by the lipid alkoxyl radicals due to their ability to donate electron or hydrogen atoms (Grace and Logan, 2000). Under oxidative stress, flavonoids and phenylpropanoids participate to neutralize the ROS by scavenging H_2O_2 to prevent cellular oxidation (Løvdal et al., 2010). Based on the structure, flavonoids are categorized into flavonols, flavones, isoflavones, and carotenoids, mainly involved in pigmentation, protection from UV light, phytopathogen defense, and seed germination. Flavonoids also protect the photosynthetic apparatus by scavenging the ROS ($^1O_2$) and neutralizing the free radicals (Agati et al., 2012). Many reports have highlighted the induction of phenolics and flavonoids under various types of stresses (Michalak, 2006; Fini et al., 2011; Choudhary and Agrawal, 2014).

14.5.6 Proline

Being an osmolyte, proline plays a crucial role in plant adaptation to different abiotic stresses because of increased synthesis and decreased degradation (Haider et al., 2017a). Besides, proline acts as a nonenzymatic antioxidant that reduces the adversity of ROS in plants, animals, and even in microbes. Free proline works as an osmoprotectant, a metal chelator, a protein stabilizer, and an LPO inhibitor by scavenging $OH^{\bullet}$ (Ashraf and Foolad, 2007). Among organic compounds, proline has better $OH^{\bullet}$ scavenging ability (Haider et al., 2017a). Hence, proline acts as a ROS quencher and redox signaling molecule in response to drought, salt, and metal stress (Lee et al., 2012). Haider et al. (2017a) tested the proline metabolic pathways and demonstrated that all gene-related to synthesis and degradation of proline were up-regulated in response to drought stress. The overexpression of P5CS (pyrroline-5-carboxylate synthase) in *Arabidopsis* enhanced the proline production, which also resulted in hastening the SOD, CAT, and GPX activities under heat stress (Lv et al., 2011).

14.6 ROS in redox signaling

The ROS-signaling mechanism is profoundly conserved between aerobic organisms and regulates the wide range of biological processes, which includes growth and development and acclimation to various biotic and abiotic stresses (Mittler et al., 2011). Previously, ROS metabolism-related studies have just focused on ROS-mediated cell toxicity and scavenging mechanism, but recently, much attention has been paid on the ROS role as a signaling molecule. In plants, NADPH oxidase (NOX) and RBOH play a vital role in the ROS production reactions (Suzuki et al., 2011). However, *Arabidopsis AtRBOH* family constitutes 10 homologs (*AtRBOHA-AtRBOHJ*), which are involved in different signaling pathways, such as stomatal closure, pollen and stigma interaction, root hair development and plant resistance, and acclimation to different environmental stresses (Alscher et al., 2002; Miller et al., 2009; Noctor et al., 2018). Various tissue-specific biological processes are also regulated by spatial-temporal coordination between ROS and other signals, such as Ca^{2+}-signaling, mitogen-activated protein kinases (MAPKs)-signaling, and protein phosphorylation in response to environmental stimuli (Petrov et al., 2013). During evolution, plants have acquired a unique acclamatory mechanism that can be activated in the primary tissues exposed directly or indirectly to stress. This acclamatory or defensive mechanism in systemic tissues known as systemically acquired acclimation (SAA) or systemic acquired resistance (SAR) and both have a crucial role in preventing plants from potential damage (Shah and Zeier, 2013; Liu and He, 2017). Furthermore, the mobility and diffusion of H_2O_2 facilitated by plasma membrane aquaporins have a significant advantage of H_2O_2 signaling in cells (Dynowski et al., 2008).

14.6.1 RBOH proteins: diversity in their regulatory mechanism

In plants, RBOHs consist of cytosolic NADPH- and FAD-binding domain at C-terminal region, six conserved transmembrane-spanning domains, and N-terminal extension, which consists of two Ca^{2+}-binding EF-hand motifs and phosphorylation activities for its activity (Suzuki et al., 2011; Oda et al., 2010; Drerup et al., 2013). The activation of RBOH proteins produces $O_2^{\bullet-}$ in the apoplast, which is further dismutated enzymatically to H_2O_2 via SOD. Due to more half-life, membrane permeable H_2O_2 can act as a signaling molecule in growth, development, and acclimation response to environmental stresses (Xia et al., 2009; Kawahara et al., 2007).

In *Arabidopsis*, homologs of RBOH proteins (e.g., *AtRBOHC*, *D*, and *F*) perform specific regulatory functions depending on various signaling components, such as Ca^{2+}, MAPK, phosphorylation, and phospholipase Dα 1 (PLDα1) (Suzuki et al., 2011; Marino et al., 2012). Cellular fractionation of plant tissues induces the Ca^{2+}_{cyt} accumulation via an influx in the apoplast across the plasma membrane. The alleviated Ca^{2+} level then stimulates the *AtRBOHC*-dependent production of ROS followed by loop-feedback between RBOHC and Ca^{2+} to regulate root hair development (Foreman et al., 2003; Carol et al., 2005). Moreover, Ca^{2+} binding and phosphorylation antagonistically regulate the ROS-producing ability of *AtRBOHD* and *ATRBOHF* in *Arabidopsis*, intimating that Cyt. Ca^{2+} accumulation is crucial for *AtRBOHD* activation, which requires structural modifications in EF-hand motifs via Ca^{2+} binding (Ogasawara et al., 2008; Kimura et al., 2012). PLDα1 and its lipid product phosphatidic acid (PA) play a vital role in ABA-mediated ROS production in guard cells of stomata via *AtRBOHD* and *AtRBOHF*. The binding PA-motifs with arginine residues 149, 150, 156, and 157 in *AtRBOHD* and *AtRBOHF* phosphorylation at Ser13 and Ser174 in *ATRBOHF* are required for ABA-dependent stomatal closure by *OPEN*

STOMATA 1 kinase (*OST1*) (Sirichandra et al., 2009; Zhang et al., 2009). Recent studies have demonstrated the positive role of *AtRBOHD* in pathogen defense shown to be phosphorylated by *calcium-dependent protein kinase 5* (*CPK5*) (Liu and He, 2017).

In rice, a homolog of mammalian RAC (*OsRac1*) shown to activate *OsRBOHB* by interacting with its N-terminal region to induce pathogen defense (Oda et al., 2010). Moreover, the phosphorylation of Ser82 by *StCDPK4* and Ser97 *by StCDPK5* at the N-terminal region of *StRBOHB* resulted in the activation of *StRBOHB*-dependent ROS production, which is required for oxidative burst during pathogen defense (Asai et al., 2013). Besides RBOH, several other pathways, such as photosynthesis (ETC, PSI, and PSII), respiration (ETC), excited chlorophyll (3Chl*), xanthine oxidase (XOD), amine oxidase (AOD), glycolate oxidase (GOX), oxalate oxidase, peroxidases (PRX), and fatty acid (FA) oxidation are responsible for cellular ROS production (Mittler, 2002). Among these, oxalate oxidase was shown to involve in ROS production in drought-responsive root cells (Voothuluru and Sharp, 2013), whereas GOX revealed nonhost pathogen defense response in *Arabidopsis* and tobacco (Rojas et al., 2012). Likewise, PRX-dependent ROS production was involved in the regulation of root growth, callose deposition, defense response, and response to potassium (K^+) deficiency. In addition, *AtPRX33* and *AtPRX34* have been characterized as the dominant ROS producers in response to bacterial pathogens in *Arabidopsis* (Wrzaczek et al., 2013).

Taken together, these findings suggested that functional diversity occur between RBOH proteins and PRXs might be attributed to the differences in the type of ROS produced by these enzymes. The $O_2^{\bullet-}$ produced by RBOH protein can regulate distinct pathways varying from H_2O_2-activated pathways. Second reason for functional diversity in both enzymes might be due to the differences in their reductants. For example, RBOH-dependent $O_2^{\bullet-}$ synthesis takes NADPH as reductant, while many organic compounds and chemicals serve as reductants in PRX-dependent H_2O_2 production (Suzuki et al., 2011; O'Brien et al., 2012).

14.6.2 Temporal coordination of ROS-signaling with other signals in plants

In response to environmental shifts, the ion fluxes across the plasma membrane, accumulation of hormones and Ca^{2+}_{cyt} level, activation of MAPKs, and production of ROS can all be activated within seconds or minutes as early signaling events in plants (Miller et al., 2009; Finka et al., 2012). For example, in photorespiratory machinery, amino acids (e.g., glycerate, glycine, and serine) accumulate in the leaves within 60 s following the high light exposure and within 15–45 min in systemic tissues of plants (Suzuki et al., 2013). The high light responses are suggested to be linked with the redox biology of PQN pool, higher ethylene and ROS production, modifications in extracellular electric potential, and reduction in photochemical efficiency and non-photochemical quenching (Rossel et al., 2007; Szechynska-Hebda et al., 2010). In tobacco plants, the production of ROS was triggered within 3 min after heat stress, which can be inhibited by NOX inhibitors, whereas the outward influx of Ca^{2+} into the cytosol by Ca^{2+} channels within 10 min (Saidi et al., 2009; Mittler et al., 2012).

Another study demonstrated that pathogen elicitor (flg22) could activate the MAPK signaling that induce the membranous phosphorylation proteins, including protein phosphatase, protein kinases, ion channels, calmodulins, auxin-signaling associated proteins along with *AtRBOHD* in *Arabidopsis* within 5–10 min. For instance, in CDPK-signaling, in vivo phosphorylation of *AtRBOHD* by *CPK5* can be persuaded within 5 min (Benschop et al., 2007; Dubiella et al., 2013).

Recent researches on mechanical wounding demonstrated that a higher level of jasmonic acid (JA) accumulates in damaged tissues and nondamaged systemic leaves within 30 s to 5 min in *Arabidopsis* (Koo et al., 2009; Glauser et al., 2009). The momentum of the long-distance signal in systemic tissues is 3.4–4.5 $cm\,min^{-1}$ that results in the de novo synthesis of JA. Miller et al. (2009) revealed that abiotic stimuli induce RBOHD (NOX homolog), which is required for the long-distance signal is also auto-propagating systemic signals that move at the rate of about 8.4 $cm\,min^{-1}$. Furthermore, the involvement of electric signals produced at comparable rates was also linked in the RBOHD-triggered prompt systemic signaling during wounding (Mittler et al., 2011; Suzuki and Mittler, 2012). Another study has revealed the essential biological role of RBOHD-dependent long-distance travel signals in SAA response of plants to high light and heat stress (Suzuki et al., 2013). SAA is known to enhance tolerance of plants against heat stress and positively correlated with the ROS production and occurred promptly within 5–10 min following the heat stress. Further validation by metabolomic analysis suggested that glycine, glycerate, and serine were accumulated in leaf tissues exposed to high light within 60 s, while in systemic tissues within 15 and 45 min, demonstrating that photorespiration portion is required in the high light intensity as an early response. During the photorespiratory process, H_2O_2 is produced by PRX, which scavenge and alter the primary responses under high light intensity.

Long-term exposure to environmental stimuli alters growth and development for plant survival. For instance, ROS signaling associated with defensive mechanism confers protection to plant for hours or days against pathogen attack (Dubiella et al., 2013). RBOH-mediated ROS accumulation was observed in correlation with gradual necrotic symptoms (Wi et al., 2012). Moreover, high light intensity may imbalance the energy dissipation in the photosystem and can modify the thylakoid membrane structure within hours and days by STN7 kinase (Pesaresi et al., 2009). Although modifications in photosynthetic rate and light intensity implicate ROS and redox signaling that can modulate the redox state of plastoquinone in chloroplast, they play a vital acclamatory role in fluctuating light conditions (Mittler et al., 2011; Li et al., 2009). As a conclusion, it has been revealed that early ROS burst is required for later ROS production, which regulates underlying pathways and plant acclimation to environmental stress (Suzuki et al., 2013).

14.6.3 Spatial coordination of ROS signaling with other signals in plants

Many studies have highlighted the multipurpose signaling role of RBOH protein in different tissues and developmental stages (Miller et al., 2009), signifying its critical regulatory role in various biological processes in coordination with ROS signals activated in different plant tissues. The transcriptomic-based comparison of local leaves, affected with high light intensity and systemic leaves revealed the 70% upregulation of transcripts in local leaves and also showed variation in their expression in systemic leaves, implying that during SAA similar signals exist in local and systemic tissues to high light intensity (Rossel et al., 2007). The relationship within local and systemic responses validated by changes in ROS and redox signals and accretion of amino acids (e.g., glycine, serine, and glycerate) with the photorespiratory pathways resulted in both local and systemic tissues to high light (Miller et al., 2009; Suzuki et al., 2013).

Generally, local and systemic tissues exhibited significant overlap in their signals; recent researches have depicted the changes in transcript and metabolites in these tissues. Ethylene accumulated in both local and systemic tissues following the high light; however, signals mediated by EIN2 are needed for the initiation of APX2 only in systemic tissues. Likewise, SID2 delays APX2 induction in leaves but not in systemic leaves following the high light (Muhlenbock et al., 2008), intimating that APX2 expression in local and systemic tissues might be regulated by the coordination of ethylene and salicylic acid signaling during SAA. Another study demonstrated the spatial diversity in high light resulted in the accumulation of ZAT10 transcripts and RRTF1 (redox responsive factor 1) in systemic leaves but vary in leaf position (Gordon et al., 2012).

Plants can alter the spatial distribution of metabolites under various abiotic stresses. The distribution of secondary metabolite was altered in the mutant lacking CAT2, proposing the ROS-scavenging mechanism play a significant role in determining the distribution of secondary metabolites upon pathogen infection (Simon et al., 2010). ROS may play an important role in propagating signals from local tissues to systemic tissues, such as wave-like patterns of APX1 correlate positively with H_2O_2 accumulation and negatively with NPQ (Karpinski et al., 2013). Recent findings on the RBOHD-dependent ROS wave revealed its linkage with the production of systemic potential, may be a link between ROS production and electrical signals in plants (Suzuki et al., 2013). Moreover, the previous study has depicted the ROS signaling to regulate the connection between different tissues, suggested by the distribution of RBOH-dependent long-distance signals from leaves to the entire plant (Suzuki et al., 2011; Miller et al., 2009).

14.6.4 ROS-mediated activation of MAPK signaling

The MAPK or MPK cascade is an intracellular pathway conserved in eukaryotes and comprised of three kinases, including MAPK, MAPK kinase (MKK), and MAPKK kinase (MEKK or MAPKKK) (Haider et al., 2017b). MAPK cascades are involved in many physiological processes, of which cell division, resistance to insects and pathogens, and response to abiotic stresses (Haider et al., 2017b; Liu and He, 2017). A recent study demonstrated that exogenous H_2O_2 activates MAPK cascades, which resulted in the initiation of ROS signaling (Pitzschke and Hirt, 2006, 2009). ROS activates MPK3, MPK4, and MPK6, but different MKKs transfer signals depending on the type of stimuli, for example, the overexpression of MKK2 that activates MPK4 and MPK6 resulted in enhanced tolerance to heat and cold stress in *Arabidopsis*

(Jaspers and Kangasjarvi, 2010; Taj et al., 2010; Ismail et al., 2014), whereas the overexpression of MKK9 (MPK3 and MPK6) revealed the enhanced sensitivity of plants to salt stress (Xu et al., 2008). The MEKK1 was shown to activate MPK4 by H_2O_2 and signaling interaction between MEKK1 and MPK4 is mediated by MKK1 and MKK2 (Nakagami et al., 2006; Qiu et al., 2008). Moreover, H_2O_2 increases the expression of nucleoside diphosphate kinase 2 (NDPK2), and overexpression of *AtNDPK2* by *Arabidopsis* showed enhanced tolerance to various abiotic stresses by reducing H_2O_2 accumulation (Moon et al., 2003). In contrast, MPK8, that is, activated by the phosphorylation of CaM in Ca^{2+} signaling has been shown to control RBOHD, negatively regulate ROS production (Marino et al., 2012; Takahashi et al., 2011).

14.6.5 ROS signaling in systemic acquired acclimation to biotic and abiotic stresses

ROS are produced in different cell organelles of the plant by the enzyme activity of NOX (plasma membrane), PRX (cell wall), and AOX (apoplast) (Haider et al., 2018). Unlike $O_2^{\bullet-}$, H_2O_2 is permeable and can diffuse into the cell and induce many plant defense responses, including PCD (Dangl and Jones, 2001). During plant-pathogen interaction, SA and nitrous oxide (NO) suppress the activity of ROS detoxifying enzymes (CAT and PRX), which disturbs the balance between ROS production and scavenging, resulting in the activation of PCD. When plants induce defense responses, including PCD in response to pathogen attack, H_2O_2 production occurs in a biphasic manner. H_2O_2 generation takes place in both local and systemic tissues, functions as a secondary messenger mediating systemic responses of specific defense-related genes in tomato plants, following wounding (Orozco-Cárdenas and Ryan, 2001). Another research has shown that *Arabidopsis* mutant lines lacking functional *rboh* genes exhibit reduced production of ROS and PCD following bacterial infection (Torres et al., 2002). In contrast, reduced expression of CAT and APX in tobacco plants infected with bacteria revealed more ROS production leading to PCD (Mittler et al., 1999).

Under abiotic stress, different antioxidant enzymes (enzymatic and nonenzymatic) are activated to reduce the intracellular ROS toxicity, while the differences in ROS signaling may depend on the chemicals, compounds, hormones, and interaction with different signaling pathways during different environmental stresses (Ashraf and Foolad, 2007). In salt-responsive root tips, an alleviated Ca^{2+}_{cyt} wave moves at 400 μm/s parallel with the ROS wave from the roots toward aerial parts of plants. An implication of Ca^{2+} channel blocker (Lanthanum; La^{3+}) can inhibit the gene expression of marker genes associated with the Ca^{2+} and ROS wave, signifying the link between ROS and Ca^{2+} wave. In contrast, plants lacking with vacuolar-ion channel (two-pore channel 1; TPC1) revealed interruption in delivering systemic Ca^{2+} wave. It was also revealed that ROS also act as a conjugate with compounds that travel systemically and can activate ROS production in distal portions of the plant (Apel and Hirt, 2004). It is still unclear whether ROS can travel long-distance because ROS are reactive and detoxified by antioxidant enzymes in the apoplast, still need considerations.

14.6.6 ROS signaling interaction with other signals

Various studies have unveiled ROS signaling with other signals, such as Ca^{2+} signaling, hormone signaling, protein phosphatase, protein kinase, and redox responses to stress (Haider et al., 2017a, 2018). Plants can launch defense responses by recognizing various pathogens via extracellular cellular receptors to decode pathogen-assisted molecular patterns (PAMPs) (Haider et al., 2017b), which are accompanied by an accumulation of SA and its glycoside (SAG) and activation of pathogenesis-related (PR) proteins. The pathogen-mediated induction of metabolic changes can lead to SAR (Dempsey and Klessig, 2012). In the infected tissues of the plants, signals are initiated and translocated by vascular tissues (phloem) to the distal portions and received in systemic tissues (Kachroo and Robin, 2013). The phloem-driven SAR signals are shown to have biologically active molecules, including JA (Truman et al., 2007); glycerol-3-phosphate (G3P) derivatives (Chanda et al., 2011), methyl salicylate (MeSA) (Park et al., 2007), azelaic acid (AzA), pipecolic acid (PCP), and a lipid transfer protein (DIR) (Maldonado et al., 2002), while few of them induce systemic resistance when applied locally. JA upregulate the *Arabidopsis AtBSMT1* that converts SA into MeSA, which can travel to systemic tissues via phloem and converted to SA by SABP2 upon arrival (Dempsey and Klessig, 2012). In addition, Pip is a potential SAR signal and alleviated by the bacterial inoculation in both local and systemic leaf tissues to induce SAR by positively regulating self-synthesis and SA priming in systemic tissues. Whereas, DA is a rapid SAR signal shown to increase SA level and expression of PR proteins to increase the pathogen resistance in *Arabidopsis*, tobacco, and tomato (Dempsey and Klessig, 2012).

The occurrence of biotic and abiotic stress could lead to the production of ROS, which can also be associated with signals generated by various plant hormones (Haider et al., 2017b). For example, ethylene biosynthesis is positively modulated by RBOH protein and negatively regulated by *constitutive triple response 1* (CTR1) (Jakubowicz et al., 2010). Ethylene plays a critical role in the regulation of SAA and SAR to high light intensity (Karpinski et al., 1999). In response to high light intensity, the redox level of the PQ pool can regulate the expression of ethylene-related genes and induce the production of ACC (*1-aminocyclopropane-1-carboxylate*) (Muhlenbock et al., 2008). ROS-mediated PCD mainly relies on the regulation of EIN2 by LSD1 (*LESION SIMULATING DISEASE 1*) (Karpinski et al., 2013). Recent studies have demonstrated the role of brassinosteroid (BR) signaling in ROS-dependent stress response and shown to enhance tolerance to oxidative stress followed by cytosolic H_2O_2 production and regulating the expression of RBOH, MPK1, and MPK3 (Xia et al., 2009). Unlike auxins and polyamines, BRs are not directly involved in long-distance signaling (Li et al., 2013). In *Arabidopsis*, the exogenous application of IAA (indole-3-acetic acid) highly activates the activity of *AtRBOHD* (Peer et al., 2013), while H_2O_2 production was inhibited in *aux1* mutants (Krishnamurthy and Rathinasabapathi, 2013).

ABA has been revealed to interact with ROS and involved in the wide range of biological functions in response to abiotic stimuli (Ma et al., 2012). In *Arabidopsis*, *AtRBOHD* and *F* regulate the stomatal closure, seed germination, root elongation, and Na^+/K^+ homeostasis following the application of salt stress (Ma et al., 2012). Also,

ABA and SA applications resulted in increased H_2O_2 production which alleviates the tolerance mechanism to drought, salt, heat, high light intensity, and oxidative stress (Xia et al., 2009). In the recent study, SAA of plants was positively correlated with the initiation of the ROS wave and transient ABA accumulation in systemic tissues following heat stress, whereas similar responses were hindered in *AtRBOHD* mutants (Suzuki et al., 2013). These results have suggested that spatial-temporal interaction between RBOHD-dependent ROS and ABA mediate SAA to high-temperature stress.

Many studies have addressed the interaction of SA and JA signaling with ABA underlying stomatal closure (Song et al., 2014). The RNAi plants deficient with *aba2-1* were failed to induce stomatal closure following exogenous SA application, however, guard cells of SA-deficient mutants (*NahG* and *sid2*) responded to ABA, implying that SA signaling functions upstream of ABA signaling (Song et al., 2014; Montillet and Hirt, 2013). Stomatal closure induced by SA was inhibited by NOX inhibitor (DPI) and plants lacking RBOHD show poor stomatal response to exogenous SA treatment (Kalachova et al., 2013). A recent study has revealed that SA-induce stomatal closure conducted by PRX-mediated extracellular production of ROS (Khokon et al., 2010). However, the underlying mechanism that involves the integration of SA signaling to ROS-producing pathways will be beneficial to elucidate in the future.

14.6.7 ROS-mediated programmed cell death (PCD) under abiotic stress

Programmed cell death (PCD) induced by higher concentration of ROS is a genetically controlled mechanism in which cells are selectively destroyed in a multistep manner mediated by particular nuclease and proteases activities, without any damage to neighboring cells (Petrov et al., 2015). Under intense drought conditions, compressed CO_2 fixation followed by eminent electron leakage to O_2 intensifies the ROS production, thus resulting in inducing PCD (Gechev et al., 2012). ROS-mediated PCD was triggered by the interaction of ABA and cytokinin under drought stress (Munn-Bosch and Alegre, 2004). Plants can make some adaptive changes, which include enhanced lateral root growth, stomatal closure, and reorientation of polyamine metabolism under drought, in which ABA plays a crucial role (Duan et al., 2010). In grapevine, ABA induces stomatal closure and regulates polyamines accumulation by AOX metabolism that results in H_2O_2 production, which leads to further stress or PCD (Haider et al., 2017a; Toumi et al., 2010).

Flooding stress-induced PCD plays a vital role in the formation of aerenchyma in selected cells to produce air channels in roots to protect plants (Gunawardena et al., 2001). Another study revealed that the exogenous application of H_2O_2 also stimulates the formation of aerenchyma in rice (Steffens et al., 2011). Moreover, the expression of RBOHD was enhanced, which plays an essential role in H_2O_2 production, which results in the formation of aerenchyma in maize under water-logged conditions (Rajhi et al., 2011).

The $O_2^{\cdot-}$ produced by NOX activity is also implicated in the PCD under osmotic and salt stress, while the signals activated by nonionic oxidative and ionic salt inducers stress do not possess overlap (Petrov et al., 2015; Monetti et al., 2014). Salt-induced PCD disturbs the ionic balance by more Na^+ influx to the cytosol and decreased K^+ level (Kim et al., 2014). The $OH^{\cdot}$ radicals in the cell alter the K^+/Na^+ ratio and regulate the PCD inducing enzymes. Moreover, the overexpression of antiapoptotic enzyme BCL2 represses the expression of vascular processing enzymes (VPEs), reduces K^+ efflux, and enhances the PCD in rice (Kim et al., 2014; Demidchik et al., 2010). Taken together, these findings intimated that upholding of proper Na^+/K^+ ratios could be helpful to enhance salt tolerance in plants.

During heat stress, mitochondrial cytochrome c is released in a ROS-dependent manner, which activates caspase-like protease that induces PCD (Abdelrahman et al., 2017). Proline is a critical osmoprotectant also implicated in the mitochondrial ROS metabolism might be a crucial component in PCD under heat stress (Miller et al., 2009). In *Arabidopsis*, exogenous application of AsA and GSH induces PCD, but CAT that detoxifies H_2O_2 suppresses PCD, implying that H_2O_2 acts as a positive regulator of PCD, while other ROS types function as a negative regulator of PCD under heat stress (Li et al., 2012). In addition, MPK6 that functions upstream to proteases and hydrolase associated with PCD under heat stress (Wituszynska et al., 2015). Under low-temperature, LSD1 (a negative regulator of PCD) interacts with CAT that reduces its activity; its deficiency leads to PCD in plants (Li et al., 2013). Recent studies have suggested that plant deficient with LSD1 exhibited more UV-C-induce PCD, however, EDS1 mutants repress the PCD following UV-C treatment, signifying that LSD1 and EDS1 are regulated by the modulation of ROS homeostasis (Wituszynska et al., 2015).

14.7 Conclusion

The overview presented in this chapter simply intimates the basis of ROS formation in different cellular compartments, different production sites, and the role of ROS as a signaling molecule in different physiological processes of the plants. In fact, ROS production is a continuous phenomenon of aerobic metabolism, but under physiological steady-state ROS is constantly scavenged by antioxidant defense systems confined in different compartments of the cell. The balance between ROS production and scavenging is frequently perturbed by various unfavorable environmental conditions, resulting in ROS accumulation and posing oxidative stress. Plants can also generate and release the ROS (oxidative burst) by activating PRXs in response to several biotic and abiotic factors, which might be chemically different or generated in different cellular compartments. ROS also play a crucial role in coordinating spatial-temporal responses in plants, resulting in the formation of ROS wave and accretion in cytosolic calcium (Ca^{2+}) level that propagates entirely in the different tissues of the plant. In chloroplast and mitochondria, ROS signaling contributes to regulating the biological pathways, such as energy metabolism and gene expression following environmental stress stimuli. These functions are the potential mediator of important pathways, including systemic and retrograde signaling and programmed cell death (PCD) under oxidative stress conditions. However, there is an overlap between chloroplast/mitochondrial retrograde signaling, RBOH-mediated systemic response, and PCD. Moreover, how these pathways are regulated by different environmental stress stimuli could be helpful by emphasizing the integration of ROS signaling and sensor of abiotic stresses.

References

Abdelrahman, M., El-Sayed, M., Jogaiah, S., Burritt, D.J., Tran, L.-S.P., 2017. The "STAY-GREEN" trait and phytohormone signaling networks in plants under heat stress. Plant Cell Rep. **36**, 1009–1025.

Abdelrahman, M., Jogaiah, S., Burritt, D.J., Tran, L.-S.P., 2018. Legume genetic resources and transcriptome dynamics under abiotic stress conditions. Plant Cell Environ. 41, 1972–1983.

Abouzari, A., Fakheri, B.A., 2015. Reactive oxygen species: generation, oxidative damage, and signal transduction. *Intl. J. Life Sci.* 9, 3–17.

Agati, G., Azzarello, E., Pollastri, S., Tattini, M., 2012. Flavonoids as antioxidants in plants: location and functional significance. *Plant Sci.* 196, 67–76.

Alscher, R.G., Erturk, N., Heath, L.S., 2002. Role of superoxide dismutases (SODs) in controlling oxidative stress in plants. J. Exp. Biol. 53, 1331–1341.

Apel, K., Hirt, H., 2004. Reactive oxygen species: metabolism, oxidative stress, and signal transduction. *Annu Rev Plant Biol.* 55, 373–399.

Asada, K., 2006. Production and scavenging of reactive oxygen species in chloroplasts and their functions. *Plant Physiol.* 141, 391–396.

Asai, S., Ichikawa, T., Nomura, H., et al., 2013. The variable domain of a plant calcium-dependent protein kinase (CDPK) confers subcellular localization and substrate recognition for NADPH oxidase. *J. Biol. Chem.* 288, 14332–14340.

Ashraf, M., Foolad, M.R., 2007. Roles of glycine betaine and proline in improving plant abiotic stress resistance. Environ. Exp. Bot. 59, 206–216.

Atkin, O.K., Macherel, D., 2009. The crucial role of plant mitochondria in orchestrating drought tolerance. *Ann. Bot.* 103, 581–597.

Baker, A., Graham, I., 2002. Plant Peroxisomes. Biochemistry, Cell Biology and Biotechnological Applications. Kluwer, Dordrecht, The Netherlands.

Barnes, J.D., Zheng, Y., Lyons, T.M., 2002. Plant resistance to ozone: the role of ascorbate. In: Omasa, K., Saji, H., Youssefian, S., Kondo, N. (Eds.), Air Pollution and Plant Biotechnology. Springer, Tokyo, pp. 235–254.

Bechtold, U., Karpinski, S., Mullineaux, P.M., 2005. The influence of the light environment and photosynthesis on oxidative signaling responses in plant-biotrophic pathogen interactions. *Plant Cell Environ.* 28, 1046–1055.

Benschop, J.J., Mohammed, S., O'Flaherty, M., Heck, A.J., Slijper, M., Menke, F.L., 2007. Quantitative phosphoproteomics of early elicitor signaling in *Arabidopsis. Mol. Cell Proteomics.* 6, 1198–1214.

Bhattacharjee, S., 2005. Reactive oxygen species and oxidative burst: roles in stress, senescence and signal transduction in plants. *Curr Sci.* 89, 1113–1121.

Bielski, B.H., Arudi, R.L., Sutherland, M.W., 1983. A study of the reactivity of HO_2/O_2^- with unsaturated fatty acids. *J. Biol. Chem.* 258, 4759–4761.

Bienert, G.P., Schjoerring, J.K., Jahn, T.P., 2006. Membrane transport of hydrogen peroxide. *BBA Biomembranes*. 1758, 994–1003.

Bindschedler, L.V., Dewdney, J., Blee, K.A., Stone, J.M., Asai, T., Plotnikov, J., Denoux, C., Hayes, T., Gerrish, C., Davies, D.R., Ausubel, F.M., Bolwell, G.P., 2006. Peroxidase-dependent apoplastic oxidative burst in Arabidopsis required for pathogen resistance. *Plant J*. 47, 851–863.

Boguszewska, D., Grudkowska, M., Zagdañska, B., 2010. Drought-responsive antioxidant enzymes in potato (*Solanum tuberosum* L.). Potato Res. 53, 373–382.

Bolwell, G.P., Bindschedler, L.V., Blee, K.A., Butt, V.S., Davies, D.R., Gardner, S.L., Gerrish, C., Minibayeva, F., 2002. The apoplastic oxidative burst in response to biotic stress in plants: a three-component system. *J. Exp. Bot*. 53, 1367–1376.

Bright, J., Desikan, R., Hancock, J.T., Weir, I.S., Neill, S.J., 2010. Aba-induced no generation and stomatal closure in Arabidopsis are dependent on h2*o2 syn*thesis. *Plant J*. 45, 113–122.

Carmody, M., Crisp, P.A., d'Alessandro, S., Ganguly, D., Gordon, M., Havaux, M., Albrecht-Borth, V., Pogson, B.J., 2016. Uncoupling high light responses from singlet oxygen retrograde signaling and spatial-temporal systemic acquired acclimation. *Plant Physiol*. 171, 1734–1749.

Carol, R.J., Takeda, S., Linstead, P., Durrant, M.C., Kakesova, H., Derbyshire, P., Drea, S., Zarsky, V., Dola, L., 2005. A RhoGDP dissociation inhibitor spatially regulates growth in root hair cells. Nature 438, 1013–1016.

Chanda, B., Xia, Y., Mandal, M.K., et al., 2011. Glycerol-3-phosphate is a critical mobile inducer of systemic immunity in plants. *Nat Genet*. 43, 421–427.

Chaves, M.M., Pereira, J.S., Maroco, J., et al., 2002. How plants cope with water stress in the field. Photosynthesis and growth. Ann. Bot. 7, 907–916.

Chen, Z., Gallie, D.R., 2006. Dehydroascorbate reductase affects leaf growth, development, and function. Plant Physiol. 142, 775–787.

Choudhary, K.K., Agrawal, S.B., 2014. Cultivar specificity of tropical mung bean (*Vigna radiata L.*) to elevated ultraviolet-B: changes in antioxidative defense system, nitrogen metabolism and accumulation of jasmonic and salicylic acids. *Environ. Exp. Bot*. 99, 122–132.

Collins, A., 2001. Carotenoids and genomic stability. *Mutat Res*. 475, 21–28.

Cosio, C., Dunand, C., 2009. Specific functions of individual class III peroxidase genes. J. Exp. Bot. 60, 391–408.

Dangl, J.L., Jones, J.D.G., 2001. Plant pathogens and integrated defence responses to infection. Nature 411, 826–833.

Das, K., Roychoudhury, A., 2014. Reactive oxygen species (ROS) and response of antioxidants as ROS-scavengers during environmental stress in plants. *Front. Environ. Sci*. 2, 53.

de Pinto, M.C., Locato, V., Sgobba, A., Romero-Puertas, M.C., Gadaleta, C., Delledonne, M., De Gara, L., 2013. *S*-nitrosylation of ascorbate peroxidase is part of programmed cell death signaling in tobacco Bright Yellow-2 cells. *Plant Physiol*. 163, 1766–1775.

del Río, L.A., 2015. ROS and RNS in plant physiology: an overview. *J. Exp. Bot*. 66, 2827–2837.

del Rio, L.A., Corpas, F.J., Sandalio, L.M., Palma, J.M., Gómez, M., Barroso, J.B., 2002. Reactive oxygen species, antioxidant systems and nitric oxide in peroxisomes. J. Exp. Bot. 53, 1255–1272.

del Rio, L.A., Sandalio, L.M., Corpas, F.J., Palma, J.M., Barroso, J.B., 2006. Reactive oxygen species and reactive nitrogen species in peroxisomes. Production, scavenging, and role in cell signalling. Plant Physiol. 141, 330–335.

Demidchik, V., Cuin, T.A., Svistunenko, D., Smith, S.J., Miller, A.J., Shabala, S., Sokolik, A., Yurin, V., 2010. Arabidopsis root K+-efflux conductance activated by hydroxyl radicals:

single-channel properties, genetic basis and involvement in stress-induced cell death. *J. Cell Sci.* 123, 1468–1479.

Dempsey, D.A., Klessig, D.F., 2012. SOS- too many signals for systemic acquired resistance? *Trends Plant Sci.* 17, 538–545.

Desel, C., Hubbermann, E.M., Schwarz, K., Krupinska, K., 2007. Nitration of γ-tocopherol in plant tissues. Planta 226, 1311–1322.

Donaldson, P.A., Anderson, T., Lane, B.G., Davidson, A.L., Simmonds, D.H., 2001. Soybean plants expressing an active oligomeric oxalate oxidase from the wheat GF-2.8 (germin) gene are resistant to the oxalate-secreting pathogen *Sclerotina sclerotiorum*. Physiol. Mol. Plant Pathol. 59, 297–307.

Drerup, M.M., Schlücking, K., Hashimoto, K., Manishankar, P., Steinhorst, L., Kuchitsu, K., Kudla, J., 2013. The calcineurin B-like calcium sensors CBL1 and CBL9 together with their interacting protein kinase CIPK26 regulate the *Arabidopsis* NADPH oxidase RBOHF. *Mol. Plant.* 6, 559–569.

Duan, Y., Zhang, W., Li, B., Wang, Y., Li, K., Sodmergen, Han, C., Zhang, Y., Li, X., 2010. An endoplasmic reticulum response pathway mediates programmed cell death of root tip induced by water stress in *Arabidopsis*. New Phytol. 186, 681–695.

Dubiella, U., Seybold, H., Durian, G., Komander, E., Lassig, R., Witte, C.P., Schulze, W.X., Romeis, T., 2013. Calcium-dependent protein kinase/NADPH oxidase activation circuit is required for rapid defense signal propagation. Proc. Natl. Acad. Sci. USA 110, 8744–8749.

Dynowski, M., Schaaf, G., Loque, D., Moran, O., Ludewig, U., 2008. Plant plasma membrane water channels conduct the signalling molecule H_2O_2. *Biochem. J.* 414, 53–61.

Elstner, E.F., 1987. Metabolism of activated oxygen species. In: Davies, D.D. (Ed.), Biochemistry of Plants. Academic Press, London, pp. 253–315.

Fini, A., Brunetti, C., DiFerdinando, M., Ferrini, F., Tattini, M., 2011. Stress-induced flavonoids biosynthesis and the antioxidant machinery of plants. *Plant Signal Behav.* 6, 709–711.

Finka, A., Cuendet, A.F., Maathuis, F.J., Saidi, Y., Goloubinoff, P., 2012. Plasma membrane cyclic nucleotide gated calcium channels control land plant thermal sensing and acquired thermotolerance. Plant Cell. 24, 3333–3348.

Flors, C., Nonell, S., 2010. Light and singlet oxygen in plant defense against pathogens: phototoxic phenalenone phytoalexins. *Acc. Chem. Res.* 37, 293–300.

Foreman, J., Demidchik, V., Bothwell, J.H., Mylona, P., Miedema, H., Torres, M.A., 2003. Reactive oxygen species produced by NADPH oxidase regulate plant cell growth. Nature 422, 442–446.

Foyer, C.H., Noctor, G., 2002. Photosynthetic nitrogen assimilation: *inter-pathway* control and signaling. In: Foyer, C.H., Noctor, G. (Eds.), *Photosynthetic Nitrogen Assimilation and Associated Carbon and Respiratory Metabolism*. Kluwer Academic Publishers, The Netherlands, pp. 1–22.

Foyer, C.H., Noctor, G., 2003. Redox sensing and signaling associated with reactive oxygen in chloroplasts, peroxisomes and mitochondria. *Physiol Plant.* 119, 355–364.

Foyer, C.H., Noctor, G., 2005. Redox homeostasis and antioxidant signaling: a metabolic interface between stress perception and physiological responses. *Plant Cell.* 17, 1866–1875.

Foyer, C.H., Noctor, G., 2011. Ascorbate and glutathione: the heart of the redox hub. *Plant Physiol.* 155, 2–18.

Fryer, M.J., Oxborough, K., Mullineaux, P.M., Baker, N.R., 2002. Imaging of photo-oxidative stress responses in leaves. *J. Exp. Bot.* 53, 1249–1254.

Gechev, T.S., Dinakar, C., Benina, M., Toneva, V., Bartels, D., 2012. Molecular mechanisms of desiccation tolerance in resurrection plants. Cell Mol Life Sci. 69, 3175–3186.

Gill, S.S., Tuteja, N., 2010. Reactive oxygen species and antioxidant machinery in abiotic stress tolerance in crop plants. *Plant Physiol. Biochem.* 48, 909–930.

Glauser, G., Dubugnon, L., Mousavi, S.A., Rudaz, S., Wolfender, J.L., Farmer, E.E., 2009. Velocity estimates for signal propagation leading to systemic jasmonic acid accumulation in wounded *Arabidopsis*. *J. Biol. Chem.* 284, 34506–34513.

Gordon, M.J., Carmody, M., Albrecht, V., Pogson, B., 2012. Systemic and local responses to repeated HL stress-induced retrograde signaling in *Arabidopsis*. Front Plant Sci. 3, 303.

Grace, S.G., Logan, B.A., 2000. Energy dissipation and radical scavenging by the plant phenylpropanoid pathway. *Philos. Trans. Roy. Soc. B.* 355, 1499–1510.

Gunawardena, A., Pearce, D.M., Jackson, M.B., Hawes, C.R., Evans, D.E., 2001. Characterisation of programmed cell death during aerenchyma formation induced by ethylene or hypoxia in roots of maize (*Zea mays* L.). Planta 212, 205–214.

Guo, T., Zhang, G., Zhou, M., Wu, F., Chen, J., 2004. Effects of aluminum and cadmium toxicity on growth and antioxidant enzyme activities of two barley genotypes with different Al resistance. *Plant Soil.* 258, 241–248.

Haider, M.S., Khan, I.A., Naqvi, S.A., Jaskani, M.J., Khan, R.W., Nafees, M., Maryam., and Pasha, I., 2013. Fruit developmental stages effects on biochemical attributes in date palm. Pak. J. Agril. Sci. 50, 577–583.

Haider, M.S., Zhang, C., Kurjogi, M.M., Pervaiz, T., Zheng, T., Zhang, C.B., Lide, C., Shangguan, L.F., Fang, J., 2017a. Insights into grapevine defense response against drought as revealed by biochemical, physiological and RNA-Seq analysis. *Sci Rep.* 7, 13134.

Haider, M.S., Kurjogi, M.M., Khalil-Ur-Rehman, M., Fiaz, M., Pervaiz, T., Jiu, S., Haifeng, J., Chen, W., Fang, J., 2017b. Grapevine immune signaling network in response to drought stress as revealed by transcriptomic analysis. *Plant Physiol. Biochem.* 121, 187–195.

Haider, M.S., Kurjogi, M.M., Khalil-Ur-Rehman, M., Pervaiz, T., Jiu, S., Fiaz, M., Jogaiah, S., Chen, W., Fang, J., 2018. Drought stress revealed physiological, biochemical and gene-expressional variations in 'Yoshihime' peach (*Prunus Persica* L) cultivar. *J. Plant Interact.* 13, 83–90.

Haider, M.S., Jogaiah, S., Pervaiz, T., Yanxue, Z., Khan, N., Fang, J., 2019. Physiological and transcriptional variations inducing complex adaptive mechanisms in 2 grapevine by salt stress. Environ. Exp. Biol. 162, 455–467.

Halliwell, B., 2006. Reactive species and antioxidants. Redox biology is a fundamental theme of aerobic life. Plant Physiol. 141, 312–322.

Hatz, S., Lambert, J.D.C., Ogilby, P.R., 2007. Measuring the lifetime of singlet oxygen in a single cell: addressing the issue of cell viability. *Photochem. Photobiol. Sci.* 6, 1106–1116.

Hayashi, M., Nishimura, M., 2006. *Arabidopsis thaliana*-a model organism to study plant peroxisomes. *Biochim. Biophys. Acta.* 1763, 1382–1391.

Hideg, E., Kálai, T., Hideg, K., 2011. Direct detection of free radicals and reactive oxygen species in thylakoids. *Methods Mol. Biol.* 684, 187–200.

Hojati, M., Modarres Sanavy, S.A.M., Karimi, M., Ghanati, F., 2010. Responses of growth and antioxidant systems in *Carthamus tinctorius* L. under water deficit stress. Acta. Physiol. Plant. 33, 105–112.

Hollander-Czytko, H., Grabowski, J., Sandorf, I., Weckermann, K., Weiler, E.W., 2005. Tocopherol content and activities of tyrosine aminotransferase and cysteine lyase in Arabidopsis under stress conditions. *J Plant Physiol.* 162, 767–770.

Horemans, N., Foyer, C., Asard, H., 2000. Transport and action of ascorbate at the plasma membrane. Trends Plant Sci. 5, 263–266.

Igamberdiev, A.U., Seregelyes, C., Manach, N., Hill, R.D., 2004. NADH-dependent metabolism of nitric oxide in alfalfa root cultures expressing barley hemoglobin. *Planta.* 219, 95–102.

Ismail, A., Takeda, S., Nick, P., 2014. Life and death under salt stress: same players, different timing? J. Exp. Bot. 65, 2963–2979.

Jakubowicz, M., Galganska, H., Nowak, W., Sadowski, J., 2010. Exogenously induced expression of ethylene biosynthesis, ethylene perception, phospholipase D, and Rboh-oxidase genes in broccoli seedlings. J. Exp. Bot. 61, 3475–3491.

Jaspers, P., Kangasjarvi, J., 2010. Reactive oxygen species in abiotic stress signaling. Physiol Plant. 138, 405–413.

Jogaiah, S., Sharathchandra, R.G., Lam-Son, P.T., 2013. Systems biology-based approaches towards understanding drought tolerance in food crops. Critical Rev. in Biotech. 33, 23–39.

Kachroo, A., Robin, G.P., 2013. Systemic signaling during plant defense. Curr. Opin. Plant Biol. 16, 527–533.

Kalachova, T., Iakovenko, O., Kretinin, S., Kravets, V., 2013. Involvement of phospholipase D and NADPH-oxidase in salicylic acid signaling cascade. *Plant Physiol Biochem.* 66, 127–133.

Karpinski, S., Reynolds, H., Karpinska, B., Wingsle, G., Creissen, G., Mullineaux, P., 1999. Systemic signaling and acclimation in response to excess excitation energy in *Arabidopsis*. Science 284, 654–657.

Karpinski, S., Szechynska-Hebda, M., Wituszynska, W., Burdiak, P., 2013. Light acclimation, retrograde signalling, cell death and immune defences in plants. Plant Cell Environ. 36, 736–744.

Kawahara, T., Quinn, M.T., Lambeth, J.D., 2007. Molecular evolution of the reactive oxygen-generating NADPH oxidase (Nox/Duox) family of enzymes. BMC Evol. Biol. 7, 109.

Khokon, M.A., Hossain, M.A., Munemasa, S., Uraji, M., Nakamura, Y., Mori, I.C., Murata, Y., 2010. Yeast elicitor-induced stomatal closure and peroxidase-mediated ROS production in *Arabidopsis*. Plant Cell Physiol. 51, 1915–1921.

Kiffin, R., Bandyopadhyay, U., Cuervo, A.M., 2006. Oxidative stress and autophagy. Antioxid Redox. Signal. 8, 152–162.

Kim, M.J., Ciani, S., Schachtman, D.P., 2010. A peroxidase contributes to ROS production during Arabidopsis root response to potassium deficiency. Mol. Plant. 3, 420–427.

Kim, Y., Wang, M., Bai, Y., Zeng, Z., Guo, F., Han, N., Bian, H., Wang, J., Pan, J., Zhu, M., 2014. Bcl-2 suppresses activation of VPEs by inhibiting cytosolic Ca(2)(+) level with elevated K(+) efflux in NaCl-induced PCD in rice. *Plant Physiol. Biochem.* 80, 168–175.

Kimura, S., Kaya, H., Kawarazaki, T., Hiraoka, G., Senzaki, E., Michikawa, M., Kuchitsu, K., 2012. Protein phosphorylation is a prerequisite for the Ca2+-dependent activation of Arabidopsis NADPH oxidases and may function as a trigger for the positive feedback regulation of Ca2+ and reactive oxygen species. Biochim. Biophys. Acta. 1823, 398–405.

Koo, A.J., Gao, X., Jones, A.D., Howe, G.A., 2009. A rapid wound signal activates the systemic synthesis of bioactive jasmonates in *Arabidopsis*. *Plant J. Cell Mol. Biol.* 59, 974–986.

Krieger-Liszkay, A., Fufezan, C., Trebst, A., 2008. Singlet oxygen production in photosystem II and related protection mechanism. *Photosynth Res.* 98, 551–564.

Krishnamurthy, A., Rathinasabapathi, B., 2013. Auxin and its transport play a role in plant tolerance to arsenite-induced oxidative stress in *Arabidopsis thaliana*. *Plant Cell Environ.* 36, 1838–1849.

Kukreja, S., Nandwal, A.S., Kumar, N., Sharma, S.K., Sharma, S.K., Unvi, V., Sharma, P.K., 2005. Plant water status, H_2O_2 scavenging enzymes ethylene evolution and membrane integrity of *Cicer arietinum* roots as affected by salinity. Biol Plant. 49, 305–308.

Lee, Y.P., Babakov, A., de Boer, B., Zuther, E., Hincha, D.K., 2012. Comparison of freezing tolerance, compatible solutes and polyamines in geographically diverse collections of Thellungiella sp. and *Arabidopsis thaliana* accessions. *BMC Plant Biol.* 12, 131.

Leng, X., Jia, H., Sun, X., Shangguan, L., Mu, Q., Wang, B., Fang, J., 2015. Comparative transcriptome analysis of grapevine in response to copper stress. *Sci Rep*. 5, 1–17.

Li, Z., Wakao, S., Fischer, B.B., Niyogi, K.K., 2009. Sensing and responding to excess light. *Ann. Rev. Plant Biol.* 60, 239–260.

Li, Z., Yue, H., Xing, D., 2012. MAP Kinase 6-mediated activation of vacuolar processing enzyme modulates heat shock-induced programmed cell death in *Arabidopsis*. *New Phytol.* 195, 85–96.

Li, Y., Chen, L., Mu, J., Zuo, J., 2013. Lesion Simulating Disease1 interacts with catalases to regulate hypersensitive cell death in *Arabidopsis*. Plant Physiol. 163, 1059–1070.

Liu, Y., He, C., 2017. A review of redox signaling and the control of map kinase pathway in plants. *Redox Biol.* 11, 192–204.

Liu, Y., Ren, D., Pike, S., Pallardy, S., Gassmann, W., Zhang, S., 2007. Chloroplast-generated reactive oxygen species are involved in hypersensitive response-like cell death mediated by a mitogen activated protein kinase cascade. Plant J. 51, 941–954.

Loew, O., 1900. A new enzyme of general occurrence in organisms. Science 11, 701–702.

Løvdal, T., Olsen, K.M., Slimestad, R., et al., 2010. Synergetic effects of nitrogen depletion, temperature, and light on the content of phenolic compounds and gene expression in leaves of tomato. Phytochemistry 71, 605–613.

Luis, A.R., Lopez-Huertas, E., 2006. ROS generation in peroxisomes and its role in cell signaling. Plant Cell Physiol. 57, 1364–1376.

Lv, W.T., Lin, B., Zhang, M., Hua, X.J., 2011. Proline accumulation is inhibitory to *Arabidopsis* seedlings during heat stress. Plant Physiol. 156, 1921–1933.

Ma, L., Zhang, H., Sun, L., Jiao, Y., Zhang, G., Miao, C., Hao, F., 2012. NADPH oxidase AtrbohD and AtrbohF function in ROS-dependent regulation of Na(+)/K(+)homeostasis in *Arabidopsis* under salt stress. *J. Exp. Bot.* 63, 305–317.

Maldonado, A.M., Doerner, P., Dixon, R.A., Lamb, C.J., Cameron, R.K., 2002. A putative lipid transfer protein involved in systemic resistance signalling in *Arabidopsis*. Nature 419, 399–403.

Mallick, N., Mohn, F.H., 2000. Reactive oxygen species: response of algal cells. J. Plant Physiol. 157, 183–193.

Marino, D., Dunand, C., Puppo, A., Pauly, N., 2012. A burst of plant NADPH oxidases. Trends Plant Sci. 17, 9–15.

Maxwell, D.P., Wang, Y., McIntosh, L., 1999. The alternative oxidase lowers mitochondrial reactive oxygen production in plant cells. Proc. Natl. Acad. Sci. USA 96, 8271–8276.

McDowell, J.M., Dangl, J.L., 2000. Signal transduction in the plant immune response. *Trends Biochem. Sci.* 25, 79–82.

Michalak, A., 2006. Phenolic compounds and their antioxidant activity in plants growing under heavy metal stress. *Pol. J. Environ. Stud.* 15, 523–530.

Millar, A.H., Mittova, V., Kiddle, G., et al., 2003. Control of ascorbate synthesis by respiration and its implications for stress responses. *Plant Physiol.* 133, 443–447.

Miller, G., Suzuki, N., Rizhsky, L., Hegie, A., Koussevitzky, S., Mittler, R., 2007. Double mutants deficient in cytosolic and thylakoid ascorbate peroxidase reveal a complex mode of interaction between reactive oxygen species, plant development, and response to abiotic stresses. *Plant Physiol.* 144, 1777–1785.

Miller, G., Shulaev, V., Mittler, R., 2008. Reactive oxygen signaling and abiotic stress. *Physiol Plant.* 133, 481–489.

Miller, G., Schlauch, K., Tam, R., Cortes, D., Torres, M.A., Shulaev, V., Dangl, J., Mittler, R., 2009. The plant NADPH oxidase RBOHD mediates rapid systemic signaling in response to diverse stimuli. Sci Signal. 2, ra45.

Mittler, R., 2002. Oxidative stress, antioxidants and stress tolerance. Trends Plant Sci. 7, 405–410.

Mittler, R., Herr, E.H., Orvar, B.L., Van, C.W., Willekens, H., Inzé, D., et al., 1999. Transgenic tobacco plants with reduced capability to detoxify reactive oxygen intermediates are hyperresponsive to pathogen infection. Proc. Nat. Acad. Sci. USA 96, 14165–14170.

Mittler, R., Vanderauwera, S., Gollery, M., Van, Breusegem. F., 2004. The reactive oxygen gene network in plants. Trends Plant Sci. 9, 490–498.

Mittler, R., Vanderauwera, S., Suzuki, N., et al., 2011. ROS signaling: the new wave? Trends Plant Sci. 6, 300–309.

Mittler, R., Finka, A., Goloubinoff, P., 2012. How do plants feel the heat? Trends Biochem. Sci. 37, 118–125.

Monetti, E., Kadono, T., Tran, D., Azzarello, E., Arbelet-Bonnin, D., Biligui, B., Briand, J., Kawano, T., Mancuso, S., Bouteau, F., 2014. Deciphering early events involved in hyperosmotic stressinduced programmed cell death in tobacco BY-2 cells. *J. Exp. Bot.* 65, 1361–1375.

Montillet, J.L., Hirt, H., 2013. New checkpoints in stomatal defense. Trends Plant Sci. 18, 295–297.

Moon, H., Lee, B., Choi, G., Shin, D., Prasad, D.T., Lee, O., et al., 2003. Ndp kinase 2 interacts with two oxidative stress-activated mapks to regulate cellular redox state and enhances multiple stress tolerance in transgenic plants. Proc. Nat. Acad. Sci. USA 100, 358–363.

Muhlenbock, P., Szechynska-Hebda, M., Plaszczyca, M., Baudo, M., Mullineaux, P.M., Parker, J.E., Karpinska, B., Karpinski, S., 2008. Chloroplast signaling and LESION SIMULATING DISEASE1 regulate crosstalk between light acclimation and immunity in *Arabidopsis*. Plant Cell 20, 2339–2356.

Mullineaux, P.M., Rausch, T., 2005. Glutathione, photosynthesis and the redox regulation of stress responsive gene expression. *Photosynth Res.* 86, 459–474.

Munn-Bosch, S., Alegre, L., 2004. Die and let live: leaf senescence contributes to plant survival under drought stress. Funct Plant Biol. 31, 203–216.

Munne-Bosch, S., Queval, G., Foyer, C.H., 2013. The impact of global change factors on redox signaling underpinning stress tolerance. Plant Physiol. 161, 5–19.

Munns, R., Tester, M., 2008. Mechanisms of salinity tolerance. *Ann. Rev. Plant Biol.* 59, 651–681.

Murphy, M.P., 2009. How mitochondria produce reactive oxygen species. *Biochem. J.* 417, 1–13.

Nakagami, H., Soukupova, H., Schikora, A., Zarsky, V., Hirt, H., 2006. A mitogen-activated protein kinase kinase kinase mediates reactive oxygen species homeostasis in *Arabidopsis*. *J. Biol. Chem.* 281, 38697–38704.

Navrot, N., Roubier, N., Gelbaye, E., Jacquot, J.P., 2007. Reactive oxygen species generation and antioxidant systems in plant mitochondria. *Physiol. Planta.* 129, 185–195.

Niyogi, K.K., Shih, C., Chow, W.S., et al., 2001. Photoprotection in a zeaxanthin-and lutein-deficient double mutant of Arabidopsis. *Photosynth Res.* 67, 139–145.

Noctor, G., Foyer, C.H., 1998. Ascorbate and glutathione: keeping active oxygen under control. Annu. Rev. Plant Phys. 49, 249–279.

Noctor, G., De Paepe, R., Foyer, C.H., 2007. Mitochondrial redox biology and homeostasis in plants. *Trends Plant Sci.* 12, 125–134.

Noctor, G., Mhamdi, A., Chaouch, S., Han, Y., Neukermans, J., Marquez-Garcia, B., Queval, G., Foyer, C.H., 2012. Glutathione in plants: an integrated overview. *Plant Cell Environ.* 35, 454–484.

Noctor, G., Reichheld, J.P., Foyer, C.H., 2018. Ros-related redox regulation and signaling in plants. Sem. Cell Develop Biol. 80, 3–12.

O'Brien, J.A., Daudi, A., Butt, V.S., Bolwell, G.P., 2012. Reactive oxygen species and their role in plant defence and cell wall metabolism. Planta 236, 765–779.

Oda, T., Hashimoto, H., Kuwabara, N., Akashi, S., Hayashi, K., Kojima, C., Wong, H.L., Kawasaki, T., Shimamoto, K., Sato, M., Shimizu, T., 2010. Structure of the N-terminal regulatory domain of a plant NADPH oxidase and its functional implications. *J. Biol. Chem.* 285, 1435–1445.

Ogasawara, Y., Kaya, H., Hiraoka, G., Yumoto, F., Kimura, S., Kadota, Y., Hishinuma, H., Senzaki, E., Yamagoe, S., Nagata, K., Nara, M., Suzuki, K., Tanokura, M., Kuchitsu, K., 2008. Synergistic activation of the arabidopsis NADPH oxidase AtrbohD by Ca2+ and Phosphorylation. *J. Biol. Chem.* 283, 8885–8892.

Okado-Matsumoto, A., Fridovich, I., 2001. Subcellular distribution of superoxide dismutases (SOD) in rat liver Cu, Zn-SOD in mitochondria. *J. Biol. Chem.* 276, 38388–38393.

Orozco-Cárdenas, M.L., Ryan, C.A., 2001. Hydrogen peroxide acts as a second messenger for the induction of defense genes in tomato plants in response to wounding, systemin, and methyl jasmonate. *Plant Cell.* 13, 179–191.

Palma, J.M., Sandalio, L.M., Corpas, F.J., Romero-Puertas, M.C., McCarthy, I., del Río, L.A., 2002. Plant proteases, protein degradation and oxidative stress: role of peroxisomes. *Plant Physiol. Biochem.* 40, 521–530.

Palma, J.M., Corpas, F.J., del Rı´o, L. A., 2009. Proteome of plant peroxisomes: new perspectives on the role of these organelles in cell biology. *Proteomics.* 9, 2301–2312.

Park, S.W., Kaimoyo, E., Kumar, D., Mosher, S., Klessig, D.F., 2007. Methyl salicylate is a critical mobile signal for plant systemic acquired resistance. Science 318, 113–116.

Passia, G., Queval, G., Bai, J., Margis-Pinherio, M., Foyer, C.H., 2014. The effects of redox controls mediated by glutathione peroxidases on root architecture in Arabidopsis thaliana. *J. Exp. Bot.* 65, 1403–1413.

Peer, W.A., Cheng, Y., Murphy, A.S., 2013. Evidence of oxidative attenuation of auxin signalling. J. Exp. Bot. 64, 2629–2639.

Peng, C.L., Zhi-Ying, O.U., Liu, N., Lin, G.Z., 2005. Response to high temperature in flag leaves of super high-yielding rice pei'ai 64s/e32 and liangyoupeijiu. Rice Sci. 12, 179–186.

Perry, J.J., Shin, D.S., Getzoff, E.D., Tainer, J.A., 2010. The structural biochemistry of the superoxide dismutases. *Biochim. Biophys. Acta.* 2, 245–262.

Pesaresi, P., Hertle, A., Pribil, M., et al., 2009. *Arabidopsis* STN7 kinase provides a link between short- and long-term photosynthetic acclimation. Plant Cell. 21, 2402–2423.

Petrov, V., Hille, J., Meuller-Roeber, B., 2013. ROS-mediated abiotic stress-induced programmed cell death in plants. *Front Plant Sci.* 6, 1–16.

Petrov, V., Hille, J., Mueller-Roeber, B., Gechev, T.S., 2015. ROS-mediated abiotic stress-induced programmed cell death in plants. *Front Plant Sci.* 6, 69.

Pitzschke, A., Hirt, H., 2006. Mitogen-activated protein kinases and reactive oxygen species signaling in plants. Plant Physiol. 141, 351.

Pitzschke, A., Hirt, H., 2009. Disentangling the complexity of mitogen-activated protein kinases and reactive oxygen species signaling. Plant Physiol. 149, 606–615.

Qiu, J.L., Zhou, L., Yun, B.W., Nielsen, H.B., Fiil, B.K., Petersen, K., Mackinlay, J., Loake, G.J., Mundy, J., Morris, P.C., 2008. Arabidopsis mitogen-activated protein kinase kinases MKK1 and MKK2 have overlapping functions in defense signaling mediated by MEKK1, MPK4, and MKS1. *Plant Physiol.* 148, 212–222.

Quan, L., Zhang, B., Shi, W., Li, H., 2008. Hydrogen peroxide in plants: a versatile molecule of the reactive oxygen species network. J. Int. Biol. 50, 2–18.

Quinlan, C.L., Perevoshchikova, I.V., Hey-Mogensen, M., Orr, A.L., Brand, M.D., 2013. Sites of reactive oxygen species generation by mitochondria oxidizing different substrates. *Redox Biol.* 1, 304–312.

Racchi, M.L., Bagnoli, F., Balla, I., Danti, S., 2001. Differential activity of catalase and superoxide dismutase in seedlings and in vitro micropropagated oak (*Quercus robur* L.). *Plant Cell Rep.* 20, 169–174.

Rajhi, I., Yamauchi, T., Takahashi, H., Nishiuchi, S., Shiono, K., Watanabe, R., Mliki, A., Nagamura, Y., Tsutsumi, N., Nishizawa, N.K., Nakazono, M., 2011. Identification of genes expressed in maize root cortical cells during lysigenous aerenchyma formation using laser microdissection and microarray analyses. *New Phytol.* 190, 351–368.

Ramel, F., Birtic, S., Ginies, C., Soubigou-Taconnat, L., Triantaphylides, C., Havaux, M., 2012. Carotenoid oxidation products are stress signals that mediate gene responses to singlet oxygen in plants. Proc. Natl. Acad. Sci. USA 109, 5535–5540.

Rasmusson, A.G., Wallström, S.V., 2010. Involvement of mitochondria in the control of plant cell NAD(P)H reduction levels. *Biochem. Soc. Trans.* 38, 661–666.

Reiter, J., Pick, A., Wiemann, L., Schieder, D., Sieber, V., 2014. A novel natural nadh and nadph dependent glutathione reductase as tool in biotechnological applications. JSM Biotech. Biomeb. Eng. 2, 1028–1035.

Rojas, C.M., Senthil-Kumar, M., Wang, K., Ryu, C.M., Kaundal, A., Mysore, K.S., 2012. Glycolate oxidase modulates reactive oxygen species-mediated signal transduction during nonhost resistance in *Nicotiana benthamiana* and *Arabidopsis*. Plant Cell. 24, 336–352.

Rossel, J.B., Wilson, P.B., Hussain, D., Woo, N.S., Gordon, M.J., Mewett, O.P., Howell, K.A., Whelan, J., Kazan, K., Pogson, B.J., 2007. Systemic and intracellular responses to photooxidative stress in *Arabidopsis*. *Plant Cell.* 19, 4091–4110.

Saidi, Y., Finka, A., Muriset, M., Bromberg, Z., Weiss, Y.G., Maathuis, F.J., Goloubinoff, P., 2009. The heat shock response in moss plants is regulated by specific calcium-permeable channels in the plasma membrane. *Plant Cell.* 21, 2829–2843.

Saxena, A., Gautam, S., Arya, K.R., Singh, R.K., 2016. Comparative study of phytochemicals, antioxidative potential and activity of enzymatic antioxidants of *Eclipta alba* and *Plumbago zeylanica* by in vitro assays. *Free Radic. Antioxidants.* 2, 139–144.

Semchuk, N.M., Lushchak, O.V., Falk, J., Krupinska, K., Lushchak, V.I., 2009. Inactivation of genes encoding tocopherol biosynthetic pathway enzymes results in oxidative stress in outdoor grown *Arabidopsis thaliana*. *Plant Physiol. Biochem.* 47, 384–390.

Shah, J., Zeier, J., 2013. Long-distance communication and signal amplification in systemic acquired resistance. *Front Plant Sci.* 4, 30.

Sharma, P., Dubey, R.S., 2005. Modulation of nitrate reductase activity in rice seedlings under aluminium toxicity and water stress: role of osmolytes as enzyme protectant. J Plant Physiol. 162, 854–864.

Sharma, P., Jha, A.B., Dubey, R.S., 2010. Oxidative stress and antioxidative defense system in plants growing under abiotic stresses. In: Pessarakli, M. (Ed.), Handbook of Plant and Crop Stress. third ed. CRC Press, Taylor & Francis, pp. 89–138.

Sharma, P., Jha, A.B., Dubey, R.S., Pessarakli, M., 2012. Reactive oxygen species, oxidative damage, and antioxidative defense mechanism in plants under stressful conditions. *J Bot.* 2012, Art. ID 217037.

Simon, C., Langlois-Meurinne, M., Bellvert, F., et al., 2010. The differential spatial distribution of secondary metabolites in *Arabidopsis* leaves reacting hypersensitively to *Pseudomonas syringae* pv. Tomato is dependent on the oxidative burst. J. Exp. Bot. 61, 3355–3370.

Singh, R., Singh, S., Parihar, P., Mishra, R.K., Tripathi, D.K., Singh, V.P., Chauhan, D.K., Prasad, S.M., 2016. Reactive oxygen species (ROS): beneficial companions of plants' developmental processes. *Front Plant Sci.* 7, 1299.

Sirichandra, C., Gu, D., Hu, H.C., Davanture, M., Lee, S., Djaoui, M., Valot, B., Zivy, M., Leung, J., Merlot, S., Kwak, J.M., 2009. Phosphorylation of the *Arabidopsis AtrbohF* NADPH oxidase by OST1 protein kinase. FEBS Lett. 583, 2982–2986.

Smirnoff, N., Wheeler, G.L., 2000. Ascorbic acid in plants: biosynthesis and function. *Crit. Rev. Biochem. Mol. Biol.* 35, 291–314.

Song, Y., Miao, Y., Song, C.P., 2014. Behind the scenes: the roles of reactive oxygen species in guard cells. *New Phytol.* 201, 1121–1140.

Steffens, B., Geske, T., Sauter, M., 2011. Aerenchyma formation in the rice stem and its promotion by H_2O_2. *New Phytol.* 190, 369–378.

Sturtz, L.A., Diekert, K., Jensen, L.T., Lill, R., Culotta, V.C., 2001. A fraction of yeast Cu/Zn-superoxide dismutase and its metallochaperone, CCS, localize to the intermembrane space of mitochondria: a physiological role for SOD1 in guarding against mitochondrial oxidative damage. *J. Biol. Chem.* 276, 38084–38089.

Suh, H.J., Kim, C.S., Jung, J., 2000. Cytochrome b6/f complex as an indigenous photodynamic generator of singlet oxygen in thylakoid membranes. Photochem. Photobiol. 71, 103–109.

Sultana, S., Khew, C.Y., Morshed, M.M., Namasivayam, P., Napis, S., Ho, C.L., 2012. Overexpression of monodehydroascorbate reductase from a mangrove plant (AeMDHAR) confers salt tolerance on rice. *J. Plant Physiol.* 169, 311–318.

Suzuki, N., Mittler, R., 2012. Reactive oxygen species-dependent wound responses in animals and plants. Free Radic. Biol. Med. 53, 2269–2276.

Suzuki, N., Miller, G., Morales, J., Shulaev, V., Torres, M.A., Mittler, R., 2011. Respiratory burst oxidases: the engines of ROS signaling. *Curr. Opin. Plant Biol.* 14, 691–699.

Suzuki, N., Miller, G., Salazar, C., Mondal, H.A., Shulaev, E., Cortes, D.F., Shuman, J.L., Luo, X., Shah, J., Schlauch, K., Shulaev, V., Mittler, R., 2013. Temporal-spatial interaction between reactive oxygen species and abscisic acid regulates rapid systemic acclimation in plants. Plant Cell. 25, 3553–3569.

Sweetlove, L.J., Foyer, C.H., 2004. Roles for reactive oxygen species and antioxidants in plant mitochondria. In: Day, D.A., Millar, A.H., Whelan, J. (Eds.), Plant Mitochondria: From Genome to Function. In: Advances in Photosynthesis and Respiration, vol. 1. Kluwer Academic Press, Dordrecht, pp. 307–320.

Szarka, A., Horemans, N., Kovacs, Z., Grof, P., Mayer, M., Banhegyi, G., 2007. Dehydroascorbate reduction in plant mitochondria is coupled to the respiratory electron transfer chain. *Physiol Planta.* 129, 225–232.

Szechynska-Hebda, M., Kruk, J., Gorecka, M., Karpinska, B., Karpinski, S., 2010. Evidence for light wavelength-specific photoelectrophysiological signaling and memory of excess light episodes in *Arabidopsis*. Plant Cell. 22, 2201–2218.

Taj, G., Agarwal, P., Grant, M., Kumar, A., 2010. MAPK machinery in plants: recognition and response to different stresses through multiple signal transduction pathways. *Plant Signal Behav.* 5, 1370–1378.

Takahashi, F., Mizoguchi, T., Yoshida, R., Ichimura, K., Shinozaki, K., 2011. Calmodulin-dependent activation of MAP kinase for ROS homeostasis in *Arabidopsis*. Mol. Cell. 41, 649–660.

Torres, M.A., Dangl, J.L., Jones, J.D., 2002. Arabidopsis gp91phox homologues atrbohd and atrbohf are required for accumulation of reactive oxygen intermediates in the plant defense response. Proc. Nat. Acad. Sci. USA 99, 517–522.

Toumi, I., Moschou, P.N., Paschalidis, K.A., Bouamama, B., Ben Salem-Fnayou, A., Ghorbel, A.W., Mliki, A., Roubelakis-Angelakis, K.A., 2010. Abscisic acid signals reorientation of polyamine metabolism to orchestrate stress responses via the polyamine exodus pathway in grapevine. J Plant Physiol. 167, 519–525.

Tretter, L., Mayer-Takacs, D., Adam-Vizi, V., 2007a. The effect of bovine serum albumin on the membrane potential and reactive oxygen species generation in succinate-supported isolated brain mitochondria. Neurochem Int. 50, 139–147.

Tretter, L., Takacs, K., Hegedus, V., Adam-Vizi, V., 2007b. Characteristics of alpha-glycerophosphate-evoked H_2O_2 generation in brain mitochondria. *J. Neurochem.* 100, 650–663.

Tripathi, D.K., Singh, S., Singh, V.P., Prasad, S.M., Chauhan, D.K., Dubey, N.K., 2016. Silicon nanoparticles more efficiently alleviate arsenate toxicity than silicon in maize cultivar and hybrid differing in arsenate tolerance. *Front. Environ. Sci.* 4, 46.

Tripathy, B.C., Oelmuller, R., 2012. Reactive oxygen species generation and signaling in plants. Plant Signal Behav. 7, 1621–1633.

Tripathy, B.C., Oelmüller, R., 2012. Reactive oxygen species generation and signaling in plants. *Plant Signaling Behav.* 7, 1621–1633.

Truman, W., Bennett, M.H., Kubigsteltig, I., Turnbull, C., Grant, M., 2007. *Arabidopsis* systemic immunity uses conserved defense signaling pathways and is mediated by jasmonates. Proc. Nat. Acad. Sci. USA 104, 1075–1080.

Turrens, J.F., 2003. Mitochondrial formation of reactive oxygen species. J. Physiol. (Lond). 552, 335–344.

Vellosillo, T., Vicente, J., Kulasekaran, S., Hamberg, M., Castresana, C., 2010. Emerging complexity in reactive oxygen species production and signaling during the response of plants to pathogens. Plant Physiol. 154, 444–448.

Voothuluru, P., Sharp, R.E., 2013. Apoplastic hydrogen peroxide in the growth zone of the maize primary root under water stress I. Increased levels are specific to the apical region of growth maintenance. J. Exp. Bot. 64, 1223–1233.

Wagner, D., Przybyla, D., Camp, R.O.D., Kim, C., Landgraf, F., Lee, K.P., Würsch, M., Laloi, C., Nater, M., Hideg, E., Apel, K., 2004. The genetic basis of singlet oxygen induced stress responses of *Arabidopsis thaliana*. Science 306, 1183–1185.

Wahid, A., Gelani, S., Ashraf, M., Foolad, M.R., 2007. Heat tolerance in plants: an overview. Environ. Exp. Bot. 61, 199–223.

Wang, Y., Chang, C.F., Chou, J., Chen, H.L., Deng, X., Harvey, B.K., Cadet, J.L., Bickford, P.C., 2005. Dietary supplementation with blueberries, spinach, or Spirulina reduces ischemic brain damage. Exp. Neurol. 1, 75–84.

Weydert, C.J., Cullen, J.J., 2010. Measurement of superoxide dismutase, catalase, and glutathione peroxidase in cultured cells and tissue. *Nat. Protoc.* 1, 51–66.

Wi, S.J., Ji, N.R., Park, K.Y., 2012. Synergistic biosynthesis of biphasic ethylene and reactive oxygen species in response to hemibiotrophic *Phytophthora parasitica* in tobacco plants. *Plant Physiol.* 159, 251–265.

Wituszynska, W., Szechynska-Hebda, M., Sobczak, M., Rusaczonek, A., Kozlowska-Makulska, A., Witon, D., Karpinski, S., 2015. Lesion simulating disease 1 and enhanced disease susceptibility 1 differentially regulate UV-C-induced photooxidative stress signalling and programmed cell death in *Arabidopsis thaliana*. Plant Cell Environ. 38, 315–330.

Wrzaczek, M., Brosche, M., Kangasjarvi, J., 2013. ROS signaling loops - production, perception, regulation. Curr. Opin. Plant Biol. 16, 575–582.

Wu, J., Liu, J., 2002. Effect of several pesticides on sod activity in different rice varieties. *Sci. Agric. Sinica.* 35, 451–456.

Wu, J., Zhang, Y., Yin, L., Qu, J., Lu, J., 2014. Linkage of cold acclimation and disease resistance through plant-pathogen interaction pathway in *Vitis amurensis* grapevine. *Func Integrat Genom.* 14, 741–755.

Xia, X.J., Wang, Y.J., Zhou, Y.H., Tao, Y., Mao, W.H., Shi, K., Asami, T., Chen, Z., Yu, J.Q., 2009. Reactive oxygen species are involved in brassinosteroid-induced stress tolerance in cucumber. *Plant Physiol.* 150, 801–814.

Xu, J., Li, Y., Wang, Y., Liu, H., Lei, L., Yang, H., Liu, G., Ren, D., 2008. Activation of MAPK kinase 9 induces ethylene and camalexin biosynthesis and enhances sensitivity to salt stress in *Arabidopsis*. *J. Biol. Chem.* 283, 26996–27006.

Xu, J., Yang, J., Duan, X.G., Jiang, Y.M., Zhang, P., 2014. Increased expression of native cytosolic Cu/Zn superoxide dismutase and ascorbate peroxidase improves tolerance to oxidative and chilling stresses in cassava (*Manihot esculenta* Crantz). *BMC Plant Biol.* 14, 208.

Yang, T., Poovaiah, B.W., 2002. Hydrogen peroxide homeostasis: activation of plant catalase by calcium/calmodulin. Proc. Nat. Acad. Sci. USA 99, 4097–4102.

Zaefyzadeh, M., Quliyev, R.A., Babayeva, S.M., Abbasov, M.A., 2009. The effect of the interaction between genotypes and drought stress on the superoxide dismutase and chlorophyll content in durum wheat landraces. *Turk. J. Biol.* 33, 1–7.

Zhai, C.Z., Zhao, L., Yin, L.J., Chen, M., Wang, Q.Y., Li, L.C., Xu, Z.S., Ma, Y.Z., 2013. Two wheat glutathione peroxidase genes whose products are located in chloroplasts improve salt and H_2O_2 tolerances in *Arabidopsis*. PLoS One. 8, e73989.

Zhang, H.J., Fang, Q., Zhang, Z.G., Wang, Y.C., Zheng, X.B., 2009. The role of respiratory burst oxidase homologues in elicitor-induced stomatal closure and hypersensitive response in *Nicotiana benthamiana*. J. Exp. Bot. 60, 3109–3122.

Zhang, S., Apel, K., Kim, C., 2014. Singlet oxygen-mediated and EXECUTER-dependent signaling and acclimation of Arabidopsis thaliana exposed to light stress. Philos. Trans. R. Soc. Lond. Ser. B Biol. Sci. 369, 20130227.

Antioxidant-mediated defense in triggering resistance against biotic stress in plants

15

Belur Satyan Kumudini and Savita Veeranagouda Patil
Department of Biotechnology, School of Sciences (Block 1), JAIN (Deemed-to-be University), Bengaluru, Karnataka, India

15.1 Introduction

Food security and public health worldwide are of great concern at present, being directed toward the third world countries, a majority of which domesticate and cultivate in semiarid and tropical conditions. The progress of the green revolution has been reversed due to abiotic factors like drought and floods, which are the key drivers. A report by the United Nations (FAO et al., 2018) showed that undernourishment resulted as a cause of climate change. It also reports that staple crops, viz., wheat, rice, and maize, are at risk due to extreme climates, encountering an increased number of stress factors (Pandey et al., 2015). Plants are influenced by both abiotic (drought, heat, salinity, heavy metal, flood, radiation) and biotic stresses at different stages of growth, which may be individual or in consortium (heat-fungal, heat-pest, salinity-fungal, etc.) (Abdelrahman et al., 2018; Haider et al., 2019). Plant stress mechanism instigated by the living organisms, including pathogens, viz., bacteria, fungi, viruses, pests, nematodes, and weeds, is biotic stress (Pandey et al., 2015; Joshi et al., 2019). When they occur in consortium, it poses a threat to the host, devastating its growth and yield (Mittler, 2006; Prasad et al., 2011; Prasch and Sonnewald, 2013; Suzuki et al., 2014; Mahalingam, 2015; Pandey et al., 2015). Due to the pivotal increase in abiotic stress, the pathogen and pest persistence in the host plant can be activated due to global warming and climate variations posing a threat to the environment (Atkinson et al., 2013; Prasch and Sonnewald, 2013; Jogaiah et al., 2013).

The defense responses by the host against phytopathogens depend on the lifestyle and the mode of nutrition acquired from the host. The different types of infections or microbial attacks to the hosts are biotrophic, necrotrophic, and hemibiotrophic (Thomma et al., 2001; Nandini et al., 2020). It has been reported by Kumudini et al. (2018) that microbes which infect, live, and continue their life cycle in the host are biotrophs (*Cladosporium fulvum*, *Blumeria graminis*) and those which strike, kill, and feed on the host for survival are termed "necrotrophs" (*Erwinia carotovora*,

Biocontrol Agents and Secondary Metabolites. https://doi.org/10.1016/B978-0-12-822919-4.00015-6

Botrytis cinerea). Hemibiotrophs are organisms which are parasitic to living tissue and further continue their life cycle on the dead cells (*Magnaporthe grisea*).

15.2 Early defense responses

Plants manage to defend against a plethora of evading pathogens by plant innate immunity as these higher organisms lack specialized immune cells as in the case of animals and human beings (Dodds and Rathjen, 2010). As a consequence, the biochemical and molecular changes induced during stress activate the signaling mechanism to evade off the pathogen (Jogaiah et al., 2018). The defense mechanisms perceive the pathogen attack through microbe-associated molecular patterns (MAMPs) or the pathogen-associated molecular patterns (PAMPs) or by their own damage-associated molecular patterns (DAMPs). The pathogen-induced damage molecules are facilitated by surface-localized pattern recognition receptors (PRRs), directing PRR-triggered immunity (PTI). These PRRs perceive the fungal attack by finding the kinase domain of chitins initiating the downstream signal cascade (Boller and Felix, 2009; Torres et al., 2006; Monaghan and Zipfel, 2012). The multifaceted signaling mechanism is highly complex due to the interwoven multiple networks overlapping with the pathogen entry, a diverse arena of phytopathogens (Knight and Knight, 2001; Jogaiah et al., 2018). Hence, with the diversified group of pathogens, the PAMPs involved in signaling depend on the group and type of pathogen. The various PAMPs include bacterial (flagellin, elongation factor-TU, peptidoglycans, lipopolysaccharide), fungal, and oomycete (chitin, ethylene-inducing xylanase). Thus, these PRRs and PAMPs determine the type of signaling cascade to be perceived during induction of resistance against the pathogen by the host's first layer of defense (Rathore and Ghosh, 2018). The membrane localized PRRs are largely categorized into receptor-like protein (RLP) and/or receptor-like kinase (RLK). Various types of PAMPs and DAMPs have been identified and characterized in many of the plant species which are activated upon pathogen response, which have been reviewed by Rathore and Ghosh (2018).

After PAMP, MAMP, and DAMP recognition, the first line of defense signaling is triggered by involvement of reactive oxygen species (free radicals), thereby regulating the defense under different stress. ROS accumulation is sensed by the involvement of Ca^{2+} and Ca^{2+}-binding proteins (calmodulin) and G-protein activation, phospholipid signaling, resulting in augmentation of phosphatidic acid and upsurge of mitogen-associated protein kinase (MAPK) pathways (Mittler et al., 2004; Opdenakker et al., 2012). MAPKs regulate cellular responses by modulating the signals formed by the different receptors/sensors. MAPK comprises three modules: MAPKK kinases (MAPKKK) activate serine/threonine kinases, phosphorylating into MAPK kinases (MAPKK or MKK). MAPKK activates MAP kinases (MAPK or MPK) by phosphorylation of tyrosine and serine/threonine residues. MAPKs regulate the signaling of different hormones, viz., salicylic acid (SA), jasmonic acid (JA), abscisic acid (ABA), ethylene, and other ROS metabolites during different abiotic and biotic stresses in various plants (Opdenakker et al., 2012).

15.3 Reactive oxygen species (ROS)

Due to diverse environmental stresses in plants, the molecular oxygen is accomplished to produce reactive species such as free radicals, nevertheless being unreactive when at ground state under normal conditions (Polidoros et al., 2005; O'Brien et al., 2012; Phaniendra et al., 2015). The free radicals, reactive oxygen intermediates such as singlet oxygen ($^1O_2$), superoxide radical ($O_2^{\cdot -}$), hydrogen peroxide, or hydroxyl radicals ($OH^{\cdot}$), are produced as a result of the excitation of the O_2 molecule (Mittler, 2002). These free radicals play a vital role in plant signaling, plant growth development, and its adaptability to various stresses (Lamb and Dixon, 1997; Shetty et al., 2008; Baxter et al., 2014; Qi et al., 2017). Recent studies (Bhatt et al., 2011; Kotapati et al., 2014; Varsha and Kumudini, 2016; Jayamohan et al., 2018) on signaling and scavenging of ROS under stress conditions have shown that a balanced ROS level has to exist to avoid cytotoxicity (Mittler et al., 2004; Sharma et al., 2012; Baxter et al., 2014; Kunstler et al., 2015; Camejo et al., 2016) in plants. The production and accumulation of free radical levels in plants are exaggerated during various stress mechanisms. This leads to increased levels of ROS and oxidative burst, thereby inducing modifications in ion fluxes, phosphorylation of proteins, and immobilization and change in extracellular pH (Bolwell et al., 1995; Wojtaszek et al., 1995; O'Brien et al., 2012).

The redox oxidative burst reacts with lipids, proteins, and DNA, thereby increasing the electrolyte leakage and reduced photosynthesis, followed by quickening of cell death (Das and Roychoudhary, 2014, Lehmann et al., 2015). Though programmed cell death (PCD) encompasses detrimental effects on the plants, oxidative burst-mediated removal of infected cells thereby inhibits the progression of the pathogen entry into the adjacent cells (Kumudini and Shetty, 2004; Mittler, 2006; Jayamohan and Kumudini, 2011; Kumudini et al., 2018; Jayamohan et al., 2018).

15.3.1 ROS and biotic stress

A successful pathogen recognition, virulent or avirulent, is eventually achieved by oxidative burst and brief production of ROS (Torres et al., 2006; Kumudini et al., 2018). The biphasic accumulation of ROS is activated by the *R*-gene of avirulent pathogens comprising a profuse, transient first phase and subsequently prolonged by a second phase with a higher degree of resistance (Kunstler et al., 2015), whereas a transient, low-amplitude phase is seen in virulent pathogens (Torres et al., 2006).

Production of ROS is a major host cellular response which can damage biological molecules during successful pathogen recognition (Kunstler et al., 2015; Camejo et al., 2016; Sewelam et al., 2016). Though earlier considered as a hazardous byproduct of O_2 metabolism, it is now evident that ROS is a cell signal which can lead to the defense-specific gene activation, apoptosis, and cell signaling cascade against various environmental stresses in plants (Das and Roychoudhary, 2014; Lehmann et al., 2015). ROS initiates oxidative homeostasis by chemical signaling and induces resistance during biotic stress (Das and Roychoudhary, 2014; Camejo et al., 2016).

The concentration of ROS ($<LD_{50}$) is sufficient for triggering the antioxidant enzymes (Baxter et al., 2014). Pathways regulating ROS homeostasis play a crucial role in mitigating ROS toxicity (Baxter et al., 2014).

15.4 ROS and reactive nitrogen species (RNS)

Nitric oxide (NO) reacts with superoxide ion, molecular oxygen, low-molecular-weight thiol groups producing derivatives, viz., nitrosoglutathione, nitrosonium cation, dinitrogen trioxide, nitroxyl, and peroxynitrite, which are referred to as reactive nitrogen species (RNS), which are known to react with lipids, proteins, and nucleic acids and induce oxidative stress. As with regard to ROS, RNS are also involved in a scaffold of reactions, amplifying the signal cascade under various stress (abiotic and biotic) conditions. Precursors of NO include L-arginine and/or *N*-hydroxy-L-arginine catalyzed by nitric oxide synthase (NOS) in all living systems (Stamler et al., 1992; Hill et al., 2010; Thomas et al., 2008; Rosnoblet et al., 2016). The impact of NO in plant immunity is well studied (Besson-Bard et al., 2008; Zipfel, 2009; Spoel and Dong, 2012; Domingos et al., 2014) since the past two decades. Studies on plant immunity and defense mechanisms signify the profound role of NO in controlling the defense system. It has been reported to act as messengers, regulates expression of defense genes and lignin biosynthesis, tempering the hypersensitive response (HR) and hormone signaling (Durner et al., 1998; Huang et al., 2002; Zeier et al., 2004; Palmieri et al., 2008; Leitner et al., 2009; Mur et al., 2008; Wendehenne et al., 2004). The perfect regulatory mechanism recruited by NO is unpredictable and modulates the protein kinase activities (MAP kinase) and synthesis of different hormones (jasmonic acid, salicylic acid, abscisic acid, and ethylene). Hence, the relative cross talk between the messengers ROS and RNS (NO) has been deduced in different pathosystems, discussed in the following section.

15.4.1 Cross talk of ROS and NO in plants

The functioning of the nonexpresser of pathogenesis-related gene1 (NPR1) during defense response depends on the signaling cross talk of NO and ROS during systemic resistance. The stability of the oligomeric state of NPR1 depends on S-nitrosylation. Also, the translocation of NPR1 to the nucleus after pathogen attack is inhibited by higher concentrations of cytoplasmic H_2O_2. Hence, NPR1 is regulated by both ROS and NO during biotic stress, thereby inhibiting the signaling cascade and highlighting its complexity. Studies (reviewed by Rosnoblet et al., 2016) show that NO reacting with superoxide ion forms peroxynitrite ($ONOO^-$), a signaling moiety for oxidation of macromolecules (DNA, proteins, lipids, and nitration of PTM tyrosine), resulting in damage of the plant cells. Correspondingly, the triggering of the signal cascade after pathogen attack by initial ROS production by the protein AtRBOHD has been constrained by S-nitrosylation (Romero-Puertas et al., 2007; Rosnoblet et al., 2016). On the contrary, a direct correlation between NO and H_2O_2 accumulation was observed in Arabidopsis leaf discs treated with oligogalacturonides, regulating the HR and programmed cell death during plant defense. This regulatory system is well ascertained in different plant-pathogen systems reviewed by Mur et al. (2008) in upstreaming the pathogenesis-related (PR) proteins (Grun et al., 2006). NOS is a well-known NO-producing enzyme regulating the increased levels of SA and PR proteins during defense in tobacco plants. In the tobacco-tobacco necrosis virus (TNV), host-pathogen interaction, TNV was inhibited by chitosan application augmenting oxidative burst, hypersensitive response, and callose deposition (Iriti et al., 2006). A similar study on the use of chitosan by seed treatment induced increased accumulation of NO, thereby regulating antioxidant production, augmenting of PR proteins, SA accumulation, and callose deposition in the pearl millet-*Sclerospora graminicola* pathosystem (Manjunatha et al., 2009).

15.5 ROS scavenging via the antioxidant system

To counteract the imbalance in the ROS system, the host cells produce antioxidants upon response by the pathogen attack (Halliwell, 2006; Sewelam et al., 2016). The antioxidants are electron donors and neutralize the ROS generated, and in the process, oxidized molecules seem to be less harmful to the cells. They are upregulated by the response at the early phase of the signaling cascade, thereby activating the enzymatic and nonenzymatic antioxidants (Jayamohan et al., 2018; Kumudini et al., 2017; Patil and Kumudini, 2018) consisting of ascorbate peroxidase (APX), guaiacol peroxidase (GPX), superoxide dismutase (SOD), catalase (CAT), and glutathione reductase (GR). The nonenzymatic antioxidants include tocopherol, ascorbic acid, glutathione, and carotenoids (Das and Roychoudhary, 2014; Lehmann et al., 2015; Yousuf et al., 2016; Qi et al., 2017). The reactions catalyzed by these antioxidants are the ascorbate-glutathione cycle, referred to as the Halliwell-Asada pathway (Foyer and Halliwell, 1976), involving the enzymatic antioxidants (Table 15.1).

Table 15.1 Sequential enzymatic antioxidants reactions mediated in redox detoxification in triggering the plant defense.

Enzymes	Enzyme code	Reaction
SOD	EC 1.15.1.1	$O_2^{\bullet -}+O_2^{\bullet -}+2H+\rightarrow H_2O_2+O_2$
CAT	EC 1.11.1.6	$2H_2O_2\rightarrow O_2+2H_2O$
APX	EC 1.11.1.11	$H_2O_2+AA\rightarrow 2H_2O+DHA$
GPX	EC 1.11.1.7	$H_2O_2+DHA\rightarrow 2H_2O+GSSG$
MDHAR	EC 1.6.5.4	$2MDHA+NADH\rightarrow 2AA+NAD^+$
DHAR	EC 1.8.5.1	$DHA+2GSH\rightarrow AA+GSSG$
GR	EC 1.6.4.2	$GSSG+NADPH\rightarrow 2GSH+NADP^+$

SOD, superoxide dismutase; *$O_2^{\bullet -}$*, superoxide radical; *H_2O_2*, hydrogen peroxide; *CAT*, catalase; *APX*, ascorbate peroxidase; *AA*, ascorbic acid; *DHA*, dehydroascorbate; *GPX*, guaiacol peroxidase; *GSSG*, oxidized glutathione; *MDHAR*, monodehydroascorbate reductase; *MDHA*, monodehydroascorbate; *DHAR*, dehydroascorbate reductase; *DHA*, dehydroascorbate; *GSH*, glutathione; *GR*, glutathione reductase.

15.5.1 Enzymatic antioxidants

The different antioxidants with an enzymatic nature are ascorbate peroxidase (APX), guaiacol peroxidase (GPX), superoxide dismutase (SOD), catalase (CAT), and glutathione reductase (GR), which facilitate ROS scavenging in plants. The minor enzymatic antioxidants are monodehydroascorbate reductase (MDR), dehydroascorbate reductase (DHAR), which are also involved in redox homeostasis.

SOD (EC 1.15.1.1) is a group of metalloproteins which are involved in catalyzing the dismutation of superoxide radical to molecular oxygen and hydrogen peroxide. Further, H_2O_2 is catalyzed by CAT, breaking it down to water and oxygen, thereby regulating the anions produced. Superoxide radicals react with ferric sulfide proteins, having a half-life of 1–4 μs, which is scavenged by SOD, and is present in different isoforms with different metallic ions consisting of copper/zinc SOD (Cu/Zn-SOD), manganese SOD (Mn-SOD), and iron SOD (Fe-SOD). They are nuclear encoded and are located in subcellular compartments than in the extracellular regions in plants. The subcellular localization includes mitochondria (Mn-SOD), chloroplasts (Fe-SOD), whereas Cu/Zn-SOD is present as three isoforms in cytosol, chloroplast, peroxisome, and mitochondria (del Rio et al., 2006; Sharma et al., 2012). Genomic studies on Arabidopsis revealed that these isoforms have three genes each for Cu/Zn-SOD and Fe-SOD, and one of Mn-SOD (Fridovich, 1989; Racchi et al., 2001). The type/isoform of SOD in plants can be detected by native PAGE, which is sensitive to KCN and H_2O_2 (Sharma et al., 2012). Thus, increased activity of SOD is positively correlated with the increased levels of tolerance toward various environmental stresses (abiotic and biotic).

CAT (EC 1.11.1.6) is present in abundant levels in peroxisomes, catalyzing the dismutation of H_2O_2, into water and oxygen molecules. This enzyme is involved in scavenging the superoxide radicals in the plant defense system along with SOD and H_2O_2. CAT is an iron porphyrin tetrameric protein, highly specific to H_2O_2 and less active against organic peroxides. Even though different enzymes catalyze H_2O_2 degradation, CAT seems to occupy the foremost position as it does not require any cellular reducing moieties (del Rio et al., 2006; Corpas et al., 2008; Sharma et al., 2012) and has lower affinity to H_2O_2. The enzymes of photorespiration, oxidizing fatty acids, and catabolizing purines are involved in H_2O_2 removal by oxidation of CAT in peroxisomes (Willekens et al., 1995). The earlier reports suggest three different isoforms of CAT: (i) Class I—regulated by light and expressed in photosynthetic tissues, (ii) Class II—present in vascular tissues (xylem and phloem), and (iii) Class III—present in excess levels in seeds and young seedlings. This classification is based on the reports of Willekens et al. (1995) on the tobacco gene profile and has also been reported in angiosperms.

GPX (EC 1.11.1.7) oxidizes H_2O_2 with an aromatic electron donor (guaiacol and pyrogallol). It is a heme-containing protein extensively present in both prokaryotes and eukaryotes. The isoforms are localized in the cell wall, cytosol, and vacuoles in all higher plants. GPX plays a crucial role in most of the plant physiological pathways under stress, hence known as a stress enzyme, quenching the reactive form of O_2 and peroxide radicals formed (Rani et al., 2014). GPX is associated with lignin biosynthesis (the phenylpropanoid pathway), IAA dissociation, ethylene synthesis, and wound response (Sharma et al., 2012; Rani et al., 2014).

APX (EC 1.1.11.1) is a principal component of the AsA-GSH cycle, controlling the intracellular ROS levels. It belongs to the super family of Class I of heme peroxidases, regulating the scavenging of H_2O_2 by reducing it to two molecules of H_2O and generation of MDHA (Patterson and Poulos, 1995; Sharma et al., 2012). The isoenzymes are reported to be present in five diverse forms which exist in subcellular regions of cytosol (cAPX), stroma (sAPX), thylakoids (tAPX), mitochondria (mitAPX), peroxisomes, and glyoxisomes as mAPX (Nakano and Asada, 1987; Mittler and Zilinskas, 1992; Madhusudhan et al., 2003; Sharma and Dubey, 2004; Sharma et al., 2012). Higher affinity (μM range) toward H_2O_2 is reported from these isoforms of APX compared to that of CAT; hence they are regarded as the most broadly distributed antioxidants of the plant kingdom during stress (Mhamdi and Breusegem, 2018).

15.5.2 Nonenzymatic antioxidants

These are nonspecific ROS scavenging metabolites, including amino acids (ascorbic acid, proline), glutathione, proline, polyamines, betaine, carotenes, flavonoids, and α-tocopherol. The present review includes the importance of tocopherol and carotenoids in defense system.

Tocopherols are lipophilic antioxidants present in all plants as molecules of its membranes (Kiffin et al., 2006), protecting lipids from auto-oxidation, protecting the alterations rendered on the structure and functioning of the PSII system (Igamberdiev et al., 2004). Different free radicals excited during PUFA oxidation (carbonyl, alkoxy, peroxy, and organic hydroperoxide radicals) react with α-tocopherols and consequently form tocopherol ions salvaging tocopherol by reacting with ascorbic acid and other antioxidants (Igamberdiev et al., 2004). In certain circumstances during oxidative stress, α-TOC genes are expressed. Oxidative stress activates the expression of genes responsible for the synthesis of α-TOC in higher plants (Havaux et al., 2005). They also deactivate superoxide anions, mediate the scavenging of ROS, and restrain the membrane fluidity (Ledford and Niyogi, 2005; Karuppanapandian et al., 2011), which is achieved by altering the permeability of glucose and protons through digalactosyldiacyl glycerol vesicles (Havaux et al., 2005; Ledford and Niyogi, 2005).

Carotenoids are lipophilic metabolites localized in plastids in all plant tissues. They absorb light (450–570nm) and convert the captured energy to chlorophyll. Carotenoids have been reported (Collins, 2001; Sharma and Dubey, 2004; Sharma et al., 2012) to detoxify the redox ions produced under stress in plants. This regulation of stress is achieved by terminating the lipid peroxides produced, dissipating heat from scavenged superoxide ions, and reacting with chlorophyll (Chl• and/or 3Chl•) molecules (Collins, 2001). The other forms like β-carotene and zeaxanthin also assist the chloroplasts in synchronizing the oxidative damage to the plants (Collins, 2001; Mortensen et al., 2001; Karuppanapandian et al., 2011).

15.6 Enhancement of ROS scavenging and plant immunity

15.6.1 *Biotic elicitors*

The elicitation of disease resistance by using microbial agents to induce the signaling cascade dates back to centuries (Kloepper and Schroth, 1978; Kumudini et al., 2017). Recent studies by Jayamohan et al. (2018) reveal the efficiency of rhizobacterial upregulation of antioxidants during *Fusarium* wilt of tomato. Disease resistance was induced in *Pseudomonas*-primed tomato plants compared to the mock plants at the early stages of disease challenge. The early (96 hpi) scavenging of H_2O_2 ruled out programmed cell death, activating the signaling cascade during disease progression, which was also observed in studies of Shaw and Long (2003) and Lohar et al. (2007). Similar results were obtained from SOD, PPO, and CAT activities along with proline and protein content showing an upsurge in primed (induced resistant) plants compared to the mock plants. These results are also supported by studies on the chickpea-*Sclerotinia sclerotiorum* host-pathogen system by Jain et al. (2012). The plants were primed using biocontrol agents like *P. aeruginosa* (PJHU15), *Trichoderma harzianum* (TNHU27), and *Bacillus subtilis* (BHHU100) individually or in consortium. The individual and consortium containing three isolates showed greater enhanced activity of the antioxidants than control at early hours of 24–48 hpi. In the *Rhizoctonia solani* and sunflower pathosystem, a rise in CAT, APX, SOD, and GPX antioxidant activities was observed by Singh et al. (2011) with *T. harzianum* treatments. In the ragi-*Magnaporthe grisea* interaction, it was observed that the role of antioxidant activities played a crucial role in induction of disease resistance. Lignin deposition resulted as an upsurge of defense at early stages of 12–36 hpi using *P. aeruginosa* for seed priming (Patil and Kumudini, 2018). The network of signal cascades involved in the mechanism of PGPR (fluorescent Pseudomonads-FP) induced disease resistance in the *M. grisea*-Ragi pathosystem has been emphasized in Fig. 15.1.

Free radical and hydroxyl radical scavenging was observed when primed with *Mesorhizobium* sp. (RL 091), fluorescent *Pseudomonas* (PHU 094), and *T. harzianum* (THU 0816) in regulating the antioxidants in edible parts of chickpea. It was evident that the application of microbes showed increased accumulation of phenolic acids, viz., gallic, tannic, p-coumaric, shikimic, and ferulic acids. It was higher when primed with consortium than with individual isolates. It was observed that accumulation of H_2O_2 in petiolar and interveinal regions of the leaf and activities of CAT, GR, and GPX were higher when primed with the same rhizobacterial species when compared to the control. Even though the ascorbic acid content and chitinase accumulation was high, APX activity was not that profound in the *Cicer arietinum*—*S. rolfsii* pathosystem (Jain et al., 2013). Use of plant extract from *Datura metel* induced disease resistance and showed HR response in pearl millet downy mildew. The study was focused on induction of HR response at the initial hours after pathogen challenge by increasing the deposition of cell wall bound molecules, viz., callose and lignin (Kumudini et al., 2001; Kumudini and Shetty, 2002a,b).

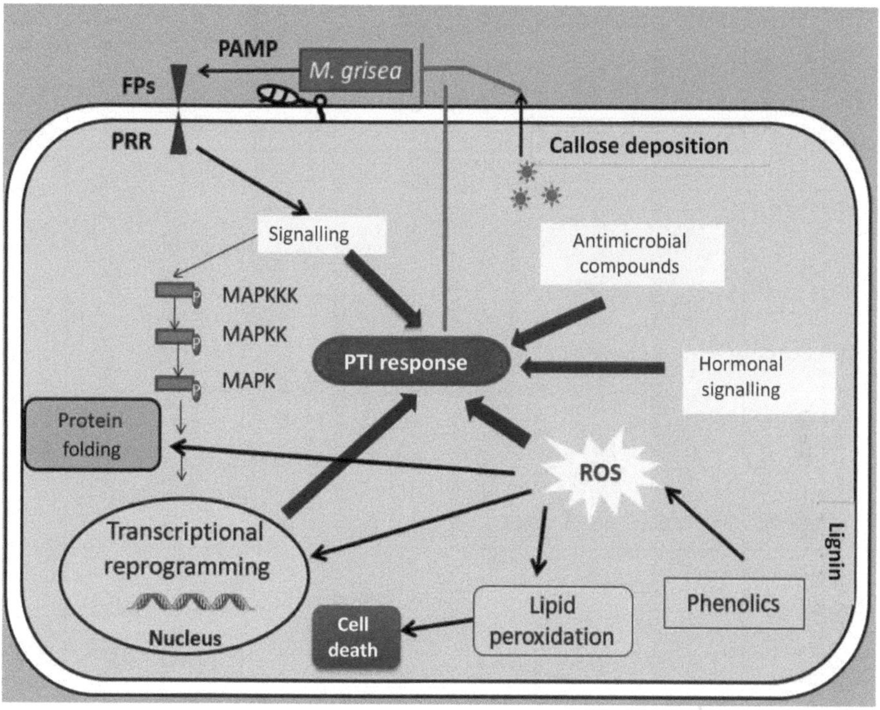

Fig. 15.1 Schematic representation of PGPR-mediated cascade of signaling mechanism upon pathogen.

15.6.2 Abiotic elicitors

Various studies have been reported in use of different abiotic (nonbiological) elicitors (Abdelrahman et al., 2017). The different elicitors are amino acids, riboflavin, 2–6 dichloro isonicotinic acid, jasmonic acid, and salicylic acid, extracts from *Datura metel* (Joe and Muthukumaran, 2008; Kumudini et al., 2001; Kumudini and Shetty, 2002a; Sudisha et al., 2011; Abhayashree et al., 2016). Therefore, the role of abiotic elicitors encompasses a specific defensive mechanism in plants which has been listed in this review (Table 15.2).

Application of β-amino butyric acid (BABA) on pearl millet seeds increased seedling vigor and induced resistance against the downy mildew pathogen *Sclerospora graminicola* with 74% of disease protection. This protection increased the accumulation of cell wall bound hydroxyproline-rich glycoproteins (HRGP), β-1,3-glucanase, phenolic acids, and peroxidase activity (Shailasree et al., 2007). Joe and Muthukumaran (2008) reported the role of both abiotic and biotic elicitors in triggering the disease. The antioxidant enzymes like polyphenol oxidase, peroxidase, lipoxygenase, and proteinase inhibitors were upregulated in the elicitor-treated plants against *Spodoptera litura*. The elicitors used were jasmonic acid (JA), salicylic acid (SA), and the biotic elicitor was also involved (*P. aeruginosa*). Abhayashree et al.

Table 15.2 Different abiotic elicitors involved in regulation of ROS scavenging in different pathosystems.

Plant	Pathogen/pest reference	Elicitor	Response	Reference
Lycopersicum esculentum	*Fusarium oxysporum* *Phytophthora capsici*	Chitosan	Produced a higher resistance	Ortega-Ortíz et al. (2003)
Lycopersicum esculentum	*Helicoverpa armigera*	SA	SAR gene expression, higher levels of H_2O_2	Peng et al. (2004)
Beta vulgaris	Tobacco necrosis virus	Benzothiadiazole	Induced chitinase and β-1,3-glucanase accumulation	Burketová et al. (1999)
Avena sativa, Oryza sativa, Raphanus sativus, Arachis hypogea, Nicotiana tabacum	a	Chitosan	Directs the lignification, HR reaction and defense systems	Maksimov et al. (2003)
Pisum sativum	*Erysiphe polygoni*	SA, 4-aminobutyric acid	Increased levels of phenolics, PPO, peroxidase, PAL	Katoch (2005)
Citrus sinensis	*Penicillium italicum*	β-Amino butyric acid	Inhibited spore germination and induced resistance	Tavallali et al. (2008)
Abelmoschus esculentus	*Erysiphe cichoracearum*	SA	PAL activity and phenolic acids	Vimala and Suriachandraselvan (2009)
Solanum melongena	*Ralstonia solanacearum*	SA, chitosan, methyl salicylate, methyl jasmonate	Lignin deposition, increase in phenolics, PAL, peroxidase, PPO, CAT activity	Mandal (2010)
Brassica juncea	*Alternaria brassicae*	Benzothiadiazole, SA	Increased activities of SOD, PAL, peroxidase	Sharma and Sohal (2010)

[a] Multiple pathogens.

(2016) studied the proficiency of different abiotic elicitors (L-isoleucine, L-leucine, L-methionine, L-phenylalanine, and L-proline) on resistance induction and growth of chili against *Colletotrichum capsici*. Significant increase in disease protection was observed in L-leucine and L-proline treatments, followed by enhanced production of phenolics, activity of PAL, and peroxidase as compared to the control.

15.7 Conclusion

Reports on different plant-pathogen systems are immense since decades. The broad-spectrum studies on biochemical, molecular, and physiological changes upon infestation have been ensured. Different omic approaches, viz., proteomics, metabolomics, and gene expression studies, have been found in deducing the role of different antioxidants/metabolomes involved in triggering plant immunity. But the precise role of ROS and/or RNS have revealed that the use of these metabolites in upregulating the signal cascade in different pathosystems needs to be understood. As every other organism (bacteria, fungi, viruses, pests) varies in its physiological, molecular makeup, the type of disease infested and signal transmitted also will vary. Hence, the use of an efficient metabolite in induction of resistance may not induce the same in another pathosystem. Therefore, extensive studies on the mode of signal cascade in induction of disease resistance in different crops have to be carried out, thereby enhancing the yield and productivity of the crop, increasing the crop production worldwide, and reducing the malnutrition and hunger prevailing.

Acknowledgments

The authors acknowledge DST-SERB (YSS/2015/001905), Government of India, New Delhi and JAIN (Deemed-to-be University), Bengaluru for their financial and infrastructural support. They apologize to all whose relevant work that could not be quoted in this chapter due to space constraints.

References

Abdelrahman, M., El-Sayed, M., Jogaiah, S., Burritt, D.J., Tran, L.-S.P., 2017. The "STAY-GREEN" trait and phytohormone signaling networks in plants under heat stress. Plant Cell Rep. 36, 1009–1025.

Abdelrahman, M., Jogaiah, S., Burritt, D.J., Tran, L.-S.P., 2018. Legume genetic resources and transcriptome dynamics under abiotic stress conditions. Plant Cell Environ. 41, 1972–1983.

Abhayashree, M.S., Murali, M., Amruthesh, K.N., 2016. Abiotic elicitors mediated resistance and enhanced defense related enzymes in *Capsicum annuum* L. against anthracnose disease. Sci. Hortic. 204, 172–178.

Atkinson, N.J., Lilley, C.J., Urwin, P.E., 2013. Identification of genes involved in the response to simultaneous biotic and abiotic stress. Plant Physiol. 162, 2028–2041.

Baxter, A., Mittler, R., Suzuki, N., 2014. ROS as key players in plant stress signalling. J. Exp. Bot. 65, 1229–1240.

Besson-Bard, A., Pugin, A., Wendehenne, D., 2008. New insights into nitric oxide signaling in plants. Annu. Rev. Plant Biol. 59, 21–39.

Bhatt, D., Negi, M., Sharma, P., Saxena, S.C., Dobriyal, A.K., Bhatt, S.A., 2011. Responses to drought induced oxidative stress in five finger millet varieties differing in their geographical distribution. Physiol. Mol. Biol. Plants 17, 347–353.

Boller, T., Felix, G., 2009. A renaissance of elicitors: perception of microbe-associated molecular patterns and danger signals by pattern-recognition receptors. Annu. Rev. Plant Biol. 60, 379–406.

Bolwell, G.P., Butt, V.S., Davies, D.R., Zimmerlin, A., 1995. The origin of the oxidative burst in plants. Free Radic. Res. 23, 517–532.

Burketová, L., Šindelářová, M., Šindelář, L., 1999. Benzothiadiazole as an inducer of β-1,3-glucanase and chitinase isozymes in sugar beet. Biol. Plantarum. 42, 279–287.

Camejo, D., Guzman-Cedeno, A., Moreno, A., 2016. Reactive oxygen species, essential molecules, during plant-pathogen interactions. Plant Physiol. Biochem. 103, 10–23.

Collins, A., 2001. Carotenoids and genomic stability. Mutat. Res. 475, 1–28.

Corpas, F.J., Palma, J.M., Sandalio, L.M., Valderrama, R., Barroso, J.B., del Rio, L.A., 2008. Peroxisomal xanthine oxidoreductase: characterization of the enzyme from pea (*Pisum sativum* L.) leaves. J. Plant Physiol. 165, 1319–1330.

Das, K., Roychoudhary, A., 2014. Reactive oxygen species (ROS) and response of antioxidants as ROS-scavengers during environmental stress in plants. Front. Environ. Sci.. https://doi.org/10.3389/fenvs.2014.00053.

Dodds, P.N., Rathjen, J.P., 2010. Plant immunity: towards an integrated view of plant-pathogen interactions. Nat. Rev. 11, 539–548. https://doi.org/10.1038/nrg2812.

Domingos, P., Prado, A.M., Wong, A., Gehring, C., Feijo, J.A., 2014. Nitric oxide: a multitasked signaling gas in plants. Mol. Plant. https://doi.org/10.1016/j.molp.2014.12.010.

Durner, J., Wendehenne, D., Klessig, D.F., 1998. Defense gene induction in tobacco by nitric oxide, cyclic GMP and cyclic ADP-ribose. Proc. Natl. Acad. Sci. U. S. A. 95, 10328–10333.

FAO, IFAD, UNICEF, WFP, WHO, 2018. The State of Food Security and Nutrition in the World 2018. Building Climate Resilience for Food Security and Nutrition. FAO, Rome.

Foyer, C.H., Halliwell, B., 1976. The presence of glutathione and glutathione reductase in chloroplasts: a proposed role in ascorbic acid metabolism. Planta 133, 21–25.

Fridovich, I., 1989. Superoxide dismutases: an adaptation to a paramagnetic gas. J. Biol. Chem. 264, 7761–7764.

Grun, S., Lindermayr, C.S., Durner, J., 2006. Nitric oxide and gene regulation in plants. J. Exp. Bot. 57, 507–516.

Haider, M.S., Jogaiah, S., Pervaiz, T., Yanxue, Z., Khan, N., Fang, J., 2019. Physiological and transcriptional variations inducing complex adaptive mechanisms in 2 grapevine by salt stress. Environ. Experiment. Biol. 162, 455–467.

Halliwell, B., 2006. Reactive species and antioxidants. Redox biology is a fundamental theme of aerobic life. Plant Physiol. 141, 312–322.

Havaux, M., Eymery, F., Porfirova, S., Rey, P., Dörmann, P., 2005. Vitamin E protects against photoinhibition and photooxidative stress in *Arabidopsis thaliana*. Plant Cell 17, 3451–3469.

Hill, B.G., Dranka, B.P., Bailey, S.M., Lancaster Jr., J.R., Darley-Usmar, V.M., 2010. What part of NO don't you understand? Some answers to the cardinal questions in nitric oxide biology. J. Biol. Chem. 285, 19699–19704.

Huang, X., Kiefer, E., von Rad, U., Ernst, D., Foissner, I., Durner, J., 2002. Nitric oxide burst and nitric oxide-dependent gene induction in plants. Plant Physiol. Biochem. 40, 625–631.

Igamberdiev, A.U., Serege'yes, C., Manac'h, N., Hill, R.D., 2004. NADH-dependent metabolism of nitric oxide in alfalfa root cultures expressing barley haemoglobin. Planta 219, 95–102.

Iriti, M., Sironi, M., Gomarasca, S., Casazza, A.P., Soave, C., Faoro, F., 2006. Cell death-mediated antiviral effect of chitosan in tobacco. Plant Physiol. Biochem. 44, 893–900.

Jain, A., Singh, S., Sarma, B.K., Singh, H.B., 2012. Microbial consortium-mediated reprogramming of defence network in pea to enhance tolerance against *Sclerotinia sclerotiorum*. J. Appl. Microbiol. 112, 537–550.

Jain, A., Singh, A., Singh, S., Singh, H.B., 2013. Microbial consortium-induced changes in oxidative stress markers in pea plants challenged with *Sclerotinia sclerotiorum*. J. Plant Growth Regul. 32, 388–398.

Jayamohan, N.S., Kumudini, B.S., 2011. Host pathogen interaction at the plant cell wall. Intl. Res. J. Pharm. Pharmacol. 1, 242–249.

Jayamohan, N.S., Patil, S.V., Kumudini, B.S., 2018. Reactive oxygen species (ROS) and antioxidative enzyme status in *Solanum lycopersicum* on priming with fluorescent *Pseudomonas* spp. against *Fusarium oxysporum*. Biologia. https://doi.org/10.2478/s11756-018-0125-3.

Joe, M.M., Muthukumaran, N., 2008. Role of certain elicitors on the chemical induction of resistance in tomato against the leaf Caterpillar *Spodoptera litura* Fab. Not. Bot. Hort. Agrobot. Cluj. 36, 71–75.

Jogaiah, S., Sharathchandra, R.G., Lam-Son, P.T., 2013. Systems biology-based approaches towards understanding drought tolerance in food crops. Critic. Rev. Biotech. 33, 23–39.

Jogaiah, S., Abdelrahman, M., Tran, L.-S.P., Ito, S.-I., 2018. Different mechanisms of *Trichoderma virens*-mediated resistance in tomato against Fusarium wilt involve the jasmonic and salicylic acid pathways. Mol. Plant Pathol. 19, 870–882.

Joshi, S.M., De Britto, S., Jogaiah, S., Ito, S., 2019. Mycogenic selenium nanoparticles as potential new generation broad Spectrum antifungal molecules. Biomol. Ther. 9 (9), 419.

Karuppanapandian, T., Moon, J., Kim, C., Manoharan, K., Kim, W., 2011. Reactive oxygen species in plants: their generation, signal transduction, and scavenging mechanisms. Aus. J. Crop. Sci. 5, 709–725.

Katoch, R., 2005. Effect of elicitors and *E. polygoni* inoculation on the activity of phenol metabolizing enzymes in garden pea (*Pisum sativum* L.). Indian J. Agricul. Biochem. 18, 87–91.

Kiffin, R., Bandyopadhyay, U., Cuervo, A.M., 2006. Oxidative stress and autophagy. Antioxid. Redox Signal. 8, 152–162.

Kloepper, J.W., Schroth, M.N., 1978. Plant growth-promoting rhizobacteria on radishes. In: Proceedings of the 4th International Conference on Plant Pathogenic Bacteria. Vol. 2. Station de Pathologie Vegetale et Phytobacteriologie, INRA, Angers, France, pp. 879–882.

Knight, H., Knight, M.R., 2001. Abiotic stress signaling pathways: specificity and cross-talk. Trends Plant Sci. 6, 262–267.

Kotapati, K.V., Palaka, B.K., Kandukuri, A., Pamuru, R.R., Ampasala, D.R., 2014. Response of antioxidative enzymes and lipoxygenase to drought stress in finger millet leaves (*Eleusine coracana* (L.) Gaertn). Int. J. Plant Ani. Environ. Sci. 4, 644–653.

Kumudini, B.S., Shetty, H.S., 2002a. Association of lignification and callose deposition with host cultivar resistance and induced systemic resistance in pearl millet to *Sclerospora graminicola*. Austr. Plant Pathol. 31, 157–164.

Kumudini, B.S., Shetty, H.S., 2002b. Induction of resistance in pearl millet and associated host defence responses—a histological perspective. J. Mycol. Plant Pathol. 32, 377.

Kumudini, B.S., Shetty, H.S., 2004. A histological illustration of host responses in induced resistant pearl millet against downy mildew. In: Proceeding of The International Joint Workshop on PR-Proteins and Induced Resistance held at Elsinore Denmark.

Kumudini, B.S., Vasanthi, N.S., Shetty, H.S., 2001. Hypersensitive response, cell death and histochemical localisation of hydrogen peroxide in host and non-host seedlings infected with the downy mildew pathogen *Sclerospora graminicola*. Ann. Appl. Biol. 139, 217–225.

Kumudini, B.S., Jayamohan, N.S., Patil, S.V., 2017. Integrated mechanisms of plant disease containment by rhizospheric bacteria: unravelling the signal crosstalk between plant and fluorescent *Pseudomonas*. In: Meena, V.S., Mishra, P.K., Bisht, J.K., Pattanayak, A. (Eds.), Agriculturally Important Microbes for Sustainable Agriculture. Springer Nature, Singapore, pp. 263–291.

Kumudini, B.S., Jayamohan, N.S., Patil, S.V., Govardhana, M., 2018. Primary plant metabolism during plant–pathogen interactions and its role in defense. In: Ahmad, P., Ahanger, M.A., Singh, V.P., Tripathi, D.K., Alam, P., Alyemeni, M.N. (Eds.), Plant Metabolites and Regulation Under Environmental Stress. Academic Press, New York, pp. 215–229.

Kunstler, A., Bacso, R., Hafez, Y.M., Kiraly, L., 2015. Reactive oxygen species and plant disease resistance. In: Gupta, D.K. et al., (Ed.), Reactive Oxygen Species and Oxidative Damage in Plants under Stress. Springer International Publishing, Switzerland, pp. 269–303.

Lamb, C., Dixon, A.R., 1997. The oxidative burst in plant disease resistance. Annu. Rev. Plant. Physiol. Plant. Mol. Biol. 48, 251–275.

Ledford, H.K., Niyogi, K.K., 2005. Singlet oxygen and photo-oxidative stress management in plants and algae. Plant Cell Environ. 28, 1037–1045.

Lehmann, S., Serrano, M., L'Haridon, F., Tjamos, S.E., Metraux, J.P., 2015. Reactive oxygen species and plant resistance to fungal pathogens. Phytochemistry 112, 54–62.

Leitner, M., Vandelle, E., Gaupels, F., Bellin, D., Delledonne, M., 2009. NO signals in the haze: nitric oxide signalling in plant defence. Curr. Opin. Plant Biol. 12, 451–458.

Lohar, D.P., Haridas, S., Gantt, J.S., Vanden Bosch, K.A., 2007. A transient decrease in reactive oxygen species in roots leads to root hair deformation in the legume-rhizobia symbiosis. New Phytol. 173, 39–49.

Madhusudhan, R., Ishikawa, T., Sawa, Y., Shigeoka, S., Shibata, H., 2003. Characterization of an ascorbate peroxidase in plastids of tobacco BY-2 cells. Physiol. Plant. 117, 550–557.

Mahalingam, R. (Ed.), 2015. Consideration of Combined Stress: A Crucial Paradigm for Improving Multiple Stress Tolerance in Plants, in Combined Stresses in Plants. Springer International Publishing, pp. 1–25.

Maksimov, I.V., Cherepanova, E.A., Khairullin, R.M., 2003. Chitin-specific peroxidases in plants. Biochemistry 68, 111–115.

Mandal, S., 2010. Induction of phenolics, lignin and key defense enzymes in eggplant (*Solanum melongena* L.) roots in response to elicitors. Afr. J. Biotechnol. 9, 8038–8047.

Manjunatha, G., Niranjan-Raj, S., Prashanth, G.N., Deepak, S., Amruthesh, K.N., Shetty, H.S., 2009. Nitric oxide is involved in chitosan-induced systemic resistance in pearl millet against downy mildew disease. Pest Manag. Sci. 65, 737–743.

Mhamdi, A., Breusegem, F.V., 2018. Reactive oxygen species in plant development. Development. https://doi.org/10.1242/dev.164376.

Mittler, R., 2002. Oxidative stress, antioxidants and stress tolerance. Trends Plant Sci. 7, 405–410.

Mittler, R., 2006. Abiotic stress, the field environment and stress combination. Trends Plant Sci. 11, 15–19.

Mittler, R., Zilinskas, B.A., 1992. Molecular cloning and characterization of a gene encoding pea cytosolic ascorbate peroxidase. J. Biol. Chem. 267, 21802–21807.

Mittler, R., Vanderauwera, S., Gollery, M., Van Breusegem, F., 2004. Reactive oxygen gene network of plants. Trends Plant Sci. 9, 490–498.

Monaghan, J., Zipfel, C., 2012. Plant pattern recognition receptor complexes at the plasma membrane. Curr. Opin. Plant Biol. 15, 349–357.

Mortensen, A., Skibsted, L.H., Truscott, T.G., 2001. The interaction of dietary carotenoids with radical species. Arch. Biochem. Biophys. 385, 13–19.

Mur, L.A.J., Kenton, P., Lloyd, A.J., Ougham, H., Prats, E., 2008. The hypersensitive response; the centenary is upon us but how much do we know? J. Exp. Bot. 59, 501–520.

Nakano, Y., Asada, K., 1987. Purification of ascorbate peroxidase in spinach chloroplasts; its inactivation in ascorbate-depleted medium and reactivation by monodehydroascorbate radical. Plant Cell Physiol. 28, 131–140.

Nandini, B., Puttaswamy, H., Prakash, H.S., Adhikari, S., Jogaiah, S., Geetha, N., 2020. Elicitation of novel trichogenic-lipid nanoemulsion signaling resistance against pearl millet downy mildew disease. Biomol. Ther. 10 (1), 25.

O'Brien, J.A., Daudi, A., Butt, V.S., Bolwell, G.P., 2012. Reactive oxygen species and their role in plant defence and cell wall metabolism. Planta 236, 765–779.

Opdenakker, K., Remans, T., Vangronsveld, J., Cuypers, A., 2012. Mitogen-activated protein (MAP) kinases in plant metal stress: regulation and responses in comparison to other biotic and abiotic stresses. Int. J. Mol. Sci. 13, 7828–7853.

Ortega-Ortíz, H., Benavides-Mendoza, A., Flores-Olivas, A., Ledezma-Pérez, A., 2003. Use of the inter polyelectrolyte complexes of poly(acrylic acid)-chitosan as inductors of tolerance against pathogenic fungi in tomato (*Lycopersicom esculentum* Mill. var. Floradade). Macromol. Biosci. 3, 566–570.

Palmieri, M.C., Sell, S., Huang, X., Scherf, M., Werner, T., Durner, J., Lindermayr, C., 2008. Nitric oxide responsive genes and promoters in *Arabidopsis thaliana*: a bioinformatics approach. J. Exp. Bot. 59, 177–186.

Pandey, P., Ramegowda, V., Senthil-Kumar, M., 2015. Shared and unique responses of plants to multiple individual stresses and stress combinations: physiological and molecular mechanisms. Front. Plant Sci.. https://doi.org/10.3389/fpls.2015.00723.

Patil, S., Kumudini, B.S., 2018. Antioxidant-mediated survival of primed finger millet plants against blast disease. Phytopathol. 108.

Patterson, W.R., Poulos, T.L., 1995. Crystal structure of recombinant pea cytosolic ascorbate peroxidase. Biochemistry 34, 4331–4341.

Peng, J., Deng, X., Huang, J., Jia, S., Miao, X., Huang, Y., 2004. Role of salicylic acid in tomato defense against cotton bollworm, *Helicoverpa armigera* Hubner. Z. Naturforsch. 59, 856–862.

Phaniendra, A., Jestadi, D.B., Periyasamy, L., 2015. Free radicals: properties, sources, targets, and their implication in various diseases. Indian J. Clin. Biochem. 30, 11–26.

Polidoros, A.N., Mylona, P.V., Pasentsis, K., Scandalios, J.G., Tsaftaris, A.S., 2005. The maize alternative oxidase 1a (Aox1a) gene is regulated by signals related to oxidative stress. Redox Rep. 10, 71–78.

Prasad, P.V.V., Pisipati, S.R., Momčilovič, I., Ristic, Z., 2011. Independent and combined effects of high temperature and drought stress during grain filling on plant yield and chloroplast EF-Tu expression in spring wheat. J. Agron. Crop Sci. 197, 430–441.

Prasch, C.M., Sonnewald, U., 2013. Simultaneous application of heat, drought, and virus to Arabidopsis plants reveals significant shifts in signaling networks. Plant Physiol. 162, 1849–1866.

Qi, J., Wang, J., Gong, Z., Zhou, J., 2017. Apoplastic ROS signaling in plant immunity. Curr. Op. Plant Biol. 38, 92–100.

Racchi, M.L., Bagnoli, F., Balla, I., Danti, S., 2001. Differential activity of catalase and superoxide dismutase in seedlings and in vitro micropropagated oak (*Quercus robur* L.). Plant Cell Rep. 20, 169–174.

Rani, T.S., Takahashi, D., Uemura, M., Podile, A.R., 2014. Proteins associated with oxidative burst and cell wall strengthening accumulate during citrus-*Xanthomonas* non-host interaction. Plant Mol. Biol. Rep.. https://doi.org/10.1007/s11105-014-0817-y.

Rathore, J.S., Ghosh, C., 2018. Pathogen-associated molecular patterns and their perception in plants. In: Singh, A., Singh, I.K. (Eds.), Molecular Aspects of Plant-Pathogen Interaction. Springer Nature, Singapore, pp. 79–113.

del Rio, L.A., Sandalio, L.M., Corpas, F.J., Palma, J.M., Barroso, J.B., 2006. Reactive oxygen species and reactive nitrogen species in peroxisomes: production, scavenging, and role in cell signalling. Plant Physiol. 141, 330–335.

Romero-Puertas, M.C., Laxa, M., Matte, A., Zaninotto, F., Finkemeier, I., Jones, A.M., Perazzolli, M., Vandelle, E., Dietz, K.J., Delledonne, M., 2007. *S*-nitrosylation of peroxiredoxin II E promotes peroxynitrite-mediated tyrosine nitration. Plant Cell 19, 4120–4130.

Rosnoblet, C., Bourque, S., Nicolas-Frances, V., Lamotte, O., Besson-Bard, A., Jeandroz, S., Wendehenne, D., 2016. NO signalling in plant immunity. In: Lamattina, L., Garcia-Mata, (Eds.), Gasotransmitters in Plants, Signaling and Communication in Plants. Springer International Publishing, Switzerlandhttps://doi.org/10.1007/978-3-319-40713-5_11.

Sewelam, N., Kazan, K., Schenk, P.M., 2016. Global plant stress signaling: reactive oxygen species at the cross-road. Front. Plant Sci. 7, 187. https://doi.org/10.3389/fpls.2016.00187.

Shailasree, S., Ramachandra, K.K., Shetty, S.H., 2007. β-Amino butyric acid-induced resistance in pearl millet to downy mildew is associated with accumulation of defence-related proteins. Aust. Plant Pathol. 36, 204–211.

Sharma, P., Dubey, R.S., 2004. Ascorbate peroxidase from rice seedlings: properties of enzyme isoforms, effects of stresses and protective roles of osmolytes. Plant Sci. 167, 541–550.

Sharma, S., Sohal, B.S., 2010. Foliar spray of benzothiadiazole and salicylic acid on *Brassica juncea* var. Rlm619 to combat *Alternaria* blight in field trials. Crop. Improv. 31, 87–92.

Sharma, P., Jha, A.B., Dubey, R.S., Pessarakli, M., 2012. Reactive oxygen species, oxidative damage, and antioxidative defense mechanism in plants under stressful conditions. J. Bot.. https://doi.org/10.1155/2012/217037.

Shaw, S.L., Long, S.R., 2003. Nod factor inhibition of reactive oxygen efflux in a host legume. Plant Physiol. 132, 2196–2204.

Shetty, N.P., Jorgensen, H.J.L., Jensen, J.D., Collinge, D.B., Shetty, H.S., 2008. Roles of reactive oxygen species in interactions between plants and pathogens. Eur. J. Plant Pathol. 121, 267–280.

Singh, B.N., Singh, A., Singh, S.P., Singh, H.B., 2011. Trichoderma harzianum -mediated reprogramming of oxidative stress response in root apoplast of sunflower enhances defence against *Rhizoctonia solani*. Eur. J. Plant Pathol. 131, 121–134.

Spoel, S.H., Dong, X., 2012. How do plants achieve immunity? Defence without specialized immune cells. Nat. Rev. Immunol. 12, 89–100.

Stamler, J.S., Singel, D.J., Loscalzo, J., 1992. Biochemistry of nitric oxide and its redox- activated forms. Science 258, 1898–1902.

Sudisha, J., Kumar, A., Amruthesh, K.N., Niranjana, S.R., Shetty, H.S., 2011. Elicitation of resistance and defense related enzymes by raw cow milk and amino acids in pearl millet against downy mildew disease caused by *Sclerospora graminicola*. Crop Prot. 30, 794–809.

Suzuki, N., Rivero, R.M., Shulaev, V., Blumwald, E., Mittler, R., 2014. Abiotic and biotic stress combinations. New Phytol. 203, 32–43.

Tavallali, V., Karimi, S., Mohammadi, S., Hojati, S., 2008. Effects of β-aminobutyric acid on the induction of resistance to *Penicillium italicum*. World Appl. Sci. J. 5, 345–351.

Thomas, D.D., Ridnour, L.A., Isenberg, J.S., Flores-Santana, W., Switzer, C.H., Donzelli, S., Hussain, P., Vecoli, C., Paolocci, N., Ambs, S., Colton, C.A., Harris, C.C., Roberts, D.D., Wink, D.A., 2008. The chemical biology of nitric oxide: implications in cellular signaling. Free Radic. Biol. Med. 45, 18–31.

Thomma, B.P.H.J., Penninckx, I.A., Broekaert, W.T., Cammue, B.P., 2001. The complexity of disease signaling in Arabidopsis. Curr. Opin. Immunol. 13, 63–68.

Torres, M.A., Jones, J.D.G., Dangl, J.L., 2006. Reactive oxygen species signaling in response to pathogens. Plant Physiol. 141, 373–378.

Varsha, T., Kumudini, B.S., 2016. Fluorescent *Pseudomonas* mediated alleviation of trivalent chromium toxicity in ragi through enhanced antioxidant activities. Proc. Natl. Acad. Sci. India Sect. B Biol. Sci.. https://doi.org/10.1007/s40011-016-0816-x.

Vimala, R., Suriachandraselvan, M., 2009. Induced resistance in bhendi against powdery mildew by foliar application of salicylic acid. J. Biopest. 2, 111–114.

Wendehenne, D., Durner, J., Klessig, D.F., 2004. Nitric oxide: a new player in plant signalling and defence responses. Curr. Opin. Plant Biol. 7, 449–455.

Willekens, H., Inze, D., Van Montagu, M., Van Camp, W., 1995. Catalases in plants. Mol. Breed. 1, 207–228.

Wojtaszek, P., Trethowan, J., Bolwell, G.P., 1995. Specificity in the immobilization of cell wall proteins in response to different elicitor molecules in suspension-cultured cells of French bean (*Phaseolus vulgaris* L.). Plant Mol. Biol. 28, 1075–1087.

Yousuf, P.Y., Ahmad, A., Ganie, A.H., Sareer, O., Krishnapriya, V., Aref, I.M., Iqbal, M., 2016. Antioxidant response and proteomic modulations in Indian mustard grown under salt stress. Plant Growth Regul.. https://doi.org/10.1007/s10725-016-0182-y.

Zeier, J., Delledonne, M., Mishina, T., Severi, E., Sonoda, M., Lamb, C., 2004. Genetic elucidation of nitric oxide signaling in incompatible plant-pathogen interactions. Plant Physiol. 136, 2875–2886.

Zipfel, C., 2009. Early molecular events in PAMP-triggered immunity. Curr. Opin. Plant Biol. 12, 414–420.

Role of terpenes in plant defense to biotic stress

16

Silvia Laura Toffolatti, Giuliana Maddalena, Alessandro Passera, Paola Casati, Piero Attilio Bianco, and Fabio Quaglino
Department of Agricultural and Environmental Sciences—Production, Landscape, Agroenergy (DISAA), University of Milan, Milano, Italy

16.1 Introduction

The role of terpenes in plant biological functions is widely established but still in evolution, as demonstrated by the increasing number of studies reported in the scientific literature. Terpenes are involved in the primary metabolism of the plant, as hormones, antioxidants, and components of the electron transfer system, but also play an important role in more specialized activities (Pichersky and Raguso, 2016). Terpenes, for example, are also involved in chemical communication with insects, both to attract beneficial insects such as pollinators or to repel herbivores (Eberhard Breitmaier, 2006). Apart from their role in plant survival, terpenes also possess biological and pharmacological properties that are beneficial to humans (Singh and Sharma, 2015). For instance, vitamin A is synthesized from β-carotene, an abundant plant terpene. Traditionally, terpene-containing plant products have been used in medicine without knowing their composition and the mode of action of the individual compounds (Cho et al., 2017). Nowadays, specific activities on components of human cells or pathogens have led to their use in the treatment of several diseases, from cancer to malaria (Pichersky and Raguso, 2016; Paduch et al., 2007; Cho et al., 2017; Jansen and Shenvi, 2014).

Terpenes are among the most widespread and chemically diverse plant products. More than 25,000 terpene structures are, in fact, produced by different plant species (Chen et al., 2011). Their basic structure consists of linked units of the five carbons (C_5) molecules dimethylallyl diphosphate (DMAPP) and isopentenyl diphosphate (IPP) (Oldfield and Lin, 2012), which are the building units of isoprenoids. According to the number of linked subunits and carbons (in brackets), terpenes are classified into hemi- (C_5) mono- (C_{10}), sesqui- (C_{15}), di- (C_{20}), sester- (C_{25}), tri- (C_{30}), tetraterpenes (C_{40}), and polyterpenes $(C_5)_n$ with $n > 8$ (Pichersky and Raguso, 2016; Eberhard Breitmaier, 2006). Terpene precursors DMAPP and IPP are synthesized through two independent metabolic pathways: the cytosolic mevalonate (MVA) pathway or the chloroplast methylerythritol phosphate (MEP) pathway (Moore et al., 2014). In general, isoprene, monoterpenes, and diterpenes are synthesized in plastids, while the biosynthesis of sesquiterpenes occurs in the cytosol. The huge structural diversity of terpenes is the result of the different linkage of the subunits to the activity of the

Biocontrol Agents and Secondary Metabolites. https://doi.org/10.1016/B978-0-12-822919-4.00016-8

terpene synthase (TPS) enzyme family and to the modification of the resulting terpene skeletons (Pichersky and Raguso, 2016). First, DMAPP and IPP units can be linked to each other through the head (the isopropyl part) or the tail (ethyl residue) and form linear chains or cyclic structures (Eberhard Breitmaier, 2006; Paduch et al., 2007; Oldfield and Lin, 2012). Second, TPSs enzymes are a vast family that can count up to 150 genes per genome; these enzymes form the basic terpene skeleton utilizing prenyl diphosphates, formed by the condensation of IPP to DMAPP (Kuzuyama, 2017), as precursors, and each TPS can produce multiple products from the same substrate (Pichersky and Raguso, 2016; Oldfield and Lin, 2012). This diversity was shown in the literature by a comparative analysis of the TPS family in different plant species, highlighting that closely related enzymes differ in their product profiles, subcellular localization, or the in planta substrates they use (Chen et al., 2011). The recent analysis of 17 genomes of flowering plants found evidence for 1904 proteins associated with terpenoid biosynthesis, among which are 840 core terpene-synthases (Hofberger et al., 2015). Finally, terpene backbones can be further modified by oxidation reactions and conjugation to other moieties, and consequently the total number of plant-synthesized terpenes is estimated to reach the tens of thousands of different molecules (Schilmiller et al., 2008).

In natural ecosystems, plants are under stress caused by biotic agents such as fungi (including Oomycetes), bacteria, viruses, nematodes, and insects that compromise plant health, survival, and offspring. Plants have evolved sophisticated strategies through time to "perceive" biotic attackers and organized a defensive response to them by a complex defense signaling network (Pieterse and Dicke, 2007; Jones and Dangl, 2006). Plants defend themselves through resistance mechanisms that are preformed or induced upon attack and consist of structural barriers (waxes, cutin, suberin, lignin, phenolic compounds, cellulose, and cell-wall proteins) and chemical compounds with antibiotic activity (Glazebrook, 2005; Panstruga et al., 2009).

Most of the studies on terpenes and biotic stress concern the interaction of the plants with insects (Aljbory and Chen, 2018; Arimura et al., 2009; War et al., 2011; Irmisch et al., 2014; Keeling and Bohlmann, 2006; Raffa et al., 2005), and less information is available on plant pathogens (fungi, bacteria, and viruses). For this reason, in this chapter particular attention will be given to the role of terpenes in plant disease resistance.

Terpenoids, a term that includes terpenes and compounds with terpene moieties linked to other moieties derived from different pathways (Pichersky and Raguso, 2016), play an important role in plant-pathogen interactions (Dudareva et al., 2004; Paschold et al., 2006). For instance, several diterpenes and sesquiterpenes belong to the class of phytoalexins (Schmelz et al., 2014; AhujaI and Bones, 2012). Phytoalexins are a heterogeneous group of compounds with low-molecular weight that show biological activity toward a variety of pathogens and are considered as markers of disease resistance (AhujaI and Bones, 2012; Harborne, 1999; Kuc, 1995; Hammerschmidt, 1999). The concept of phytoalexins emerged in 1940 from the observation that potato tuber slices inoculated with avirulent races of the late blight pathogen *Phytophthora infestans* were immune to virulent races and it was hypothesized that this effect could be due to the synthesis of compounds that inhibited fungal growth in the inoculated tubers (Widmark, 2010). However, it took almost 30 years to isolate the first phytoalexin compound, the bicyclic sesquiterpene rishitin, from tuber tissues (Tomiyama et al., 1968; Jadhav et al., 1991). Since then, numerous studies, reviewed in Refs. (AhujaI and Bones, 2012; Schmelz et al., 2014), have been carried out on the role of phytoalexins in plant-pathogen interaction and genetic manipulation of phytoalexins has been investigated to increase the disease resistance of plants (Jeandet et al., 2013).

Essential oils are volatile aromatic oily liquids obtained from different plant portions, ranging from flowers and leaves to roots (Burt, 2004). The main constituents of essential oils are mono and sesquiterpenes including carbohydrates (such as the unsaturated hydrocarbon limonene), phenols, alcohols (menthol), ethers, aldehydes, and ketones (myrtenal, carvone) responsible for the biological activity of the molecules (Santos et al., 2011). Monoterpenes can be acyclic (myrcene, linalool, geraniol), monocyclic (α-terpineol and terpinolene), and bicyclic (α-pinene, thujone, camphor, fenchone). In general, the inhibitory action of monoterpenes on microorganism cells involves cytoplasm granulation, cytoplasmic membrane rupturing, and inactivation and/or synthesis inhibition of intracellular and extracellular enzymes (Sukanya et al., 2011). Many essential oils have been reported to have antifungal activities (Nuzhat and Vidyasagar, 2013) and their properties have been employed in plant disease management: formulates including geraniol, eugenol, and menthol are currently employed in European grapevine fields to control important plant pathogens such as *Botrytis cinerea*, the causal agent of grey mold (European Food Safety Authority (EFSA), 2012), thanks to their activity against spore germination and mycelial growth (Tsao and Zhou, 2000).

Due to the high diversity of terpenes and the very specific role that some of them play in plant-pathogen interaction, in many cases mediated by biotic vectors, it is difficult to make generalizations about their mode of action. In the following paragraphs, therefore, specific examples on the role of terpenes in plant defense mechanisms will be described.

16.2 Role of terpenes in resistance to fungal diseases

Fungi are the major agents of plant disease in terms of number of diseases caused and economic losses (Doehlemann et al., 2017). According to the opinion of the scientific community, among the most important phytopathogenic fungi are *Magnaporthe oryzae* (agent of rice blast), *B. cinerea* (agent of grey mold on numerous crops), *Puccinia* spp. (rust disease agents), *Fusarium graminearum*, and *Fusarium oxysporum* (vascular wilt agents) (Dean et al., 2012). Fungal species of the genera *Aspergillus*, *Penicillium*, and *Alternaria* are also considered to be major plant pathogens worldwide, and their threat is also linked to the production of mycotoxins (Moretti et al., 2017). To this list can be added important Oomycetes pathogens, with *P. infestans* being the most representative example because of the terrible crop failures this pathogen can cause, which led to the Irish famine in 1845–1846 and great economic losses caused to potato and tomato production at present (Kamoun et al., 2015). Among other important Oomycetes is *Plasmopara viticola*, the agent of grapevine downy mildew, a disease that affects in particular the *Vitis vinifera* species, a crop of great economic impact, worldwide.

The role of terpenes in plant resistance to fungal pathogens has been analyzed through the evaluation of the direct activity of the molecules against the pathogens or the involvement of terpene synthase genes in the resistance response of the plant to the pathogen or a combination of these approaches.

Several studies were carried out through comparative transcriptome analyses, demonstrating that resistant cultivars show an overexpression of genes associated with terpene biosynthesis in comparison with susceptible ones upon pathogen inoculation. For instance, based on gene ontology and pathway analyses, the biosynthesis of terpenes was found to be activated shortly after infection in the rice variety Digu, resistant to *M. oryzae* (Li et al., 2016). The role of terpenes in rice defenses is supported by other studies, where jasmonic acid-dependent terpene synthase (AK071447) showed a role in the rice defense system toward *M. oryzae* in the rice cultivar Hinohikari by producing the volatile sesquiterpene β-elemene, which showed a direct action on the pathogen by significantly reducing hyphal growth in in vitro tests (Taniguchi et al., 2014). Furthermore, a terpene synthase gene (*OsTPS19*) upregulated in a rice line with enhanced resistance to *M. oryzae* was reported to synthesize (S)-limonene, a monoterpene inducing significant reduction of the spore germination rate (Chen et al., 2018).

Genes encoding for enzymes involved in terpene biosynthesis (valencene synthase and geraniol 8-hydroxylase) have been found to be overexpressed in a *V. vinifera* variety resistant to the downy mildew agent *P. viticola* (Toffolatti et al., 2018). The involvement of terpenes in grapevine resistance to *P. viticola* has been previously demonstrated in non-*vinifera* species where volatile monoterpenes and sesquiterpenes have been produced following pathogen inoculation (Algarra Alarcon et al., 2015). Volatile organic compounds are important secondary metabolites acting, among the other functions, as airborne defense signals that prime defense response to pathogens in distant plants (Chalal et al., 2015).

Association genetics, combining RNAseq and genome wide association studies, showed that genes of the mevalonate pathway (hydroxy-3-methylglutaryl coenzyme A reductase) and terpene biosynthesis (cytochrome P450 protein) are strong candidates for controlling quantitative resistance to *P. infestans* in potato together with a lipoxygenase gene involved in the jasmonate pathway (Mosquera et al., 2016). Root metabolomic profiling showed that the volatile sesquiterpene β-selinene and in particular its nonvolatile antibiotic derivative β-costic acid, produced by terpene synthase21 (ZmTps21), are highly produced in maize following fungal elicitation (Ding et al., 2017).

Numerous studies highlighted the role of terpene phytoalexins in plant defense to fungi, linked to a direct inhibition of the pathogen growth. Nearly 50 years ago, pathogen-inducible diterpenoid production was described in rice, and these compounds were shown to function as antimicrobial phytoalexins. Fourteen diterpenoid phytoalexins, grouped into four structurally distinct types (momilactones A and B, oryzalexin S, phytocassanes A–E), have been identified in response to inoculation with *M. grisea* (Prisic et al., 2004). The accumulation of these phytoalexins in resistant rice cultivars was associated with hypersensitive response following *M. grisea* inoculation. Moreover, Momilactone A completely suppressed mycelial growth for 24 hours following inoculation on potato dextrose agar medium (PDA) (Hasegawa et al., 2010). Capsidiol, a bicyclic sesquiterpene, is the major phytoalexin produced by many *Solanaceae* species, among which are pepper fruit and tobacco, following inoculation with pathogenic fungi (Ahujal and Bones, 2012). It prevents the germination and growth of several important fungal species such as *B. cinerea* and *Colletotrichum gloeosporioides*, the anthracnose agent (Park et al., 2014). Gene expression profiling in roots of *Zea mays* inoculated with *Phytophthora cinnamomi* (causing root rot in numerous plant species) showed that maize, which is naturally resistant to the pathogen, rapidly responds to it by overexpressing two genes (*terpene synthase 11* and *kaurene synthase 2*) involved in biosynthesis of the terpenoid phytoalexins zealexin and kauralexin, providing confirmation of the important role in plant defense mechanism of these phytoalexins (Allardyce et al., 2013). Genes encoding for these two phytoalexins were shown to be furthermore involved in response to *Cercospora zeina* (the maize gray leaf spot agent) in resistant subtropical maize lines (Meyer et al., 2017). Kauralexin B3 and Zealexin A1 inhibited the growth of *Rhizopus microsporus* and *Colletotrichum graminicola* (kauralexin B3) and *Aspergillus flavus* and *F. graminearum* (zealexin A1) in in vitro studies (Schmelz et al., 2014, 2011).

Experimental evidence that terpenoids with antifungal activity interfere with plant-insect interaction and, as a consequence, with fungal pathogenesis has been found (Block et al., 2018). An example of this is described. *Ostrinia nubilalis*, the European corn borer, creates stem wounds in maize that are a suitable environment for important fungal pathogens such as *Fusarium* spp. Stem surfaces treated with kauralexins A3 and B3 significantly reduced the tissue consumed by *O. nubilalis* compared with the untreated control stem sections, suggesting that these phytoalexins, despite having no direct effect on insect growth, act as local antifeedants (Schmelz et al., 2011).

A number of scientific investigations have highlighted the importance and the contribution of the essential oils produced by plants belonging to several different families (i.e., *Apocynaceae, Asteraceae, Caesalpinaceae, Liliaceae, Piperaceae, Rutaceae, Sapotaceae, Solanaceae,* etc.) against fungal pathogens, confirming the involvement of essential oils in defense response. The antifungal activity of essential oil of chrysanthemum (*Chrysanthemum coronarium* L.) was evaluated against different plant pathogens such as *Alternaria* spp., *A. flavus, Pythium ultimum, Fusarium* spp., *B. cinerea, Sclerotinia sclerotiorum, Rhizoctonia solani,* and *Penicillium digitatum* (Alvarez-Castellanos et al., 2001). The main compounds in the oil were camphor, α- and β-pinene, and lyratyl acetate. The essential oil was active both in contact and headspace in vitro assays by inhibiting hyphal growth (Nuzhat and Vidyasagar, 2013).

Another example of the importance of essential oils can be taken from coniferous plants. Conifers are ancient plants with great ability to adapt and diversify (Farjon, 2018), which enabled them to inhabit 80% of the available niches (Scagel et al., 1965). Part of this success can be attributed to their highly evolved defense system, which allowed them to withstand interspecific competition as well as herbivore and pathogen damages. The most familiar of these defenses is the viscous, odoriferous secretion mobilized at wound and infection sites called oleoresin, or pitch. Oleoresin is a complex mixture of terpenoids, consisting of two fractions, diterpene rosin and turpentine, a fraction containing mono and sesquiterpenes. The biologically active agents contained in turpentine, such as limonene and 3-carene, show activity against insect and microbes, often acting in synergy (Raffa and Berryman, 1983). Turpentine has a role in transporting the rosin fraction of oleoresin to the site of injury and when evaporating upon exposure to the atmosphere, it creates a semicrystalline mass of resin that acts as a physical barrier to pathogen infection and insects trap (Gijzen et al., 1993). In spite of these effective defense mechanisms, the most common agents of tree death for coniferous plants appear to be pathogenic fungi (often of the class *Ophiostomataceae*), which are able to kill conifers through a combination of toxin production and invasive growth in the conduction elements. For instance, the causal agents of Dutch elm disease (*Ophiostoma ulmi* and *Ophiostoma novo-ulmi*) have killed millions of elm trees in the Northern Hemisphere during the past (Brasier, 1990). Mixtures of mono- and sesquiterpenes have been associated with increased resistance against fungal pathogens in host trees and numerous terpenes synthesized by conifers have shown an inhibitory activity toward fungal pathogens (Cates, 1996). In particular, limonene, β-pinene, and myrcene (both individually and in mixture) were found to affect adversely fungal growth across a wide range of hosts and fungi. An in vitro assay on three terpenes produced by conifers showed that abietic and isopimaric acid inhibit spore germination and mycelial growth (only in the case of abietic acid) of *Ophiostoma ips*, a conifer pathogen symbiotically associated with the pine engraver *Ips pini* (Kopper et al., 2005). The production of resins in conifers can be constitutive and inducible: while members of the genus *Pinus* constitutively produce and store copious amounts of oleoresin, the true firs (*Abies* genus) store only small amounts of primary resin and produce oleoresin in response to wounding, targeting both insects and their vectored fungal pathogens (Phillips and Croteau, 1999).

16.3 Role of terpenes in interaction with bacteria

As terpenes and terpenoids are involved in a very wide array of molecular functions and interactions, it comes as no surprise that they are also part of the plant-bacteria interactions. The bacteria that plants come in contact with are not only pathogens, and therefore the plants may produce antimicrobial terpenes in response to pathogen attack, or may be stimulated by beneficial bacteria in producing defense terpenes to protect themselves against future pathogen attack. One particular difficulty that plants may face in carrying out their defense responses against bacteria can come from their ability to organize themselves in a sessile, highly resistant state called biofilm (Flemming et al., 2016). While organized in a biofilm, protected by layers of extracellular polymeric substances (EPS; mostly excreted lipopolysaccharides, lipids, and proteins), pathogenic bacteria can become resistant to several antimicrobial compounds and escape detection from the plant host, effectively raising their chances of surviving inside a host and of successfully expressing their pathogenic potential. Another phenomenon related to both expression of pathogenicity and formation of biofilm is bacterial quorum sensing (QS) in which bacteria, by the secretion and detection of specific extracellular signal molecules, can sense the presence and quantity of other bacteria and modulate their metabolism accordingly, leading even to a complete change of lifestyle once the quorum has been reached. QS is known to be a crucial step in the infection process by several bacterial pathogens such as *Agrobacterium tumefaciens*, *Pectobacterium carotovorum*, *Pseudomonas syringae*, *Ralstonia solanacearum*, and *Xylella fastidiosa*, just to cite a few (von Bodman et al., 2003; Simionato et al., 2007). Plant terpenoids can be an effective weapon against both biofilm formation and QS and can leave pathogenic bacteria more susceptible to the whole array of defenses that the plant can activate against pathogens. For example, the monoterpene flavonoid carvacrol, found to be a main determinant of the aroma of oregano essential oil, has been characterized to reduce the production of QS signal molecules (Joshi et al., 2016) and inhibit the formation of biofilm (Tapia-Rodriguez et al., 2017) in treated bacteria. These two effects could be related to inhibition of QS, leading to the impossibility of organizing a biofilm, but recent studies attributed this effect also to a more direct effect of carvacrol on the bacteria cells: it has been reported that cells of *P. carotovorum* treated with this terpenoid had lower motility, adhesion potential, and produced less EPS compared to the nontreated control (Gutierrez-Pacheco et al., 2018). All these effects could be explained by the chemical interaction of carvacrol, a highly lipophilic molecule, with the bacterial cell wall: the terpenoid can easily be incorporated in the cell wall and make it more hydrophobic overall, changing its physical properties and reducing the bacteria's ability to move and adhere, and therefore to form a functional biofilm. Similar results and conclusions were obtained with another molecule, the sesquiterpene farnesol, for which the change in the cell wall structure was hypothesized to be the main biofilm-related mechanism (Feng et al., 2014), even if in this case the terpene was necessary for biofilm formation rather than inhibiting it.

While these experiments were mostly conducted in vitro, the importance of terpenes in defenses against plant pathogenic bacteria was also demonstrated through in planta experiments. For example, the transgenic expression of two terpene synthase genes from *Panax ginseng* in the model plant *Arabidopsis thaliana* could confer tolerance toward the *P. syringae* pathogen to the otherwise susceptible host plant (Yoon et al., 2016). In these plants, as expected, a higher production of terpenes and terpenoids could be detected compared to the wild-type control, especially after treatment with salicylic acid or methyl-jasmonate, the two main hormones responsible for the activation of plant defense mechanisms known as systemic acquired resistance (SAR) and induced systemic resistance (ISR), respectively. The fact that terpenes can be produced in higher quantities once the plant's defense mechanisms have been activated leads to the other side of the plant-bacteria interaction regarding terpenes: how bacteria can promote the production of these terpenoids. It is well known that one of the key features of many beneficial bacteria is to promote plant defenses by operating a "priming" effect that is defined as ISR, in which the plant is not under attack by a pathogen but still readies a whole array of defenses against biotic threats, becoming more capable of reacting against pathogens and predators (Farag et al., 2013). This phenomenon, which is accompanied by an increase in jasmonates, can ultimately lead to the production of different terpenes and terpenoids. For example, experiments carried out on specific bacterial strains isolated from the roots of grapevine plants, and reinoculated as pure cultures in micropropagated grapevine plants, showed how the introduction of specific bacteria could induce the production of mono- and sesquiterpenes that were not produced in nontreated plants (Salomon et al., 2016). Among these terpenes were strong antioxidants, such as α-pinene and nerolidol. Furthermore, the presence of these bacteria induced a way higher production of other terpenes, such as α- and γ-tocopherol, compared to control plants. Whether these changes in plant metabolites are simply the byproduct of metabolism stimulation from the bacteria is still a matter of debate: while there is no doubt that the presence of different microorganisms can induce different responses in the host, it cannot be ignored that bacteria themselves produce and metabolize a wide array of complex molecules, including terpenes and isoprenoids. It has been demonstrated that applying a surface sterilization to the leaves and flowers of *Sambucus nigra* plants, therefore eliminating all the epiphytic community, can lead to a drastic change in the whole metabolome of the plants, suggesting that the degradation of plant-originated compounds and the de novo synthesis of molecules from the microorganisms living in association with the host can greatly contribute to the host's ability to produce some compounds, which may in turn be essential for the host's health (Gargallo-Garriga et al., 2016).

Phytoplasmas are pleomorphic, cell wall-less bacteria of the class *Mollicutes* that exist as obligate plant pathogens. Based on molecular and other biological features, phytoplasma strains have been classified into 40 species within the provisional genus "*Candidatus* Phytoplasma" (Marcone, 2014). Phytoplasmas have been associated with several hundred diseases affecting economically important crops, such as ornamentals, vegetables, fruit trees, and grapevine (Bertaccini et al., 2014). In diseased plants, phytoplasmas are restricted to the phloem sieve tubes and are transmitted between plants by phloem-sap-feeding leafhoppers, planthoppers, or psyllids in a persistent manner (Weintraub and Beanland, 2006). Recent studies showed that phytoplasmas secrete effector proteins (i.e., SAP11, SAP54, TENGU) to induce alterations in the morphology and phytohormone-mediated defense system of the plant hosts (Bai et al., 2009; Hoshi et al., 2009; Sugio et al., 2011a; Maclean et al., 2014; Marcone et al., 2016).

Changes in the terpenoids profile have been observed in response to infection by phytoplasma and in recovered plants. A comprehensive NMR-based metabolomic profiling of periwinkle (*Catharanthus roseus*) infected by strains of ten mutually distinct "*Ca.* Phytoplasma" species identified the metabolites that were present in different levels in phytoplasma-infected plants than in healthy ones. Phytoplasma infection increased the levels of metabolites related to the biosynthetic pathways of phenylpropanoids or terpenoid indole alkaloids: chlorogenic acid, loganic acid, secologanin, and vindoline. Principal component analysis of 1H-NMR signals of healthy and phytoplasma-infected plants shows that these metabolites are major discriminating factors to characterize the phytoplasma-infected plants from healthy ones. Based on the NMR and PCA analysis, it might be suggested that the biosynthetic pathway of terpenoid indole alkaloids, together with that of phenylpropanoids, is stimulated by the infection of phytoplasma (Choi et al., 2004). Further gene expression studies showed a positive correlation between the expression of terpenoid pathway genes and the accumulation of related alkaloids in phytoplasma-infected plants. The phytoplasma modifies the gene expression system under the control of cytochrome P450 reductase, which results in upregulation of the defense pathway through the accumulation and secretion of catharanthine by the root system (Srivastava et al., 2014).

In some phytoplasma-infected plant hosts, recovery implies the complete remission of symptoms in previously symptomatic plants (Osler et al., 1993). This phenomenon is linked with the disappearance of phytoplasmas from the crowns of infected trees. Cytochemical analyses have shown that recovery is associated with biochemical changes in the phloem. Although recovered plants show overproduced hydrogen peroxide in phloem tissue, the physiological causes of the recovery process have still not been elucidated (Bianco et al., 2011). In grapevines infected by "*Ca.* Phytoplasma solani," the etiological agent of Bois noir disease (Quaglino et al., 2013), some important terpene synthases and the genes involved in phenylpropanoids were enhanced. Compared to the control healthy status, the analysis showed that recovery status was associated with the induction of several secondary metabolism categories, such as terpenoids, phenylpropanoids, lipid biosynthesis, sulfur-containing compounds (glucosinolates), flavonoids, and simple phenols (Punelli et al., 2016). Such upregulation of the secondary metabolism genes could allow plants to be less susceptible to further infection (Fang et al., 2018). The roles suggested for phenylpropanoid compounds in plant defenses have been traditionally based on biological activities in vitro, and on correlations between accumulation rates and reduced susceptibility in vivo. Yet, as plant defense responses are multicomponent, it is not easy to define which components are both necessary and sufficient to confer protection. Regarding the biotic stress-related pathways, it is noteworthy that very few key genes, such as HSC70, terpene synthase 14 (TPS14), a protein kinase family protein, were upregulated by recovery (Punelli et al., 2016).

Terpenes were also able to suppress the symptom development induced by aster yellows phytoplasma in grafted periwinkles, although the plants were not tested for phytoplasma presence after the treatment (Ćurković Perica et al., 2007).

On the other side, emission of volatile terpenes can attract vector insects, thereby increasing the spread of the disease in field. Phytoplasmas can also change the composition of plant-emitted volatile compounds, and their ability to attract insect vectors (Sugio et al., 2015). "*Ca.* Phytoplasma mali," the causal agent of apple proliferation disease (Seemüller and Schneider, 2007), changes the apple volatile emission, increasing the sesquiterpene β-caryophyllene which lures the psyllid vector *Cacopsylla picta*. The higher emission of VOCs in plants infected with "*Ca.* Phytoplasma mali" could be caused by an increased amount of vascular tissue and leaf surface due to the proliferation associated with the production of witches'-broom symptoms (Mayer et al., 2008). Moreover, a further study reported that genetically distinct strains of "*Ca*. Phytoplasma mali," differentiated by their virulence, possess specific abilities to modulate the emission of volatiles (Rid et al., 2016). This could be due to the differential presence of phytoplasma effectors which can interfere with the expression of terpene synthases/cyclase genes of infected plants (Hoshi et al., 2009; Sugio et al., 2011b; Maclean et al., 2014).

16.4 Role of terpenes in interaction with viruses

The major terpene components of plant essential oils also show antiviral activities. Citronellal, limonene, 1,8-cineole, and α-zingiberene from lemongrass, tangerine peel, lemon, tea tree, and artemisia essential oils had in vitro and in vivo inhibitory activity against Tobacco Mosaic Virus (TMV). These compounds, more effective when used individually, could have a direct effect on virus replication cycles (Min et al., 2013). Other studies demonstrated that essential oil of *Satureja montana* and *Plectranthus tenuiflorus* have antiphytoviral activity against TMV, Cucumber Mosaic Virus (CMV), Tobacco Necrosis Virus (TNV), and Tomato Spotted Wilt Virus (TSWV) (Othman and Shoman, 2004; Dunkić et al., 2010). In detail, Dunkić and coauthors (2010) identified carvacrol and thymol as the most abundant compounds in the essential oil able to reduce CMV local lesion on *Chenopodium amaranticolor* and *C. quinoa,* respectively. On the other hand, Othman and Shoman (2004) did not identify the oil component responsible for antiviral activities; anyway, the low content of plant proteins suggested the involvement of diterpenoids because of their antimicrobial and antiviral activities.

As other obligate plant pathogens, for example, liberibacters, viruses can induce changes in infected plants, which emit volatile blends enriched in monoterpenes, aldehydes, and sesquiterpenes compared to those of uninfected plants. These changes in volatile emissions also vary over the course of disease progression in ways that are conducive to transmission (Mann et al., 2012; Rajabaskar et al., 2014; Mauck et al., 2016).

16.5 Conclusion

The role of terpenoids in plant defenses is gaining in importance over time, mainly thanks to the availability of omics studies (in particular transcriptomics and metabolomics); however, the direct involvement of terpenes in the resistance mechanism is not always easily established. In some cases, in fact, terpenoids show a contradictory role in defense to pathogens. For example, a preformed unidentified terpenoid compound (T1) was associated with moderate resistance to *P. infestans* in tissues of a potato cultivar (Henriquez et al., 2012). However, the discovery that the accumulation of T1 occurred at high levels following inoculation with an aggressive *P. infestans* isolate, also in a susceptible potato cultivar, suggested that this compound alone is not able to inhibit infection by the pathogen but that more likely it could function in additive or synergistic ways with different plant secondary metabolites, such as phenolics and terpenoids, in resistance establishment (Henriquez et al., 2012). Analogously, the co-occurrence of the phytoalexin zealexins with a number of antimicrobial proteins, including chitinases and pathogenesis-related proteins, in transcript profiling of *F. graminearum*-infected maize tissues indicates a cooperative role of terpenoid phytoalexins in biochemical defense to fungi (Huffaker et al., 2011).

Terpenes and terpenoids show a direct role in the inhibition of plant pathogens and an indirect role in the plant defense mechanism through the interaction with the pathogen vectors or interruption of signal communications among pathogen cells that need to be better elucidated in the future. The identification and characterization of resistance traits within this context could help in obtaining less susceptible varieties to be cultivated worldwide for a more sustainable agriculture.

References

Ahujal, K.R., Bones, A.M., 2012. Phytoalexins in defense against pathogens. Trends Plant Sci. 17, 73–90.

Algarra Alarcon, A., Lazazzara, V., Cappellin, L., Bianchedi, P.L., Schuhmacher, R., Wohlfahrt, G., Pertot, I., Biasioli, F., Perazzolli, M., 2015. Emission of volatile sesquiterpenes and monoterpenes in grapevine genotypes following *Plasmopara viticola* inoculation *in vitro*. J. Mass Spectrom. 50, 1013–1022.

Aljbory, Z., Chen, M.S., 2018. Indirect plant defense against insect herbivores: a review. Insect Sci. 25, 2–23.

Allardyce, J.A., Rookes, J.E., Hussain, H.I., Cahill, D.M., 2013. Transcriptional profiling of *Zea mays* roots reveals roles for jasmonic acid and terpenoids in resistance against *Phytophthora cinnamomi*. Funct. Integr. Genomics 13, 217–228.

Alvarez-Castellanos, P.P., Bishop, C.D., Pascual-Villalobos, M.J., 2001. Antifungal activity of the essential oil of flowerheads of garland chrysanthemum (*Chrysanthemum coronarium*) against agricultural pathogens. Phytochemistry 57, 99–102.

Arimura, G., Matsui, K., Takabayashi, J., 2009. Chemical and molecular ecology of herbivore-induced plant volatiles: proximate factors and their ultimate functions. Plant Cell Physiol. 50, 911–923.

Bai, X.D., Correa, V.R., Toruno, T.Y., Ammar, E.D., Kamoun, S., Hogenhout, S.A., 2009. AY-WB phytoplasma secretes a protein that targets plant cell nuclei. Mol. Plant Microbe Interact. 22, 18–30.

Bertaccini, A., Duduk, B., Paltrinieri, S., Contaldo, N., 2014. Phytoplasmas and phytoplasma diseases: a severe threat to agriculture. Am. J. Plant Sci. 5, 1763–1788.

Bianco, P.A., Bulgari, D., Casati, P., Quaglino, F., 2011. Conventional and novel strategies for the phytoplasma diseases containment. Phytopathogenic Mollicutes 1, 77–82.

Block, A.K., Vaughan, M.M., Schmelz, E.A., Christensen, S.A., 2018. Biosynthesis and function of terpenoid defense compounds in maize (*Zea mays*). Planta. https://doi.org/10.1007/s00425-018-2999-2.

Brasier, C.M., 1990. China and the origins of Dutch elm disease: an appraisal. Plant Pathol. 39, 5–16.

Burt, S., 2004. Essential oils: their antibacterial properties and potential applications in foods—a review. Int. J. Food Microbiol. 94, 223–253.

Cates, R.G., 1996. The role of mixtures and variation in the production of terpenoids in conifer-insect-pathogen interactions. In: Romeo, J., Saunders, J.A., Barbosa, P. (Eds.),

Phytochemical Diversity and Redundancy in Ecological Interactions. Plenum Press, New York, pp. 179–216.

Chalal, M., Winkler, J.B., Gourrat, K., Trouvelot, S., Adrian, M., Schnitzler, J.-P., Jamois, F., Daire, X., 2015. Sesquiterpene volatile organic compounds (VOCs) are markers of elicitation by sulfated laminarine in grapevine. Front. Plant Sci. 6, 350.

Chen, F., Tholl, D., Bohlmann, J., Pichersky, E., 2011. The family of terpene synthases in plants: a mid-size family of genes for specialized metabolism that is highly diversified throughout the kingdom. Plant J. 66, 212–229.

Chen, X., Chen, H., Yuan, J.S., Köllner, T.G., Chen, Y., Guo, Y., Zhuang, X., Chen, X., Zhang, Y.J., Fu, J., Nebenführ, A., Guo, Z., Chen, F., 2018. The rice terpene synthase gene *OsTPS19* functions as an (*S*)-limonene synthase *in planta*, and its overexpression leads to enhanced resistance to the blast fungus *Magnaporthe oryzae*. Plant Biotechnol. J. 16, 1778–1787.

Cho, K.S., Lim, Y.R., Lee, K., Lee, J., Lee, J.H., Lee, I.S., 2017. Terpenes from forests and human health. Toxicol. Res. 33, 97–106.

Choi, Y.H., Casas Tapias, E., Kim, H.K., Lefeber, A.W.M., Erkelens, C., Verhoeven, J.T.J., Brzin, J., Zel, J., Verpoorte, R., 2004. Metabolic discrimination of *Catharanthus roseus* leaves infected by phytoplasma using 1H-NMR spectroscopy and multivariate data analysis. Plant Physiol. 135, 2398–2410.

Ćurković Perica, M., Lepeduš, H., Šeruga Musić, M., 2007. Effect of indole-3-butyricacid on phytoplasmas in infected *Catharanthus roseus* shoots grown *in vitro*. FEMS Microbiol. Lett. 286, 171–177.

Dean, R., van Kan, J.A.L., Pretorius, Z.A., KE, H.-.K., Di Pietro, A., Spanu, P.D., Rudd, J.J., Dickman, M., Kahmann, R., Ellis, J., Foster, G.D., 2012. The top 10 fungal pathogens in molecular plant pathology. Mol. Plant Pathol. 13, 414–430.

Ding, Y., Huffaker, A., Köllner, T.G., Weckwerth, P., Robert, C.A.M., Spencer, J.L., Lipka, A.E., Schmelz, E.A., 2017. Selinene volatiles are essential precursors for maize defense promoting fungal pathogen resistance. Plant Physiol. 175, 1455–1468.

Doehlemann, G., Ökmen, B., Zhu, W., Sharon, A., 2017. Plant Pathogenic Fungi. Microbiol. Spectr. 5 (1) FUNK-0023-2016.

Dudareva, N., Pichersky, E., Gershenzon, J., 2004. Biochemistry of plant volatiles. Plant Physiol. 135, 1893–1902.

Dunkić, V., Bezić, N., Vuko, E., Cukrov, D., 2010. Antiphytoviral activity of *Satureja montana* L. ssp. *variegata* (Host) P. W. Ball essential oil and phenol compounds on CMV and TMV. Molecules 15, 6713–6721.

Eberhard Breitmaier, E., 2006. Terpenes Flavors, Fragrances, Pharmaca, Pheromones. WILEY-VCH Verlag GmbH & Co. KGaA, Weinheim.

European Food Safety Authority (EFSA), 2012. Conclusion on the peer review of the pesticide risk assessment of the active substance eugenol. EFSA J. 10, 2914.

Fang, J., Jogaiah, S., Guan, L., Sun, X., Abdelrahman, M., 2018. Coloring biology in grape skin: a prospective strategy for molecular farming. Physiol. Plant. 164, 429–441.

Farag, M.A., Zhang, H.M., Ryu, C.M., 2013. Dynamic chemical communication between plants and bacteria through airborne signals: induced resistance by bacterial volatiles. J. Chem. Ecol. 39, 1007–1018.

Farjon, A., 2018. Conifers of the World. Kew Bull. 73, 8.

Feng, X., Hu, Y., Zheng, Y., Zhu, W., Li, K., Huang, C.-H., Ko, T.-P., Ren, F., Chan, H.-C., Nega, M., Bogue, S., López, D., Kolter, R., Götz, F., Guo, R.-T., Oldfield, E., 2014. Structural and functional analysis of *Bacillus subtilis* YisP reveal a role of its product in biofilm production. Chem. Biol. 21, 1557–1563.

Flemming, H.C., Wingender, J., Szewyk, U., Steinberg, P., Rice, S.A., Kjelleberg, S., 2016. Bacterial adhesion and biofilms on surfaces. Prog. Nat. Sci. 18, 1049–1056.
Gargallo-Garriga, A., Sardans, J., Pérez-Trujillo, M., Guenther, A., Llusià, J., Rico, L., Terradas, J., Farrè-Armengol, G., Filella, I., Parella, T., Peñuelas, J., 2016. Shifts in plant foliar and floral metabolomes in response to the suppression of the associated microbiota. BMC Plant Biol. 16, 78.
Gijzen, M., Lewinsohn, E., Savage, T.J., Croteau, R.B., 1993. Conifer monoterpenes: biochemistry and bark beetle chemical ecology. In: Bioactive Volatile Compounds from Plants. Symposium Series 525, American Chemical Society. Washington, DC.
Glazebrook, J., 2005. Contrasting mechanisms of defense against biotrophic and necrotrophic pathogens. Annu. Rev. Phytopathol. 43, 205–227.
Gutierrez-Pacheco, M.M., Gonzales-Aguilar, G.A., Martinez-Tellez, M.A., Lizardi-Mendoza, J., Madera-Santana, T.J., Bernal-Mercado, A.T., Vazquez-Armenta, F.J., Ayala-Zavala, J.F., 2018. Carvacrol inhibits biofilm formation and production of extracellular polymeric substances of *Pectobacterium carotovorum* subsp. *carotovorum*. Food Control 89, 210–218.
Hammerschmidt, R., 1999. Phytoalexins: what have we learned after 60 years? Annu. Rev. Phytopathol. 37, 285–306.
Harborne, J.B., 1999. Classes and functions of secondary products from plants. In: Walton, N.J., Brown, D.E. (Eds.), Chemicals from Plants. Imperial College Press, London, pp. 1–25.
Hasegawa, M., Mitsuhara, I., Seo, S., Imai, T., Koga, J., Okada, K., Yamane, H., Ohashi, Y., 2010. Phytoalexin accumulation in the interaction between rice and the blast fungus. Mol. Plant Microbe Interact. 23, 1000–1011.
Henriquez, M.A., Adam, L.R., Daayf, F., 2012. Alteration of secondary metabolites' profiles in potato leaves in response to weakly and highly aggressive isolates of *Phytophthora infestans*. Plant Physiol. Biochem. 57, 8–14.
Hofberger, J.A., Ramirez, A.M., van den Bergh, E., Zhu, X., Bouwmeester, H.J., Schuurink, R.C., Schranz, M.E., 2015. Large-scale evolutionary analysis of genes and supergene clusters from terpenoid modular pathways provides insights into metabolic diversification in flowering plants. PLoS One. 10, e0128808.
Hoshi, A., Oshima, K., Kakizawa, S., Ishii, Y., Ozeki, J., Hashimoto, M., Komatsu, K., Kagiwada, S., Yamaji, Y., Namba, S., 2009. A unique virulence factor for proliferation and dwarfism in plants identified from a phytopathogenic bacterium. Proc. Natl. Acad. Sci. U S A 106, 6416–6421.
Huffaker, A., Kaplan, F., Vaughan, M.M., Dafoe, N.J., Ni, X., Rocca, J.R., Alborn, H.T., Teal, P.E., Schmelz, E.A., 2011. Novel acidic sesquiterpenoids constitute a dominant class of pathogen-induced phytoalexins in maize. Plant Physiol. 156, 2082–2097.
Irmisch, S., Jiang, Y., Chen, F., Gershenzon, J., Köllner, T.G., 2014. Terpene synthases and their contribution to herbivore-induced volatile emission in western balsam poplar (*Populus trichocarpa*). BMC Plant Biol. 14, 270.
Jadhav, S.J., Mazza, G., Salunkhe, D.K., 1991. Terpenoid phytoalexins in potatoes: a review. Food Chem. 41, 195–217.
Jansen, D.J., Shenvi, R.A., 2014. Synthesis of medicinally relevant terpenes: reducing the cost and time of drug discovery. Future Med. Chem. 6, 1127–1148.
Jeandet, P., Clément, C., Courot, E., Cordelier, S., 2013. Modulation of phytoalexin biosynthesis in engineered plants for disease resistance. Int. J. Mol. Sci. 14, 14136–14170.
Jones, J.D., Dangl, J.L., 2006. The plant immune system. Nature 444, 323–329.

Joshi, J.R., Khazanov, N., Senderowitz, H., Burdman, S., Lipsky, A., Yedidia, I., 2016. Plant phenolic volatiles inhibit quorum sensing in pectobacteria and reduce their virulence by potential binding to ExpI and ExpR proteins. Sci. Rep. 6, 1–15.

Kamoun, S., Furzer, O., Jones, J.D.G., Judelson, H.S., Ali, G.S., Dalio, R.J.D., Roy, S.G., Schena, L., Zambounis, A., Panabières, F., Cahill, D., Ruocco, M., Figueiredo, A., Chen, X.-R., Hulvey, J., Stam, R., Lamour, K., Gijzen, M., Tyler, B.M., Grünwald, N.J., Mukhtar, M.S., Tomé, D.F.A., Tör, M., Van den Ackerveken, G., McDowell, J., Daayf, F., Fry, W.E., Lindqvist-Kreuze, H., Meijer, H.J.G., Petre, B., Ristaino, J., Yoshida, K., Birch, P.R.J., Govers, F., 2015. The top 10 oomycete pathogens in molecular plant pathology. Mol. Plant Pathol. 16, 413–434.

Keeling, C.I., Bohlmann, J., 2006. Genes, enzymes and chemicals of terpenoid diversity in the constitutive and induced defence of conifers against insects and pathogens. New Phytol. 70, 657–675.

Kopper, B.J., Illman, B.L., Kersten, P.J., Klepzdig, K.D., Raffa, K.F., 2005. Effects of diterpene acids on components of a conifer bark beetle–fungal interaction: tolerance by *Ips pini* and sensitivity by its associate *Ophiostoma ips*. Environ. Entomol. 34, 486–493.

Kuc, J., 1995. Phytoalexins, stress, metabolism and disease resistance in plants. Annu. Rev. Phytopathol. 33, 275–297.

Kuzuyama, T., 2017. Biosynthetic studies on terpenoids produced by *Streptomyces*. J. Antibiot. 70, 811–818.

Li, W., Liu, Y., Wang, J., HeM, Z.X., Yang, C., Yuan, C., Wang, J., Chern, M., Yin, J., Chen, W., Ma, B., Wang, Y., Qin, P., Li, S., Ronald, P., Chen, X., 2016. The durably resistant rice cultivar Digu activates defence gene expression before the full maturation of *Magnaporthe oryzae* appressorium. Mol. Plant Pathol. 17, 354–368.

Maclean, A.M., Orlovskis, Z., Kowitwanich, K., Zdziarska, A.M., Angenent, G.C., Immink, R.G.H., Hogenhout, S.A., 2014. Phytoplasma effector SAP54 hijacks plant reproduction by degrading mads-box proteins and promotes insect colonization in a Rad23-dependent manner. PLoS Biol. 12, e1001835.

Mann, R.S., Ali, J.G., Hermann, S.L., Tiwari, S., Pelz-Stelinski, K.S., Alborn, H.T., Stelinski, L.L., 2012. Induced release of a plant-defensevolatile 'deceptively' attracts insect vectors to plants infected with a bacterial pathogen. PLoS Pathog. 8, e1002610.

Marcone, C., 2014. Molecular biology and pathogenicity of phytoplasmas. Ann. Appl. Biol. 165, 199–221.

Marcone, C., Bellardi, M.G., Bertaccini, A., 2016. Phytoplasma diseases of medicinal and aromatic plants. J. Plant Pathol. 98, 379–404.

Mauck, K.E., De Moraes, C.M., Mescher, M.C., 2016. Effects of pathogens on sensory-mediated interactionsbetween plants and insect vectors. Curr. Opin. Plant Biol. 32, 53–61.

Mayer, C.J., Vilcinskas, A., Gross, J., 2008. Phytopathogen lures its insect vector by altering host plant odor. J. Chem. Ecol. 34, 1045–1049.

Meyer, J., Berger, D.K., Christensen, S.A., Murray, S.L., 2017. RNA-Seq analysis of resistant and susceptible sub-tropical maize lines reveals a role for kauralexins in resistance to grey leaf spot disease, caused by *Cercospora zeina*. BMC Plant Biol. 17, 197.

Min, L., Han, Z., Xu, Y., Yao, L., 2013. In vitro and in vivo anti-Tobacco Mosaic Virus activities of essential oils and individual compounds. J. Microbiol. Biotechnol. 23, 771–778.

Moore, B.D., Andrew, R.L., Kulheim, C., Foley, W.J., 2014. Explaining intraspecific diversity in plant secondary metabolites in an ecological context. New Phytol. 201, 733–775.

Moretti, A., Logrieco, A.F., Susca, A., 2017. Mycotoxins: an underhand food problem. Methods Mol. Biol. 1542, 3–12.

Mosquera, T., Alvarez, M.F., Jiménez-Gómez, J.M., Muktar, M.S., Paulo, M.J., Steinemann, S., Li, J., Draffehn, A., Hofmann, A., Lübeck, J., Strahwald, J., Tacke, E., Hofferbert, H.-R.,

Walkemeier, B., Gebhardt, C., 2016. Targeted and untargeted approaches unravel novel candidate genes and diagnostic SNPs for quantitative resistance of the potato (*Solanum tuberosum* L.) to *Phytophthora infestans* causing the late blight disease. PLoS One. 11, e0156254.

Nuzhat, T., Vidyasagar, G.M., 2013. Antifungal investigations on plant essential oils. A review. Int. J. Pharm. Pharm. Sci. 5, 2–9.

Oldfield, E., Lin, F.Y., 2012. Terpene biosynthesis: modularity rules. Angew. Chem. 51, 1124–1137.

Osler, R., Carraro, L., Loi, N., Refatti, E., 1993. Symptom expression and disease occurrence of a yellows disease of grapevine in Northeastern Italy. Plant Dis. 77, 496–498.

Othman, B.A., Shoman, S.A., 2004. Antiphytoviral activity of the *Plectranthus tenuiflorus* on someimportant viruses. Int. J. Agric. Biol. 6, 844–849.

Paduch, R., Kandefer-Szerszen, M., Trytek, M., Fiedurek, J., 2007. Terpenes: substances useful in human healthcare. Arch. Immunol. Ther. Exp. (Warsz.) 55, 315–327.

Panstruga, R., Parker, J.E., Schulze-Lefert, P., 2009. SnapShot: plant immune response pathways. Cell 136, 978.e1–978.e3.

Park, S., Park, A.R., Im, S., Han, Y.-J., Lee, S., Back, K., Kim, J.-I., Kim, Y.S., 2014. Developmentally regulated sesquiterpene production confers resistance to *Colletotrichum gloeosporioides* in ripe pepper fruits. PLoS One. 9, e109453.

Paschold, A., Halitschke, R., Baldwin, I.T., 2006. Using 'mute' plants to translate volatile signals. Plant J. 45, 275–291.

Phillips, M.A., Croteau, R.B., 1999. Resin-based defenses in conifers. Trends Plant Sci. 4, 184–190.

Pichersky, E., Raguso, R.A., 2016. Why do plants produce so many terpenoid compounds? New Phytologist. https://doi.org/10.1111/nph.14178.

Pieterse, C.M.J., Dicke, M., 2007. Plant interactions with microbes and insects: from molecular mechanisms to ecology. Trends Plant Sci. 12, 564–569.

Prisic, S., Xu, M.M., Wilderman, P.R., Peters, R.J., 2004. Rice contains two disparate *ent*-copalyl diphosphate synthases with distinct metabolic functions. Plant Physiol. 136, 4228–4236.

Punelli, F., Al Hassan, M., Fileccia, V., Uva, P., Pasquini, G., Martinelli, F., 2016. A microarray analysis highlights the role of tetrapyrrole pathways in grapevine responses to "stolbur" phytoplasma, phloem virusinfections and recovered status. Physiol. Mol. Plant Pathol. 93, 129–137.

Quaglino, F., Zhao, Y., Casati, P., Bulgari, D., Bianco, P.A., Wei, W., Davis, R.E., 2013. '*Candidatus* Phytoplasma solani', a novel taxon associated with stolbur and bois noir related diseases of plants. Int. J. Syst. Evol. Microbiol. 63, 2879–2894.

Raffa, K.F., Berryman, A.A., 1983. The role of host plant resistance in the colonization behavior and ecology of bark beetles (*Coleoptera: Scolytidae*). Ecol. Monogr. 53, 27–49.

Raffa, K.F., Aukema, B.H., Erbilgin, N., Klepzig, K.D., Wallin, K.F., 2005. Interactions among conifer terpenoids and bark beetles across multiple levels of scale: an attempt to understand links between population patterns and physiological processes. Recent Adv. Phytochem. 39, 79–118.

Rajabaskar, D., Bosque-Pérez, N.A., Eigenbrode, S.D., 2014. Preference bya virus vector for infected plants is reversed after virus acquisition. Virus Res. 186, 32–37.

Rid, M., Mesca, C., Ayasse, M., Gross, J., 2016. Apple proliferation phytoplasma influences the pattern of plant volatiles emitted depending on pathogen virulence. Front. Ecol. Evol. 3, 152.

Salomon, M.A., Purpora, R., Bottini, R., Piccoli, P., 2016. Rhizosphere associated bacteria trigger accumulation of terpenes in leaves of *Vitis vinifera* L. cv. Malbec that protect cells against reactive oxygen species. Plant Physiol. Biochem. 106, 295–304.

Santos, R.B., Capanema, E.A., Balakshin, M.Y., Chang, H.-M., Jameel, H., 2011. Effect of hardwoods characteristics on kraft pulping process: emphasis on lignin structure. Bio-Resources 6, 3623–3637.

Scagel, R.F., Bandoni, R.J., Rouse, G.E., Schofield, W.B., Stein, J.R., Taylor, T.M.C., 1965. An Evolutionary Survey of the Plant Kingdom. Wadsworth, Belmont, CA.

Schilmiller, A.L., Last, R.L., Pichersky, E., 2008. Harnessing plant trichome biochemistry for the production of useful compounds. Plant J. 54, 702–711.

Schmelz, E.A., Kaplan, F., Huffaker, A., Dafoe, N.J., Vaughan, M.M., Ni, X., Rocca, J.R., Alborn, H.T., Teal, P.E., 2011. Identity, regulation, and activity of inducible diterpenoid phytoalexins in maize. Proc. Natl. Acad. Sci. U. S. A. 108, 5455–5460.

Schmelz, E.A., Huffaker, A., Sims, J.W., Christensen, S.A., Lu, X., Okada, K., Peters, R.J., 2014. Biosynthesis, elicitation and roles of monocot terpenoid phytoalexins. Plant J 79, 659–678.

Seemüller, E., Schneider, B., 2007. Differences in virulence and genomic features of strains of '*Candidatus* Phytoplasma mali', the apple proliferation agent. Phytopathology 97, 964–970.

Simionato, A.V.C., de Silva, D.S., Lambais, M.R., Carrilho, E., 2007. Characterization of a putative *Xylella fastidiosa* diffusible signal factor by HRGG-EI-MS. J. Mass Spectrom. 42, 490–496.

Singh, B., Sharma, R.A., 2015. Plant terpenes: defense responses, phylogenetic analysis, regulation and clinical applications. 3 Biotech 5, 129–151.

Srivastava, S., Pandey, R., Kumar, S., Nautiyal, C.S., 2014. Correspondence between flowers and leaves in terpenoid indole alkaloid metabolism of the phytoplasma-infected *Catharanthus roseus* plants. Protoplasma 251, 1307–1320.

Sugio, A., Kingdom, H.N., MacLean, A.M., Grieve, V.M., Hogenhout, S.A., 2011a. Phytoplasma protein effector SAP11 enhances insect vector reproduction by manipulating plant development and defense hormone biosynthesis. Proc. Natl Acad. Sci. U S A 108, E1254–E1263.

Sugio, A., MacLean, A.M., Kingdom, H.N., Grieve, V.M., Manimekalai, R., Hogenhout, S.A., 2011b. Diverse targets of phytoplasma effectors: from plant development to defense against insects. Annu. Rev. Phytopathol. 49, 175–195.

Sugio, A., Dubreuil, G., Giron, D., Simon, J.-C., 2015. Plant–insect interactions under bacterial influence: ecological implications and underlying mechanisms. J. Exp. Bot. 66, 467–478.

Sukanya, S.L., Sudisha, J., Prakash, H.S., Fathima, S.K., 2011. Isolation and characterization of antimicrobial compound from *Chromolaena odorata*. J. Phytology 3, 26–32.

Taniguchi, S., Miyoshi, S., Tamaoki, D., Yamada, S., Tanaka, K., Uji, Y., Tanaka, S., Akimitsu, K., Gomi, K., 2014. Isolation of jasmonate-induced sesquiterpene synthase of rice: product of which has an antifungal activity against *Magnaporthe oryzae*. J. Plant Physiol. 171, 625–632.

Tapia-Rodriguez, M.R., Hernandez-Mendoza, A., Gonzales-Aguilar, G.A., Martinez-Tellez, M.A., Martins, C.M., Ayala-Zavala, J.F., 2017. Carvacrol as potential quorum sensing inhibitor of *Pseudomonas aeruginosa* and biofilm production on stainless steel surfaces. Food Control 75, 255–261.

Toffolatti, S.L., De Lorenzis, G., Costa, A., Maddalena, G., Passera, A., Bonza, M.C., Pindo, M., Stefani, E., Cestaro, A., Casati, P., Failla, O., Bianco, P.A., Maghradze, D., Quaglino, F., 2018. Unique resistance traits against downy mildew from the center of origin of grapevine (*Vitis vinifera*). Sci. Rep. 8, 12523.

Tomiyama, K., Sakuma, T., Ishizaka, N., Sato, N., Takasugi, M., Katsui, T., 1968. A new antifungal substance isolated from potato tuber tissue infected by pathogens. Phytopathology 58, 115–116.

Tsao, R., Zhou, T., 2000. Antifungal Activity of Monoterpenoids against Postharvest Pathogens *Botrytis cinerea* and *Monilinia fructicola*. J. Essent. Oil Res. 12, 113–121.

von Bodman, S.B., Bauer, W.D., Colpin, D.L., 2003. Quorum sensing in plant-pathogenic bacteria. Annu. Rev. Phytopathol. 41, 455–482.

War, A.R., Sharma, H.C., Paulraj, M.G., War, M.Y., Ignacimuthu, S., 2011. Herbivore induced plant volatiles: their role in plant defense for pest management. Plant Signal. Behav. 6, 1973–1978.

Weintraub, P.G., Beanland, L., 2006. Insect vectors of phytoplasmas. Annu. Rev. Entomol. 51, 91–111.

Widmark, A.-K., 2010. *Phytophthora infestans*-interaction with the potato plant and inoculum sources. Doctoral Thesis, Swedish University of Agricultural Sciences, Uppsala, p. 67.

Yoon, S.-J., Sukweenadhi, J., Khorolragchaa, A., Mathiyalagan, R., Subramaniyam, S., Kim, Y.-J., Kim, H.-B., Kim, M.-J., Kim, Y.-J., 2016. Overexpression of *Panax ginseng* sesquiterpene synthase gene confers tolerance against *Pseudomonas syringae* pv. *tomato* in *Arabidopsis thaliana*. Physiol. Mol. Biol. Plants 22, 485–495.

Role of phenols and polyphenols in plant defense response to biotic and abiotic stresses

17

Palistha Tuladhar, Santanu Sasidharan, and Prakash Saudagar
Department of Biotechnology, National Institute of Technology, Warangal, Telangana, India

17.1 Introduction

Plants are frequently exposed to a wide range of stresses, both biotic and abiotic. The study of the stresses and the defense mechanism employed by the plants is highly being pursued in order to accommodate the increase in the stresses due to global warming (Di Ferdinando et al., 2014). Weeds, insect pests, fungi, and other microorganisms are considered to be biotic stress and the remaining physical and environmental conditions such as salinity, drought, ultraviolet radiation, extreme temperatures, and heavy metals are known as abiotic stresses (Atkinson and Urwin, 2012). Plants are immobile and hence they are constantly exposed to these stresses, therefore, they require a highly efficient mechanism to combat them. Some studies have shown that these stresses reduce the overall crop production and crop yield by 50%–70%, respectively, which may affect the productivity and growth of the plant (Sharma et al., 2019a). These stresses cause a rapid change in the cellular redox homeostasis with the abundant generation of reactive oxygen species which impair the cell organelles and cause obstructions in the ROS-promoted pathways.

There are two types of metabolites produced by plants. Primary metabolites and secondary metabolites. These metabolites have different roles in various functions and systems in the plant (Sasidharan and Saudagar, 2019). Among these metabolites, a great proportion is devoted to its defense mechanisms adopted by plants to protect themselves from the plethora of stresses that they face frequently. Primary metabolites are produced by plants in order to promote growth, reproduction, and development (Dewick, 2002). However, secondary metabolites are produced by different other pathways that are not necessarily considered essential for the functioning of the plant, but they play a huge role in the survival of the plant in their environment. Few examples of the primary metabolites are vitamins, hormones, and carbohydrates, organic and inorganic amino acids. The secondary metabolites include compounds such as terpenes, steroids, saponins, lignins, coumarins, etc. (Wink, 1999).

Phenols and polyphenols are secondary metabolites that comprise a vast portion of the secondary metabolites repertoire of plants. Polyphenols are very commonly present in the plant kingdom and are mainly found in the bark, fruits, and leaves with concentrations as high as 50% of the dry (Pandey and Rizvi, 2009). They are involved

Biocontrol Agents and Secondary Metabolites. https://doi.org/10.1016/B978-0-12-822919-4.00017-X

in the defense mechanism of the plant system, however, it has been found that they perform multiple functions such as cell division, photosynthetic activity, reproduction, hormonal regulation, and nutrient mineralization, when there are drastic environmental constraints (Sharma et al., 2019a). They also have dynamic antioxidant properties which neutralize the consequences produced by oxidative stress. Few phenolic compounds also display an ability to chelate heavy metal ions. Phenolic compounds such as the phytoalexins have antifungal and antibiotic activity whereas, other phenolic compounds such as coumarins and tannins reduce stress on plants by repelling herbivores (Lattanzio, 2013). Flavonoids are one of the largest categories of secondary metabolites in plants. Phenolic compounds are typically distributed into two groups, simple phenols, and complex phenol derivatives which usually contain many aromatic rings attached together. A detailed classification of phenols and their derivatives is given below.

17.2 Phenols and polyphenols in crops

Phenols and polyphenols are the secondary metabolites produced by plants which play a vital role in resistance produced by plants. This group contains at least one aromatic ring which is bonded to one or more hydroxyl groups and is generally formed from the amino acid, phenylalanine which gets converted into cinnamic acid (Neish, 1960). They are involved in the regulation of seed germination and also helps in the regulation of the growth of plants, along with active participation in the defense responses produced by the plants as a result of stresses faced during various biotic and abiotic conditions. One of the key features possessed by these phenolic compounds is the antioxidant activity which is largely attributed to its chemical structure. Furthermore, they may exhibit chelating properties also (Kaurinovic and Vastag, 2019). These diverse functions and its necessity for plant metabolism makes it one of the most widely studied groups of substances produced by the plants. It is one of the largest and the varied groups among the active plant substances.

17.2.1 Classification of phenols and polyphenols

The categorization of phenols is based on complexity. There are simple phenols with aromatic rings whereas polyphenols, which are complex derivatives of phenols, have several aromatic rings attached to each other (Kubalt, 2016). The number and the characteristics are the determinants of their distinct chemical, biological, and physical properties of the members in each category. The structures of the major types of phenolic compounds have been shown in Table 17.1. The section below discusses the classification of phenols based on their structure and biological function.

Table 17.1 Major classification of phenolic compounds identified in plants.

Number of carbon atoms	Structural skeleton	Category of phenolic compound
6	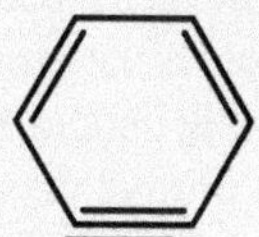	Basic phenol and benzoquinones
7	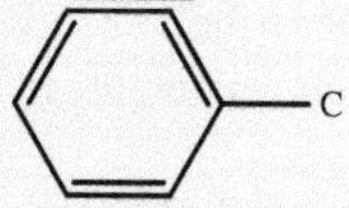	Phenolic acids
8	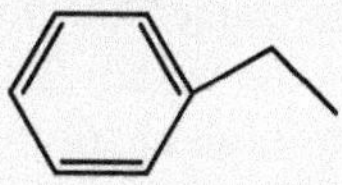	Tyrosine derivatives; phenylacetic acids; acetophenones
9	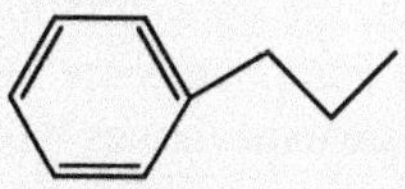	Isocoumarins; chromones; coumarins; phenylpropenes; hydroxycinnamic acids
10	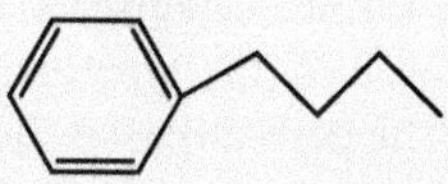	Naphthoquinones
13	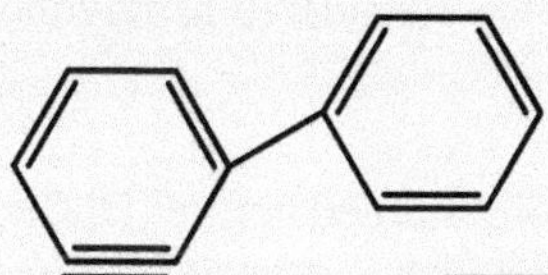	Xanthones
14	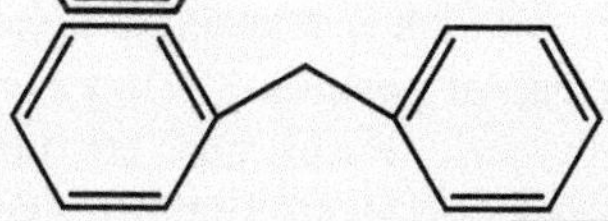	Anthraquinones; stilbenes
15	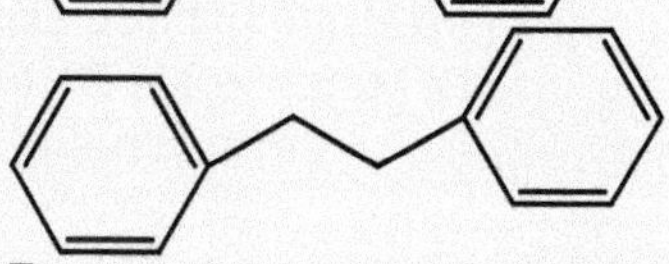	Flavonoids; isoflavonoids
18	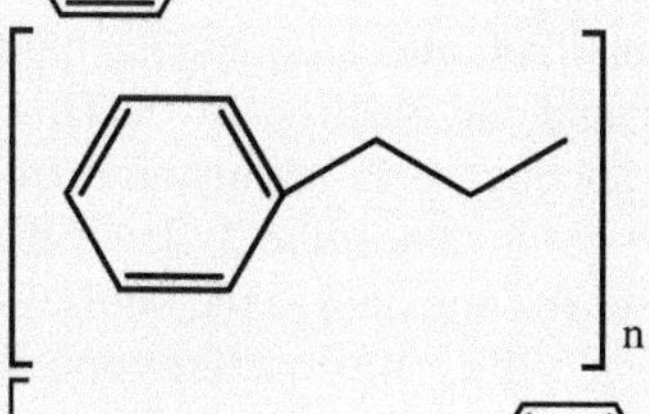	Neolignans; lignans
30	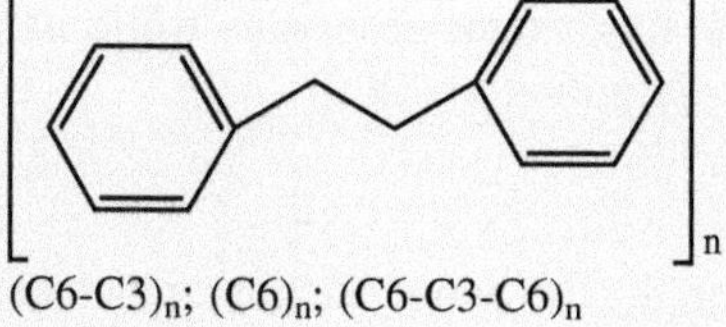	Biflavonoids
n	$(C6\text{-}C3)_n$; $(C6)_n$; $(C6\text{-}C3\text{-}C6)_n$	Lignins; catechol; melanins; flavolans

17.2.1.1 *Flavonoids*

They are the derivatives of simple phenols and are considered to be one of the biggest groups of secondary metabolites produced by plants. As with all the phenolic compounds, flavonoids are biosynthesized by the phenylpropanoid pathway which is followed by the shikimic acid pathway and the malonic acid pathway (Kubalt, 2016). The structure of flavonoid consists of a 15-carbon based skeleton with two aromatic rings joined by three-carbon-bridge. Its construction is generally based on the flavone skeleton. Few of the flavonoids have the ability to protect the plants from the stress produced due to the presence of heavy metals due to their ability to chelate transition metals such as Fe, Cu, Ni, and Zn, which generate hydroxyl radical (Michalak, 2006).

Around 6000 flavonoid compounds have been recognized till now and most of them are known to have antimicrobial properties (Ferrer et al., 2008). Flavonoids are well known for their ability to absorb radiation of high energy when the plants are exposed to UV light. On exposure to UV light, the concentration of flavonoids was observed to have increased in the plants. The compounds are known to be collected in vacuoles of the epithelial cells, in order to save the tissues lying beneath from the dangerous UV radiation. Flavonoids can be further categorized into chalcones, flavandiols, flavonols, and proanthocyanidins based on the type and the number of the substituents (Falcone Ferreyra et al., 2012). The derivatives of these compounds, anthocyanidins and condensed tannins are also considered as flavonoids.

Catechins

This group of flavan-3-ols has been extensively studied before (Di Chen et al., 2011). The structure contains a benzopyran skeleton which is attached to a phenyl group, that is, bound at position-2 and a hydroxyl group at position-3. They can exist in three forms, monomeric, oligomeric, and polymeric forms and are not glycosylated (Fantini et al., 2015). Catechins are strong antioxidants and also exhibit the ability to induce H_2O_2 molecules (Ramos, 2008; Brglez Mojzer et al., 2016).

Flavonols, flavones, and flavanones

Flavonols are a diverse group with around 200 quercetin and kaempferol glycosides and 380 flavonol glycosides (Weichselbaum and Buttriss, 2010). This group of flavonoids with flavones, flavonols, and flavanones are all mainly antioxidative in nature. But, because of the absence of hydroxyl group at position 3, flavones and flavanones tend to have less antioxidant properties when compared to flavonols whereas, the double bonds at the position 2 and 3 bond results in a more reactive compound and for this very reason, apigenin has moderate antioxidant properties while naringenin has no activity against the superoxide ion (Tripoli et al., 2007).

Isoflavones and chalcones

Isoflavones are the phytochemicals in the flavonoid category which are also known as "phytoestrogens" due to their estrogen-like effect. Isoflavones are predominately found in legumes, to be more specific, they are isolated from soy products. Another important category of flavonoids is the chalcones. They also act as the metabolic precursors of certain isoflavonoid and flavonoid compounds (Brglez Mojzer et al., 2016).

Anthocyanidins

Anthocyanidins are the universal colorants and they represent the highest percentage of bioflavonoid (Konczak and Zhang, 2004). They are naturally present as glycosides called anthocyanins. More than 600 anthocyanidins have been recognized to date. They mostly perform the function of providing colors to the plant parts such as red, blue, and purple shades (Asensi et al., 2011). The color that they provide have been understood to be pH-dependent (Dai and Mumper, 2010). Anthocyanins may be present as different chemicals in plant tissues which makes them unique amidst plant phenolics (Brglez Mojzer et al., 2016).

17.2.1.2 Coumarins

The cyclization of coumaric acid, which is produced as one of the products of the shikimic acid pathway, results in an important compound, coumarins. Coumaric acid is produced as the result of deamination of tyrosine by tyrosine-ammonia lyase enzyme or by hydroxylation of cinnamic acid. Coumarins have been known to be toxic for the herbivores (Kubalt, 2016). One of the coumarins, dicumarol exhibits anticoagulant properties and because of its bitterness property, it also repels the animals from consuming the plants. These compounds are present in high concentrations making it very repulsive for the animals. Few of the compounds have also been reported to be carcinogenic and hepatotoxic in nature (Lake, 1999).

17.2.1.3 Phytoalexins

Phytoalexins are commonly found to possess protective properties. They belong to the flavonoid and isoflavonoid family of the phenolic compounds. Some of them have structures similar to that of stilbenes or terpenoids. They are responsible for the protection against infection by nematodes and microorganisms like bacteria, fungi, and viruses. They act similar to antibiotics by expressing bacteriostatic properties, at times even bactericidal. These compounds can also inhibit sporulation, hyphal growth, and spore germination of phytopathogenic fungi. The antimicrobial efficiency of phytoalexins is dependent on their concentration and the rate of synthesis in the plant tissues. They are produced de novo also in response to various abiotic stresses such as heavy metals, UV radiation, and mechanical damage. Flavonoid phytoalexins are commonly found in Fabaceae and the terpenoidphytoalexins are characteristic for Solanaceae (Kubalt, 2016). Although they are common for microbial protection, some pathogens are found to be resistant to them. These pathogens have the ability to demethylate or hydroxylate the aromatic rings in the phytoalexins and hence inactivate them. These compounds are also more susceptible to oxidation and are soluble in water (Kubalt, 2016).

17.2.1.4 Tannins

Tannins protect plants from herbivores. Unlike the other polyphenols, they have an intermediate to high molecular weight, which contains oligomers and polymers of other polyphenols (Bravo, 1998). Tannins have two or three phenolic groups in their structure and are highly hydroxylated. They can be divided into two types: easily hydrolyzable tannins and condensed tannins. Hydrolyzable tannins are produced by the polymerization of gallic acid or other phenolic acids and some sugars which contain a central core of glucose or another polyol whereas the condensed tannins are produced by a combination of multiple flavonoid units through an inter-flavan carbon bond. The bark of the trees and the leaves have the highest concentration of tannins. The herbivores-repulsive property is brought about by their signature unpleasant and bitter taste. They are capable of binding to and denaturing proteins, which also makes them toxic in nature. Due to the mentioned properties of tannins, they also provide protection from a variety of insects (Kubalt, 2016).

17.2.1.5 Other polyphenols

Phenolic acid

Phenolic acids represent one-third of the total phenolic compounds consumed. They can be divided into two classes: hydroxycinnamic acids and hydroxybenzoic acids along with their derivatives. Phenolic acids are highly present in leguminous plants. They have antioxidant properties and also function as metal chelators and scavengers of free radicals, especially hydroxyl, peroxyl, peroxynitrites, and superoxide anion radicals (Brglez Mojzer et al., 2016). In hydroxybenzoic acids, gallic and ellagic acids comprise the major proportion and are the precursors of tannins. Ellagic acid is commonly found in flowering plant families. It is present as a part of hydrolyzable tannins or in glucoside form. The product of the shikimic acid pathway, cinnamic acid, is the precursor of hydroxycinnamic acids. The most abundantly present hydrocinnamic acids are ferulic acids and caffeic acid (Brglez Mojzer et al., 2016).

Stilbenes and lignans

The stilbene class belongs to phytoalexins. Pterostilbene, piceatannol, and resveratrol are the most important representatives of stilbenes. They are produced in the plant as a response to fungal infection or any other environmental stress (Brglez Mojzer et al., 2016). Lignans are dimeric in nature and contain a 2,3-dibenzylbutane structure which is formed as a result of dimerization of two sinapyl or coniferyl alcohol units. They are derivatives of cinnamic acid. They exist in less quantity in plants and is prominently used for the formation of plant cell wall formation (Brglez Mojzer et al., 2016).

17.2.2 Biosynthesis of phenols and polyphenols

As we saw from the classification, phenols vary from simple aromatic rings like flavones to highly complex compounds like lignin. For all these phenol compounds, phenylalanine is the common precursor and hence they are also known as phenylpropanoids (Nanda et al., 2016). The phenylpropanoid pathway present in the plant is present in Fig. 17.1. Polyphenols have multiples of these phenolic structural units. The aromatic ring of these phenolic compounds is synthesized through the shikimic acid pathway from the precursor amino acid. Initially, the erythrose 4-phosphate combines with phosphoenolpyruvate to form the amino acid, phenylalanine (Santos-Sánchez et al., 2019). With phenylalanine ammonia lyase (PAL) as the enzyme, in the initial step, a simple fragmentation of the amine group from the amino acid, phenylalanine, converts it into trans-cinnamic acid. Numerous other phenolic compounds such as flavonoids, monolignols, lignins, lignans, coumarins, and hydrolyzable tannins are synthesized through this pathway. This cinnamic acid is also used as the initial compound for the biosynthesis of many other phenolic compounds (Sharma et al., 2019a). Under stressful conditions, the biosynthesis of the phenolics is dependent on the enzymes, like PAL and CHS. Enhanced performance of these phenolics is often associated with the increase in the transcript levels of the genes coding for key enzymes such as C4H, 4CL, PAL, CHI, CHS, DFR, IFR, IFS, etc. (Sharma et al., 2019a).

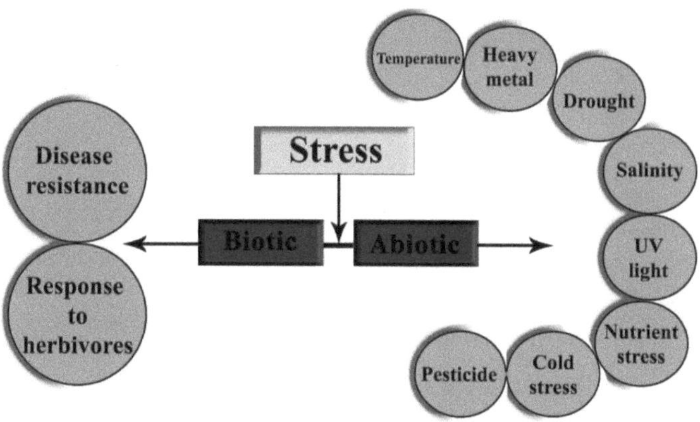

Fig. 17.1 The figure above describes the biosynthesis of three monolignols and anthocyanin. ρ-Coumaroyl CoA is the node where the metabolic routes branch into anthocyanin synthesis or lignin production. Enzymes in each step: *AGT*, anthocyanin glycosyltransferase; *ANS*, anthocyanidin synthase; *C3H*, ρ-coumarate 3-hydroxylase; *C4H*, cinnamic acid 4-hydroxylase; *4CL*, ρ-coumaroyl-CoA synthase; *CAD*, cinnamyl alcohol dehydrogenase; *CCoAOMT*, caffeoyl-CoA 3-*O*-methyltransferase; *CCR*, cinnamoyl-CoA reductase; *CHI*, chalcone-flavanoneisomerase; *CHS*, chalcone synthase; *COMT*, caffeic acid *O*-methyltransferase; *DRF*, dihydroflavonol 4-reductase; *F3H*, flavanone 3-hydroxylase; *F3′H*, flavanone 3′-hydroxylase; *HCT*, hydroxycinnamoyl-CoA: skimimate/quinatehydroxycinnamoyltransferase; *PAL*, phenylalanine ammonia lyase.

17.3 Systemic protection toward biotic and abiotic stresses

When a plant is subjected to a wide variety of stresses, a series of biochemical modifications occur in plants (Gull et al., 2019). These defense mechanisms are activated in the span of a very short time or moreover, spontaneous and as a response to the hazardous agents. Among these reactions, the increase in the concentration of the reactive oxygen species is one of the first lines of reaction to be activated which ultimately damages the cell organelles. At normal times, a threshold level of ROS has been found to be essential for the functioning of the plants, however, an increase beyond the threshold takes place when plants face various stresses which leads to peroxidation and destabilization of cellular membranes (Sharma et al., 2019a; Suzuki et al., 2012). Chloroplasts are the chief sites for ROS production in plants (Asada, 2006). The increase in ROS also disrupts the CO_2 fixation in chloroplasts. These stresses tend to cause modification in physiological and biochemical systems of plants which decreases its growth and results in poor yield (Wani and Sah, 2014). This ROS production is caused by various biotic and abiotic factors. It is important to understand the type of stress and the phenols produced in response to the stress. Fig. 17.2 provides a concise representation of the abiotic and biotic factors discussed in this review. The section below gives an insight into the types of stresses faced by a plant.

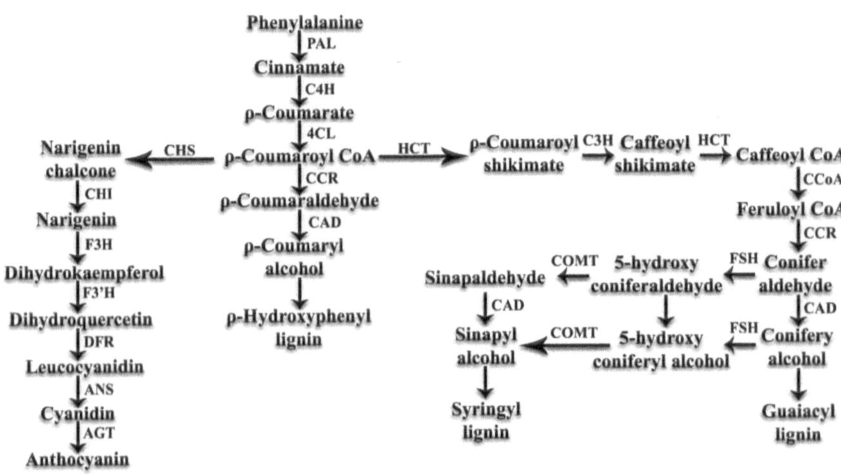

Fig. 17.2 The image above briefly illustrates the different types of biotic and abiotic stresses faced by plants.

17.3.1 Biotic stress

17.3.1.1 Disease resistance

Many plant phenols, the secondary metabolites produced by plants, are produced in increased amounts and act as phytoanticipins, phytoalexins, and nematicides against a number of soil pathogens and phytophagous insect. Most of the polyphenols act negatively on these pathogens and microbes. In order to shield the plants from numerous disease-causing pathogens, the plants produce phytoalexins like hydroxycinnamate conjugates and hydroxycoumarins. Initially, there is an appearance of "reaction zones" in plant saps when they suffer an injury or an attack from a microorganisms (Kemp and Burden, 1986). This phenomenon has been observed in a variety of plants including soybean, cucumber, peach, wheat, tobacco, etc. (Chalker-Scott and Fuchigami, 2018). Phenolic compounds are synthesized when plant pattern recognition receptors identify potential pathogens by their conserved pathogen-associated molecular patterns which lead to the PAMP triggered immunity. Hence, the progression of infection is stopped before the pathogen is able to gain complete hold of the plant.

In some cases, it was observed that the plants gained immunity from a previous infection or wounding. In the study performed by Ostrofsky et al. (1984) on the beech plants, it was illustrated that the bark of the beech trees which suffered prior injuries has more accumulation of phenolic compounds when compared to the susceptible ones. Few other plants which are inherently resistant have higher phenolic concentration than the rest. For example, the roots of rot-resistant plants of passionfruit, pecan, and persimmon had a higher concentration of phenolic compounds when compared with the susceptible species such as peach, almond, and apple. Phenolic compounds such as p-coumaric acid, o-coumaric acid, ferulic acid, protocatechuic acids inhibit the fungal growth by a similar mechanism (Perradin et al., 1983). Other factors such as the concentration of the enzymes required for the biosynthesis of the phenolic compounds also might be responsible for these disease resistance as it has been observed that treatments that reduce such enzymes or inhibition of such enzymes resulted in lower resistance (Chalker-Scott and Fuchigami, 2018).

17.3.1.2 Response to herbivores

The phenolic compounds defend the plants from herbivores by attacking the gut flora of ruminants (Schultz et al., 1992). Unlike insects, which feed on a specific plant only, herbivores have the capacity to ingest a wide variety of plants (Bernays et al., 1989). Studies have been conducted in the mammals to see the effect of polyphenols as a defense mechanism for plants and it has been determined that deterrence seems to be a common reaction, however, the extent varies. Tannins can precipitate the proteins and may reduce the activity of many enzymes. Tannins, as we have seen in insects, act by reducing the availability of the protein in the diet, therefore, they are known as digestibility reducing compounds (Bernays et al., 1989). The condensed tannins can inhibit the extracellular enzymes produced by these bacteria and the presence of tannin also adds an unpleasant and bitter taste to the leaves of plants, eventually, repel the herbivores (Kubalt, 2016). The absorption of the phenolic products of digested hydrolyzable tannins can also have detrimental effects such as overloading and damage to the liver and kidneys in vertebrates which is a consequence of detoxification and excretory processes failing to keep up (Bernays et al., 1989).

17.3.2 Abiotic stress

17.3.2.1 Heavy metal

The presence of metals and minerals in excess or nonessentially can provoke a lot of stressful reactions in plants. Generally, metals are known to cause oxidative stress by promoting the generation of harmful ROSs which leads to toxic results and retarded growth in the plants (Reichman, 2002). However, the heightened synthesis of phenolic compounds during the stressful period aids in the protection of plants from the oxidative stresses. As an example, flavonoids can reduce the levels of hydroxyl radical which is known to be harmful to the plant cells by augmenting the metal chelation process. Hence, the concentration of flavonoids has been observed to be enhanced by the presence of excess metal (Kumar and Pandey, 2013). The concentration of specific flavonoids such as flavonols and anthocyanins, which play a vital role in aiding the plant's defense system, have also been found to be enhanced when plants are exposed to metal toxicity (Sharma et al., 2019a). Flavonoids also have the ability to scavenge H_2O_2 and are known to play an essential role in the phenolic/ascorbate-peroxidase cycle. This accumulation of the phenolics during heavy metal stress has been attributed to the increased biosynthesis of phenylpropanoid enzymes, which include phenylalanine ammonia lyase, cinnamyl alcohol dehydrogenase, polyphenol oxidase, shikimate dehydrogenase, and chalcone synthase (Deng and Lu, 2017). The increase in biosynthesis of phenylpropanoids relies on the variation of transcription levels of the genes which encode the enzymes when exposed to heavy metals.

Apart from phenols, polyphenols are polyfunctional, that contain many chelating o-dihydroxyphenyl groups in each of its molecules. This property of the polyphenols allows it to remove metals like iron from the other iron compounds with greater efficiency than the monofunctional low weight phenols (Cherrak et al., 2016). The effect of heavy metals has been studied in many plants. For instance, in *Brassicajuncea*, the effect of various heavy metals such as copper, chromium, and lead has been extensively studied. From these investigations, it has been observed that in the presence of chromium, there is an increase in the total contents of phenols, anthocyanins, and flavonoids along with the increase in expressions of CHS and PAL (Handa et al., 2019). Similarly in another study involving *Kandeliaobovate*, the presence of zinc and cadmium in the surroundings, upregulated the levels of phenolics and also increased the concentrations as well as the activities of the enzymes required for the metabolism of phenols such as cinnamyl alcohol dehydrogenase, polyphenol oxidase, and shikimate dehydrogenase (Chen et al., 2019a).

17.3.2.2 Drought

The upregulation of the phenolic compounds is extremely crucial for the protection against the harmful impacts of the drought conditions in the plants (Naikoo et al., 2019). In a study performed on *Arabidopsis* sp., it was confirmed that the increase in the accumulation of phenolic compounds such as flavonoids under the drought conditions was found to be highly helpful in order to provide resistance (Nakabayashi et al., 2014). Accumulation and the biosynthesis of the flavonols were enhanced under the water deficiency which was also concluded to have provided the resistance against the drought conditions (Kirakosyan et al., 2003). Drought has also been shown to cause the activation of biosynthetic pathways of flavonoids and phenolic acids which results in their accumulation (Gharibi et al., 2019; Li et al., 2018). Once their concentration is increased enough, they act as antioxidants and protect plants from the severe effects of the stress caused by water paucity (Nichols et al., 2015). Another investigation on the tomato plant established that the presence of flavonoids such as quercetin and kaempferol enhanced drought tolerance (Sánchez-Rodríguez et al., 2011). When flavonoids are accumulated in the cytoplasm of a plant cell, they are able to detoxify the deleterious H_2O_2 molecules which are generated as a consequence of water scarcity. Also, after the complete oxidation of flavonoids, ascorbic acid facilitates the reconversion of flavonoids into primary metabolites (Hernández et al., 2009). During droughts, an increase in the concentration of phenolic compounds take place because of the modulation of the phenylpropanoid pathway. The drought stress regulates many major genes that encode the enzymes required for the phenylpropanoid pathway and this, in turn, stimulates the synthesis of phenolic compounds (Sharma et al., 2019a). Research conducted on *Chrysanthemum morifolium* also concluded that there was an increase in the concentration of flavonoids and flavonols during water scarcity conditions. Also, the expression of the enzymes involved in the synthesis of the phenolic compounds, PAL, F3H, and CHI was increased when in drought conditions (Hodaei et al., 2018).

17.3.2.3 Salinity

The stress caused by excess salinity requires activation of well-organized and fine-tuned antioxidant systems in the plant to counteract the generation of ROS like superoxide anions, hydroxyl ions, and hydrogen peroxide when under the salt stress (Taïbi et al., 2016; Martinez et al., 2016). Phenolic compounds are equipped with powerful antioxidant properties and are known to help in scavenging the detrimental ROS in plants (Chen et al., 2019b; Valifard et al., 2014). Due to the salt stress, the phenylpropanoid pathway gets stimulated and produces a range of phenolic compounds which possess a strong antioxidative potential (Bistgani et al., 2019).

Genes like VvbHLH1 are associated with the enhancement of the production of flavonoids by monitoring the genes in the biosynthetic pathways and also provide the tolerance to salt stress (Golkar and Taghizadeh, 2018). The biosynthesis of flavones was upregulated when under salt stress, also the genes which encode for flavone synthase, GmFNSII-1 and GmFNSII-2 were highly transcribed (Yan et al., 2014).

The NtCHS1 gene controls the biosynthesis of flavonoids under saline conditions in tobacco plants. These flavonoids are directly involved with the scavenging of ROS (Chen et al., 2019b). Some other phenolic compounds which are accumulated in plants under the increased salinity are cinnamylmalic acid, caffeic acid, gallic acid, caftaric acid, ferulic acid, anthocyanins, and vanillic acid (Sharma et al., 2019a). In *Amaranthus tricolor*, there was an increase in the total phenolics, hydroxybenzoic acids, flavonoids and hydroxycinnamic acids (Sarker and Oba, 2018). In a study conducted on the *Thymus* sp., it was found that there was an increase in the concentration of phenolics like quercetin, gallic acid, caffeic acid, cinnamic acid, syringic acid, apigenin, luteolin, rosemarnic acid, naringenin, rutin, and vanillic acid (Bistgani et al., 2019).

17.3.2.4 Ultraviolet rays

Light stress is one of the common stresses faced by plants in Mediterranean, tropical, and subtropical regions. Most of the polyphenols, which include flavonoids and hydroxycinnamates [absorbers of both UV-A (320–400 nm) and UV-B (280–320 nm)], have the ability to absorb the short solar wavelength. This is further supported by the fact that the UV light stress induces the accumulation of phenolics. Ellagitannins and poly-galloyl derivatives are also suspected to absorb a wide range of solar wavelengths including the UV-B radiation (Di Ferdinando et al., 2014).

The constant exposure to UV-B radiations causes protein damage and mutations in the DNA of the plant cells and generates the injurious ROS. The phenolics, which are produced to counteract the effects of UV-B, do so by establishing a shield under the epidermal layer (Sharma et al., 2019a). Moreover, the phenolic compounds reduce the damage caused by UV-B on the DNA as they can prevent the dimerization of thymine and also reduce the photodamage by NAD/NADP (Naikoo et al., 2019). Flavonoids, in particular, protect the plants from these harmful radiations by acting as light screens because of their ability to absorb both UV radiations and visible radiation (Ramawat and Mérillon, 2013). This information was reinforced by the studies where the researchers observed an enhanced biosynthesis of flavonoids in the plants under UV radiations along with an increased UV absorption and tolerance to the radiations (Agati and Tattini, 2010). Also, it was seen that the antioxidant capacity of the plants was also increased (Agati et al., 2012). The central genes that were transcripted increasingly in case of exposure to UV radiation includes CHI (chalconeisomerase); PAL (phenylalanine ammonia lyase); CHS (chalcone synthase); DFR (dihydroflavonol 4-reductase); FLS (flavonol synthase); FGT (flavonoid glycosyltransferases); and FHT (flavanone 3-hydroxylase). UV light has also been found to use the jasmonate-dependent and independent pathways in order to trigger the synthesis of different phenols in plants (Demkura et al., 2010). Moreover, abscisic acid also regulates the phenolic biosynthetic pathways in the presence of UV radiation (Berli et al., 2011). In *Arbutus unedo* it was found that there was an increase in compounds like thogallin, juglanin, and avicularin which belong to the phenolic compounds (Nenadis et al., 2015). In another plant, *Tricticumaestivum*, there was an increase in the accumulation of all the phenolic compounds including ferulic acid, vanillic acid, and p-hydroxybenzoic acid. Also, alterations were observed in the transcript levels of enzymes C4H, 4CL, PAL, and COMT (Chen et al., 2019c).

17.3.2.5 Cold stress

When plants are exposed to nonfreezing cold temperatures, the accumulation of phenolic compounds has been observed to have increased. Phenolic metabolism has been observed at the cold temperatures at which the chilling injuries take place. Additionally, these cold temperatures also result in the stimulation of enzymes such as PAL which are vital in phenolic metabolism. This increase has been observed when plants of apples, potatoes, and *Pinus* were exposed to cold climates (Chalker-Scott and Fuchigami, 2018; Naikoo et al., 2019). One of the prominent changes observed in the plants due to cold stress is the increase in the concentration of suberin and lignin which could protect the plants from the adverse conditions by strengthening their cell walls. Apple trees, grapevines, and sugarcanes have also been found to adopt lignification as a frost tolerance mechanism (Chalker-Scott and Fuchigami, 2018).

17.3.2.6 Nutrient stress

Like all living organisms, a constant supply of nutrients is crucial for the proper growth and development of a plant. The nutrients are gathered from the environment and from the soil after mineralization by the soil microbes (Naikoo et al., 2019). Many factors such as decomposer organisms, substrate quality, and climate, influence the regulation of nutrients. Polyphenols are involved in the supply and flow of organic and inorganic soil nutrients for the plants. The composition and activity of decomposers are controlled by polyphenols which in turn affects the rates of decomposition and nutrient cycling. The concentration of phenolic compounds is increased when there is a deficiency of nutrients like sulfur, potassium, phosphorous, and nitrogen (Gershenzon, 1984). In one of the studies, it has been reported that the concentration of anthocyanins was increased in *Vitisvinifera* due to the osmotic stress exerted by sucrose on the plant (Tuteja and Mahajan, 2007).

17.3.2.7 Other abiotic factors

Apart from the stresses discussed above, there are other abiotic factors such as temperature, pesticides, and nanoparticles which can also encourage the biosynthesis of plant phenols and help them by providing resistance against various phytotoxic effects produced by these abiotic stresses (Ghasemi et al., 2019). When the phenolic compounds are increased as a result of pesticide stress, they provide resistance from the pesticide toxicity by phenol production, for the plant to be able to survive (Sharma et al., 2016). This upregulation of the synthesis of phenolic compounds is due to the activation of the major enzymes required for the biosynthetic pathways and upregulation of the main genes in the phenylpropanoid pathway (Sharma et al., 2019b). On the other hand, the excessive application of insecticides on the plant increases the accumulation of anthocyanins, thereby, helping in the recovery of the plant's photosynthetic efficiency. *Oryza sativa* has been known to produce ferulic acid, phenylalanine, and p-hydroxybenzoic acid in higher concentrations when exposed to insecticides (Mahdavi et al., 2015). Nanoparticles of metals like titanium, copper, zinc, silver, and silicon are also known to cause an increase in the accumulation of phenolic compounds. For example, in the plant *Vitisvinifera*, the contents of total phenolics which included kaempferol derivatives, caftaric acid, and quercetin derivatives were relatively more when exposed to titanium nanoparticles than to the normal conditions (Sharma et al., 2019a).

Temperature is another vital factor that all the plants have to get acclimatized to, in general. But, plants tend to produce phenolic acids, flavonols, flavonoids, and anthocyanins in extreme temperatures to protect the plant cells (Ancillotti et al., 2015). In many research studies performed on the plant *Festucatrachyphylla*, it was observed that, under the heat stress, there was an enhancement in the synthesis of phenolic compounds like benzoic acid, 4-hydroxybenzoic acid, coumaric acid, caffeic acid, cinnamic acid, homovanillic acid, gallic acid, ferulic acid, vanillic acid, and salicylic acid. This increase in the accumulation of phenolic compounds resulted in the enhancement of heat tolerance in *F. trachyphylla* plants (Wang et al., 2019). Carrots, according to another study, produce phenolics such as anthocyanins, coumaric acid, and caffeic acid and accumulate them, in order to protect the plant from the oxidative damage induced by heat (Sharma et al., 2019a). Furthermore, phenolics like salicylic acid has also been reported to act as a stimulant for phenol synthesis in heat conditions. Because of this, there is an increased accumulation of the phenolic compounds which allows them to detoxify the excessive ROS and provide heat tolerance to the plants. When the conditions are adversely cold and chilly, there is an accumulation of phenolic compounds like lignin or suberin which protects the plant cells from the cold stress and prevents the collapse of cells and the injuries caused by the cold conditions (Sharma et al., 2019a). The effect of cold stress has been discussed earlier in Section 17.3.2.5.

Apart from the role of phenols in the protection of plants under stress conditions, phenols also play an important stand in plant growth. The following section discusses the diverse plant growth roles of phenols and polyphenols in detail.

17.4 Role of phenols and polyphenols in plant growth

As seen before in this chapter, there is a wide range of phenolics available which perform highly diverse functions involving key metabolic and physiological processes in the plants (Boudet, 2007; Kumar et al., 2019). They have been found to impact various physiological processes regarding the development and growth of plants including cell division, synthesis of photosynthetic pigments, and seed germination (Tanase et al., 2019). These compounds play vital roles in many physiological processes involving the improvement of adaptability and tolerance of plants under the suboptimal conditions (Andersen, 2003; Lattanzio et al., 2009; Dixon and Paiva, 1995). Particularly, the phenolics possessing the antioxidant properties have been found to enhance and maintain the plant performances under the stressful environment (Sharma et al., 2019a). Few of the metabolic roles of phenols are addressed below.

17.4.1 Physiological roles

Plants interact with their surrounding using phenolic compounds. Polyphenols are also critical for signal transduction from the root to the shoot and in nutrient mobilization as discussed earlier. The relationship with the symbionts is said to be maintained with the help of polyphenols. The soil microbes alter them into the substances which aid in the humus formation and nitrogen mineralization (Halvorson et al., 2009). Moreover, they also enhance the uptake of nutrients by chelating the metallic ions, modulate soil porosity, thereby, accelerating mobilization of elements like magnesium, potassium, iron, calcium, zinc, and manganese and enhance the active absorption sites (Seneviratne and Jayasinghearachchi, 2003).

Plant phenolics also act as the physiological regulator/chemical messengers which tend to inhibit the catabolism of IAA or limit the synthesis of IAA (Mathesius, 2001). Flavonoids have a crucial role in the development of functional pollens. It has been shown through studies that the addition of small doses of aglycones, quercetin, flavonol, or kaempferol restores the fertility of mature pollens during the process of pollination (Taylor and Grotewold, 2005; Van Der Meer et al., 1992). Certain phenolics such as trans-cinnamic acid, p-hydroxybenzoic acid, benzoic acid, and coumarin might be phytotoxic when accumulated in huge quantities and may also inhibit the seedling growth and germination by disrupting the cellular functioning of the enzymes and impairing the cell division (Sharma et al., 2019a; Baleroni et al., 2000). Some phenolic compounds are known to inhibit the prolyl aminopeptidase and phosphatase enzyme which are involved in the seed germination of the legumes (Shankar et al., 2009). However, it has also been reported that the presence of phenolic compounds in high concentrations exerts a positive effect on seed germination (Chen et al., 2016). Further research is required in this regard for substantiating both the claims. In a focused study, wherein spruce bark which had polyphenols stimulated the germination rate of *Lycopersicon esculentum* but, it simultaneously inhibited the process of root elongation (Balas and Popa, 2008). Phenolics also increases the seed tegument porosity and the thickness which aids in water imbibition and in turn boost the germination rate (Tobe et al., 2001). Another interesting role of phenolic compounds is to reduce the energy required for the ion transfer by altering the structure of mitochondrial and thylakoid membranes (Moreland and Novitzky, 1987).

17.4.2 Symbiotic relation formation

The interaction of plants with the microorganisms in the soil and the surrounding is maintained by a wide range of phenolics as well as other chemicals produced by the roots of the plant (Bais et al., 2006). Phenolics have been found to play a crucial role in the nitrogen fixation by legumes. The secondary metabolites released by the roots of leguminous plants, mainly flavonoids, are essential for the synthesis of Nod factors and the formation of the infection column (Zhang et al., 2009). The collection of functions performed by the phenolic compounds in the plant rhizosphere are appropriately termed as "rhizosphere effect" (Andersen, 2003; Dakora, 2003; Bhattacharya et al., 2010). The type of phenolic compound released may differ according to the species of plant, time, location, and space. For monocotyledonous plants, the concentration of phenolic compounds in soil range from 2.1% to 4.4% whereas it ranges from 0.1% to 0.6% in the dicotyledons (Hartley and Harris, 1981). Phenolic compounds initiate redox reactions in the soil surrounding and it also selectively influence the growth of soil microbes which develop symbiotic relationships with the plants. The availability of essential phytonutrients, enzymatic activity, hormonal balance, and competition with the neighboring plants are all governed by the phenolic compounds (Kraus et al., 2003). As a result of these complex and ever-changing interactions, the chemical composition and the structure of soil vary depending on the identity and quantity of the phenolic compound produced by the plants. Because of these dynamic conditions, the composition of the microorganism species found in different root areas is also constantly altered and shaped. Furthermore, as phenolics seeps into the soil, they bind to the organic matter and then gets metabolized by the bacterial community in the soil (Kefeli et al., 2003). There are several mechanisms of gene regulations in *Agrobacterium* and *Rhizobium* that are modified by phenols and they have been summarized in Table 17.2.

17.4.2.1 Chemotaxis

For the soil microbial community, the phenolic compounds play a major role involving the chemotaxis of symbionts such as *Agrobacterium* and *Rhizobium,* including their means of obtaining the vital substrates and hosts. These are excellent

Table 17.2 Symbiotic relation formation between plants and bacterial species like *Agrobacterium* sp. and *Rhizobium* sp. and the genes regulated by the compounds.

Role	**Phenolic compounds**	**Factors regulated in *Agrobacterium* sp. and *Rhizobium* sp.**
Chemotaxis	Hydroxyacetosyringone; syringone; flavonoids; luteolin; vanillyl alcohol; p-hydroxybenzoic acid; naringenin; gallate	VirA; NodD
Gene inducer	Catechol; p-resorcylate; vanillin; gallic acid; pyrogallic acid; flavonoids, flavones; flavonols; genistein; daidzein; glycitin; apigenin	VirA; VirG; NodD
Detoxification	Flavonoids; quercetin; daidzein; genistein	VirH2; Enzymes cleaving C-ring
Quorum signaling and quenching	Flavonoids; opines; salicylic acid	CinR/CinI; RhiI; RhiA; RhiB; RhiC; RhiR; RaiI; RaiR; TraI; TraB-I; BisR; TraR, repABC; BisR; TraM; AttM; AttL; AttK

representations of plant-microbe interactions and signal transduction (Palmer et al., 2004). Different phenolic compounds generated by plants have a variety of substitutions that are responsible for the chemotactic movement of rhizobium or agrobacterium toward that of higher concentration of possible nutrients and lower levels of inhibitors. Many phenolic compounds such as 3,4-dihydroxybenzoic acid, vanillyl alcohol, acetosyringone, umbelliferone, and p-hydroxybenzoic acids provoke a strong chemotactic response in *Rhizobium leguminosarum*. Some rhizobium strains also respond to apigenin and luteolin (Abdelrahman et al., 2017).

17.4.2.2 Quorum sensing

Quorum sensing is an important mechanism by which the members of rhizobium observe their surroundings (Bjarnsholt and Givskov, 2007). It allows the bacteria to perform cell to cell communication and induces a fruitful lifestyle for both the maintenance and survival of symbiotic relationships in their environment (Joint et al., 2002). The cell density-dependent regulation of the gene expression allows the bacteria to coordinate and conduct processes which cannot be done by a single bacterium in quorum sensing. In many cases, quorum sensing helps the symbionts to synchronize with the phenolic signals. This helps them operate as multicellular organisms and establish a successful symbiosis (Bhattacharya et al., 2010).

17.4.2.3 Activation of nodulation and virulence genes

A variety of phenols produced by plants influence the microbial gene expression after chemotaxis takes place. These signals generated by are known as "xenognosins" or "host recognition factors" (Campbell et al., 2000). Phenolics, which function as a chemoattractant, act in the same manner in both pathogenic *Agrobacterium* sp. and symbiotic *Rhizobium* sp. They are also known to regulate the expression of *nod* and *vir* genes (Djordjevic et al., 1987). Flavonoids, isoflavonoids, flavones, and flavonols are found to be responsible for their nod inducing ability and also a variety of phenolics, especially, acetosyringone induces *vir* gene expression (Bekkara et al., 1998; Perret et al., 2000).

17.4.2.4 Detoxification

As phenolic compounds disrupt the structural integrity of bacterial membranes and also inhibit the bacterial enzymes crucial for the electron transfer system, they can be used as antimicrobial compounds (Hirsch et al., 2003). In the course of evolution, few bacteria have gained tolerance to these toxic compounds by degrading and utilizing the toxic phenolics as their carbon source (Dua et al., 2002). *Agrobacterium* and *Rhizobium*, both have the ability to remove the toxic flavonoids levels with the help of their hydrophobic efflux pumps (Burse et al., 2004). Even when *rhizobia* are not threatened by the plant-based phenolics, they have an inbuilt mechanism to detoxify other xenobiotics present in the rhizosphere. The ability of these microbes to utilize and convert the phenols into a nontoxic substance, which provides energy to them, is the way these organisms have gained natural tolerance to these compounds and allows them to flourish in the soil and inside the host plant (Bhattacharya et al., 2010).

17.5 Conclusion

The world of phenol and polyphenols is vast and exploration of these compounds and their function is still incomplete. The phenolic compound producing pathway, i.e., the phenylpropanoid pathway, is the most researched and evaluated pathway of the secondary metabolite production in the Plantae kingdom. These compounds are produced in both optimal and suboptimal conditions, however, their concentration is increased when plants face some sort of stress either of biotic origin or abiotic origin. This protective role that the polyphenols perform is attributed to their ability to scavenge ROS along with their ability to protect the plants from excessive light. They also act as information carriers and play a vital part in the initiation and integration of the symbionts with the plants. Because of the plethora of advantages provided by phenolic compounds, they have been proposed to serve as the alternatives to the chemicals used for control of pathogens in the agriculture sector. Also, due to these reasons, these compounds have been used in various fields including allelochemicals, bioremediation, antioxidants as food additives and promotion of plant growth and development (Bujor et al., 2015). Plant polyphenols also have the potential to be used as an indicator for various stresses that the plant could be facing. Because of their influence on the nutrient cycle, they could also be used to diagnose nutrient deficiencies before the appearance of the visible symptoms. Concluding, phenols and polyphenols does play an important role in the diagnosis and protection of plants against biotic and abiotic stresses as well as being vital for the various metabolic functions in the plant systems.

References

Abdelrahman, M., El-Sayed, M., Jogaiah, S., Burritt, D.J., Tran, L.-S.P., 2017. The "STAY-GREEN" trait and phytohormone signaling networks in plants under heat stress. Plant Cell Rep. 36, 1009–1025.

Agati, G., Tattini, M., 2010. Multiple functional roles of flavonoids in photoprotection. New Phytol. 186 (4), 786–793.

Agati, G., et al., 2012. Flavonoids as antioxidants in plants: location and functional significance. Plant Sci. 196, 67–76.

Ancillotti, C., et al., 2015. Changes in polyphenol and sugar concentrations in wild type and genetically modified *Nicotiana langsdorffii* Weinmann in response to water and heat stress. Plant Physiol. Biochem. 97, 52–61.

Andersen, C.P., 2003. Source–sink balance and carbon allocation below ground in plants exposed to ozone. New Phytol. 157 (2), 213–228.

Asada, K., 2006. Production and scavenging of reactive oxygen species in chloroplasts and their functions. Plant Physiol. 141 (2), 391–396.

Asensi, M., et al., 2011. Natural polyphenols in cancer therapy. Crit. Rev. Clin. Lab. Sci. 48 (5–6), 197–216.

Atkinson, N.J., Urwin, P.E., 2012. The interaction of plant biotic and abiotic stresses: from genes to the field. J. Exp. Bot. 63 (10), 3523–3543.

Bais, H.P., et al., 2006. The role of root exudates in rhizosphere interactions with plants and other organisms. Annu. Rev. Plant Biol. 57, 233–266.

Balas, A., Popa, V., 2008. Bioactive compounds extracted from *Picea abies* bark. In: Proceedings of 10th European Workshop on Lignocellulosics and Pulp, Stock-holm.

Baleroni, C., et al., 2000. Lipid accumulation during canola seed germination in response to cinnamic acid derivatives. Biol. Plant. 43 (2), 313–316.

Bekkara, F., et al., 1998. Distribution of phenolic compounds within seed and seedlings of two *Vicia faba* cvs differing in their seed tannin content, and study of their seed and root phenolic exudations. Plant Soil 203 (1), 27–36.

Berli, F.J., et al., 2011. Solar UV-B and ABA are involved in phenol metabolism of *Vitis vinifera* L. increasing biosynthesis of berry skin polyphenols. J. Agric. Food Chem. 59 (9), 4874–4884.

Bernays, E.A., Driver, G.C., Bilgener, M., 1989. Herbivores and plant tannins. In: Advances in Ecological Research. Elsevier, pp. 263–302.

Bhattacharya, A., Sood, P., Citovsky, V., 2010. The roles of plant phenolics in defence and communication during agrobacterium and rhizobium infection. Mol. Plant Pathol. 11 (5), 705–719.

Bistgani, Z.E., et al., 2019. Effect of salinity stress on the physiological characteristics, phenolic compounds and antioxidant activity of *Thymus vulgaris* L. and *Thymus daenensis* Celak. Ind. Crop. Prod. 135, 311–320.

Bjarnsholt, T., Givskov, M., 2007. Quorum-sensing blockade as a strategy for enhancing host defences against bacterial pathogens. Philos. Trans. Royal Soc. B Biol. Sci. 362 (1483), 1213–1222.

Boudet, A.-M., 2007. Evolution and current status of research in phenolic compounds. Phytochemistry 68 (22–24), 2722–2735.

Bravo, L., 1998. Polyphenols: chemistry, dietary sources, metabolism, and nutritional significance. Nutr. Rev. 56 (11), 317–333.

Brglez Mojzer, E., et al., 2016. Polyphenols: extraction methods, antioxidative action, bioavailability and anticarcinogenic effects. Molecules 21 (7), 901.

Bujor, O.-C., et al., 2015. Biorefining to recover aromatic compounds with biological properties. TAPPI J. 14 (3), 187–193.

Burse, A., Weingart, H., Ullrich, M.S., 2004. The phytoalexin-inducible multidrug efflux pump AcrAB contributes to virulence in the fire blight pathogen, *Erwinia amylovora*. Mol. Plant-Microbe Interact. 17 (1), 43–54.

Campbell, A., et al., 2000. Xenognosin sensing in virulence: is there a phenol receptor in agrobacterium tumefaciens? Chem. Biol. 7 (1), 65–76.

Chalker-Scott, L., Fuchigami, L., 2018. The role of phenolic compounds in plant stress responses. In: Low Temperature Stress Physiology in Crops. CRC Press., pp. 67–80.

Chen, Z., et al., 2016. Changes of phenolic profiles and antioxidant activity in canaryseed (*Phalaris canariensis* L.) during germination. Food Chem. 194, 608–618.

Chen, S., et al., 2019a. Phenolic metabolism and related heavy metal tolerance mechanism in Kandelia Obovata under Cd and Zn stress. Ecotoxicol. Environ. Saf. 169, 134–143.

Chen, S., et al., 2019b. NtMYB4 and NtCHS1 are critical factors in the regulation of flavonoid biosynthesis and are involved in salinity responsiveness. Front. Plant Sci. 10, 178.

Chen, Z., et al., 2019c. Effects of exogenous Ca^{2+} on phenolic accumulation and physiological changes in germinated wheat (*Triticum aestivum* L.) under UV-B radiation. Food Chem. 288, 368–376.

Cherrak, S.A., et al., 2016. In vitro antioxidant versus metal ion chelating properties of flavonoids: a structure-activity investigation. PLoS One **11**(10).

Dai, J., Mumper, R.J., 2010. Plant phenolics: extraction, analysis and their antioxidant and anticancer properties. Molecules 15 (10), 7313–7352.

Dakora, F.D., 2003. Defining new roles for plant and rhizobial molecules in sole and mixed plant cultures involving symbiotic legumes. New Phytol. 158 (1), 39–49.

Demkura, P.V., et al., 2010. Jasmonate-dependent and-independent pathways mediate specific effects of solar ultraviolet B radiation on leaf phenolics and antiherbivore defense. Plant Physiol. 152 (2), 1084–1095.

Deng, Y., Lu, S., 2017. Biosynthesis and regulation of phenylpropanoids in plants. Crit. Rev. Plant Sci. 36 (4), 257–290.

Dewick, P.M., 2002. The biosynthesis of C 5–C 25 terpenoid compounds. Nat. Prod. Rep. 19 (2), 181–222.

Di Chen, S.B.W., et al., 2011. EGCG, green tea polyphenols and their synthetic analogs and prodrugs for human cancer prevention and treatment. Adv. Clin. Chem. 53, 155.

Di Ferdinando, M., et al., 2014. Multiple functions of polyphenols in plants inhabiting unfavorable Mediterranean areas. Environ. Exp. Bot. 103, 107–116.

Dixon, R.A., Paiva, N.L., 1995. Stress-induced phenylpropanoid metabolism. Plant Cell 7 (7), 1085.

Djordjevic, M.A., Gabriel, D.W., Rolfe, B.G., 1987. Rhizobium-the refined parasite of legumes. Annu. Rev. Phytopathol. 25 (1), 145–168.

Dua, M., et al., 2002. Biotechnology and bioremediation: successes and limitations. Appl. Microbiol. Biotechnol. 59 (2–3), 143–152.

Falcone Ferreyra, M.L., Rius, S., Casati, P., 2012. Flavonoids: biosynthesis, biological functions, and biotechnological applications. Front. Plant Sci. 3, 222.

Fantini, M., et al., 2015. In vitro and in vivo antitumoral effects of combinations of polyphenols, or polyphenols and anticancer drugs: perspectives on cancer treatment. Int. J. Mol. Sci. 16 (5), 9236–9282.

Ferrer, J.-L., et al., 2008. Structure and function of enzymes involved in the biosynthesis of phenylpropanoids. Plant Physiol. Biochem. 46 (3), 356–370.

Gershenzon, J., 1984. Changes in the levels of plant secondary metabolites under water and nutrient stress. In: Phytochemical Adaptations to Stress. Springer, pp. 273–320.

Gharibi, S., et al., 2019. The effect of drought stress on polyphenolic compounds and expression of flavonoid biosynthesis related genes in *Achillea pachycephala* Rech. f. Phytochemistry 162, 90–98.

Ghasemi, S., Kumleh, H.H., Kordrostami, M., 2019. Changes in the expression of some genes involved in the biosynthesis of secondary metabolites in *Cuminum cyminum* L. under UV stress. Protoplasma 256 (1), 279–290.

Golkar, P., Taghizadeh, M., 2018. In vitro evaluation of phenolic and osmolite compounds, ionic content, and antioxidant activity in safflower (*Carthamus tinctorius* L.) under salinity stress. Plant Cell, Tissue and Organ Culture (PCTOC) 134 (3), 357–368.

Gull, A., Lone, A.A., Wani, N.U.I., 2019. Biotic and abiotic stresses in plants. In: Abiotic and Biotic Stress in Plants. IntechOpen.

Halvorson, J.J., et al., 2009. Sorption of tannin and related phenolic compounds and effects on soluble-N in soil. Soil Biol. Biochem. 41 (9), 2002–2010.

Handa, N., et al., 2019. Selenium modulates dynamics of antioxidative defence expression, photosynthetic attributes and secondary metabolites to mitigate chromium toxicity in *Brassica juncea* L. plants. Environ. Exp. Bot. 161, 180–192.

Hartley, R.D., Harris, P.J., 1981. Phenolic constituents of the cell walls of dicotyledons. Biochem. Syst. Ecol. 9 (2–3), 189–203.

Hernández, I., et al., 2009. How relevant are flavonoids as antioxidants in plants? Trends Plant Sci. 14 (3), 125–132.

Hirsch, A.M., et al., 2003. Molecular signals and receptors: controlling rhizosphere interactions between plants and other organisms. Ecology 84 (4), 858–868.

Hodaei, M., et al., 2018. The effect of water stress on phytochemical accumulation, bioactive compounds and expression of key genes involved in flavonoid biosynthesis in *Chrysanthemum morifolium* L. Ind. Crop. Prod. 120, 295–304.

Joint, I., et al., 2002. Cell-to-cell communication across the prokaryote-eukaryote boundary. Science 298 (5596), 1207.

Kaurinovic, B., Vastag, D., 2019. Flavonoids and phenolic acids as potential natural antioxidants. In: Antioxidants. IntechOpen.

Kefeli, V.I., Kalevitch, M.V., Borsari, B., 2003. Phenolic cycle in plants and environment. J. Cell Mol. Biol 2 (1), 13–18.

Kemp, M., Burden, R., 1986. Phytoalexins and stress metabolites in the sapwood of trees. Phytochemistry 25 (6), 1261–1269.

Kirakosyan, A., et al., 2003. Antioxidant capacity of polyphenolic extracts from leaves of *Crataegus laevigata* and *Crataegus monogyna* (Hawthorn) subjected to drought and cold stress. J. Agric. Food Chem. 51 (14), 3973–3976.

Konczak, I., Zhang, W., 2004. Anthocyanins—more than nature's colours. Biomed. Res. Int. 2004 (5), 239–240.

Kraus, T.E., Dahlgren, R.A., Zasoski, R.J., 2003. Tannins in nutrient dynamics of forest ecosystems-a review. Plant Soil 256 (1), 41–66.

Kubalt, K., 2016. The role of phenolic compounds in plant resistance. Biotechnology and Food Sciences 80 (2), 97–108.

Kumar, S., Pandey, A.K., 2013. Chemistry and biological activities of flavonoids: an overview. Sci. World J.. 2013.

Kumar, V., et al., 2019. Differential Distribution of Polyphenols in Plants Using Multivariate Techniques. Biotechnology Research and Innovation.

Lake, B., 1999. Coumarin metabolism, toxicity and carcinogenicity: relevance for human risk assessment. Food Chem. Toxicol. 37 (4), 423–453.

Lattanzio, V., 2013. Phenolic compounds: introduction. In: Ramawat, K., Mérillon, J.M. (Eds.), Natural Products. Springer, Berlin, Heidelberg.

Lattanzio, V., et al., 2009. Relationship of secondary metabolism to growth in oregano (*Origanum vulgare* L.) shoot cultures under nutritional stress. Environ. Exp. Bot. 65 (1), 54–62.

Li, M., et al., 2018. Metabolomics analysis reveals that elevated atmospheric CO_2 alleviates drought stress in cucumber seedling leaves. Anal. Biochem. 559, 71–85.

Mahdavi, V., et al., 2015. A targeted metabolomics approach toward understanding metabolic variations in rice under pesticide stress. Anal. Biochem. 478, 65–72.

Martinez, V., et al., 2016. Accumulation of flavonols over hydroxycinnamic acids favors oxidative damage protection under abiotic stress. Front. Plant Sci. 7, 838.

Mathesius, U., 2001. Flavonoids induced in cells undergoing nodule organogenesis in white clover are regulators of auxin breakdown by peroxidase. J. Exp. Bot. 52 (suppl_1), 419–426.

Michalak, A., 2006. Phenolic compounds and their antioxidant activity in plants growing under heavy metal stress. Pol. J. Environ. Stud.. 15(4).

Moreland, D.E., Novitzky, W.P., 1987. Effects of Phenolic Acids, Coumarins, and Flavonoids on Isolated Chloroplasts and Mitochondria. ACS Publications.

Naikoo, M.I., et al., 2019. Role and regulation of plants phenolics in abiotic stress tolerance: an overview. In: Plant Signaling Molecules. Elsevier, pp. 157–168.

Nakabayashi, R., et al., 2014. Enhancement of oxidative and drought tolerance in Arabidopsis by overaccumulation of antioxidant flavonoids. Plant J. 77 (3), 367–379.

Nanda, S., et al., 2016. Metabolic Engineering of Phenyl Propanoids in Plants. Transgenesis and Secondary Metabolism: Part of the Series Reference Series in Phytochemistry. Springer, New York, pp. 1–26.

Neish, A., 1960. Biosynthetic pathways of aromatic compounds. Annu. Rev. Plant Physiol. 11 (1), 55–80.

Nenadis, N., et al., 2015. Interactive effects of UV radiation and reduced precipitation on the seasonal leaf phenolic content/composition and the antioxidant activity of naturally growing *Arbutus unedo* plants. J. Photochem. Photobiol. B Biol. 153, 435–444.

Nichols, S.N., Hofmann, R.W., Williams, W.M., 2015. Physiological drought resistance and accumulation of leaf phenolics in white clover interspecific hybrids. Environ. Exp. Bot. 119, 40–47.

Ostrofsky, W., Shortle, W., Blanchard, R., 1984. Bark phenolics of American beech (*Fagus grandifolia*) in relation to the beech bark disease 1. Eur. J. For. Pathol. 14 (1), 52–59.

Palmer, M., et al., 2004. Ecology for a Crowded Planet. American Association for the Advancement of Science.

Pandey, K.B., Rizvi, S.I., 2009. Plant polyphenols as dietary antioxidants in human health and disease. Oxidative Med. Cell. Longev. 2 (5), 270–278.

Perradin, Y., Mottet, M., Lalonde, M., 1983. Influence of phenolics on in vitro growth of Frankia strains. Can. J. Bot. 61 (11), 2807–2814.

Perret, X., Staehelin, C., Broughton, W.J., 2000. Molecular basis of symbiotic promiscuity. Microbiol. Mol. Biol. Rev. 64 (1), 180–201.

Ramawat, K.G., Mérillon, J.-M., 2013. Natural Products: Phytochemistry, Botany and Metabolism of Alkaloids, Phenolics and Terpenes. Springer.

Ramos, S., 2008. Cancer chemoprevention and chemotherapy: dietary polyphenols and signalling pathways. Mol. Nutr. Food Res. 52 (5), 507–526.

Reichman, S., 2002. The Responses of Plants to Metal Toxicity: A Review Forusing on Copper, Manganese & Zinc. Australian Minerals & Energy Environment Foundation, Melbourne.

Sánchez-Rodríguez, E., et al., 2011. Differential responses of five cherry tomato varieties to water stress: changes on phenolic metabolites and related enzymes. Phytochemistry 72 (8), 723–729.

Santos-Sánchez, N.F., et al., 2019. Shikimic acid pathway in biosynthesis of phenolic compounds. In: Plant Physiological Aspects of Phenolic Compounds. IntechOpen.

Sarker, U., Oba, S., 2018. Augmentation of leaf color parameters, pigments, vitamins, phenolic acids, flavonoids and antioxidant activity in selected Amaranthus tricolor under s alinity stress. Sci. Rep. 8 (1), 1–9.

Sasidharan, S., Saudagar, P., 2019. Plant metabolites as new leads to drug discovery. In: Natural Bio-active Compounds. Springer, pp. 275–295.

Schultz, J.C., Hunter, M.D., Appel, H.M., 1992. Antimicrobial activity of polyphenols mediates plant-herbivore interactions. In: Plant Polyphenols. Springer, pp. 621–637.

Seneviratne, G., Jayasinghearachchi, H., 2003. Mycelial colonization by bradyrhizobia and azorhizobia. J. Biosci. 28 (2), 243–247.

Shankar, S., et al., 2009. Allelopathic effects of phenolics and terpenoids extracted from Gimelina arborea on germination of Black gram (*Vigna mungo*) and Green gram (*Vigna radiata*). Allelopath. J. **23**(2).

Sharma, A., et al., 2016. Epibrassinolide-imidacloprid interaction enhances non-enzymatic antioxidants in *Brassica juncea* L. Indian J. Plant Physiol. 21 (1), 70–75.
Sharma, A., et al., 2019a. Response of phenylpropanoid pathway and the role of polyphenols in plants under abiotic stress. Molecules 24 (13), 2452.
Sharma, A., et al., 2019b. Castasterone attenuates insecticide induced phytotoxicity in mustard. Ecotoxicol. Environ. Saf. 179, 50–61.
Suzuki, N., et al., 2012. ROS and redox signalling in the response of plants to abiotic stress. Plant Cell Environ. 35 (2), 259–270.
Taïbi, K., et al., 2016. Effect of salt stress on growth, chlorophyll content, lipid peroxidation and antioxidant defence systems in *Phaseolus vulgaris* L. S. Afr. J. Bot. 105, 306–312.
Tanase, C., Bujor, O.-C., Popa, V.I., 2019. Phenolic natural compounds and their influence on physiological processes in plants. In: Polyphenols in Plants. Elsevier, pp. 45–58.
Taylor, L.P., Grotewold, E., 2005. Flavonoids as developmental regulators. Curr. Opin. Plant Biol. 8 (3), 317–323.
Tobe, K., et al., 2001. Characteristics of seed germination in five non-halophytic Chinese desert shrub species. J. Arid Environ. 47 (2), 191–201.
Tripoli, E., et al., 2007. Citrus flavonoids: molecular structure, biological activity and nutritional properties: a review. Food Chem. 104 (2), 466–479.
Tuteja, N., Mahajan, S., 2007. Calcium signaling network in plants: an overview. Plant Signal. Behav. 2 (2), 79–85.
Valifard, M., et al., 2014. Effects of salt stress on volatile compounds, total phenolic content and antioxidant activities of *Salvia mirzayanii*. S. Afr. J. Bot. 93, 92–97.
Van Der Meer, I.M., et al., 1992. Antisense inhibition of flavonoid biosynthesis in petunia anthers results in male sterility. Plant Cell 4 (3), 253–262.
Wang, J., Yuan, B., Huang, B., 2019. Differential heat-induced changes in phenolic acids associated with genotypic variations in heat tolerance for hard fescue. Crop Sci. 59 (2), 667–674.
Wani, S.H., Sah, S., 2014. Biotechnology and abiotic stress tolerance in rice. J Rice Res. 2, e105.
Weichselbaum, E., Buttriss, J., 2010. Polyphenols in the diet. Nutr. Bull. 35 (2), 157–164.
Wink, M., 1999. Functions of Plant Secondary Metabolites and Their Exploitation in Biotechnology. Vol. 3. Taylor & Francis.
Yan, J., et al., 2014. GmFNSII-controlled soybean flavone metabolism responds to abiotic stresses and regulates plant salt tolerance. Plant Cell Physiol. 55 (1), 74–86.
Zhang, J., et al., 2009. Flavones and flavonols play distinct critical roles during nodulation of *Medicago truncatula* by *Sinorhizobium meliloti*. Plant J. 57 (1), 171–183.

Terpenoid indole alkaloids, a secondary metabolite in plant defense response

18

M. Thippeswamy[a], *V. Rajasreelatha*[b], *Raju Krishna Chalannavar*[c], *and Sudisha Jogaiah*[d]

[a]Department of Botany, Davangere University, Davanagere, Karnataka, India, [b]Department of Biochemistry, Indian Institute of Science, Bangalore, Karnataka, India, [c]Department of Applied Botany, Mangalore University, Mangalagangotri, Karnataka, India, [d]Laboratory of Plant Healthcare and Diagnostics, P.G. Department of Biotechnology and Microbiology, Karnatak University, Dharwad, Karnataka, India

18.1 Introduction

Secondary metabolites play an important role in the way plants interact with their environment and are usually produced in select cell types within the plant. These compounds play an active role in defense against herbivores, fungi, bacteria, viruses, and other plants competing for resources. They are also used as signal molecules, as attractants of pollinators, and as protection against physical stress (Verpoorte, 2000). One class of secondary metabolites is the alkaloids which are found in 20% of plant species. They are characterized as low-molecular weight, nitrogen-containing compounds. Many of these compounds have biological activity and are being used in the pharmaceutical industry as analgesics, anticancer, and antiarrhythmic drugs (Facchini, 2001). Now, a number of examples of alkaloids, cyanogenic glycosides, glucosinolates, terpenes, saponins, tannins, and anthraquinones serving as defense molecules are known. Secondary metabolites may occur in organisms either in the active state or as a prodrug that becomes active upon infection by a pathogen or upon wounding as in the case of cyanogenic glycosides. Synthesis of certain secondary metabolites like phytoalexins are initiated only upon infection or wounding (Ahuja et al., 2012).

Secondary metabolites can also serve as key players in the interaction between plants and their environment. For example, anthocyanins and betulins, which contribute to the flower color in plants, attract pollinators. Fragrant monoterpenes also serve as attractants in certain plants. It was shown that a flavone, luteolin exuded from roots of alfalfa, serves as a signal that activates rhizobial nodulation genes and therefore plays a major role in its root colonization (Kent and Long, 1988). Secondary metabolites are also known to function as signaling molecules. Methyl jasmonate, a secondary metabolite derived from the membrane associated fatty acid-linolenic acid is involved in defense gene activation, tuber formation, and plant senescence (WeilerElmar et al., 1999). Certain alkaloids and peptides mediate the transport

Biocontrol Agents and Secondary Metabolites. https://doi.org/10.1016/B978-0-12-822919-4.00018-1

and storage of toxic nitrogen compounds and certain phenolic compounds such as avanoids function as UV-protectants. Secondary metabolites, although being complex in their chemical structure and conformations, are mostly, if not always, derived from primary metabolites or their intermediates. They have been categorized using a variety of criteria such as chemical characteristics/complexity, plant origin, or biosynthetic origin. A classification scheme, largely based either on the primary metabolite, the metabolic pathway through which they are derived, or on the basis of a common structural element between them, is described here (Schmidt et al., 2007a).

18.2 Secondary metabolites classification

Secondary metabolites are chemically diverse and heterogenous. They are classified based on the primary metabolite from which they are derived or the biosynthetic pathway through which they are derived or the common structural elements present in them (Fig. 18.1). A simple classification of secondary metabolites includes three main groups: terpenoids, phenolics, and nitrogen-containing compounds.

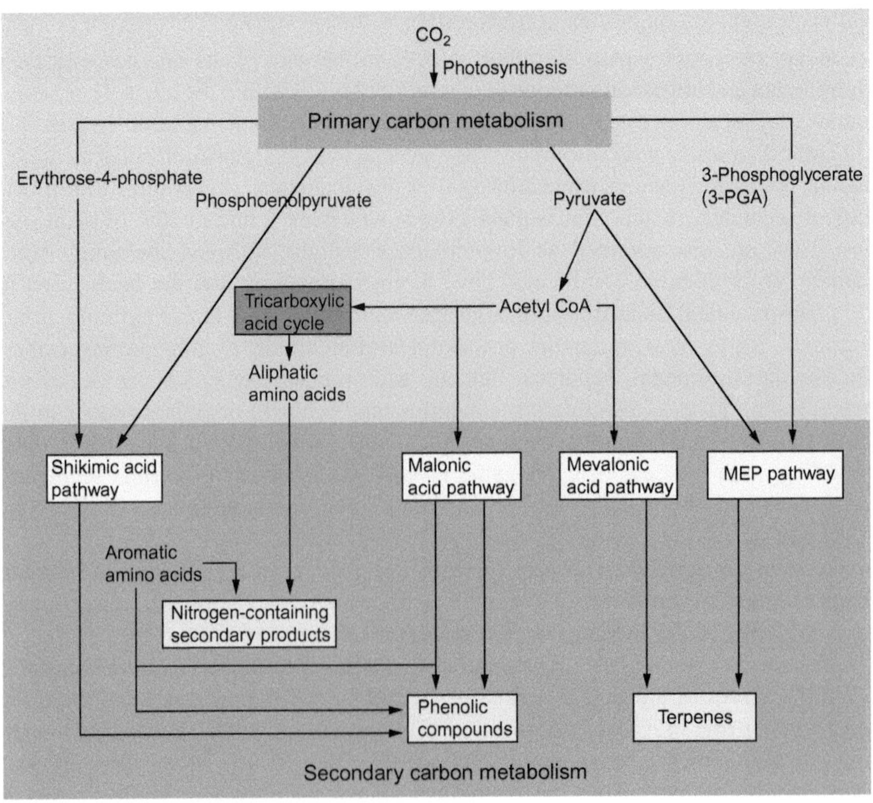

Fig. 18.1 A simplified scheme of secondary metabolite biosynthesis and their interrelationship with primary metabolism (Taiz and Zeiger, 2006).

18.2.1 Terpenoids

Terpenoids form the largest group of secondary metabolites (Croteau et al., 2000). The basic unit of these compounds is a five carbon (5C) building block known as isoprene. Based on the number of such isoprenoid units (5C) which constitute the final structure, they are grouped into mono-(10C), sesqui-(15C), di-(20C), tri-(30C) or tetra-terpenes (40C), respectively (Ashour et al., 2010). These primarily act as deterrents to herbivory. A few of them, such as gibberellic acid, abscisic acid, phytol, and carotenoids, also work in plant growth and metabolism. Monoterpenes (pyrethroid, insecticides, and essential oils of plants), sesquiterpenes (Herbivory deterrents in sunflower), diterpenes (toxins and irritants, plants resins which help to seal wounds in certain plants), triterpenes (Saponins and vertebrate heart toxins), and tetraterpenes (lycopene, β-carotene, etc.) are a few of the innumerable terpenoids produced by different plants.

18.2.2 Phenolic compounds

This group consists of thousands of diverse compounds having a central phenolic group in common, as the name suggests. Most of these are derived from the amino acids, mainly phenylalanine or sometimes tyrosine. The major subgroups are phenyl propanoids like caffeic acid (an allelopathic compound) and furanocoumarins, which are again toxic to herbivores, and some may act as UV activated toxins. Benzoic acid derivatives like salicylic acid act as a systemic resistance signal in plants. Phytoalexins are a chemically diverse group of compounds, some of which are phenolics and act in immune response in plants. Lignins and tannins are phenolic compounds and form part of the primary level of defense in plants. Flavonoids constitute a major subgroup encompassing anthocyanins (pigmentation), flavones, flavonols (UV protection and bee attractant), and isoflavones (antimicrobials).

18.2.3 Nitrogen-containing compounds

This group comprises a diverse group of chemicals which have a nitrogen atom in the molecule. Some of them have medicinal value and therefore are well studied. The following are the major subgroups of this type.

18.2.3.1 Cyanogenic glycosides

These are a large group of compounds which release cyanide gas upon herbivory. It has been shown that cyanogenic glycosides have protective function against herbivory.

18.2.3.2 Glucosinolates

These compounds are found mainly in family Brassicaceae (mustard oil glycosides). They form volatile defensive compounds that are released on herbivory through enzymatic action. In some cases, they also act as attractants to pollinators.

18.2.3.3 Nonprotein amino acids

Many plants contain unusual amino acids that are generally not incorporated into proteins, but are present in the free form and, being toxic to animals, act as protective substances (canavanine). Upon digestion and assimilation by the herbivore, they are incorporated in place of the standard amino acids in the proteins of the animals, thus leading to the formation of nonfunctional proteins.

Over the decades, with the advent of improved detection methods, a large amount of novel secondary metabolites are being discovered each day and being added to the already huge repertoire of metabolites.

18.2.3.4 Alkaloids

Alkaloids were one of the first natural products to be isolated from medicinal plants. They are basic compounds containing nitrogen atom(s) as a part of a heterocyclic ring. All of these compounds contain a heterocyclic ring which has a nitrogen atom. They can be further subdivided into pyrrolidine alkaloids and tropane alkaloids (derived from ornithine or arginine, e.g., scopolamine), isoquinoline alkaloids (derived from tyrosine), indole alkaloids (tryptophan derived, e.g., strychnine), piperidine alkaloids (lysine derived, e.g., coniine), etc. Alkaloids are classified based on the structure of the ring system that contains a nitrogen atom, which in turn reflects the precursor amino acid from which they are derived (Table 18.1.). Several pharmacologically active compounds like morphine, quinine, ephedrine, and pilocarpine belong to the alkaloid family.

Table 18.1 Major types of alkaloids, their amino acid precursor, and well-known examples.

Alkaloid class	Biosynthetic precursor	Examples	Human uses
Pyrrolidine	Ornithine (Aspartate)	Nicotine	Stimulant, depressant, tranquilizer
Tropane	Ornithine	Atropine	Prevention of intestinal spasms, antidote to other poisons, dilation of pupils for examination
		Cocaine	Stimulant of the central nervous system, local anesthetic
Piperidine	Lysine (or acetate)	Coniline	Poison (paralyzes motor neurons)
Pyrrolizidine	Ornithine	Retrosine	None
Quinilizidine	Lysine	Lupinine	Restoration of heart rhythm
Isoquinoline	Tyrosine	Codeine	Analgesic (pain relief), treatment of coughs
		Morphine	Analgesic
Indole	Tryptophan	Psilocybin	Halucinogen
		Reserpine	Treatment of hypertension, treatment of psychoses
		Strychnine	Rat poison, treatment of eye disorders

18.2.3.5 Tropane alkaloids

Tropane alkaloids are valued because of their effects on the parasympathetic nervous system. They are antagonists of acetylcholine and other muscarinic agonists and thus are anticholinergics. Tropane alkaloids are used in ophthalmology, anesthesia, and in the treatment of cardiac and gastrointestinal diseases. In additional to peripheral anticholinergic effects, they also act on the central nervous system and thus are used to relieve the symptoms of Parkinson's disease and antidotes. Scopolamine is a more preferred drug than hyoscyamine because of its stronger effect on the central nervous system and reduced side effects. It is used in treating motion sickness and gastric disorders. Both hyoscyamine and scopolamine alkaloids have dextro (+) and levo rotator (−) forms, and the (−) forms are biologically active, while atropine is a racemic mixture. Plants are the only source of these alkaloids, as chemical synthesis has proven to be difficult and economically not feasible. The demand for scopolamine is 10-fold higher than that for hyoscyamine and atropine combined. Although inexpensive, their demand is quite high as they cannot be substituted by any other compounds (Oksman-Caldentey and Arroo, 2000).

18.2.3.6 Terpenoidindole alkaloids (TIAs)

These are secondary metabolites produced by plants like the Madagascar periwinkle (*Catharanthusroseus*). Of the 130 known terpenoid secondary metabolites produced by *Catharanthus* sp., the highly functionalized vincristine and vinblastine are used as antineoplastic drugs. Originally, these TIAs were obtained by extraction from leaves of *C. roseus* and are a constituent agent in the treatment of breast, throat, nonsmall cell lung, ovarian, and prostate cancer. The TIAs are used against different cancers and in combination treatments with other chemotherapeutic agents. These metabolites seem to be involved in plant defense processes, which have enabled the survivability of this long-lived plant genus (Geerlings et al., 2000). This compound is a potent antifungal agent and serves the plant by protecting it against potential fungal pathogens. A brief summary of the different structural classes of plant secondary metabolites is shown in Fig. 18.2.

18.3 Terpenoidindole alkaloid pathway

TIAs are one of the largest groups of alkaloids with more than 3000 representatives (Schmidt et al., 2007b). The biosynthesis of TIAs involves over 20 enzymatic steps and begins with the convergence of the shikimate pathway (of tryptophan biosynthesis) and the isopentenyl diphosphate pathway (Fig. 18.3). The indole moiety in the TIAs is provided by the amino acid derivative tryptamine, and the terpenoid moiety is provided by secologanin, which are derived via the shikimate and mevalonate pathways, respectively.

Tryptophan decarboxylase (TDC) catalyzes the formation of tryptamine from L-tryptophan. Tryptamine is a key precursor of a wide range of terpenoid-derived indole alkaloids in plants. The decarboxylation of L-tryptophan may be viewed as a

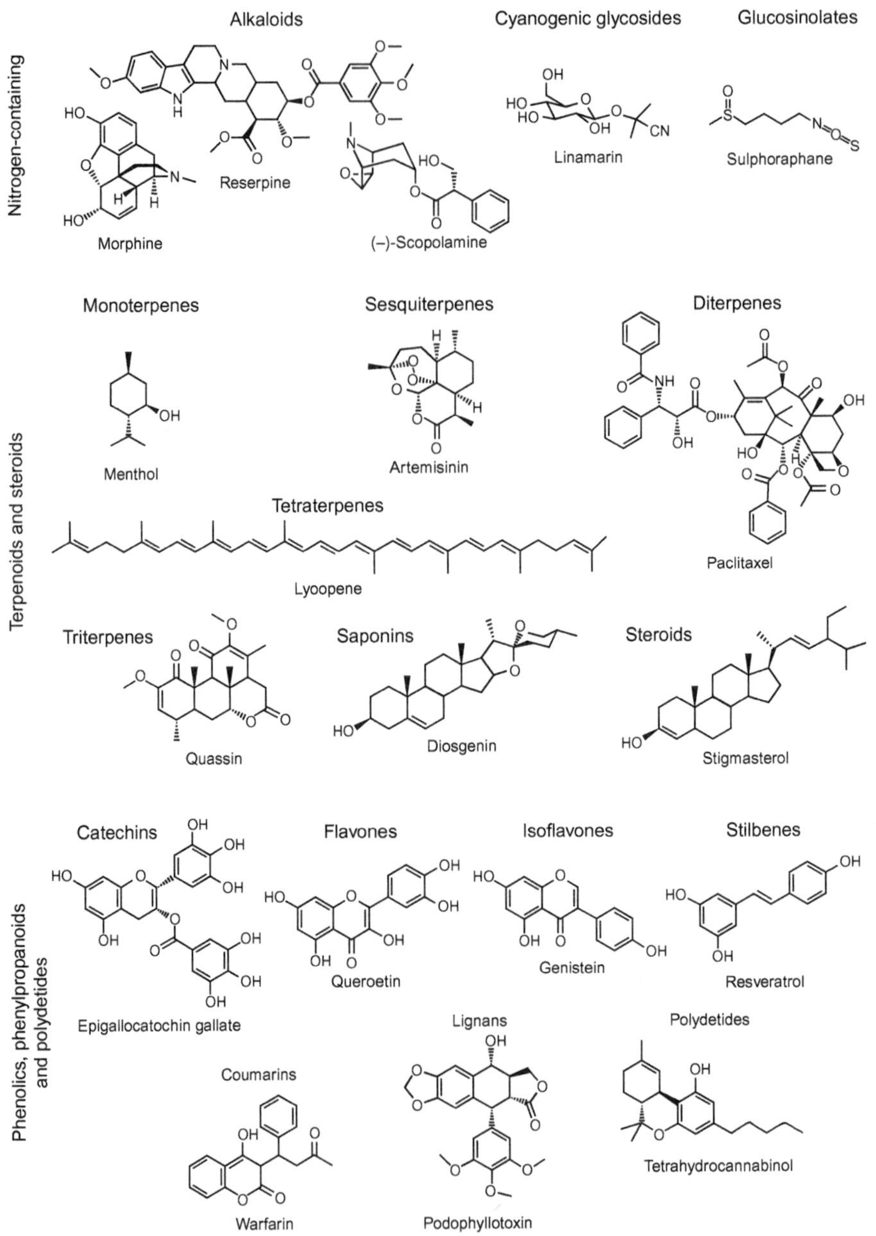

Fig. 18.2 Classification of plant secondary metabolites. A brief outline of the classification of secondary metabolites based on the carbon skeleton and the side groups/derivatives is illustrated with examples. The list of secondary metabolites discovered to date is nonexhaustive with a new compound being added to the database each day.
Adapted from Ref. Schmidt, B.M., Ribnicky, D.M., Lipsky, P.E., Raskin, I. 2007. Revisiting the ancient concept of botanical therapeutics. Nat. Chem. Biol. 3, 360–366.

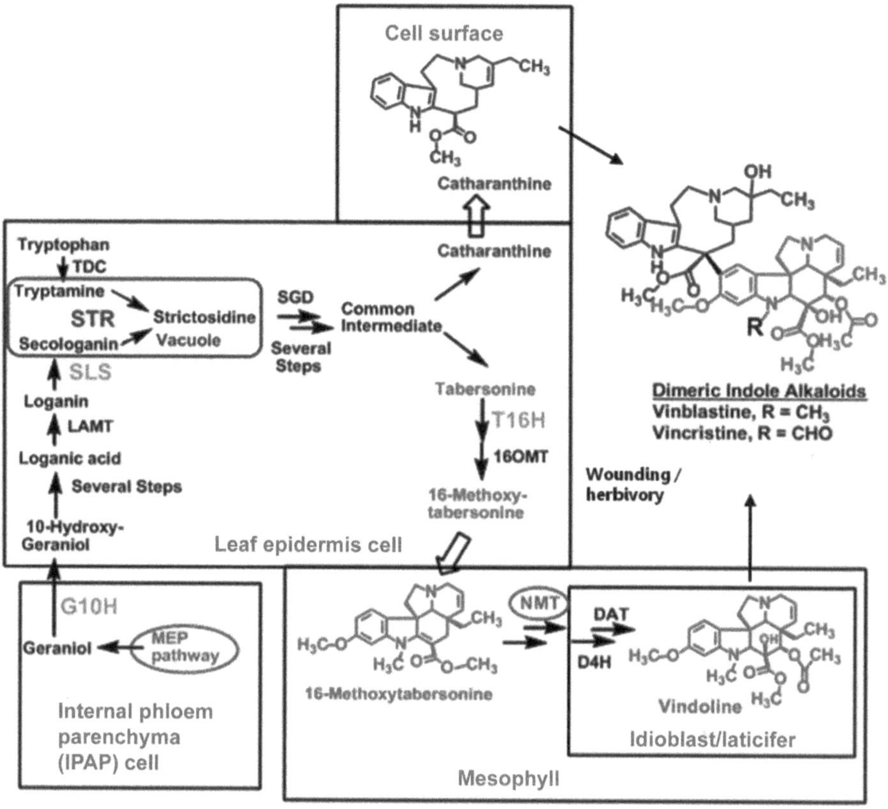

Fig. 18.3 Overview of biosynthesis and compartmentalization of TIAs. The TIA pathway essentially begins with the amino acid tryptophan and the terpene intermediate geraniol, both derived from the primary metabolism of a plant. These undergo various reactions involving multiple enzymes which are compartmentalized into various cell types (leaf epidermal cells, parenchymal cells, mesophyll cells, and specialized cells called idioblasts) and cellular organelles (chloroplast, vacuole, cytoplasm), finally producing the dimeric TIAs like vincristine upon wounding. *MEP pathway*: methyl erythritol phosphate pathway, *G10H*: Geraniol 10-hydroxylase, *LAMT*: Loganic acid-*O*-methyl transferase, *TDC*: Tryptophan decarboxylase, *SLS*: Secologanin synthase, *STR*: Strictosidine synthase, *SGD*: Strictosidine β glucosidase, *T16H*: Tabersonine 16-hydroxylase, *16OMT*: 16-hydroxy tabersonine-*O*-methyl transferase, *NMT*: *N*-methyl transferase, *D4H*: dihydrovindoline 4-hydroxylase, *DAT*: Desacetoxyvindoline acetyl transferase.
Adapted from Ref. Roepke, J., Salim, V., Wu, M., Thamm, A.M.K., Murata, J., Ploss, K., Boland, W., De Luca, V. 2010. Vinca drug components accumulate exclusively in leaf exudates of Madagascar periwinkle. Proc. Natl. Acad. Sci. USA 107, 15287–15292.

branching point from primary to secondary metabolism. Strictosidine synthase (STR) is the first enzyme of the TIA pathway catalyzing the condensation of the amino acid derivative tryptamine and the terpenoidsecologanin to form strictosidine. Strictosidine is the universal precursor for therapeutically valuable TIAs, which are produced by four plant families (Apocyanaceae, Loganiaceae, Rubiaceae, and Nyssaceae). Strictosidine occupies a central role in the biosynthesis of all major classes of monoterpenoidindole alkaloids of *C. roseus* as well as in the other members that produce these compounds. Both TDC and STR could be viewed as the important regulatory enzymes, which regulate the influx of metabolites into the TIA pathway. Strictosidine can undergo various conversion reactions to finally yield catharanthine and in a separate reaction in the chloroplast to tabersonine. Tabersonine in the chloroplast is converted to vindoline in a light-dependent and differentiation-dependent manner. Finally, catharanthine and vindoline are shuttled into the vacuole to yield vinblastine or vincristine and stored in specialized cells called idioblasts and laticifers.

18.4 Localization of the TIA pathway

The enzymes of the TIA pathway are located in different compartments of the cell and in different tissues of the plant. This compartmentalization helps to keep toxic compounds from killing the plant cell (Facchini, 2001). The shikimate pathway leading to tryptophan and the non-mevalonate pathway leading to geraniol occur within the chloroplast of the cell TDC, OMT, D4H, and DAT that are located in the cytosol. NMT activity is associated with the thylakoid located in the chloroplast. Along with the enzymes that combine to form anhydrovinblastine, STR is located within the vacuole with G10H occurring in the provacuolar membrane (Facchini, 2001; Meijer et al., 1993; Sottomayor and Barcelo, 2005).

In addition to the compartmentalization of the different reactions, another degree of complexity comes from the fact that the enzymes are located within different tissues and organs within the plant. TDC and STR are abundant in the protoderm and cortical cells around the apical meristem of the root tips and can also be found at low levels in the epidermis of stems, leaves, and flower buds (St-Pierre et al., 1999). DXS, DXR, MECS, and G10H were reported to be expressed in the internal phloem parenchyma and were present in roots, flower buds, and leaves (Burlat et al., 2004). SLS, SGD, T16H, and OMT are found in the epidermis (Irmler et al., 2000; Murata and De Luca, 2005). G10H expression was also found in epidermal cells (Murata and De Luca, 2005). D4H and DAT are found in young leaves and shoot organs; specifically, D4H and DAT are found in the laticifers and edioblasts in the mesophyll of the leaves which are several tissue layers away from the epidermis (St-Pierre et al., 1999) (Fig. 18.4).

18.5 Regulation of the TIA pathway

The TIA pathway is regulated by a number of different signaling events within *C. roseus*. Expression of TIA enzymes is regulated by development cues, by light, and by abiotic and biotic stress responses. Studies with germinating seedlings and with mature plants show that alkaloid biosynthesis and accumulation are under developmental control (Aerts et al., 1994; Westekemper et al., 1980). Vindoline biosynthesis

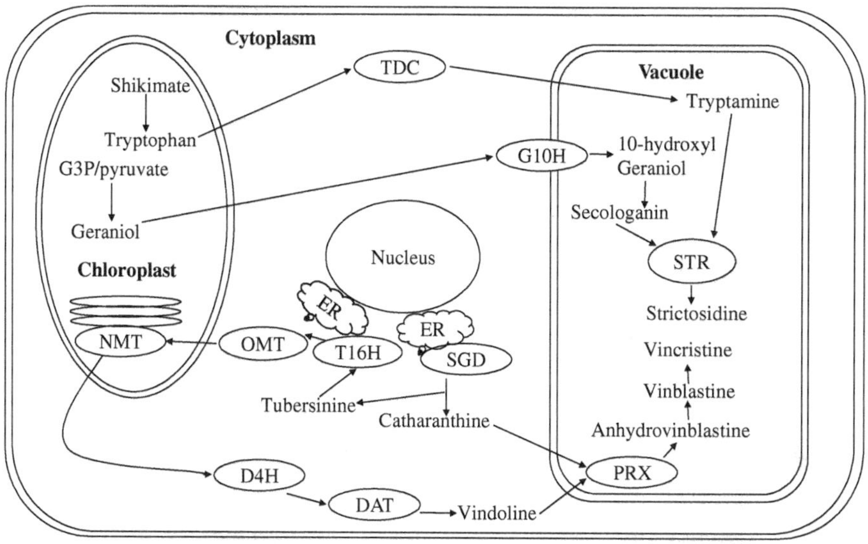

Fig. 18.4 Compartmentalization of terpenoidindole alkaloid biosynthesis in *Catharanthusroseus*. The *dashed arrows* represent hypothetical steps. All of these reactions do not occur within the same plant cell. Some reactions are localized to specific tissues within the plant. *TDC*, tryptophan decarboxylase; *G3P*, glyceraldehyde-3-phosphate; *G10H*, geraniol 10-hydroxylase; *STR*, strictosidine synthase; *SGD*, strictosidine β-D-glucosidase; *T16H*, tabersonine 16-hydroxylase; *OMT*, 16-hydroxytabersonine-16-*O*-methyltransferase; *NMT*, *N*-methyl-transferase; *D4H*, desacetoxyvindoline 4-hydroxylase; *DAT*, deacetylvindolineacetyltransferase; *PRX*, peroxidase.
The figure was adapted from the following Facchini, P.J. 2001. Alkaloid biosynthesis in plants; biochemistry, cell biology, molecular regulation and metabolic engineering applications. Annu. Rev. Plant Physiol. Plant Mol. Biol. 52, 29–66; Meijer, A.H., Verpoorte, R., Hoge, J.H.C. 1993. Regulation of enzymes and genes involved in terpenoidindole alkaloid biosynthesis in Catharanthusroseus. J. Plant Res. 2, 145–164; Sottomayor, M., Barcelo, A.R. 2005. The vincaalkaloids: from biosynthesis and accumulation in plant cells, to uptake, activity and metabolism in animal cells. Studies Nat. Prod. Chem. 33, 813–857.

is also activated by light. Light exposure of etiolated seedlings activates T16H (St-Pierre and De Luca, 1995), D4H (Vazquez-Flota and De Luca, 1998b), and DAT (St-Pierre et al., 1998). In addition, the light activation of D4H seems to be induced by a photoreversible phytochrome. Major isoforms of this enzyme occur in both etiolated and light grown seedlings, with the light grown seedlings showing the greatest activity of the enzyme (Vazquez-Flota and De Luca, 1998a).

In recent years, researchers are beginning to develop a clearer picture of the control network of the TIA pathway in response to abiotic and biotic stresses. Feeding exogenous jasmonic acid (JA) or methyl jasmonate, important plant growth hormones, to *C. roseus* causes an increase in TIA metabolites and an increase in mRNA transcripts of all TIA genes tested (Aerts et al., 1994; Rijhwani and Shanks, 1998; Van der Fits and Memelink, 2000). ORCA2 and ORCA3 are jasmonate responsive AP2 domain transcription factors that promote transcription of TIA genes. In *C. roseus* cell suspension cultures, ORCA3 overexpression led to an increase in ASα, TDC, DXS, CPR, STR, SGD, and D4H mRNA transcripts. G10H and DAT mRNA transcripts were not affected by ORCA3 overexpression (Van der Fits and Memelink, 2000). This points to other signaling mechanisms that control the TIA pathway under JA signaling.

The zinc finger-binding proteins ZCT1, ZCT2, and ZCT3 (members of the transcription factor III-type zinc finger family) were found to bind to the promoters of STR and TDC. This interaction repressed the activity of STR and TDC. The binding of the ZCTs to the STR promoter has been suggested to counteract the activation of STR by ORCA2 or ORCA3 (Pauw et al., 2004). The STR and TDC promoters contain a G-box or G-boxlike sequence. Two G-box binding factors, Crgbf1 and Crgbf2, were identified in *C. roseus*. These factors were shown to repress the transcription of STR by binding to the G-box sequence (Siberil et al., 2001). The signal that stimulates the G-box binding factors is not known. A box P binding factor 1 (CrBPF-1) was shown to bind to the BA promoter region of the STR promoter. CrBPF1 mRNA accumulated in elicitor-treated cells but not JA-treated cells, indicating that factors play a role in an elicitor-responsive, jasmonate-independent signal transduction pathway (Van der Fits et al., 2000).

Recently, promoter analysis of G10H led to the identification of three unique regions that could act as potential enhancers to G10H transcript abundance. These regions were different from the elements found in the promoter of STR and TDC. These unique regions point to the possibility of another signaling cascade that regulates the activity of G10H in response to jasmonic acid, thus complicating the regulatory landscape of TIA synthesis (Suttipanta et al., 2007). G10H is also upregulated through the addition of cytokinin, ethylene, or both cytokinin and ethylene, whereas DXS, DXR, and MECS are upregulated only by the addition of both cytokinin and ethylene (Papon et al., 2005). The activity of CaaX-prenyltransferases is necessary for the expression of DXS, DXR, and G10H (Courdavault et al., 2005).

18.5.1 Posttranscriptional regulation

The TDC and STR enzyme activity is reportedly regulated both at the translational (Pasquali et al., 1992) and posttranslational levels (Aerts et al., 1992; Fernandez et al., 1989). However, the mechanism by which enzymes of the TIA pathway are regulated is still unclear.

18.5.2 Regulation by conditions of growth and environmental factors

Chemical synthesis of bioactive secondary metabolites having the same stereochemistry, activity, and efficacy as the natural product is difficult because of their complex structure and high cost of production (Balandrin et al., 1985; Abdelrahman et al., 2018).

Since the production of secondary metabolites has reportedly been induced in response to both biotic and abiotic stresses, it is possible to simulate their production by subjecting plant tissues to similar stress conditions *in vitro*. Various strategies (optimization of medium composition, supply of biosynthetic precursors, treatment with plant growth regulators and elicitors) have been used to increase alkaloid production in cell suspension cultures (Moreno et al., 1993; Verpoorte et al., 1993).

Studies in cell suspension cultures of *C. roseus* have established that indole alkaloid biosynthesis can be transiently induced by changes in nutrient composition (Bhadra and Shanks, 1997). Auxins stimulate cell proliferation while repressing the transcriptional activation of the *TDC* gene (Goddijn et al., 1992). In the cell lines overexpressing the *STR* gene, addition of indole and iridoid precursors resulted in enhanced accumulation of TIAs (Whitmer et al., 2002).

The TIA pathway is known to respond to pathogens, osmotic stress (Godoy-Hernandez and Loyola-Vargas, 1991), salt stress (Zhao et al., 2000), fungal elicitors (Zhao et al., 2001b), yeast elicitors (Menke et al., 1999a), wounding, plant growth regulators (Zhao et al., 2000, 2001b; El-Sayed and Verpoorte, 2002), chemicals (Choudhury, 1999; Contin et al., 1999; Zhao et al., 2001a), methyl jasmonate (Menke et al., 1999b), and UV light (Ouwerkerk et al., 1999a,b). Despite the reports of enhanced TIA accumulation using biotic and abiotic elicitors, the information on the mechanism by which this increase is brought about, and the early signal transduction events involved, is very limited.

18.6 Defense responses of TIAs in plants

The plant kingdom has direct and indirect defense responses when they come in contact with microbial pathogens. The direct mode of defense mechanism includes physical structures such as trichomes, thorns as well as the accumulation of phytochemicals that have antibiotic activities. Airborne terpenoids are also critical components of plant defense responses to abiotic and biotic stresses (Unsicker et al., 2009; Vickers et al., 2009). In leaves of *C. roseus*, UV light induced the accumulation of mRNAs encoding the key TIA biosynthetic enzymes TDC and STR, leading to accumulation of the dimeric alkaloids (Hirata et al., 1992). Several aspects of elicitor signal transduction leading to production of catharanthine were studied, including elicitor signal perception by various receptors of plants, Ca^{2+} signaling, medium alkalinization, generation of reactive oxygen species (ROS), and involvement of protein kinases (Ramani et al., 2010). The diverse lines of research explored to address the caveats in TIA production through varied approaches ranging from chemical synthesis, plant tissue culture, bioreactor studies on cell suspension cultures, various elicitation studies, and metabolic engineering approaches, among others, proved to be expensive and offered a very poor yield. Bioprospecting for endophytes is another attractive alternative, with active research being focused on production of many such secondary metabolites, and is deemed to be a promising arena for TIA production (Dhayanithy et al., 2019).

References

Abdelrahman, M., Jogaiah, S., Burritt, D.J., Tran, L.-S.P., 2018. Legume genetic resources and transcriptome dynamics under abiotic stress conditions. Plant Cell Environ. 41, 1972–1983.

Aerts, R.J., Alarco, A.M., De Luca, V., 1992. Auxins induce tryptophan decarboxylase activity in radicles of *Catharanthus* seedlings. Plant Physiol. 100, 1014–1019.

Aerts, R.J., Gisi, D., De Carolis, E., De Luca, V., Baumann, T.W., 1994. Methyl jasmonate vapor increases the developmentally controlled synthesis of alkaloids in *Catharanthus* and *Cinchona* seedlings. Plant J. 5, 635–643.

Ahuja, I., Kissen, R., Bones, A.M., 2012. Phytoalexins in defense against pathogens. Trends Plant Sci. 17, 73–90.

Ashour, M., Wink, M., Gershenzon, J., 2010. Biochemistry of terpenoids: Monoterpenes, sesquiterpenes and diterpenes. In: Wink, M. (Ed.), Annual Plant Reviews: Biochemistry of Plant Secondary Metabolism, Vol 40, second ed. Wiley, New York.

Balandrin, M.F., Klocke, J.A., Wurtele, E.S., Bollinger, W.H., 1985. Natural plant chemicals: sources of industrial and medicinal materials. Science 228, 1154–1160.

Bhadra, R., Shanks, J.V., 1997. Transient studies of nutrient uptake, growth, and indole alkaloid accumulation in heterotrophic cultures of hairy roots of *Catharanthus roseus*. Biotechnol. Bioeng. 55, 527–534.

Burlat, V., Oudin, A., Courtois, M., Rideau, M., Saint-Pierre, B., 2004. Co-expression of three MEP pathway genes and geraniol 10-hydroxylase in internal phloem parenchyma of *Catnaranthusroseus* implicates multicellular translocation of intermediates during the biosynthesis of monoterpeneindole alkaloids and isoprenoid – derived primary metabolites. Plant J. 38, 131–141.

Choudhury, S., Gupta, K., 1999. Effect of dikegulac and on biomass and production in *Catharanthus roseus* (L) G Don under *in vitro* condition. Indian J. Exp. Bot. 37, 594–598.

Contin, A., Collu, G., van der Heijden, R., Verpoorte, R., 1999. The effects of phenobarbital and ketoconazole on the alkaloid biosynthesis in *Catharanthus roseus* cell suspension cultures. Plant Physiol. Biochem. 37, 139–144.

Courdavault, V., Thiesault, M., Courtois, M., Gantet, P., Oudin, A., Dioreau, P., St-Pierre, B., Giglioli-Guivarc'h, N., 2005. CaaX-prenyltransferases are essential for expression of genes involvedinthe early stages of monoterpenoid biosynthetic pathway in *Catharanthusroseus* cells. Plant Mol. Biol. 57, 855–870.

Croteau, R., Kutchan, T.M., Lewis, N.G., 2000. Natural products (secondary metabolites). Biochemmolbiol. Plants 24, 1250–1319.

Dhayanithy, G., Subban, K., Chelliah, J., 2019. Diversity and biological activities of endophytic fungi associated with *Catharanthusroseus*. BMC Microbiol. 19, 8–22.

El-Sayed, M., Verpoorte, R., 2002. Effect of phytohormones on growth and alkaloid accumulation by *Catharanthus roseus* cell suspension cultures fed with alkaloid precursors tryptamine and loganin. Plant Cell Tissue Organ Cult. 68, 265–270.

Facchini, P.J., 2001. Alkaloid biosynthesis in plants; biochemistry, cell biology, molecular regulation and metabolic engineering applications. Annu. Rev. Plant Physiol. Plant Mol. Biol. 52, 29–66.

Fernandez, J.A., Owen, T.G., Kurz, W.G., De Luca, V., 1989. Immunological detection and quantitation of tryptophan decarboxylase in developing *Catharanthus roseus* seedlings. Plant Physiol. 91, 79–84.

Geerlings, A., Ibanez, M.M., Memelink, J., van Der Heijden, R., Verpoorte, R., 2000. Molecular cloning and analysis of strictosidine beta-D-glucosidase, an enzyme in terpenoidindole alkaloid biosynthesis in Catharanthusroseus. J. Biol. Chem. 275, 3051–3056.

Goddijn, O.J., de Kam, R.J., Zanetti, A., Schilperoort, R.A., Hoge, J.H., 1992. Auxin rapidly down-regulates transcription of the tryptophan decarboxylase gene from *Catharanthus roseus*. Plant Mol. Biol. 18, 1113–1120.

Godoy-Hernandez, G., Loyola-Vargas, V.M., 1991. Effect of fungal homogenate, enzyme inhibitors and osmotic stress on alkaloid content of *Catharanthus roseus* cell suspension cultures. Plant Cell Rep. 10, 537–540.

Hirata, K., Horiuchi, M., Asada, M., Ando, T., Miyamota, K., Miura, Y., 1992. Stimulation of dimeric alkaloid production by nearultraviolet light in multiple shoot cultures of Catharanthusroseus. Ferment Bioeng. 74, 222–225.

Irmler, S., Schroder, G., St-Pierre, B., Crouch, N.P., Hotze, M., Schmidt, J., Strack, D., Matern, U., Schroder, J., 2000. Indole alkaloid biosynthesis in *Catharanthusroseus*: new enzyme activities and identification of cytochrome P450 CYP72A1 as secologanin synthase. Plant J. 24, 797–804.

Kent, P.N., Long, S.R., 1988. Alfalfa root exudates and compounds which promote or inhibit induction of rhizobium meliloti nodulation genes. Plant Physiol. 88, 396–400.

Meijer, A.H., Verpoorte, R., Hoge, J.H.C., 1993. Regulation of enzymes and genes involved in terpenoidindole alkaloid biosynthesis in *Catharanthusroseus*. J. Plant Res. 2, 145–164.

Menke, F.L., Champion, A., Kijne, J.W., Memelink, J., 1999a. A novel jasmonate- and elicitor-responsive element in the periwinkle secondary metabolite biosynthetic gene Str interacts with a jasmonate- and elicitor-inducible AP2-domain transcription factor, ORCA2. EMBO J. 18, 4455–4463.

Menke, F.L., Champion, A., Kijne, J.W., Memelink, J., 1999b. A novel jasmonate- and elicitor-responsive element in the periwinkle secondary metabolite biosynthetic gene Str interacts with a jasmonate- and elicitor-inducible AP2-domain transcription factor, ORCA2. EMBO J. 18, 4455–4463.

Moreno, P.R., van der Heijden, R., Verpoorte, R., 1993. Effect of terpenoid precursor feeding and elicitation on formation of indole alkaloids in cell suspension cultures of *Catharanthus roseus*. Plant Cell Rep. 12, 702–705.

Murata, J., De Luca, V., 2005. Localization of tabersonine 16-hydroxylase and 16-OH-tabersonine −16-O-methyltransferase to leaf epidermal cells define them as a major site of precursor biosynthesis in the vindoline pathway in *Catharanthusroseus*. Plant J. 44, 581–594.

Oksman-Caldentey, K.M., Arroo, R., 2000. Regulation of tropane alkaloid metabolism in plants and plant cell cultures. In: Verpoorte, R., Alfermann, A.W. (Eds.), Metabolic Engineering of Plant Secondary Metabolism. Kluwer Academic Publishers, Netherlands, pp. 253–281.

Ouwerkerk, P.B., Hallard, D., Verpoorte, R., Memelink, J., 1999a. Identification of UV-B light-responsive regions in the promoter of the tryptophan decarboxylase gene from *Catharanthus roseus*. Plant Mol. Biol. 41, 491–503.

Ouwerkerk, P.B., Trimborn TO, Hilliou, F., Memelink, J., 1999b. Nuclear factors GT-1 and 3AF1 interact with multiple sequences within the promoter of the Tdc gene from Madagascar periwinkle: GT-1 is involved in UV light-induced expression. Mol. Gen. Genet. 261, 610–622.

Papon, N., Bremer, J., Vansiri, A., Andreu, F., Rideau, M., Creche, J., 2005. Cytokinin and ethylene control indole alkaloid production at the level of the MEP/terpenoid pathway in *Catharanthusroseus* suspension cells. Planta Med. 71, 572–574.

Pasquali, G., Goddijn, O.J., de Waal, A., Verpoorte, R., Schilperoort, R.A., Hoge, J.H., Memelink, J., 1992. Coordinated regulation of two indole alkaloid biosynthetic genes from *Catharanthus roseus* by auxin and elicitors. Plant Mol. Biol. 18, 1121–1131.

Pauw, B., Hilliou, F.A., Martin, V.S., Chatel, G., de Wolf, C.J., Champion, A., Pre, M., van Duijn, B., Kijne, J.W., van der Fits, L., et al., 2004. Zinc finger oriteins act as transcriptional repressors of alkaloid biosynthesis genes in *Catharanthusroseus*. J. Biol. Chem. 279, 52940–52948.

Ramani, S., Patil, N., Jayabaskaran, C., 2010. UV-B induced transcript accumulation of DAHP synthase in suspension-cultured *Catharanthus roseus* cells. J. Mol. Signal. 5, 13–20.

Rijhwani, S.K., Shanks, J.V., 1998. Effect of elicitor dosage and exposure time on biosynthesis of indole alkaloids by *Catharanthusroseus* hairy root cultures. Biotehnol. Prog. 14, 442–449.

Schmidt, B.M., Ribnicky, D.M., Lipsky, P.E., Raskin, I., 2007a. Revisiting the ancient concept of botanical therapeutics. Nat. Chem. Biol. 3, 360–366.

Schmidt, B.M., Ribnicky, D.M., Lipsky, P.E., Raskin, I., 2007b. Revisiting the ancient concept of botanical therapeutics. Nat. Chem. Biol. 3, 360–366.

Siberil, Y., Benhamron, S., Memelink, J., Giglioli-Guivarc'h, N., Thiersault, M., Biosson, B., Doireau, P., Gantet, P., 2001. *Catharanthusroseus* G-box binding factors 1 and 2 act as repressors of strictosidine synthase gene expression in cell cultures. Plant Mol. Biol. 45, 477–488.

Sottomayor, M., Barcelo, A.R., 2005. The vincaalkaloids: from biosynthesis and accumulation in plant cells, to uptake, activity and metabolism in animal cells. Studies Nat. Prod. Chem. 33, 813–857.

St-Pierre, B., De Luca, V., 1995. A cytochrome P-450 monooxygenase catalyzes the first step in the conversion of tabersonine to vindoline in *Catharanthusroseus*. Plant Physiol. 109, 131–139.

St-Pierre, B., Laflamme, P., Alarca, A.M., De Luca, V., 1998. The terminal O-acetyltransferase involved in vindoline biosynthesis defines a new class of proteins responsible for coenzyme A-dependent acyl transfer. Plant J. 14, 703–713.

St-Pierre, B., Vazquez-Flota, F.A., De Luca, V., 1999. Multicellular compartmentation of *Catharanthusroseus* alkaloid biosynthesis predicts intercellular translocation of a pathway intermediate. Plant Cell 11, 887–900.

Suttipanta, N., Pattanaik, S., Gunjan, S., Xie, C.H., Littleton, J., Yuan, L., 2007. Promoter analysis of the *Catharanthus roseus* geraniol 10-hydroxylase gene involved in terpenoidindole alkaloid biosynthesis. Biochim. Biophys. Acta 1769, 139–148.

Taiz, L., Zeiger, E., 2006. Fisiologia Vegetal. vol. 10. Universitat Jaume I.

Unsicker, S.B., Kunret, G., Gershenzon, J., 2009. Protective perfumes: the role of vegetative volatiles in plant defense against herbivores. Curr. Opin. Plant Biol. 12, 479–485.

Van der Fits, L., Memelink, J., 2000. ORCA3, a jasmonate-responsive transcriptional regulator of plant primary and secondary metabolism. Science 289, 295–297.

Van der Fits, L., Zang, H., MenkeFL, D.M., Memelink, J., 2000. A *Catharanthusroseus*BPF-1 homologue interacts with an elicitor-responsive region of the secondary metabolite biosynthetic gene STR and induced by elicitor via a JA – independent signal transduction pathway. Plant Mol. Biol. 44, 675–685.

Vazquez-Flota, F.A., De Luca, V., 1998a. Developmental and light regulation of desacetoxyvindoline 4-hydroxylase in *Catharanthusroseus*. (L) G. Don. Plant Physiol. 117, 1351–1361.

Vazquez-Flota, F.A., De Luca, V., 1998b. Developmental and light regulation of desacetoxyvindoline 4-hydroxylase in *Catharanthusroseus* (L) G. Don. Plant Physiol. 117, 1351–1361.

Verpoorte, R., 2000. Secondary Metabolism. Kluwer Academic Publishers, Dordrecht, The Netherlands.

Verpoorte, R., Heijden, R.V.D., Schripsema, J., Hoge, J.H.C., Hoopen, H.J.G.T., 1993. Plant cell biotechnology for the production of alkaloids: present status and prospects. J. Nat. Prod. 56, 186–207.

Vickers, C.E., Gershenzon, J., Lerdan, M.T., Loreto, F., 2009. A unified mechanism of action for volatile isoprenoids in plant abiotic stress. Nat. Chem. Biol. 5, 283–291.

WeilerElmar, W., Dietmar, L., Stelmach Boguslawa, A., Peter, H., Ines, B.C.K., 1999. Octadecanoid and hexadecanoid signalling in plant defence. In: Pages 191-204 of: Novartis Foundation Symposium 223-Insect-Plant Interactions and Induced Plant Defence. Wiley Online Library.

Westekemper, P., Wieczorek, U., Gueritte, F., Langlois, N., Langlois, Y., Potier, P., Zenk, M.H., 1980. Radioimmunoassay for the determination of the indole alkaloid vindoline in *Catharanthus*. Planta Med. 39, 24–37.

Whitmer, S., van der Heijden, R., Verpoorte, R., 2002. Effect of precursor feeding on alkaloid accumulation by a tryptophan decarboxylase over-expressing transgenic cell line T22 of *Catharanthus roseus*. J. Biotechnol. 96, 193–203.

Zhao, J., Zhu, W., Hu, Q., Xing, W.H., 2000. Improved alkaloid production in *Catharanthus roseus* suspension cell cultures by various chemicals. Biotechnol. Lett. 22, 1221–1226.

Zhao, J., Zhu, W., Hu, Q., 2001a. Effects of light and plant growth regulators on the biosynthesis of vindoline and other indole alkaloids in *Catharanthus roseus* callus cultures. Plant Growth Regul. 33, 43–49.

Zhao, J., Zhu, W., Hu, Q., 2001b. Selection of fungal elicitors to increase indole alkaloid accumulation in *Catharanthus roseus* suspension cell culture. Enzym. Microb. Technol. 28, 666–672.

Exploring plant volatile compounds in sustainable crop improvement

19

Younes M. Rashad
Plant Protection and Biomolecular Diagnosis Department, Arid Lands Cultivation Research Institute, City of Scientific Research and Technological Applications, Alexandria, Egypt

19.1 Introduction

Plant volatile compounds (PVCs) are bioactive small molecules produced by a large number of plant species and released in a gaseous form. To date, more than 2000 PVCs of different odors are known to be released by more than 90 plant families (Park et al., 2016; Altindal and Altindal, 2017; Jing et al., 2020; Zouaouia et al., 2020). PVCs are emitted into the surrounding air from different plant parts such as leaves, flowers, or fruits and released into the rhizosphere from the roots. They are implicated in a large array of vital physiological and ecological functions including attraction of pollinators, seed dispersers, and natural enemies (Amo et al., 2013; Schiestl, 2015), defense against herbivorous insects, predators/parasitoids, and microbial attack (Dong et al., 2016), airborne communication signals in plant-plant interactions (Karban et al., 2014; Effaha et al., 2019), and interactions between plants and other organisms, which is called plant language (Simpraga et al., 2016). Moreover, PVCs emitted by a plant organ can determine which type(s) of microorganisms can live in their phyllosphere or rhizosphere at the same time; these microorganisms release their own VCs which affect the plant growth and resistance in a bidirectional manner (Jogaiah et al., 2013; Farré-Armengol et al., 2016). The emission, composition, and release duration of PVCs depend on the type of the emitter plant species (Jaeger et al., 2016; Malik et al., 2018), and the inducer whether a biotic such as single or continuous wounding, low or high temperature, lack or excess of light, or biotic such as herbivorous insects, egg deposition, attacking predators or microbes. In other words, PVCs reflect the physiological status of the emitter plant (Sharifi et al., 2018). Moreover, recent researches have shown that emission of PVCs depends on the richness of plant species and the type of neighboring plant community (Kigathi et al., 2019).

Based on their biochemical origin, PVCs are classified into four groups: (A) terpenes derivatives (isoprenoids) including monoterpenes (e.g., α-pinene, limonene, and α-terpineol), sesquiterpenes (e.g., α-farnesene, γ-bisabolene, and α-guaiene), diterpenes (e.g., sclarene) or irregular terpenes (e.g., geranylacetone and dihydro-β-ionone), (B) fatty acid derivatives (e.g., (Z)-3-hexenyl acetate and 2*E*,6*Z*-nonadienol), (C) phenylpropanoids/benzenoids (e.g., carvacrol and (*E*)-anethole), or (D) amino acids

Biocontrol Agents and Secondary Metabolites. https://doi.org/10.1016/B978-0-12-822919-4.00019-3

derivatives (e.g., methanethiol, dimethyl disulfide, and thioesters) (Dudareva et al., 2013; Bialecki and Smadja, 2014). In this chapter, we will explore the functional roles of PVCs in sustainable crop improvement including their application for protecting crops against different biotic stresses, improving of plant growth and productivity, and their application in smart agricultural practices.

19.2 PVCs in protection against pathogens

Emission of PVCs by many plant species represents an adaptive mechanism to overcome different stressful conditions including attacking pathogens. In this regard, plenty of PVCs-emission cases have been reported as plant defense responses against various phytopathogens (Spinelli et al., 2011; Iakimova et al., 2013; Cellini et al., 2016). In a recent study on apple plant, emission of an array of PVCs including 3-hexenal, 2-hexenal, 2-Methyl-1-propanol, and a-farnesene was reported to be induced by infection with *Erwinia amylovora* or *Pseudomonas syringae* pv. *syringae*. Moreover, emission of disease-specific PVCs such as hexenal isomers and 2,3-butanediol, 3-hydroxy-2-butanone, 1,2-propanediol, and phenylethyl alcohol was also detected (Cellini et al., 2018). López et al. (2015) reported also pathosystem-specific emissions of (Z)-3-hexenyl 2-methylbutanoate from apple plants infected with *Penicillium expansum* and *Rhizopus stolonifer*, and 2-butanone and a-pinene from pear plants infected with *R. stolonifer*. These PVCs may promote plant growth or elicit their resistance against the attacking pathogens. The stimulated plant exhibits earlier, higher, and faster responses upon further stress occurrence (Conrath et al., 2015), which has been considered as a kind of green vaccination (Luna-Diez, 2016). In this regard, exposure of *Arabidopsis thaliana* to a mixture of α-pinene and β-pinene stimulated the plant defense, and triggered transcriptional expression of salicylic acid-related genes. The volatile emissions from these resistance-elicited plants stimulated plant resistance in the neighboring plants (Riedlmeier et al., 2017). The defensive mechanisms involved in the PVCs-induced protection may affect the host plant itself by triggering their defense responses, and/or the invading pathogen by inhibiting their viability through reducing the membrane permeability, disturbing the nutrient transport or interfering with the essential metabolic reactions (Spinelli et al., 2012; Abdelrahman et al., 2016; Gong et al., 2019). In addition, emission of PVCs can be induced also by beneficial microbes such as plant growth-promoting rhizobacteria (Cappellari et al., 2020). A threefold increase was recorded in *Mentha piperita* plants inoculated with plant growth-promoting rhizobacteria (Cappellari et al., 2017). However, PVCs emission patterns of a plant cannot be determined based on their interactions with microbial monocultures. Richness and diversity of microbiota communities in the plant phyllosphere and rhizosphere drive the balance between PVCs-mediated microbe-pathogen and microbe-plant interactions (Raza et al., 2020). In general, variability in PVCs emission between plant species/cultivars may reflect their susceptibility/resistance during plant-microbe interactions (Louw and Korsten, 2014).

Applications of PVCs as gaseous treatments, in a process called biofumigation, to control postharvest decay have received growing interest in the last decade

(Sivakumar and Bautista-Baños, 2014; Pekmezovic et al., 2015; Mari et al., 2016). Owing to their known antimicrobial, preservative, and biodegradability properties, PVCs can play an important role in the postharvest systems (Ugolini et al., 2014; Sun et al., 2014). In a recent study, application of two essential oils, caraway and spearmint, was investigated to provide a preservation method for Jerusalem artichoke tubers against *Sclerotium* tuber rot fungus (*Sclerotium rolfsii*). Tubers infected with *S. rolfsii* and treated with caraway oil at 2% and kept in peat moss exhibited lower disease severity, sprouting percentage, weight loss, and higher quality parameters and storage life potential (Ghoneem et al., 2016). The antifungal activities of essential oils from different citrus species were evaluated in vitro against the postharvest pathogens *P. digitatum*, *Trichoderma viride,* and *Botrytis cinerea.* With exception to *P. digitatum*, all the tested PVCs significantly inhibited mycelial growth of *B. cinerea* and *T. viride* (Simas et al., 2017). In another study, Wood et al. (2013) tested the PVCs acetaldehyde and 2*E*-hexenal to control three postharvest diseases of potato (black dot, silver scurf, and bacterial soft rot). 2*E*-hexenal at $5\,\mu L\,L^{-1}$ showed a complete growth suppression of all tested pathogens. All of these promising results suggest that PVCs represent a potential and eco-safe approach to enhance productivity of crops in the future sustainable agriculture.

19.3 PVCs in protection against herbivores

Many PVCs are released by several plants as a defense response to attack of herbivorous arthropods (Villamar-Torres et al., 2018). Different direct and indirect mechanisms were described in this regard. A direct toxic effect of gossypol, a highly toxic sesquiterpenoid, released by emitting glands in cotton leaves was reported as a defense response against lepidopteran pests (Krempl et al., 2016; Hagenbucher et al., 2017). On the other hand, different indirect mechanisms were also reported such as the triggered release of PVCs (semiochemicals) by several plant species in response to herbivore arthropods attack, with a repellent effect on these invading arthropods (Krokene, 2015), other arthropods (Danielsson et al., 2019), or with an attractive effect on their predators (Aartsma et al., 2017). Zhang et al. (2014) found that the PVCs: eugenol, *l*-carvone, *p*/*l*-methone, and methyl salicylate and the essential oils: clove, lemongrass, spearmint, and ylang-ylang have a strong repellent effect on the brown marmorated stink bug *Halyomorpha halys*. Attraction and recruitment of arthropod natural enemies (predators and parasitoids) by induced emission of PVCs from arthropod-infected plants have also a considerable potential in their biological control strategies. In this regard, Xiu et al. (2019) investigated the attractiveness of PVCs from various aphid-infected plants combinations to adults of the multicolored Asian lady beetle (*Harmonia axyridis* Pallas). Five electrophysiologically active PVCs significantly attracted lady beetles including *p*-diethylbenzene, 3-ethylacetophenone, 1,2-diethylbenzene, α-pinene, and butyl acrylate. These results have great contribution in the establishment of integrated pest management programs to control aphid or developing volatile-based traps for their natural enemies, lady beetles. More interesting, numerous herbivore-induced PVCs have the ability for selective attraction or repulsion of a particular natural enemy such as *H. halys* (Weber et al., 2017). Moreover, many studies have elucidated the exploitation of egg parasitoids for herbivore-induced PVCs to find their oviposited eggs at an early developmental stage (Dicke and Baldwin, 2010; Michereff et al., 2011). In addition, oviposition-induced emissions of PVCs have been exploited also by early larval parasitoids (Fatouros et al., 2012). Rondoni et al. (2017) reported the attraction of the parasitoids (*Anastatus bifasciatus* and *Ooencyrtus telenomicida*) to PVCs released by faba bean plants oviposited upon by *H. halys*. In another study, Milonas et al. (2019) showed that changes in the profile of the PVCs (3-(Z)-hexen-1-ol and β-myrcene), released by tomato plants, were qualitatively and quantitatively induced by oviposition of *Tuta absoluta* females leading to attraction of their egg parasitoid *Trichogramma cordubense*. In other words, herbivore-induced PVCs have a potential role as cues for foraging parasitoids.

19.4 PVC-mediated weed control

Weed control represents one of the crucial problems in organic agricultural systems worldwide (Peigné et al., 2016). PVCs have attained great interest as eco-friendly and fast biodegradable herbicides (Benvenuti et al., 2017; Brilli et al., 2019). Various PVCs with a phytotoxic effect on germination and growth of weeds have been reported by many researchers (Araniti et al., 2013; Puig et al., 2018; Souza-Alonso et al., 2018). In this regard, Pardo-Muras et al. (2018) studied the phytotoxic activity of PVCs emitted by the legume shrubs *Ulex europaeus* and *Cytisus scoparius* and the possibility to be used as sources of natural products with bioherbicide potential in sustainable agriculture systems. PVCs released by both shrubs, particularly linalool, α-terpineol, aspirane, verbenone, and eugenol, exhibited a phytotoxic effect against the growth of two agricultural weeds: *Amaranthus retroflexus* and *Digitaria sanguinalis* up to 80% inhibition. In a recent study, Pardo-Muras et al. (2019) investigated the herbicidal effect of PVCs, namely eugenol, verbenone, terpinen-4-ol, α-terpineol, and linalool, released from *U. europaeus* and *C. scoparius*, individually and in binary mixtures. Strong, irreversible, and synergistic inhibitory effects by the tested PVCs were achieved against germination and early growth of the agricultural weed species *A. retroflexus* and *D. sanguinalis*. The reported phytotoxicity of the paired mixtures indicated that PVCs act synergistically, not only additively. Thus, interactions of different PVCs with allelopathic potential have an advantageous property owing to their variability of chemical composition and mechanisms of action which guarantee an increased efficiency and lowered herbicide resistance. However, the phytotoxicity of PVCs is affected by various factors such as physiological plant stage, climate, flow rate, and season, and their release is a function of the surrounding conditions such as temperature, light, and irrigation (Holopainen and Gershenzon, 2010).

19.5 PVCs in improving/suppressing plant growth and productivity

PVCs have another valuable role in improving plant growth and productivity by reducing the plant-plant competitions through their allelopathic effects on the surrounding weeds, allowing the emitter plant to obtain more nutrients, water, and light (Puig et al., 2018). In general, physiological mechanisms utilized by PVCs in plant-plant competition may be directly through allelopathic potential affecting seed germination or growth of the neighboring plant, or indirectly through suppressing the plant's competitive ability such as associational resistance or chemical camouflage (Effaha et al., 2019). Senescence regulation in plant tissues has been also reported for some PVCs which synergistically work with plant hormones such as cytokinins. Synergy of isoprenoids with hormones biosynthesis leads to improved antioxidant potential in the plant tissue which inhibits the cell degradation and prolongs the lifespan of the plant tissues, in addition, it improves the whole plant production (Dani et al., 2016). In contrast, it is well accepted that some PVCs (i.e., terpenes) have a crucial role in inducing plant senescence symptoms such as membrane destruction and cell deaths. In this regard, microtubules disruption was observed in *A. thaliana* plants exposed to citral vapor in a dose- and time-dependent manner, suggesting their interaction with tubulin (Chaimovitsh et al., 2010). Inhibition of root growth, triggering oxidative damage by inducing reactive oxygen species, enhanced lipid peroxidation, disintegration of the cell membrane, and/or suppression of antioxidant enzymes were also reported as a result of exposure of roots of *Celtis occidentalis* to α-pinene (Singh et al., 2009).

19.6 PVCs in smart agriculture practices

To reduce the economic losses in global crop yields due to invading pathogens and pests, there is an increased demand for developing a fast and accurate method for early detection of the plant diseases. Monitoring PVCs emission of a plant represents an advanced approach for early diagnosis of their invaders pathogens and pests. Specificity of PVCs emission of a plant can serve as an indicator of their real-time physiological health status and can be utilized in developing a PVCs-specific electrochemical biosensor, which is also known as electronic nose. These gas sensors include three different categories: conductivity sensors, gravimetric sensors, and optical sensors, and are characterized by portability, sensitivity, nondestructive, rapid detection, and reduced cost (Cui et al., 2018). Aksenov et al. (2014) developed a novel biomarker, with 90% accuracy, based on the chemical analysis of the emitted PVCs from the citrus trees infected with citrus greening (huanglongbing), caused by *Candidatus Liberibacter*, using gas chromatography/mass spectrometry and gas chromatography/differential mobility spectrometry. Emission of PVCs including hexanol, tetradecene, linalool, and phenylacetaldehyde was found to be induced in the diseased citrus trees. In another study, electrical biosensor arrays were also developed for the detection of

PVCs emitted by huanglongbing disease-infected citrus trees based on single-walled carbon nanotubes decorated with single-stranded DNA for early detection of hexanol, linalool, tetradecene, and phenylacetaldehyde release over a wide concentration range from 5% to 100% (Wang et al., 2019). However, most of these developed technologies (electronic noses) have several limitations, including low sensitivity at a very low concentration of PVCs, short specificity to PVCs with similar chemical structures, and severe inconsistency due to environmental conditions. Alternatively, a new chemical sensor array has been developed with high sensitivity, multiplexity, and chemical selectivity. Portable smartphone-based PVCs fingerprinting platform was integrated for early diagnosis of tomato late blight, caused by *Phytophthora infestans*, using a paper-based colorimetric sensor array that incorporates plasmonic cysteine-functionalized gold nanoparticles or nanorods and chemo-responsive organic dyes (Li et al., 2019). This method is based on specific detection of the green leafy aldehyde (*E*)-2-hexenal-emission, which is one of the main PVCs released by late blight-diseased tomato leaves.

References

Aartsma, Y., Bianchi, F.A., Werf, W.V.D., Poelman, E.H., Dicke, M., 2017. Herbivore-induced plant volatiles and tritrophic interactions across spatial scales. New Phytol. 216, 1054–1063.

Abdelrahman, M., Abdel-Motaal, F., El-Sayed, M., Jogaiah, S., Shigyo, M., Ito, S.-i., Tran, L.-S.P., 2016. Dissection of *Trichoderma longibrachiatum*-induced defense in onion (*Allium cepa* L.) against *Fusarium oxysporum* f. sp. *Cepa* by target metabolite profiling. Plant Sci. 246, 128–138.

Aksenov, A.A., Pasamontes, A., Peirano, D.J., Zhao, W.X., Dandekar, A.M., Fiehn, O., Ehsani, R., Davis, C.E., 2014. Detection of Huanglongbing disease using differential mobility spectrometry. Anal. Chem. 86, 2481–2488.

Altindal, D., Altindal, N., 2017. Plant volatile compounds in growth. In: Choudhary, D., Sharma, A., Agarwal, P., Varma, A., Tuteja, N. (Eds.), Volatiles and Food Security. Springer, Singapore.

Amo, L., Jansen, J.J., Dam, N.M., Dicke, M., Visser, M.E., 2013. Birds exploit herbivore-induced plant volatiles to locate herbivorous prey. Ecol. Lett. 16, 1348–1355.

Araniti, F., Lupini, A., Sorgonà, A., Statti, G.A., Abenavoli, M.R., 2013. Phytotoxic activity of foliar volatiles and essential oils of*Calamintha nepeta* (L.) Savi. Nat. Prod. Res. 27, 1651–1656.

Benvenuti, S., Cioni, P.L., Flamini, G., Pardossi, A., 2017. Weeds for weed control: Asteraceae essential oils as natural herbicides. Weed Res. 57, 342–353.

Bialecki, A., Smadja, J., 2014. Identification of volatile compounds from flowers and aromatic plants: how and why? In: Gupta Bhowon, M., Jhaumeer-Laulloo, S., Li Kam Wah, H., Ramasami, P. (Eds.), Chemistry: The Key to our Sustainable Future. Springer, Dordrecht.

Brilli, F., Loreto, F., Baccelli, I., 2019. Exploiting plant volatile organic compounds (VOCs) in agriculture to improve sustainable defense strategies and productivity of crops. Front. Plant Sci. 10, 1–8.

Cappellari, L.R., Chiappero, J., Santoro, M., Giordano, W., Banchio, E., 2017. Inducing phenolic production and volatile organic compounds emission by inoculating*Mentha piperita* with plant growth-promoting rhizobacteria. Sci. Hortic. 220, 193–198.

Cappellari, L.R., Santoro, M.V., Schmidt, A., Gershenzon, J., Banchio, E., 2020. Improving phenolic total content and monoterpene in*Mentha* x*piperita* by using salicylic acid or methyl jasmonate combined with rhizobacteria inoculation. Int. J. Mol. Sci. 21 (1), 50.

Cellini, A., Biondi, E., Buriani, G., Farneti, B., Rodriguez Estrada, M.T., Braschi, I., Savioli, S., Blasioli, S., Rocchi, L., Biasioli, F., Costa, G., Spinelli, F., 2016. Characterization of volatile organic compounds emitted by kiwifruit plants infected with *Pseudomonas syringae* pv. *actinidiae* and their effects on host defences. Trees 30, 795–806.

Cellini, A., Buriani, G., Rocchi, L., Rondelli, E., Savioli, S., Rodriguez Estrada, M.T., Cristescu, S.M., Costa, G., Spinelli, F., 2018. Biological relevance of volatile organic compounds emitted during the pathogenic interactions between apple plants and *Erwinia amylovora*. Mol. Plant Pathol. 19 (1), 158–168.

Chaimovitsh, D., Abu-Abied, M., Belausov, E., Rubin, B., Dudai, N., 2010. Microtubules are an intracellular target of the plant terpene citral. Plant J. 61, 399–408.

Conrath, U., Beckers, G.J., Langenbach, C.J., Jaskiewicz, M.R., 2015. Priming for enhanced defense. Annu. Rev. Phytopathol. 53, 97–119.

Cui, S., Ling, P., Zhu, H., Keener, H.M., 2018. Plant pest detection using an artificial nose system: a review. Sensors (Basel). 18 (2) E378.

Dani, K.G.S., Fineschi, S., Michelozzi, M., Loreto, F., 2016. Do cytokinins, volatile isoprenoids and carotenoids synergically delay leaf senescence? Plant Cell Environ. 39, 1103–1111.

Danielsson, M., Zhao, T., Borg-Karlson, A., 2019. Arthropod infestation sites and induced defence can be traced by emission from single spruce needles. Arthropod Plant Interact. 13, 253–259.

Dicke, M., Baldwin, I.T., 2010. The evolutionary context for herbivore-induced plant volatiles: beyond the 'cry for help'. Trends Plant Sci. 15 (3), 167–175.

Dong, F., Fu, X., Watanabe, N., Su, X., Yang, Z., 2016. Recent advances in the emission and functions of plant vegetative volatiles. Molecules 21 (2), 124.

Dudareva, N., Klempien, A., Muhlemann, J.K., Kaplan, I., 2013. Biosynthesis, function and metabolic engineering of plant volatile organic compounds. New Phytol. 198, 16–32.

Effaha, E., Holopainenb, J.K., McCormicka, A.C., 2019. Potential roles of volatile organic compounds in plant competition. Perspect. Plant Ecol. Evol. Syst. 38, 58–63.

Farré-Armengol, G., Filella, I., Llusia, J., Peñuelas, J., 2016. Bidirectional interaction between phyllospheric microbiotas and plant volatile emissions. Trends Plant Sci. 21 (10), 854–860.

Fatouros, N.E., Lucas-Barbosa, D., Weldegergis, B.T., Pashalidou, F.G., van Loon, J.J., Dicke, M., Harvey, J.A., Gols, R., Huigens, M.E., 2012. Plant volatiles induced by herbivore egg deposition affect insects of different trophic levels. PLoS One. 7(8), e43607.

Ghoneem, K.M., Saber, W.I.A., El-Awady, A.A., Rashad, Y.M., Al-Askar, A.A., 2016. Alternative preservation method against *Sclerotium* tuber rot of Jerusalem artichoke using natural essential oils. Phytoparasitica 44 (3), 341–352.

Gong, D., Bi, Y., Li, Y., Zong, Y., Han, Y., Prusky, D., 2019. Both*Penicillium expansum* and*Trichothecim roseum* infections promote the ripening of apples and release specific volatile compounds. Front. Plant Sci. 10, 338.

Hagenbucher, S., Eisenring, M., Meissle, M., Romeis, J., 2017. Interaction of transgenic and natural insect resistance mechanisms against *Spodoptera littoralis* in cotton. Pest Manag. Sci. 73, 1670–1678.

Holopainen, J.K., Gershenzon, J., 2010. Multiple stress factors and the emission of plant VOCs. Trends Plant Sci. 15 (3), 176–184.

Iakimova, E.T., Sobiczewski, P., Michalczuk, L., Wegrzynowicz-Lesiak, E., Mikicinski, A., Woltering, E.J., 2013. Morphological and biochemical characterization of *Erwinia amylovora*-induced hypersensitive cell death in apple leaves. Plant Physiol. Biochem. 63, 292–305.

Jaeger, D.M., Runyon, J.B., Richardson, B.A., 2016. Signals of speciation: volatile organic compounds resolve closely related sagebrush taxa, suggesting their importance in evolution. New Phytol. 211, 1393–1401.

Jing, X., Lun, X., Fan, C., Ma, W., 2020. Emission patterns of biogenic volatile organic compounds from dominant forest species in Beijing, China. J. Environ. Sci. https://doi.org/10.1016/j.jes.2020.03.049.

Jogaiah, S., Mostafa, A., Tran, L.S.P., Shin-ichi, I., 2013. Characterization of rhizosphere fungi that mediate resistance in tomato against bacterial wilt disease. J. Exp. Bot. 64, 3829–3842.

Karban, R., Yang, L.H., Edwards, K.F., 2014. Volatile communication between plants that affects herbivory: a meta-analysis. Ecol. Lett. 17, 44–52.

Kigathi, R.N., Weisser, W.W., Reichelt, M., Gershenzon, J., Unsicker, S.B., 2019. Plant volatile emission depends on the species composition of the neighboring plant community. BMC Plant Biol. 19, 58.

Krempl, C., Heidel-Fischer, H.M., Jiménez-Alemán, G.H., Reichelt, M., Menezes, R.C., Boland, W., Vogel, H., Heckel, D.G., Joußen, N., 2016. Gossypol toxicity and detoxification in *Helicoverpa armigera* and *Heliothis virescens*. Insect Biochem. Mol. Biol. 78, 69–77.

Krokene, P., 2015. Conifer defense and resistance to bark beetles. In: Vega, F.E., Hofstetter, R.W. (Eds.), Bark Beetles: Biology and Ecology of Native and Invasive Species. Elsevier Academic Press, San Diego, pp. 177–207.

Li, Z., Paul, R., Ba Tis, T., Saville, A.C., Hansel, J.C., Yu, T., Ristaino, J.B., Wei, Q., 2019. Non-invasive plant disease diagnostics enabled by smartphone-based fingerprinting of leaf volatiles. Nat. Plants 5 (8), 856–866.

López, L., Echeverria, G., Usall, J., Teixidó, N., 2015. The detection of fungal diseases in the "Golden smoothee" apple and "Blanquilla" pear based on the volatile profile. Postharvest Biol. Technol. 99, 120–130.

Louw, J.P., Korsten, L., 2014. Pathogenic *Penicillium* spp. on apple and pear. Plant Dis. 98, 590–598.

Luna-Diez, E., 2016. Using green vaccination to brighten the agronomic future. Outlooks Pest Manage. 27, 136–140.

Malik, T.G., Gajbhiye, T., Pandey, S.K., 2018. Plant specific emission pattern of biogenic volatile organic compounds (BVOCs) from common plant species of Central India. Environ. Monit. Assess. 190, 631.

Mari, M., Bautista-Baños, S., Sivakumar, D., 2016. Decay control in the postharvest system: role of microbial and plant volatile organic compounds. Postharvest Biol. Technol. 122, 70–81.

Michereff, M.F.F., Laumann, R.A., Borges, M., Michereff, F.M., Diniz, I.R., Faria, N.A., Moraes, M.C.B., 2011. Volatiles mediating plant-herbivory-natural enemy interaction in resistant and susceptible soybean cultivars. J. Chem. Ecol. 37, 273–285.

Milonas, P.G., Anastasaki, E., Partsinevelos, G., 2019. Oviposition-induced volatiles affect electrophysiological and behavioral responses of egg parasitoids. Insects 10 (12), 437.

Pardo-Muras, M., Puig, C.G., López-Nogueira, A., Cavaleiro, C., Pedrol, N., 2018. On the bioherbicide potential of *Ulex europaeus* and *Cytisus scoparius*: profiles of volatile organic compounds and their phytotoxic effects. PLoS One. 13(10), e0205997.

Pardo-Muras, M., Puig, C.G., Pedrol, N., 2019. *Cytisus scoparius* and *Ulex europaeus* produce volatile organic compounds with powerful synergistic herbicidal effects. Molecules 24, 4539.

Park, Y.J., Baskar, T.B., Yeo, S.K., Arasu, M.V., Al-Dhabi, N.A., Lim, S.S., Park, S.U., 2016. Composition of volatile compounds and in vitro antimicrobial activity of nine *Mentha* spp. Springerplus 5 (1), 1628.

Peigné, J., Casagrande, M., Payet, V., David, C., Sans, F.X., Blanco-Moreno, J.M., Cooper, J., Gascoyne, K., Antichi, D., Bàrberi, P., Bigongiali, F., Surböck, A., Kranzler, A., Beeckman, A., Willekens, K., Luik, A., Matt, D., Grosse, M., Heß, J., Clerc, M., Dierauer, H., Mäder, P., 2016. How organic farmers practice conservation agriculture in Europe. Renewable Agric. Food Syst. 31 (1), 72–85.

Pekmezovic, M., Rajkovic, K., Barac, A., Senerovic, L., Arsic, A.V., 2015. Development of kinetic model for testing antifungal effect of *Thymus vulgaris* L. and *Cinnamomum cassia* L. essential oils on *Aspergillus flavus* spores and application for optimization of synergistic effect. Biochem. Eng. J. 99, 131–137.

Puig, C.G., Gonçalves, R.F., Valentão, P., Andrade, P.B., Reigosa, M.J., Pedrol, N., 2018. The consistency between phytotoxic effects and the dynamics of allelochemicals release from *Eucalyptus globulus* leaves used as bioherbicide green manure. J. Chem. Ecol. 44, 658–670.

Raza, W., Wang, J., Jousset, A., Friman, V.-P., Mei, X., Wang, S., Wei, Z., Shen, Q., 2020. Bacterial community richness shifts the balance between volatile organic compound-mediated microbe–pathogen and microbe–plant interactions. Proc. R. Soc. B 287, 20200403.

Riedlmeier, M., Ghirardo, A., Wenig, M., Knappe, C., Koch, K., Georgii, E., Dey, S., Parker, J.E., Schnitzler, J.-P., Vlot, A.C., 2017. Monoterpenes support systemic acquired resistance within and between plants. Plant Cell 29, 1440–1459.

Rondoni, G., Bertoldi, V., Malek, R., Foti, M.C., Peri, E., Maistrello, L., Haye, T., Conti, E., 2017. Native egg parasitoids recorded from the invasive *Halyomorpha halys* successfully exploit volatiles emitted by the plant-herbivore complex. J. Pest. Sci. 90, 1087–1095.

Schiestl, F.P., 2015. Ecology and evolution of floral volatile- mediated information transfer in plants. New Phytol. 206, 571–577.

Sharifi, R., Lee, S.M., Ryu, C.M., 2018. Microbe-induced plant volatiles. New Phytol. 220, 684–691.

Simas, D.L.R., de Amorim, S.H.B.M., Goulart, F.R.V., Alviano, C.S., Alviano, D.S., da Silva, A.J.R., 2017. Citrus species essential oils and their components can inhibit or stimulate fungal growth in fruit. Ind. Crop. Prod. 98, 108–115.

Simpraga, M., Takabayashi, J., Holopainen, J.K., 2016. Language of plants: where is the word? J. Integr. Plant Biol. 58, 343–349.

Singh, H.P., Kaur, S., Mittal, S., Batish, D.R., Kohli, R.K., 2009. Essential oil of *Artemisia scoparia* inhibits plant growth by generating reactive oxygen species and causing oxidative damage. J. Chem. Ecol. 35, 154–162.

Sivakumar, D., Bautista-Baños, S., 2014. A review on the use of essential oils for postharvest decay control and maintenance of fruit quality during storage. Crop Prot. 64, 27–37.

Souza-Alonso, P., González, L., López-Nogueira, A., Cavaleiro, C., Pedrol, N., 2018. Volatile organic compounds of *Acacia longifolia* and their effects on germination and early growth of species from invaded habitats. Chem. Ecol. 34, 126–145.

Spinelli, F., Costa, G., Rondelli, E., Busi, S., Vanneste, J.L., Rodriguez, E.M.T., Savioli, S., Harren, F.J.M., Crespo, E., Cristescu, S.M., 2011. Emission of volatiles during the pathogenic interaction between *Erwinia amylovora* and *Malus domestica*. Acta Hortic. 896, 55–63.

Spinelli, F., Cellini, A., Vanneste, J.L., Rodriguez-Estrada, M.T., Costa, G., Savioli, S., Harren, F.J.M., Cristescu, S.M., 2012. Emission of volatile compounds by *Erwinia amylovora*: biological activity in vitro and possible exploitation for bacterial identification. Trees Struct. Funct. 26, 141–152.

Sun, X., Narciso, J., Wang, Z., Ference, C., Bai, J., Zhou, K., 2014. Effect of chitosan-essential oil coatings on safety and quality of fresh blueberries. J. Food Sci. 79, 955–960.

Ugolini, L., Martini, C., Lazzeri, L., D'Avino, L., Mari, M., 2014. Control of postharvest grey mould (*Botrytis cinerea* Per.: Fr.) on strawberries by glucosinolate-derived allyl-isothiocyanate treatments. Postharvest Biol. Technol. 90, 34–39.

Villamar-Torres, R., Seyed Mehdi Jazayeri, S.M., Liuba-Delfini, G., Cruzaty, L.G., Viot, C.-R., 2018. Volatile organic compounds: plant natural defense mechanisms against herbivorous arthropods and an opportunity for plant breeding of cotton. Sci. Agropecu. 9 (2), 287–297.

Wang, H., Ramnani, P., Pham, T., Villarreal, C.C., Yu, X., Liu, G., Mulchandani, A., 2019. Gas biosensor arrays based on single-stranded DNA-functionalized single-walled carbon nanotubes for the detection of volatile organic compound biomarkers released by huanglongbing disease-infected citrus trees. Sensors 19, 4795.

Weber, D.C., Morrison, W.R., Khrimian, A., Rice, K.B., Leskey, T.C., Rodriguez-Saona, C., Blaauw, B.R., 2017. Chemical ecology of *Halyomorpha halys*: discoveries and applications. J. Pest. Sci. 90, 989–1008.

Wood, E.M., Miles, T.D., Wharton, P.S., 2013. The use of natural plant volatile compounds for the control of the potato postharvest diseases, black dot, silver scurf and soft rot. Biol. Control 64, 152–159.

Xiu, C., Zhanga, W., Xua, B., Wyckhuysa, K.A.G., Caic, X., Sub, H., Lu, Y., 2019. Volatiles from aphid-infested plants attract adults of the multicolored Asian lady beetle *Harmonia axyridis*. Biol. Control 129, 1–11.

Zhang, Q.-H., Schneidmiller, R.G., Hoover, D.R., Zhou, G., Margaryan, A., Bryant, P., 2014. Essential oils as spatial repellents for the brown marmorated stink bug, *Halyomorpha halys* (Stål) (Hemiptera: Pentatomidae). J. Appl. Entomol. 138, 490–499.

Zouaouia, N., Chenchounic, H., Bouguerrab, A., Massourase, T., Barkatb, M., 2020. Characterization of volatile organic compounds from six aromatic and medicinal plant species growing wild in north African drylands. NFS J. 18, 19–28.

Biostimulants: Promising probiotics for plant health

20

S.A. Belorkar
Department of Microbiology and Bioinformatics, Atal Bihari Vajpayee University, Bilaspur, India

20.1 Introduction

Biostimulants are the edge of agro-technology focusing on the overall health of Plants. They empower the host plant with attributes like improved nutrient absorption efficacy, tolerance against stresses, and bestows virtues of faster growth, health, and productivity. The derived edibles also exhibit increased nutritional content and shelf life with a better fresh look. Although the term biostimulants encompasses the word "Bio," yet it includes many other nonbiological ingredients. Although all the components do not exert a direct effect on the plant growth, they stimulate the biologicals of the microbiome to trigger the probiotic effect.

20.2 Biostimulant: A changing perspective

In the past, biostimulants existed persistently as an agricultural aid in the form of use of waste materials, ashes, microbes for soil amendment. Attempts were made for categorization of biostimulants since 1951. Scientists provided various grounds for their classification and ended up with different approaches. The review by Yakhin et al. (2017) provides deep insight into the various classifications of biostimulants. Du Jarden had an explicit work on Biostimulants and categorized them in seven classes considering microbial hormone source and biopesticides as a separate entity. Bulgari et al. in 2015 suggested mode of action as the basis of biostimulant classification.

20.3 Active components of biostimulant

The composition of a Biostimulant is very critical since its impact is dependent on the interplay of its components. Every component exerts diverse effects on Plant parts where it is applied and transverse effects on other cocomponents are also seen. These components modulate the plant response, the microbes in the rhizosphere, and interaction of all components of the microbiome. Fig. 20.1 provides insight into the components which add up to make up a Biostimulant formulation. As is clear from Fig. 20.1, the biostimulant is made up of a variety of Biological and Abiological,

Biocontrol Agents and Secondary Metabolites. https://doi.org/10.1016/B978-0-12-822919-4.00020-X

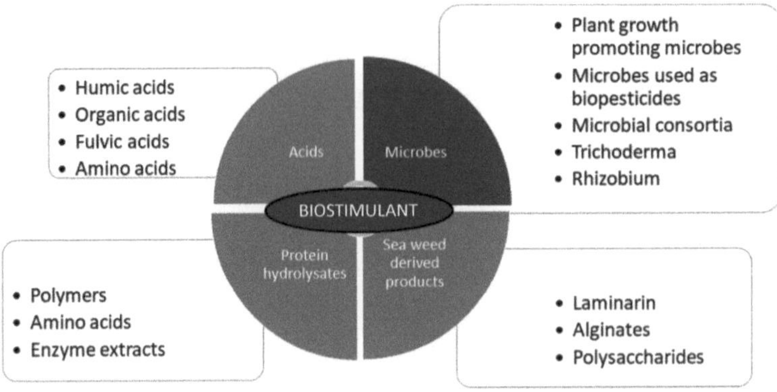

Fig. 20.1 Active components of biostimulant.

organic, inorganic components which are interdependent and exhibit a concerted positive effect on plant growth and yield. These components are environment friendly. The types of materials used for every component in Fig. 20.1 are representative. There are numerous active molecules which can be used under each category.

20.3.1 Acids

Humic acid as a part of Humus is a natural ingredient of soil and contributes toward improved biological properties of soil (Pukalchik et al., 2019). It is a well-established fact that one of the most important constituents of BS is Humic acid (HA). Even Fulvic acid and similar compounds have a coordinated similar effect on the overall health improvement of the plant. Past research reveals that HA, by nature being polyionic, favors the ion exchanges carried out in soil, resulting in blocking of free calcium. The calcium is now not available for reaction with phosphates, and hence P availability increases for plants.

Humic acid also improves the primary and secondary metabolism of plants by inducing certain necessary changes (Canellas et al., 2015). Jindo et al. (2012) reported the influence of HA on the ATPases and electromembrane potential of the roots. This results in cell enlargement, cell growth, increasing nutrient uptake and ion uptake. HA also has an impact on nitrate absorption and improved invertase activity. Even stress management by plants has exhibited modulation by production of phenolic compounds under the influence of Humic acid (Olivares et al., 2015). Drobek et al. (2019) reported the studies on Apricots representing an overall increase in the yields of Apricots under the treatment of Humic acid. Table 20.1 provides the information regarding the major effects of all the nonbiological components of Biostimulants.

Most of the work directed toward the influence of Humic acid/Humus concludes that HA is a natural soil health improver and enhances the overall absorption capacity of roots, resulting in improved growth.

Novel approaches for studies on HA effect are being successfully accomplished. A new technological approach conducted in vitro on isolated Cuticles was reported

Table 20.1 Role of different non-microbial components of biostimulants.

Components of biostimulants	Probiotic effect on plants
Humic acids	Improves biological properties of soil Improves natural interaction between soil components (Pukalchik et al., 2019) Stress response modulation by production n of phenolic compounds (Olivares et al., 2015) Free phosphate availability increases Increases invertase activity by increasing free C availability Improves NO_3 absorption Overall increase in yield (Drobek et al., 2019) Increases electromembrane potential leading increased ATP are activity (Jindo et al., 2012) Results in cell enlargement, cell growth Increased ion/nutrient uptake
Protein hydrolysates	Improves plants stress management against salinity, drought and heavy metals (Rouphael et al., 2017; Colla et al., 2013) Enhances microbial activity (Colla et al., 2015a; Du Jardin, 2015) Increases nutrient availability (Colla et al., 2015b) Auxin and Giberellin like activity-promotes root and shoot growth (Colla et al., 2014) Stimulates C and N metabolism (Nardi et al., 2016) Uptake of macro and micro nutrients (Halpern et al., 2015)
Seaweed-derived additives	Enhancing plant water and nutrients use efficiency (Mattner et al., 2018) Increase in Chlorophyll content (Al-Ghamdi and Elansary, 2018) Antioxidant defense mechanism (Debnath et al., 2011) Stimulatory effect on the microbiome (Ji et al., 2017) ROS Scavenging activity (Zou et al., 2019) Stress resistance due to salinity (Liu et al., 2019) Priming (Jisha et al., 2013) Improves absorption efficiency of Ca, Mg, S, Fe, Cu, Mn, Mo, Zn, and B (Ertani et al., 2018) Enhanced uptake of NPK (Rathore et al., 2009)
(Biopolymer) Chitosan	Empowers plant for Abiotic and Biotic stress management (Rahman et al., 2018) Topical applications prevents postharvest fruit infections (Xu et al., 2007; Pillai et al., 2009) Chelation with metal ions provides strong antimicrobial activity (Cárdenas and Ramírez, 2004) Improves plant microbe interactions, improves metabolic functions and seed germination (Hirano, 1997)

to be successful by Smilkova et al. (2019). The studies were conducted on the transportation of HA-based biostimulants through the cuticle. Such techniques, if developed further, can help in answering mechanisms which are unexplored till now.

20.3.2 Protein hydrolysates

Biostimulants based on (PDPH) plant-derived protein hydrolysates are exhibiting enormous beneficial effects on the plants. They are reported to develop stress resistance to salinity, drought, and heavy metals (Colla et al., 2013; Rouphael et al., 2017).

PH increases the nutrient availability, uptake of macro and micronutrients, and also stimulates the carbon and nitrogen metabolism.

In a review by Colla et al. (2017), the Protein Hydrolysate was highlighted as a key factor to have an impact on the stimulation of plant microbiomes. The natural inhabitants of the rhizosphere are in complete harmony with the soil and the plant root, creating an environment favorable for plant roots to perform better. The microbiome signifies the presence of an enormous and wide variety of microbial growth and activity around the roots. Although research reports direct growth enhancement to the credit of Protein Hydrolysatein Biostimulant, it is now realized that Protein Hydrolysates are assimilated by the microbes, which in turn release metabolites in plant assimilable form as amino acids, and that they cannot directly be absorbed by plants.

In nature, the rhizosphere microflora is a distinct feature of every plant species. Scientists have a concept that the nature, type, and number of microorganisms encountered in the rhizoshere indicate a watermark of the interaction between the root, soil, and microbes. The number and types of organisms vary at different points of contact with the root. The variation is also noticed with the distance of the microbe from the plant root. The variations are found both vertically and laterally in relation to the proximity of the roots. Kawasaki et al. (2016) reported the impact of rhizosphere microbes on the increase in yield of *Brachypodium distachyon* as a model for Wheat species. They reported the exuberant presence of Burkholderiales. The study reported the variation of species with the degree of binding with the root. The genetic analysis revealed that different genes are activated in species found in root tips and bases of the roots.

The roots of plants act as a biochemical expression center signaling and governing the microbiome existing around it for maximization of the exchange reactions occurring in the rhizosphere and the root favored microbes. Badri and Vivanco (2009) reported the excretion of root comprising micro and macromolecules like amino acids, organic acids, sugars, phenolics and oligosaccharides, polysaccharides, and proteins, respectively. The roots not only impart strength and anchorage to the plant shoot and are mediators of absorption of water and minerals, but are also synthesizers of biochemical signaling molecules expressed in the rhizosphere. These expressed molecules play a pivotal role in the Plant microbiome ecology and productivity. These biochemical communicating molecules are recognized as root exudates (Nazir et al., 2016). The root exudates are rich in a variety of amino acids and are an excellent N source for the rhizospheric microbes favoring their growth in the rhizosphere.

The studies of Moe in 2013 reported that 80% of the microbes are capable of utilizing these amino acid exudates as an excellent N source for self-growth.

Hu et al. (2018) reported that the production of secondary metabolites by roots of maize and wheat has an effect on microbe root association, decreases the growth, and increases the defense mechanism due to the product 6-methoxy-benzoxazolin-2-one (MBOA), which is registered to exhibit the above effect. These secondary metabolites have been reported to be of class benzoxazinoid. The protein hydrolysates therefore, if added as a biostimulant component, aggregates the beneficial microbial population and provides assimilable N compounds and other metabolites stimulating its growth.

20.3.3 Seaweed-derived additives (SWDA)

Algae has been identified as a rich nutritional source for a long time. The enriched nutrients of marine algae make it exploitable as a raw material for various industrial processes directed toward medicine and agricultural products. It has been extensively used in nutrient amendment of agricultural soils. Although SWDA exerts a stimulatory effect on the overall growth of plants, yet the mechanism has not been unraveled. Seaweeds extracts biochemical composition is complex containing polysaccharides, minerals, vitamins, oils, fats, acids, antioxidants, pigments, and hormones (Michalak and Chojnacka, 2014).

The seaweeds, predominantly macroalgae, are of three types: Brown, Red, and Green algae in accord of their pigmentation; biologically classified as Phaeophyta, Rhodophyta, and Chlorophyta; or the brown, red, and green algae, respectively. The extracts are derived after the treatment of Natural raw materials generating variation in the final product composition. The common active elements invariably present in the extract are bioactive multidimensionally effective molecules. Apart from minerals and polysaccharides, selective processing methods may impart the boon of the presence of phytohormones, cytokinins, vitamins, polyphenols, antimicrobial agents, and other biostimulating molecules (Rasyid, 2017; Zerrifi et al., 2018).

The preliminary challenge faced in the use of SWDA is its extraction as the product composition is process dependent. New technological approaches are explored for maximum extraction of bioactive components of SWDA. Godlewska et al. (2016) investigated the efficiency of water as a solvent for extraction of Phenolic compounds, micro- and macroelements, lipids content, and antibacterial properties. They used two approaches of boiling and soaking method. The boiled extract was successful in inhibiting *E. coli*. The investigation proved that an improvement in shoot length, more nutrient absorption increased Chlorophyll but the morphology remained moderately affected.

Another novel marine-derived nutrient source having prospects of soil amelioration is the Fish industry wastes. These novel nutrient-rich sources are reported to exhibit stimulatory effects like improved plant nutrient uptake, nutritional efficiency, plant yields, and the quality of products. The ever-increasing demand for fertilizers, which resulted in usage of chemical fertilizers compromising environmental safety, can be satisfied by this nutrient-rich marine-derived substitute. The use of these substitutes is giving promising results in Horticulture. The application of smaller quantities of fish protein hydrolysates in horticultural practices has resulted in increased crop yields and fruit and vegetable quality compared to chemical fertilizers in several studies (Madende and Hayes, 2020).

Looking forward to the superiority of seaweed-derived fertilizers over conventional chemical fertilizers due to the high level of organic matter, micro and macroelements, vitamins, and fatty acids, this resulted in the approval of SWDA to be used as fertilizers in organic food production where chemical fertilizer usage is facing restriction (Nagaral et al., 2018). The most popularly used alga as a biostimulant in commercial formulations is *Ascophyllum nodosum* as a seaweed component for biostimulation. Some products are launched in the market using *Durvillea antarctica*, Ecklonia maxima, *Macrocystis pyrifera*, and *Lithothamnium calcareum* for the desired effect.

20.3.4 Biopolymers

Chitosan is a modification polymer obtained from Chitin. Chitosan is a Nitrogen-rich polymer. The variable length polymers are economically important; they find application in many fields, including Agriculture, since they are is eco-friendly, nontoxic, and degradable; hence, they act as substrates for various degradative enzymes (Nandeesh Kumar et al., 2008).

The general probiotic effects of the biopolymer include natural defense responses in plants, controls of pre- and postharvest pathogenic diseases, antimicrobial activities against phytopathogens, which are some of the important attributes (Rahman et al., 2018). A novel approach using UHPLC/QTOF-MS metabolomics revealed the pivotal role of brassinosteroids and their interaction with other hormones. Root application of the biopolymer-based biostimulant induced changes of root development. This is found to be associated with the accumulation of abscisic acid, cytokines, and gibberellin-related compounds. The treatment also evidenced enhancement of flavonoids, carotenoids, and glucosinolates marking the activation of Abiotic and biotic stress management defense (Luigi et al., 2018). An experiment involving spraying of the polymer on strawberry plants revealed improved growth and fruit yield (Rahman et al., 2018). Ramírez et al. (2010) focused on the following main activities of biostimulation by Chitin and its derivatives that triggers resistance against pest and diseases.

20.3.5 The microbial component

As described earlier, biostimulants have many components of which microorganisms are critically important. The microbes in biostimulants activate a plethora of natural interrelations and reactions which provide a competing edge to the biostimulant effect. The microbes used are plant growth promoting fungi and bacteria, a parasite of plant pathogens. Table 20.2 enlists the important microbes exerting a probiotic effect in plant biostimulants. The microbes are filamentous fungi, yeasts, and bacteria. The mechanism of growth promotion is dependent on the type of host-microbe interrelationship. The plant benefit is derived due to one

Table 20.2 Important classes of microbes exhibiting probiotic effect on plant health.

Probiotic microbe	Probiotic effect
PGP Rhizobacteria (nonpathogenic) *Pseudomonas* spp. *Bacillus* spp., *Azotobacter* spp., *Serratia* spp., *Azospirillum* spp.	Capable of improving nutrient availability in soil Improves plant nutrient uptake and assimilation Improves nitrogen cycling Capable of induced resistance (Sudisha et al., 2006; Raaijmakers et al., 2009; Berg et al., 2014; Lugtenberg, 2015)
Lactobacillus plantarum	Biopesticide Activity through active compounds, for instance bacteriocins or organic acids (Reis et al., 2012; Daranas et al., 2018)
AMF and VAM Mycorrhizal fungi *Gigaspora* spp. *Funneliformis* spp. *Rhizophagus (Glomus) Laccaria* spp.	Mutualistic symbiosis with most of vascular plant species. Promotes carbon exchange, and augment the capacity of the plant to absorb water, nutrients, thus counteracting negative effects of biotic and abiotic stresses Active ingredient in commercialized products worldwide (Rouphael et al., 2015; Bulgari et al., 2019)
Pseudomonas fluorescens	Produces siderophores (growth factors) and phenazines (stimulator) (Pierson and Pierson, 2010; Babu et al., 2015)
Trichoderma spp.	Antagonism Multiple beneficial effects on plants Used extensively in biological and integrated pest management (Nagaraju et al., 2012; Jogaiah et al., 2013, 2018; Frac, 2015; Panek and Frac, 2018)
Pseudomonas fluorescens	Bioactive compounds such as Actofunicone, deoxy-funicone and vermistatin (Proksa, 2010; Pylak et al., 2019)
Aureobasidium pullulans	Parasitism (Pylak et al., 2019) Efficient in strawberry crops as Biocontrol agent scientists (Lima et al., 1997; Prokkola and Kivijarvi, 2007; Sylla et al., 2013; Wagner and Hetman, 2016)

or combined effects of natural relationships like symbiosis, N-fixation, P solubilization, parasitism against plant pathogen, antagonism, competition, synthesis of antimicrobial agents like bacteriocins, etc.

These microbes impart varied benefits independently to the plant, but microbial consortium is one of the promising approaches to optimize the biostimulant impact exuberantly.

20.4 Biofilms: A natural consortium

In nature, the ability of microbes to indulge in biofilm formation is not a new concept. This ability is imparted due to the virtue of amino acid metabolism. Biofilms formation is a microbial barrier created to sustain environmental stress like change in pH, metabolites produced by plants and other microbes, predation by protozoa, and helps in Conjugation. These biofilms were noticed to play a key role in sustaining critical mass for duration needed to initiate consortial metabolism that a single microbe fails to achieve efficiently. The consortial metabolism results in expression of degradative metabolites and exoenzymes, biocontrol agents (Kolokodin-Gal et al., 2010).

20.4.1 Microbial consortia means of wonderful soil remediation

The crop domestication, repeated crop cultivation, pesticides, and use of chemical fertilizers for improved yield have degraded the agricultural soil quality extensively. These activities have a deleterious effect on the soil health and also create environmental pollution in other abiotic factors. These activities have silently affected the soil vitals like microbiome. On the other hand, the crop plants are facing different environmental biotic and abiotic stresses which result in decreased yield.

To satisfy the yield expectations of the agricultural market, it is the need of the present day to rejuvenate the soil and make it nutrient rich for healthy crop production. The crops should also be imparted the potential of improved nutrient uptake, stress sustenance, pest resistance, longer shelf life for perishable products, which would confer an elevated yield.

A synergistic approach was proposed by Woo and Pepe in 2018 to achieve the above goal of empowering the plant and soil together. The benefits of coculture use are depicted in Fig. 20.2. None of the abovementioned components can achieve the desired impact single-handedly. A holistic approach would cater to the formulation of Biostimulants embedding the nonmicrobial components, active molecules like metabolites, and microbial consortia like bacteria-fungi consortia.

20.5 Future prospects

Although the results of Biostimulant usage are very promising, yet there are many unexplored mechanisms to be unraveled. Every component has an intricate biochemical plethora of reactions for execution of its effect. Multidisciplinary approaches, use of technologies and methodologies will have to be sought to elaborate the metabolic and genetic level impacts. It will be a promising and sustainable probiotic for plant health.

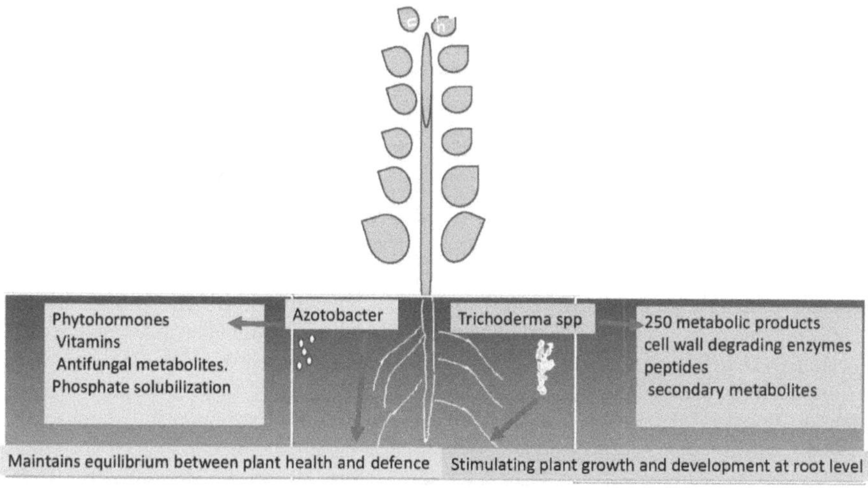

Fig. 20.2 Trichoderma and Azotobacter consortia-A promising probiotic additive to biostimulant.

References

Al-Ghamdi, A.A., Elansary, H.O., 2018. Synergetic effects of 5-aminolevulinicacid and *Ascophyllum nodosum* seaweed extraction *Asparagus phenolics* and stress related genes under saline irrigation. Plant Physiol. Biochem. 2018 (129), 273–284.

Babu, A.N., Jogaiah, S., Ito, S., Nagaraj, A.K., Tran, L.P., 2015. Improvement of growth, fruit weight and early blight disease protection of tomato plants by rhizosphere bacteria is correlated with their beneficial traits and induced biosynthesis of antioxidant peroxidase and polyphenol oxidase. Plant Sci. 231, 62–73.

Badri, D.V., Vivanco, J.M., 2009. Regulation and function of root exaudates. Plant Cell Environ. 32, 666–681.

Berg, G., Grube, M., Schloter, M., Smalla, K., 2014. Unraveling the plant microbiome: looking back and future perspectives. Front. Microbiol. 5, 148. https://doi.org/10.3389/fmicb.2014.00148.

Bulgari, R., Cocetta, G., Trivellini, A., Vernieri, P., Ferrante, A., 2015. Biostimulants and crop responses: a review. Biol. Agric. Hortic. 31, 1–17. https://doi.org/10.1080/01448765.2014.964649.

Bulgari, R., Franzoni, G., Ferrante, A., 2019. Biostimulants application in horticultural crops under abiotic stress conditions. Agronomy 2019 (9), 306. https://doi.org/10.3390/agronomy9060306.

Canellas, L.P., Olivaries, F.L., Aguiar, N.O., Jones, D.L., Pieluigi, A.N., Piccolo, M.A., 2015. Humic and fulvic acids as biostimulants in horticulture. Sci. Hortic. 196, 15–27. https://doi.org/10.1016/j.scienta.2015.09.013.

Cárdenas, R., Ramírez, M., 2004. Efecto de los derivados de quitina y su combinación con sulfato de cobre en el comportamiento del crecimiento micelial y esporulación de un aislamiento monospórico del hongo *Pyricularia grisea*, Sacc. Cultiv. Tropic. 25 (4), 89–93.

Colla, G., Svecova, E., Rouphael, Y., Cardarelli, M., Reynaud, H., Canaguier, R., 2013. Effectiveness of a plant–derived protein hydrolysate to improve crop performances under different growing conditions. Acta Hortic. 1009, 175–179. https://doi.org/10.17660/ActaHortic.2013.1009.21.

Colla, G., Rouphael, Y., Canaguier, R., Svecova, E., Cardarelli, M., 2014. Biostimulant action of a plant-derived protein hydrolysate produced through enzymatic hydrolysis. Front. Plant Sci. 5, 448. https://doi.org/10.3389/fpls.2014.00448.

Colla, G., Nardi, S., Cardarelli, M., Ertani, A., Lucini, L., Canaguier, R., 2015a. Protein hydrolysates as biostimulants in horticulture. Sci. Hortic. 96, 28–38. https://doi.org/10.1016/j.scienta.2015.08.037.

Colla, G., Rouphael, Y., Di Mattia, E., El-Nakhel, C., Cardarelli, M., 2015b. Co-inoculation of *Glomus intraradices* and *Trichoderma atroviride* acts as a biostimulant to promote growth, yield and nutrient uptake of vegetable crops. J. Sci. Food Agric. 95, 1706–1715. https://doi.org/10.1002/jsfa.6875doi:10.1002/jsfa.6875.

Colla, G., Hoagland, L., Ruzzi, M., Cardarelli, M., Bonini, P., Canaguier, R., Rouphael, Y., 2017. Biostimulant action of protein hydrolysates: unraveling their effects on plant physiology and microbiome. Front. Plant Sci. 8, 2202. https://doi.org/10.3389/fpls.2017.02202.

Daranas, N., Roselló, G., Cabrefiga, J., Donati, I., Francés, J., Badosa, E., Spinelli, F., Montesinos, E., 2018. Biological control of bacterial plant diseases with *Lactobacillus plantarum* strains selected for their broad-spectrum activity. Ann. Appl. Biol. 174 (1), 92–105. https://doi.org/10.1111/aab.12476.

Debnath, M., Pandey, M., Bisen, P.S., 2011. An omics approach to understand the plant abiotic stress. Omi. J. Integr. Biol. 2011 (15), 739–762.

Drobek, M., Frac, M., Cybulska, J., 2019. Plant biostimulants: importance of the quality and yield of horticultural crops and the improvement of plant tolerance to abiotic stress—a review. Agronomy 9 (6), 335. https://doi.org/10.3390/agronomy9060335.

Du Jardin, P., 2015. Plant biostimulants: definition, concept, main categories and regulation. Sci. Hortic. 196, 3–14. https://doi.org/10.1016/j.scienta.2015.09.021.

Ertani, A., Francioso, O., Tinti, A., Schiavon, M., Pizzeghello, D., Nardi, S., 2018. Evaluation of seaweed extracts from laminaria and *Ascophyllum nodosum* spp. as biostimulants in Zea mays L. using a combination of chemical, biochemical and morphological approaches. Front. Plant Sci. 9, 428.

Frac, M., 2015. Occurrence, detection, and molecular and metabolic characterization of heat-resistant fungi in soils and plants and their risk to human health. Adv. Agron. 132, 161–204. https://doi.org/10.1016/bs.agron.2015.02.003.

Godlewska, K., Michalak, I., Tuhy, U., Chojnacka, K., 2016. Plant growth biostimulants based on different methods of seaweed extraction with water. Biomed. Res. Int. 2016, 5973760. https://doi.org/10.1155/2016/5973760.

Halpern, M., Bar-Tal, A., Ofek, M., Minz, D., Muller, T., Yermiyahu, U., 2015. The use of biostimulants for enhancing nutrient uptake. Adv. Agron. 130, 141–174. https://doi.org/10.3389/fpls.2017.00597.

Hirano, S., 1997. MFA, G. (Ed.), Applications of chitin and chitosan in the ecological and environmental fields. Technomic, Lancaster, PA, pp. 31–54.

Hu, L., Robert, C.A.M., Cadot, S., 2018. Root exudate metabolites drive plant-soil feedbacks on growth and defense by shaping the rhizosphere microbiota. Nat. Commun. 9, 2738. https://doi.org/10.1038/s41467-018-05122-7.

Ji, R., Dong, G., Shi, W., Min, J., 2017. Effects of liquid organic fertilizers on plant growth and Rhizosphere soil characteristics of chrysanthemum. Sustainability 2017 (9), 841.

Jindo, K., Martim, S.A., Navarro, E.C., Aguiar, N.O., Canellas, L.P., 2012. Root growthpromotion by humic acids from composted and non-composted urban organicwastes. Plant Soil 353, 209–220.

Jisha, K.C., Vijayakumari, K., Puthur, J.T., 2013. Seed priming for abiotic stress tolerance: an overview. Acta Physiol. Plant. 35, 1381–1396.

Jogaiah, S., Mostafa, A., Tran, L.S.P., Shin-ichi, I., 2013. Characterization of rhizosphere fungi that mediate resistance in tomato against bacterial wilt disease. J. Exp. Bot. 64, 3829–3842.

Jogaiah, S., Abdelrahman, M., Tran, L.-S.P., Ito, S.-I., 2018. Different mechanisms of *Trichoderma virens*-mediated resistance in tomato against Fusarium wilt involve the jasmonic and salicylic acid pathways. Mol. Plant Pathol. 19, 870–882.

Kawasaki, A., Donn, S., Ryan, P.R., Mathesius, U., Devilla, R., Jones, A., Watt, M., 2016. Microbiome and exudates of the root and rhizosphere of *Brachypodium distachyon*, a model for wheat. PLoS One. 11(10), e0164533. https://doi.org/10.1371/journal.pone.0164533.

Kolokodin-Gal, I., Romero, D., Cao, S., Clardy, J., Kolter, R., Losick, R., 2010. D-amino acids trigger biofilm disassembly. Science 328 (5978), 627–629. https://doi.org/10.1126/science.1188628.

Lima, G., Ippolito, A., Nigro, F., Salerno, M., 1997. Effectiveness of *Aureobasidium pullulans* and Candida oleophila against postharvest strawberry rots. Postharvest Biol. Technol. 10, 169–178. https://doi.org/10.1016/S09255214(96)01302-6.

Liu, H., Chen, X., Song, L., Li, K., Zhang, X., Liu, S., Qin, Y., Li, P., 2019. Polysaccharides from Grateloupia filicina enhance tolerance of rice seeds (*Oryza sativa* L.) under salt stress. Int. J. Biol. Macromol. 124, 1197–1204.

Lugtenberg, B., 2015. Principles of plant-microbe interactions: Microbes for sustainable agriculture. Springer International Publishing, Cham, p. 448.

Luigi, L., Rouphael, Y., Cardarelli, M., Bonini, P., Baffi, C., Colla, G., 2018. A vegetal biopolymer-based biostimulant promoted root growth in melon while triggering brassinosteroids and stress-related compounds. Front. Plant Sci. 9, 472. https://doi.org/10.3389/fpls.2018.00472.

Madende, M., Hayes, M., 2020. A review fish by-product use as biostimulants: an overview of the current state of the art, including relevant legislation and regulations within the EU and USA. Molecules 2020 (25), 1122. https://doi.org/10.3390/molecules25051122.

Mattner, S.W., Milinkovic, M., Arioli, T., 2018. Increased growth response of strawberry roots to a commercial extract from *Durvillaea potatorum* and *Ascophyllum nodosum*. J. Appl. Phycol. 30, 2943–2951.

Michalak, I., Chojnacka, K., 2014. Algal extracts: technology and advances. Eng. Life Sci. 2014 (14), 581–591.

Nagaraju, A., Jogaiah, S., Mahadeva Murthy, S., Shin-ichi, I., 2012. Seed priming with *Trichoderma harzianum* isolates enhances plant growth and induces resistance against *Plasmopara halstedii*, an incitant of sunflower downy mildew disease. Aust. J. Plant Path. 41, 609–620.

Nagaral, K.S., Kisan, B., Haleypyati, A.S., 2018. Sea weed based liquid fertilizer as a source of plant nutrition—a review. Int. J. Curr. Microbiol. App. Sci. 7 (4), 1984–1989.

Nandeesh Kumar, P.K., Sudisha, J., Kini, K.R., Prakash, H.S., Niranjana, S.R., Shekar Shetty, H., 2008. Chitosan induced resistance to downy mildew in sunflower caused by *Plasmopara halstedii*. Physiol. Mol. Plant Path. 7, 188–194.

Nardi, S., Pizzeghello, D., Schiavon, M., Ertani, A., 2016. Plant biostimulants: physiological responses induced by protein hydrolyzed-based products and humic substances in plant metabolism. Sci. Agric. 73, 18–23. https://doi.org/10.1590/01039016-2015-0006.

Nazir, N., Kamili, A., Zargar, M., Khan, I., Shah, D., Parray, J., Tyub, S., 2016. Effect of root exudates on rhizosphere soil microbial communities. J. Res. Dev. 16, 88–94.

Olivares, F.L., Aguiar, N.O., Rosa, R.C.C., Canellas, L.P., 2015. Substrate biofortification in combination with foliar sprays of plant growth promoting bacteria and humic substances boosts production of organic tomatoes. Sci. Hortic. 183, 100–108.

Panek, J., Frac, M., 2018. Development of a qPCR assay for the detection of heat-resistant *Talaromyces flavus*. Int. J. Food Microbiol. 270, 44–51. https://doi.org/10.1016/j.ijfoodmicro.2018.02.010.

Pierson, L.S., Pierson, E.A., 2010. Metabolism and function of phenazines in bacteria: impacts on the behavior of bacteria in the environment and biotechnological processes. Appl. Microbiol. Biotechnol. 86, 1659–1670. https://doi.org/10.1007/s00253-010-2509-3.

Pillai, C.K.S., Paul, W., Sharma, C.P., 2009. Chitin and chitosan polymers: chemistry, solubility and fiber formation. Prog. Polym. Sci. 34 (7), 641–678.

Prokkola, S., Kivijarvi, P., 2007. Effect of biological sprays on the incidence of grey mould, fruit yield and fruit quality in organic strawberry production. Agric. Food Sci. 16, 25–33. https://doi.org/10.2137/145960607781635886.

Proksa, B., 2010. *Talaromyces flavus* and its metabolites. Chem. Pap. 64, 696–714. https://doi.org/10.2478/s11696-010-0073-z.

Pukalchik, M., Kydralieva, K., Yakimenko, O., Fedoseeva, E., Terekhova, V., 2019. Outlining the potential role of humic products in modifying biological properties of the soil—a review. Front. Environ. Sci. 7, 80. https://doi.org/10.3389/fenvs.2019.00080.

Pylak, M., Oszust, K., Frąc, M., 2019. Review report on the role of bioproducts, biopreparations, biostimulants and microbial inoculants in organic production of fruit. Rev. Environ. Sci. Biotechnol. 18, 597–616. https://doi.org/10.1007/s11157-019-09500-5.

Raaijmakers, J.M., Paulitz, T.C., Steinberg, C., Alabouvette, C., Moënne-Loccoz, Y., 2009. The rhizosphere: a playground and battlefield for soilborne pathogens and beneficial microorganisms. Plant Soil 321, 341–361. https://doi.org/10.1007/s11104-008-9568-6.

Rahman, M., Mukta, J.A., Sabir, A.A., Gupta, D.R., Mohi-Ud-Din, M., Hasanuzzaman, M., Miah, M.G., Rahman, M., Islam, M.T., 2018. Chitosan biopolymer promotes yield and stimulates accumulation of antioxidants in strawberry fruit. PLoS One. https://doi.org/10.1371/journal.pone.0203.

Ramírez, M.A., Rodríguez, A.T., Alfonso Land Peniche, C., 2010. Chitin and its derivatives as biopolymers with potential agricultural applications. Biotecnología Aplicada *27*, *4*.

Rasyid, A., 2017. Evaluation of nutritional composition of the dried seaweed Ulva lactuca from Pameungpeuk waters, Indonesia. Trop. Life Sci. Res. 28, 119.

Rathore, S.S., Chaudhary, D.R., Boricha, G.N., Ghosh, A., Bhatt, B.P., Zodape, S.T., Patolia, J.S., 2009. Effect of seaweed extract on the growth, yield and nutrient uptake of Soybean (Glycine max) under rainfed conditions. S. Afr. J. Bot. 75, 351–355.

Reis, J.A., Paula, A.T., Casarotti, S.N., Penna, A.L.B., 2012. Lactic acid bacteria antimicrobial compounds: characteristics and applications. Food Eng. Rev. 4, 124–140. https://doi.org/10.1007/s12393-012-9051-2.

Rouphael, Y., Franken, P., Schneider, C., Schwarz, D., Giovannetti, M., Agnolucci, M., De Pascale, S., Bonini, P., Colla, G., 2015. Arbuscular mycorrhizal fungi act as biostimulants in horticultural crops. Sci. Hortic. 196, 91–108. https://doi.org/10.1016/j.scienta.2015.09.002.

Rouphael, Y., Cardarelli, M., Bonini, P., Colla, G., 2017. Synergistic action of a microbial-based biostimulant and a plant derived-protein hydrolysate enhances lettuce tolerance to alkalinity and salinity. Front. Plant Sci. 8, 131. https://doi.org/10.3389/fpls.2017.00131.

Smilkova, M., Smilek, J., Kalina, M., 2019. A simple technique for assessing the cuticular diffusion of humic acid biostimulants. Plant Methods 15, 83. https://doi.org/10.1186/s13007-019-0469-x.

Sudisha, J., Niranjana, S.R., Umesha, S., Prakash, H.S., Shekar Shetty, H., 2006. Transmission of seed-borne infection of muskmelon by *Didymella bryoniae* and effect of seed treatments on disease incidence and fruit yield. Biol. Control 37, 196–205.

Sylla, J., Alsanius, B., Kruger, E., Reineke, A., Bischoff-Schaefer, M., Wohanka, W., 2013. Introduction of *Aureobasidium pullulans* to the phyllosphere of organically grown strawberries with focus on its establishment and interactions with the resident microbiome. Agronomy 3, 704–731. https://doi.org/10.3390/agronomy3040704.

Wagner, A., Hetman, B., 2016. Effect of some biopreparations on health status of strawberry (*Fragaria ananassa* Duch.). J. Agric. Sci. Technol. B 6, 295–302. https://doi.org/10.17265/2161-6264/2016.05.002.

Woo, S.L., Pepe, O., 2018. Microbial consortia: Promising probiotics as plant biostimulants for sustainable agriculture. Front. Plant Sci. 9, Article 1801www.frontiersin.org.

Xu, W.T., Huang, K.L., Guo, F., Qu, W., Yang, J.J., Liang, Z.H., Luo, Y.B., 2007. Postharvest grapefruit seed extract and chitosan treatments of table grapes to control Botrytis cinerea. Postharvest Biol. Technol. 46, 86–94. https://doi.org/10.1016/j.postharvbio.2007. 03.019.

Yakhin, O.I., Lubyanov, A.A., Yakin, I.A., Brown, P.H., 2017. Biostimulants in plant science: a global perspective. Front. Plant Sci. https://doi.org/10.3389/fpls.2016.02049.

Zerrifi, S.E.A., El Khalloufi, F., Oudra, B., Vasconcelos, V., 2018. Seaweed bioactive compounds against pathogens and microalgae: potential uses on pharmacology and harmful algae bloom control. Mar. Drugs 2018 (16), 55.

Zou, P., Lu, X., Zhao, H., Yuan, Y., Meng, L., Zhang, C., Li, Y., 2019. Polysaccharides derived from the brown algae *Lessonia nigrescens* enhance salt stress tolerance to wheat seedlings by enhancing the antioxidant system and modulating intracellular ion concentration. Front. Plant Sci. 10, 48.

Explorations of fungal diversity in extreme environmental conditions for sustainable agriculture applications

21

H.V. Pavan[a], *S. Mahadeva Murthy*[a], *and Sudisha Jogaiah*[b]
[a]Department of Microbiology, Yuvaraja's College, University of Mysore, Mysore, Karnataka, India, [b]Laboratory of Plant Healthcare and Diagnostics, P.G. Department of Biotechnology and Microbiology, Karnatak University, Dharwad, Karnataka, India

21.1 Introduction

The fungi occupy an extraordinarily diverse range of habitats. They play a very important role in different terrestrial ecosystems as decomposers, mutualists, and pathogens, and are one of the most diverse groups of the eukarya (Mueller et al., 2007). To study the ecological factors responsible for the growth and development of fungal communities is a big challenge because of its high taxonomic and ecological diversity. They occupy a wide range of niches in various ecosystems. They are present in association with different organisms like algae, higher plants and even in the guts of some insects. In plants, they may be associated with roots in the form of mycorrhizal association or with other parts of the plant body like leaves forming nonmycorrhizal associations also. Fungi can be found at various environmental conditions starting from tropical forest regions to places with extreme environmental conditions like deserts and hot springs. New habitats which are rich sources of undiscovered fungal biodiversity continue to be discovered.

21.2 Explorations of fungal diversity

21.2.1 Fungal diversity in insect gut

Insect guts are found to be quite rich in fungal biodiversity. The interactions between fungi and insects may be transient or obligate. In some cases, fungi may kill insects, but in a majority of the cases it is beneficial for the insects. They are the home for many of the fungal species belonging to Trichomycetes. A study conducted in Norway where aquatic insect larvae collected over a one to 40-day period yielded 25 species, including one new genus and nine new species (White and Lichtwardt, 2004).

Biocontrol Agents and Secondary Metabolites. https://doi.org/10.1016/B978-0-12-822919-4.00021-1

Trichomycetes (Zygomycota) are a cosmopolitan class of filamentous fungi. Many of them usually live as obligate symbionts in the guts of various arthropods (Lichtwardt, 1986; Lichtwardt et al., 2001; Abdelrahman et al., 2016). Larval black flies (Diptera: Simuliidae) harbor eight genera and approximately 33 species of Harpellales (Lichtwardt, 1986; Labeyrie et al., 1996; Abdelrahman and Jogaiah, 2019; Lichtwardt and Arenas, 1996; Lichtwardt and Williams, 1990; Williams and Lichtwardt, 1990). The presence of trichomycetes in a particular host can vary seasonally (Labeyrie et al., 1996; Beard and Adler, 2002; Lichtwardt and Williams, 1988). *Harpella* is a ubiquitous genus of gut fungus in black fly populations, the prevalence of which has been noted to vary seasonally in black fly larvae (Beard et al., 2003).

The termite gut harbors various microorganisms that are symbiotically associated. The cellulosic degradation is found to be effectively carried out with the help of protozoan symbionts in lower termites, whereas in higher termites, gut microbiota and the host cellulose genes are mainly responsible for cellulosic and hemicellulosic degradation (Yang et al., 2004; Huang et al., 2008). In the macrotermitine termite *Odontotermes formosanus*, symbiosis between the gut symbionts, termite, and the ectosymbiotic fungi Termitomyces facilitate the plant cellulolytic degradation (Rouland-Lefèvre et al., 2006; Mathew et al., 2012).

The number of Trichomycetes reported in the tropical regions is quite low, and is considered to be a rich source of new unexplored and unusual Trichomycetes. One such example is the discovery of a new unusual genus *Gauthieromyces* (*G. indicus*) that lives in mayfly nymphs (Misra and Tiwari, 2008).

21.2.2 Nematophagous fungi

Fungi can also be utilized as a biological control agent in controlling nematodes in the environment. One of the biological phenomena by which nematodes are controlled is by egg parasitism (Lysek, 1982; Larsen, 1999). Nematophagous fungi can be endoparasites, predators, and parasites of nematode eggs (Araujo et al., 2008; Liu et al., 2009). Nematophagous fungi are ubiquitous in nature. They occur in natural and agricultural soil and in all types of organic matter in decomposition. More than 150 fungal species have been found to show inhibitory action on various parasitic helminths such as *Toxocara canis*, *Ascarissuum*, *Ascaris lumbricoides*, *Trichuris trichiura*, and *T. vulpis* (Lysek, 1982; Araujo et al., 1995; Ferreira et al., 2011). The mechanism by which the nematophagous fungi degrade the eggshell of helminths could be associated with the penetration of the eggshell by a combination of mechanical and enzymatic activity on its chitin and protein structure (Bonants et al., 1995; Lopez-Llorca et al., 2002; Tikhonov et al., 2002).

Nematophagous fungi, such as *Paecilomyces chlamydosporia*, *P. lilacinus*, and *Dactylella oviparasitica*, have been shown to display a very high ovicidal activity. Many fungal species, including *Penicillium frequentans*, *Stachybotrys chartarum*, *Fusarium culmorum*, *F. pallidoroseum*, *Paecilomyces fumosrseus*, *P. lilicinus*, *Metahizium flavoviride*, *Aspergillus versicolor*, *A. niger*, *Trichoderma viride*, and *Verticillium chlamydiosporum*, have been shown to display inhibitory activity on the development of *Ascaris* eggs (Lysek and Sterba, 1991; Kuzna-Grygiel et al., 2001; Jaborowska, 2006; Jaborowska et al., 2006).

Taenia taeniaeformis is one of the widely distributed cestodes which parasitizes the small intestine of cats (*Felis catus*) and other carnivores. The occurrence of this parasite is high in cats and has not been recorded to pose any risk to human beings. Two fungal species belonging to the *Monacrosporium* species, *Monacrosporium sinense* and *M. thaumasium*, are found to be potentbiological control agents against gastrointestinal nematodes in domestic animals (Araujo et al., 2004; Campos et al., 2007). *Pochonia chlamydosporia* has been shown to be ovicidal against various helminth eggs like *Ascaris lumbricoides*, *Fasciola hepatica*, and *Schistosoma mansoni* and has been used in the *in vitro* control of helminth eggs (Braga et al., 2007, 2008). *Pochonia chlamydosporia* has also been shown to have ovicidal activity on *Dipylidium caninum* egg capsules (Braga et al., 2007, 2008).

21.2.3 Fungal association with orchids

The family *Orchidaceae* is one of the largest plant families with more than 24,000 species (Jones, 2006). Many of the orchid species are quite rare and either endangered or threatened throughout the world. Orchids are very much dependent on fungi for nutrition, growth, and development after seed germination (Smith and Read, 1997; Bonnardeaux et al., 2007). All orchids need to establish a relationship with mycorrhizal fungi for seed germination and subsequent growth and development (Rasmussen, 1995; Kull and Arditti, 2002). Achlorophyllous orchids are dependent on their fungal partners throughout their life for the purpose of nutrition, which is referred to as mycoheterotrophy (Leake, 1994). Even though the majority of orchid species are photosynthetic, they still depend on their fungal associates for carbon nutrition in particular living conditions, like situations when the availability of light is very low (Gebauer and Meyer, 2003). The orchids use fungi as a source of carbon, vitamins, hormones, and amino acids for their growth and development. The absence of fungal association with orchid not only reduces the growth and development of orchid, but also results in high mortality of germinating seedlings during the early stages of development (Zettler et al., 2007).

Various morphological as well as molecular studies show the association of mycorrhizal fungi with the orchids to be extremely specific. Usually, fungi from a single species are found to associate in many of the orchid species (Shefferson et al., 2007). Although mycorrhizal association is a very important phenomenon in orchid biology, the knowledge about the diversity of mycorrhizal fungi associated with orchids in nature is very little (Dearnaley, 2007; Bidartondo and Read, 2008). The seeds of orchids are very minute and contain few stored reserve food material. For the germination and early development of the seedling on the substrate, colonization by a compatible fungus is very much essential (Smith and Read, 1997). The fungal hyphae grow into orchid tissues where they form elaborate coiled structures within cortical cells called pelotons.

Orchids also receive compounds other than carbon from their fungal partners. Studies showed that mycorrhizal *Goodyera repens* acquired 100 times more phosphorus than nonmycorrhizal controls (Alexander et al., 1984). The majority of fungi isolated from orchids belong to the anamorphic form genus *Rhizoctonia* (Filipello-Marchisio and Berta, 1985). The nonphotosynthetic orchids have been increasingly shown to associate with non-Rhizoctonia fungi (Dearnaley, 2007). Rhizoctonia-forming fungi are the most frequently encountered fungi belonging to the Ceratobasidiaceae and Tulasnellaceae mainly associated with photosynthetic orchids. Ectomycorrhizal Basidiomycetes like Thelephoraceae and Russulaceae usually occur in mycoheterotrophic and mixotrophic orchids (McCormick et al., 2004; Rasmussen, 2002). Even though orchidaceae is one of the largest plant families, mycorrhizal association with orchids has received greater attention from investigators when compared to association of orchids with nonmycorrhizal fungi (Bonnardeaux et al., 2007; Shefferson et al., 2005; Kaliamoorthy, 2007; Yuan et al., 2009; Bagyalakshmi et al., 2010). Although orchids host many nonmycorrhizal fungi in leaves, studies on such fungi, especially in tropical and subtropical orchids, are lacking (Dearnaley, 2007; Yuan et al., 2009; Bayman and Otero, 2006).

Like mycorrhizal fungi, the nonmycorrhizal fungi of orchids are equally important due to their multiple ecological roles (Yuan et al., 2009; Bayman and Otero, 2006). Few studies are available on endophytic fungal taxa of different tissues of tropical orchids (Bayman and Otero, 2006). Tropical forests have been considered to be the hot spots of endophytic fungi on leaves (Babu et al., 2015). Endophytic fungi inhabiting leaves are under high selective pressure, compared to those of mycorrhizal fungi and fungi associated with persistent tissues (Arnold, 2007). One of the most commonly found mycorrhizal associations with the roots of many species of orchids is *Epulorhiza*. *Epulorhiza* has a very important role in the germination of seeds of orchids. It is observed that *Epulorhiza* not only promotes seed germination of orchid species in which it inhabits, but also promotes seed germination when introduced into the orchids in which it does not inhabit naturally. Hence, the use of mycorrhiza fungi would substantially reduce the cost during the artificial propagation of orchids and greatly increase the germination efficiency (Nontachaiyapoom et al., 2010). Various reports suggest the occurrence as well as the beneficial role of *Fusarium* in many orchids. *Fusarium* species have been shown to have a role in stimulation of germination of seeds of *Cypripedium reginae* (Vujanovic et al., 2000).

Nearly 125 species of the epiphytic or lithophytic genus *Coelogyne* have been found which are distributed from India and China through South East Asia and Indonesia, the Philippines, to the Pacific and Fiji islands (Abraham and Vatsala, 1981). Eight species of *Coelogyne* are found to be distributed across various parts of South India. The epiphytic *Coelogyne nervosa* is one of the endemic orchid species found in southern India, particularly in the Southern Western Ghats of Kerala and Tamil Nadu (Murugesan, 2005).

21.2.4 Fungi in desert ecosystem

Many species of fungi are found in the desert ecosystem that evade those extremely harsh environmental conditions. Atacama desert is one of the driest locations on earth where some regions have no recorded rainfall for decades. Even in such extreme environmental conditions, 12 genera of fungi, namely: *Cladophialophora, Cladosporium, Leptosphaerulina, Alternaria, Ulocladium, Eupenicillium, Aspergillus, Penicillium, Ascobolus, Monodictys, Periconia,* and *Giberella,* were isolated (Conley et al., 2006). Other species, including *Protophysarium phleogenum, Badhamnia gracilis,* and *Physarum straminipes,* have been isolated from the cortex tissues of the giant cactus in deserts of North America (Sonora and Mojave desert) and in dry regions of Europe and Russia (Moreno et al., 1998).

Rocks in desert regions are one of the most stressful environments on Earth. The temperatures on surfaces of exposed rock reach from 45°C up to 60°C and infrared and ultraviolet radiation is very high. The availability of organic carbon is rare and dewfall might be the only available water (Gorbushina et al., 2002; Gorbushina, 2007; Omelon, 2008). Black fungi are the rock inhabiting fungi, which, together with some lichens, are assumed to be the most stress resistant eukaryotic organisms known on Earth (de Hoog and Grube, 2008). Recent experiments showed that their stress resistance against solar radiation, radioactivity, desiccation, and oligotrophic conditions even allows them to survive space and Martian conditions (Onofri et al., 2008). For this reason, black fungi are now model organisms for Astrobiology (Onofri et al., 2004) and for gamma radiation experiments. Black fungi are not only resistant to high levels of radioactivity, they actually are able to use radioactivity as a source of energy for ATP generation (Dadachova and Casadevall, 2008).

21.2.5 Fungi in denitrification

Denitrification, a sequential reduction of NO_3- to NO_2-, NO, N_2O, and N_2, has been regarded to be the major biological process for terrestrial N_2O production under O_2 limited conditions (Davidson, 1991). A number of fungal isolates have shown N_2O-producing capability (Yanai et al., 2007; Lavrent'ev et al., 2008; Prendergast-Miller et al., 2011). Denitrifying fungi are found to be quite different physiologically and biochemically when compared to denitrifying bacteria. Denitrifying bacteria require strict anaerobic conditions for NO_3- or NO_2- reduction, whereas fungi need the presence of small amounts of O_2 for denitrification function (Zumft, 1997; Zhou et al., 2001). In addition, denitrifying fungi generally lack N_2O reductase, and thus their reactions end with N_2O production (Zhou et al., 2001; Jogaiah et al., 2013). Fungal contribution to soil N_2O production was found to be more significant in acidic woody plant-dominated plantation forestry than in other herbaceous plant-dominated ecosystems (Nagaraju et al., 2012).

21.2.6 Fungi in marine ecosystem

The marine ecosystem is a huge reservoir of fungi. Fungi live in the marine environment mainly as saprophytes, parasites, and symbionts with algae and other marine organisms. Many of these fungi are rich sources of various bioactive compounds, and now with the increasing quest for the synthesis of newer bioactive compounds, interest in the investigations on marine fungi has increased enormously. The presence of fungi in a particular marine location is influenced by various factors including water temperature, its salinity, the water movement, the presence of suitable substrates for colonization, pollution, and the oxygen content of the water (Gareth, 2000). The distribution of marine fungi are quite distinct and distributed based on the substratum on which they grow freely (Gareth et al., 2006). Fungi isolated from the marine ecosystem have been widely studied for their bioactive metabolites. Many of these have proven to be a rich and promising source of novel anticancer, antibacterial, antiplasmodial, antiinflammatory, and antiviral agents (Bhadury et al., 2006; Sudisha and Shekar Shetty, 2009).

Some of the metabolites isolated from marine fungi have been reported to have antiprotozoal activities against *Trypanosoma cruzi*, *T. brucei*, and *Plasmodium falciparum* (Kasettrathat et al., 2008; Pontius et al., 2008; Watts et al., 2010). Recently, a number of fungal species have been isolated from marine environments and have been recognized as a rich source of biologically active metabolites. Hence, these fungi can be a rich source for finding new drugs (Shang et al., 2012).

21.2.7 Radiotrophic fungi

These are the fungi which are able to harvest energy from ionizing radiations such as gamma radiations emitted from the nuclear reactors. The phenomenon is termed radiotropism. Black molds that were observed growing in and around the Chernobyl Nuclear Power Plant in Ukraine led to the discovery of these radiotrophic fungi (De Britto et al., 2020). The radiation dose in these regions was actually three to five orders greater than that of the normal radioactivity. But these fungal species are able to inhabit even in such extreme conditions (Belozerskaya et al., 2010). Radiotrophic fungi have been observed to inhabit the Arctic and Antarctic regions, as well as high altitude terrains, where naturally high levels of radiation occur (Robinson, 2001). One of the most interesting facts is that these radiotrophic fungi are even seen in the spacecrafts orbiting outer space. Beyond the protective shield of the Earth's atmosphere, there is the presence of high levels of ionizing radiations and fungi are able to grow in such extreme conditions (Dadachova and Casadevall, 2008).

21.2.8 Fungi in Antarctica

Psychrophiles are the organisms capable of surviving at extremely low temperatures. Antarctica is a continent present at the south pole consisting of extremely low temperatures. Even in such extreme environments, many of the bacterial and fungal species do survive. The predominance of sterile mycelia could be one of the physiological adaptations of these fungi to survive the extreme cold conditions of Antarctica. Many of these fungi also produce melanin, which may be one of the protective mechanisms to survive the subzero temperatures (Dadachova and Casadevall, 2008).

Nine species of fungi from the Antarctica region including *Arthrobotrys ferox* on moss, *Torulopsis psychrophila,* and *Phoma herbarum* grow on bird excreta, *P. herbarum* on skeletal remains, *Acremonium antarcticum*, and *A. psychrophilum* usually grow on lichens (Sharma, 2000). Some of the common fungal genera like *Acremonium, Aspergillus, Cladosporium, Fusarium,* and *Trichoderma* were also observed in Antarctic soils. However, it was found that they were quite different from their mesophilic counterparts when their mycelial characters were compared. Most of the fungi recovered from the Antarctic soils were found to show unique mycelia, in that they consistently had abundant intercalary, swollen, thick-walled cells that sometimes formed conspicuous mycelial cords besides chlamydospores. These are definitely the adaptations of these fungi to overcome the harsh cold conditions (Singh et al., 2006).

21.2.9 Thermophilic fungi

These are the fungi which comprise thermophilic and thermotolerant forms. The thermophilic fungi have maximum growth at or above 50°C, whereas the thermotolerant forms grow at a temperature range of 20–55°C. Thermophilic bacteria grow near or above 100°C in thermal springs, solfatara fields, or hydrothermal vents, whereas thermophilic fungi are not as extreme as in eubacteria or archaea (Blochl et al., 1997; Brock, 1995). Hot springs and geothermal vents are found in many parts of the earth. The largest single concentration is found in Yellowstone National Park, USA. Many of the thermophilic fungi belong to the Ascomycetes or zygomycetes class (Morgenstern et al., 2012). *Dichanthelium lanuginosum* (panic grass) is one of the common plants found on the hot ground near the geothermal areas of Yellowstone National Park. The fungus *Curvularia protuberate* is found to live as an endophyte in the roots of the panic grass.

21.3 Conclusion

The number of fungal species present is estimated to be 1.5 million and they occupy a wide range of ecosystems, much of which are yet to be explored. With the increasing interest to find newer fungal species, investigators are looking for the various habitats starting from tropical forests to extreme environmental habitats like deserts, hot springs, and marine environment. Even though these fungi contribute a greater number in the fungal diversity, very little is known about them. Exploring these fungi inhabiting various extreme and unique habitats may result in the finding of new fungal species which could have a better perspective in the field of medicine, agriculture, and industrial applications. Newer fungal species from various habitats which are rich sources of undiscovered fungal diversity continue to be discovered.

References

Abdelrahman, M., Jogaiah, S., 2019. Defense mechanism and diverse actions of fungal biological control agents against plant biotic stresses. In: Khurana, S., Gaur, R. (Eds.), Plant Biotechnology: Progress in Genomic Era. Springer, Singapore, pp. 461–478.

Abdelrahman, M., Abdel-Motaal, F., El-Sayed, M., Jogaiah, S., Shigyo, M., Ito, S.-I., Tran, L.S.P., 2016. Dissection of *Trichoderma longibrachiatum*-induced defense in onion (*Allium cepa* L.) against *Fusarium oxysporum* f. sp. *Cepa* by target metabolite profiling. Plant Sci. 246, 128–138.

Abraham, A., Vatsala, P., 1981. Introduction to Orchids. Tropical Botanic Garden and Research Institute. Trivandrum, India.

Alexander, C., Alexander, I.J., Hadley, G., 1984. Phosphate uptake by *Goodyera repens* in relation to mycorrhizal infection. New Phytol. 97, 401–411.

Araujo, J.V., Santos, M.A., Ferraz, S., 1995. Ovicidal effect of nematofagous fungi on embryonate eggs of *Toxocara canis*. Arq. Bras. Med. Vet. Zootec. 47, 37–42.

Araujo, J.V., Mota, M.A., Campos, A.K., 2004. Controle biologico de helmintos parasitos de animais por fungos nemato´ fagos. Rev. Bras. Parasitol. Vet. 13, 165–170.

Araujo, J.V., Braga, F.R., Araujo, J.M., Silva, A.R., Tavela, A.O., 2008. *In vitro* evaluation of the effect of the nematophagous fungi *Duddingtonia flagrans*, *Monacrosporium sinense* and *Pochonia chlamydosporia* on *Ascaris suum* eggs. Parasitol. Res. 102, 787–790.

Arnold, A.E., 2007. Understanding the diversity of foliar endophytic fungi: progress, challenges, and frontiers. Fungal Biol. Rev. 21, 51–66.

Babu, N.A., Jogaiah, S., Ito, S.-I., Nagraj, K.A., Tran, L.-S.P., 2015. Improvement of growth, fruit weight and early blight disease protection of tomato plants by rhizosphere bacteria is correlated with their beneficial traits and induced biosynthesis of antioxidant peroxidase and polyphenol oxidase. Plant Sci. 231, 62–73.

Bagyalakshmi, G., Muthukumar, T., Sathiyadash, K., Muniappan, V., 2010. Mycorrhizal and dark septate fungal associations in shola species of WesternGhats, Southern India. Mycoscience 9, 297–313.

Bayman, P., Otero, J.T., 2006. Microbial endophytes of orchid roots. In: Schulz, B., Boyle, C., Sieber, T. (Eds.), Microbial Root Endophytes. Springer, New York, pp. 153–173.

Beard, C.E., Adler, P.H., 2002. Seasonality of trichomycetes in larval black flies from South Carolina, USA. Mycologia 94, 200–209.

Beard, C.E., McCreadie, J.W., Adler, P.H., 2003. Prevalence of the trichomycete fungus *Harpella melusinae* (Harpellales: Harpellaceae) in larval black flies (Diptera: Simuliidae) across a heterogeneous environment. Mycologia 95, 577–583.

Belozerskaya, T., Aslanidi, K., Ivanova, A., Gessler, N.E., Karpenko, Y., Olishevskaya, S., 2010. Characteristics of extremophylic fungi from Chernobyl Nuclear Power Plant. In: Mendez-Vilas, A. (Ed.), Current Research, Technology and Education Topics in Applied Microbiology and Microbial Biotechnology. In: vol. 1. Formatex Research Centre, pp. 88–94.

Bhadury, P., Mohammad, B.T., Wright, P.C., 2006. The current status of natural products from marine fungi and their potential as anti-infective agents. J. Ind. Microbiol. Biotechnol. 33, 325–337.

Bidartondo, M.I., Read, D.J., 2008. Fungal specificity bottlenecks during orchid germination and development. Mol. Ecol. 17, 3707–3716.

Blochl, E., Rachel, R., Burggraf, S., Hafenbradl, D., Jannasch, H.W., Stetter, K.O., 1997. *Pyrolobus fumarii*, gen. And sp. nov., represents a novel group of archaea, extending the upper temperature limit for life to 113°C. Extremophiles 1, 14–21.

Bonants, P.J.M., Fitters, P.F.L., Thijs, H., den Belder, E., Waalwijk, C., Henfling, J.W.D.M., 1995. A basic serine protease from *Paecilomyces lilacinus* with biological activity against *Meloidogyne hapla* eggs. Microbiology 141, 775–784.

Bonnardeaux, Y., Brundrett, M., Batty, A., Dixon, K., Koch, J., Sivasithamparam, K., 2007. Diversity of mycorrhizal fungi of terrestrial orchids: compatibility webs, brief encounters, lasting relationships and alien invasions. Mycol. Res. 111, 51–61.

Braga, F.R., Arau'jo, J.V., Campos, A.K., Carvalho, R.O., Silva, A.R., Tavela, A.O., Maciel, A.S., 2007. Observação in vitro da ação dos isolados fu'ngicos *Duddingtonia flagrans, Monacrosporium thaumasium e Verticillium chlamydosporium* sobre ovos de *Ascaris lumbricoides* (Lineu, 1758). Rev. Soc. Bras. Med. Trop. 40, 356–358.

Braga, F.R., Arau'jo, J.V., Campos, A.K., Arau'jo, J.M., Silva, A.R., Carvalho, R.O., Tavela, A.O., 2008. *In vitro* evaluation of the action of the nematophagous fungi *Duddingtonia flagrans, Monacrosporium sinense* and *Pochonia chlamydosporia* on *Fasciola hepatica* eggs. World J. Microbiol. Biotechnol. 24, 1573–1972.

Brock, T.D., 1995. The road to Yellowstone—and beyond. Ann. Rev. Microbiol. 49, 1–28.

Campos, A.K., Arau'jo, J.V., RCL, A., Gandra, J.R., Guimaraes, M.P., 2007. Viabilidade de formulação peletizada do fungo nemato'fago *Monacrosporium sinense*, no controle biolo'gico de nemato'ides parasitos gastrintestinais de bezerros. Arq. Bras. Med. Vet. Zootec. 59, 14–20.

Conley, C.A., Ishkahnova, G., McKay, C.P., Cullings, K., 2006. A preliminary survey of non-lichenized fungi cultured from the hyperarid Atacama Desert of Chile. Astrobiology 6, 521–526.

Dadachova, K., Casadevall, A., 2008. Ionizing radiation: how fungi cope, adapt, and exploit with the help of melanin. Curr. Opin. Microbiol. 11, 525–531.

Davidson, E.A., 1991. Fluxes of nitrous oxide and nitric oxide from terrestrial ecosystems. In: Rogers, J.E., Whitman, W.E. (Eds.), Microbial Production and Consumption of Greenhouse Gases. American Society for Microbiology, Washington, DC, pp. 219–235.

De Britto, S., Tanzeembanu, D.G., Praveen, S., Lalitha, S., Ramachandra, Y.L., Jogaiah, S., Ito, S., 2020. Isolation and characterization of nutrient dependent pyocyanin from *Pseudomonas aeruginosa* and its dye and agrochemical properties. Sci. Rep. 10, 1542.

de Hoog, G.S., Grube, M. (Eds.), 2008. Black fungal extremes. Stud. Mycol. 61, 198.

Dearnaley, J.D.W., 2007. Further advances in orchids mycorrhizal research. Mycorrhiza 17, 475–486.

Ferreira, S.R., Araujo, J.V., Braga, F.R., Araujo, J.M., Frassy, L.N., Ferreira, A.S., 2011. Biological control of *Ascaris suum* eggs by *Pochonia chlamydosporia* fungus. Vet. Res. Commun. 35, 553–558.

Filipello-Marchisio, V., Berta, G., 1985. Endophytes of wild orchids native to Italy: their morphology, caryology, ultrastructure and cytochemical characterisation. New Phytol. 100, 623–641.

Gareth, J.E.B., 2000. Marine fungi: some factors influencing biodiversity. Fungal Diver. 4, 53–73.

Gareth, J.E.B., Pilantanapak, A., Chatmala, I., Sakayaroj, J., Phongpaichit, S., Choeyklin, R., 2006. Thai marine fungal diversity. Songklanakarin J. Sci. Technol. 28 (4), 687–708.

Gebauer, G., Meyer, M., 2003. 15N and 13C natural abundance of autotrophic and mycoheterotrophic orchids provides insight into nitrogen and carbon gain from fungal association. New Phytol. 160, 209–223.

Gorbushina, A., 2007. Life on the rocks. Environ. Microbiol. 9, 1613–1631.

Gorbushina, A.A., Krumbein, W.E., Volkmann, M., 2002. Rock surfaces as life indicators: new ways to demonstrate life and traces of former life. Astrobiology 3, 543–554.

Huang, Q.Y., Wang, W.P., Mo, R.Y., Lei, C.L., 2008. Studies on feeding and trophallaxis in subterranean termite *Odontotermes formosanus* using Rubidium chloride. Entomol. Exp. Appl. 129, 210–215.

Jaborowska, M., 2006. Limitation in the population of parasitic geohelminths by saprotrophic soil fungi and their secretions. Ann. Acad. Med. Stetin. 52, 37–46.

Jaborowska, M., Kuzna-Grygiel, W., Mazurkiewicz-Zapalowicz, K., Kolodziejczyk, L., 2006. Comparative studies on the antagonistic effects of mould fungi on the embryogenesis of *Ascaris suum*. Adv. Agric. Sci. 10, 45–49.

Jogaiah, S., Mostafa, A., Tran, L.S.P., Shin-ichi, I., 2013. Characterization of rhizosphere fungi that mediate resistance in tomato against bacterial wilt disease. J. Exp. Bot. 64, 3829–3842.

Jones, D.L., 2006. A Complete Guide to Native Orchids of Australia Including the Island Territories. Reed New Holland, Sydney.

Kaliamoorthy, S., 2007. Pattern of mycorrhizal infections in the roots of *Aerides maculosum* Lindl. and *Calanthe triplicate* (Willem.) Ames. Mycorrhiza News 19, 14–18.

Kasettrathat, C., Ngamrojanavanich, N., Wiyakrutta, S., Mahidol, C., Ruchirawat, S., Kittakoop, P., 2008. Cytotoxic and antiplasmodial substances from marine derived fungi, *Nodulisporium* sp. and CRI247-01. Phytochemistry 69, 2621–2626.

Kull, T., Arditti, J., 2002. Orchid Biology: Reviews and Perspectives. vol. VIII. Kluwer Academic Publishers, Dordrecht.

Kuzna-Grygiel, W., Kolodziejczyk, L., Janowicz, K., Mazurkiewicz-Zapalowicz, K., 2001. Effect of some saprotrophic soil fungi on the embryonic development of *Ascaris suum*. Acta Mycol. 36, 283–291.

Labeyrie, E.S., Molloy, D.P., Lichtwardt, R.W., 1996. An investigation of Harpellales (Trichomycetes) in New York state blackflies (Diptera:Simuliidae). J. Invertebr. Pathol. 68, 293–298.

Larsen, M., 1999. Biological control of helminths. Int. J. Parasitol. 29, 139–146.

Lavrent'ev, R., Zaitsev, S., Sudnitsyn, I., Kurakov, A., 2008. Nitrous oxide productionby fungi in soils under different moisture levels. Mosc. Univ. Soil Sc. Bull 63, 178–183.

Leake, J.R., 1994. The biology of mycoheterotrophic ('saprophytic') plants, Tansley review no. 69. New Phytol. 127, 171–216.

Lichtwardt, R.W., 1986. The Trichomycetes: Fungal Associates of Arthropods. Springer-Verlag, New York.

Lichtwardt, R.W., Arenas, M.J., 1996. Trichomycetes in aquatic insects from Southern Chile. Mycologia 88, 844–857.

Lichtwardt, R.W., Williams, M.C., 1988. Distribution and species diversity of trichomycete gut fungi in aquatic insect larvae in two Rocky Mountain streams. Can. J. Bot. 66, 1259–1263.

Lichtwardt, R.W., Williams, M.C., 1990. Trichomycete gut fungi in Australian aquatic insect larvae. Can. J. Bot. 68, 1057–1974.

Lichtwardt, R.W., Cafaro, M.J., White, M.M., 2001. The Trichomycetes: Fungal Associates of Arthropods. Revised ed. published on the Internet: Available from: www.nhm.ukans.edu/~fungi.

Liu, X., Xiang, M., Che, Y., 2009. The living strategy of nematophagous fungi. Mycoscience 50, 20–25.

Lopez-Llorca, L.V., Olivares-Bernabeu, C., Salinas, J., Jansson, H.B., Kolattukudy, P.E., 2002. Pre-penetration events in fungal parasitism of nematode eggs. Mycol. Res. 106, 499–506.

Lysek, H., 1982. The problem of human geohelminthoses and the prospects for their biological control. Acta Univ. Palacki. Olomuc. Fac. Med. 103, 315–328.

Lysek, H., Sterba, J., 1991. Colonization of *Ascaris lumbricoides* eggs by the fungus *Verticillium chlamydosporium* Goddard. Folia Parasitol. (Praha) 38, 255–259.

Mathew, G.M., Ju, Y.M., Lai, C.Y., Mathew, D.C., Huang, C.C., 2012. Microbial community analysis in the termite gut and fungus comb of *Odontotermes formosanus*: the implication of *Bacillus* as mutualists. FEMS Microbial. Ecol. 79, 504–517.

McCormick, M.K., Whigham, D.F., O'Neill, J., 2004. Mycorrhizal diversity in photosynthetic terrestrial orchids. New Phytol. 163, 425–438.

Misra, J.K., Tiwari, V.K., 2008. A new species of Gauthieromyces and range extensions for other Harpellales in India. Mycologia 100, 94–98.

Moreno, G., Illana, C., Lizarraga, M., 1998. *Protophysarum phloiogenum* and a new family in the Physarales. Mycol. Res. 102, 838–842.

Morgenstern, I., Powlowski, J., Ishmael, N., Darmond, C., Marqueteau, S., Moisan, M.C., Quenneville, G., Tsang, A., 2012. A molecular phylogeny of thermophilic fungi. Fungal Biol. 116, 489–502.

Mueller, G.M., Schmit, J.P., Leacock, P.R., Buyck, B., Cifuentes, J., Desjardin, D.E., Halling, R.E., Hjortstman, K., Iturriaga, T., Larsson, K., 2007. Global diversity and distribution of macrofungi. Biodivers. Conserv. 16, 37–48.

Murugesan, M., 2005. Floristic diversity and ethnobotanical studies in Velliangiri Hills, the Western Ghats of Coimbatore district, Tamil Nadu, India. PhD thesisBharathiar University, Coimbatore, Tamil Nadu, India.

Nagaraju, A., Murali, M., Jogaiah, S., Amruthesh, K.N., Mahadeva Murthy, S., 2012. Beneficial microbes promote plant growth and induce systemic resistance in sunflower against downy mildew disease caused by *Plasmopara halstedii*. Curr. Bot. 3, 12–18.

Nontachaiyapoom, S., Sasirat, S., Manoch, L., 2010. Isolation and identification of Rhizoctonia-like fungi from roots of three orchid genera, *Paphiopedilum*, *Dendrobium*, and *Cymbidium*, collected in Chiang Rai and Chiang Mai provinces of Thailand. Mycorrhiza 20, 459–471.

Omelon, C.R., 2008. Endolithic microbial communities in the polar desert habitats. Geomicrobiol J. 25, 404–414.

Onofri, S., Selbmann, L., Zucconi, L., Pagano, S., 2004. Antarctic microfungi as models for exobiology. Planet Space Sci. 52, 229–237.

Onofri, S., Barreca, D., Selbmann, L., Isola, D., Rabbow, E., Horneck, G., de Vera, J.P.P., Hatton, J., Zucconi, L., 2008. Resistance of Antarcticblack fungi cryptoendolithic communities to simulated space Mars conditions. Stud. Mycol. 61, 99–109.

Pontius, A., Krick, A., Kehraus, S., Brun, R., König, G.M., 2008. Antiprotozoal activities of heterocyclic substituted xanthones from the marine-derived fungus *Chaetomium* sp. J. Nat. Prod. 71, 1579–1584.

Prendergast-Miller, M.T., Baggs, E.M., Johnson, D., 2011. Nitrous oxide productionby the ectomycorrhizal fungi *Paxillus involutus* and *Tylospora fibrillosa*. FEMS Microbiol. Lett. 316, 31–35.

Rasmussen, H.N., 1995. Terrestrial Orchids: From Seed to Mycotrophic Plant. Cambridge University Press, Cambridge, UK.

Rasmussen, H.N., 2002. Recent developments in the study of orchid mycorrhiza. Plant Soil 244, 149–163.

Robinson, C.H., 2001. Cold adaptation in Arctic and Antarctic fungi. New Phytol. 151, 341.

Rouland-Lefèvre, C., Inoue, T., Johjima, T., 2006. Termite/termitomyces interactions. In: König, H., Varma, A. (Eds.), Soil Biology: Intestinal Microorganisms of Soil Invertebrates. Springer-Verlag, Berlin, pp. 335–347.

Shang, Z., Li, X., Li, M., 2012. Chemical profile of the secondary metabolites produced by a deep-sea sediment derived fungus *Penicillium commune* SD-118. Chin. J. Oceanol. Limnol. 30 (2), 305–314.

Sharma, J.R., 2000. Mycological studies at Antarctica. In: Seventeenth Indian Expedition to Antarctica, Scientific Report 2000, Department of Ocean Development, Technical Publication, No. 15, pp. 165–168.

Shefferson, R.P., Weib, M., Kull, T., Taylor, D.L., 2005. High specificity generally characterizes mycorrhizal association in rare lady's slipper orchids, genus *Cypripedium*. Mol. Ecol. 14, 613–626.

Shefferson, R.P., Taylor, D.L., Weiß, M., Garnica, S., McCormick, M.K., Adams, S., Gray, H.M., McFarland, J.W., Kull, T., Tali, K., Yukawa, T., Kawahara, T., Miyoshi, K., Lee, Y.I., 2007. The evolutionary history of mycorrhizal specificity among lady's slipper orchids. Evolution 61, 1380–1390.

Singh, S.M., Puja, G., Bhat, D.J., 2006. Psychrophilic fungi from Schirmacher oasis, East Antarctica. Curr. Sci. 90 (10), 1388–1392.

Smith, S.E., Read, D.J., 1997. Mycorrhizal Symbiosis. Academic Press, Cambridge, p. 312.

Sudisha, J., Shekar Shetty, H., 2009. Anti-oomycete compounds from *Ganoderma appalantum*, a wood rot basidiomycete. Nat. Prod. Res. 23, 737–753.

Tikhonov, V.E., Lopez-Llorca, L.V., Salinas, J., Jansson, H.B., 2002. Purification and characterization of chitinases from the nematophagous fungi *Verticillium chlamydosporium* and *V. suchlasporium*. Fungal Genet. Biol. 35, 67–78.

Vujanovic, V., St-Arnaud, M., Barabe, D., Thibeault, G., 2000. Viability testing of orchid seed and the promotion of colouration and germination. Ann. Bot. 86, 79–86.

Watts, K.R., Ratnam, J., Ang, K.H., Tenney, K., Compton, J.E., McKerrow, J., Crews, P., 2010. Assessing the trypanocidal potential of natural and semi-synthetic diketopiperazines from two deep water marine-derived fungi. Bioorg. Med. Chem. 18, 2566–2574.

White, M.M., Lichtwardt, R.W., 2004. Fungal symbionts (Harpellales) in Norwegian aquatic insect larvae. Mycologia 96, 891–910.

Williams, M.C., Lichtwardt, R.W., 1990. Trichomycete gut fungi in New Zealand aquatic insect larvae. Can. J. Bot. 68, 1045–1956.

Yanai, Y., Toyota, K., Morishita, T., Takakai, F., Hatano, R., Limin, S.H., Darung, U., Dohong, S., 2007. Fungal N_2O production in an arablepeat soil in Central Kalimantan, Indonesia. Soil Sci. Plant Nutr. 53, 806–811.

Yang, T.C., Mo, J.C., Cheng, J.A., 2004. Purification and some properties of cellulase from *Odontotermes formosanus* (Isoptera: Termitidae). Entomol. Sin. 11, 1–10.

Yuan, Z.-l., Chen, Y.-c., Yang, Y., 2009. Diverse non-mycorrhizal fungal endophytes inhabiting an epiphytic, medicinal orchid (*Dendrobium nobile*): estimation and characterization. World J. Microbiol. Biotechnol. 25, 295–303.

Zettler, L.W., Pouter, S.B., Mc Donald, K.I., Stewart, S.L., 2007. Conservation driven propagation of an epiphytic orchid (*Epidendrum nocturnum*) with a mycorrhizal fungus. Hort. Sci. 42, 135–139.

Zhou, Z., Takaya, N., Sakairi, M.A.C., Shoun, H., 2001. Oxygen requirement for denitrification by the fungus *Fusarium oxysporum*. Arch. Microbiol. 175, 19–25.

Zumft, W.G., 1997. Cell biology and molecular basis of denitrification. Microbiol. Mol. Biol. R. 61, 533–616.

Diversity and functions of secondary metabolites secreted by epi-endophytic microbes and their interaction with phytopathogens

22

G. Karthikeyan, L. Rajendran, V. Sendhilvel, K. Prabakar, and T. Raguchander
Department of Plant Pathology, Centre for Plant Protection Studies, Tamil Nadu Agricultural University, Coimbatore, Tamil Nadu, India

22.1 Introduction

In India, agriculture is the largest private enterprise, contributes nearly one-fourth of the national GDP, and is considered as the backbone of the agro-based industry. Through the update of modern agricultural technologies, India has moved from an era of food shortages to a status of food self-sufficiency and even food exports. In achieving a green revolution, no doubt that chemicals have played a significant role in improving production and productivity (Sudisha et al., 2010). In contrast, plant diseases caused by fungi, bacteria, and viruses cause significant damage and economic losses in agri-horti ecosystem every year. As a consequence, management strategies including the use of chemical pesticides are often applied inappropriately, at the wrong time, or when they are not required (Satapute et al., 2019). Hence, the effective and sustained control of phytopathogens is an important issue. Global losses caused by pathogens are estimated to be 10–15%, despite the continuous release of new resistant cultivars and pesticides. Furthermore, pathogens are continually developing resistance against the existing chemicals and a few of the pesticides are being withdrawn from the market for environmental reasons (Jogaiah et al., 2007). In addition to the reduction of crop yield, plant pathogens often lower crop quality by producing toxins which affect the human being. At this juncture, an ecologically sustainable method for the management of crop diseases is highly desirable. This paper deals with eco-friendly approaches involving some of the beneficial microbes (epi and endophytes) for the management of diseases in important agri-horti crops. Plants serve as a reservoir for an untold number of microbes known as endophytes which are colonizing the plant internally without doing substantial harm to the host (Wilson, 1995). They are found in numerous plant species and different parts of plants (seed, root, stem, leaf, and flower) with the most being members of common soil bacterial genera namely *Pseudomonas*, *Bacillus*, *Azospirillum*, and many strains can promote plant growth

Biocontrol Agents and Secondary Metabolites. https://doi.org/10.1016/B978-0-12-822919-4.00022-3

includes nitrogen fixation and increased resistance to phytopathogens (Babu et al., 2015). In the case of epiphytes, it is isolated from the surface of any plant parts whereas endophytic bacteria are commonly isolated from internal plant tissue either directly by centrifugation or pressure bomb extraction as well as indirectly following the disinfestation of the plant surface of different parts. These endophytic bacteria colonize a broad spectrum of plant species, plant parts, and it has often been stated that all plants in nature harbor endophytic bacteria. The total population densities depend on plant species, plant genotype, plant tissue, growth stage, and environmental conditions (Sturz et al., 1997). Overall population densities of endophytic bacteria are found to be higher in the root tissue with approximately log 5 cfu g^{-1} fresh weight compared to aerial parts (around log 4 cfu g^{-1} fresh stem weight and log 3 cfu g^{-1} in the leaf tissue) (Quadt-Hallmann and Kloepper, 1996; Hallmann et al., 1997), while they are significantly lower (or) generally absent in organs like flowers, fruits, and seeds. Among the different plant parts, the root is thought to be a preferred site of bacterial entrance and colonizes more bacteria which further move through the stomata, hydathodes, and micropores of aboveground parts. Intensive studies regarding the population dynamics of endophytic bacteria in the roots and stems of cotton and sweet corn over two successive cropping seasons were reported by McInroy and Kloepper (1995a,b). The main source of endophytic colonization is believed to be rhizosphere and phyllosphere, where high similarity was found between the bacteria present internally in the root and rhizosphere or phyllosphere (Jogaiah et al., 2010). In greenhouse experiments with cotton, 82% of the endophytic species identified from root tissue were also isolated from the rhizosphere (Hallmann et al., 1998) and 94% similarity was found between endorhiza and rhizosphere microbial communities in field-grown cucumber (Mahaffee and Kloepper, 1997). Sturz et al. (1997) reported the presence of nonpathogenic isolates of the genera *Pasteurella*, *Escherichia*, *Curtobacterium*, and *Xanthomonas* in leaves; however, since the root is exposed to bacteria long before the seedling emerges, the rhizosphere is the predominant source of endophytic colonization. Successful endophytic colonization includes the stages namely host finding, recognition, colonization, and entrance into internal plant tissue. Precolonization interactions include bacterial movement toward the root, bacterial attachment to the surface, plant-bacterial recognition processes at the root surface, and finally root penetration by the bacterium, whereas post colonization involves bacterial multiplication and localization within the root tissue and this endophytic bacteria probably find their host by chemotaxis, accidental encounter, or a combination of both. Bashan (1986) reported that the migration of the beneficial plant-associated bacteria, *P. fluorescens* and *Azospirillum brasilense* toward wheat roots was stimulated by various wheat genotypes and synthetic attractants. The attachment of endophytic bacteria with the roots may be due to the interaction of bacterial surface polysaccharides with root surface lectins and only limited data support this hypothesis. Duijiff et al. (1997) reported that o-antigenic side chain of the outer membrane lipopolysaccharides of *P. fluorescens* WCS417r may play a significant role in attachment. The main routes of entry are by natural opening such as hydathodes, stomata, and lenticels and wounds by abrasion with soil particles, pathogen damage, and formation of lateral roots, micropores, and mechanical damage. However, the most important entry of endophytic bacteria is thought to be through wounds and micropores present in the early stage of root development. Mahaffee and Kloepper (1997) reported increased endophytic bacterial populations as a consequence of *Rhizoctonia solani* infection of *Phaseolus vulgaris*. Nematode infestation also serves as a source of entry for endophytic bacteria (Hallmann et al., 1998). Following penetration of the plant tissue, endophytic bacteria have to multiply and colonize the plant tissue to establish a successful plant endophytic bacteria association. Hurek et al. (1994) reported inter and intracellular multiplication of *Azoarcus* sp. in the root cortex of rice. Lamb et al. (1996) reported the persistence of *P. aureofaciens* in maize plants due to continual bacterial invasion of newly formed lateral and crown roots. *B. subtilis* naturally colonizes plant roots using biofilm formation which is important for plant root colonization and protection. The biofilm formation by *B. subtilis*, embedded in a matrix composed of exopolysaccharides (EPS), helps to hold the bacterial community together. Quadt-Hallmann and Kloepper (1996) reported that intercellular spaces or epidermal cells colonized by endophytic bacteria are usually packed with bacteria. Bacterial endophytes promote plant growth and yield, suppress phytopathogens, may help to remove contaminants, solubilize phosphate, or contribute assimilable nitrogen to plants (Rosenblueth and Martínez-Romero, 2006). In plant tissues, bacterial endophytes may originate from seeds (McInroy and Kloepper, 1995a), vegetative material (Sturz, 1995), soil (McInroy and Kloepper, 1995b), and the phylloplane (Raaijmakers et al., 1995). A unique characteristic of *B. subtilis* is its ability to produce endospores when environmental conditions are stressful. Further, *B. subtilis* with an average of 4%–5% of its genome devoted to antibiotic synthesis, has the potential to produce more than two dozen structurally diverse antimicrobial compounds (Stein, 2005). The antibiotics belong to the families of surfactin, iturin, and fengycin and have well-recognized potential uses in biotechnology applications because of their surfactant properties. Peptide antibiotics or small molecules secreted by *Bacillus* species contribute their activity against pathogenic root-knot nematodes (Cadena et al., 2008; Kavitha et al., 2012). The production of surfactin and other lipopeptides by *Bacillus* cells is one of the main mechanisms for plant biocontrol since these molecules can induce systemic resistance (ISR) as well as strongly inhibit the growth of common plant pathogen such as *F. oxysporum* (Ramyabharathi and Raguchander, 2014b). PGPR includes many well-known genera *Rhizobia*, *Azospirillum*, *Klebsiella*, *Bacillus*, *Burkholderia*, *Azotobacter*, *Enterobacter*, and *Pseudomonas*, etc., but some of these genera include endophytic species as well. The best-characterized endophytic bacteria include *Azoarcus* spp., *Gluconacetobacter diazotrophicus*, *Herbaspirillum seropedicae*, *Bacillus subtilis*, etc. (Suyal et al., 2016).

22.2 Biocontrol agents (BCAs)

Alternative strategies to a chemical pesticide which are frequently used in the field of disease management are necessary which mainly include the conservation of microbial biodiversity and the introduction of antagonistic fungi and bacteria which show beneficial effects on plants by reducing disease incidence and increasing the crop yield significantly (Jogaiah et al., 2013). The goal of biological control is to use naturally occurring organisms to maintain the pathogen population at a level that does not cause any disease or yield loss. The conservation of biological control aims to conserve existing populations of naturally occurring organisms by adopting agricultural practices that favor the populations of these beneficial microbes. Antagonist bacteria are considered as an important functional group of beneficial bacteria used for biological control as well as for promoting plant growth. Recently biological control has become an important approach to suppress many plant pathogens and nematodes (Siddiqui and Shaukat, 2004; Jogaiah et al., 2018; Joshi et al., 2019). The strain, *Bacillus subtilis* Cohn. is considered to be an excellent biocontrol agent due to its ability to not only inducing plant systemic resistance but also producing various hydrolytic enzymes and antibiotics. From the large quantities of antimicrobials produced, lipopeptides stand among the most representative (Nandini et al., 2020). These antifungal peptides have been proved safe to people and no pollution to environment. So, they have a high potential for being used in biological control against pathogens and nematodes (Chen et al., 2012). Fluorescent pseudomonads and many *Bacillus* spp. are saprophytic root-colonizing bacteria. Fluorescent pseudomonads associated with plants include *P. fluorescens*, *P. putida*, *P. aeruginosa*, and *P. aureofaciens*. The species of fluorescent pseudomonads are grouped in different biovars and subgroups based on similarity in biochemical tests (Barrett et al., 1986). The use of FAME and UP-PCR fingerprinting profiles was generally helpful in the identification of *Bacillus* spp., making these features useful for the classification of genus at the species level. A total of 51 *Bacillus* isolates were characterized by universal primer polymerase chain reaction fingerprinting and were clustered into three different groups viz., *Bacillus amyloliquefaciens*, *B. subtilis*, and *B. pumilus* (Wulff et al., 2002). Daffonchio et al. (2003) investigated the molecular nature of polymorphism in ITS PCR fingerprinting of spore-forming bacteria belonging to the genera *Bacillus*, *Brevibacillus*, *Geobacillus*, and *Paenibacillus*. Hill et al. (2004) characterized more than 300 *Bacillus* isolates by fluorescent amplified fragment length polymorphism (AFLP) which revealed extensive diversity within *B. thuringiensis* and *B. cereus* compared to *B. anthracis*. Dickinson et al. (2004) demonstrated the versatility of matrix-assisted laser desorption ionization time-of-flight mass spectrometry (MALDI-TOFMS) protein profiling for the species differentiation of diverse suite of *Bacillus* spores. Various soil microorganisms which are capable of exerting beneficial effects on plant pests and diseases either in culture or protected environment have the potential of using in both agriculture and horticulture crops which result in enhanced yield. The modes of action are nitrogen fixation, phosphate solubilization and mobilization, plant growth regulation, siderophores production, lytic enzyme production, and antibiotics production and induced systemic resistance (Babu et al., 2015; Abdelrahman et al., 2016). It has been found that microbes have immense potential for plant growth promotion, as well as the degradation of environmental pollutants and can be used as bio-inoculant (bio-fungicides and bioinsecticides) (Nagaraju et al., 2012). Development cost, time, and ease of registration and potential growing market demand in contrast to chemical pesticides make biopesticides an attractive option to investigate. They suppress pests either by producing toxic metabolites specific to the pest, pathogenic infection of the pest, preventing the establishment of other microorganisms through competition for space or through various other modes of action (Singh et al., 2019).

22.3 Epi/endophytes

Rhizosphere bacteria that favorably affect plant growth and yield of commercially important crops are denominated by plant growth-promoting rhizobacteria (PGPR) which includes bacteria belonging to the genera *Azotobacter*, *Azospirillum*, *Arthrobacter*, *Bacillus*, *Burkholderia*, *Enterobacter*, *Klebsiella*, *Pseudomonas*, *Xanthomonas*, and *Serratia*. The PGPR affect plant growth directly by producing and secreting plant growth-promoting substances such as auxins, gibberellins, and cytokinins, by eliciting root metabolic activities with bacterial surface components, and/or by supplying biologically fixed nitrogen. Other PGPR affects plant growth by indirect mechanisms such as suppression of bacterial, fungal, and nematode pathogens (biocontrol) by the production of various metabolites, induced systemic resistance, and/or by competing with the pathogen for nutrients or for the colonization of space. Further, plant endophytes were reported to possess the capability of inhibiting host plant pathogens, accelerating seedling emergence, and promoting plant growth and yield, thus they were considered as more suitable biocontrol agents (Eljounaidi et al., 2016; Santoyo et al., 2016). Bacteria inhabiting seeds form an important group among the plant-associated bacteria. They are presumably present in seeds of all plant species and numerous bacterial genera of different phyla were reported as seed endophytes. They can play important roles during germination, seedling development, and plant growth (Sudisha et al., 2006). As they possess plant-growth-promoting and biocontrol properties, the study of their application in diverse processes such as biofertilization, bioenergy production, and bioremediation should be encouraged. Most seed endophytes isolated until now are cultivable bacteria and several of them are accidentally discovered during a sidetrack of the main research issue. A large fraction of the bacterial endophytic population in seeds probably has unknown cultivation conditions or is in a viable but noncultivable state. Metagenomic studies will shed more light on this part of the seed endophyte populations and provide a better understanding of the total bacterial populations present in seeds, both concerning the genera that are present as well as their phenotypic characteristics and possible role(s) in germination and plant development. Until now, the studies suggesting vertical transmission of bacterial endophytes have used 16S rRNA gene sequencing, leading to identification at genus or species level. To confirm the existence of vertical transmission, techniques allowing us to generate strain-level identification, such as repetitive element palindromic PCR, single nucleotide polymorphism analysis, or pulsed-field gel electrophoresis, should be applied in order to detect the presence of the same bacterial strains in consecutive plant generations (Lopez-Velasco et al., 2013).

It has been suggested that microorganisms isolated from the surface of the plant or as endophyte from specific crop may be better adapted to that crop and may provide better control of diseases than organisms originally isolated from other plant species. Such plant-associated microorganisms may serve as better biocontrol agents because they are already closely associated with and adapted to the plant or plant part as well as the particular environmental conditions in which they are supposed to function. The screening of such locally adapted strains has yielded improved biocontrol in some cases and was reported to possess the capability of inhibiting host plant pathogens, accelerating seedling emergence, and promoting plant growth and yield (Eljounaidi et al., 2016; Santoyo et al., 2016). The inoculation of plant growth-promoting endophytic bacteria may reduce the application of potentially polluting chemical fertilizers and pesticides (Ambrosini et al., 2016). The microorganisms present in the plants compared to those in the rhizosphere have a closer interaction with their host due to their ability to colonize the plant tissues. The endophytic microbes besides having superiority over rhizosphere microbes due to stability of endosphere have plants' preference (Fig. 22.1). Consequently, endophytic bacteria are comparatively less prone to environmental adversities, xenobiotics, and nutrient limitations, which are

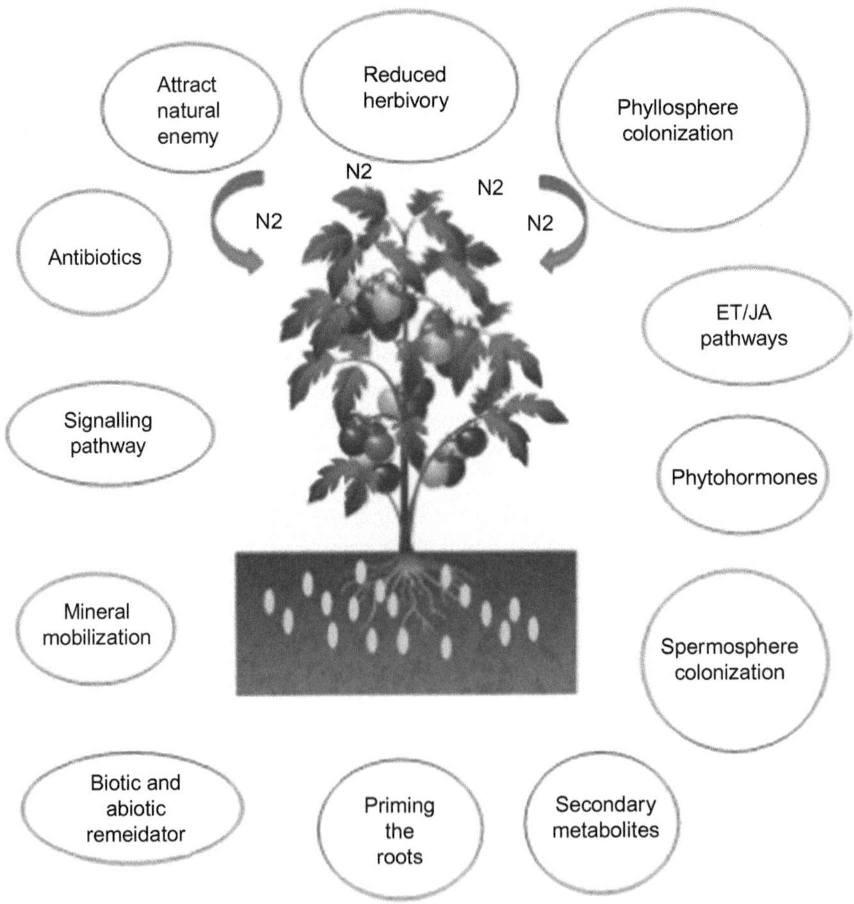

Fig. 22.1 Multidimensional interaction of endophytes with host plants.

quite prevalent in soil. These characteristics make endophytes preferable candidates (Mahmood et al., 2019). Supplementation with plant growth-promoting rhizobacteria (PGPR) *Pseudomonas aeruginosa* and *Burkholderia gladioli* alleviates cadmium toxicity in *Solanum lycopersicum* by modulating the expression of secondary metabolites [CHS (chalcone synthase; 138.4%), PAL (phenylalanine ammonia lyase; 206.7%), CS (citrate synthase; 61.3%), SUCLG1 (succinyl Co-A ligase; 33.6%), SDH (succinate dehydrogenase; 23.2%), FH (fumarate hydratase; 12.4%), and MS (malate synthase; 41.2%)] (Khanna et al., 2019).

22.4 Secondary metabolites

The communication of endophytic communities with the host plant significantly influences the physiological processes of the plant such as the activation of silent gene clusters leading to the synthesis of novel secondary metabolites (Nakkeeran et al., 2019). Microbial secondary metabolites are low-molecular mass products usually produced during the late growth phase (idiophase) of a relatively small sort of microorganisms. PGPR have been classified as biofertilizers, phytostimulators, rhizomediators, and biopesticides, depending on their adapted functional strategies under various physiological conditions (Ahemad and Kibret, 2014). To perform each of the above roles, PGPR produces a variety of primary or low-molecular-weight secondary metabolites (Table 22.1). The ability of PGPR, as antagonists, to inhibit pathogen growth and to produce secondary metabolites has been claimed to be important for biological control (El-Fawy et al., 2018). Many bacterial strains are known to suppress fungal growth in vitro by producing one or more antifungal antibiotics that may also have activity in vivo. Many endophytes also secrete specialized metabolites or biologically active compounds. Bioactive compounds like alkaloids, steroids, terpenoids, peptides, polyketones, flavonoids, quinols and phenols, and the natural insecticide azadirachtin produced by endophytic bacteria (Molina et al., 2012) (Fig. 22.2) that help host plant to develop systemic resistance against pathogens are also used in the pharmaceutical industries. Several strains of *Pseudomonas* spp. and *Bacillus* spp. have been shown to produce a wide array of antibiotics which includes ammonia, butyrolactones, 2-4 diacetyl phloroglucinol, HCN, kanosamine, oligomycin A, oomycin A, phenazine-1-carboxylic acid, pyoluteorin, pyrrolnitrin, tropolone, pyocyanin, iturin, surfactin, viscosinamide, zwittermicin A, agrocin 84 as well as several other uncharacterized moieties. The production of antimicrobial metabolites by *Bacillus* species determines the ability to control plant diseases (Silo-suh et al., 1994). Antibiotics from iturin family show strong antifungal and hemolytic activities with limited antibacterial activity. Iturin has a broad antifungal spectrum and serve as a potential agent for biocontrol of plant diseases (Maget-Dana and Peypoux, 1994). Damodaran et al. (2013) reported that two stress-tolerant rhizobacteria *B. pumilus* B-1 and *B. subtilis* B-3 had extensive zone formation for indole-3-acetic acid (>1 cm) and siderophore production with higher zone ranging from 0.6 to 0.9 cm. Ramyabharathi and Raguchander (2014a) reported the presence of lipopeptide genes viz., Iturin-ItuC gene (506 bp), ItuD gene (800 bp), Bacillomycin A-BmyA gene (344 bp), Bacillomycin D-BacD gene (482 bp), Bacilysin-BacAB gene

Table 22.1 Secondary metabolites produced by microbes and its effect on phytopathogens.

Crop	Epi/endophyte	Source	Effective against	Secondary metabolites	References
–	*Streptomyces igroscopicus*	–	–	ACC deaminase	Nascimento et al. (2014)
Wheat	*Streptomyces rochei, S. carpinensis, S. thermolilacinus*	Rhizosphere	–	Production of siderophore, IAA synthesis, and phosphate solubilization	Jog et al. (2012)
Blackgram, *Triticum aestivum*	*Pseudomonas* spp.	Rhizosphere	Soilborne pathogens, *Gaeumanomyces graminis* var. *tritici* Bacterial canker of tomato caused by *Clavibacter michiganensis* subsp. *michiganensis*	Cyclic lipopeptide amphysin, 2,4-diacetylphloroglucinol (DAPG), oomycin A, the aromatic polyketide pyoluteorin, pyrrolnitrin, the antibacterial compound tropolone Phenazine-1-carboxylic acid (PCA) Hydrogen cyanide-nematicidal activity	Défago (1993), De Souza et al. (2003), Lanteigne et al. (2012), Ju et al. (2018), and Kang et al. (2018)
	Bacillus, Streptomyces, Stenotrophomonas spp.,			Oligomycin A, kanosamine, the linear aminopolyol zwittermicin A, and xanthobactin	Kim et al. (1999) and Compant et al. (2005)
Betelvine	*Bacillus subtilis* Bbv57	Rhizosphere	Fusarium wilt nematode complex in Gerbera	Butanedioic acid, Hexadecanoic acid ethyl ester, Pentanedioic acid 2-oxo-dimethyl ester, Pyrrolo [1,2-a]pyrazine-1,4-dione, hexahydro-3-(2-methylpropyl) and Pyrrolo [1,2-a]pyrazine-1,4-dione, hexahydro-3-(phenylmethyl) ester]	Ramyabharathi and Raguchander (2014a,b) antifungal, antibacterial and antinematicidal activity

Citrus	*Bacillus subtilis*	Unknown	Citrus black spot (CBS) *Phyllosticta citricarpa*	Iturin and surfactin antibiotics	Kupper et al. (2020)
Eleusine coracana	*Enterobacter* sp.	Unknown	Suppressing *Fusarium graminearum* in plant tissues	Reduction of deoxynivalenol mycotoxin	Mousa et al. (2015)
Beta vulgaris	*Pseudomonas poae* RE*1-1-14	Roots	Suppressing *Phytophthora capsici* and *P. infestans* zoospores	Production of novel lipopeptide Poaeamide	Zachow et al. (2015)

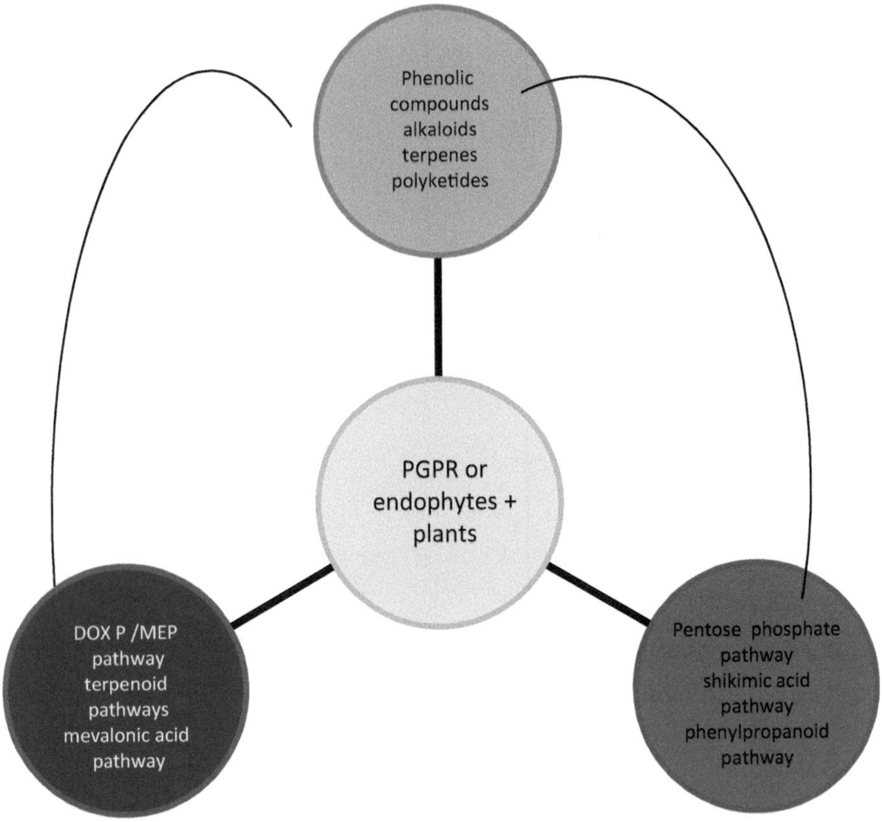

Fig. 22.2 Pathways for the synthesis of some secondary metabolites from epi/endophytes.

(815 bp), and Fengycin-fenD gene (220 bp) in *B. subtilis* strain EPCO 16. Applications of biological control agents (BCAs) and their secondary metabolites are important strategies in the management of plant diseases. Metabolomic profiling of PGPR *Pantoea agglomerans* strain Pa under one strain many compounds (OSMAC) conditions revealed a wide diversity of secondary metabolites (SM) with interesting salt stress alleviation and PGP activities.

22.5 Synthesis pathway and diversity

Endophytes synthesize secondary metabolites via a variety of pathways, e.g., polyketide, isoprenoid, or amino acid derivation (Jalgaonwala, 2013) and the biosynthetic pathways are responsible for the production of both primary and secondary metabolites (Nicolaou et al., 2011) that act as biofilm, toxins, and virulence factors (Raaijmakers and Mazzola, 2012). The aliphatic hydrocarbon-based compounds, namely decane, 2,3,5,8-tetramethyl (Sathyaprabha et al., 2010), benzoic acid,

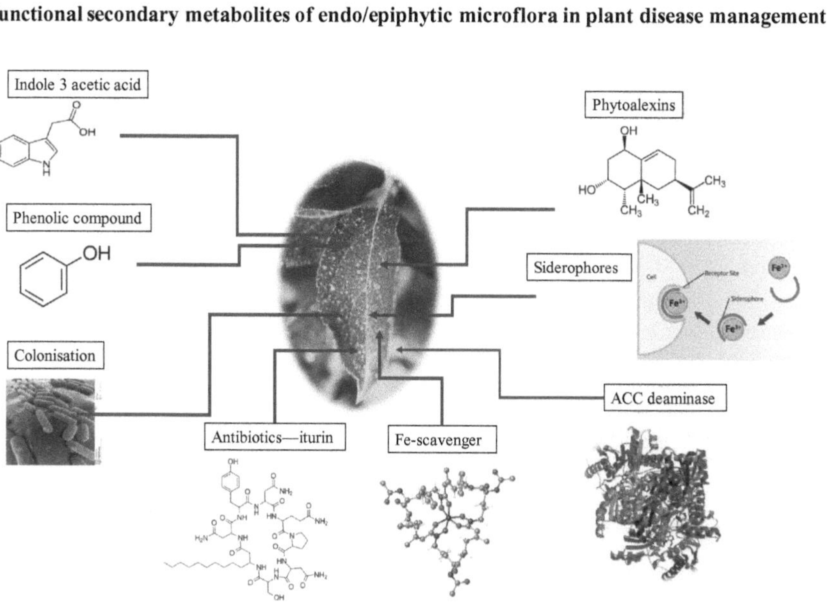

Fig. 22.3 Multifunctional secondary metabolites of endo/epiphytic microflora in plant disease management.

4-ethoxy-ethyl ester (Patel et al., 2006), and plasticizer compound diisooctyl phthalate have both antibacterial and antifungal properties (Helal et al., 2006).

The compounds associated with *B. cereus* are 3-methyldec-3-ene, an aliphatic hydrocarbon possess antibacterial activity, 2,5-piperazinedione, 3,6-bis (2-methylpropyl)—the aliphatic hydrocarbon that has antibacterial activity (Dehpour et al., 2011) and the aliphatic hydrocarbon namely dihydroergotamine possess both antibacterial and antifungal activity. Tridecyl trichloroacetate and hexadecan1-ol a lipopeptide-based main polar compound of surfactin, possess antimicrobial activity (Ongena et al., 2007), and tetracosyl heptafluorobutyrate, erycilamide, and pentatriacontene belonging to amide group had antibacterial and antifungal activity (Ashwanikumar et al., 2011). Pentadecanoic acid, 14-methyl ester belonging to fatty acid has antibacterial and antifungal activity (Ghazala et al., 2004) (Fig. 22.3).

22.6 Interaction in spermosphere

Endophytic microbes, especially seed endophytic bacteria, are considered to be excellent biofertilizers and biocontrol agents in agriculture not only associated with plant growth promotion and health but also transmitted these benefits from one generation to other through gynophores and ovules. *Bacillus* spp. are known to promote plant growth and are associated with N2-fixation in peanut and other plants. Gene clusters responsible for antifungal metabolites (fengycin, surfactin, bacilysin) and antibacterial metabolites (butirosin, bacillaene, difficidin, macrolactin, surfactin, bacilysin) were identified in peanut endophyte *Bacillus velezensis* LDO2 (Chen et al., 2019). Several *Paenibacillus* spp. were isolated from *Lupinus albus* nodules, indicating that they play a role in nodulation (Carro et al., 2014). Bacteria and fungi were determined with approximately 10^{10} gene copy numbers g^{-1} seed of native alpine plants as abundant inhabitants. Archaea, which were newly discovered as seed endophytes, are less and represent only 1.1% of the signatures. The seed microbiome was highly diversified, and all seeds showed a species-specific, highly unique microbial signature (Wassermann et al., 2019). Truyens et al. (2015) reported that *Bacillus*, *Pseudomonas*, *Paenibacillus*, *Micrococcus*, *Staphylococcus*, *Pantoea*, and *Acinetobacter*, in ascending order, are the most common bacteria within seeds of very different crop species. Hexanoate, succinate, and jasmonic acid (JA) accumulated in peanut root tips after incubation with *Paenibacillus glycanilyticus* YMR3 which increases the JA content (14.93-fold change) and modulates the metabolism of peanut to facilitate nodule formation and growth (Li et al., 2019).

22.7 Interaction in rhizosphere

The rhizosphere is a reservoir for plant endophytes and represents the belowground interface with the highly diverse soil microbiota. Yung (2010) reported that bacteria synthesize auxins in order to perturb host physiological processes for their own benefit. The microorganisms isolated from the rhizosphere region of the various crop can produce indole-acetic acid as secondary metabolites due to the rich supply of substrates. Indole acetic acid helps in the production of longer roots with an increased number of root hairs and root laterals which are involved in nutrient uptake (Datta and Basu, 2000). IAA stimulates cell elongation by modifying certain conditions like, increase in osmotic contents of the cell, increase in permeability of water into cell, decrease in wall pressure, and an increase in cell wall synthesis and protein synthesis (Zhao, 2010). Further, Cadena et al. (2008) reported that *B. amyloliquefaciens* strain FZB42 produced lipopeptides, surfactins, bacillomycin D, and fengycins, which are secondary metabolites with mainly antifungal activity, also decreased gall formation, egg mass count, and juvenile counts of *M. incognita* extracted from roots of tomatoes. Plant growth-promoting actinobacteria reduce the severity of soilborne diseases and benefit the host plant growth (Qin et al., 2015; Jogaiah et al., 2016). Such actinobacteria favor the plant development directly by providing nutrients (e.g., iron, phosphorus, and nitrogen) and synthesizing hormones (e.g., indole-3-acetic acid and ethylene) (Alekhya and Gopalakrishnan, 2017), as well as indirectly by controlling the development of phytopathogens and pests, and protecting against heavy metals and salt stress (Jacob and Sudini, 2016).

22.8 Interaction with postharvest pathogens

The endophytic Actinobacterium *Streptomyces griseocarneus* R132 controls phytopathogens namely *Fusarium oxysporum*, *Botryosphaeria dothidea*, *Colletotrichum gloeosporioides* and promotes the growth of pepper (*Capsicum annuum*)

(Liotti et al., 2019). The Gram-positive bacterium *Bacillus* spp. is one of the most studied antagonistic microorganisms for biological control in agriculture, being used in citrus crops for the prevention and control of post bloom fruit drop (*Colletotrichum acutatum*), a fungal disease that occurs in the citrus phylloplane (Kupper et al., 2012; Klein et al., 2016).

PGPB are a group of nonpathogenic beneficial bacteria that can directly and/or indirectly promote plant growth, disease resistance, and abiotic stress tolerance. They may live autonomously in the soil or colonize the rhizosphere, phyllosphere (epiphytes), and plants' interior tissues (endophytes) (Baez-Rogelio et al., 2016; Lastochkina et al., 2017; Seifikalhor et al., 2018). Particularly, interesting PGPB, belonging to the genus *Bacillus* spp. viz. *B. subtilis*, is one of the most attractive agents for the development of natural plant protection products, as recommended by United States Food and Drug Administration. *Bacillus* spp. are generally recognized as safe microorganisms for application in the food industry. *Bacillus* spp. occupy the same niche as many pathogens and have the capacity to produce a wide range of bioactive substances with antibiotic activity (Lastochkina et al., 2019). These substances induce various physiological features in host plant metabolism without causing adverse effects on the environment and human health (Sarma et al., 2012; Maksimov et al., 2015).

The ability of *B. subtilis* to suppress the development of postharvest pathogens causing gray mold (*Botrytis cinerea* and *B. mali*) has been demonstrated in strawberry, pear, apple, and tomato (Lastochkina et al., 2019; Kim et al., 2016; Zhao et al., 2007; Kilani-Feki et al., 2016). More in-depth studies suggest that microbial antagonists from the *Bacillus* genus possess the substantial potential to increase vegetables/fruit sets, quality, postharvest disease resistance, and tolerance under temperature fluctuations, mechanical injury associated with loading of product for transportation, unloading, packaging, and storage (Jiang et al., 2001; Miller, 2003).

22.9 Interaction in phyllosphere

In control of phyllosphere diseases, a contribution of both iturins and fengycins was shown in the antagonism of *B. subtilis* toward *Podosphaera fusca* infecting melon leaves (Romero et al., 2007). In the phyllosphere, endophytes, host, and pathogens virulence are involved to contain the foliar diseases. There are two hypotheses for how endophytes reduce the activity of the pathogen to express the symptom severity in their host plants: direct interaction between endophytes and pathogens, or local induction of quantitative genetic resistance. Systemic induction is not consistent with observed patterns of local resistance conferred by endophytes in the aforementioned study of Populus and in the studies of foliar endophytes of the tropical tree *Theobroma cacao* (Arnold et al., 2002). While there have been no direct tests of these two hypotheses, the results of Raghavendra and Newcombe (2013) are more consistent with direct interaction than with local induction. More specifically, they argue that the consistent ranking of endophytes across plant genotype-rust combinations is compatible with direct interaction between endophytes and rust (Raghavendra and Newcombe, 2013). An alternative mechanistic explanation for leaf endophyte effects on disease severity is direct interaction with pathogens. Leaf endophytes could reduce disease severity via direct competition for limited nutrient resources, by parasitizing pathogens, or by producing mycotoxins. For example, the fungus *Eudarluca caricis* has been shown to parasitize rust fungi on the outer surfaces of leaves (Nischwitz et al., 2005), and the fungal leaf pathogen *Stachybotrys cylindrospora* produces mycotoxins that can negatively affect other pathogens (Ayer and Miao, 1993). Direct effects could also occur in the opposite direction (Kurose et al., 2012). For example, leaf endophytes could increase disease severity by producing a compound that would otherwise be limited to pathogen growth. The extraneous greenhouse fungi could have influenced the direct or indirect effects between plants, endophytes, and pathogens. Indeed, some of the plants were infected with such fungi, although any incidental colonization would have been constant across all treatments.

22.10 Epiphytic microflora for plant disease management

Epiphytic bacteria have been defined as populations that can survive and multiply on the surface of plants (Hirano et al., 1982). They occupy ecological niches on the phylloplane that could be occupied by pathogens (Monier and Lindow, 2005) and to their broad antagonistic effect against pathogens. Biosurfactants, pyocyanin, antibiotics, bacteriocins, and volatile organic compounds (VOCs) synthesis, siderophores, and competition for space and nutrients are related to the antagonistic effects of epiphytic bacteria on the phytopathogen growth (Beattie and Lindow, 1999; Lindow and Brandl, 2003; De Britto et al., 2020). The phenological and V10 stages of maize plants sprayed with *Bacillus* spp. (isolate 8) caused the greatest effect on reducing the severity of northern leaf blight. Moreover, isolate 8 was the potential biocontrol agent that showed more stability in the phyllosphere all potential epiphytic biocontrol agents had a significant effect on controlling the disease caused by *Exserohilum turcicum* in maize (Melina et al., 2017).

The epiphytic bacteria such as *Paenibacillus macerans* and *Bacillus pumilus* epiphytic bacteria reduced *Xanthomonas vesicatoria* and *Alternaria solani* disease severity in tomato plants (Filho et al., 2010). They found that epiphytic bacteria can inhibit the growth of tested phytopathogens in vitro and efficiently colonize the phylloplane of tomato plants. Halfeld-Vieira et al. (2008) proved the efficiency of the epiphyte *Bacillus cereus* UFV-IEA6 against *Phytophthora infestans*. Gnanamanickam and Immanuel (2006) reported the importance of epiphyte *Bacillus* sp. proved for plant disease control. The epiphytic growth of biocontrol agents in the host plant is determined by the physical environment (availability of nutrients, weather conditions) and the microbiological environment (indigenous microbial community). The ability of antagonists to colonize on apple blossoms under field conditions is crucial to ensure effective biological control of fire blight (Johnson and Stockwell, 1998a,b). The epiphytic bacteria *P. fluorescens* EPS62e was able to reach very high population levels on both apple and pear blossoms, between 10^7 and 10^8 CFU per blossom after a single treatment during bloom against fire blight pathogen *E. amylovora* (Pujol et al., 2007). The epiphytic microflora of *Bacillus* and *Pantoea* isolates from leaves were potentially control the *Exserohilum turcicum* disease in maize (Sartoria et al., 2015). The use of microorganisms that antagonize foliar pathogens is the risk free when these organisms come from the same ecosystem. The inhabitants of the phyllosphere are termed epiphytes and may consist of a variety of bacteria, yeasts, or filamentous fungi (Lindow and Brandl, 2003). Microorganisms within the phyllosphere can include those that are pathogenic to the plant but can also include nonpathogenic organisms that prevent the colonization of leaf by pathogens (Kishore et al., 2005). The pear and apple disease caused by the bacterium *Erwinia amylovora* was controlled using *Pantoea agglomerans* strain whose mechanism of action is not the synthesis of antibiotics (Johnson and Stockwell, 1998a,b). The most epiphytic population consisted of bacteria which were found in the order of 6 log CFU per gram of maize leaf (Yadav et al., 2005; Jurkevitch and Shapira, 2000). Therefore, the interaction of phyllosphere microorganisms can play an important role for plant health and protection (Andrews and Harris, 2000). Antagonistic effect was observed with *Bacillus subtilis* and *Pseudomonas fluorescens* against *E. turcicum* (Harlapur et al., 2007). The epiphytic bacteria, especially the *pseudomonads* and *bacilli* have been shown to play a key role in the suppression of plant pathogens in different cropping systems (Ramarathnam and Dilantha, 2006).

22.11 Challenges and future perspectives for upscaling the secondary metabolites

The present scenario in the plant disease management is aiming to explore the biological originated secondary metabolites. The large-scale production paves the way for the natural products to contain the plant diseases which support for the organic farming concepts. The multi-scale method is widely applied for the optimization of industrial bioprocesses for penicillin, erythromycin, chlortetracycline, inosine, and guanosine productions (Zhang et al., 2004). But, owing to the complexity of the fermentation system, the multi-scale method is unable to deduce unequivocal relationships between bioprocess variables, product formation, and microbial growth.

The development of the fermentation process and its optimization is seriously challenged by a lack of knowledge about scale-up and other issues such as the influence of morphology on broth rheology and mass transfer. Microbial products are extensively used for pest and disease management. This paper is a precursor for an idea to exploit the endophytic bacteria for the interdisciplinary approach to explore the secondary metabolites, to utilize in agriculture, and to produce either novel compounds or enhance the existing important secondary metabolites.

References

Abdelrahman, S., Abdel-Motaal, F., El-Sayed, M., Jogaiah, S., Shigyo, M., Ito, S.-I., Tran, L.-S.P., 2016. Dissection of *Trichoderma longibrachiatum*-induced defense in onion (*Allium cepa* L.) against *Fusarium oxysporum* f. sp. Cepa by target metabolite profiling. Plant Sci. 246, 128–138.

Ahemad, M., Kibret, M., 2014. Mechanisms and applications of plant growth promoting rhizobacteria: current perspective. J. King Saud Univ. Sci. 26, 1–20.

Alekhya, G., Gopalakrishnan, S., 2017. Biological control and plant growth-promotion traits of Streptomyces species under greenhouse and field conditions in chickpea. Agric. Res. 6, 410–420. https://doi.org/10.1007/s40003-017-c0278-2.

Ambrosini, A., de Souza, R., Passaglia, L.M.P., 2016. Ecological role of bacterial inoculants and their potential impact on soil microbial diversity. Plant Soil 400, 193–207. https://doi.org/10.1007/s11104-015-2727-7.

Andrews, J., Harris, R., 2000. The ecology and biogeography of microorganisms on plant surfaces. Annu. Rev. Phytopathol. 38, 145–180.

Arnold, A.E., Maynard, Z., Gilbert, G.S., Coley, P.D., Kursar, T.A., 2002. Are tropical fungal endophytes hyperdiverse? Ecol. Lett. 3, 267–274.

Ashwanikumar, S., Surender, S., Sandeep, J., Parvin, K., 2011. Synthesis, antimicrobial evaluation, qsar and in silico admet studies of decanoic acid derivatives. Acta Pol. Pharm. Drug Res. 68, 191–204.

Ayer, W.A., Miao, S., 1993. Sccondary metabolites of the aspen fungus *Stachybotrys cylindrospora*. Can. J. Chem. 71, 487–493.

Babu, N.A., Jogaiah, S., Ito, S.-i., Kestur, A., Tran, N.L.-S.P., 2015. Improvement of growth, fruit weight and early blight disease protection of tomato plants by rhizosphere bacteria is correlated with their beneficial traits and induced biosynthesis of antioxidant peroxidase and polyphenol oxidase. Plant Sci. 231, 62–73.

Baez-Rogelio, A., Morales-García, Y.E., Quintero-Hernández, V., Muñoz-Rojas, J., 2016. Next generation of microbial inoculants for agriculture and bioremediation. Microb. Biotechnol. 10, 19–21.

Barrett, E.L., Solanes, R.E., Tang, J.S., Palleroni, N.J., 1986. *Pseudomonas fluorescens* biovar V its resolution into distinct component groups and relationship of these to other *P. fluorescens* biovars to *P. putida* and to psychrophilic pseudomonads associated with food spoilage. J. Gen. Microbiol. 132, 2709–2721.

Bashan, Y., 1986. Alginate beads as synthetic inoculant carriers for slow release of bacteria that affect plant growth. Appl. Environ. Microbiol. 51, 1089–1098.

Beattie, G.A., Lindow, S.E., 1999. Bacterial colonization of leaves: a spectrum of strategies. Phytopathology 89, 53–359.

Cadena, M.B., Burelle, N.K., Kathy, S.L., Santen, E.V., Kloepper, J.W., 2008. Suppressiveness of root-knot nematodes mediated by rhizobacteria. Biol. Control 47, 55–59.

Carro, L., Flores-Felix, J.D., Ramı'rez-Bahena, M.H., Garcı'a-Fraile, P., Martı'nez-Hidalgo, P., Igual, J.M., Tejedor, C., Peix, A., Velazquez, E., 2014. *Paenibacillus lupini* sp. nov., isolated from nodules of *Lupinus albus*. Int. J. Syst. Evol. Microbiol. 64, 3028–3033. https://doi.org/10.1099/ijs.0.060830.

Chen, F., Han, D., Wang, M., 2012. Production and identification of antifungal compounds produced by *Bacillus subtilis* B579. In: International Conference on "Bioinformatics and Biomedical Engineering"vol 4. pp. 87–90.

Chen, L., Shi, H., Heng, J., Wang, D., Bian, K., 2019. Antimicrobial, plant growth-promoting and genomic properties of the peanut endophyte *Bacillus velezensis* LDO2. Microbiol. Res. 218, 41–48. https://doi.org/10.1016/j.micres.2018.10.002.

Compant, S., Duffy, B., Nowak, J., Clement, C., Barka, E.A., 2005. Use of plant growth promoting bacteria for biocontrol of plant diseases: principles, mechanisms of action, and future prospects. Appl. Environ. Microbiol. 71, 4951–4959.

Daffonchio, D., Cherif, A., Brusetti, L., Rizzi, A., Mora, D., Boudabous, A., Borin, S., 2003. Nature of polymorphisms in 16s-23s rRNA gene intergenic transcribed spacer fingerprinting of *Bacillus* and related genera. Appl. Environ. Microbiol. 69 (9), 5128–5137.

Damodaran, T., Sah, V., Rai, R.B., Sharma, D.K., Mishra, V.K., Jha, S.K., Kannan, R., 2013. Isolation of salt tolerant endophytic and rhizospheric bacteria by natural selection and screening for promising plant growth-promoting rhizobacteria (PGPR) and growth vigour in tomato under sodic environment. Afr. J. Microbiol. Res. 7 (44), 5082–5089.

Datta, C., Basu, P., 2000. Lndole acetic acid production by a rhizobium species from root nodules of a leguminous shrub *Cajanus cajan*. Microbiol. Res. 155, 123–127.

De Britto, S., Tanzeembanu, D.G., Praveen, S., Lalitha, S., Ramachandra, Y.L., Jogaiah, S., Ito, S., 2020. Isolation and characterization of nutrient dependent pyocyanin from *Pseudomonas aeruginosa* and its dye and agrochemical properties. Sci. Rep. 10, 1542.

de Souza, J.T., Weller, D.M., Raaijmakers, J.M., 2003. Frequency, diversity, and activity of 2,4-diacetylphloroglucinol-producing fluorescent *Pseudomonas* spp. in Dutch take-all decline soils. Phytopathology 93 (1), 54–63. https://doi.org/10.1094/PHYTO.2003.93.1.54.

Défago, G., 1993. 2,4-Diacetylphloroglucinol, a promising compound in biocontrol. Plant Pathol. 42, 311–312.

Dehpour, A.A., Mohammad, E., Roudgar, A., Nabavi, S.F., Nabavi, S.M., 2011. Antioxidant and antibacterial activity of *Consolida orientalis*. World Acad. Sci. Eng. Technol. 73, 162–166.

Dickinson, D.N., La Duc, M.T., Haskins, W.E., Gornushkin, I., Winefordner, J.D., Powell, D.H., Venkateswaran, K., 2004. Species differentiation of a diverse suite of *Bacillus* spores by mass spectrometry-based protein profiling. Appl. Environ. Microbiol. 70 (1), 475–482.

Duijiff, B.J., Gilaninazzi-Pearson, V., Lemancelu, P., 1997. Involvement of the outer membrane lipopolysaccharides in the endophytic colonization of tomato roots by biocontrol *Pseudomonas fluorescens* strain WCS417r. New Phytol. 135, 325–334.

El-Fawy, M.M., El-Sharkawy, R.M., Abo-Elyousr, K.A.M., 2018. Evaluation of certain*Penicillium frequentans* isolates against Cercospora leaf spot disease of sugar beet. Egypt J. Biolog. Pest. Cont. 28, 49. https://doi.org/10.1186/s41938-018-0053-0.

Eljounaidi, K., Lee, S.K., Bae, H., 2016. Bacterial endophytes as potential biocontrol agents of vascular wilt diseases-review and future prospects. Biol. Control 103, 62–68.

Filho, R.L., da Silva Romeiro, R., Alves, E., 2010. Bacterial spot and early blightbiocontrol by epiphytic bacteria in tomato plants. Pesq. Agropec. Bras. 45(12). https://doi.org/10.1590/S0100-204X2010001200007.

Ghazala, B., Shameel, M., Choudhary, M.I., Shahzad, S., Leghari, S.M., 2004. Phytochemistry and bioactivity of Tetraspora (VOLVOCOPHYTA) from Sindh. Pak. J. Bot. 36, 531–547.

Gnanamanickam, S.S., Immanuel, J.E., 2006. Epiphytic bacteria, their ecology and functions. In: Gnanamanickam, S.S. (Ed.), Plant-Associated Bacteria. Springer, Dordrecht, pp. 131–154.

Halfeld-Vieira, B.A., Romeiro, R.S., Mounteer, A., Mizubuti, E.S.G., 2008. Efficiency of phylloplane bacteria in controlling aerial tomato diseases under field conditions. Summa Phytopathologica 34, 86–87.

Hallmann, J., Quadt-Hallmann, A., Mahaffee, W.F., Kloepper, J.W., 1997. Bacterial endophytcs in agricultural crops. Can. J. Microbiol. 43, 895–914.

Hallmann, J., Quadt-Hallmann, A., Rodriguez-Kabana, R., Kloepper, J.W., 1998. Interactions between *Meloidogyne incognita* and endophytic bacteria in cotton and cucumber. Soil Biol. Biochem. 30, 925–937.

Harlapur, S., Kulkarni, M., Wali, M., Kulkarni, S., 2007. Evaluation of plant extract, bio-agents and fungicides against *Exserohilum turcicum* (Pass.) Leonard and Suggs causing turcicum leaf bligh to maize. Karnataka J. Agric. Sci. 20, 541–544.

Helal, G.A., Sarhan, M.M., Abu Shahla, A.N.K., Abou El-Khair, E.K., 2006. Effects of *Cymbopogon citratus* L. essential oil on the growth, lipid content and morphogenesis of

Aspergillus niger ML2-strain. J. Basic Microbiol. 46, 456–469. https://doi.org/10.1002/jobm.200510106.

Hill, K.H., Ticknor, L.O., Okinaka, R.T., Asay, M., Blair, H., Bliss, K.A., Laker, M., Pardington, P.E., Richardson, A.P., Tonks, M., Beecher, D.J., Kemp, J.D., Kolst, A., Lee Wong, A.C., Keim, P., Jackson, P.J., 2004. Fluorescent amplified fragment length polymorphism analysis of *Bacillus anthracis*, *Bacillus cereus* and *Bacillus thuringiensis* isolates. Appl. Environ. Microbiol. 70 (2), 1068–1080.

Hirano, S.S., Nordheim, M.E.V., Arny, D.C., Upper, C.D., 1982. Lognormal distribution of epiphytic bacterial populations on leaf surfaces. Appl. Environ. Microbiol. 44, 695–700.

Hurek, T., Reinhold-Hurek, B., Van Montagu, M., Kellenberger, E., 1994. Root colonization and systemic spreading of *Azoarcus* sp. Strain BH72 in grasses. J. Bacteriol. 176, 1913–1923.

Jacob, S., Sudini, H.K., 2016. Indirect plant growth promotion in grain legumes: role of actinobacteria. In: Plant Growth Promoting Actinobacteria. Springer Singapore, Singapore, pp. 17–32. https://doi.org/10.1007/978-981-10-0707-1_2.

Jalgaonwala, R.E., 2013. Bioprospecting for microbial endophytes and their natural products. Ph.D ThesisNorth Maharastra University, Jalgaon, Maharastra, India.

Jiang, Y.M., Chen, F., Li, Y.B., Liu, S.X., 2001. A preliminary study on the biological control of postharvest diseases of Litchi fruit. J. Fruit Sci. 14, 185–186.

Jog, R., Nareshkumar, G., Rajkumar, S., 2012. Plant growth promoting potential and soil enzyme production of the most abundant *Streptomyces* spp. from wheat rhizosphere. J. Appl. Microbiol. 113, 1154–1164.

Jogaiah, S., Abdelrahman, M., Tran, L.-S.P., Ito, S.-I., 2018. Different mechanisms of *Trichoderma virens*-mediated resistance in tomato against Fusarium wilt involve the jasmonic and salicylic acid pathways. Mol. Plant Pathol. 19, 870–882.

Jogaiah, S., Mahantesh, K., Sharathchnadra, R.G., Shetty, H.S., Vedamurthy, A.B., Tran, L.-S.P., 2016. Isolation and evaluation of proteolytic actinomycete isolates as novel inducers of pearl millet downy mildew disease protection. Sci. Rep. 6, 30789. https://doi.org/10.1038/srep30789.

Jogaiah, S., Mitani, S., Amruthesh, K.N., Shekar Shetty, H., 2007. Activity of cyazofamid against *Sclerospora graminicola*, a downy mildew disease of pearl millet. Pest Manag. Sci. 63, 722–727.

Jogaiah, S., Mostafa, A., Tran, L.S.P., Shin-ichi, I., 2013. Characterization of rhizosphere fungi that mediate resistance in tomato against bacterial wilt disease. J. Exp. Bot. 64, 3829–3842.

Jogaiah, S., Roopa, K.S., Pushpalatha, H.G., Shekar Shetty, H., 2010. Evaluation of plant growth-promoting rhizobacteria for their efficiency to promote growth and induce systemic resistance in pearl millet against downy mildew disease. Arch. Phytopath. Plant Prot. 43, 368–378.

Johnson, K., Stockwell, V., 1998a. Management of fire blight: a case study in microbial ecology. Annu. Rev. Phytopathol. 36, 227–248.

Johnson, K.B., Stockwell, V.O., 1998b. Management of fire blight: a case study in microbial ecology. *Annu. Rev. Phytopatho*. 136, 227–248.

Joshi, S.M., De Britto, S., Jogaiah, S., Ito, S., 2019. Mycogenic selenium nanoparticles as potential new generation broad Spectrum antifungal molecules. Biomol. Ther. 9 (9), 419.

Ju, M., Wang, D., Pierson, L., Pierson, E., 2018. Disruption of MiaA provides insights into the regulation of phenazine biosynthesis under suboptimal growth conditions in Pseudomonas chlororaphis 30-84. Microbiology 163, 94–108.

Jurkevitch, E.J., Shapira, G., 2000. Structure and colonization dynamics of epiphytic bacterial communities and of selected component strains on tomato (*Lycopersicon esculentum*) leaves. Microb. Ecol. 40, 300–308. https://doi.org/10.1007/s002480000023.

Kang, B.R., Anderson, A.J., Kim, Y.C., 2018. Hydrogen cyanide produced by Pseudomonas chlororaphis o6 exhibits nematicidal activity against *Meloidogyne hapla*. Plant Pathol. J. 34, 35–43.

Kavitha, P.G., Jonathan, E.I., Nakkeeran, S., 2012. Effects of crude antibiotic of Bacillus subtilis on hatching of eggs and mortality of juveniles of *Meloidogyne incognita*. Nematol. Medit. 40, 203–206.

Khanna, K., Vijay Lakshmi, J., Sharma, A., Gandhi, S.G., Ohri, P., Bhardwaj, R., Al-Huqail, A.A., Siddiqui, M.H., Ali, H.M., Ahmad, P., 2019. Supplementation with plant growth promoting rhizobacteria (PGPR)alleviates cadmium toxicity in Solanum lycopersicum by modulating the expression of secondary metabolites. Chemosphere 230 (628), 639.

Kilani-Feki, O., Ben, K.S., Dammak, M., Kamoun, A., Jabnoun-Khiareddine, H., Daami-Remadi, M., Touns, S., 2016. Improvement of antifungal metabolites productionby Bacillus subtilis V26 for biocontrol of tomato postharvest disease. Biol. Control. 95, 73–82.

Kim, B.S., Moon, S.S., Hwang, B.K., 1999. Isolation, identification, and antifungal activity of a macrolide antibiotic, oligomycin A, produced by *Streptomyces libani*. Can. J. Bot. 77, 850–858.

Kim, Y.S., Balaraju, K., Jeon, Y., 2016. Effects of rhizobacteriaPaenibacillus polymyxaAPEC136 and*Bacillus subtilis* APEC170 on biocontrol of postharvest pathogens of apple fruits. J. Zhejiang Univ. Sci. B. 17, 931–940.

Kishore, G., Pande, S., Podile, A., 2005. Biological control of late leaf spot of peanut (Arachishypogaea) with chitinolytic bacteria. Biol. Control 95, 1157–1165.

Klein, M.N., Silva, A.C., Kupper, K.C., 2016. *Bacillus subtilis* based-for-mulation for the control of postbloom fruit drop of citrus. World J. Microbiol. Biotechnol. 32, 205.

Kupper, K.C., Corrêa, F.E., Azevedo, F.A., Silva, A.C., 2012. *Bacillus subtilis* to biological control of postbloom fruit drop caused by *Colletotrichum acutatum* under feld conditions. Sci. Hortic. 134, 139–143.

Kupper, K.C., Moretto, R.K., Fujimoto, A., 2020. Production ofantifungal compounds byBacillus spp. isolates andits capacity forcontrolling citrus black spot underfeld conditions. World J. Microbiol. Biotechnol. 36, 7.

Kurose, D., Furuya, N., Tsuchiya, K., Tsushima, S., 2012. Endophytic fungi associated with *Fallopia japonica* (Polygonaceae) in Japan and their interactions with Puccinia polygoni-amphibii var. tovariae, a candidate for classical biological control. Fungal Biol. 116, 785–791.

Lamb, T.G., Tonkyn, D.W., Kloeplel, D.A., 1996. Movement of *Pseudomonas aureofaciens* from the rhizosphere to aerial plant tissue. Can. J. Microiol. 42, 1112–1120.

Lanteigne, C., Gadkar, V.J., Wallon, T., Novinscak, A., Filion, M., 2012. Production of DAPG and HCN by *Pseudomonas* sp. LBUM300 contributes to the biological control of bacterial canker of tomato. Phyto. Pathol. 102, 967–973.

Lastochkina, O., Pusenkova, L., Yuldashev, R., Babaev, M., Garipova, S., Blagova, D., Khairullin, R., Aliniaeifard, S., 2017. Effects ofBacillus subtilison some physiological and biochemical parameters of*Triticum aestivum* L. (wheat) under salinity. Plant Physiol. Biochem. 121, 80–88. https://doi.org/10.1016/j.plaphy.2017.10.020.

Lastochkina, O., Seifikalhor, M., Aliniaeifard, S., Baymiev, A., Pusenkova, L., Garipova, S., Kulabuhova, D., Maksimov, I., 2019. *Bacillus* Spp.: efficient biotic strategy to control postharvest diseases of fruits and vegetables. Plants (Basel, Switzerland) 8 (4), 97.

Li, L., Zhang, Z., Pan, S., Li, L., Li, X., 2019. Characterization and metabolism effect of seed endophytic bacteria associated with peanut grown in South China. Front. Microbiol. 10, 2659. https://doi.org/10.3389/fmicb.2019.02659.

Lindow, S.E., Brandl, M.T., 2003. Microbiology of the phyllosphere. Appl. Environ. Microbiol. 69, 1875–1883.

Liotti, R.G., Figueiredoa, M.I.S., Soares, M.A., 2019. Streptomyces griseocarneus R132 controls phytopathogens and promotes growth of pepper (*Capsicum annuum*). Biol. Control 138, 104065.

Lopez-Velasco, G., Carder, P.A., Welbaum, G.E., Ponder, M.A., 2013. Diversity of the spinach (*Spinacia oleracea*) spermosphere and phyllosphere bacterial communities. FEMS Microbiol. Lett. 346, 146–154.

Maget-Dana, R., Peypoux, F., 1994. Iturins, a special class of pore-forming lipopeptides: biological and physicochemical properties. Toxicology 87, 151–174.

Mahaffee, W.F., Kloepper, J.W., 1997. Temporal changes in the bacterial communities of soil, rhizosphere and endorhiza associated with field-grown cucumber (*Cucumis sativus* L.). Microbial. Ecol. 34, 210–223. https://doi.org/10.1007/s002489900050.

Mahmood, A., Takagi, K., Ito, K., Kataoka, R., 2019. Changes in endophytic bacterial communities during diferent growth stages of cucumber (*Cucumis sativus* L.). World J. Microbiol. Biotechnol. 35, 104.

Maksimov, I.V., Veselova, S.V., Nuzhnaya, T.V., Sarvarova, E.R., Khairullin, R.M., 2015. Plant growth promoting bacteria in regulation of plant resistance to stress factors. Rus. J. Plant Physiol. 62, 715–726.

Pujol, M., Badosa, E., Montesinos, E., 2007. Epiphytic fitness of a biological control agent of fire blight in apple and pear orchards under Mediterranean weather conditions. FEMS Microbiol. Ecol. 59 (1), 186–193. https://doi.org/10.1111/j.1574-6941.2006.00227.x.

McInroy, J.A., Kloepper, J.W., 1995a. Survey of indigenous bacterial endophytes from cotton and sweet corn. Plant Soil 173, 337–342.

McInroy, J.A., Kloepper, J.W., 1995b. Population dynamics of endophytic bacteria in field-grown sweet corn and cotton. Can. J. Microbiol. 41, 895–901.

Miller, A.R., 2003. Harvest and handling injury: Physiology, biochemistry, and detection. In: Postharvest Physiology and Pathology of Vegetables. Marcel Dekker Inc, New York, NY.

Molina, G., Pimentel, M.R., Bertucci, T.C.P., Pastore, G.M., 2012. Application of fungal endophytes in biotechnological processes. Chem. Eng. Trans. 27, 289–294.

Monier, J.M., Lindow, S.E., 2005. Aggregates of resident bacteria facilitate survival of immigrant bacteria on leaf surfaces. Microbial. Ecol. 49, 343–352.

Mousa, W.K., Schwan, A., Davidson, J., Strange, P., Liu, H., Zhou, T., et al., 2015. An endophytic fungus isolated from finger millet (*Eleusine coracana*) produces anti-fungal natural products. Front. *Microbiol.* 6, 1157. https://doi.org/10.3389/fmicb.2015. 01157.

Nagaraju, A., Jogaiah, S., Mahadeva Murthy, S., Shin-ichi, I., 2012. Seed priming with *Trichoderma harzianum* isolates enhances plant growth and induces resistance against *Plasmopara halstedii*, an incitant of sunflower downy mildew disease. Aust. J. Plant Path. 41, 609–620.

Nakkeeran, S., Vinodkumar, S., Renukadevi, P., Rajamanickam, S., Jogaiah, S., 2019. Bioactive molecules from *Bacillus* spp.: an effective tool for plant stress management. In: Jogaiah, S., Abdelrahman, M. (Eds.), Bioactive Molecules in Plant Defense. Springer, Berlin, pp. 1–23.

Nandini, B., Puttaswamy, H., Prakash, H.S., Adhikari, S., Jogaiah, S., Geetha, N., 2020. Elicitation of novel trichogenic-lipid nanoemulsion signaling resistance against pearl millet downy mildew disease. Biomol. Ther. 10 (1), 25.

Nascimento, F.X., Rossi, M.J., Soares, C.R., McConkey, B.J., Glick, B.R., 2014. New insights into 1-aminocyclopropane1-carboxylate (ACC) deaminase phylogeny, evolution and ecological significance. PLoS One. 6, e99168.

Nicolaou, K.C., Chen, J.S., Corey, E.J., 2011. Classics in Total Synthesis, Further Targets, Strategies, Methods III. Wiley, Weinheim, pp. 1–770.

Nischwitz, C., Newcombe, G., Anderson, C.L., 2005. Host specialization of the mycoparasite *Eudarluca caricis* and its evolutionary relationship to *Ampelomyces*. Mycol. Res. 109, 421–428.

Ongena, M., Adam, A., Jourdan, E., Paquot, M., Brans, A., Joris, B., Arpigny, J.L., Thonart, P., 2007. Surfactin and fengycin lipopeptides of *Bacillus subtilis* as elicitors of induced systemic resistance in plants. Environ. Microbiol. 9, 1084–1090. https://doi.org/10.1111/j.1462-2920.2006.01202.x.

Patel, B.B., Patel, R.G., Pate, M.P., 2006. Synthesis and studies of the biological activity of novel pyrimido fused acridine derivatives. J. Serb. Chem. Soc. 71, 1015–1023.

Qin, S., Miao, Q., Feng, W.W., Wang, Y., Zhu, X., Xing, K., Jiang, J.H., 2015. Biodiversity and plant growth promoting traits of culturable endophytic actinobacteria associated with *Jatropha curcas* L. growing in Panxi dryhot valley soil. Appl. Soil Ecol. 93, 47–55.

Quadt-Hallmann, A., Kloepper, J.W., 1996. Immunogical detection and localization of cotton endophyte *Enterobacter arburiae* JM22 in different plant species. Can. J. Microbiol. 42, 1144–1154.

Raaijmakers, J.M., Mazzola, M., 2012. Diversity and natural functions of antibiotics produced by beneficial and plant pathogenic bacteria. Annu. Rev. Phytopathol. 50, 403–424. https://doi.org/10.1146/annurev-phyto-081211-172908.

Raaijmakers, J.M., der Sluis, V., Van den Hout, I., Bakker, M., P. A. H. M. and Schippers, B., 1995. Dispersal of wild-type and genetically-modified *Pseudomonas* spp. from treated seeds or soil to aerial parts of radish plants. Soil. Biol. Biochem. 27, 1473–1478.

Raghavendra, A.K.H., Newcombe, G., 2013. The contribution of foliar endophytes to quantitative resistanceto Melampsora rust. New Phytol. 197, 909–918.

Ramarathnam, R., Dilantha, F.W., 2006. Preliminary phenotypic and molecular screening for potential bacterial biocontrol agentes of *Leptosphaeria maculans*, the black leg pathogen of canola. Biocontrol Sci. Tech. 16, 567–582.

Ramyabharathi, S.A., Raguchander, T., 2014a. Characterization of antifungal antibiotic synthesis genes from different strains of *Bacillus subtilis*. J. Pure Appl. Microbiol. 8 (3), 2337–2344.

Ramyabharathi, S.A., Raguchander, T., 2014b. Efficacy of secondary metabolites produced by Bacillus subtilis EPCO16 against tomato wilt pathogen *Fusarium oxysporum* f. sp. *lycopersici*. *J. Mycol. Plant Pathol.* 44 (2), 148–153.

Romero, D., de Vicente, A., Rakotoaly, R.H., Dufour, S.E., Veening, J.W., Arrebola, E., Cazorla, F.M., Kuipers, O.P., Paquot, M., Perez-Garcia, A., 2007. The iturin and fengycin families of lipopeptides are key factors in antagonism of *Bacillus subtilis* towards *Podosphaera fusca*. Mol. Plant-Microbe Interact. 20, 430–440.

Rosenblueth, M., Martínez-Romero, E., 2006. Bacterial endophytes and their interactions with hosts. Mol. Pant-Microbe Int. 19 (8), 827–837.

Santoyo, G., Moreno-Hagelsieb, G., Orozco-Mosqueda Mdel, C., Glick, B.R., 2016. Plant growth-promoting bacterial endophytes. Microbiol. Res. 183, 92–99.

Sarma, B.K., Yadav, K.S., Singh, D.P., Singh, H.B., 2012. Rhizobacteria mediated induced systemic tolerance in plants: Prospects for abiotic stress management. In: Bacteria in Agrobiology: Stress Management. Springer, Berlin/Heidelberg, pp. 225–238.

Melina, S., Nesci, A., García, J., Passone, M., Montemarani, A., Etcheverry, M., 2017. Efficacy of epiphytic bacteria to prevent northern leaf blight caused by *Exserohilum turcicum* in maize. Revista Argentina de Microbiología 49, 75–82.

Sartoria, M., Nescia, A., Formentoc, A., Etcheverry M., 2015. Selection of potential biological control of *Exserohilum turcicum* with the epiphytic microorganism from maize. Rev. Argent. Microbiol. 47 (1), 62–71.

Satapute, P., Milan, V.K., Shivakantkumar, S.A., Jogaiah, S., 2019. Influence of triazole pesticides on tillage soil microbial populations and metabolic changes. Sci. Total Environ. 651, 2334–2344.

Sathyaprabha, G., Kumaravel, S., Ruffina, D., Praveenkumar, P.A., 2010. Comparative study on antioxidant, proximate analysis, antimicrobial activity and phytochemical analysis of *Aloe vera* and *C. quadrangularis* by GC-MS. J. Pharm. Res. 3, 2970–2973.

Seifikalhor, M.S., Aliniaeifard, S., Self, M., Javadi, F., Bernard, F., Li, T., Lastochkina, O., 2018. Rhisobacteria *Bacillus subtilis* reduces toxic effects of high electrical conductivity insoilless culture of lettuce. ActaHortic. 1227(59).

Siddiqui, I.A., Shaukat, S.S., 2004. Systemic resistance in tomato induced by biocontrol bacteria against the root-knot nematode, Meloidogyne javanica is independent of salicylic acid production. Phytopathology 152, 48–54.

Silo-suh, L.A., Lethbridge, B.J., Raffel, S.J., He, H., Clardy, J., Handelsman, J., 1994. Biological activities of two fungistatic antibiotics produced by *Bacillus cereus* UW 85. Appl. Environ. Microbiol. 60, 2023–2030.

Singh, H.B., Keswani, C., Reddy, M.S., Sansinenea, E., García-Estrada, C., 2019. Secondary Metabolites of Plant Growth Promoting. Springer Nature Singapore Pte, Ltd., Singapore.

Stein, T., 2005. *Bacillus subtilis* antibiotics: structures, syntheses and specific functions. Mol. Microbiol. 56, 845–857.

Sturz, A.V., 1995. The role of endophytic bacteria during seed piece dacy and potato tuberization. Plant Soil 175, 257–265.

Sturz, A.V., Christie, H.R., Matheson, B.G., Arsenault, W.J., Buchman, N.A., 1997. Biodiversity of endophytic bacteria which colonize red clover nodules, roots, stems and foliage and their influence on host growth. Biol. Fert. Soils 25, 13–19.

Sudisha, J., Niranjana, S.R., Umesha, S., Prakash, H.S., Shekar Shetty, H., 2006. Transmission of seed-borne infection of muskmelon by *Didymella bryoniae* and effect of seed treatments on disease incidence and fruit yield. Biol. Control 37, 196–205.

Sudisha, J., Niranjana, S.R., Sukanya, S.L., Girijamba, R., Lakshmidevi, N., Shekar Shetty, H., 2010. Relative efficacy of strobilurin formulations in the control of downy mildew of sunflower. J. Pest. Sci. 83, 461–470.

Suyal, D.V., Soni, R., Sai, S., Goel, R., 2016. Microbial inoculants as biofertilizer. In: Singh, D.P. et al., (Ed.), Microbial Inoculants in Sustainable Agricultural Productivity.https://doi.org/10.1007/978-81-322-2647-5_18.

Truyens, S., Weyens, N., Cuypers, A., Vangronsveld, J., 2015. Bacterial seed endophytes: genera, vertical transmission and interaction with plants. Environ. Microbiol. Rep. 7, 40–50.

Wassermann, B., Cernava, T., Müller, H., Berg, C., Berg, G., 2019. Seeds of native alpine plants host unique microbial communities embedded in cross-kingdom networks. Microbiome 7, 108.

Wilson, D., 1995. Endophyte-the evolution of a term, and clarification of its use and definition. Oikos 73, 274–276.

Wulff, E.G., Mguni, C.M., Mansfeld-Giese, K., Fels, J., Lubeck, M., Hockenhull, J., 2002. Biochemical and molecular characterization of *Bacillus amyloliquefaciens*, *B. subtilis* and *B. pumilus* isolates with distinct antagonistic potential against *Xanthomonas campestris* pv. *campestris*. Plant Pathol. 51, 574–584.

Yadav, R.K.P., Karamanoli, K., Vokou, D., 2005. Bacterial colonization of the Phyllosphere of Mediterranean perennial species as influenced by leaf structural and chemical features. Microb. Ecol. 50, 185–196. https://doi.org/10.1007/s00248-004-0171-y.

Yung, S.H., 2010. IAA Production by *Streptomyces scabies* and its Role in Plant Microbe Interaction. M.Sc. ThesisCornell University, New York.

Zachow, C., Jahanshah, G., de Bruijn, I., Song, C., Ianni, F., Pataj, Z., et al., 2015. The novel lipopeptide poaeamide of the endophyte Pseudomonas poae RE*1-1-14 is involved in pathogen suppression and root colonization. Mol. Plant-Microbe Interact. 28, 800–810. https://doi.org/10.1094/mpmi-12-14-0406-r.

Zhang, S.L., Chu, J., Zhuang, Y.P., 2004. A Multi-Scale Study of Industrial Fermentation Processes and their Optimization. Biomanufacturing. Springer, Berlin, pp. 97–150.

Zhao, Y., Shao, X.F., Tu, K., Chen, J.K., 2007. Inhibitory effect of*Bacillus subtilis* B10 on the diseases of postharvest strawberry. J. Fruit Sci. 24, 339–343.

Zhao, Y., 2010. Auxin biosynthesis and its role in plant development. Annu. Rev. Plant Biol. 61, 49–64.

Fungal diversity and its role in sustainable agriculture

23

Kushal Raj[a], *Leela Wati*[b], *and Anil Kumar*[a]
[a]Department of Plant Pathology, CCS HAU, Hisar, Haryana, India, [b]Department of Microbiology, CCS HAU, Hisar, Haryana, India

23.1 Introduction

Fungi are a group of organisms devoid of chlorophyll and therefore differ from the green plants in the essential ways which result from this deficiency. The fungal cells contain chitin, unlike the cell wall of plants and some protists, which contain cellulose. Transpiration, respiration, and true assimilation are the same as with the green plants, but photosynthesis or starch manufacture cannot be accomplished by them. Sunlight being thus useless to them directly as they can live in the dark as well as in the light. Having no ability to elaborate their own foods from inorganic matter these organisms are limited to such nutriment as they can obtain from plants or animals which have elaborated it; that is, they must have organic foods for their sustenance. The fungi have acquired various food habits and adapted themselves to different modes of nutrition (Nagaraju et al., 2012). Some are nearly omnivorous and can subsist upon almost any decaying tissue or upon soils or solutions rich with organic debris. Others thrive only upon special substances, for example, some particular plant or animal, or their particular parts.

Morphologically they vary from microscopic yeasts to gigantic mushrooms which can grow to weigh up to 316kg. A fungus also holds the record for the biggest living organism ever recorded. *Armillaria bulbosa* was found in forest soil in Oregon which spread for over 40ha. Fungi appeal to the senses by their beauty and strangeness. They are often delicate and short-lived, sudden in appearance above ground. They can be colorful. A few are smelly too. Fungi excluding yeasts and some zoosporic taxa consist of filaments (hyphae) that increase in length by the deposition of cell wall material from growing tips. As these tips expand and produce new growing tips network termed the mycelium develops. Once they establish, fungal mycelia are capable of essentially unlimited growth and persistence. This form of indeterminate body structure differs from the determinant body structure of most animal and plant species. As mycelium continues to expand it may come to occupy heterogenous suite of micro and macro environments at spatial scales that range from several millimeters to entire landscapes.

23.2 Classification of fungi

Fungi are a group of organisms united by their mode of nutrition, namely absorption. The idea that fungi form a kingdom distinct from plants and animals gradually became accepted only after Whittaker (1969). Presently, the "fungi" as a mega-diverse group

Biocontrol Agents and Secondary Metabolites. https://doi.org/10.1016/B978-0-12-822919-4.00023-5

span three kingdoms, most belonging to the Fungi (Eumycota), while others are classified in the Protozoa and Chromista (Stramenophila) (Cavalier-Smith, 1998; Hawksworth, 2004; James et al., 2006). It might be expected that the predicted number of fungi on earth would have been considerably greater than the 1.5 M as suggested by Hawksworth (1991), which is currently accepted as a working figure although recognized as conservative. The 10th edition of Ainsworth and Bisby's Dictionary of the Fungi provided a total of 98,998 for the number of fungal species accepted to date (excluding taxa treated under Chromista and Protozoa). Kirk et al. (2008) reported 1039 species as chromistan fungal analogues and 1165 as protozoan in which 1038 are regarded as protozoan fungal analogues: Percolozoa (Acrasida), Amoebozoa (Dictyostelia, Myxogastria, and Protostelia), Cercozoa (Plasmodiophorida) which were previously treated as Myxomycota and Plasmodiophoromycota.

The major groups of fungi are listed in three categories: (1) those groups that are relatively well known, for which 70%–90% of the species have been described; (2) those groups that are moderately well known, for which 40%–70% of the species have been described; and (3) those groups that are poorly known for which only 10%–40% of the species have been described.

23.3 Well-known groups

Under this group keys to genera and species exist at least for some parts of the world and for some genera, relatively few new species expected, and about 70%–90% are described.

23.3.1 Macrolichens including most foliose and fruticose species

The group of fungi discussed as macrolichens is an artificial assemblage of lichens that form a foliose or fruticose thallus. These lichens are perennial, varying in size from very large to inconspicuous, and occur on all kinds of surfaces (Galloway, 1992). Identification of species is based partially on lichen acids present in the thallus but the genetic basis for acid production is not well known. Species concepts may change in the next 10 years as molecular approaches are applied to an understanding of species and evolution within the lichens.

23.3.2 Polyporaceae sensu lato including Corticiaceae, Stereaceae, etc.

Members of this group of fungi generally form large, conspicuous, perennial fruiting bodies that develop primarily on wood. The species are relatively well-known with floristic treatments available for some areas of the world, primarily in temperate regions (Ginns and Lefebvre, 1993) but also parts of the tropics. For some genera, computerized keys are available suggesting that parataxonomists could be trained to collect and identify this group of fungi.

23.4 Moderately well-known groups

Under this group identification of the genus is possible, over 50% of the species are described but many new species are expected. Only three moderately well-known groups of fungi are discussed here.

23.4.1 *Agaricales (mushrooms including secotioid and hypogeous relatives)*

The Agaricales includes species that function as ectomycorrhizae associated with the roots of conifer and dicotyledonous trees and as saprophytes on woody tissues, leaves, herbaceous debris, dung, and associated with ants.

23.4.2 *Uredinales (rusts)*

The Uredinales or rusts are a group of virulent plant pathogens that are obligately parasitic primarily on vascular plant hosts. Most rusts are host specific at the generic or species level. Monographs exist for two host groups, namely members of the Asteraceae and Fabaceae with emphasis on North America (Cummins and Hiratsuka, 1983). Based on well-studied temperate and tropical regions, about one out of every 4–25 vascular plant species is host to unique rust, thus the worldwide number of species in the Uredinales is estimated at 10,000–60,000 species.

23.4.3 *Hypocreales and Xylariales*

The perithecial Euascomycetes in these two orders, the Hypocreales and Xylariales, have received considerable attention in the last 20 years in the New World. With increased study, it is now possible to identify and characterize many of the species collected from these areas. Although primarily saprophytic, these fungi include some virulent facultative plant pathogens as well as insect-associated and fungicolous species.

23.5 Poorly known groups

In this group identification of genus is possible but many new species are encountered and useful keys to species are lacking.

23.5.1 *Perithecial Euascomycetes and Loculoascomycetes (excluding the Erysiphales, Hypocreales, and Xylariales)*

The perithecial Euascomycetes and Loculoascomycetes with their asexual states listed as Deuteromycetes-Hyphomycetes and Deuteromycetes-Coelomyetes are by far the largest and least known groups of fungi. Considering their numbers and importance, relatively few systematists are working on them. A few orders within this group are considered moderately well-known, namely the Erysiphales, Hypocreales, and Xylariales. With a few exceptions, most genera or subclass taxa within the perithecial Euascomycetes and Loculoascomycetes are poorly defined. Few specimens in these groups can be identified even to a meaningful generic level and certainly not to species. These fungi occur as parasites, endophytes, and saprophytes on all substrates including insects, living leaves, dead wood, twigs, leaves, dung, and lichens.

23.5.2 Nondematiaceous hyphomycetes

Most anamorph fungi do not have any known sexual states but are known to have been derived from or related to Ascomycetes. Based on the known sexual and asexual states of the same species reported by Farr et al. (1989), only 9% of the Deuteromycetes have described sexual states in the United States, a part of the world where the fungi are relatively well known. Thus the overlap between Ascomycetes and Deuteromycetes is much lower than one might expect. With the exception of a few well-studied genera such as *Aspergillus, Fusarium* and *Penicillium*, and the aquatic hyphomycetes, the nondematiaceous hyphomycetes represent a large group of taxonomically difficult species with few systematists working on them. Although many species are saprophytic or endophytic, the group also includes plant-parasitic species on living leaves or causing cankers on woody plants. They can be isolated from all substrates.

23.5.3 Endogonales and Glomales (vesicular mycorrhizae)

Most species are obligately endomycorrhizal with many kinds of herbaceous plants. Morphologically based species concepts may be questionable and, as yet, molecular approaches have been difficult to apply to this group. Minimal morphology exists on which to describe species with the details of the vesicle wall structure providing an important character. Species can adapt to changing environmental conditions and vary with edaphic factors such as soil pH and aluminium concentration rather than being vascular plant host specific. These fungi can be dispersed in the dung of large animals (McGee and Baczocha, 1994).

23.6 Fungi and ecosystems

Fungi are important components in every ecosystem, intimately associated with crucial processes like the decomposition, recycling, and transportation of nutrients in different environments. It has been estimated that there may be over a million different fungal species on this Earth, of which only a small fraction (approx. 5%) has been identified.

As a consequence of their absorptive mode of nutrition, fungi are able to exploit an almost infinite diversity of nutritional microniches. Fungi are known to colonize, multiply, and survive in diversified habitats, viz., water, soil, air, litter, dung, foam, etc. Fungi are ubiquitous and cosmopolitan in distribution covering tropics to poles and mountain tops to the deep oceans. Many of the described species are known only as dead herbarium material and around 5% of species are isolated as pure cultures. Geographic location, climatic conditions, microhabitat, substrate type, distribution of fauna and flora are all important factors contributing to fungal distribution around the world.

For fungi as in other groups of organisms, those species for which the fruiting bodies or, in the case of lichenized fungi, the thallus is relatively large in size and persistent are more well-known than those that are small and ephemeral. Thus, fungi that form large, perennial conks on dead tree trunks or that form extensive foliose mats in old-growth forests, that is, macrolichens, are better known than small, diaphanous mushrooms on rapidly decomposing tropical leaf litter or endophytic ascomycetes that occur in living plant tissue and are isolated after long incubation periods of the host material.

23.6.1 Fungi and animals

Many fungi have important symbiotic relationships with organisms from most if not all Kingdoms. These interactions can be mutualistic or antagonistic in nature, or in the case of commensal fungi are of no apparent benefit or detriment to the host (Nandini et al., 2020). Fungi provide food and living space for hundreds of species of animals. Many toadstools will be found that have been partly eaten by mammals and mollusks. These browsers may benefit the fungus by dispersing the fungal spores in their droppings. There are also many hundreds of insect species that are completely dependent on fungi, breeding in toadstools or bracket fungi. Some, especially beetles using bracket fungi, are restricted to a particular fungus species Many of the invertebrates which are classed as decomposers, such as woodlice and freshwater shrimps, get their nourishment not directly from dead leaves but from the bacteria and fungi growing on the dead leaves. The fungal hyphae must convert the indigestible leaf chemicals into more readily assimilated material before the invertebrates can use it. Among palaearctic beetles, 349 species in 57 families are known to be dependent on fungi (Anderson, 2001).

Many insects also engage in mutualistic relationships with fungi (Hackman and Meinander, 1979). Several groups of ants cultivate fungi in the order Agaricales as their primary food source, while ambrosia beetles cultivate various species of fungi in the bark of trees that they infest. Likewise, females of several wood wasp species (genus *Sirex*) inject their eggs together with spores of the wood-rotting fungus, *Amylostereum areolatum* into the sapwood of pine trees; the growth of the fungus provides ideal nutritional conditions for the development of the wasp larvae. Termites on the African savanna hare also known to cultivate fungi and yeasts of the genera *Candida* and *Lachancea* inhabit the gut of a wide range of insects, including neuropterans, beetles, and cockroaches; it is not known whether these fungi benefit their hosts. The larvae of many families of fungicolous flies, particularly those within the superfamily Sciaroidea such as the Mycetophilidae and some Keroplatidae feed on fungal fruiting bodies and sterile mycorrhizae.

23.6.2 Fungi and plants

Mycorrhizal symbiosis between plants and fungi is one of the most well-known plant-fungus associations and is of significant importance for plant growth and persistence in many ecosystems; over 90% of all plant species engage in mycorrhizal relationships with fungi and are dependent on this relationship for survival. The mycorrhizal symbiosis is ancient, dating to at least 400 million years ago. It often increases the plant's uptake of inorganic compounds, such as nitrate and phosphate from soils having low concentrations of these key plant nutrients. The fungal partners may also mediate plant-to-plant transfer of carbohydrates and other nutrients. Such mycorrhizal communities are called "common mycorrhizal networks." A special case of mycorrhiza is myco-heterotrophy, whereby the plant parasitizes the fungus, obtaining all of its nutrients from its fungal symbiont. Some fungal species inhabit the tissues inside roots, stems, and leaves, in which case they are called endophytes similar to mycorrhiza, endophytic colonization by fungi may benefit both symbionts; for example, endophytes of grasses impart to their host increased resistance to herbivores and other environmental stresses and receive food and shelter from the plant in return.

23.6.3 Fungi and algae/cyanobacteria

Lichens are formed by a symbiotic relationship between algae or cyanobacteria (referred to in lichen terminology as "photobionts") and fungi (mostly various species of Ascomycetes and a few Basidiomycetes) in which individual photobiont cells are embedded in a tissue formed by the fungus. Lichens occur in every ecosystem on all continents, play a key role in soil formation and the initiation of biological succession, and are the dominating life forms in extreme environments, including polar, alpine, and semiarid desert regions. They are able to grow on inhospitable surfaces, including bare soil, rocks, tree bark, wood, shells, barnacles, and leaves. As in mycorrhizas, the photobiont provides sugars and other carbohydrates via photosynthesis, while the fungus provides minerals and water. The functions of both symbiotic organisms are so closely intertwined that they function almost as a single organism; in most cases, the resulting organism differs greatly from the individual components. Lichenization is a common mode of nutrition; around 20% of fungi—between 17,500 and 20,000 described species are lichenized. Characteristics common to most lichens include obtaining organic carbon by photosynthesis, slow growth, small size, long-life, long-lasting (seasonal) vegetative reproductive structures, mineral nutrition obtained largely from airborne sources, and greater tolerance of desiccation than most other photosynthetic organisms in the same habitat.

23.7 Economic value of fungi

Fungi are important because they are:

- Agents of biodegradation and biodeterioration
- Responsible for the majority of plant diseases and several diseases of animals (including humans)

- Used in industrial fermentation processes
- Used in the commercial production of many biochemicals
- Cultured commercially to provide us with a direct source of food
- Used in bioremediation
- Beneficial in agriculture, horticulture, and forestry.

Fungi, particularly as the asexual states of Ascomycetes, are important producers of biologically active molecules such as penicillin produced by *Penicillium chrysogenum*. Cyclosporin A, the immune-suppressant drug used in organ transplant operations is produced by *Beauveria nivea* and the cholesterol-reducing lovastatin derives from *Aspergillus terreus* Stierle et al. (1993) have shown that a previously undescribed fungus isolated from Pacific yew tree, *Taxus brevifolia* Nutt., produces the drug taxol effective in suppressing breast cancer.

Fungi serve as the source of commercially important enzymes and natural products ranging from abscisic acid to zymosterol that result in a billion-dollar industry (Holler et al., 2000) They are increasingly used to ferment solid organic waste substrates into useable products such as methane and fertilizers and are invaluable as substitutes for chemicals in the pulp and paper industry. New classes and sources of natural products are being discovered from fungi (Dreyfuss and Chapela, 1994), the search for organisms that produce useful secondary metabolites has been limited in scope. Fungal strains of *T. reesei* from tropical countries produce the industrially important enzyme cellulase in much greater quantity than the strain currently used in commercial production. The fungi as a biological resource of commercially important products are virtually untapped. Fungal species screened for secondary metabolites using modern techniques are less than 1% of those that may exist. Thus, the potential is enormous for the discovery of valuable natural products resulting from a directed search and screening of fungi from unexplored habitats.

In the production of food both in agriculture and in the food-processing industry, fungi play influential roles. The filamentous fungi cause billions of dollars damage to agricultural crops, yet they hold the potential to serve as beneficial organisms in the biological control of insects, pathogenic fungi, and weeds (Kluth et al., 2003) and reduce the use of chemical fertilizers through an enhanced formation of endomycorrhizae in crop plants. The fungi associated with roots are essential for the growth of over 90% of all vascular plants as both endomycorrhizae in crops and ectomycorrhizae developing in most woody plants. The remarkable beneficial effect of inoculating tree seedlings with selected ectomycorrhizal fungi has been studied by many workers. The fungus-root association can be enhanced through the effective use of both ecto- and endomycorrhizae to reduce the need for fertilizer in working toward sustainable agricultural and forest systems.

23.8 Biodiversity of fungi

Although mycologists are often overwhelmed by the vast diversity of fungi, literature to document this diversity is highly fragmented or lacking. In attempting to estimate the total number of fungi, both described and undescribed, Hawksworth (1991) compared the vascular plant-fungus ratios in areas that had been relatively well-studied.

He arrived at an estimate of 1.5 million species based partly on an estimated number of unique fungal species to vascular plant species of 6:1. Even if one accepts a conservative estimate of a 1:1 ratio as in Farr et al. (1989), the total number of fungal species on plants would be about 250,000. Many fungi occur on substrates for which the plant-derived host taxon cannot be identified such as soil, litter, and rotten wood. Another enormous group of fungi to be considered in discussing numbers of fungal species are those associated with insects of which estimates range from 500,000 to 1.5 million. Other substrates harboring fungi include other fungi, animals, and animal parts such as hair, bones, horn, and the digestive tract of ruminants. Additional evidence to support the worldwide estimate of 1.5 million fungal species is based on the rates of newly described species of 50%–66% in some fungal groups over the last 20 years or by examining countries or specific groups that fungi have been relatively well studied (Manoharachary et al., 2005). The off-repeated number of described fungal species is 47,000–69,000. This number is an extremely conservative estimate derived by totaling the number of species listed in each genus known to the compilers of the *Dictionary of the Fungi* (Hawksworth et al., 2011). In terms of biomass, diversity, and their ecological functions, fungi are the most important group of organisms in forest ecosystems after the trees themselves. It is estimated that the fungal biomass in soils exceeds the biomass of all other soil organisms combined, except plant roots.

23.9 Fungi in sustainable agriculture

It is believed that agriculture and biodiversity are essential elements of life on planet earth. Biodiversity and agriculture are connected and can benefit greatly from each other. The concept of "sustainability" is becoming ever more prominent in almost every area of human affairs, from individual households to the planet Earth itself. A study modeled the removal of different biotic components of the soil found that the removal of only two component groups, bacteria, and saprophytic (decay) fungi caused drastic changes in net primary productivity. This suggests that as a functional group (decomposers in the latter case), fungi are essential for soil processes. The biodiversity of fungi in an ecosystem has been shown to affect plant diversity (van der Heijden et al., 1998a,b) and thus primary productivity in terrestrial ecosystems.

Fungi can play a significant role in the pursuit of sustainability. Fungi are vitally important for the good growth of most plants, including crops, through the development of associations between the fungi and the roots of the plants (Jogaiah et al., 2018). On the other hand, they can be extremely injurious, causing disease and death or reduced fitness. Fungal pathogens can have enormous negative consequences for crop production. Some fungi are parasites of plants. Most of our common crop plants are susceptible to the fungal attack of one kind or another. Fungal diseases can on occasion result in the loss of entire crops if they are not managed correctly.

In general, major taxonomic groups reflect ecological functions. For example, all rusts and most smuts are obligate parasites of vascular plants. One does not find rust on leaf litter or associated with insects nor would one find a member of the Entomophthorales, a group of obligate insect-associated fungi, on plant tissues as a parasite, saprophyte, or forming ectomycorrhizae with forest trees. Most endophytic and litter-decomposing fungi associated with vascular plants are Ascomycetes and their anamorphs, although some "lower" Basidiomycetes have this capability (Jogaiah et al., 2016a). Many exceptions exist, for example, among the fungi forming ectomycorrhizae with vascular plants, most are members of particular genera in the Agaricales, and however, a few Ascomycetes are known to form ectomycorrhizae. It is true that in agriculture, we fertilize crops, in woodlands, fungi do most of this job.

The role of fungi in sustainable agriculture is listed as:

1. Vital importance in ecosystem maintenance-saprophytic fungi as natural recyclers–mycorrhizal fungi as important symbionts with plants
2. Use in biological control of insects, nematodes, pathogenic fungi, and weeds–*Ampelomyces quisqualis*, *Cordyceps sinensis*, *Nematophthora gynophila*, *Verticillium chlamydosporium*, Phlebiagigantea, *Trichoderma* sp., *Fusarium* sp.
3. Recycling organic matter (plant debris) by saprophytic fungi mycorrhizal fungi associated with roots of 90% of all vascular plants such as arbuscular mycorrhizae in crops (*Glomus* sp., *Gigaspora* sp.), and ectomycorrhizae in most woody plants (*Lactarius* sp., *Laccaria* sp.)
4. Edible fungi: edible mushrooms, cultivated or wild (*Boletus edulis*, *Morchella* sp., Tuber sp., *Lentinula edodes*, *Agaricus bisporus*)
5. Use in ecology as indicators of vegetational or atmospherical changes-Lichens (association of a fungus and an alga)

23.10 Nutrient recycling

23.10.1 Decomposition of organic matter

Comprising interactions that range from mutualism to antagonism, fungal symbioses with plants are key determinants of biomass, nutrient cycling, and ecosystem productivity in terrestrial habitats from the poles to the equator (Hawksworth, 2001). Fungi also have a vital role in the decomposition of waste. Along with bacteria, worms, woodlice, and other invertebrates, fungi break down wood, leaves, dung, and help to release the nutrients back into the soil.

Most of the fungi comprise a network of hyphae, unseen within the wood or other substrates. Few organisms other than fungi have the enzymes able to decompose such tough materials as lignin and keratin (Dix and Webster, 1995). Fungi are critical biodiversity in maintaining soil fertility (Isaac, 1998). They drive major nutrient cycling processes in woodlands by decomposing dead plant and animal material, supplying nutrients to ensure the healthy growth of plants, and providing food for insects, small mammals and soil microbes. A diverse array of native fungi exists in woodland remnants but these fungi take a very long time, if they ever do, to self-establish in revegetation sites. The fungi need to be assisted back to promote sustained soil functional values for the greater long-term sustainability of woodland revegetation, increased nature conservation, and improved landscapes.

Fungi hold key roles in the maintenance of forest ecosystems in particular listed as:

- Nutrient cycling, retention, and formation of soil structure.
- Food in detritivore food webs in forests and forest streams.
- Microhabitat creation in forests by fungal pathogens.
- Mycorrhizal mutualisms.

23.11 Mycorrhiza

It has been discovered that most plant roots can form links with fungi, to the benefit of both plant and fungus (Allen, 1996; van der Heijden, 2002; Klironomos, 2003). The association between a fungus and a plant root is called a mycorrhiza. The fungal mycelium collects water and dissolved minerals more efficiently from a much wider area than the plant's roots could cover, some of these nutrients being transferred to the plant; and the fungus gains sugars produced by the plant's leaves (Hart et al., 2003). Plants that fail to establish a mycorrhiza usually grow weak if they grow at all. As well as improving the plant's nutrition, mycorrhizae may protect the plant against diseases and toxic metals (Lonsdale and Gibbs, 1996; Colpaert and Van Tichelen, 1996a,b). Most remarkable of all, mycorrhizae link different trees, even different species of tree, and they transfer nutrients between trees from those with an excess to those with a shortage. By these various mechanisms, mycorrhizae help maintain tree diversity in natural woodlands (Perry et al., 1992; Colpaert and Van Tichelen, 1996a) and the species of tree present in natural woodland may be partly determined by the types of mycorrhizae in the soil (Allen et al., 1995). This partly explains why recent secondary woodland is so obviously different from ancient woodland (Merryweather, 2001). There is evidence that mycorrhizae are equally influential in grasslands and can even determine the growth form of a plant (Streitwolf-Engel et al., 1997; van der Heijden et al., 1998a,b). The influence of mycorrhizae can also show higher up the food web: caterpillars of the common blue butterfly fail to mature when feeding on plants that have no mycorrhiza (Goverde et al., 2000).

Mycorrhizal fungi are widespread in agricultural systems and are especially relevant for organic agriculture because they can act as natural fertilizers, enhancing plant yield. The 400 million-year-old symbiosis between the majority of land plants and arbuscular mycorrhizal (AM) fungi is one of the most ancient and abundant mutualisms on Earth (Streitwolf-Engel et al., 1997a). AM fungi form extensive hyphal networks in soil and provide plants with nutrients in return for assimilates (Smith and Read, 1996). AM fungi can act as support systems for seedling establishment; provide resistance against drought and some pathogens, and AM fungi can enhance biological diversity in grassland (van der Heijden et al., 1998a). Several studies have shown that AM fungi contribute to up to 90% of plant P demand (van der Heijden et al., 2006). AM fungi are especially important for sustainable farming systems because AM fungi are efficient when nutrient availability is low and when nutrients are bound to organic matter and soil particles. Many important agricultural crops can benefit from AM fungi, including maize, potato, sunflower, wheat, onion, leek, and soybean, especially under conditions where nutrient availability is limiting plant growth (Kernighan, 2005). Moreover, AM fungi not only can promote via directs effects, but there are also a number of indirect effects such as stimulation of soil quality and the suppression of organisms that reduce crop productivity (van der Heijden et al., 2008).

Arbuscular vesicular (AM) fungi play a role in the uptake of minerals by plants, especially zinc and phosphorus (Abbott and Robson, 1992). In addition, AM fungi modify plant response to the environment, including pathogens, and are associated with the formation of microaggregates and are therefore central to soil structure and stability. AM fungi not only exist for long periods in soil, but they are now known to produce a glycoprotein called glomalin, especially in soils of fine texture (Rillig and Steinberg, 2002). Glomalin may exist for long periods in the absence of the fungus (Steinberg and Rillig, 2002). Glomalin is important because it contains carbon and thus may be a carbon store. Glomalin also attaches to the surface of and binds micro-aggregates (Wright and Upadhyaya, 1998). Thus it also plays a secondary role in soil structure. The overall effects of mycorrhizal fungi in sustainable agriculture are:

Direct effects on crops

1. Stimulation of plant productivity of various crops
2. Nutrient acquisition (P, N, Cu, Fe, Zn)
3. Enhanced seedling establishment
4. Drought resistance
5. Heavy metal resistance

Indirect effects

1. Weed suppression
2. Stimulation of nitrogen fixation by legumes (green manure)
3. Stimulation of soil aggregation and soil structure
4. Suppression of some soil pathogens
5. Stimulation of soil biological activity
6. Increased soil carbon storage
7. Reduction of nutrient leaching

23.12 Endophytic fungi

Endophytes are microorganisms colonizing healthy plants tissue without causing any apparent symptoms and noticeable injury to the host. Both fungi and bacteria are the most common microbes existing as endophytes and are to be found in virtually every plant on earth (Padhi et al., 2013; Saikkonen et al., 1998). It is also suspected that other types of microorganisms, viz., archaebacteria and mycoplasmas can undoubtedly exist in plants as endophytes, but no such evidence for them has yet been explored. Recent studies have revealed the ubiquity of these fungi, with an estimate of 1 million species of endophytic fungi residing in plants and even lichen. Endophytic fungi represent an important and quantifiable component of fungal biodiversity and are known to affect plant community diversity and structure. In recent year's special attention have been made to endophytic fungi because of its ability to produce a good number of new and interesting bioactive secondary metabolites, which are of pharmaceutical, industrial, and agricultural importance (Jalgaonwala et al., 2011; Joshi et al., 2019). The existence of endophytes has been known for over 100 years. They live as imperfect fungi most of the time and have been described as benign parasites or true symbionts. It has been suggested that they can influence the distribution, ecology, physiology, and biochemistry of the host plants (Sridhar and Raviraja, 1995).

A variety of relationships exist between fungal endophytes and their host plants, ranging from mutualistic or symbiotic to antagonistic or slightly pathogenic. Endophytes may produce overabundance of substances of potential use to agriculture, industry, and modern medicine such as novel antibiotics, antimycotics, immunosuppressants, and anticancer compounds (Patel et al., 2013). In addition, the studies of endophytic fungi and their relationships with host plants will shed light on the ecology and evolution of both the endophytes and their hosts: the evolution of endophyte plant symbioses; the ecological factors that influence the direction and strength of the endophyte host plant interaction. Since natural products are likely adapted to a specific function in nature, so search for novel secondary metabolites should concentrate on organisms that inhabit novel biotopes. Schulz et al. (2002) isolated about 6500 endophytic fungi from herbaceous plants and trees and screened them for biologically active compounds. They found a correlation between biological activity and biotope. They also got a higher proportion of the fungal endophytes, in contrast to the soil isolates, suppressed at least one of the test organisms for antialgal and herbicidal activities. Medicinal plants have been recognized as a repository of fungal endophytes with novel metabolites of pharmaceutical importance. The various natural products produced by endophytic fungi possess unique structures and great bioactivities, representing a huge reservoir which offers enormous potential for exploitation for medicinal, agricultural, and industrial uses.

The endophytic fungi are of biotechnological importance as new pharmaceutical compounds, secondary metabolites, agents of biological control, and other useful characteristics would be found by further exploration of endophytes. Dreyfuss and Chapela (1994) estimated that there may be at least one million species of endophytic fungi alone. Recently they have received considerable attention after they were found to protect their host against insects, pests, pathogens, and even domestic herbivores.

23.12.1 Mushroom cultivation

The most significant virtue of mushroom cultivation is that mushrooms can perform the alchemy of transforming agricultural and other organic waste into a nutritious and marketable product (Sudisha and Shekar Shetty, 2009). Oyster mushrooms can grow on cottonseed hulls, cocoa hulls, banana leaves, coffee waste, straw, and even newspaper and cardboard. Shiitake mushrooms grow well on many different kinds of wood and forest waste materials. But, that is not the end, once the mushroom harvest is over, the spent mushroom substrate has all the nutrients, protein, and medicinal compounds found in the mushrooms themselves. This makes it an ideal feed product for livestock, being both nutritional and medicinal (Adamovic et al., 1998). Alternatively, it can be used as excellent compost for other plant or vegetable crops, again turning a waste product into a valuable resource.

Mushrooms are quite nutritious, and they can be a potential food source as well as a marketable product in impoverished areas. In addition, mushrooms have significant medicinal properties, which make them a potential health food commodity (Jogaiah et al., 2016). Shiitake mushrooms, for example, are a source of the compound Lentinan, which is being evaluated as an anticancer drug. Even the ubiquitous polypore, *Trametes versicolor* ("Turkey Tail") is a source of "PSK," another substance with anticancer potential (Hobbs, 1995). Health food stores and upscale "whole foods" markets now have whole lines of "mycomedicinals."

Most people encounter fungi as edible mushrooms produced under cultivation or gathered from the wild. Mushroom cultivation may be integrated into schemes for recycling the agricultural waste as well as providing nutrition and income for people living in developing nations. In 1992 the value of this crop was about $700 million, just in the United States. There is increasing interest in expanding the range of species produced as edible fungi and in growing edible fungi on composted plant biomass as a means of converting waste material into a cash commodity. In addition to a gourmet food item, fungi serve as a low-cost protein source made from *Fusarium graminearum* Schwäbe sold as Quorn with a market value of 25 million British pounds.

23.13 Bioremediation

Certain fungi, in particular "white rot" fungi, can degrade insecticides, herbicides, pentachlorophenol, creosote, coal tars, and heavy fuels and turn them into carbon dioxide, water, and basic elements. Fungi have been shown to biomineralize uranium oxides, suggesting they may have application in the bioremediation of radioactively polluted sites (Raghukumar, 2000; Jogaiah et al., 2019).

23.14 Fungi as biocontrol agents

In agriculture, fungi may be useful if they actively compete for nutrients and space with pathogenic microorganisms such as bacteria or other fungi via the competitive exclusion principle or if they are parasites of these pathogens. Certain fungal species may be used to eliminate or suppress the growth of harmful plant pathogens, such as insects, mites, weeds, nematodes, and other fungi that cause diseases of important crop plants (Murali et al., 2013; Jogaiah et al., 2013). This has generated strong interest in practical applications that use these fungi in the biological control of these agricultural pests (Babu et al., 2015). Entomopathogenic fungi can be used as biopesticides, as they actively kill insects. Fungi that have been used as biological insecticides are *Beauveria bassiana*, *Metarhizium* spp., *Hirsutella* spp., *Paecilomyces* (Isaria) spp., and *Lecanicillium lecanii*. Endophytic fungi of grasses of the genus Neotyphodium, such as *N. coenophialum*, produce alkaloids that are toxic to a range of invertebrate and vertebrate herbivores. These alkaloids protect grass plants from herbivory, but several endophyte alkaloids can poison grazing animals, such as cattle and sheep. Infecting cultivars of pasture or forage grasses with Neotyphodium endophytes is one approach being used in grass breeding programs; the fungal strains are selected for producing only alkaloids that increase resistance to herbivores such as insects while being nontoxic to livestock. Antagonistic strains of the genus *Trichoderma* sp. are able to produce various secondary metabolites that may play a role in the mechanism of action of their biological activity. *Trichoderma harzianum* strain 1295–22 (commercial product

T-22/TRIANUM-G) has currently the greatest ability to colonize plant roots and to inhibit pathogens like *Pythium*, *Rhizoctonia*, *Fusarium* (Alexandru et al., 2013). AM fungi have a negative impact on several ruderal plants. Many important weeds have a ruderal lifestyle, suggesting that AM fungi have the potential to suppress weed growth.

23.15 Conclusion

On considering sustainability on a larger perspective, namely on the level of ecosystems, and ultimately the entire planet the importance of fungal biodiversity in sustaining agriculture becomes increasingly clear. The fungi in the soil of terrestrial ecosystems comprise a vast network of hyphae in complex relationships with many organisms including bacteria, nematodes, and arthropods which is absolutely essential for the well-being and resilience of the entire ecosystem Fungi are intimately associated with crucial processes like the decomposition, recycling, and transportation of nutrients. As time goes on, it becomes clearer how vital wild fungi are in sustainable agriculture. It is realized that fungi can contribute to soil fertility and weed control are sources of drugs and other useful chemicals may be luxury food items, and form a large proportion of the world's biodiversity. Mushroom cultivation can play a role in the transformation of agricultural waste into a delicious, nutritious and marketable product, and as a tool for restoring and sustaining forest communities. About 90% of all plants have a symbiotic relationship with fungi via mycorrhizae. VAMs account for a considerable amount of the fungal biodiversity in soils and performing vital functions. Increased knowledge of the diversity and characteristics of the fungi would contribute directly and indirectly to sustainable agriculture and would strengthen the initiative to preserve fungal diversity.

References

Abbott, L.K., Robson, A.D., 1992. External hyphae of vesicular–arbuscular mycorrhizal fungi associated with *Trifolium subterraneum* L I. Spread of hyphae and phosphorus inflow into roots. *New Phytol.* 120, 371–380.

Adamovic, M.G., Grubic, I., Milenkovic, R., Jovanovic, R., Protic, R., Strenovic, L., Stoicevic, L., 1998. The biodegradation of wheat straw by *Pleurotus ostreatus* mushrooms and its use in cattle feeding. Anim. Feed Sci. Technol. 71930, 357–362.

Alexandru, P., Cristian Iulian, E.N.E., Stefan, I. A., 2013. Fungal biodiversity and climate change on corn: a key tool in building an innovative and sustainable agriculture on dobrogea area. Sci. Papers. Series A. Agon. 56, 406–411.

Allen, M.F., 1996. The ecology of arbuscular mycorrhizas: a look back into the 20th century and a peek into the 21st. Mycol. Res. 100, 769–782.

Allen, E.B., Allen, M.F., Helm, D.J., Trappe, J.M., Molina, R., Rincon, E., 1995. Patterns and regulation of mycorrhizal plant and fungal diversity. Plant Soil 170, 47–62.

Anderson, R., 2001. Fungi and beetles: diversity within diversity. Field Mycol. 2, 82–87.

Babu, N.A., Jogaiah, S., Ito, S.-I., Nagraj, K.A., Tran, L.-S.P., 2015. Improvement of growth, fruit weight and early blight disease protection of tomato plants by rhizosphere bacteria is

correlated with their beneficial traits and induced biosynthesis of antioxidant peroxidase and polyphenol oxidase. Plant Sci. 231, 62–73.
Cavalier-Smith, T., 1998. A revised six-kingdom system of life. Biol. Rev. 73, 203–266.
Colpaert, J.V., Van Tichelen, K.K., 1996a. Mycorrhizas and environmental stress. In: -Frankland, J.C., Magan, N., Gadd, G.M. (Eds.), Fungi and Environmental Change. Cambridge University Press, for the British Mycological Society, Cambridge, pp. 109–128.
Colpaert, J.V., van Tichelen, K.K., 1996b. Decomposition, nitrogen and phosphorus mineralization from beech leaf litter colonized by ectomycorrhizal or litter decomposing basidiomycetes. New Phytol. 134, 123–132.
Cummins, G.B., Hiratsuka, Y., 1983. Illustrated Genera of Rust Fungi, second ed. p. 120.
Dix, N.J., Webster, J., 1995. Höller, U. (Ed.), Fungal Ecology. Chapman and Hall, London, p. 38.
Dreyfuss, M.M., Chapela, I.H., 1994. Potential of fungi in the discovery of novel, low molecular weight pharmaceuticals. In: Gullo, V.P. (Ed.), The Discovery of Natural Products with Therapeutic Potential. Butterworth-Heinemann, Boston, pp. 49–80.
Farr, D.F., Bills, G.F., Chamuris, G.P., Rossman, A.Y., 1989. Fungi on Plants and Plant Products in the United States. The American Phytopathological Society, Minnesota, USA, p. 1252.
Galloway, D.J., 1992. Biodiversity: a lichenological perspective. Biodivers. Conserv. 312–323.
Ginns, J., Lefebvre, M.N.L., 1993. Lignicolous corticioid fungi (Basidiomycota) of North America. Mycologia Memoirs. 19, 1–247.
Goverde, M., van der Heijden, M.G.A., Wiemken, A., Sanders, I.R., Erhardt, A., 2000. Arbuscular mycorrhizal fungi influence life history traits of a lepidopteran herbivore. Oecologia 125, 362–369.
Hackman, W., Meinander, M., 1979. Diptera feeding as larvae on macrofungi in Finland. Annales Zoologica Fennica 16, 50–83.
Hart, M.M., Richard, J.R., Klironomos, J.N., 2003. Plant coexistence mediated by arbuscular mycorrhizal fungi. Trends Ecol. Evol. 18(8).
Hawksworth, D.L., 1991. The fungal dimension of biodiversity, magnitude, significance and conservation. Mycol. Res. 95, 641–655.
Hawksworth, D.L., 2001. The magnitude of fungal diversity: the 1.5 million species estimate revisited. Mycol. Res. 105, 1422–1432.
Hawksworth, D.L., 2004. Fungal diversity and its implications for genetic resource collections. Stud. Mycol. 50, 9–18.
Hawksworth, D.L., Crous, P.W., Redhead, S.A., Reynolds, D.R., Samson, R.A., 2011. The Amsterdam declaration on fungal nomenclature. IMA Fungus 2, 105–112.
Hobbs, C., 1995. Medicinal Mushrooms: An Exploration of Tradition, Healing and Culture. Botanica Press, Santa Cruz, CA.
Holler, U., Wright, A.D., Matthée, G.F., Konig, G.M., Draeger, S., Aust, H.J., Schulz, B., 2000. Fungi from marine sponges: diversity, biological activity and secondary metabolites. Mycol. Res. 104, 1354–1365.
Isaac, S., 1998. To what extent does fungal activity contribute to the processes of decomposition in soils and in composts? The Mycol. 12, 185–186.
Jalgaonwala, R.E., Vishwas, M.B., Mahajan, R.T., 2011. Natural products from plant associated endophytic fungi. J. Microbiol. Biotech. Res. 1 (2), 21–32.
James, T.Y., Kauff, F., Schoch, C.L., 2006. Reconstructing the early evolution of the fungi using a six gene phylogeny. Nature 443, 818–822.
Jogaiah, S., Mostafa, A., Tran, L.S.P., Shin-ichi, I., 2013. Characterization of rhizosphere fungi that mediate resistance in tomato against bacterial wilt disease. J. Exp. Bot. 64, 3829–3842.

Jogaiah, S., Shetty, H.S., Ito, S.-I., Tran, L.-S.P., 2016. Enhancement of downy mildew disease resistance in pearl millet by the G_app7 bioactive compound produced by *Ganoderma applanatum*. Plant Physiol. Biochem. 105, 109–117.

Jogaiah, S., Mahantesh, K., Sharathchnadra, R.G., Shetty, H.S., Vedamurthy, A.B., Tran, L.-S.P., 2016a. Isolation and evaluation of proteolytic actinomycete isolates as novel inducers of pearl millet downy mildew disease protection. Sci. Rep. 6, 30789. https://doi.org/10.1038/srep30789.

Jogaiah, S., Abdelrahman, M., Tran, L.-S.P., Ito, S.-I., 2018. Different mechanisms of *Trichoderma virens*-mediated resistance in tomato against Fusarium wilt involve the jasmonic and salicylic acid pathways. Mol. Plant Pathol. 19, 870–882.

Jogaiah, S., Kurjogi, M., Abdelrahman, M., Nagabhushana, H., Tran, L.S.P., 2019. *Ganoderma applanatum*-mediated green synthesis of silver nanoparticles: structural characterization and in vitro and in vivo biomedical and agrochemical properties. Arabian J. Chem. 12, 1108–1120.

Joshi, S.M., De Britto, S., Jogaiah, S., Ito, S., 2019. Mycogenic selenium nanoparticles as potential new generation broad Spectrum antifungal molecules. Biomol. Ther. 9 (9), 419.

Kernighan, G., 2005. Mycorrhizal diversity: cause and effect? Pedobiologia 49, 511–520.

Kirk, P., Cannon, P.F., Minter, D.W., Stalpers, J.A., 2008. Ainsworth & Bisby's Dictionary of the Fungi, tenth edn CAB International, Wallingford, UK.

Klironomos, J.N., 2003. Variation in plant response to native and exotic arbuscular mycorrhizal fungi. Ecology 84, 2292–2301.

Kluth, S., Kruess, A., Tscharntke, T., 2003. Influence of mechanical cutting and pathogen application on the performance and nutrient storage of *Cirsium arvense*. J. Appl. Ecol. 40, 334–343.

Lonsdale, D., Gibbs, J.N., 1996. Effects of climate change on fungal diseases of trees. In: Frankland, J.C., Magan, N., Gadd, G.M. (Eds.), Fungi and Environmental Change. Cambridge University Press, for the British Mycological Society, pp. 1–19.

Manoharachary, C., Sridhar, K., Singh, R., Adholeya, A., Suryanarayanan, T.S., Rawat, S., Johri, B.N., 2005. Fungal biodiversity: distribution, conservation and prospecting of fungi from India. Curr. Sci. 89 (1), 58–71.

McGee, P.A., Baczocha, N., 1994. Sporocarpic endogonales and Glomales in the scats of *Rattus* and *Perametes*. Mycol. Res. 98, 246–249.

Merryweather, J., 2001. Meet the Glomales – the ecology of mycorrhiza. British Wildlife 13, 86–93.

Murali, M., Jogaiah, S., Amruthesh, K.N., Shin-ichi, I., Shekar Shetty, H., 2013. Rhizosphere fungus *Penicillium chrysogenum* promotes growth and induces defense-related genes and downy mildew disease resistance in pearl millet. Plant Biol. 15, 111–118.

Nagaraju, A., Jogaiah, S., Mahadeva Murthy, S., Ito, S.-I., 2012. Seed priming with *Trichoderma harzianum* isolates enhances plant growth and induces resistance against *Plasmopara halstedii*, an incitant of sunflower downy mildew disease. Aust. J. Plant Path. 41, 609–620.

Nandini, B., Puttaswamy, H., Prakash, H.S., Adhikari, S., Jogaiah, S., Geetha, N., 2020. Elicitation of novel trichogenic-lipid nanoemulsion signaling resistance against pearl millet downy mildew disease. Biomol. Ther. 10 (1), 25.

Padhi, L., Mohanta, Y.K., Panda, S.K., 2013. Endophytic fungi with great promises: A review. J. Adv. Pharm. Edu. & Res. **3**, 152–170.

Patel, C., Yadav, S., Rahi, S., Dave, A., 2013. Studies on biodiversity of fungal endophytes of indigenous monocotaceous and dicotaceous plants and evaluation of their enzymatic potentialities. Int. J. Sci. Res. 3 (7), 1–5.

Perry, D.A., Bell, T., Amaranthus, M.P., 1992. Mycorrhizal fungi in mixed-species forests and other tales of positive feedback, redundancy and stability. In: Cannell, M.G.R., Malcolm, D.C., Robertson, P.A. (Eds.), The Ecology of Mixed-Species Stands of Trees. Oxford, Blackwell Scientific Publications, pp. 151–179.
Raghukumar, C., 2000. Fungi from marine habitats: an application in bioremediation. Mycol. Res. 104, 1222–1226.
Rillig, M.C., Steinberg, P.D., 2002. Differential decomposition of arbuscular mycorrhizal fungal hyphae and glomalin. Soil Biol. Biochem. 35, 191–194.
Saikkonen, K., Faeth, S.H., Helander, M., Sullivan, T.J., 1998. Fungal endophytes, A continuum of interactions with host plants. Ann. Rev. Ecol. Syst. 29, 319–343.
Schulz, B., Boyle, C., Draeger, S., Rommert, A.K., 2002. Endophytic fungi, a source of novel biologically active secondary metabolites. Mycol. Res. 106, 996–1004.
Smith, S.E., Read, D.J., 1996. Mycorrhizal Symbiosis. Academic Press, San Diego.
Sridhar, K.R., Raviraja, N.S., 1995. Endophytes-A crucial issue. Curr. Sc. 69 (7), 570–574.
Steinberg, P.D., Rillig, M.C., 2002. Glomalin production by an arbuscular mycorrhizal fungus: a mechanism of habitat modification. Soil Biol. Biochem. 34, 1371–1374.
Stierle, A., Strobel, G.A., Stierle, D., 1993. Taxol and taxane production by *Taxomyces andreanae*, an endophytic fungus of Pacific yew. Science 260, 214–216.
Streitwolf-Engel, R., Boller, T., Wiemken, A., Sanders, I.R., 1997. Clonal growth traits of two *Prunella* species are determined by co-occurring arbuscular mycorrhizal fungi from a calcareous grassland. J. Ecol. 85, 181–191.
Streitwolf-Engel, R., van der Heijden, M.G.A., Wiemken, A., Sanders, I.R., 1997a. The ecological significance of arbuscular mycorrhizal fungal effects on clonal reproductionin plants. Ecology 82, 2846–2859.
Sudisha, J., Shekar Shetty, H., 2009. Anti-oomycete compounds from *Ganoderma appalantum*, a wood rot basidiomycete. Nat. Prod. Res. 23, 737–753.
van der Heijden, M., 2002. Arbuscular mycorrhizal fungi as a determinant of plant diversity: In search for underlying mechanisms and general principals. In: Van der Heijden, M., Sanders, I. (Eds.), Mycorrhizal Ecology. Ecological Studies. vol. 157. Springer, Berlin, pp. 243–261.
van der Heijden, M., Boller, T., Wiemken, A., Sanders, I., 1998a. Different arbuscular mycorrhizal fungal species are potential determinants of plant community structure. Ecology 79, 2082–2091.
van der Heijden, M., Klironomos, J., Ursic, M., Moutoglis, P., Streitwolf-Engel, R., Boller, T., Wiemken, A., Sanders, I., 1998b. Mycorrhizal fungal diversity determines plant biodiversity, ecosystem variability and productivity. Nature 396, 69–72.
van der Heijden, M.G.A., Streitwolf-Engel, R., Riedl, R., Siegrist, S., Neudecker, A., Ineichen, K., Boller, T., Wiemken, A., Sanders, I.R., 2006. The mycorrhizal contribution to plant productivity, plant nutrition and soil structure in experimental grassland. New Phytol. 172, 739–752.
van der Heijden, M.G.A., Rinaudo, V., Verbruggen, E., Scherrer, C., Bàrberi, P., Giovannetti, M., 2008. The significance of mycorrhizal fungi for crop productivity and ecosystem sustainability in organic farming systems. In: 16th IFOAM Organic World Congress, Modena, Italy, June 16–20,.
Whittaker, R.H., 1969. New concepts of kingdoms of organisms. Science 163, 150–160.
Wright, S.F., Upadhyaya, A., 1998. A survey of soils for aggregate stability and glomalin, a glycoprotein produced by hyphae of arbuscular mycorrhizal fungi. Plant Soil 198, 97–107.

Exploring the biogeographical diversity of *Trichoderma* for plant health

24

S. Nakkeeran[a], T. Marimuthu[b], P. Renukadevi[c], S. Brindhadevi[c], and Sudisha Jogaiah[d]
[a]Department of Plant Pathology, Centre for Plant Protection Studies, Tamil Nadu Agricultural University, Coimbatore, India, [b]World Noni Research Centre, Chennai, India, [c]Department of Plant Pathology, Centre for Plant Protection Studies, Tamil Nadu Agricultural University, Coimbatore, India, [d]Laboratory of Plant Healthcare and Diagnostics, P.G. Department of Biotechnology and Microbiology, Karnatak University, Dharwad, Karnataka, India

24.1 Introduction

A thorough knowledge of the systematic of *Trichoderma* is highly imperative to discriminate between the beneficial and pathogenic species of *Trichoderma* and to protect the diversity (Abdelrahman et al., 2016). Hence, in-depth knowledge of *Trichoderma* on the present existence of the diversity, where do they exist? and any threat to them has to be well documented to explore their potential for human welfare (Nagaraju et al., 2012b). Though morphological taxonomy helps in the differentiation of strains, the era of omics adds value to the phenotypic description of the strains and helps in documenting and differentiating the diversity. Hence, the systematic of *Trichoderma* should be carried in combination with morphological, biochemical, and omic approaches in a polyphasic manner. Biodiversity studies on *Trichoderma* conducted in European countries, Russia, and China reflect the presence of 62 teleomorphs of *Trichoderma* (Hyphocrea) and 25 species of *Trichoderma* (Anamorphic of Hypocrea). The biodiversity of *Trichoderma* from different habitats plays a crucial role in the management of plant diseases, bioremediation and thus improves human welfare.

24.2 Is *Trichoderma* important?

The need for the agricultural community was the force of motivation for the development of broad-spectrum pesticides to manage pests and diseases of both agricultural and horticultural crops (Jogaiah et al., 2013). Advances in the area of pesticide development were measured by the assessment of its efficacy in the management alone. However, today the food production has ceased to be a major concern, due to stable population levels and the success of innovative agricultural technologies. At this juncture, the agrochemical industry that primarily recognized the farmers as their customers have failed to acknowledge the public as an important client. The perception of the public reflects that the negative aspects of pesticides seem to

Biocontrol Agents and Secondary Metabolites. https://doi.org/10.1016/B978-0-12-822919-4.00024-7

outnumber their benefits. The increased reflection on environmental concern over pesticide use has been instrumental in a large upsurge of bio-agents for both pest and disease control (Jogaiah et al., 2018). The development of fungicide resistance among the pathogens, groundwater, and foodstuff pollution and the development of oncogenic risks have further encouraged the exploitation of antagonistic microflora in disease management (Satapute et al., 2019).

The decades of laboratory experiments have led to the explosion in the field of biocontrol and introduced more than 50 commercial formulations comprising of fungi and bacteria (Whipps, 1997; Murali et al., 2013). Among the antagonists, *Trichoderma* plays a vital role in plant disease management and has opened new vistas for the commercialization. *Trichoderma* spp. are cosmopolitan soil fungi, remarkable for their rapid growth, capable of utilizing diverse substrates, and resistance to noxious chemicals (Nagaraju et al., 2012b). It is a predominant component among the mycoflora in various soils in all climatic zones. The frequent association of *Trichoderma* (*Hypocreales*, *Ascomycota*) with decaying wood produces more enzymes of industrial value (*Trichoderma reesei*=*Hypocrea jecorina*) (Kubicek and Penttilä, 1998; Sudisha and Shekar Shetty, 2009) and antibiotics (Sivasithamparam and Ghisalberti, 1998) or has applied as biocontrol agents against plant pathogens (i.e., *T. harzianum*=*H. lixii*; *T. atroviride*=*H. atroviridis*; *T. asperellum*, *T. virens*) (Hjeljord and Tronsmo, 1998; Jogaiah et al., 2018). More recently, *T. longibrachiatum* has also known as an opportunistic pathogen of immunocompromised mammals including humans (Kredics et al., 2003). *Trichoderma* has been recognized to comprise a significant amount of fungal biomass in soil (Nelson, 1982) and is frequently present as an indoor contaminant (Thrane et al., 2001). These diverse implications of *Trichoderma*/*Hypocrea* with human society render an accurate species and strain identification as paramount importance (Abdelrahman and Jogaiah, 2019). However, classical approaches based on the use of morphological criteria are difficult to apply to *Trichoderma* due to the plasticity of characters. Hence, the knowledge on the diversity and distribution pattern of *Trichoderma* is highly imperative to explore the best among the diverse species of *Trichoderma*. The analysis of the *Trichoderma* products registered with CIB, Faridabad, and Haryana revealed that only *T. viride*, *T. harzianum*, and *T. hamatum* have been explored to a maximum extent.

24.3 Attributes of *Trichoderma* as a successful biocontrol organism

Trichoderma is the most widely used biocontrol agent. Among the various isolates of *Trichoderma*, *T. viride*, *T. harzianum*, *T. virens*, and *T. hamatum* are used for the management of various diseases of crop plants (Nagaraju et al., 2012a,b; Jogaiah et al., 2013).

It has many attributes to call it as a best suited biocontrol agent, which are:

a. High rhizosphere competence
b. High-competitive saprophytic ability

c. Enhance plant growth
d. Great arsenal of inducible polysaccharide—degrading enzymes
e. Ease for mass multiplication
f. Broad spectrum of action against various pathogens
g. Provide reliable control
h. Environmentally safe

24.4 Ecology of *Trichoderma*

The significance of genetic variation is a key factor to evaluate biodiversity. The genetic variations in *Trichoderma* are mainly attributed to the adaptation of different species suited to different ecological conditions. The hymenomycetous genus, *Trichoderma* is a predominant cosmopolitan soil fungus, widely distributed all over the world, well known for its rapid growth and has the potential to use the varied substrates and often they are resistant to toxic chemicals. *Trichoderma* occurs in all soils and other natural habitats that are rich in organic matter. *Trichoderma* is also found on root surfaces of various plants, decayed bark, mainly after the decay by other fungi (Esposito and da Silva, 1998). The population of *Trichoderma* decreases when it is exposed to dry conditions for a long period of time. Some strains of *T. hamatum* and *T. pseudokoningii* are adapted to excessive soil moisture and those of *T. viride* and *T. polysporum* are not found to be restricted to areas of low temperature. *T. harzianum* is predominant in warm climatic regions, while *T. hamatum* and *T. koningii* are widely distributed in the areas of diverse climatic conditions (Papavizas, 1985; Grondona et al., 1997). It is also distributed in different soils including agricultural, forest, prairie, salt marsh, and desert soils distributed in different agroclimatic zones. The propagules of *Trichoderma* were abundant in forest soils and they proliferated well in the forest soils rich in carbon and nitrogen (Danielson and Davey, 1973). Some of the species of *Trichoderma* survive as decomposers of woody and herbaceous materials and are also necrotrophic against the primary wood decomposers (Rossman, 1996).

The population dynamics of *Trichoderma* are also altered due to the variations in the iron content of the soil (Xu, 1996). Certain strains of *Trichoderma* have evolved and have occupied the niches of higher elevations. Analysis of soil samples from higher elevations of about 2700m in the Himalayas reflected the presence of section *Longibrachiatum*. Based on the parsimony analysis the same was grouped under *H. audinensis*. The same was also isolated from 2300m elevation of Venezuelan Andes (Samuels et al., 1998). Another member of the section *Longibrachiatum* namely, *T. konilangbra* was isolated from locations between 1750 and 3400m in Uganda.

24.4.1 Ecology-based diversity of Trichoderma

The diversity of the species of *Trichoderma* varies based on locations and with the fauna associated with the microsites. The soils of Taiwan were rich in *T. asperellum*, *T.* sp. *novum* was found in association with tree barks of Taiwan, and tree bark with lichen in Chitan forest of Taiwan had *Hypocrea jecorina*, *T. viride*, *T. atroviride*, and *T. harzianum*, but the tree barks that were in association with moss were found to contain *T. harzianum* alone.

Similarly, the tropical forest soil of Cambodia was rich in *T. harzianum*, indicating that it prefers warm soil temperature for its survival and multiplication. Analyses of the samples from the ruined walls of the temple indicate that they were predominant with *T. harzianum* and to a certain extent the ruined walls were also found to be associated with *T. virens*. The soils near the seashore of Thailand at Coral Island and Koh Samet Island was colonized by *T.* sp. *novum* and *T. harzianum*. The wet soils of Kanchanaburi province of Burma were colonized by *T. harzianum* and *T. ghanense*. The tropical rain forest soils of Malaysia had *T. virens*, *T. spirale*, *T. harzianum*, *T. asperellum*, and *T. sp. novum*. The lake garden soils of Kuala Lumpur were associated with *T. virens*, *T. spirale*, *T. harzianum*, *T. sp. novum*, *T. koningii*, and *T. atroviride*. The forest soil of Singapore was colonized by *T. koningii*, *T. asperellum*, *T. virens*, *T. spirale*, and *T. harzianum*. Though there was a huge variation in the distribution pattern of the *Trichoderma* spp., the cultivated garden soils of South East Asia were predominant with *T. virens* which was used as biocontrol fungi for the management of soilborne diseases. This vividly explains that the ecological requirements are different from the species colonizing tree bark, forest soils, and cultivated garden soils (Kubicek et al., 2003).

Isolates of the *T. spirale* was found to be more in South East Asia than the cosmopolitan species *T. koningii*. Similarly, several isolates of *T. spirale* were observed in the soils of Canada. However, the abundance of the same was more in Central America, Columbia, and South Africa which suggests that this species is more abundant in tropical climates (Kubicek et al., 2003). *H. jecorina* a pantropical ascomycete is abundant in South-East Asia, The Pacific Isles, and South America, but the same was previously reported to be restricted to the regions around equator (Samuels et al., 1998). Soils from Central Russia, the Urals, Siberia, and the Himalayan mountains were predominant with *T. atroviride*, *T. hamatum*, *T. virens*, *T. asperellum*, *T. koningii*, *T. harzianum*, and *T. oblongisporum* (Kullnig et al., 2000).

24.5 Systematics of *Trichoderma* and its significance in biodiversity

Knowledge of *Trichoderma* systematics is highly essential to explore the biodiversity of *Trichoderma*. *Trichoderma* is a complex group of fungi that are used in the production of cellulolytic and hemicellulolytic enzymes, biocontrol of plant diseases, and biodegradation of chlorophenolic compounds and soil bioremediation. Since *Trichoderma* is being used as a very common beneficial fungus for several applications (Nandini et al., 2020), and the systematics of *Trichoderma* becomes very imperative in the identification and utilization of various strains. For this purpose, several molecular and morphological approaches are available for the identification of various isolates.

Trichoderma was introduced by Persoon (1794). The taxonomy and identification of *Trichoderma* remain inconclusive. All strains of *Trichoderma* were identified as "*T. viride*" till 1969 (including all the cellulase-producing strains of *T. reesei*) owing to Bisby's (1939) concept that *Trichoderma* consists of a single species. This led to the erroneous statement even in textbooks that "*T. viride* is an industrial cellulose producer." Hence, most of the taxa sampled and determined before 1969 were misidentified. Rifai (1969) adopted the concept of 'aggregate' species, distinguished nine 'species aggregates,' and admitted that some of them (particularly *T. hamatum*) likely contain two or more morphologically indistinguishable species. Bissett (1984, 1991a,b,c, 1992) and Gams and Bissett (1998) revised Rifai's aggregate species by recognizing the continuities of morphological characters, expanded the morphological criteria to accommodate the wider range of morphological variation expressed by some anamorphs of *Hypocrea*, and adopted some forms previously included in *Gliocladium*. While dissecting several of Rifai's species aggregates into several defined taxa, Bissett (1991a) established a subdivision of the genus into five sections: *Longibrachiatum*, *Pachybasium*, *Trichoderma*, *Saturnisporum*, and *Hypocreanum*. The advent of molecular tools for investigations in fungal taxonomy prompted research in the mid-1990s to reassess the morphology-based taxonomy in *Trichoderma*. The laboratories of G.J. Samuels (Beltsville, MD, USA), T. Börner (Berlin, FRG), and C.P. Kubicek (Vienna, Austria) collaboratively pioneered a revision of Bissett's section *Longibrachiatum*. They combined the use of molecular markers (ITS1 and ITS2 sequence analysis, RAPD), physiological (isoenzyme analysis), and phenetic characters, and also for the first time included an analysis of potential teleomorphs of the *Trichoderma* spp. from this section (Samuels et al., 1998). As a result, section *Longibrachiatum* was recognized to be monophyletic and to contain 10 taxa, within which four pairs displayed teleomorph-anamorph relationships: *H. schweinitzii*/*T. citrinoviride*; *H. pseudokoningii*/*T. pseudokoningii*; *H. jecorina*/*T. reesei*; and *H. orientalis*/*T. longibrachiatum*.

Further merged section *Saturnisporum* included only two species, *T. saturnisporum* and *T. ghanense* (Doi et al., 1987), with section *Longibrachiatum*, and recognized the synonymy of *T. ghanense* with *T. parceramosum*. Yet, as a whole, the concept of section *Longibrachiatum* as defined by Bissett (1984) was largely confirmed, suggesting a degree of correlation between morphological and molecular approaches to taxonomy in *Trichoderma*. Section *Longibrachiatum* is a comparably small section of *Trichoderma*, and phylogenetically most distant from the other sections. The relationship between morphological characters and molecular phylogeny became more complex, however, when larger sections of *Trichoderma* were investigated. Kindermann et al. (1998) attempted a first phylogenetic analysis of the whole genus. Using sequence analysis of the ITS1 region of rDNA, they found that the largest section, *Pachybasium*, is actually paraphyletic. Although the use of a single gene fragment alone is insufficient by today's standards, this finding has been confirmed by phylogenetic analysis of several other genes (Kullnig-Gradinger et al., 2002; Chaverri et al., 2004). In this context the nomenclatural-type strain of sect. *Pachybasium*, *Pachybasium hamatum* (Bonord.) = *T. hamatum*, is not a member of the major one of the two *Pachybasium* clades (clade B) (Kindermann et al., 1998), but clusters together with *T. pubescens* and *T. strigosum* in a clade otherwise containing almost all taxa (i.e., with the exception of *T. aureoviride*) from section *Trichoderma*. Because of the lack of a distinctive morphological hiatus between clades A and B of *Pachybasium* (Kindermann et al., 1998), researchers have so far maintained the name *Pachybasium* for both clades, but it is clear that this is an unsatisfactory situation from a taxonomic point of view. *Pachybasium* B contains all taxa attributed to this section by Bissett (1991b) with the exception of the three mentioned above. In addition, *Pachybasium* B also poses the problem of strong genetic variability of several of its species (e.g., *T. harzianum*). Since its species (both ana- and teleomorphs) account for the majority of taxa found in field investigations (Gherbawy et al., 2004), this raises the question of how to identify species in this large heterogeneous group.

24.5.1 Morphological taxonomy of Trichoderma

Morphological taxonomy was based on colony morphology, characters of conidiophore, phialides, phialospores, and chlamydospores.

24.5.1.1 Key to *Trichoderma* genus

Colony

- Colonies rapidly growing
- Mycelium mostly submerged
- Mycelium may be matte or floccose or woolly or arachnoid aerial mycelium
- The reverse side of the plate may be buff or yellowish or amber or reddish or yellow-green or colorless
- Conidiation effuse/tufted/compact pustules; white at first, remaining as such or turns green, gray, or brown

Chlamydospores

- Abundant in submerged mycelium; intercalary or terminal on short lateral branches of vegetative hyphae
- Globose to ellipsoid, smooth or thick walled, colorless to pale yellowish or greenish

Conidiophores

- The main axis is broad, straight, or flexuous
- Primary branches arise at regular intervals and produce smaller secondary branches of branching nature
- Branches are more or less divergent, solitary, paired, or in verticils
- Repeated verticillate branching results in the pyramidal structure
- In some species, the main axis of the conidiophore ends in sterile structures which might be simple, branched, or coiled

Phialides

- Flask shaped, slightly narrower at the base, bulged at the middle and with narrow conical or subcylindrical neck
- Phialides are disposed of in verticils or whorls beneath the septa of the conidiophore

Phialospores

- Conidia single celled, colorless/grayish/green or brown
- Cell wall may be smooth/roughened or with wing-like projection on the outer wall
- Conidial shape may be subglobose, obovoid, ellipsoid, oblong, or short cylindrical Bissett's (1991a,b,c)

24.5.2 *Sections of* Trichoderma

The genus *Trichoderma* is divided into five sections. They are

- Trichoderma
- Pachybasium
- Longibrachiatum
- Saturnisporum
- Hypocreanum

24.5.2.1 *Trichoderma*

- The species in section *Trichoderma* have narrow and flexuous conidiophores
- Branches of conidiophore and phialides are uncrowded
- Branches are frequently paired and never in verticils of more than three.
- (*T. koningii, T. aureoviride, T. viride, T. atroviride*)

24.5.2.2 *Pachybasium*

- Highly ramified colonies
- Conidiophores are broad, arranged in compact pustules, or fascicles
- Branches and phialides are broad or inflated, relatively short
- Phialides on the branches are distributed in crowded verticils
- Few species have the extension of sterile conidiophore
- Many of them produce compact conidiogenous pustules by anatomizing with adjacent conidiophores (Rifai, 1969)

  (*T. hamatum, T. harzianum, T. piluliferum, T. polysporum*)

24.5.2.3 *Longibrachiatum (Bissett, 1984)*

- Distinct greenish-yellow pigments are produced on the reverse of the cultures

  (*T. pseudokoningii, T. longibrachiatum*)

24.5.2.4 *Saturnisporum*

- Branching system is characterized by *T. viride*
- Uncrowded branches and phialides
- Frequently paired branches
- Compact conidiogenous pustules
- Wing like or bullate conidial ornamentation

  (*T. saturnisporum*)

24.5.2.5 *Hypocreanum (teleomorph of* Trichoderma*)*

Isolates from soil and wood pertain to this section

- Conidiation is effuse and sparse
- Conidiophores are sparingly branched
- Phialides are cylindrical to subulate
- They are borne in divergent verticils like verticillium Bissett's (1991a,b,c)

24.6 Global diversity of *Trichoderma*—An overview

Trichoderma being cosmopolitan fungi found to be widely distributed in different agroclimatic zones and habitats in both tropical and subtropical countries. However, the biogeographical distribution of *Trichoderma* is not concentrated except in European countries, Russia, and China. Since *Trichoderma* has a wide scope and potential to improve human life from different angles, attention must be focused to identify the diversity and distribution pattern of *Trichoderma*. About 35 *Trichoderma* species are recognized on the basis of morphological and molecular characters. Besides the role of a few of these species in biotechnology, several seem to play prominent roles in soil ecosystems. The diversity of *Trichoderma* spp. in Russia, Siberia, and the Himalayan Mountains were identified through molecular approaches using ITS 1 and 2 sequence of the rDNA cluster of the 75 isolates obtained. The comparison of sequences confirmed the presence of *T. atroviride*, *T. virens*, *T. hamatum*, *T. asperellum*, *T. koningii*, and *T. oblongisporum*, *T. harzianum*, and *T. inhamatum*. Section *Longibrachiatum* and *T. stromaticum* were found distributed at an elevation of 2700 m in the Himalayas (Kullnig et al., 2000).

Parsimony and distance analyses of DNA sequences from multiple genetic loci of *Trichoderma* with 18S rDNA sequence analysis suggested that the genus *Trichoderma* might have evolved 110 million years ago as *Hypomyces* and *Fusarium*. 28S rDNA sequence analysis shows that the genus *Trichoderma* is part of a monophyletic branch within the *Hypocreaceae*. Gene trees inferred from a combined analysis of the nuclear ribosomal internal transcribed spacer provided strong statistical support for a phylogeny consistent with the existence of four clades. Clade A comprises species of Bissett's (1991) sect. *Trichoderma* in addition to *T. hamatum*, *T. pubescens*, *T. asperellum*, and *T. strigosum*; clade C comprises all the species contained in section *Longibrachiatum* as revised by Samuels et al. (1998), and clade D contains only *T. aureoviride* which is genetically most distant to all other species. Clade B, on the other hand, contains a large and taxonomically heterogeneous mixture of species, among which several subclades could be identified: subclade B1 containing *H. lactea*, *H. citrina*, *H. citrina* var. *americana*, *H. lutea*, and an unnamed *T.* sp. 1; subclade B2 containing *T. stromaticum*, and an unnamed *T.* sp. PPRI3559; subclade B3 containing *T. fertile*, *T. oblongisporum*, and *H. hunua*; subclade B5 containing *T. polysporum*, *T. roceum*, *Hypocrea pilulifera*, and *T. minutisporum*; and a larger subclade (B4), in which three strain clusters could be distinguished: one comprising *T. harzianum*, *T. inhamatum*, *H. vinosa*, and *T. aggressivum*, another one containing *T. fasciculatum*,

T. longipile, and *T. strictipile*; and a third containing *T. virens*, *T. avofuscum*, and *T. crassum*. The position of the remaining species of subclade B4 (*T. spirale*, *T. cfr aureoviride*, *H. tawa*, and *T. tomentosum*) was not resolved (Kullnig-Gradinger et al., 2002).

The variation in *Trichoderma* in South East Asia (Taiwan and Indonesia) was identified through morphological, biochemical, and through sequence analysis. A total of 96 strains were isolated. In all, 78 isolates were identified as *Trichoderma harzianum*/*Trichoderma inhamatum* (37 strains) *Trichoderma virens* (16 strains), *Trichoderma spirale* (8 strains), *Trichoderma koningii* (3 strains), *Trichoderma atroviride* (3 strains), *Trichoderma asperellum* (4 strains), *Hypocrea jecorina* (anamorph: *Trichoderma reesei*; 2 strains), *Trichoderma viride* (2 strains), *Trichoderma hamatum* (1 strain), and *Trichoderma ghanense* (1 strain). Analysis of biochemical characters revealed that *T. virens*, *T. spirale*, *T. asperellum*, *T. koningii*, *H. jecorina*, and *T. ghanense* formed well-defined clusters, thus exhibiting species-specific metabolic properties. In addition, two new taxa of *Trichoderma* were identified (Kubicek et al., 2003).

The biodiversity and biogeography of *Trichoderma* in four areas of China: North (Hebei province), South-East (Zhejiang province), West (Himalayan, Tibet), and South-West (Yunnan province) include 135 isolates. They were grouped tentatively based on morphological characters which showed the existence of 64 isolates. But, all were grouped into 11 species as *Trichoderma asperellum*, *T. koningii*, *T. atroviride*, *T. viride*, *T. velutinum*, *T. cerinum*, *T. virens*, *T. harzianum*, *T. sinensis*, *T. citrinoviride*, *T. longibrachiatum*, and two putative new species (T. sp. C1, and T. sp. C2) by both morphological characters and phylogenetic analysis. *T. harzianum* accounted for almost half of the biodiversity and was predominant in North China. In addition, a unique haplotype of *T. harzianum* which is rarely found elsewhere was identified (Zhang et al., 2005).

24.7 Species diversity of *Trichoderma*

Trichoderma being cosmopolitan fungi, they are often the predominant components of the soil mycoflora in various soils, such as agricultural, prairie, forest, salt marsh, and desert soils in all climatic zones where they are involved in a number of processes like humic acid synthesis and degradation of xenobiotics (Klein and Eveleigh, 1998). Some are also used as good biocontrol agents to improve plant health (Joshi et al., 2019). There is a huge diversity of *Trichoderma* species in different soils and habitats. The diversity of *Trichoderma* from different geographical locations will be ascertained on the basis of morphology, physiological characteristics, and ITS1 and 2 sequences. Nucleotide variation in ITS has recently been used to identify most species and clades of *Trichoderma* that have been recognized by combined analyses of nuclear and mitochondrial rDNA plus fragments of two genes, translation elongation factor and endochitinase 42 (Kullnig-Gradinger et al., 2002). Various species of identified *Trichoderma* include:

Trichoderma viride, *T. harzianum*, *T. hamatum*, *T. cerinum*, *T. citrinoviride*, *T. asperellum*, *T. aggressivum f. aggressivum*, *T. cerinum.*, *T. atroviride*, *T. koningii*, *T. konilangbra*, *T. pseudokoningii*, *T. inhamatum*, *T. virens*, *T. velutinum.*, *T. spirale*, *T. reesei*, *T. ghanense*, *T. spirale*, *T. sinensis.*, *T. rossicum*, *T. longibrachcyatum*, *T. pubescens*, *T. tomentosum.*, *T. strigosum*, *T. saturnisporum*, *T. erinaceum*, *T. effusum*, *T. atroviride*, *T. polysporum*, etc. The anamorphs of *Trichoderma species* are *Hypocrea aureoviridis*, *H. andinensis*, *H. jecorina*, *H. orientalis*, *H. novazelandia*, *H. tawa.*, *H. lixli.*, *H. cinnamomea.*, *H. catoptron.*, *H. thailandica.*, etc.

24.8 Ecological significance of *Trichoderma*

Trichoderma plays an immense role in creating a pollution-free environment through multivarious action. Based on the well-established enzyme profile in *Trichoderma*, it is used for

- Plant growth promotion and disease control
- Improve soil health
- Bioremediation of polluted soils
- Industrial applications
- Phytobial remediation

24.8.1 Trichoderma *and plant health*

Trichoderma species have been widely exploited for the management of seed and soil-borne diseases of agricultural and horticultural crops (Tables 24.1 and 24.2). Soilborne diseases due to *Rhizoctonia solani*, *Macrophomina phaseolina*, and *Fusarium* spp. causes a high percentage of mortality of the plants and consequent yield losses in

Table 24.1 Effect of *Trichoderma* spp. against plant diseases (Kannaiyan et al., 2001).

Trichoderma	Pathogen/disease	Crop	Delivery system
Trichoderma sp. C62	*Sclerotium cepivorum*	Onion	Soil application as bran sand inoculums
Trichoderma spp.	*G. graminis* var. *tritici*	Wheat	Spores added to soil
		Lettuce	Soil
	R. solani	Lettuce	Soil
T. harzianum	*Fusarium graminearum*	Radish	Soil
T. harzianum T95I	*Pythium ulktimum*	Cucumber	Soil
T.koningii	*S. rolfsii*	Tomato	Soil
T.koningii	*Pythium* spp.	Chickpea	Seed
	Pythium spp.	Pea	Seed
	P. ultimum	Pea	Soil
T. harzianum	*Pythium ulktimum*	Cotton	Seed
		Maize	Seed
		Pea	Seed

Table 24.1 Continued

Trichoderma	Pathogen/disease	Crop	Delivery system
T. harzianum +Rhizobium	*S. rolfsii*	Groundnut	Seed and soil
T. viride *T. harzianum* *T. longibrachiatum*	*M. phaseolina*	Sesamum	Seed
T. viride	*M. phaseolina*	Groundnut Sesamum Blackgram Greengram Pigeonpea Sunflower	Seed
T. hamatum	*R .solani*	Cotton Sugarbeet Radish	Soil-as mycelial preparations
T. viride	*M. phaseolina*	Blackgram	Seed
T. viride	*M. phaseolina*	Greengram	Furrow application
T. viride	*F. udum* *M. phaseolina*	Pigeon-pea	Seed and Soil
T. viride	*F. oxysporum f.* sp. *cubense* *R. solani* *M. phaseolina* *Pythium ultimum*	Banana Cotton Green gram Cowpea Chilies, Tomato	*Trichoderma* capsules Seed Seed Seed +Soil Solid matrix priming
Trichoderma spp.	*M. phaseolina*	Sesamum Blackgram	Seed Seed and soil
T. viride	*R. solani* *S. rolfsii*	Cotton Brinjal Sunflower Bhendi	Seed

rain-fed crops especially in pulses and oilseeds. Soil incorporation of *T. harzianum* in the field reduced the inoculum density of *R. solani* by 85% and the fruit rot of tomato by 27%–51%. Soil application of *T.* harzianum was found effective in reducing the incidence of *Macrophomina* root rot of beans by 37%–74%. Coating of melon seeds with conidia of *T. harzianum* reduced the disease caused by *Macrophomina* to the extent of 46.3%. Seed coating of green gram and cowpea at the rate of 10 g of *T. viride* isolate MG6 combined with soil application reduced dry root rot incidence and increased the yield (Nakkeeran and Doraisamy, 2001). In addition to disease reduction, dry matter production was also enhanced which was attributed to the production

Table 24.2 Commercial bio-products of *Trichoderma* spp. (Kannaiyan et al., 2001).

Product	Target pathogens	Crops recommended	Manufacturer
Bio-fungus-*Trichoderma* spp.	*Sclerotinia*, *Phytophthora*, *R. solani*, *Pythium* spp., *Fusarium*, *Verticillium*	Flowers, Strawberries, Trees, Vegetables	De Cuester, Belgium
Binab T–*T. harzianum* and *T. polysporum*	Pathogens causing wilt, take all, root rot, and internal decay of wood products	Flower, fruits, ornamentals, turf and vegetables	Bio-innovation Lab-UK
Root-Pro-*T. harzianum*	*R. solani*, *Pythium* spp., *Fusarium* spp., *and S. rolfsii*	Flowers and vegetables	Mycontrol Ltd., Israel
Root shield-*T. harzianum*	*R. solani*, *Pythium* spp., *Fusarium* spp.	Trees, Shrubs, cabbage, tomato, cucumber, and all ornamentals	Bioworks, Inc., USA
Supresivit-*T. harzianum*	Several fungal pathogens	Broad spectrum	Borregaard BioPlant, Denmark
T22G, T22 planter box-*T. harzianum*	*R. solani*, *Pythium* spp., *Fusarium* spp., *and Sclerotinia homeocarpa*	Bean, cabbage, corn, cotton, cucumber, peanut, potato, sorghum, soybean, sugarbeet, tomato, turf, and ornamentals	Bioworks, Inc., USA
Trieco-*T. viride*	*R. solani*, *Pythium* spp., *Fusarium* spp.	Cotton, chillies, tomato, sugarcane, citrus, grapes, sunflower, cereals, and vegetables	Ecosense labs (1) Pvt. ltd., Mumbai, India
Trichodex-*T. harzianum*	*B. cinerea*, *Colletotrichum* spp., *Fulvia fulva*, *Monilia laxa*, *Plasmopara viticola*, *Pseudoperonospora cubensis*, *Rhizopus stolonifer*, *S. sclerotiorum*	Cucumber, grape, Nectarine, Soybean Strawberry, sunflower, tomato	Makhteshim Chemical Works Ltd., New York

Table 24.2 Continued

Product	Target pathogens	Crops recommended	Manufacturer
Trichopel, Trichoject, Trichodowel, Trichoseal-*T. harzianum and T. viride*	*Armillaria, Botryosphaeria, Chondrostereum, Fusarium, Nectria, Phytophthora, Pythium, Rhizoctonia*	Broad spectrum	Agrimm Technologies Ltd., Newzealand
Trichoderma 2000-*Trichoderma* spp.	*R. solani, Pythium* spp., *Fusarium* spp., *and S. rolfsii*	Nursery and field crops	Mycontrol Ltd., Israel

of growth hormones. Treatment of sunflower seeds with *T. viride* at the rate of 4 g/kg reduced the root rot incidence to 18.15 as against 37.1% in control. It increased the yield of 21%.

Seed treatment with *T. viride* @ 4 g/kg reduced the root rot incidence of sesame to 12.8% and increased the root, shoot length, and oil content. It recorded a yield of 677 kg/ha as against 301 kg/ha in control. Results of the field demonstrations conducted in the farmer's fields for the control of root rot in groundnut, sesamum, urdbean, and chickpea against dry root rot through *T. viride* was found to be promising. *T. harzianum* was found effective in the management of *Fusarium* wilt diseases. Strain T35 checked the *Fusarium* wilt of cotton and melon under natural soil conditions. Incorporation of *T. harzianum* as wheat bran-peat preparation into the soil under greenhouse conditions reduced the incidence of crown rot of tomato up to 80%. Soil application increased the yield of tomato by 18.8%.

Soil application of *T. koningii* at the rate of 2 g per vine of black pepper at the time of transplanting was highly effective than metalaxyl compounds against *Phytophthora* foot rot. The disease incidence was 23%–35% lower and yield was 30%–40% higher than control. Pre- and postplanting application of *T. viride* and *T. harzianum* was effective against foot rot disease of the beetle vine. Amendment of nursery soils with *T. harzianum* prior to the colonization of damping off pathogen in nursery beds resulted in the reduction of pre and postemergence damping-off of cardamom. Delivery of *Trichoderma* strains through solid matrix priming in tomato and chillies reduced the incidence of damping-off under field conditions than dry seed treatments (Nakkeeran and Doraisamy, 2001; Kannaiyan et al., 2001).

24.9 Factors influencing bioefficacy of *Trichoderma* in maintaining plant health

The soil physicochemical factors like pH, moisture, temperature, soil type, components of soil atmosphere, inorganic and organic constituents of soil, and the quantity and type of pesticides applied to soil largely influence the bio-efficacy of *Trichoderma*.

24.9.1 pH

Hydrogen ion concentration played a major role in the adaptability of biocontrol agents to their nutrient environment. The activities of *Trichoderma* spp. were more pronounced at low pH levels and the efficacy was influenced by acidic conditions (Chet and Baker, 1980). The activity of *T. lignorum* was poor in neutral and alkaline soils. Acidic pH favored disease suppression induced by *T. harzianum* (Chet and Baker, 1980) and *T. hamatum*. The spore germination was enhanced under acidic conditions (Chet and Baker, 1980), mycelial growth, conidiophore production (Chet and Baker, 1980), antibiotics production (Dennis and Webster (1971), or the activity of lytic enzymes (Chet and Baker, 1980). Jeyarajan et al. (1994) found that the isolates of *T. viride* performed on a wide range of pH varying from 5.0 to 9.0. Since the isolates of *T. viride* perform in a wide range of soil pH, a composite, compatible formulation of *T. viride* could be developed and used in disease management programs.

24.9.2 Moisture

The soil moisture content is very important compared with soil reactions in altering the population dynamics of the soil. It affects the growth of the antagonist. The bacterial population dropped rapidly as the soil moisture was falling below 15%–20% while fungi and actinomycetes population were not affected. The minimum relative humidity required for the growth of *Trichoderma* was 91%. The maximum growth of *Trichoderma* on ground pine needles occurred at 144% and 201% moisture. Growth was good at 103% moisture but was depressed at 46% moisture. Growth of *T. viride* was more at 40%–60% moisture than at higher soil moisture levels. Jeyarajan et al. (1994) observed that all the isolates of *T. viride* and *T. harzianum* survived uniformly well at 40% followed by 60% moisture-holding capacity.

24.10 Mode of action

Antagonist interacts with pathogen and host in soils either directly or indirectly. In indirect interactions, plant responds to the presence of the antagonist, resulting in induced resistance or plant growth promotion. Based on the interaction, the modes of action may be direct or indirect.

24.10.1 Direct mode of action

24.10.1.1 Competition

Most of the soilborne pathogens are controlled by the competition for space or court of infection on roots and seeds. Biological control of silver leaf disease caused by *Chondrostereum purpureum* was controlled by *T. viride through* the early colonization of fresh wounds (Corke and Hunter, 1979). Competition also seems to be the most potent mechanism of *T. harzianum* T-35 in the control of *F. oxysporum*. f. sp. *vasinfectum* and *F. oxysporum* f. sp. *melonis* in cotton and melon, respectively (Sivan and Chet, 1989).

24.10.1.2 Antibiosis

Antibiotics are low-molecular-weight (<1 kDa) secondary metabolites produced during nutrient limiting conditions. Gliotoxin produced by *G. virens* G20 is a key factor in reducing damping-off of Zinnia (Wilhitte et al., 1994). *G. virens* pertaining to 'Q' group produced gliotoxin and dimethyl gliotoxin, but 'P' group produced gliovirin (diketopiperazine) and heptelidic acid. Gliovirin was inhibitory to *P. ultimum* which causes damping-off in cotton compared with 'Q' strains, which was effective against *R. solani* (Howell et al., 1993).

Trichoderma produced several kinds of antibiotics like Trichodermain, dermadin, trichoviridin, and sesquiterpene heptalic acid. The UV mutants of *T. harzianum* produced trichozianines, the hydrophobic peptides, which interact with phospholipid membrane thus inducing membrane permeability. *T. hamatum* produced trichoviridin 3-(3-isocyano-6-oxabicyclo (3,10) hex-2-ch-5yl) and acrylic acid 3-(3-isocyano cyclopet-2-enzylidne). In addition, it produced three isonitriles among which isonitrin A had broad-spectrum activity. *T. longibrachiatum* produced Trichodermain and trichobrachin (Jeyarajan and Nakkeeran, 2000).

24.10.1.3 Hydrolytic enzymes

Cellulases

Trichoderma species were potential biocontrol agents and producers of powerful cellulases (Papavizas, 1985). Mandels et al. (1971) irradiated *T. viride* and produced a new strain that secreted twice as much cellulase as its wild parent. Improved strains of *T. reesei* were isolated through mutation and selection. It produced more cellulase than did wild strains (Montenecourt and Eveleigh, 1979). Baker (1991) found a perfect correlation between increased cellulase production and rhizosphere competence by mutants of *T. harzianum.* It is suggested that mutants use the mucigel on roots as a substrate. Mutants with rhizosphere competence were a more efficient biocontrol agent and induced increased growth responses from seed treatment than wild-type parents. The quality and quantity of cellulases from *Trichoderma* spp. were improved through mutagenesis (Cuevas et al., 1994).

Proteases

Trichoderma species are well-known producers of proteases. Proteolytic activity of *T. viride* was claimed to be involved in biocontrol of *Sclerotium rolfsii* (Rodriguez-Kabana et al., 1978). Serine proteases produced by *T. harzianum* played an important role in mycoparasitism of plant pathogens (Geremia et al., 1993). Autoclaved mycelium, fungal cell wall, or chitin induced this enzyme, and its production was repressed by glucose. Proteases are suggested to degrade cell wall, membranes, and proteins released by the lysis of the pathogen, thus making the nutrients available for the mycoparasite (Goldman et al., 1994).

24.10.1.4 α-glucosidases

It is an *exo*-carbohydrase that catalyzes the splitting of α-glucosyl residues from the nonreducing terminus of substrates to liberate α-glucose. α-glucosidase is a key enzyme used in the food industries. But its role in the inactivation of phytotoxins has been recently exploited for disease management. *T. viride* produces 67 kDa protein with a pI value of 5.0. *Rhizoctonia solani* causing rice sheath blight produced a glycoproteinaceous host-specific RS toxin. α-glucosidase of high molecular weight (110 kDa) from *T. viride* isolate (Tv-MNT7) was reported to degrade the RS toxin (Sriram et al., 2000).

24.10.1.5 *β-1,3 glucanases*

Lytic enzymes play a vital role in mycoparasitism and degrade the pathogen cell wall. The enzymes like chitinases and β-1,3 glucanases lyse the host wall and lead to the leakage of protoplasmic contents, which are in turn used as food material for the multiplication of antagonist (Cherif and Benhamou, 1990; Elad et al., 1983). Strains of *T. harzianum* produced a mixture of lytic enzymes comprising of chitinolytic and gluconolytic enzymes, which involved in mycoparasitism (Sivan and Chet, 1989). The excretion of β-1,3 glucanase into growth medium and to the soil by *T. harzianum* was observed (Elad et al., 1982). Carbon sources like laminarin, pustulan, cell walls of *R. solani*, and mycelia of several fungi incorporated in the enzyme production medium induced the secretion of β-1,3 glucanase by *T. harzianum* (Jacobs et al., 1991). The purified endochitinase, chitobiosidase, *N*-acetyl-(-glucosaminidase and glucan, 1,3-β-glucosidase, and combinations inhibited the spore germination and germ tube elongation of *Botrytis cinerea*. Combinations of the purified enzymes resulted in the synergistic increase in antifungal activity (Lorito et al., 1994a,b). High synergistic activity was observed by combining 1,3-β-glucosidases and chitinolytic enzymes. This corresponds to the effect observed with the combination of β-1,3 glucanase and endochitinase isolated from plants (Mauch et al., 1988). *T. harzianum* also produced *N*-acetylglucosaminidase, endochitinase, chitobiosidase, and *endo*-β-1,3 glucanases in soybean rhizosphere active against *R. solani* (Dalsoglio et al., 1998).

Chitinases

Chitinase, the basic protein catalyzes the hydrolysis of chitin. Chitin a linear polymer of β-1,4-*N*-acetyl glucosamine is often considered to be the second most abundant polysaccharide in nature (Deshpande, 1986; Nicol, 1991). Relationship between mycolytic enzymes, chitinases, and β-1,3 glucanases produced by mycoparasitic fungi and their significance in fungal cell wall lysis and degradation have been well established (Elad et al., 1980, 1983). The assay of chitinase could be used as a basis for screening potential biocontrol agents (Elad et al., 1982). *Trichoderma* spp. attack several plant pathogenic fungi by excreting lytic enzymes including β-1, 3 glucanase, proteinases, and chitinases enabling them to degrade host cell walls and thus reduce disease incidence (Geremia et al., 1993; Chet et al., 1993; Lorito et al., 1993; Goldman et al., 1994). *T. harzianum* strain Tm had six distinct chitinolytic enzymes. Two were identified as β-1, 4-*N*-acetyl glucosaminidases (Ec. 3.2.1.30). It hydrolyzes chitin to *N*-acetyl glucosamine (Glc NAc) monomers. Four were endochitinases (Ec 3.2.1.14) that randomly cleave at internal sites over the entire length of the chitin microfibril (CHIT 52, CHIT 42, CHIT 33, and CHIT31). *T. harzianum*—P1 released chimeric units from chitin and termed as chitobiosidase (40 kDa) (Harman et al., 1993).

The involvement of *Trichoderma* chitinases in mycoparasitism was tested in liquid cultures supplemented with different carbon sources, e.g., chitin, purified fungal cell walls, glucose, or GlcNAc (Garcia et al., 1994). Complete degradation of mycelial walls of pathogenic fungi depends to major extent on chitinolysis and hence the degrading ability to chitin by *Trichoderma* spp. assumes a greater significance in antagonistic process than the other mechanisms/enzymes involved in the process (Cherif and Benhamou, 1990). Chitinases and glucanases were produced by *Trichoderma* isolates, in soil, which had live mycelium of *S. rolfsii* and *R. solani.* The high level of β-1,3 glucanase and chitinase activities were also detected by dual agar cultures of *T. harzianum* with *S. rolfsii* and *R. solani* (Elad et al., 1983). Direct interaction of *T. harzianum* with *R. solani*, strongly enhanced the endochitinase CHIT 42 (Carsolio et al., 1994). The spore germination and hyphal elongation of *B. cinerea, F. solani, F. graminearum, Ustilago avenae*, and *Uncinula necator* were inhibited substantially by the combination of chitinolytic and gluconolytic enzymes purified from *T. harzianum.* The chitinolytic enzymes from *T. harzianum* were more active than that of the enzyme from other sources like plant-bacteria and were effective against a wider range of fungi (Harman et al., 1993).

24.10.2 Spatial distribution of glucanases and chitinases

β-1,3 glucanase degrades the substrate β-1,3 glucan, which is the major component of the fungal cell wall next to chitin. The physiological concentration of these fungi effectively inhibits the growth of many potential pathogenic fungi. Though this enzyme plays a major role in cell wall lysis, it is imperative to understand the in vitro production and subcellular localization of the constitutive expression of the enzyme, β-1,3-glucanase. Expression of β-1,3 glucanase by *T. harzianum* was observed in extracellular, periplasmic, cell wall-bound internal fractions during submerged fermentation in the absence of lactose and glucose as carbon source. In submerged culturing of *T. harzianum* with lactose, β-1,3 glucanase was found in all extracellular, cell wall bound, and in periplasmic fractions. The relative distribution of the enzyme in the cell is independent of the nature of carbon source and its concentration.

Chitinases are extracellular and intracellular. Chitinases are localized either outside or inside the cell and it is localized at subcellular level in various fractions including cell wall, periplasmic space, vacuoles, and membranes. Extracellular chitinases are more important than the intracellular chitinases, since the extracellular chitinases provide the first line of defense against plant pathogens.

24.10.3 Mycoparasitism

Mycoparasitism relies on the production of lytic enzymes by the mycoparasite for the degradation of cell walls of the host fungus. Isolates of *Trichoderma* spp. coiled around the hyphae of *R. solani*, which consequently lost their cell contents and collapsed (Chu and Wu, 1981). *Trichoderma* spp. attached to *S. rolfsii* or *R. solani* by hyphal coils, hooks, or appressoria. Lysed sites and penetration holes were found in hyphae of the pathogen following the removal of parasitic hyphae (Elad et al., 1983). *Trichoderma* coiled over *Pythium* resulted in cell wall lysis, vacuolation, and coagulation of protoplasm (Lifshitz et al., 1986).

24.10.4 Indirect modes of action

24.10.4.1 Induced resistance

The induction of resistance in plants by *Trichoderma* spp. has been poorly studied. The first demonstration of induced resistance by *Trichoderma* was published by Bigirimana (1997). Soil treatment with *T. harzianum* strain T-39 induced resistance against foliar pathogens *B. cinerea* and *Colletotrichum lindemuthianum*, even though T-39 was present only on the roots. *T. harzianum* strain T-22 induced systemic resistance to pathogens in maize. Therefore, the capacity to induce resistance to a range of diseases in a variety of plants seems to be widespread in *T. harzianum*.

24.10.4.2 Plant growth promotion

Growth promotion may occur either directly or indirectly. Promotion of growth may occur either through biocontrol of deleterious pathogens (Baker, 1989) or through the production of plant hormones like auxin, cytokinin, or gibberellins (Arshad and Frankenberger, 1991). Baker et al. (1984) found that *T. harzianum* can also stimulate plant growth, including floricultural and horticultural plants. Pepper seed germinated 2 days earlier in raw soils containing the fungus than in untreated controls. The flowering of periwinkle was accelerated, the number of blooms per plant on chrysanthemum was increased, and the height and weight of other plants were greater in either steamed or raw soil infested with *T. harzianum*. These responses have occurred consistently with population densities of *T. harzianum* higher than 10^8 cfu/g of soil, whether the biocontrol agent was applied as conidial suspension or as mycelium and spore in a peat bran medium.

Growth stimulation by *T. harzianum* could be the result of the production of plant hormones by antagonist, increased uptake of nutrients by plant, through nutrient mobilization, or the control of one or more subclinical pathogens. *Pythium* spp. are the capable of stunting plants and hinder the nutrient uptake of roots (Backer and Cook, 1988). Hence the suppression of *Pythium* spp. could increase plant growth (Salt, 1978). Increased growth was also induced by a diffusible growth-regulating factor produced by *Trichoderma* spp. (Windham et al., 1986).

24.11 Commercial production and formulations

It is the formulation that can transform a potential biological control agent from a laboratory curiosity to a commercially successful venture in terms of disease management. A successful biocontrol formulation is one, which is economical to produce, safe, stable in the environment, and easily delivered and should enable the biocontrol agent to act effectively and consistently under varied environmental conditions. The nutrient status of the substrate or additives is one of the important factors influencing the efficacy of the formulation. Introduction of antagonists through organic carriers alleviate the competition from autochthonous microorganisms since the organic carriers serve as a protectant and a food base during the establishment of the antagonist. Delivering of *Trichoderma* spp. to the site of infection via the organic food bases like diatomaceous earth granule with molasses (Backman and Rodríguez-Kabana, 1975), wheat bran, wheat bran + sawdust (Elad et al., 1980), wheat bran + peat (Sivan et al., 1984), molasses yeast (Papavizas et al., 1984), alginate pellets (Fravel et al., 1985), pyrax (Papavizas and Lewis, 1989), vermiculite + wheat bran (Lewis et al., 1991; Nakkeeran et al., 1997), and processed manure pellets (Kok et al., 1996) allows the establishment and multiplication of the antagonist in the soil resulting in high population levels during an extended period of time.

24.11.1 Talc-based formulation

The Tamil Nadu Agricultural University, Coimbatore, has developed the technology of mass production of talc-based formulation of *Trichoderma viride* and *Pseudomonas fluorescens* for seed treatment and soil application (Sudisha et al., 2006). Several private industries in Coimbatore, Bangalore, Kerala, Maharashtra, Chethali, Delhi, and Chennai produce in large quantities. However, demand exceeds the present supply. The annual requirement of *Trichoderma* has been estimated as 5000 tons to cover 50% area in India. Indirectly, it also creates self-employment opportunities for the unemployed youths. The first commercial seed treatment formulation of the biocontrol agent in India was developed by Jeyarajan et al. (1994). *T. viride* was grown in molasses yeast medium for 10 days in flasks. Then, 2.5 L of the culture broth was inoculated to 50 L of sterilized molasses yeast medium in fermentor. It was incubated for 10 days with 4–8 h aeration/day. The fungal biomass and the broth were mixed with 100-kg talc (super white) powder and 500 g of carboxymethylcellulose as a sticker. It was dried in shade for 72 h and packed in alkathene bags. The initial population of *Trichoderma* in the product was 300×10^6 cfu/g. The product should contain a minimum of 20×10^6 cfu/g at the time of use. This is applied at the rate of 4 g per kg of seed. The shelf life was 4 months. Angappan (1992) found that in chickpea treated with this product, the rhizosphere population was maintained at $11–13 \times 10^3$ cfu per gram throughout the crop growth stage.

24.11.2 Gypsum

Ranganathan et al. (1995) found that gypsum was a good substitute for talc but was much cheaper than talc.

24.11.3 Industrial wastes

Nakkeeran and Jeyarajan (1996) tested two industrial wastes namely precipitated silica and calcium silicate in the place of talc and found them to give a population of 99, 104×10^6 cfu/g, respectively, compared to 143×10^6 cfu/g in talc substrate after 4 months of storage. These two substrates were also much cheaper than talc.

24.11.4 Diatomaceous earth granules

Diatomaceous earth granules were added into a broth consisting of 100-mL molasses, 900 mL of water, 3 g each of KNO_3 and KH_2PO_4 till the level of saturation. It was autoclaved at 121°C for 15 min and spread in shallow pans to a height of 3–5 cm and autoclaved again. *T. harzianum* (3-day-old culture) was homogenized in a blender for 3 s and mixed with sterilized granules in shallow pans and incubated at 25°C for 4–7 days. Clumps were broken up and granules air dried with frequent stirring and used as an inoculum. The inoculum was mixed at 1.1 (*v*/v) with sterilized diatomaceous earth granules impregnated with 10% molasses solution. It was applied to peanut at 140 kg/ha on 70 and 100 days after sowing to control *Sclerotium rolfsii*. The disease was reduced by 42% over control and yield increased by 13.5% (Backman and Rodríguez-Kabana, 1975).

24.11.5 Wheat bran: Sawdust formulation

Wheat bran:Sawdust:tap water mixture (3:1:4 *v*/v) is taken in polypropylene bags and autoclaved for 1 h at 121°C for two successive days. The bags were inoculated with *T. harzianum* and incubated in illuminated chambers for 14 days at 30°C. It was applied at the time of sowing and mixed with the soil to a depth of 7–10 cm with a rotary hoe. It increased yield of beans (1500 kg/ha), tomato (300 kg/ha), cotton (500 kg/ha), and potato (40–600 kg/ha) and controlled *S. rolfsii* and *R. solani* (Elad et al., 1986).

24.11.6 Wheat bran: Peat

Wheat bran:peat mixture (1:1 *v*/v) was autoclaved for 1 h. Substrate moisture was adjusted to 50% (*w*/w) with sterile water, medium was inoculated with 0.1 mL of a conidial suspension containing 2×10^4 conidia/mL and incubated for 7 days at 30°C, and it was mixed with rooting mixture for a tomato at 10% *v*/v to control crown rot (*Fusarium oxysporum f.* sp. *radicis lycopersici*). The disease was reduced by 29.7% over control and yield increased by 7% (Sivan et al., 1984).

24.11.7 Vermiculite—Wheat bran

Trichoderma was multiplied in molasses—yeast medium for 10 days. Vermiculite (Grade 4) and milled wheat bran (250 mesh) were heated in a hot air oven at 70°C for 3 days using metal pans. Vermiculite (100 g), wheat bran (3.3 g), liquid culture (14 mL), and 0.05 N HCl (17.5 mL) were mixed and packed. This was immediately used for soil application (Lewis et al., 1991). Vidhya (1995) applied this formulation at 250 kg/ha to mung bean and found 41% reduction in root rot (*M. phaseolina*) and 91% increase in yield.

24.11.8 Alginate pellets

Sodium alginate (20 g) was dissolved in 750 mL water at 40°C on a stirring hot plate. Wheat bran ground to pass through a 0.425 mm mesh screen was placed in a glass blender container with distilled water (50 g/250 mL). Kaolin can also be used in place of bran. It was autoclaved for 30 min and cooled. Fermentor biomass (16–21 g/L) was added to provide 7×10^6 conidia and chlamydospores. The mixture containing fungus, alginate, and bran or kaolin was added dropwise into 500-mL gellant solution (0.25 M $CaCl_3$, pH 5.4). As it entered, each droplet gelled, and a distinct spherical bead formed. After 20 min in the gellant, beads were separated from the solution by gentle filtration, washed, and dried for 24 h in a stream of air at 25°C (Lewis and Papavizas, 1986).

24.11.9 Other substrates

Farmyard manure, gobar gas slurry, press mud, paddy chaff, rice bran, and groundnut shell were also found to support the growth of *T. viride* and *T. harzianum* (Kousalya and Jeyarajan, 1988).

24.12 Shelf life

The shelf life of a biocontrol agent plays a crucial role in storage. In general, the antagonists multiplied in an organic food base has a greater shelf life than the inert or inorganic food bases. Viability of the encapsulated alginate pellets of *T. harzianum* isolate (Thr IDI) remained viable for at least 6 months when stored at 5°C and were reduced significantly when stored at 2°C (Dandurand and Knudsen, 1993). Jeyarajan et al. (1994) developed talc, peat, lignite, and kaolin-based formulation of *T. viride*, which had a shelf life of 4 months. The shelf life of the same in gypsum-based formulation was also 4 months but was cheaper than talc-based formulations (Ranganathan et al., 1995). Vermiculite bran acid fermentor biomass (VBA-FB) of *T. viride* recorded the highest mean population of 20.5×10^7 cfu/g up to 75 days of storage at room temperature (Nakkeeran et al., 1997).

24.13 Delivery system

Biocontrol agents can be applied to various plant parts like seed, root, foliage, inflorescence, and into the soil. Several effective methods of delivery have been developed which enhanced the yield of crops considerably (Nakkeeran et al., 1995; Nakkeeran and Renukadevi, 1997).

24.13.1 Seed treatment

Seed treatment is one of the methods for introducing biocontrol agents into soil–plant environment. The efficacy of seed treatment depends on the ability of the antagonist to multiply in the rhizosphere region. Seed treatment with *T. harzianum* reduced the disease caused by *R. solani* on cotton under field conditions (Elad et al., 1982). Treatment of black gram, green gram, pigeon pea, cowpea, groundnut, sunflower, sesamum, and cotton with *T. viride* at the rate of 4 g/kg reduced the incidence of root rot and wilt (Dinakaran et al., 1995; Jeyarajan et al., 1994; Jeyarajan and Nakkeeran, 1996; Muthamilan and Jeyarajan, 1996; Nakkeeran and Renukadevi, 1997).

24.13.2 Solid matrix priming

It is a process in which moistened seeds are mixed with an organic carrier and the moisture content of the mixture is brought to a level just below that required for seed sprouting. Solid matrix priming can further enhance the efficacy of antagonists in treated seeds as it grows on seed surface during the priming process and thus the propagule numbers increase. The slurry of *Trichoderma* is first added to seeds followed by the addition of lignite and water. *Trichoderma* required less moisture for growth. The primed seeds were incubated for 4 days at 80%–100% RH. During this process, *Trichoderma* profusely sporulates on the seed surface. In *Pythium ultimum* infested soil, the stand was increased to 70%–80% as against 10% in control. In tomato seed, the population of *Trichoderma* increased by 10-fold due to solid matrix priming (Harman et al., 1989).

24.13.3 Soil

The high population of the antagonist in the soil is necessary for the effective management of soilborne pathogens. Adams (1990) defined the efficiency of biocontrol agents as the ratio of the number of propagules of mycoparasites required to obtain disease control to the typical inoculum density of a plant pathogen. Therefore, attempts were made to develop a suitable delivery system for soil. The addition of wheat bran-based inoculum to soil gave 80% reduction of root rot over control in chickpea and bean (Elad et al., 1986). The incidence of urdbean root rot was reduced by 91.3% by adding *T. viride*+*T. harzianum* to soil (Kousalya and Jeyarajan, 1988). Delivering *T. harzianum* through soil during sowing increased the survival of peanut (90%), while in control none of the plants survived (Muthamilan and Jeyarajan, 1996).

24.13.4 Cut stump application

Ricard (1981) developed a special shear to apply *T. viride* simultaneously with pruning to protect fruit trees against silver leaf disease caused by *Chondrostereum purpureum*. The shears have a reservoir for the suspension of *T. viride*. Each pruning cut was covered by spore suspension.

24.13.5 Hive insert

Thomson et al. (1992) developed an innovative method of application of biocontrol agents right in the infection court at the exact time of susceptibility. *Botrytis cinerea* (fruit rot of strawberry) was effectively controlled by *T. virens* through honeybees which acted as a vehicle to carry the spores of *Trichoderma*. Peng et al. (1992) developed a novel and more efficient method of applying *T. virens* on talc and cornmeal (5×10^8 cfug/g), which was placed inside the beehive in a dispenser. Bees acquired the conidia on their legs and bodies as they crawled. Each bee had a spore load of $88–1800 \times 10^3$. Each flower got 1.6 to 27×10^3 conidia of the antagonist. It suppressed *B. cinerea* on stamens (54%) and petals (47%) and controlled fruit rot.

24.14 Population dynamics of *Trichoderma*

24.14.1 Soil

The soil ecosystem is a hostile environment for introduced microorganisms, which often cannot establish and eventually disappear. The biocontrol efficacy of *Trichoderma* isolates seems to depend on the population size, reached after introduction into the soil. A successful antagonist, to survive better in soil and rhizosphere, should have the ability to proliferate well in the introduced environment (Baker and Cook, 1974). Population densities of *T. viride* (T-1-R4) and *T. harzianum* (WT-6-24) increased to 10^4- and 10^3-fold, respectively, in natural soil during the first 3 weeks of incubation (Lewis and Papavizas, 1984). The decline in the population of *T. viride* in the soil might be due to lysis and disintegration of conidia. The stable population densities of *T. viride* from 9 to 36 weeks of incubation may be due to the survival of chlamydospores. In no case, conidia of *Trichoderma* and *Gliocladium* did not proliferated in soil without the addition of a food base (Papavizas, 1985). Pyrax (Clay powder), alginate pellets, vermiculite-bran mixture-based formulations allowed establishment, and multiplication of the antagonist in soil, resulting in high population levels during the extended period of time. The activity and proliferation of *Trichoderma* spp. in the protection of radish seeds against *R. solani* was effective only in acidic soils (Chet and Baker, 1980). A food base such as ground wheat bran enables the fungal propagule in a given preparation to germinate and to colonize the soil. In sterile conditions, the propagation of *T. harzianum* was much greater than under unsterile soils. The incorporation of *T. harzianum* conidia through processed manure pellets helped colonization of nonsterile soils which indicated the necessity of food base for establishment. The soil pH also influenced population dynamics (Kok et al., 1996).

24.14.2 Seeds

Biocontrol agents must proliferate rapidly and become established in the place where they are applied. The first 12–24 h after planting is a crucial period for the bioprotectant applied as seed treatment against soilborne pathogens and they need to be grown and established on the seed coat (Hubbard et al., 1983; Taylor and Harman, 1990; Nagaraju et al., 2012a,b). Solid matrix priming is the process, in which seeds are moistened and mixed with a finely ground organic carrier, and sufficient water was added to achieve the appropriate moisture potential for priming and then incubated for a given duration at a constant temperature. Organic carriers have been shown to enhance the efficacy of *Trichoderma* spp., the bio-agent proliferate during the incubation period and become established, more active, or have colonized the spermosphere thoroughly before planting (Harman et al., 1981). Primed seeds consistently emerge more rapidly than the untreated seeds and provide better stands under adverse environmental conditions than do untreated seeds. Solid matrix priming of tomato seeds increased the number of *Trichoderma* propagules by one order of magnitude in the spermosphere. The effectiveness of biocontrol agents may depend partially on their ability to proliferate during a short period of favorable environmental conditions before they encounter plant pathogens. The number of colony-forming units (cfu) of *Trichoderma* increased on seeds during the matrix priming in Agro-lig by about 10-fold. The number of *Trichoderma* on matrix priming was 10^4 cfu per seed after 4 days of incubation, while on seeds treated with *Trichoderma* alone were about 10^3 cfu per seed. Hence the increase in the cfu of *Trichoderma* provided better emergence of tomato.

24.15 Strain improvement of *Trichoderma*

The existing strains can be further improved for achieving maximum disease control and for developing superior strains suited for industrial applications.

24.15.1 Genetic engineering

Genetic engineering is considered as one of the major scientific revolutions of the century. A successful biocontrol agent requires many characteristics. All of them may not be present in one strain. Since many biocontrol agents have infrequent/unknown sexual stages, crossing and progeny testing by conventional breeding is not possible. Genetic engineering is a useful tool for:

- **i.** Transfer, ability to produce antibiotics from one species to another.
- **ii.** Produce location-specific strains
- **iii.** Improve rhizosphere competence of the antagonist and
- **iv.** Improve its biocontrol potential

By these techniques, important attributes like tolerance to pesticides, adverse environmental conditions like extremes of moisture, temperature and pH, vigorous growth, sporulation in deep tank fermentation, and improved shelf life can be incorporated in one organism.

24.15.1.1 Methods of induction of new biotypes

New biotypes of biocontrol organisms can be developed by

(i) Exposure to fungicides
(ii) Exposure to chemical and physical mutagens
(iii) Protoplast fusion and
(iv) Protoplast transformation

24.15.1.2 Exposure to fungicides

T. harzianum when exposed to chlorothalonil, chlorothalonil-tolerant strains developed. They differed in morphological characters, growth habit, and sporulation from the wild type. Some showed enhanced biocontrol activity. This may be due to mutation, chromosomal aberration, or mitotic recombination. Prolonged and repeated exposure to fungicides like chlorothalonil, iprodione, procymidone, and vinclozolin has resulted in the selection of several strains of *T. harzianum* resistant to above fungicides.

24.15.1.3 Exposure to mutagens

Trichoderma has 109 alleles with great genetic variability and could be explored to induce improved strains from the existing wild ones. Evidence in the late 1970s showed that it was possible by physical and chemical mutagenesis to improve the yield and quality of enzymes by *Trichoderma* spp. (Montenecourt and Eveleigh, 1979). Improved strains of *T. reesei* were obtained by mutations and produced more cellulase than the wild parents (Mandels et al., 1971), but this technology was not imported to enhance the biocontrol potentialities of antagonistic fungi until 1980s.

24.15.1.4 Gamma irradiation

Mutants of *T. viride* resistant to benomyl were obtained through gamma irradiation, but its biocontrol properties were not reported (Papavizas and Lewis, 1983). Stable mutants of *G. virens* were obtained by exposing the conidia to gamma rays. It showed variations in phenotype, growth rate, sporulation, and antagonistic potential against *Rhizoctonia solani* and *Sclerotium rolfsii* from the parent strain. One mutant produced more chlamydospores but no conidia and was more aggressive than the wild type in antagonizing the pathogens in vitro. However, its performance under field conditions was not tested.

24.15.1.5 UV irradiation

The biotypes of *Trichoderma* differed considerably from wild type (WT-6) in appearance, growth habit, fungi toxic metabolite production against *Sclerotium cepivorum* and ability to survive in soil. Several genetic variants from wild-type strains of *T. harzianum* and *T. viride* were developed. Several biotypes tolerated high concentrations (500 μg/mL of the active ingredient) of benomyl and other fungicides like thiabendazole and carbendazim. A few benomyl tolerant biotypes of *T. viride* and *T. harzianum* differed in their growth characters, ability to sporulate, and survival in soil and the suppression of saprophytic growth of *R. solani* in soil. Several

UV-induced biotypes were consistently more effective than the wild-type strains in quenching the activity of *P. ultimum* on peas, *R. solani* on cotton and radish, *S. rolfsii* on beans, and *S. cepivorum* on onion. The isolate (T-1-R9) was also effective against *Fusarium* wilt of chrysanthemum and *Rhizoctonia* scurf of potato (Beagle-Ristaino and Papavizas, 1985). Biotype T-1-R9 was able to recolonize the soil even in the presence of benomyl. It prevented the recolonization of pathogen and brought control similar to that of the integrated control that includes the use of benomyl, lime, and nitrate nitrogen. It was the most effective biocontrol agent against white rot of onion. The wild-type strain did not produce heat-labile metabolite to the assay organism (*S. cepivorum*).

24.15.1.6 Chemical mutagenesis

Exposing the conidia to *N*-methyl-*N*-nitro-N-nitrosoguanidine produced benomyl tolerant mutants of *T. harzianum* (T-95, T-12B). It produced more cellulase than the wild-type strain (Ahmad and Baker, 1987). They also found that the mutants utilized cellulose as a sole carbon source for growth and were able to colonize plant rhizosphere, whereas the wild-type strains did not. NTG induced mutants of *Trichoderma* spp. were tolerant to benomyl, but not rhizosphere competent (Ahmad and Baker, 1987).

Advantages

- **(i)** The development of such fungicide resistant/tolerant strains would encourage the industry to take up pilot testing and commercial production
- **(ii)** Could be combined with a minimal amount of fungicide for disease control.
- **(iii)** Such mutants can also be used in genetic recombination program by the introduction of the specific gene encoding for increased production of antibiotics and enzymes into a recipient biocontrol fungus that possesses other desirable traits.

24.15.1.7 Protoplast fusion

Protoplast fusion is an excellent approach to the construction of fungal strains with novel gene combinations. It is possible to combine desirable traits from various parental strains to produce superior biocontrol strains. Interspecific crosses have been accomplished in *T. viride*.

24.15.1.8 Protoplast transformation

It is the process by which naked DNA is introduced into a cell resulting in a heritage change. It involves two basic steps.

1. Partial or complete elimination of cell barriers to allow uptake of DNA into the cell
2. Stable integration, maintenance, and self-replication of DNA into cell. Isolated protoplasts are excellent materials for the introduction of foreign DNA. This can be done in three ways
   - **i.** Fusion of whole protoplasts from two strains, species, or genera.
   - **ii.** Transfer of mitochondrial plasmids-plasmid-mediated transformation. Hybrids contain genetic material only from the naturally inbreeding microbes from which crossed strains were derived.
   - **iii.** Transfer of DNA segments (autonomously replicating and integrated vectors)—recombinant DNA technique.

DNA segments are separated from the donor and inserted or attached to other cells. Eukaryotic genes cloned into *E. coli*, sequence is rearranged and introduced into the new organism. Resistance to benomyl was constructed by transformation (Ossanna and Mischke, 1990). This was accomplished by transferring to Gl 21, a wild-type strain of *T. virens*, very sensitive to benomyl. The β-tubulin gene from *Neurospora crassa* that codes for resistance to benomyl was successfully expressed in *T. virens* strain GT-21. It suggests that there are tremendous possibilities for desirable transformations in biocontrol fungi provided that the genes coding for certain desirable traits have been identified.

24.16 Industrial application of *Trichoderma*

Trichoderma has been well explored for industrial applications such as bioremediation and enzyme production.

24.16.1 *Bioremediation by* Trichoderma

The variety of organic pollutants released by the ever-increasing number of industries is the direct cause of environmental and health-related problems that have detrimental effects on living beings. Conventional approaches (e.g., landfilling, recycling, pyrolysis, and incineration) to the remediation of contaminated sites are inefficient and costly and can also lead to the formation of toxic intermediates. Thus, biological decontamination methods are preferable to conventional approaches because, in general, microorganisms degrade numerous environmental pollutants without producing toxic intermediates. Some environmental pollutants, however, persist in the environment.

The many different organic pollutants that are present in the environment persist for varying periods of time. Depending on their half-life, some pollutants either disappear within a short period of time or transformed into nontoxic end products. Many others, however, such as polychlorinated dibenzodioxyfurans (PCDDF/Fs), dioxins, dichlorodiphenyltrichloroethane (DDT), chlordane, and hexachlorocyclohexane (HCH), have long half-lives in the environment and tend to enter the food web, where they are subsequently "biomagnified." On entering mammalian systems, most of these pollutants are often metabolized to generate more toxic and harmful intermediates owing to the nonspecific activity of different mammalian enzymes. Bioremediation is the process by which living organisms degrade or transform hazardous organic contaminants. *Trichoderma* plays a vital role in the bioremediation of heavy metals, toxic chemicals such as xenobiotic compounds, cobalt, lignite, cyanide, nitrocellulose, etc.

Trichoderma is recognized as "keystone or controller organism" in forest soil due to their involvement in decomposition, nutrient cycling, and their regulation of associated mycobiota. Recent evidence have proved their potential to decompose xenobiotics. *T. harzianum* and *T. viride* are capable of degrading organochlorides under in vitro. The persistence of organochloride contaminants in forest soil may be influenced by the predominant association of *T. harzianum* and *T. viride*. *T. harzianum* was capable of degrading chlorophenols and adsorbable organic halogens. It also could degrade ciliatine, glyphosate, and phosphonic acid (Lupicka et al., 1997).

24.16.2 Biodegradation of noxious chemicals

Trichoderma also has the potential to degrade noxious chemicals like nitrocellulose, cobalt 60, cyanides, and heavy metals. *Trichoderma reesei* could degrade nitrocellulose which is commonly used in explosives and pharmaceutical industries. The fungus was able to degrade nitrocellulose completely after 22 days of application (Auer et al., 2005).

24.16.3 Biodegradation of pesticide

T. harzianum degraded DDT, dieldrin, endosulfan, PCNB, and pentachlorophenol. *T. harzianum* CCT-4790 degraded 60% of the herbicide diuron in soil and opens a good potential for the bioremediation of herbicide-contaminated soils. *T. viride* degraded pesticide photodieldrin into water-soluble and noninsecticidal compounds after 4–5 weeks. Thus, *Trichoderma* strains play a crucial role in the bioremediation process. *Trichoderma harzianum*, a biocontrol species colonize plant rhizosphere and it metabolizes several pesticides. Pentachlorophenol (PCP) was metabolized to pentachloroanisole (Rigout and Matsumura, 2002).

24.16.4 Phytobiation

Physical and chemical techniques are the most commonly used methods for the remediation of pollutants. The high cost of these methods has stimulated the interest in alternative biologically based systems, in particular, bioremediation and phytoremediation. Bioremediation consists of adding microorganisms to a polluted system, in the presence of nutrients, for degrading toxic compounds. Phytoremediation uses plants to degrade or remove toxicants from soil lacking a defined or managed root-microbial population. The combined application of both remediations is termed as phytobiation. The cultivation of cotton, beans, or corn as well as ferns on a polluted site rich in heavy metals, arsenic, cyanide, and metallocyanides, nitrates, etc. in association with *Trichoderma* spp. increased rapid uptake of the toxic elements by the plants. In addition, there was a strong reduction of toxicity caused by polycyclic aromatic hydrocarbons and phenolic compounds in water and soil by *Trichoderma* (Vinale et al., 2003).

24.17 Conclusion

The investigation of the biogeographical distribution of fungi is a challenging task. Firstly, it is generally regarded that only a small proportion of fungi (approximately 70,000 species) have been identified to date, with the estimates of the numbers of fungi at 1.5 million. This estimate is not universally accepted, as it is based on calculations from temperate regions indicating an approximately 6:1 ratio of fungal species to plant species. However, for tropical regions, there are conflicting estimates, with this ratio either considered too high or too low or proportions of up to 33:1 considered more realistic. Consequently, the incomplete knowledge of the diversity and distribution of fungi have to be taken into account in any fungal biogeography study. Owing to the significant contribution of *Trichoderma* to human welfare and due to the richness in the biodiversity of India, the following researchable issues have to be addressed to conserve and to claim the IPR rights of our microbial wealth. They are:

- Adoption of polyphasic systematics to map the distribution of *Trichoderma* from different agroclimatic zones
- Identifying the habitat preference of individual taxa
- Exploring the potential of all the strains of *Trichoderma* for human welfare
- Research on bioremediation, phytoremediation, and phytobiation has to be strengthened to select the potential strains of *Trichoderma*
- Transformation and gene manipulation strategies have to be strengthened for exploring the potential of *Trichoderma* other than *T. reesei.*
- Strengthening proteomic and genomic research of *Trichoderma* for better pest and disease control of plants.
- Improving the knowledge of the ecology of *Trichoderma*-plant and pathogen interaction.
- Assessment of human and plant health risks.
- Human Resource Development on bioinformatics, systems biology, digital biology with molecular techniques for the documentation, and exploitation of the full potential of *Trichoderma* for the welfare of the country and mankind.

References

Abdelrahman, M., Jogaiah, S., 2019. Defense mechanism and diverse actions of fungal biological control agents against plant biotic stresses. In: Khurana, S., Gaur, R. (Eds.), Plant Biotechnology: Progress in Genomic Era. Springer, Singapore, pp. 461–478.

Abdelrahman, M., Abdel-Motaal, F., El-Sayed, M., Jogaiah, S., Shigyo, M., Ito, S.-i., Tran, L.S.P., 2016. Dissection of *Trichoderma longibrachiatum*-induced defense in onion (*Allium cepa* L.) against *Fusarium oxysporum* f. sp. *Cepa* by target metabolite profiling. Plant Sci. 246, 128–138.

Adams, P.B., 1990. The potential of microparasites for biological control of plant diseases. Ann. Rev. Phytopathol. 28, 59–72.

Ahmad, J.S., Baker, R., 1987. Competitive saprophytic ability and cellulolytic activity of rhizosphere-competent mutants of *Trichoderma harzianum*. Phytopathology 77, 358–362.

Angappan, K., 1992. Biological control of chickpea root rot caused by *Macrophomina phaseolina* (Tassi) Goid. M.Sc. (Ag). Thesis,TNAU, Coimbatore, p. 114.

Arshad, M., Frankenberger, W.T., 1991. Microbial production of plant hormones. In: Keister, D.L., Cregan, P.B. (Eds.), The rhizosphere and plant growth. Kluwer Academic Publisher, Dodrecht, The Netherlands, pp. 324–327.

Auer, N., Hedger, J.N., Evans, C.S., 2005. Degradation of nitrocellulose by fungi. Biodegradation 16 (3), 229–236.

Backer, J.O., Cook, R.J., 1988. The role of siderophores in suppression of *Pythium* spp. and production of increased growth response of wheat by fluorescent pseudomonads. Phytopathology 78, 778–782.

Backman, P.A., Rodríguez-Kabana, R., 1975. A system for the growth and delivery of biological control agents to the soil. Phytopathology 65, 819–821.

Baker, R., 1989. Improved *Trichoderma* spp. for promoting crop productivity. Trends Biotechnol. 7, 34–38.

Baker, R., 1991. Induction of rhizosphere competence in the biocontrol fungus, *Trichoderma*. In: Keister, D.L., Cregan, P.B. (Eds.), Beltsville symposia in Agricultural Research. The Rhizosphere and Plant Growth.pp. 221–228.

Baker, K.F., Cook, R.J., 1974. Biological Control of Plant Pathogens. W.H. Freeman and Co, San Francisco, CA, p. 433.

Baker, R., Elad, Y., Chet, I., 1984. The controlled experiment in the scientific method with special emphasis on biological control. Phytopathology 74, 1919–1984.

Bigirimana, 1997. Induction of systemic resistance on bean (*Phaseolus vulgaris*) by *Trichoderma harzianum*. Med. Fac. Landbouww. Univ. Gent. 62, 1001–1007.

Bisby, G.R., 1939. *Trichoderma viride* Pers. ex fries, and notes on *Hypocrea*. Trans. Brit. Mycol. Soc. 23, 149–168.

Bissett, J., 1984. A revision of the genus, *Trichoderma*. I. Sect. *Longibrachiatum* sect. nov. Can. J. Bot. 62, 924–931.

Bissett, J., 1991a. A revision of the genus *Trichoderma*. II. Infrageneric classification. Can. J. Bot. 69, 2357–2372.

Bissett, J., 1991b. A revision of the genus *Trichoderma*. III. Sect. *Pachybasium*. Can. J. Bot. 69, 2373–2417.

Bissett, J., 1991c. A revision of the genus *Trichoderma*. IV. Additional notes on section *Longibrachiatum*. Can. J. Bot. 69, 2418–2420.

Bissett, J., 1992. Trichoderma atroviride. Can. J. Bot. 70, 639–641.

Carsolio, C., Gutierrez, A., Jimenez, B., Van Montagu, M., Herrera-Estrella, A., 1994. Characterization of ech-42, a *Trichoderma harzianum* endochitinase gene expressed during mycroparasitism. *Proc. Natl. Acad. Sci.* USA 91, 10903–10907.

Chaverri, P., Candoussau, F., Samuels, G.J., 2004. *Hypocrea phyllostachydis* and its *Trichoderma* anamorph, a new bambusicolous species from France. Mycol. Progr. 3, 29–36.

Cherif, M., Benhamou, N., 1990. Cytochemical aspects of chitin breakdown during the parasitic action of a *Trichoderma* sp. on *Fusarium oxysporum f.sp. radicis lycopersici*. Phytopathology 80, 1406–1414.

Chet, I., Baker, R., 1980. Induction of suppressiveness to *Rhizoctonia solani* in soil. Phytopathology 70, 994–998.

Chet, I., Barak, Z., Oppenheim, A., 1993. Genetic engineering of microorganisms for improved biocontrol activity. In: Chet, I. (Ed.), Biotechnological Prospects of Plant Disease Control. Wiley-Liss, New York, pp. 211–235.

Chu, F.F., Wu, W.S., 1981. Biological and chemical control of *Rhizoctonia solani* by pea seed treatment. Memoris College Agric. 21, 19–28 National Taiwan University.

Corke, A.T.K., Hunter, T., 1979. Biocontrol of *Nectria galligera* infection of purning wounds on apple shoots. J. Hort. Sci. 54, 47–55.

Cuevas, V.C., Culiat, C.T., Manaligod, R.L., Sajise, N.E., 1994. Induction of fungicide resistance for enhanced cellulase production of *Trichoderma harzianum* Rifai. Philipp. J. Sci. 123, 177–183.

Dalsoglio, F.K., Bertagnolli, B.L., Sinclair, J.B., Eastbrn, D.M., 1998. Production of chitinolytic enzymes and endoglucanases in the soybean rhizosphere in the presence of *T. harzianum* and *R. solani*. Biol. Cont. 12, 111–117.

Dandurand, L.M., Knudsen, G.R., 1993. Influence of *Pseudomonas fluorescens* on hyphal growth and biocontrol activity of *Trichoderma harzianum* in the spermosphere and rhizosphere of pea. Phytopathology 83, 265–270.

Danielson, R.M., Davey, C.B., 1973. The abundance of *Trichoderma* propagules and the distribution of species in forest soils. *Soil Biol. Biochem.* **5**, 486–494.

Dennis, C., Webster, J., 1971. Antagonistic properties of species-groups of *Trichoderma*: II. Production of volatile antibiotics. Trans. Br. Mycol. Soc. 57 (1), 41–48.

Deshpande, M.V., 1986. Enzymatic degradation of chitin and its biological applications. J. Sci. Indus. Res. 45, 273–287.

Dinakaran, D., Ramakrishnan, G., Sridar, R., Jeyarajan, R., 1995. Management of sesamum root rot with biocontrol agents. J. Oilseeds Res. 12, 262–263.

Doi, Y., Abe, Y., Sugyama, J., 1987. *Trichoderma* sect. *Saturnisporum*, sect. Nov. and *Trichoderma ghanense*, sp. nov. Bull. Nat. Sci. Museum, Ser B (Botany) 12, 1–9.

Elad, Y., Chet, I., Katan, P., 1980. Trichoderma harzianum. A biocontrol agent effective against *Sclerotium rolfsii* and *Rhizoctonia solani*. Phytopathology 70, 119–121.

Elad, Y., Chet, I., Henis, Y., 1982. Degradation of plant pathogenis fungi by *Trichoderma harzianum*. Can. J. Microbiol. 28, 719–725.

Elad, Y., Chet, I., Boyle, P., Henis, Y., 1983. Parasitism of *Trichodermaia* spp. on *Rhizoctonia solani* and *Sclerotium rolfsii*. Scanning electron microscopy and fluorescenct microscopy. Phytopathology 73, 85–88.

Elad, Y., Zuiel, Y., Chet, I., 1986. Biological control of *Macrophomina phaseolina* (Tassi) Goid. by *Trichoderma harzianum*. Crop. Prot. 5, 288–292.

Esposito, E., da Silva, M., 1998. Systematics and environmental application of the genus *Trichoderma*. Crit. Rev. Microbiol. 24, 89–98.

Fravel, D.R., Marois, J.J., Lumsden, R.D., Connick, W.J., 1985. Encapsulation of potential biocontrol agents in an alginate-clay matrix. Phytopathology 75, 774–777.

Gams, W., Bissett, J., 1998. Morphology and Identification of *Trichoderma*. In: Kubicek, CP, Harman, GE (Eds.), Trichoderma and Gliocladium. In: Basic Biology, Taxonomy and GeneticsVol. 1. Taylor &amp Francis Ltd., London, pp. 3–34.

Garcia, I., Lora, J.M., De La Cruz, J., Benitez, T., Lobell, A., Pintor-Toro, J.A., 1994. Cloning and characterization of chitinase CHIT 42 cDNA from the mycoparasitic fungus *Trichoderma harzianum*. Curr. Genet. 27, 83–89.

Geremia, R., Goldman, G.H., Jacobs, D., Ardiles, S., Vila, B., Van Montagu, M., Herera, E., 1993. Moelcular characterization of the proteinase-encoding gene, Prb1, related to mycoparasitism by *Trichoderma harzianum*. Mol. Microbiol. 8, 603–613.

Gherbawy, Y., Druzhinina, I., Shaban, G.M., Wuczkowsky, M., Yaser, M., El-Naghy, M.A., Prillinger, H.J., Kubicek, C.P., 2004. *Trichoderma* populations from alkaline agricultural soil in the Nile valley, Egypt, consist of only two species. Mycol. Prog. 3, 211–218.

Goldman, H.G., Hayes, C., Harman, G.E., 1994. Molecular and cellular biology of biocontrol by *Trichoderma* spp. Trends Biotechnol. 12, 478–482.

Grondona, I., Hermosa, R., Tejada, M., Gomis, M.D., Mateos, P.F., Bridge, P.D., Monte, E., Garcia-Acha, I., 1997. Physiological and biochemical characterization of *Trichoderma harzianum*, a biological control agent against soilborne fungal plant pathogens. Appl. Environ. Microbiol. 63, 3189.

Harman, G.E., Chet, I., Baker, R., 1981. Factors affecting *Trichoderma hamatum* applied to seed as a biocontrol agent. Phytopathology 71, 569–572.

Harman, G.E., Taylor, A.G., Stasz, T.E., 1989. Combining effective strains of *Trichoderma harzianum* and solid martix priming to improve biological seed treatment. Plant Dis 73, 631–637.

Harman, G.E., Hayes, C.K., Lorito, M., Broadway, R.M., Di-Pietro, A., Petebauer, C., Tronsmo, A., 1993. Chitinolytic enzymes of *Trichoderma harzianum*. Purification of chitobiosidase and endochitinase. Phytopathology 83, 313–318.

Hjeljord, L., Tronsmo, A., 1998. *Trichoderma* and *Gliocladium*in biological control: An overview. In: Harman, G.E., Kubicek, C.P. (Eds.), *Trichoderma* and *Gliocladium.* In: Enzymes, Biological Control and Commercial Applications, vol. 2. Taylor and Francis Ltd., London, pp. 131–151.

Howell, C.R., Stipanovic, R.D., Lumsden, R.D., 1993. Antibiotic production by strains of *Glicaldium virens* and its relation to the biocontrol of cotton seedling diseases. Biocontrol Sci. Tech. 3, 435–441.

Hubbard, J.P., Harman, G.E., Hadar, Y., 1983. Effect of soil-borne *Pseudomonas* spp. on the biological control agent, *Trichoderma hamatum*, on pea seeds. Phytopathology 73, 655–659.

Jacobs, D., Geremia, R.A., Goldman, G.H., Kamoen, O., Van Montagu, M., Herrera-Estrella, A., 1991. Study of the expression of β-1, 3 – glucanase of *Trichoderma harzianum.* Petria 1, 125.

Jeyarajan, R., Nakkeeran, S., 1996. Exploitation of biocontrol potential of *Trichoderma* for field use. In: Manibhushan Rao, K, Mahadevan, A (Eds.), Current Trend in Life Sciences. In: -vol. XXI. Today and Tomorrow's Printers and Publishers, New Delhi, India, pp. 61–66.

Jeyarajan, R., Ramakrishnan, G., Dinakaran, D., Sridar, R., 1994. Development of products of *Trichoderma viride* and *Bacillus subtilis* for biocontrol of root rot diseases. In: Dwivedi, (Ed.), Biotechnology in India. Bioved Research Society, Allahabad, pp. 25–36.

Jeyarajan, R., Nakkeeran, S., 2000. Exploitation of microorganisms and viruses as biocontrol agents for crop disease management. In: Upadhyay, et al., (Ed.), Biocontrol Potential and Their Exploitation in Crop Disease Management. Kluwer Academic/Plenum Publishers, USA, pp. 95–116.

Jogaiah, S., Mostafa, A., Tran, L.S.P., Shin-ichi, I., 2013. Characterization of rhizosphere fungi that mediate resistance in tomato against bacterial wilt disease. J. Exp. Bot. 64, 3829–3842.

Jogaiah, S., Abdelrahman, M., Tran, L.-S.P., Ito, S.-I., 2018. Different mechanisms of *Trichoderma virens*-mediated resistance in tomato against Fusarium wilt involve the jasmonic and salicylic acid pathways. Mol. Plant Pathol. 19, 870–882.

Joshi, S.M., De Britto, S., Jogaiah, S., Ito, S., 2019. Mycogenic selenium nanoparticles as potential new generation broad spectrum antifungal molecules. Biomol. Ther. 9 (9), 419.

Kannaiyan, S., Doraiswamy, S., Ramakrishnan, G., Rabindra, R.J., Samiyappan, R., Udayasurian, Gunasekaran, K., Kennedy, J.S., Nakkeeran, S., Sathiah, N., 2001. Recent Advances in Eco-Friendly Approaches for the Management of Pests and Diseases in Crops. Tamil Nadu Agricultural University, Coimbatore, pp. 1–389.

Kindermann, J., El-Ayouti, Y., Samuels, G.J., Kubicek, C.P., 1998. Phylogeny of the genus *Trichoderma* based on sequence analysis of the internal transcribed spacer region 1 of the rDNA clade. Fungal Genet. Biol. 24, 298–309.

Klein, D., Eveleigh, DE., 1998. Ecology of Trichoderma. In: Kubicek, CP, Harman, GE (Eds.), Trichoderma and Gliocladium: Basic Biology, Taxonomy and Genetics. In: 1, Taylor & Francis, London, pp. 57–74.

Kok, C.J., Hageman, P.E.J., Maas, P.W.T., Postma, J., Roozen, N.J.M., Van Lvurde, J.W., 1996. Processed manure as carrier to introduce *Trichoderma harzianum*: Population dynamics and biocontrol effect on *Rhizoctonia solani*. Biocontrol Sci. Technol. 6 (2), 147–162.

Kousalya, G., Jeyarajan, R., 1988. Techniques for mass multiplication of *Trichoderma viride* Pers. Fr and *T. harzianum Rifai*. National Seminar on management of crop diseases with plant products/biological agents. Agricultural College and Research Institute, Madurai, pp. 32–33.

Kredics, L., Antal, Z., Doczi, I., Manczinger, L., Kevei, F., Nagy, E., 2003. Clinical importance of the genus *Trichoderma*. A review. Acta Microbiol. Immunol. Hung. 50, 105–117.

Kubicek, C.P., Penttilä, M.E., 1998. Regulation of production of plant polysaccharide degrading enzymes by *Trichoderma*. In: Harman, G.E., Kubicek, C.P. (Eds.), *Trichoderma* and

Gliocladium. In: Enzymes, Biological Control and Commercial Applications, vol. 2. Taylor and Francis Ltd, London, pp. 49–71.

Kubicek, C.P., Bissett, J., Druzhinina, I., Kullnig-Gradinger, C.M., Szakacs, G., 2003. Genetic and metabolic diversity of *Trichoderma*: a case study on South East Asian isolates. Fungal Genet. Biol. 38, 310–319.

Kullnig, C., Szakacs, G., Kubicek, C.P., 2000. Molecular identification of *Trichoderma* from Russia, Siberiya and the Himalya. *Mycol. Res.* **104** (9), 1117–1125.

Kullnig-Gradinger, C.M., Szakacs, G., Kubicek, C.P., 2002. Phylogeny and evolution of the fungal genus *Trichoderma* – a multigene approach. Mycol. Res. 106, 757–767.

Lewis, J.A., Papavizas, G.C., 1984. Chlamydospore formation by *Trichoderma* spp. in natural substrates. Can. J. Microbiol. 30, 1–7.

Lewis, J.A., Papavizas, G.C., 1986. Reduced incidence of *Rhizoctonia* damping off of cotton seedlings with preparations of biocontrol fungi. Biol. Cult. Tests 1, 48.

Lewis, J.A., Papavizas, G.C., Lumsden, R.D., 1991. A new formulation system for the application of biocontrol fungi to soil. Biocontrol. Sci. Technol. 1, 59–69.

Lifshitz, R., Windham, M.T., Baker, R., 1986. Mechanism of biological control of preemergence damping-off by use of integrated chemical and biological controls. Phytopathology 69, 431–434.

Lorito, M., Harman, G.E., Hayes, C.K., Broadway, R.M., Tronsmo, A., Woo, S.L., Di-pietro, A., 1993. Chitinolytic enzymes produced by *Trichoderma harzianum*; antifungal activity of purified endochitinase and chitobiosidase. Phytopathology **83**, 302–307.

Lorito, M., Hayes, C.K., Di Pietro, A., Woo, S.L., Harman, G.E., 1994a. Purification, characterization and synergistic activity of a glucan 1,3- β-glucosidase and an *N*-acetyl-β-glucosaminidase from *Trichoderma harzianum*. Phytopathology 84, 398–405.

Lorito, M., Peterbauer, C., Hayes, C.K., Harman, G.E., 1994b. Synergistic interaction between fungal cell wall degrading enzymes and different antifungal compounds enhances inhibition of spore germination. Microbiology 140, 623–629.

Lupicka, K., Strof, W., Kubs, K., Skorupa, M., Wieczorek, P., Lejczak, B., Kafarski, P., 1997. The ability of soil–borne fungi to degrade organophosphanate carbon to phosphorus bounds. Appl. Microbiol. Biotechnol. 48, 549.

Mandels, M., Weber, J., Parizek, R., 1971. Enhanced cellulase production by a mutant of *Trichoderma viride*. Appl. Microbiol. 21, 152–154.

Mauch, F., Mauch-Mani, B., Boller, T., 1988. Antifungal hydrolases in pea tissue. II. Inhibition of fungal growth by combinations of chitinase and β-1,3-*glucanase*. *Plant Physiol.* 88, 936–942.

Montenecourt, B.S., Eveleigh, D.E., 1979. Production and characterization of high yielding cellulase mutants of *Trichoderma reesei*. In: *Tech. Assoc. Pulp Paper Ind. Annu. Meet*pp. 101–108.

Murali, M., Sudisha, J., Amruthesh, K.N., Ito, S.I., Shekar, Shetty H., 2013. Rhizosphere fungus *Penicillium chrysogenum* promotes growth and induces defence-related genes and downy mildew disease resistance in pearl millet. Plant Biol. 15, 111–118.

Muthamilan, M., Jeyarajan, R., 1996. Integrated management of *Sclerotium* root rot of groundnut involving *Trichoderma harzianum* Rhizobium and Carbendazim. Indian J. Mycol. Plant Pathol. 26, 204–209.

Nagaraju, A., Jogaiah, S., Mahadeva Murthy, S., Shin-ichi, I., 2012a. Seed priming with *Trichoderma harzianum* isolates enhances plant growth and induces resistance against *Plasmopara halstedii*, an incitant of sunflower downy mildew disease. *Aust. J.* Plant Path. 41, 609–620.

Nagaraju, A., Murali, M., Jogaiah, S., Amruthesh, K.N., Mahadeva Murthy, S., 2012b. Beneficial microbes promote plant growth and induce systemic resistance in sunflower against downy mildew disease caused by *Plasmopara halstedii*. Curr. Bot. 3, 12–18.

Nakkeeran, S., Doraisamy, S., 2001. Genetic improvement of *Trichoderma viride* and assessment of its efficacy against soil borne pathogens. In: National Seminar on Emerging Trends in Pests and Disease Management. Center for Plant Protection Studies. TNAU, Coimbatore, p. 130.

Nakkeeran, S., Jeyarajan, R., 1996. Exploitation of antagonistic potential of *Trichoderma* for field use. Paper presented in National Symposium on disease of plantation crops and their management at Institute of Agriculture. Sriniketan. .

Nakkeeran, S., Renukadevi, P., 1997. Seed borne microflora of pigeonpea and their management. Plant Dis. Res. 12, 103–107.

Nakkeeran, S., Gangadharan, Kousalya, Renukadevi, P., 1995. Efficacy of organic amendments and antagonists on root rot and wilt of pigeonpea. In: Proceedings of the National Symposium on Organic Farming. TNAU, Madurai, pp. 141–142.

Nakkeeran, S., Sankar, P., Jeyarajan, R., 1997. Standardization of storage conditions to increase the shelf life of *Trichoderma* formulations. J. Mycol. Plant Pathol. 27, 60–63.

Nandini, B., Puttaswamy, H., Prakash, H.S., Adhikari, S., Jogaiah, S., Geetha, N., 2020. Elicitation of novel trichogenic-lipid nanoemulsion signaling resistance against pearl millet downy mildew disease. Biomol. Ther. 10 (1), 25.

Nelson, E.E., 1982. Occurrence of *Trichoderma* in a Douglas-fir soil. Mycologia 74, 280–284.

Nicol, S., 1991. Life after death for empty shells. New Sci. 129, 46–48.

Ossanna, N., Mischke, S., 1990. Genetic transformation of the biocontrol fungus *Gliocladium virens* to benomyl resistance. Appl. Environ. Microbiol. 56 (10), 3052–3056.

Papavizas, G.C., 1985. *Trichoderma* and Gliocladium: biology, ecology and potential for biocontrol. Annu. Rev. Phytopathol. 23, 23.

Papavizas, G.C., Lewis, J.A., 1983. Physiological and biocontrol characteristics of stable mutants of *Trichoderma viride* resistant to MBC fungicides. Phytopathology 73 (3), 407–411.

Papavizas, G.C., Lewis, J.A., 1989. Effect of *Trichoderma* and *Gliocladium* on damping-off and blight of snapbean caused by *Sclerotium rolfsii*. Plant Pathol. 38, 277–286.

Papavizas, G.C., Dunn, M.T., Lewis, J.A., Beagle-Ristaino, J., 1984. Liquid fermentation technology for experimental production of biocontrol fungi. Phytopathology 74, 1171–1175.

Peng, G., Sutton, J.C., Kevan, P.G., 1992. Effectiveness of honey bees for applying the biocontrol agents *Gliocladium rosdeum* to strawberry flowers to suppress *Botrytis cinerea*. Can. J. Pathol. 14, 117–129.

Ranganathan, K., Sridar, R., Jeyarajan, R., 1995. Evaluation of gypsum as a carrier in the formulation of *Trichoderma viride*. J. Biol. Control 9, 61–62.

Ricard, J.L., 1981. Commercialization of *Trichoderma* based myco-fungicide. Some problems and solutions. Biocontrol News Inform. 2, 95–98.

Persoon, C.H., 1794. Disposito methodica fungorum in classes, ordines, Familias et genera. In: - Römer, J.J. (Ed.), Neues Magazin für Botanik. Ziegler und Söhne, Zürich, pp. 63–128.

Rifai, M.A., 1969. A revision of the genus *Trichoderma*. Mycol. Pap. 116, 1–116.

Rigout, J., Matsumura, F., 2002. Assessment of the rhizosphere competency and pentachlorophenol—metabolizing activity of a pesticide-degrading strain of *Trichoderma* harzianum introduced into the root zone of corn seedlings. J. Env. Sci. Health 37, 201–210.

Rodriguez-Kabana, R., Kelley, D., Curl, E.A., 1978. Proteolytic activity of *Trichoderma viride* with mixed culture in *Sclerotium rolfsii* in soil. Can. J. Microbiol. 24, 487–490.

Rossman, A.Y., 1996. Morphological and molecular perspectives on systematics of Hypocreales. Mycologia 88, 1–19.

Salt, G.H., 1978. The increasing interest in minor pathogens. In: Schippers, B., Gams, W. (Eds.), Soil-Borne Plant Pathogens. Academic Press, New York, pp. 289–312.

Samuels, G.J., Petrini, O., Kuhls, K., Lieckfeldt, E., Kubicek, C.P., 1998. The *Hypocrea schweinitzii* complex and *Trichoderma* sect. *Longibrachiatum*. Stud. Mycol. 41, 1–54.

Satapute, P., Milan, V.K., Shivakantkumar, S.A., Jogaiah, S., 2019. Influence of triazole pesticides on tillage soil microbial populations and metabolic changes. Sci. Total Environ. 651, 2334–2344.

Sivan, A., Chet, I., 1989. The possible role of competition between *Trichoderma harzianum* and *Fusarium oxysporium* on rhizosphere colonization. Phytopathology 79, 198–203.

Sivan, A., Elad, Y., Chet, I., 1984. Biological control effects of a new isolate of *Trichoderma harzianum* on *Pythium aphanidermatum*. Phytopathology 74, 498–501.

Sivasithamparam, K., Ghisalberti, E.L., 1998. Secondary metabolism in *Trichoderma* and *Gliocladium*. In: Kubicek, C.P., Harman, G.E. (Eds.), *Trichoderma* and *Gliocladium*. In: Basic Biology, Taxonomy and GeneticsVol. 1. Taylor and Francis Ltd., London, pp. 139–191.

Sriram, S., Raguchander, T., Babu, S., Nandakumar, R., Shanmugam, V.V., Balasubramanian, P., Samiyappan, R., 2000. Inactivation of rice sheath blight pathogen *Rhizoctonia solani*. Can. J. Microbiol. 46, 520–524.

Sudisha, J., Shekar Shetty, H., 2009. Anti-oomycete compounds from *Ganoderma appalantum*, a wood rot basidiomycete. Nat. Prod. Res. 23, 737–753.

Sudisha, J., Niranjana, S.R., Umesha, S., Prakash, H.S., Shekar Shetty, H., 2006. Transmission of seed-borne infection of muskmelon by *Didymella bryoniae* and effect of seed treatments on disease incidence and fruit yield. Biol. Control 37, 196–205.

Taylor, A.G., Harman, G.E., 1990. Concepts and technologies of selected seed treatments. Ann. Rev. Phytopathol. 28, 321–329.

Thomson, S.V., Hansen, D.R., Flint, K., Vandenberg, S.D., 1992. Dissemination of bacteria antagonistic to *Erwinia amylovora* by honey bees. Plant Dis. 76, 1052–1056.

Thrane, U., Poulsen, S.B., Nirenberg, H.I., Lieckfeldt, E., 2001. Identification of *Trichoderma* strains by image analysis of HPLC chromatograms. FEMS Microbiol. Lett. 203, 249–255.

Vidhya, R., 1995. Studies on biological control of mungbean root rot (*Macrophomina phaseolina* (Tessi) Goid by *Trichoderma viride* Pers. M.Sc. (Ag.) Thesis. TNAU. Coimbatore, India 143.

Vinale, F., Abadi, K., Ruocco, M., Marra, R., Scala, F., Zoina, A., Woo, S., Lorito, M., 2003. Remediation of pollution by using biological systems based on beneficial plant-micro-organ-ISMS interactions. J. Plant Pathol. **85**, 301.

Whipps, J.M., 1997. Developments in the biological control of soil-borne plant pathogens. Adv. Bot. Res. 26, 1–134.

Wilhitte, S.E., Lumsden, R.D., Skraney, D.C., 1994. Mutational analysis of gliotoxin production by the biological fungus *Glicladium virens* in relation to suppression of Pythium damping-off. Phytopathology 84, 816–821.

Windham, M.T., Elad, Y., Baker, R., 1986. A mechanism for increased plant growth induced by *Trichoderma* spp. Phytopathology 76, 518–521.

Xu, T., 1996. Advances of molecular biology of Trichoderma spp. Zhenjun Xuebao 15, 143.

Zhang, C.L., Druzhinina, I.S., Kubicek, P.C., Xu, T., 2005. *Trichoderma* biodiversity in China: evidence for a North to South distribution of species in East Asia. FEMS Microbiol. Lett. 251, 251–257.

Pathogenesis-related proteins: Role in plant defense

25

Veenu Joshi[a], *Neelu Joshi*[b], *Amber Vyas*[c], *and S.K. Jadhav*[d]
[a]Center for Basic Sciences, Pt. Ravishankar Shukla University, Raipur, Chhattisgarh, India, [b]School of Biotechnology & Bioinformatics, D.Y. Patil Deemed To Be University, Navi Mumbai, Maharashtra, India, [c]University Institute of Pharmacy, Pt. Ravishankar Shukla University, Raipur, Chhattisgarh, India, [d]School of Studies in Biotechnology, Pt. Ravishankar Shukla University, Raipur, Chhattisgarh, India

25.1 Introduction

Plants are sessile in nature; they cannot move around to avoid adverse environmental conditions. Thus, they are constantly exposed to infection and invasion by several biotic (bacteria, virus, fungi, insect, and herbivores) (Murali et al., 2013; Babu et al., 2015) as well as abiotic agents (drought, salinity) that affect their growth and survival (Cramer et al., 2011; Abdelrahman et al., 2017; Haider et al., 2018). Plants do not have an acquired immune system and specialized immune cells like animals to fight with the pathogens. But nature has bestowed them with the innate defense system, which has a much larger recognition range to combat and defend themselves from the rigors of nature (Joshi et al., 2019). This defense system of plants is highly complex called plant immunity (PI). PI involves a cascade of signaling pathways leading to the expression of a number of genes and production of several protein factors (Jogaiah et al., 2018).

There are two lines of immune response in plants, the first line of defense includes cutin, waxes, rigid lignin, and callose depositions on the cell walls, which prevents the invasion of pathogens (Van Etten et al., 1994; Osbourn, 1996). The second line is inducible defense that comprises the production of reactive oxygen species leading to hypersensitive response, synthesis of antimicrobial molecules, phytoalexins, and ultimately the production of pathogenesis-related (PR) proteins (Van Loon et al., 1994; Van Loon and Van Strien, 1999; Van Baarlen et al., 2007; Sudisha et al., 2012; Ali et al., 2018).

The plant responds at two different levels in the cell (Fig. 25.1). The first level response is at the cell surface with preformed structures like special plant receptors called pattern recognition receptors (PRR) that interacts with pathogen-associated molecular patterns resulting in the activation of pathogen-associated molecular pattern-triggered immunity (PTI) (Zipfel and Felix, 2005). Pathogen-associated molecular patterns (PAMP) are the nonself-pathogen molecules that are highly conserved representing the features of a whole class of that pathogen. These PAMPs majorly include outer structural components of the pathogens like flagellins of

Biocontrol Agents and Secondary Metabolites. https://doi.org/10.1016/B978-0-12-822919-4.00025-9

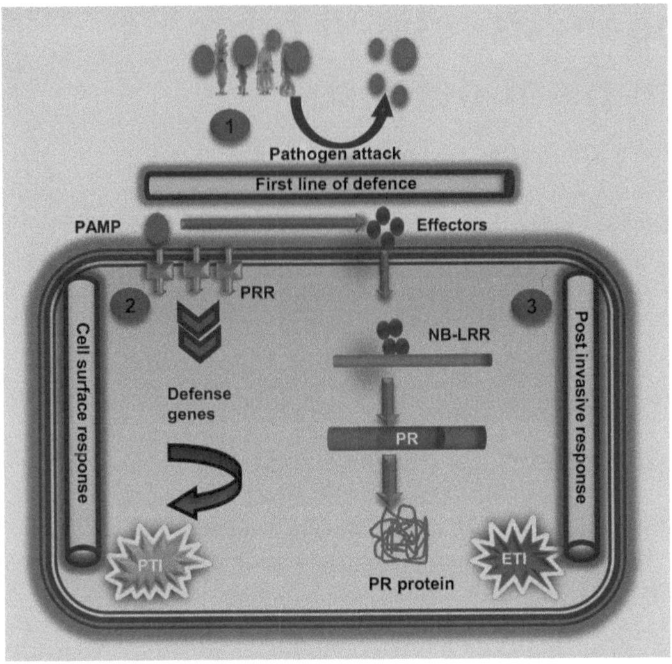

Fig. 25.1 Levels of defense response in plants. (1) First line of defense including layers of cutin and waxes that stop the pathogen attack (Alexander et al., 1993). (2) Cell surface response–PAMP recognized by PRR present in the membrane triggers a cascade of reactions to activate defense genes leading to PTI (Ali et al., 2018). (3) Postinvasive response-effector molecules released by pathogens upon entering into the cell cytoplasm interacts with R proteins like NB-LRR activates PR protein synthesis leading to ETI.

bacteria, lipopolysaccharides of membrane, and chitins and glucans of the fungal cell wall (Ali et al., 2018). The second level response operates at the post-invasive level where effector proteins which are the key elements produced by pathogens for their virulence (Sonah et al., 2016) triggers a particular group of resistance proteins inside the cell. These resistance proteins such as the nod-like receptor or nucleotide-binding leucine-rich repeat proteins results in the activation of induced defense response called as effector-triggered immunity (ETI) (Dangl and Jones, 2001).

However, the defense response efficiency of a plant against any pathogen differs from species to species (Jones and Dangl, 2006). But the plant response is generally mediated through two signaling pathways, salicylic acid (SA) and jasmonic acid (JA) pathway, that are activated upon encountering the biotrophic and the necrotrophic type of pathogens, respectively. SA pathway stimulates the transcription of nonexpressor pathogen-related gene 1 (NPR1) leading to the expression and accumulation of pathogenesis-related proteins locally as well as systemically resulting in systemic acquired resistance (SAR), providing resistance to a variety of pathogens (Sticher et al., 1997; Van Loon et al., 2006; Fu and Dong, 2013), whereas JA pathway induces transcription of the genes and synthesis of protein products accumulating locally providing only local acquired resistance (LAR) (Ali et al., 2017). It has also been reported that plant immune response is also regulated by plant growth regulators like ethylene (ET) and ABA (Shigenaga and Argueso, 2016). These hormones have antagonistic or synergistic interactions among themselves, thus generating complex signaling networks (Tsuda and Katagiri, 2010). SA has negative or positive interactions with JA and ET in regulating plant defenses against microbes (Robert-Seilaniantz et al., 2011).

Among the various inducible immune responses production of PR proteins is the most important component that prevents the spread of infection to healthy parts of the host (Jogaiah et al., 2013). In this chapter, we have reviewed the discovery, properties, classification, role, and mechanism of action of PR proteins (PR14–PR19).

25.2 PR proteins

Plants synthesize special proteins on exposure to biotic or abiotic stress conditions. These proteins that are induced by bacterial, viral, fungal pathogens, drought, salinity, heavy metals etc. as well as signaling molecules like salicylic acid and jasmonic acid are called as PR proteins.

25.2.1 Discovery

PR proteins were discovered in 1970 for the first time, where these were identified in leaf extracts of tobacco plants infected with tobacco mosaic virus (Van Loon, 1985; Van Loon, 1999). Five distinct bands of new protein components I, II, III, and IV with increasing electrophoretic mobilities were identified in inoculated plants that were absent in uninoculated plants. Since then, several researchers have worked on different plant species and identified the presence of many such proteins that are induced in pathological conditions. A different protein appeared upon TMV infection in *N. glutinosa* (Van Loon and Van Kammen, 1970; Kassanis et al., 1974), indicating that these new proteins were host specific. Aided by selective extraction methods (Van Loon, 1983; Parent and Asselin, 1984), it was established that not only viruses, but also fungi and bacteria were able to induce similar new protein components in various plant species, particularly in incompatible combinations resulting in hypersensitive necrosis. PR proteins are also known to be induced in plants during flowering (Fraser, 1981; Pierpoint, 1986), senescence (Camacho-Henriquez and Singer, 1982; Fraser, 1981), and plasmolysis (Wagih and Coutts, 1982). High levels of phytohormones (Antoniw et al., 1981; Memelink et al., 1987) or cell wall fractions from fungi (Somssich et al., 1986) can also induce their production in cell cultures (Kauffmann and Legrand, 1990). In order to unify the naming of these proteins, to describe the full set of novel proteins in the reaction of a plant to a pathogen, and to facilitate their study and explore their role in pathology and host defense, the term pathogenesis-related proteins (PRs) was introduced in 1980 to designate "proteins coded for by the host plant but induced only in pathological or related situations" (Antoniw et al., 1980; Jain and Khurana, 2018). However, some members of the different PR families have been found to be induced upon treatments with chemicals such as polyacrylic acid (Dumas et al., 1987), amino acid derivatives (Asselin et al., 1985), heavy metal salts (White, 1979; Nasser et al., 1990) such as aspirin and salicylic acid (Bol et al., 1990; Van de Rhee et al., 1990), and air pollutants (Ernst et al., 1992; Didierjean et al., 1993; Stintz et al., 1993). Some PRs are induced under osmotic stress or salt stress conditions (King et al., 1986; Grosset et al., 1990) with a possible involvement of abscisic acid (Nelson et al., 1992). PR production has also been shown to be developmentally regulated in healthy plants since high levels of some PRs have been reported in roots and senescent leaves (King et al., 1986; Castresana et al., 1990) and during flowering (Fraser, 1981).

25.2.2 Properties

PR proteins are generally monomers with few exceptions with low-molecular weight in the range of 6–43 kDa. These are grouped into two categories namely acidic and basic proteins based on their isoelectric points (Fig. 25.2). Acidic proteins are secreted in the extracellular space or apoplast, where these are stable at a very low pH and resistant to proteases enzymes (Van Loon et al., 1994; Dzhavakhiya et al., 2007). While basic proteins are usually transported to the vacuole by a specific signal sequence located at their carboxyl end (Takeda et al., 1991) and are also sensitive to proteolytic enzymes as in case of potato and tomato PR proteins (Dzhavakhiya et al., 2007). But apoplastic or vacuolar localization of acidic and basic proteins may vary in different PR protein families. Basic proteins are triggered by jasmonic acid and ethylene pathways, whereas acidic proteins are triggered by salicylic acid-mediated pathways (Jain and Khurana, 2018; Sinha et al., 2014; Xu et al., 1994). PRs are not limited to photosynthetic tissues but they have been also reported in all the plant organs like leaves, flowers, roots, stems, and seedlings (Van Loon, 1999).

25.2.3 Classification

On the basis of similarities in molecular weights, amino acid composition, serological properties, and confirmed by nucleotide sequencing of corresponding cDNAs, the 10 major acidic PRs of tobacco were grouped into five families, designated PR-1 to -5 (Van Loon, 1999; Linthorst, 1991). This categorization became a criterion to put PRs from other plants homologous to the tobacco PRs in the same family. Later many other pathogen-induced proteins reported in different plant varieties were also clustered into families based on the same criterion. A common nomenclature system was adopted in which the families are numbered and the different members within

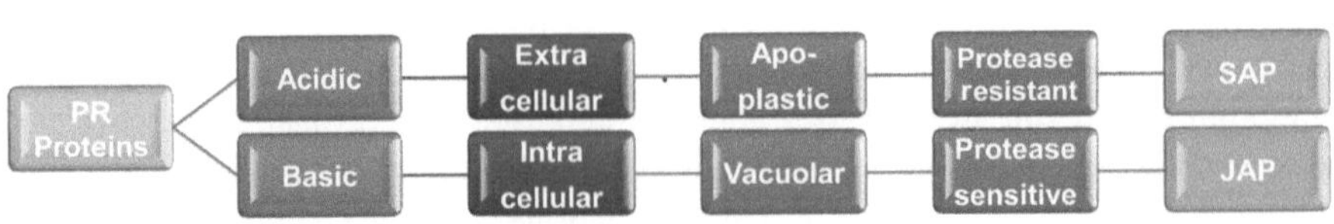

Fig. 25.2 Characteristics of PR proteins.

each family are assigned letters according to the order in which they are described and the lettering reflects the number of proteins of that family that had been identified within those plant species previous to their discovery (Van Loon, 1999). PRs are reported in several plant species like *Arabidopsis, Hordeum vulgare, Cucumis sativus, Nicotiana tabacum, Oryza sativa, Piper longum, Petroselinum crispum, Raphanus raphanistrum, Helianthus annuus, Pinus sylvestris, Triticum, Lactuca sativa, Zea mays, Solanum lycopersicum,* etc. Up till now, 17 confirmed and two putative PR families have been recognized and classified. The details of the different PR protein families, plant source, experimental material or tissue and the inducing factor used in the study in which they were first described along with their class, molecular weight, localization, and properties are presented in Table 25.1.

25.2.4 Role and mechanism of action in plant defense

The defensive action of PR proteins against pathogens can be accrued to a number of their characteristics like antibacterial, antifungal, membrane-permeabilizing ability, proteinase, peroxidase, ribonuclease, and lysozyme activities (Jeun and Buchenauer, 2001; Pushpalatha et al., 2013). Communication between overlapping stress response pathways and hormone signaling has been a challenge in unraveling the clear mechanism of action of PR proteins. However, studies on PR gene expression have given some insight on the role of PR proteins. Differential expression of PR genes in pathogen-resistant and susceptible plants has been reported in tomato plants infected with *Cladosporium fulvum* (Wubben et al., 1996), *Alternaria solani*-infected tomato (Lawrence et al., 2000), *Phytophthora infestans*-infected potato (Tonón et al., 2002), *Venturia inaequalis*-infected apple (Poupard et al., 2003), *Pseudomonas syringae*-infected grapevine (Robert et al., 2001), *Xanthomonas campestris* pv. *vesicatoria*, TMV-Po-infected hot pepper (Park et al., 2004a,b), *Venturia inaequalis*-infected apple (Gau et al., 2004), *F. circinalum*-infected *P.tecunumanii* (Visser et al., 2018), tomato spotted wilt tospovirus-infected tomato *Sw*-7 (Padmanabhan et al., 2019), *P. capsici*-infected melon (Wang et al., 2020), and whitefly-infected cassava (Irigoyen et al., 2020). In all these examples an upregulation of PR genes was recorded in a resistant variety suggesting their role in defense.

Another approach informing the role of PR proteins is through the production of transgenic plants. PR genes have emerged as a promising candidate to engineer biotic stress tolerance in plants (Sudisha et al., 2011; Ali et al., 2018). Such engineered tolerant plants can serve as a system to understand the signaling pathways involved in the activation and regulation of these genes. There are many evidences on the induction of resistance upon overexpression of PR genes in transgenic plants like tolerance to *Peronospora tabacina* and *Phytophthora parasitica* var. *nicotianae* by overexpression of PR1 in tobacco (Alexander et al., 1993), tolerance to *Rhizoctonia solani* and *Phytophthora citrophthora* upon overexpression of PR-5 in transgenic rice and orange plants, respectively (Datta et al., 1999; Fagoaga et al., 2001), and tolerance to *Phytophthora infestans* in potato overexpressing PR-2 and PR-3 (Bachmann et al., 1998). Similarly, transgenic carrot and tomato overexpressing PR-2 and PR-3 genes showed increased resistance to several fungal pathogens (Melchers et al., 1998).

Table 25.1 Types and properties of PR proteins.

Family	Source	Experimental material	Pathogen/ Inducer	Class	MW (kDa)	Localization	Properties	Reference
PR1	*Nicotiana tabacum*	Leaves	Tobacco Mosaic Virus (TMV)	PR-1a	15	Extracellular	Antifungal	Antoniw et al. (1980)
2	*Nicotiana tabacum*	Leaves	TMV	PR-2a	30	Intracellular	β-1,3-Glucanase	Antoniw et al. (1980)
3	*Nicotiana tabacum*	Leaves	TMV	P, Q	25–30	Extracellular+ Intracelluar	Chitinase type I, II,IV, V, VI, VII	Van Loon (1982)
4	*Nicotiana tabacum*	nf	TMV	R	15–20	Extracellular	Chitinase type I, II	Van Loon (1982)
5	*Nicotiana tabacum*	nf	TMV	S	25	Extracellular	Thaumatin-like proteins	Van Loon (1982)
6	*Solanum lycopersicum*	Leaves	Colorado Potato Beetles	Inhibitor I	8	nf	Proteinase-inhibitor	Green and Ryan (1972)
7	*Solanum lycopersicum*	Leaves	Citrus Exocortis viroid	P69	69	nf	Endoproteinase	Vera and Conejero (1988)
8	*Cucumis sativus*	nf	nf	Chitinase	28	nf	Chitinase type III	Metraux et al. (1988)
9	*Nicotiana tabacum*	Leaves	nf	Lignin-forming peroxidise	35	Cell wall	Peroxidase	Lagrimini et al. (1987)
10	*Petroselinum crispum*	Cell culture	*Phytophthora megasperma*	Pc PR1	17	Extracellular	"Ribonuclease-like"	Somssich et al. (1986)
11	*Nicotiana. tabacum*	Leaves	TMV	NtClass V chitinase	42	Vacuole	Chitinase typeI	Melchers et al. (1994)
12	*Raphanus raphanistrum*	Germinating seeds	*Pyrenophora tritici*-repentis, *Fusarium culmorum*, and *Pyricularia oryzae*	RsAFP3	5	Cell wall	Defensin	Terras et al. (1995)

13	*Arabidopsis thaliana*	Seedlings	*Fusarium oxysporum*, silver nitrate, and methyl jasmonate	Ath2.1	5	Membrane	Thionin	Epple et al. (1995)
14	*Hordeum vulgare*	Leaves	*Clavibacter michiganensis*	HvLTP4	9	Membrane	Lipid-transfer protein	Garcıa-Olmedo et al. (1995)
15	*Hordeum vulgare*	Leaves	*Blumeria graminis*	HvOxOa	20	Membrane	Oxalate oxidase	Zhang et al. (1995)
16	*Hordeum vulgare*	Leaves	*Blumeria graminis*-Powdery mildew fungus	HvOxOLP	25	Extracellular	Oxalate oxidase like	Wei et al. (1998)
17	*Nicotiana tabacum*	Leaves	TMV, mechanical wounding, and drought treatment	NtPRp27	27	nf	Antiviral and antifungal	Okushima et al. (2000)
18 Putative	Helianthus annus and *Lactuca sativa*	Leaves	Salicylic acid	HaCHOX, LsCHOX	60	nf	Carbohydrate oxidases	Custers et al. (2004)
19 Putative	*Pinus sylvestris* and *Macadamia integrifolia*	Root Nut Kernels	*Heterobasidion annosum*	SpAMP MiAMP1	8	nf	Antifungal, glucan binding	Asiegbu et al. (2003)

nf-data not found.

More investigations based on omics technology need to be carried out to further understand the molecular mechanism underlying PR protein-induced pathogen tolerance. In this review we have focused on the insight gained on the mode of action of the following classes of PR proteins.

25.2.4.1 PR-13

The best characterized activity for thionins is their ability to inhibit a broad spectrum of fungi and bacteria (Cammue et al., 1992) by increasing the membrane permeability (Stec, 2006). Synergistic effect of PR-13 and PR-14 against fungi has been reported, indicating their cooperation in membrane permeabilization (Molina et al., 1993). Induction of pathogen resistance upon overexpression of PR-13 gene has been reported in several plant species. For example, transgenic tomato overexpressed with the *A. thaliana* Thi2.1 gene showed resistance against bacterial wilt and *Fusarium wilt* (Chan et al., 2005). Also, overexpression of a hordothionin gene from barley in transgenic tobacco conferred resistance to *Pseudomonas syringae* (Carmona et al., 1993). Iwai et al. (2002) transferred an oat cell-wall-bound thionin gene in rice and noted enhanced resistance to bacterial diseases. Interestingly, Epple et al. (1997) described that overexpression of the endogenous thionin Thi2.1 gene enhanced resistance of *A. thaliana* against *Fusarium oxysporum*. All these reports inform the role of PR-13 in defense. However, no investigations on the mechanism underlying resistance have been conducted.

25.2.4.2 PR-14

This class of protein possesses antibacterial and antifungal properties (Zottich et al., 2011; Sun et al., 2008) and their protective role is attributed to their ability to permeabilize membranes (Regente et al., 2005) and participation in lipid catabolism.

Enhanced pathogen tolerance due to the overexpression of LTP genes in transgenic plants has been observed by several workers. For example, protection against *Pseudomonas syringae pv. tabaci* by expression of barley PR-14 gene in tobacco (Molina and Garcia-Olmedo, 1997) enhanced resistance against *Pseudomonas syringae* pv. tomato and *B. cinerea* in *A.thaliana* by the PR-14 gene (Jung et al., 2005) and tolerance against *B. cinerea* in *A.thaliana* by the expression of endogenous LTP-like genes (Chassot et al., 2007). Attempts were made to investigate the mode of action of LTPs and it was found that these are calmodulin-binding protein. Studies have shown that the calmodulin-binding site is conserved among LTPs, indicating the role of Ca^{2+} and calmodulin signaling in functions of LTPs (Wang et al., 2004). According to Maldonado et al. (2002), putative LTP (gene locus At5g48485) was found to be involved in SAR (systemic acquired resistance) in *A. thaliana*. They identified the DIR-1 gene encoding a putative apoplastic LTP and proposed its interaction with a lipid-derived molecule to promote long-distance signaling involved in SAR. In parallel, they discovered a dir1-1 mutant which failed to show SAR against virulent *Pseudomonas syringae* or *Peronospora parasitica*. The dir1-1 mutant exhibited wild-type resistance to *P. syringae*, so a direct effect of the knockout of the LTP gene on plant resistance was not shown.

25.2.4.3 PR-15 and PR-16

Germins (oxalate oxidase) and germin-like (oxalate oxidase like) proteins have been classified as PR-15 and PR-16, respectively. Oxalate oxidase was first isolated and characterized in *Hordeum vulgare* and *Triticum aestivum* and it shares sequence similarity with wheat germins. In Barley, six different oxalate oxidase and oxalate oxidase-like proteins have been characterized. These proteins are reported to possess heteropentameric structures that are mainly secreted into the extracellular spaces (Xu Bidney et al., 2003). Oxalate oxidase catalyzes the aerobic oxidation of oxalic acid into CO_2 and H_2O_2, the later in turn triggers signal transduction cascade which activates the plant defense mechanism (Mittler, 2002). Defensive action of these proteins became evident when constitutive expression of PR-15 in barley conferred enhanced resistance to oxalic acid-generating pathogens like *Sclerotinia sclerotiorum, Cristulariella pyramidalis,* and *Septoria musiva* (Singh et al., 2014). PR-16 proteins were identified in hot pepper infected with viral and bacterial pathogens (Park et al., 2004a). Seven germin-like protein (GLP) cDNA clones were isolated from *Vitis vinifera* (Godfrey et al., 2007). VvGLP3 was strongly induced by *Erysiphe necator* infection while VvGLP1 and VvGLP7 were induced by *B. cinerea* and *P. viticola* infection and not by *E. necator*. He et al. (2013) showed that transgenic *Colocasia esculenta* (Taro) transformed with oxalate oxidase gene gf2.8 from wheat confer increased resistance to Taro pathogen *Phytophthora colocasiae*. Another activity reported for PR-16 is superoxide dismutase (SOD)-type (Bernier and Berna, 2001).

25.2.4.4 PR-17

This class of protein is not much characterized compared with other PR proteins. However, investigations carried out by few workers have helped gain insight on its presence and role in plant defense.

Sequence analysis of PR-17 detected in barley showed similarity to the active sites of zinc metalloproteinase, exopeptidase 'aminopeptidase N' from eukaryotes and endopeptidase 'thermolysin' from bacteria (Christensen et al., 2002). This gives a hint toward proteinase-like activity of PR-17. They are supposed to be involved in cell wall degradation or signal transduction releasing signal molecules in a manner analogous to the release of certain carbohydrate elicitor. They may also exhibit antibiotic-like activity against an invading pathogen (Christensen et al., 2002).

The PR-17 protein has been found in tobacco, wheat, barley, and potato. Gorlach et al. (1996) reported benzothiadiazole-mediated induction of a PR member WCI-5 in wheat in response to fungal infection. NtPRp27 was detected in tobacco exposed to tobacco mosaic virus (Okushima et al., 2000). HvPR-17a and HvPR-17b were identified in barley upon infection with powdery mildew fungi *Blumeria graminis* f.sp. *hordei* (Christensen et al., 2002).

Comparison of pathogen-resistant and susceptible variety in few crops confirmed the defensive role of PR-17. For example, a resistant (var. Lince) variety of barley exposed to aphid *Rhopalosiphum padi* showed the upregulation of PR-17a compared with susceptible var. Golf (Gronberg, 2006). Elvira et al. (2008) carried out proteomic-based investigations in *Capsicum chinense* PI159236 plants harboring the L3 gene exposed to two strains of pepper mild mottle virus (PMMoV-S, PMMov-I, the Spanish and the Italian strain, respectively). The results showed the presence of PRp27 (PR-17) in addition to other PR proteins. Potato plant infected with *Phytophthora infestans* showed resistance and that was attributed to the expression of StPRp27(PR-17), a potato pathogenesis-related protein gene (Shi et al., 2011).

Application of transgenic technology provided more evidences on the defensive role of PR-17. For example, overexpression of PR17 protein in wheat (WCI-5) conferred resistance against powdery mildew fungi B. *graminis* f.sp. *tritici* (Schweizer et al., 1999).

25.2.4.5 PR-18 and PR-19

Two putative novel families of PR proteins have been discovered (Van Loon et al., 2006; Jaber et al., 2018). Sunflower and lettuce plants when treated with fungus and salicylic acid showed the presence of novel antimicrobial proteins PR-18. Enzymatic assay revealed them to be carbohydrate oxidases (Custers et al., 2004). Sooriyaarchchi et al. (2011) reported a new class of antimicrobial protein (PR-19) in *Pinus sylvestris*. Their proposed mode of action is through interaction with glucan, the major component of fungal cell wall.

25.3 Conclusion

In conclusion, PR proteins are of various types, which are classified into 17 confirmed and two putative families. These families differ in their chemistry, activation pathway, and function. They have been identified in various plant species, viz., tobacco, *A. thaliana*, potato, wheat, barley, and pepper.

Comparison of transcripts expressed in pathogen resistance and susceptible plants has led to the identification of differentially activated PR genes in tomato, potato, apple, melon, cassava and barley. Similarly, comparison at the proteome level helped in the identification of PR-17 in *Capsicum chinense*. There are reports on the activation of PR genes in plants upon exposure to pathogens and inducers together and that informed about the role of inducers and the signaling pathways involved in PR gene expression, for example, benzothiadiazole induced activation of PR-17 in wheat against fungal pathogens. In several plant species overexpression of various PR genes have resulted in the development of pathogen tolerance confirming their role in defense.

Despite these efforts, further investigations of natural and engineered pathogen-tolerant plants using the tools of functional genomics and in silico expression analysis needs to be carried out to improve the current understanding on the mechanism of action of PR proteins.

Acknowledgment

Authors are thankful to Center for Basic Sciences, School of Studies in Biotechnology & UIOP Pt. Ravishankar Shukla University Raipur, DST (New Delhi) for FIST scheme and DST Nanomission (DST/NM/NT/2018/20).

References

Abdelrahman, M., El-Sayed, M., Jogaiah, S., Burritt, D.J., Tran, L.-S.P., 2017. The "STAY-GREEN" trait and phytohormone signaling networks in plants under heat stress. Plant Cell Rep. 36, 1009–1025.

Alexander, D., Goodman, R.M., Gut-Rella, M., Glascock, C., Weymann, K., Friedrich, L., et al., 1993. Increased tolerance to two oomycete pathogens in transgenic tobacco expressing pathogenesis-related protein 1a. Proc. Natl. Acad. Sci. 90 (15), 7327–7331.

Ali, S., Mir, Z.A., Bhat, J.A., Chandrashekar, N., Papolu, P.K., Rawat, S., et al., 2017. Identification and comparative analysis of *Brassica juncea* pathogenesis-related genes in response to hormonal, biotic and abiotic stresses. Acta Physiol. Plant. 39, 1–15.

Ali, S., Ganaib, B.A., Kamili, A.N., Bhat, A.A., Mir, Z.A., Bhat, J.A., et al., 2018. Pathogenesis-related proteins and peptides as promising tools for engineering plants with multiple stress tolerance. Microbiol. Res. 212–213, 29–37.

Antoniw, J.F., Ritter, C.E., Pierpoint, W.S., Van Loon, L.C., 1980. Comparison of three pathogenesis-related proteins from plants of two cultivars of tobacco infected with TMV. J. Gen. Virol. 47, 79–87.

Antoniw, J.F., Kueh, J.S.H., Walkey, D.G.A., White, R.F., 1981. The presence of pathogenesis-related proteins in callus of Xanthi nc tobacco. Phytopathol. Z. I01, 179–184.

Asiegbu, F.O., Choi, W., Li, G., Nahalkova, J., Dean, R.A., 2003. Isolation of a novel antimicrobial peptide gene (Sp-AMP) homologue from *Pinus sylvestris* (Scots pine) following infection with the root rot fungus *Heterobasidion annosum*. FEMS Microbiol. Lett. 228 (1), 27–31.

Asselin, A., Grenier, J., Cote, F., 1985. Light-influenced extracellular accumulation of b (pathogenesis-related) proteins in Nicotiana green tissue induced by various chemicals or prolonged floating on water. Cart. J. Bot. 63, 1276–1283.

Babu, N.A., Jogaiah, S., Ito, S.-i., Nagaraj, A.K., Tran, L.-S.P., 2015. Improvement of growth, fruit weight and early blight disease protection of tomato plants by rhizosphere bacteria is correlated with their beneficial traits and induced biosynthesis of antioxidant peroxidase and polyphenol oxidase. Plant Sci. 231, 62–73.

Bachmann, D., Rezzonico, E., Retelska, D., Chetelat, A., Schaerer, S., Beffa, R., 1998. Improvement of potato resistance to *Phytophthora infestans* overexpressing antifungal hydrolases. In: 5th International workshop on pathogenesis-related proteins. Signaling pathways and biological activities, Aussois, France, March 29–April 2, 1998 (Abstract P-57).

Bernier, F., Berna, A., 2001. Germins and germin-like proteins: plant do-all proteins. But what do they do exactly? Plant Physiol. Biochem. 39, 545–554.

Bol, J.F., Linthorst, H.J.M., Comelissen, B.J.C., 1990. Plant pathogenesis-related proteins induced by virus infection. Phytopathology 28, 113–138.

Camacho-Henriquez, A., Singer, H.L., 1982. Analysis of acid extractable tomato leaf proteins after infection with a viroid, two viruses and a fungus and partial purification of the 'pathogenesis-related' protein. Arch. Virol. 74, 181–196.

Cammue, B.P., De Bolle, M.F., Terras, F.R.P., Van Damme, P., Rees, J., et al., 1992. Isolation and characterization of a novel class of plant antimicrobial peptides from *Mirabilis jalapa* L. Seeds. J. Biol. Chem. 267, 2228–2233.

Carmona, M.J., Molina, A., Fernandez, J.A., Lopez-Fando, J.J., Garcia-Olmedo, F., 1993. Expression of the alpha-thionin gene from barley in tobacco confers enhanced resistance to bacterial pathogens. Plant J. 3, 457–462.

Castresana, C., De Carvalho, F., Gheysen, G., Habets, M., Inz, C.D., Van Montagu, M., 1990. Tissue-specific andpathogen-induced regulation of a *Nicotiana plumhaginifolk* 1,3-glucanase gene. Plant Cell 2, 1131–1143.

Chan, Y.L., Prasad, V., Sanjaya, K., Chen, K.H., Liu, P.C., Chan, M.T., Cheng, C.P., 2005. Transgenic tomato plants expressing an *Arabidopsis thionin* (Thi2.1) driven by fruit-inactive promoter battle against phytopathogenic attack. Planta 221, 386–393.

Chassot, C., Nawrath, C., Mctraux, J.P., 2007. Cuticular defects lead to full immunity to a major plant pathogen. Plant J. 49, 972–980.

Christensen, A.B., Cho, B.H., Naesby, M., Gregersen, P.L., et al., 2002. The molecular characterization of two barley proteins establishes the novel PR-17 family of pathogenesis related proteins. Mol. Plant Pathol. 3, 135–144.

Cramer, G.R., Urano, K., Delrot, S., Pezzotti, M., Shinozaki, K., 2011. Effects of abiotic stress on plants: a systems biology perspective. BMC Plant Biol. 11–163.

Custers, J.H.H.V., Harrison, S.J., Sela-Buurlage, M.B., Van Deventer, E., Lageweg, W., et al., 2004. Isolation and characterization of a class of carbohydrate oxidases from higher plants, with a role in active defence. Plant J. 39, 147–160. https://doi.org/10.1111/j.1365-313X.2004.02117.

Dangl, J.L., Jones, J.D.G., 2001. Plant pathogens and integrated defence responses to infection. Nature 411, 826–833.

Datta, K., Velazhahan, R., Oliva, N., Ona, I., Mew, T., Khush, G.S., et al., 1999. Over-expression of the cloned rice thaumatin-like protein (PR-5) gene in transgenic rice plants enhances environmental friendly resistance to *Rhizoctonia solani* causing sheath blight disease. Theor. Appl. Genet. 98, 1138–1145. https://doi.org/10.1007/s001220051178.

Didierjean, L., Frendo, P., Nasser, W., Marivet, J., Genot, G., Margis-Pinheiro, M., et al., 1993. Plant genes induced by chemicals and pollutants. In: Fritig, B., Legrand, M. (Eds.), Mechanisms of Plant Defense Responses. Kluwer Academic Publishers, Dordrecht, pp. 276–285.

Dumas, E., Gianinazzi, S., Comu, A., 1987. Genetic aspects of polyacrylic-acid-induced resistance to tobacco mosaic virus and tobacco necrosis virus in Nicotiana plants. Plant Pathol. 36, 544–550.

Dzhavakhiya, V.G., Ozeretskovskaya, O.L., Zinovyeva, S.V., 2007. Immune response. In: Dyakov Yu, T., Dzhavakhiya, V.G., Korpela, T. (Eds.), Comprehensive and Molecular Phytopathology. Elsevier, pp. 265–314.

Elvira, M.I., Galdeano, M.M., Gilardi, P., Garcı'a-Luque, I., Serra, M.T., 2008. Proteomic analysis of pathogenesis-related proteins (PRs) induced by compatible and incompatible interactions of pepper mild mottle virus (PMMoV) in Capsicum chinense L3 plants. J. Exp. Bot. 59 (6), 1253–1265.

Epple, P., Apel, K., Bohlmann, H., 1995. An Arabidopsis thaliana thionin gene is inducible via a signal transduction pathway different from that for pathogenesis-related proteins. Plant Physiol. 109, 813–820.

Epple, P., Apel, K., Bohlmann, H., 1997. Overexpression of an endogenous thionin enhances resistance of Arabidopsis against *Fusarium oxysporum*. Plant Cell 9, 509–520.

Ernst, D., Schraudner, M., Langebartels, C., Sandermann Jr., H., 1992. Ozone induced changes of mRNA levels of p-1,3-glucanase, chitinase and 'pathogenesis-related' protein 1 bin tobacco plants. Plant Mol. Bio. 20, 673–682.

Fagoaga, C., Rodrigo, I., Conejero, V., et al., 2001. Increased tolerance to *Phytopthora citophthora* in transgenic ornge plants constitutively expressing a tomato pathogenesis related protein PR5. Mol. Breed. 7, 175–185.

Fraser, R.S.S., 1981. Evidence for the occurrence of the 'pathogenesis- related' proteins in leaves of healthy tobacco plants during flowering. Physiol. Plant Pathol. 19, 69–76.

Fu, Z.Q., Dong, X., 2013. Systemic acquired resistance: turning local infection into global defense. Annu. Rev. Plant Biol. 64, 839–863.

Garcıa-Olmedo, F., Molina, A., Segura, A., Moreno, M., 1995. The defensive role of non-specific lipid-transfer proteins in plants. Trends Microbiol. 3, 72–74.

Gau, A.E., Koutb, M., Piotrowski, M., Kloppstech, K., 2004. Accumulation of pathogenesis-related proteins in the apoplast of a susceptible cultivar of apple (*Malus domestica* cv. Elstar) after infection by *Venturia inaequalis* and constitutive expression of PR genes in the resistant cultivar Remo. Eur. J. Plant Pathol. 110 (7), 703–711.

Godfrey, D., Able, A.J., Dry, I.B., 2007. Induction of a grapevine germin-like protein (VvGLP3) gen is closely linked to the site of *Erysiphe necator* infection: a possible role in defense? Mol. Plant-Microbe Interact. 20, 1112.

Gorlach, J., Volrath, S., Knauf-Beiter, G., Hengy, G., Beckhove, U., Kogel, K.H., et al., 1996. Benzothiadiazole, a nove l class of inducers of systemic acquired resistance, activates gene expression and disease resistance in wheat. Plant Cell 8, 629–643.

Green, T.R., Ryan, C.A., 1972. Wound-induced proteinase inhibitor in plant leaves: a possible defense mechanism against insects. Science 175, 776–777.

Gronberg, N., 2006. Induction of pathogenesis-related genes, PR-17a and N-methyl transferase, in barley infested by the aphid *Rhopalosiphum padi*. Master thesis, Södertörns högskola, University College.

Grosset, J., Meyer, Y., Chattier, Y., Kauffmann, S., Legrand, M., Friti, B., 1990. Tobacco mesophyll protoplasts synthesize1,3- *if* glucanase, chitinases, and 'osmotins' during *in vitro* culture. Plant Physiol. 92, 520–527.

Haider, M.S., Kurjogi, M., Fiaz, M., Jogaiah, S., Wang, C., Fang, J., 2018. Drought stress revealed physiological, biochemical and gene-expressional variations in 'Yoshihime' peach (*Prunus persica* L) cultivar. J. Plant Interact. 13, 83–90.

He, X., Miyasaka, S.C., Fitch, M.M.M., Khuri, S., Zhu, Y.J., 2013. Taro (*Colocasia esculenta*) transformed with a wheat oxalate oxidase gene for improved resistance to Taro pathogen *Phytophthora colocasiae*. Hort. Sci. 48, 22–24.

Irigoyen, M.L., Garceau, D.C., Bohorquez-Chaux, A., et al., 2020. Genome-wide analyses of cassava pathogenesis-related (PR) gene families reveal core transcriptome responses to whitefly infestation, salicylic acid and jasmonic acid. BMC Genomics 21, 93.

Iwai, T., Kaku, H., Honkura, R., Nakamura, S., Ochiai, H., Sasaki, T., Ohashi, Y., 2002. Enhanced resistance to seed-transmitted bacterial diseases in transgenic rice plants over-producing an oat cell-wall-bound thionin. Mol. Plant-Microbe Interact. 15, 515–521.

Jaber, E., Kovalchuk, A., Raffaello, T., Keriö, S., Teeri, T., Asiegbu, F.O., 2018. A gene encoding scots pine antimicrobial protein Sp-AMP2 (PR-19) confers increased tolerance against Botrytis cinerea in transgenic tobacco. Forests 9, 10.

Jain, D., Khurana, J.P., 2018. Role of pathogenesis-related proteins in plant defense mechanism. In: Singh, A., Singh, I. (Eds.), Molecular Aspects of Plant-Pathogen Interaction. Springer, Singapore.

Jeun, Y.C.H., Buchenauer, H., 2001. Infection structures and localization of the pathogenesis-related protein AP 24 in leaves of tomato plants exhibiting systemic acquired resistance against *Phytophthora infestans* after pre-treatment with 3-aminobutyric acid or tobacco necrosis virus. J. Phytopathol. 149 (3–4), 141–154.

Jogaiah, S., Mostafa, A., Tran, L.S.P., Shin-ichi, I., 2013. Characterization of rhizosphere fungi that mediate resistance in tomato against bacterial wilt disease. J. Exp. Bot. 64, 3829–3842.

Jogaiah, S., Abdelrahman, M., Tran, L.-S.P., Ito, S.-I., 2018. Different mechanisms of *Trichoderma virens*-mediated resistance in tomato against Fusarium wilt involve the jasmonic and salicylic acid pathways. Mol. Plant Pathol. 19, 870–882.

Jones, J.D.G., Dangl, J.L., 2006. The plant immune system. Nature 444, 323–329.

Joshi, S.M., De Britto, S., Jogaiah, S., Ito, S., 2019. Mycogenic selenium nanoparticles as potential new generation broad spectrum antifungal molecules. Biomol. Ther. 9 (9), 419.

Jung, H.W., Kim, K.D., Hwang, B.K., 2005. Identification of pathogen-responsive regions in the promoter of a pepper lipid transfer protein gene (CALTPI) and the enhanced resistance of the CALTPI transgenic Arabidopsis against pathogen and environmental stresses. Planta 221, 361–373.

Kassanis, B., Gianinazzi, S., White, R.F., 1974. A possible explanation of the resistance of virus-infected tobacco to second infection. J. Gen. Virol. 23, 11.

Kauffmann, S., Legrand, M., 1990. Isolation and characterization of six pathogenesis-related (PR) proteins of Samsun NN tobacco. Plant Mol. Biol. 14, 381–390.

King, G.J., Husscy, C.E., Turner, V.A., 1986. A protein induced by NaCi in suspension cultures of *Nicotiana tabacum* accumulates in whole plant roots. Planta 186, 367–375.

Lagrimini, L.M., Burkhart, W., Moyer, M., Rothstein, S., 1987. Molecular cloning of complementary DNA encoding the lignin forming peroxidase from tobacco: molecular analysis and tissue-specific expression. Proc. Natl. Acad. Sci. U. S. A. 84, 7542–7546.

Lawrence, C.B., Singh, N.P., Qui, J., Gardner, R.G., Tuzun, S., 2000. Constitutive hydrolytic enzymes are associated with polygenic resistance of tomato to *Alternaria solani* and may function as an elicitor release mechanism. Physiol. Mol. Plant Pathol. 57 (5), 211–220.

Linthorst, H.J.M., 1991. Pathogenesis-related proteins of plants. Crit. Rev. Plant Sci. 10, 123.

Maldonado, A.M., Doerner, P., Dixon, R.A., Lamb, C.J., Cameron, R.K., 2002. A putative lipid transfer protein involved in systemic resistance signalling in Arabidopsis. Nature 419, 399–403.

Melchers, L.S., Potheker-de Groot, M.A., Van der Knaap, J.A., Ponstein, A.S., Sela-Buurlage, M.B., Bol, J.F., et al., 1994. A new class of tobacco chitinases homologous to bacterial exo-chitinases displays antifungal activity. Plant J. 5, 469–480.

Melchers, L.S., et al., 1998. The utility of PR genes to develop disease resistance in transgenic crops. In: Proceedings of the 5th International Workshop on Pathogenesis-Related Proteins. Signalling Pathways and Biological Activities.

Memelink, J., Hoge, J.H.C., Sehilperoort, R.A., 1987. Cytokininstress changes the developmental regulation of several defenee-related genes in tobacco. EMBO J. 6, 3579–3583.

Metraux, J.P., Streit, L., Staub, T., 1988. A pathogenesis-related protein in cucumber is a chitinase. Physiol. Mol. Plant Pathol. 33, 1–9.

Mittler, R., 2002. Oxidative stress, antioxidants and stress tolerance. Trends Plant Sci. 7, 405–410.

Molina, F., Garcia-Olmedo, 1997. Enhanced tolerance to bacterial pathogens caused by the transgenic expression of barley lipid transfer protein LTP2. Plant J. 12, 669–675.

Molina, A., Segura, F., Garcia-Olmedo, F., 1993. Lipid transfer proteins (nsLTPs) from barley and maize leaves are potent inhibitors of bacterial and fungal plant pathogens. FEBS Lett. 316, 119–122.

Murali, M., Jogaiah, S., Amruthesh, K.N., Shin-ichi, I., Shekar Shetty, H., 2013. Rhizosphere fungus *Penicillium chrysogenum* promotes growth and induces defense-related genes and downy mildew disease resistance in pearl millet. Plant Biol. 15, 111–118.

Nasser, W., De 'l' dpia, M., Burkard, G., 1990. Maize pathogenesis-related proteins: characterization and cellular distribution of' P-I,3glucutlitses and chitinases induced by brome mosaic virus infection or mercuric chloride treatment. Physiol Mel Pkrnt Putho!. 36, I–I4.

Nelson, D.E., Raghothama, K.G., Singh, N.K., Hasegawa, P.M., Bressan, R.A., 1992. Analysis of structure and transcripti'analactivation of an osmotin gene. Plant Mol. Biol. 19, 577–588.

Okushima, Y., Koizumi, N., Kusano, T., Sano, H., 2000. Secreted proteins of tobacco cultured BY2 cells: identification of a new member of pathogenesis-related proteins. Plant Mol. Biol. 42, 479–488.

Osbourn, A.E., 1996. Preformed antimicrobial compounds and plant defense against fungal attack. Plant Cell 8, 1821–1831.

Padmanabhan, C., Ma, Q., Shekasteband, R., et al., 2019. Comprehensive transcriptome analysis and functional characterization of PR-5 for its involvement in tomato *Sw*-7 resistance to tomato spotted wilt tospovirus. Sci. Rep. 9, 7673.

Parent, J.G., Asselin, A., 1984. Detection of pathogenesis-related proteins (PR or *b*) and of other proteins in the intercellular fluid of hypersensitive plants infected with tobacco mosaic virus. Can. J. Bot. 62, 564.

Park, C.J., An, J.M., Shin, Y.C., Kim, K.J., Lee, B.J., Paek, K.H., 2004a. Molecular characterization of pepper germin-like protein as the novel PR-16 family of pathogenesis-related proteins isolated during the resistance response to viral and bacterial infection. Planta 219, 797–806.

Park, C.J., Kim, K.J., Shin, R., Park, J.M., Shin, Y.C., Paek, K.H., 2004b. Pathogenesis-related protein 10 isolated from hot pepper functions as a ribonuclease in an antiviral pathway. Plant J. 37 (2), 186–198.

Pierpoint, W.S., 1986. The pathogenesis-related proteins of tobacco leaves. Phytochemistry 25, 1595–1601.

Poupard, P., Parisi, L., Campion, C., Ziadi, S., Simoneau, P., 2003. A wound-and ethephon-inducible PR-10 gene subclass from apple is differentially expressed during infection with a compatible and an incompatible race of *Venturia inaequalis*. Physiol. Mol. Plant Pathol. 62 (1), 3–12.

Pushpalatha, H.G., Jogaiah, S., Ashok, S., Geetha, N.P., Kini, K.R., Shekar Shetty, H., 2013. Association between accumulation of allene oxide synthase activity and development of resistance against downy mildew disease of pearl millet. Mol. Biol. Rep. 40, 6821–6829.

Regente, M.C., Giudici, A.M., Villalaín, J., de la Canal, L., 2005. The cytotoxic properties of a plant lipid transfer protein involve membrane permeabilization of target cells. Lett. Appl. Microbiol. 40 (3), 183–189.

Robert, N., Ferran, J., Breda, C., Coutos-Thevenot, P., Boulay, M., et al., 2001. Molecular characterization of the incompatible interaction of *Vitis vinifera* leaves with *Pseudomonas syringae* pv. pisi: expression of genes coding for stilbene synthase and class 10 PR protein. Eur. J. Plant Pathol. 107 (2), 249–261.

Robert-Seilaniantz, A., Grant, M., Jones, J.D.G., 2011. Hormone crosstalk in plant disease and defense: more than just jasmonate-salicylate antagonism. Annu. Rev. Phytopathol. 49 (1), 317–343. https://doi.org/10.1146/annurev-phyto-073009-114447.

Schweizer, P., Pokorny, J., Abderhalden, O., Dudler, R., 1999. A transient assay system for the functional assessment of defense-related genes in wheat. Mol. Plant-Microbe Interact. 12, 647–654.

Shi, X., Tian, Z., Liu, J., et al., 2011. A potato pathogenesis-related protein gene, StPRp27, contributes to race-nonspecific resistance against *Phytophthora infestans*. Mol. Biol. Rep. 39 (2), 1909–1916.

Shigenaga, A.M., Argueso, C.T., 2016. No hormone to rule them all: interactions of plant hormones during the responses of plants to pathogens. Semin. Cell Dev. Biol. 56, 174–189.

Singh, R., Tiwari, J.K., Sharma, V., Singh, B.P., Rawat, S., 2014. Role of pathogen related protein families in defence mechanism with potential role in applied biotechnology. Int. J. Adv. Res. 2 (8), 210–226.

Sinha, M., et al., 2014. Current overview of allergens of plant pathogenesis related protein families. Sci. World J., 543195.

Somssich, I.E., Schmelzer, E., Bollmann, J., Hahlbrock, K., 1986. Rapid activation by fungal elicitor of genes encoding "pathogenesis- related" proteins in cultured parsley cells. Proc. Natl. Acad. Sci. U. S. A. 83, 2427–2430.

Sonah, H., Zhang, X., Deshmukh, R.K., Borhan, M.H., Fernando, W.G., Bélanger, R.R., 2016. Comparative transcriptomic analysis of virulence factors in *Leptosphaeria maculans* during compatible and incompatible interactions with canola. Front. Plant Sci. 7, 1784.

Sooriyaarchchi, S., Jaber, E., Covarrubias, A.S., Abhayasekera, W., Asiegbu, F.O., Mowbray, S.L., 2011. Expression and β-glucan binding properties of scots pine (*Pinus sylvestris* L.) antimicrobial protein (Sp-AMP). Plant Mol. Biol. 77, 33–45. https://doi.org/10.1007/s11103-011-9791-z.

Stec, B., 2006. Plant thionins-the structural perspective. Cell. Mol. Life Sci. 63, 1370–1385.

Sticher, L., Mauch-Mani, B., Métraux, J.P., 1997. Systemic acquired resistance. Annu. Rev. Phytopathol. 35, 235–270.

Stintzi, A., Heitza, T., Prasadb, V., 1993. Plant 'pathogenesis-related' proteins and their role in defense against pathogens. Biochimie 75, 687–706.

Sudisha, J., Kumar, A., Amruthesh, K.N., Niranjana, S.R., Shekar Shetty, H., 2011. Elicitation of resistance and defense related enzymes by raw cow milk and amino acids in pearl millet against downy mildew disease caused by *Sclerospora graminicola*. Crop Prot. 30, 794–801.

Sudisha, J., Sharathchandra, R.G., Amruthesh, K.N., Kumar, A., Shetty, H.S., 2012. Pathogenesis related proteins in plant defense response. Plant Def. Biol. Control 379–403.

Sun, J.Y., Gaudet, D.A., Lu, Z.X., Frick, M., Puchalski, B., Laroche, A., 2008. Characterization and antifungal properties of wheat nonspecific lipid transfer proteins. Mol. Plant-Microbe Interact. 21 (3), 346–360.

Takeda, S., Sato, F., Ida, K., Yamada, Y., 1991. Nucleotide sequence of a cDNA for osmotin-like protein from cultured tobacco cells. Plant Physiol. 97, 844–846.

Terras, F.R., Eggermont, K., Kovaleva, V., Raikhel, N.V., Osborn, R.W., Kester, A., et al., 1995. Small cysteine-rich antifungal proteins from radish: their role in host defense. Plant Cell 7, 573–588.

Tonón, C., Guevara, G., Oliva, C., Daleo, G., 2002. Isolation of a potato acidic 39 kDa β-1, 3-glucanase with antifungal activity against *Phytophthora infestans* and analysis of its expression in potato cultivars differing in their degrees of field resistance. J. Phytopathol. 150 (4–5), 189–195.

Tsuda, K., Katagiri, F., 2010. Comparing signalling mechanisms engaged in pattern triggered and effector triggered immunity. Curr. Opin. Plant Biol. 13, 459–565.

Van Baarlen, P., Van Belkum, A., Summerbell, R.C., Crous, P.W., Bart, P., et al., 2007. Molecular mechanisms of pathogenicity: how do pathogenic microorganisms developcross-kingdom host jumps? FEMS Micro. Rev. 3, 239–277.

Van de Rhee, M.D., Van Kan, J.A.L., Gonzales-Jaen, M.T., Bol, J.F., 1990. Analysis of regulatory elements involved in the induction of two tobacco genes by salicylate treatment and virus infection. Plant Cell 2, 357–366.

Van Etten, H.D., Mansfield, J.W., Bailey, J.A., Farmer, E.E., 1994. Two classes of plant antibiotics: phytoalexins versus phytoanticipins. Plant Cell 6, 1191–1192.

Van Loon, L.C., 1982. Regulation of changes in proteins and enzymes associated with activeDefense against virus infection. In: Wood, R.K.S. (Ed.), Active Defense Mechanisms in Plants. Plenum Press, New York, pp. 247–273.

Van Loon, L.C., 1983. Mechanisms of resistance in virus-infected plants. In: Bailey, J.A., Deverall, B.J. (Eds.), The Dynamics of Host Defense. Academic Press, North Ryde, NSW, p. 123.

Van Loon, L.C., 1985. Pathogenesis-related proteins. Plant Mol. Biol. 4, 111–116.

Van Loon, L.C., 1999. Occurrence and properties of plant pathogenesis-related proteins. In: Datta, S.K., Muthukrishnan, S. (Eds.), Pathogenesis-Related Proteins in Plants. CRC Press LLC, Boca Raton, pp. 1–19.

Van Loon, L.C., Van Kammen, A., 1970. Polyacrylamide disc electrophoresis of the changes in protein constitution after infection with tobacco mosaic virus. Virology 40, 199.

Van Loon, L.C., Van Strien, E.A., 1999. The families of pathogenesis-related proteins, their activities, and comparative analysis of PR-1 type proteins. Physiol. Mol. Plant Pathol. 55, 85–297.

Van Loon, L.C., Pierpont, W.S., Boller, T., Conejero, V., 1994. Recommendations for naming plant pathogenesis-related proteins. Plant Mol. Biol. Rep. 12, 245–264.

Van Loon, L.C., Rep, M., Pieterse, C.M., 2006. Significance of inducible defense-related proteins in infected plants. Annu. Rev. Phytopathol. 44, 135–162.

Vera, P., Conejero, V., 1988. Pathogenesis-related proteins of tomato: p-69 as an alkaline endoproteinase. Plant Physiol. 87, 58–63.

Visser, E.A., Wegrzyn, J.L., Myburg, A.A., Naidoo, S., 2018. Defence transcriptome assembly and pathogenesis related gene family analysis in *Pinus tecunumanii* (low elevation). BMC Genomics 19, 632.

Wagih, E.E., Coutts, R.H.A., 1982. Comparison of virus elicited and other stresses on the soluble protein fraction of cucumber cotyledons. Phytopathol. Z. 104, 364–374.

Wang, S.Y., Wu, J.H., Ng, T.B., Ye, X.Y., Rao, P.F., 2004. A non-specific lipid transfer protein with antifungal and antibacterial activities from the mung bean. Peptides 25, 1235–1242.

Wang, P., Wu, H., Zhao, G., He, Y., Kong, W., Zhang, J., et al., 2020. Transcriptome analysis clarified genes involved in resistance to*Phytophthora capsici* in melon. PLoS One. 15(2), e0227284.

Wei, Y.D., Zhang, Z.G., Andersen, C.H., Schmelzer, E., Gregersen, P.L., Collinge, D.B., 1998. An epidermis/papilla-specific oxidase-like protein in the defence response of barley attacked by the powdery mildew fungus. Plant Mol. Biol. 36, 101–112.

White, R.F., 1979. Acetylsalicylic acid (aspirin) induces resistance to tobacco mosaic virus in tobacco. Virology 99, 410–412.

Wubben, J.P., Lawrence, C.B., DeWitt, P.J.G.M., 1996. Differential induction of chitinase and 1, 3-β-glucanase gene expression in tomato by *Cladosporium fulvum* and its race-specific elicitors. Physiol. Mol. Plant Pathol. 48 (2), 105–116.

Xu Bidney, D.L., Yalpani, N., Duvick, J.P., Crasta, O., Folker, O., Lu, G., 2003. Overexpression of a gene encoding H2O2 generating oxalate-oxidase evokes defence response in sunflower. Plant Physiol. 133, 170–181.

Xu, Y., et al., 1994. Plant defense genes are synergistically induced by ethylene and methyl jasmonate. Plant Cell 6 (8), 1077–1085.

Zhang, Z., Collinge, D.B., Thordal-Christensen, H., 1995. Germin-like oxalate oxidase, a H2O2-producing enzyme, accumulates in barley attacked by the powdery mildew fungus. Plant J. 8, 139–145.

Zipfel, C., Felix, G., 2005. Plants and animals: a different taste for microbes? Curr. Opin. Plant Biol. 8, 353–360.

Zottich, U., Da Cunha, M., Carvalho, A.O., Dias, G.B., 2011. Purification, biochemical characterization and antifungal activity of a new lipid transfer protein (LTP) from Coffea canephora seeds with α-amylase inhibitor properties. Biochim. Biophys. Acta 1810, 375–383.

Different mechanisms of signaling pathways for plant protection from diseases by fungi

26

Swapan Kumar Ghosh and Atanu Panja
Molecular Mycopathology Laboratory, Biocontrol Unit, PG Department of Botany, Ramakrishna Mission Vivekananda Centenary College (Autonomous), Kolkata, India

26.1 Introduction

Plant diseases in each year cause almost 10%–16% of the global harvest loss costing of an estimated US$220 billion. All crop pests (bacteria, fungi, nematodes, insects, and weeds) cause 42% preharvest loss and 10% postharvest loss (Fletcher et al., 2006). To save the crop loss, application of chemical fungicides and pesticides is currently the main strategy. But nonjudicial use of these chemicals is creating unintended negative effects on the environment (Ghosh and Pal, 2017; Ward et al., 2005; Satapute et al., 2019a,b). Including nontarget species and soil and water quality (Tjamos et al., 2013; Hossain et al., 2017) and the diversity of organisms including beneficial fungi, bacteria, and animals are vanishing from this green planet. At the same time, continuous use of pesticides over a long period results in developing resistance of the pest (Aktar et al., 2009; Sudisha et al., 2010) and also develop the heavy metal contamination (Satapute et al., 2019a,b). Recently reported that the chemical fungicides like benomyl, carbamates, and mancozeb are carcinogenic. Male (Goetz et al., 2006) and female mice (Rockett et al., 2006) are being affected in their reproductive systems by triazole fungicides (Satapute et al., 2019a,b). Now we carry a body burden of persistent pesticides, many of which are linked to chronic health effects (Schaefer et al., 2004; PAN, 2007). Some pesticides have been reported to cause some human cancers, e.g., leukemia and non-Hodgkin's lymphoma in childhood (Meinert et al., 2000), breast cancer (Abdalla et al., 2003), and colorectal cancer (Lee et al., 2007). Hence to ban these chemical pesticides or to reduce their application, alternative approaches are being searched by scientists. Biological control of plant diseases by beneficial microbes may be an alternative to save the crop loss and keep the environment unpolluted. Recent findings of the potentiality of fungi and fungal products to manage different diseases and pest of crops indicated that fungi can be used as the substitute for synthetic fungicides (Murali et al., 2013). Fungi play a major role in biocontrol of diseases in the plant and have contributed to the plant protection against diseases called fungal biocontrol agents (FBCA) (Jogaiah et al., 2013). Near about 75,000 species of fungi exist in soil of the world (Finley, 2007) and several genera of fungi, e.g., *Trichoderma*, *Penicillium*, *Aspergillus*, *Fusarium*, mycorrhizal fungi (MF),

Biocontrol Agents and Secondary Metabolites. https://doi.org/10.1016/B978-0-12-822919-4.00026-0

yeast, etc., have been widely studied and reported as highly potent for killing or arresting plant pathogens and they are also able to survive by tolerating several environmental hazardous chemical groups (Ghosh et al., 2013; Morley and Gadd, 1995; Nagaraju et al., 2012; Murali et al., 2013; Nandini et al., 2020). As well as killing plant pathogens, fungi induce the plant immunity despite their own immune system. There are different physical and chemical barriers inside the plant body, which are responsible for immunity like lignin and callose synthesis, antimicrobial compound production, induction of specific signaling pathways (Kachroo and Kachroo, 2009; Pushpalatha et al., 2011). The immune system is further divided into two parts such as primary immune and secondary immune system. With the primary immune system, plants recognize microbe-associated molecular patterns of potential pathogens through pattern recognition receptors that mediate a basal defense response which is also called pattern-triggered immunity (de Wit, 2007). Secondary immunity involves effectors triggered response. These effectors trigger pathway governed by R gene (resistance gene) and its product which is a species-specific response (Nürnberger et al., 2004). Another mode of immune system is induced systematic response (van Loon et al., 2006) which is activated by rhizospheric nonpathogenic fungus (Khan et al., 2010; Jogaiah et al., 2018). For identification of signaling pathways regarding plant disease response, various genetic approaches have been targeted. Out of them, the approach of resistant gene mutation is widely used. In the case of *Arabidopsis* the different disease resistance genes like ndr1, eds1-5, pbs1, pbs2, have been identified through mutation approach. A maximum number of genes are involved in multiple disease resistance pathways. Another mutant-like Arabidopsis pad (phytoalexin accumulation deficient) loci pad1-4 have been recognized for a defect in pathogen-induced accumulation of phytoalexin, camalexin, moreover, another group of mutants like npr1 (nonexpression of PR) has been identified as a result of defects in salicylic acid signaling pathways. After a number of research regarding the molecular approach of disease resistance, different genes have been identified which are involved in upstream and downstream signaling pathways. Further research work is carried on due to the finding of the mammalian disease-resistant gene homolog like Bcl-2, Bcl-$x_{L,}$ which are involved in plant disease response (Dickinson, 2003).

Peptaibol, linear peptide antibiotic consists of 5–20 amino acid residues originated by non-ribosomal peptide synthetase activity from *Trichoderma,* is now a hot topic of research. *T. viride* secrets Alamethicin, a 20 mer peptaibol which elicits JA and SA biosynthesis in lima bean (Engelberth et al., 2001), similarly 18 mer peptaibols generating from *T. virens* elicit systemic defenses against *Pseudomonas syringae* pv. *lachrymans* in cucumber against (Viterbo et al., 2007). Multiple defense signaling pathways were found to be involved in the protection of tobacco plants against TMV (tobacco mosaic virus) by *Trichoderma* peptaibols (Luo et al., 2010). Some fungi such as *Trichoderma* secrete certain secondary metabolites exhibit, at the high dose as an antimicrobial but act as MAMPs (microbe-associated molecular patterns) and as auxin-like compounds at low concentrations (Abdelrahman and Jogaiah, 2019). The first recognized *Trichoderma* MAMP was identified as an ET inducing xylanase (Xyn2/Eix), produced by *Trichoderma* as a potent elicitor of plant defense responses in specific tobacco and tomato cultivars (Rotblat et al., 2002; Jogaiah et al., 2013).

The metabolites like 6-pentyl-a-pyrone, harzianolide, and harzianopyridone at 1 ppm activate plant defense mechanisms in pea, tomato, and canola (Vinale et al., 2008).

In this book chapter, two-way interaction such as FBCA (fungal biocontrol agents) plant pathogens and FBCA-plant for protection, and some of the signaling pathways that are to be induced through resistance gene recognition of pathogen elicitors or others, and also signaling pathways triggered by fungi for plant defense will be discussed.

26.2 Plant defense mechanism by utilization of fungi

26.2.1 Plant protection mechanisms by FBCAs (fungal biocontrol agents) against pathogens

Various fungal biocontrol agents against different nature of plant pathogens/pests (bacteria, fungi virus, insect, etc.) for disease management in vitro and agrifields have been documented by several workers (Murali et al., 2013; Babu et al., 2015). Some of the important FBCAs are *Trichoderma* spp., *Gliocladium*, *Beuveria* spp. *Penicillium* spp., *Aspergillus* sp., *Metarhizium anisopliae, Pythium oligandrum, Verticillium lecanii, Aureobasidium pullulans,* etc. Out of them, *Trichoderma* are excellent for this purpose as they have a better capacity to control the phytopathogens in broad range bacteria, virus, ascomycetous, basidiomycetous, deuteromycetous fungi, soilborne as well as airborne pathogens (Monte, 2001; Harman et al., 2004; Jogaiah et al., 2018; Joshi et al., 2019) in singly or in combination with moderate-to-low dose fungicide formula as they are compatible with fungicides and they are very efficient in acidic soils. For instances, strains of *T. virens* is compatible with metalaxyl against *Pythium ultimum* infecting cotton (Chet et al., 1997) and Fusarium wilt (Jogaiah et al., 2018), *T. harzianum* with captan against *Verticillium dahliae* infecting potato (Chet and Inbar, 1994), *T. virens* with thiram against *Rhizoctonia solani* infecting tobacco, and others (Chet et al., 1997). The mechanism of BCA-mediated control of plant pathogen or pest or interactions are broadly categorized as by indirect mechanism or direct mechanism. Under indirect are competition, antibiosis, and inactivation of pathogens' enzymes by BCA. Induction of plant resistance by BCA also is under an indirect mechanism for pathogen elimination; it will be discussed in a separate section. The direct mechanism includes mycoparasitism or hyperparasitism and hypovirulence. They have summarized below.

26.2.1.1 Competition

Win of BCA from the competition in the rhizospheric zone against other microbes for root colonization is important because a biocontrol agent cannot compete for space and nutrients if it is unable to grow in the rhizosphere. *Trichoderma* species, either added to the soil or applied as seed treatments, grow readily along with the developing root system of the treated plant. Efficient BCA *Trichoderma* has high competitive saprophytic ability (CSA) and quickly occupies the dead organic matter in crop zone, and root zone as a result plant pathogen get starved and eliminated. The most successful biological control of root rot of pine caused by *Heterbasidiun annosus* (*Fomes annosus*) with saprophytic or weak pathogenic BCA *Peniophora gigantea* as shown by (Rishbeth, 1963) in England. This BCA has high CSA and grows very rapidly and colonized the left out stumps of pine which are the primary source of the pathogenic fungus. As a result, *H. annosus* get starved and cannot spread in fresh pine stem.

Another thing is that FBCA produces Fe chelator which is known as a siderophore. By siderophore, BCAs scavenge Fe element near crop zone in the soil. So subsequence Fe starvation is another cause for elimination of crop pathogen (Chet et al., 1997). Eisendle et al. (2004) recorded that in *Aspergillus fumigatus* and *Aspergillus nidulans*, siderophore biosynthesis is negatively regulated by carbon source. In *Ustilago maydis*, gene products related to iron uptake affect the development of plant disease (McIntyre et al., 2004). Some *Trichoderma* BCAs produce highly efficient siderophores that chelate iron and stop the growth of other fungi (Chet et al., 1997; Jogaiah et al., 2013). For this reason, soil composition influences the biocontrol effectiveness of *Pythium* by *Trichoderma* according to iron availability. In addition, *T. harzianum* T35 controls *Fusarium oxysporum* by competing for both rhizosphere colonization and nutrients, with biocontrol becoming more effective as the nutrient concentration decreases (Tjamos et al., 2013). Competition has proved to be particularly important for the biocontrol of phytopathogens such as *Botrytis cinerea*, the main pathogenic fungi. In our experiment (Ghosh and Santra, 2016), we tried to record the capacity of colonization and dominancy over pathogen (*Penicillium digitatum*) of *Candida famata* on the fruit of orange by SEM (Scanning Electron microscopy) study where we inoculated both *P. digitatum* and *Candida famata* yeast at the same time in injured fruit surface. SEM (Fig. 26.1) showed that within 24 h, cells of *C. famata* quickly multiplied and colonized on the infected zone and intermingled with the hyphae of *P. digitatum*. The cells

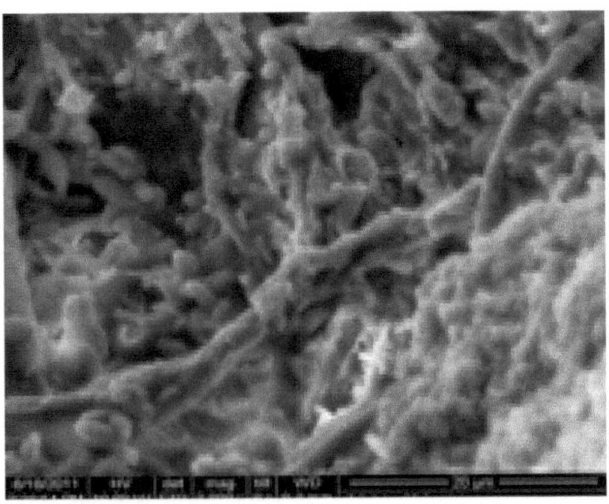

Fig. 26.1 Scanning electron microscopy (SEM) shows quick multiplication of *C. famata* on inoculated zone of citrus fruit and adhere to mycelium of *P. digitatum* after 24 h inoculation (Ghosh and Santra, 2016).

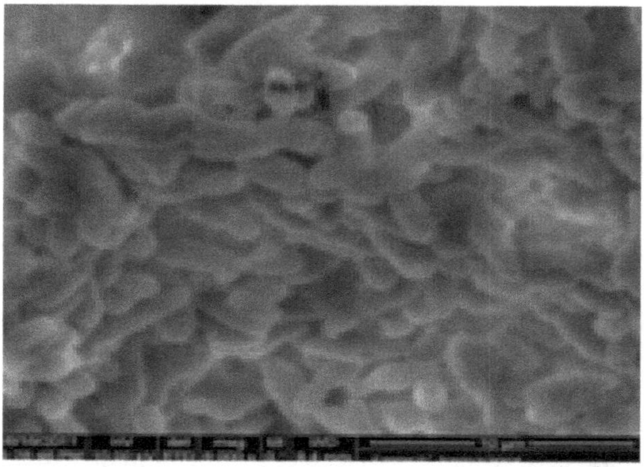

Fig. 26.2 Scanning electron microscopy (SEM) shows quick multiplication of *C. famata* and over-dominant over *P. digitatum* on inoculated zone of citrus fruit and no mycelium of pathogen after 48 h inoculation (Ghosh and Santra, 2016).

of *C. famata* adhered on the cell wall of the pathogen. SEM (Fig. 26.2) showed that after 48 h, *C. famata* highly multiplied, colonized, and dominated the zone and hyphae of *P. digitatum* are not seen.

26.2.1.2 Antibiosis

Secretion of chemical compounds from FBCA which inhibits plant pathogens is another mechanism. The phenomenon of inhibition of pathogens by these chemicals is known as antibiosis. The chemical secreted by BCA maybe antibiotics, toxins, enzymes, etc. We first came to know that *T. lignorum* produce toxic metabolite from the work of Weindling (1934). Later this fungus was identified as *G. frimbiatum* and the metabolite was named as gliotoxin. Antibiosis occurs during interactions involving low molecular weight diffusible compounds or antibiotics produced by *Trichoderma* strains that inhibit the growth of other microorganisms (Howell, 1998). Most *Trichoderma* strains secret metabolites and they are classified into three categories such as (i) volatile substances, e.g., 6-pentyl-α-pyrone (6-PP) and isocyanide derivates; (ii) water-soluble compounds, i.e., heptelidic acid or koningic acid; and (iii) peptaibols, which are linear oligopeptides of 12–22 amino acids rich in α-aminoisobutyric, *N*-acetylated at the *N*-terminus and containing amino alcohol at the *C*-terminus (Ghisalberti and Sivasithamparam, 1991). Among these metabolites, some are harmonic acid, tricholin, peptaibols, alamethicins, antibiotics, 6-pentyl-a-pyrone, viridian, massoilactone, glisoprenins, gliovirin, heptelidic acid, and others (Vey et al., 2001). Glucanases and or chitinases enzymes produced by the BCAs are responsible for killing or suppression of the plant pathogens (Jogaiah et al., 2013). These enzymes act by breaking down the polysaccharides, chitin, and glucans that are responsible for the rigidity of fungal cell walls, thereby, destroying cell wall integrity. Metcalf and Wilson (2001) showed the colonization of onion roots by *T. koningii* (Tr5) and protection from *Sclerotium cepivorum.*

Hyphae of the BCA penetrated infected epidermal and cortical tissue of the root to destroy the hyphae of the pathogen, without damaging plant tissue. Actually, here *T. koningii* produced endo- and exo-chitinases by which it killed the pathogen. Baek et al. (1999) disrupted or overexpressed the gene coding for chitinase (cht42) in *T. virens* (Gv29-8). In vitro studies have revealed that purified endochitinase, chitobiosidase, n-acetyl-β-glucosidase, and glucan 1,3-β-glucosidase and combinations of them greatly suppressed the spore germination and germ tube elongation in nine different fungal species (Lorito et al., 1993, 1994; Di Pietro et al., 1993). *T. harzianum* TM transformants overexpressing chit36 chitinase inhibited *F. oxysporum* and *S. rolfsii* more strongly than the wild type. Moreover, culture filtrates inhibited the germination of *B. cinerea* almost completely (Viterbo et al., 2001). HCN (hydrogen cyanide) production by biocontrol agents, during the biological control of pathogens, has been reported by many authors (Laville et al., 1998) and a positive correlation between HCN production in vitro and plant protection has been reported (Ramette et al., 2003). In our recent work (Ghosh et al., 2018) it was shown that among the five fungal antagonists, only *T. harzianum* (Fig. 26.3) produced HCN, whereas *T. viride*, *T. asperellum*, *T. longibrachiatum*, and *B. bassiana* were unable to produce HCN. This HCN was able to inhibit the growth of *A. brassica* in dual reverse culture (Fig. 26.4). The direct inhibition of fungi by HCN was thought to be one of the main mechanisms of action (Voisard et al., 1989).

26.2.1.3 Mycoparasitism or hyperparasitism

One of the most salient characteristics of FBCA is their ability to parasitize other microbes or fungi (Fig. 26.5). *Trichoderma* sp. may exert direct biocontrol by parasitizing a range of fungi detecting other fungi and growing toward them.

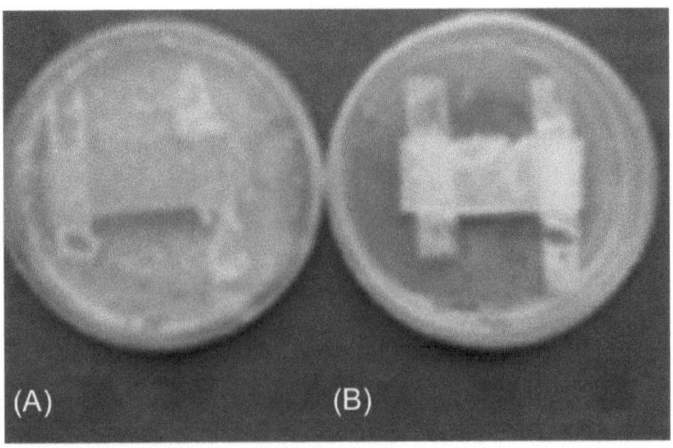

Fig. 26.3 HCN production by *T. harzianum* in sealed Petri plate pair. Panel (A) contains culture of *T. harzianum* and picric acid dipped sterile blotting paper which turned to orange from yellow indicating production of HCN by this BCA. Panel (B) contains no fungal culture but has picric acid dipped sterile blotting paper which remained yellow indicating no HCN (control plate) (Ghosh et al., 2018).

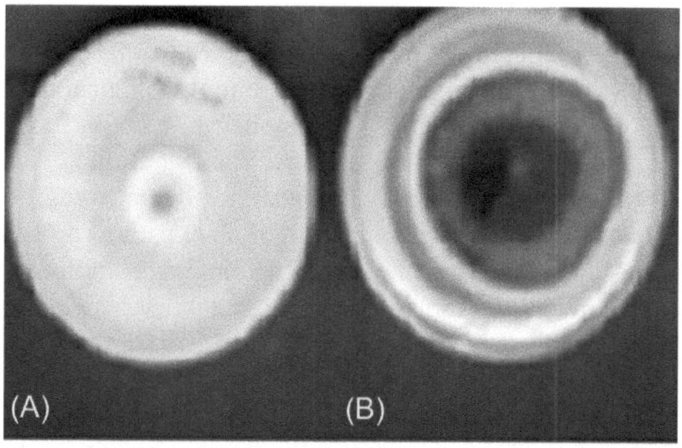

Fig. 26.4 Growth inhibition of pathogen by *Trichoderma*. (A) Sealed Petri plate pair contains culture of *T. harzianum* at lower and pathogen *A. brassisea* on upper plate indicating very poor or no growth of pathogen. Panel (B) contains no *T. harzianum* on lower plate but profuse growth of *A. brassisea* on upper plate (control) (Ghosh et al., 2018).

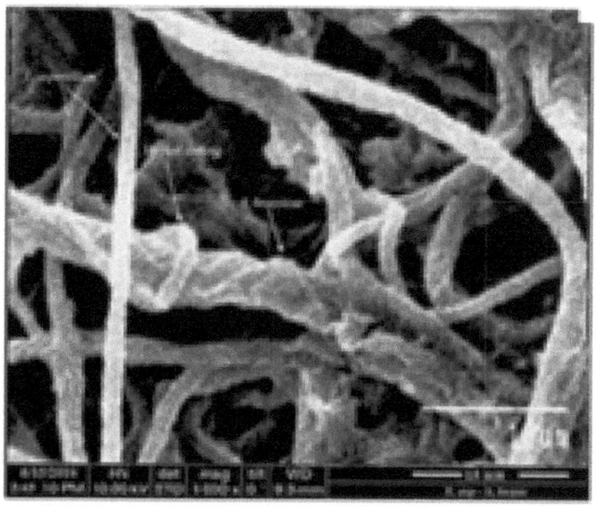

Fig. 26.5 Hyperparasitism nature of *Trichoderma* on pathogen. Scanning electron microscopy (SEM) shows *T. asperellum* hyphae coils around the pathogen *A brassicae* (Ghosh et al., 2018).

Weindling (1934) documented the mycoparasitism of *R. solani* hyphae by the hyphae of the biocontrol agent, many instances of successful biocontrol with *Trichoderma* species have been ascribed to the mechanisms of mycoparasitism and antibiosis. *Trichoderma* attaches to the pathogen with cell wall carbohydrates that bind to pathogen lectins. Once *Trichoderma* is attached, it coils around the pathogen and forms the appressoria. The step consists of the production of cell wall degrading enzymes and peptaibols (Howell, 2003) which facilitate both the entry of *Trichoderma* hypha into the lumen of the parasitized fungus and the assimilation of the cell wall content. *Trichoderma* sp. reacts violently with hyphae of the *Fusarium* species. The hyphae of *Trichoderma* sp. when near to pathogen induce morphological deformities in the host hyphae. Many a time bursting of hyphae and vacuolation has been observed (Komatsu, 1968; Gao et al., 2001). The classical work of Deacon (2006) and Laing and Deacon (1991) mycoparasitism by *Pythium oligandrum* has been studied in details by video microscopy where a hyphal tip of plant pathogen of *B. cinerea* contacted a hyphae *P. oligandrum* and then lysed within 3 min, spilling the contents of *Botrytis* hyphae and in addition, granulation, coagulation, disintegration, and finally lysis of the pathogen occurs (Lim and Teh, 1990; Elad et al., 1980; Nigam et al., 1997; Gao et al., 2001). Electron microscope observations have led to the hypothesis that during the interaction of *Trichoderma* spp. with either *Sclerotium rolfsii* or *Rhizoctonia solani*, the host cell walls are enzymatically digested by the parasite (Cherif and Benhamou, 1990; Elad et al., 1983a,b). The support for this hypothesis includes the finding that *Trichoderma* spp. produce extracellular β-(1,3)-glucanases, chitinases, lipases, and proteases when they are grown on cell walls of pathogenic fungi. In addition, several pieces of evidence have shown that the production of some lytic enzymes is induced during the parasitic interaction between *Trichoderma* spp. and some pathogenic fungi. The most interesting individual enzyme of the complex is the 42-kDa endochitinase (Ech42), which can hydrolyze in vitro *Botrytis cinerea* cell walls and inhibits spore germination and germ tube elongation of various fungi (Schirmböck et al., 1994). The corresponding gene (*ech42*) is strongly induced during fungus interactions and when the fungus is grown in the presence of autoclaved mycelia of several fungi or with colloidal chitin as the sole carbon source. Ech42 expression is repressed by glucose and is increased by exposure to light and by nutritional stress conditions (Carsolio et al., 1994; García et al., 1994). Later the role of the *Trichoderma harzianum* endochitinase (Ech42) in mycoparasitism was studied by genetically manipulating the gene that encodes Ech42 (Carsolio et al., 1999). In our recent work, an SEM (Fig. 26.5) study confirmed the hyperparasitic nature of *T. asperellum* through hyphal coiling around *Alternaria brassicae*, leaf spot pathogen of *Aloe vera* (Ghosh et al., 2018) and vacuolation and distortion of *B. bassiana* hyphae after coming in contact with *T. viride* T12 hyphae (Banerjee et al., 2016). Although at least 750 species of fungi are known to be entomopathogenic (Copping, 2004), and among them, several were used as BCA against phytopathogens. *Metarhizium anisopliae* and *Beauveria bassiana* have been mainly developed for commercial applications. Recently, it has been first reported by us that *Trichoderma longibrachiatum* can parasitize shoot and fruit borer (*Leucinodes orbonalis*) pest and also its ability to satisfactorily control the pest amount in field trial has been documented. It penetrates the pest body through ostiole of the body surface and also blocks the ostiolesas shown in SEM study (Fig. 26.6) (Ghosh and Pal, 2016).

Mycoparasitism of *Trichoderma* is regulated by cAMP signaling pathway and MAPkinase pathway as documented from *T. atroviride* IMI206040 where, Tga1 and Tga3, two G-protein α-subunits, monitor coiling (Rocha-Ramírez et al., 2002; McIntyre et al., 2004; Omero et al., 1999, 2004) of hyphae around pathogen and

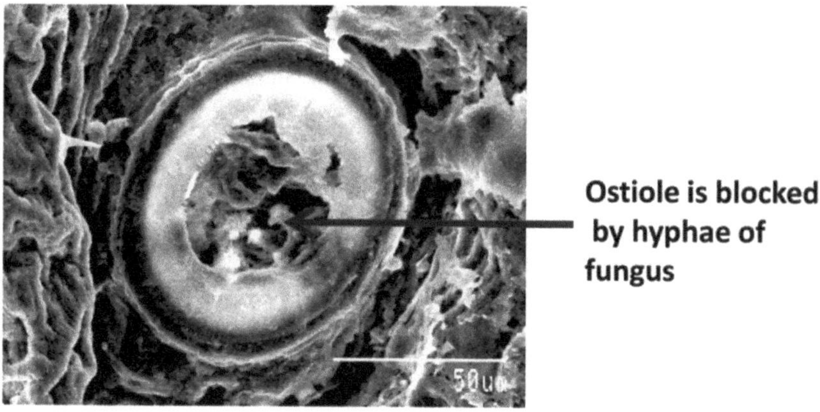

Fig. 26.6 *T. longibrachiatum* fungal hyphae plugged the ostiole of shoot and root borer of brinjal.

Tga1 also regulates the secretion of lytic enzymes and biosynthesis of antifungal chemicals (Reithner et al., 2005). Zeilinger et al. (2016) added that Tga3 triggers the synthesis of cell wall degrading enzymes. G-α gene (*tga1*) has been reported to express either under the monitoring of its own promoter or under the control of the promoter of the basic proteinase *prb1* in *T. atroviride* (Rocha-Ramírez et al., 2002). Signaling genes responsible for mycoparasitism are the kinase Tvk1/TmkA from *T. virens* and Tmk1 from *T. atroviride*, which are negative regulators of hydrolytic enzymes and antibiotics. The corresponding gene mutations were done and found they were more potential in managing plant pathogen *R. solani* than the chemical fungicides in beans (Reithner et al., 2007; Mukherjee et al., 2013). During the searching of Promoter of genes for the antagonism in *Trichoderma* strains it was found that transcription factors responsible for mycoparasitism (MYC) regulation along with other transcription factors of carbon (CreaA), nitrogen (AreA), Stress (Msn2/Msn4), pH (PacC), etc. (Mach and Zeilinger, 2003). More information is available in the article of Mach and Zeilinger (2003). So, appropriate manipulation of these regulators may serve us more potent BCAs (Benítez, 1998).

26.2.1.4 Hypervirulence

One unique phenomenon is that infection of some virus may affect the virulence of fungal pathogen of the plant. A classical example is the hypovirulence caused by *Cryphonecteria hypovirus* (CHV) on *Cryphonecteria parasitica*, the dangerous causal agent of Chestnut blight (Robin et al., 2000). It was noted that hypovirulent (with virus) results in superficial canker disease while virus-free strains of *C. parasitica* causes deep and lethal canker on both European (*Castanea sativa*) and American (*C. dentata*) chestnut trees. Hypovirulent strain of *C. parasitica* can transfer the virus to virus-free strain of *C. parasitica* by hyphal anastomosis mechanism (Robin et al., 2000) as a result virus-free strain getting virus became avirulent or less virulent. Applying hypovirulence of this BCA, chestnut blight disease in European and American countries has been managed successfully (Robin et al., 2000; Bissegger et al., 1997).

26.2.1.5 Inactivation of pathogens enzymes by BCA

T. harzianum strain T39 and NCIM 1185 by secreting serine protease inactivate the enzymes produced by a pathogen (*Botrytis cinerea*) as a result disease become combated (Elad and Kapat, 1999)

26.2.1.6 Induction of plant resistance by BCA

Its details are given in the next section.

26.2.2 Induction of plant resistance/plant protection mechanisms by FBCA (fungal biocontrol agents)-plant interaction

The plant has its own immune system, but upon pathogen attack sometime gets weaken and

requires reinforcement to fight back. At the same time, BCA interacts with plant and fulfills the need of host plant and provides with support in many ways by induction of resistance for combating against phytopathogens. Similarly, FBCA interacts with the plant to boost up plant immunity and as a result, pathogens are unable to infect plant or suppressed for time being. Research works and application in agrifield are mainly concentrated on *Trichoderma* mediated and mycorrhiza mediated induction of resistance in crops or plants. *Trichoderma* has been found in many ecosystems and some spp. have the potentiality to reduce the severity of plant diseases by inducing host defense mechanism (Viterbo and Horwitz, 2010; Jogaiah et al., 2018). The induction of plant defense responses mediated by *Trichoderma* BCA has been well studied (De Meyer et al., 1998; Yedidia et al., 1999; Hanson and Howell, 2004; Harman et al., 2004). Both mono- and dicotyledonous species when treated with *Trichoderma,* exhibited enhanced resistance to many pathogens (Harman et al., 2004). Plant colonization by *Trichoderma* spp. Induced localized acquired resistance (LAR), as well as induced systemic resistance or ISR. The control of a broad range of plant pathogens/diseases through elicitation by *Trichoderma* of ISR or localized resistance has been observed by Harman et al. (2004). Recently, an increasing number of research works has directed to unveil the molecular basis of *Trichoderma* induced defense in plants. The pathogen- or microbe-associated molecular patterns (PAMPs or MAMPs) as we found from the reviewed work of Hermosa et al. (2012) MAMP-triggered plant responses are elicited rapidly and transiently. Firstly MAMP responses include ion fluxes across the plasma membrane, the generation of reactive oxygen species (ROS), nitric oxide, ET (ethylene) and also callose formation and the generation of antibiotics. Many MAMPs have been identified for PGPR, such as flagellin or lipopolysaccharides, but also secreted compounds including antibiotics, biosurfactants, and volatile organic compounds have been shown to elicit systemic resistance. Potential *Trichoderma* spp. produce a variety of MAMPs (Lorito et al., 2010). The first recognized *Trichoderma* MAMP was identified as an ET inducing xylanase (Xyn2/Eix), produced by *Trichoderma* as a potent elicitor of plant defense responses in specific tobacco and tomato cultivars (Rotblat et al., 2002). Others that were reported subsequently were cellulases from *T. longibrachiatum* (Martinez et al., 2001), endopolygalacturonase ThPG1 from *T. harzianum* (Morán-Diez et al., 2009), Trichomonas (20mer peptaibol) from *T. pseudokoningii* (Luo et al., 2010), etc.

26.2.2.1 General induced defense in the plant body

After the infection of the pathogen in the plant body, a chain of defense reactions occurs, which involves the activation of the respective genes like the gene encoding for the enzyme involved in the phytoalexin synthesis (Dixon and Harrison, 1990; Tagu and Martin, 1996). Plant immune response involves MAPK signaling, activation of the protein kinase, ROS production pathway, etc. Many of these pathways have a mammalian similarity. In the case of the plants, local responses involve the production of the salicylic acid, and different protein molecules which help in inducing the plant systemic defense response. Necrotic pathogen attachment involves activation of jasmonic acid (JA) pathways, which is referred to as induced systemic resistance (ISR). In the plant cell, MAPK pathway involves in the disease resistance pathway, which is similar to the mammalian system. This pathway involves the activation of the serine/threonine MAP enzyme system through phosphorylation of the tyrosine or threonine residue. Recent researches revealed that *Arabidopsis* immune system involves analogous MAP kinase cascade. FLS2 receptor gene was identified in this cascade system. FLS2 encodes leucine-rich repeat trans-membrane domain-kinase protein which is a similar structure to Xa21 of Rice. MAPK is now identified in many plant species. *Arabidopsis* genome contains about 20 MAPK—like genes. Another pathway involved calcium homeostasis pathway, in this pathway after the entry of pathogen-derived product, plasma membrane ion fluxes are changed. Recently it has been found that Phospholipase C, an important protein of G protein signaling cascade, works through the Diacyl Glycerol (DAG) to activate Protein Kinase C and Calcium influx (Dickinson, 2003). Although DAG is converted to phosphatidic acid which is a signal molecule in the plant defense system. IP_3 is also generated in plant and reported to release Ca^{2+}. Calcium is involved in the oxidative burst and Nitric Oxide (NO) production. ROSs are weakly induced in compatible and incompatible interactions. Primarily O_2^- and H_2O_2, etc., like ROS, are a sequential reduction of molecular oxygen. Hydrogen Peroxide reported to work through the activation of the PAL (phenylalanine ammonia-lyase) and glutathione-*S*-transferase helps in the SA pathway since PAL is the first enzyme for SA pathways. In Fig. 26.7, a general defense response of the plant body has been presented (Dickinson, 2003). Now the question is how far these plant immune systems are boosted/response with BCA or mycorrhizal fungi.

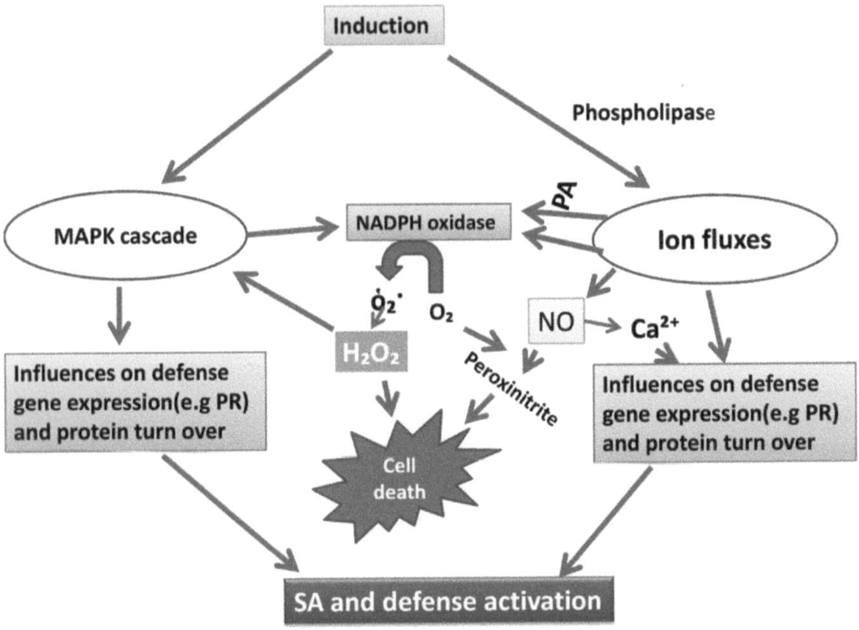

Fig. 26.7 A generalized diagram of induction of plant defense response pathways (Dickinson, 2003). *PLC*, phospholipase; *PA*, phosphatidic acid; *NO*, nitric oxide; *SA*, salicylic acid.

26.3 Signaling pathways during induced resistance (ISR and SAR)

In plant systems, two important signal transduction pathways were operated during induced resistance against pathogens whether they are bacteria or fungi or virus. They are Salicylic acid (SA) (Vlot et al., 2009) (Fig. 26.8), jasmonic acid (JA) (Fig. 26.9), and Ethylene (ET) signaling pathway. Induced resistance responses in plants are subdivided into two major categories: systemic acquired resistance (SAR) and induced systemic resistance (ISR). SAR is associated with both local and systemic increases in Salicylic acid levels, and the expression of pathogenesis-related (PR) genes (Sudisha et al., 2012) is mediated via the SA pathway. ISR does not depend on the accumulation of SA or pathogenesis-related proteins, but on pathways regulated by JA and ET (Elsharkawy et al., 2012).

Strains of *Trichoderma* in the rhizosphere zone trigger resistance mechanisms similar to the hypersensitive response (HR), systemic acquired resistance (SAR), and induced systemic resistance (ISR) in plants (Harman et al., 2004). The molecular study shows an increase in the concentration of enzymes and metabolites responsible for defensive mechanisms, e.g., phenyl-alanine ammonia-lyase (PAL) and chalcone synthase (CHS), involved in the biosynthesis of phytoalexins (HR response), chitinases, and glucanases. These compounds consist of pathogenesis-related proteins (PR) (SAR response) and enzymes involved in response to oxidative stress (Stacey

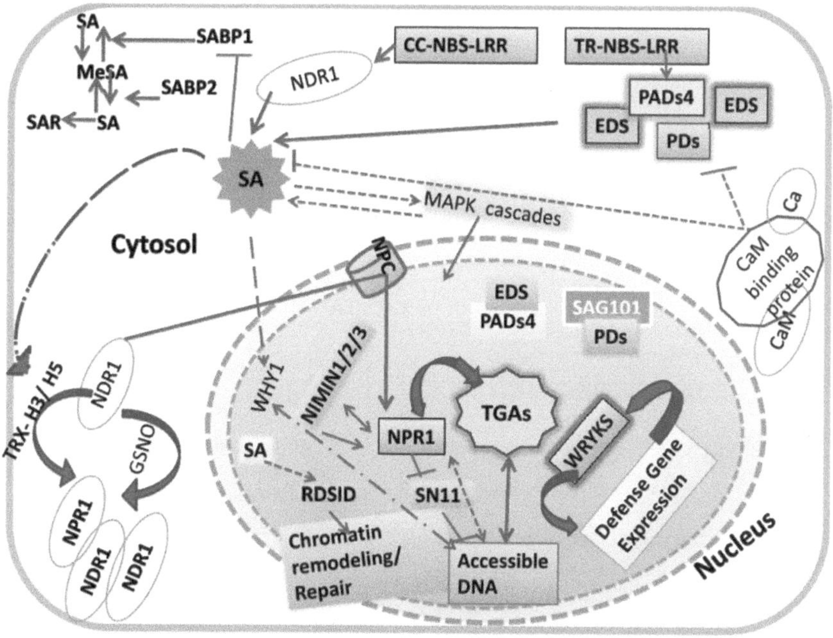

Fig. 26.8 Salicylic acid (SA) pathway (Vlot et al., 2009). *CC-NBS-LRR*, coiled coil nucleotide bindings leucine rich repeat; *TR-NBS-LRR*, toll receptor nucleotide bindings leucine rich repeat; *PAD4*, phytoalexin accumulation deficient 4; *NDR1*, nonspecific disease resistance; *EDS1*, enhanced disease susceptibility 1; *NPR1*, nonexpressor of PR1; *SA*, salicylic acid; *SAG*, SA O-β-glucoside; *LRR*, leucine rich repeat; *TGA*, TGACGTCA *cis*-element-binding protein; *WRKY*, transcription factor; *SABP2*, SA-binding protein 2; *TRX-H3/H5*, thioredoxins H3 and H5; *MeSA*, methyl salicylate; *SAG101*, senescence-associated gene 101; NPR1, also known as NIM1; NIMIN1, 2, and 3 (NIM1-interacting 1, 2 and 3); *SNI1*, suppressor of npr1-1 inducible 1.

and Keen, 1996). Evidence indicating the partial involvement of the SA-dependent signaling pathway in ISR has been reported (Elsharkawy et al., 2012). In an experiment, when *T. harzianum* T22 was applied before or simultaneously with CMV (PD and PE), it triggered concomitant marked increases in the foliar JA/ET ratio and SA levels, compared to healthy controls, suggesting that ISR and SAR have occurred. Therefore, regarding the role of plant hormones in the regulation of the tomato-CMV-*Trichoderma* complex, workers concluded that in all plants treated with *T. harzianum* and inoculated with CMV the responsiveness to ABA, JA, ET, and SA had happened to develop lower levels of disease, and also crosstalk among different hormone-related signaling pathways took place. The significant increase in the foliar JA/ET ratio, as well as the SA levels observed in both leaves and roots, suggest that *T. harzianum* was able to induce systemic resistance against CMV, requiring both JA/ET and the SA signaling pathway. In the study of Contreras-Cornejo et al. (2011), they used transgenic Arabidopsis lines expressing β-glucuronidase (GUS) fusions to elucidate the signal transduction pathway by which *T. virens* and *T. atroviride* elicit plant defense responses. Their results suggested that SA and JA were likely involved

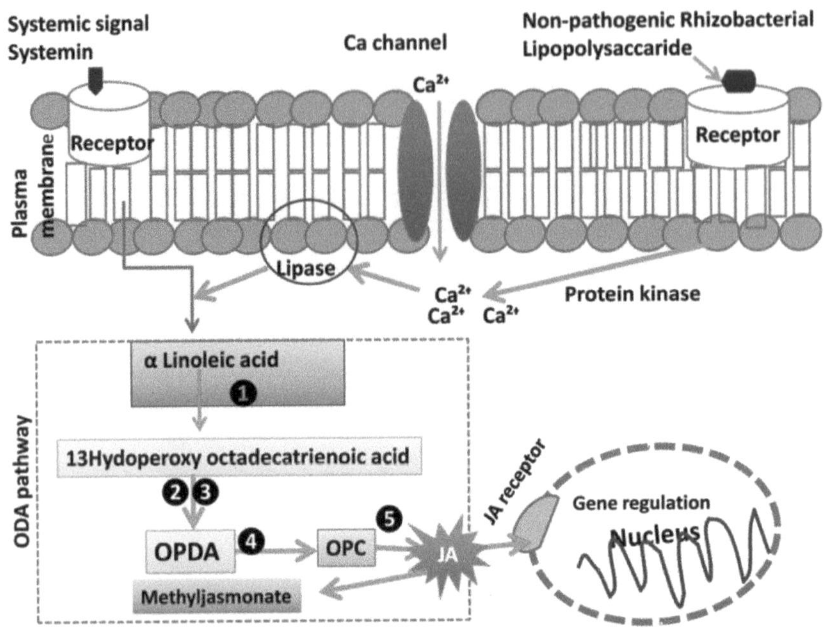

Fig. 26.9 Jasmonic acid (JA) biosynthesis and pathway in *Arabidopsis* (Dickinson, 2003). ❶, lipoxygenase; ❷, allene oxide synthase; ❸, allene oxide cyclase; ❹, OPDA reductase; ❺, beta-oxidase; *OPDA*, 12-oxophytodienoic acid; *OPC*, 12-oxophtoenoic acid; *ODA*, octadecanoic acid.

in Arabidopsis responses to *Trichoderma* inoculation as *T. virens* and *T. atroviride* induced both SA and JA accumulation.

Moreover, they showed that *Trichoderma* induced phytoalexin camalexin production and accumulation in Arabidopsis and produced indole-3-carboxaldehyde. Segarra et al. (2007) exhibited that both SA and JA changes have been taken place in cucumber plants inoculated with *Trichoderma asperellum* strain T34 indicating that hormonal defense protection of plants due to *Trichoderma* species. In the case of maize, *T. virens* induces the expression of *PAL* through a JA-dependent response (Djonović et al., 2006, 2007). De Meyer et al. (1998) found that bean root colonization by *T. harzianum* was effective in inducing defense responses and Yedidia et al. (1999) also noted that penetration of *T. asperellum* in the root system triggered ISR in cucumber seedlings. Shoresh et al. (2006) recorded involvement of jasmonic acid/ethylene signaling pathway in the systemic resistance induced in cucumber by *Trichoderma asperellum* T203 against the leaf pathogen *Pseudomonas syringae* pv. *lachrymans* and higher systemic expression of the pathogenesis-related genes encoding for chitinase 1, β-1,3-glucanase, and peroxidase in inoculated leaf than to noninoculated were recorded (Shoresh et al., 2005), as monitored by several workers. Djonović et al. (2007) recorded that cerato-platanin Sm1 was required for *T. virens*-mediated ISR against *Colletotrichum graminicola* in maize. Korolev et al. (2008) found that *Trichoderma* treatment of JA/ET-deficient Arabidopsis genotypes leads to enhanced susceptibility to *Botrytis cinerea*. ISR triggered by PGPR and *Trichoderma* converges upstream from MYB72, a crucial early component of the onset of ISR (Segarra et al., 2009). The defense response of *T. harzianum* challenged potato to *Rhizoctonia solani* was dependent on JA/ET and SA (Gallou et al., 2009). Up to ninefold increase in the concentration of jasmonic acid (JA) was observed in mycorrhizal tomato plants compared to *Fusarium oxysporum lycopersi* infected control plants. The increased JA concentration in mycorrhizal plants was concomitant to systemic enhanced defense response measured regarding increase in Phenylalanine Ammonia-Lyase (PAL) activity, the concentration of phenols and trichome density in leaves. The above parameters showed a strong positive correlation with the endogenous level of JA. The results suggest the implication of JA in AMF-induced systemic resistance (Kapoor, 2008). Increase in levels of endogenous jasmonates is followed by activation of genes involved in plant defense responses such as those coding for enzymes of phytoalexins synthesis, vegetative storage proteins (Hause et al., 2007; Lorenzo and Solano, 2005). The finding of (Abeysinghe, 2009) supported the fact that *T. harzianum* RU01 actively induces systemic resistance against bean rust caused by *Uromyces appendiculatus*.

Information is not available on changes of plant physiology under treatment with *Trichoderma* spp. in the plant-virus pathosystem except the work of Vitti et al. (2016). The effects of *T. harzianum* on the induction of defense against CMV in *Solanum lycopersicum* var. *cerasiforme* (tomato) have been previously investigated by analyzing various biochemical and molecular aspects involved in the plant-virus interactions and demonstrating the apparent involvement of ROS (Vitti et al., 2016). The analysis of plant hormones shown that treating with *Trichoderma* before or simultaneously to CMV infection, led to a systemic resistance by jasmonic acid/ethylene and salicylic acid signaling pathways. Conversely, systemic resistance was abscisic acid-dependent when *T. harzianum* treatment was administered after the CMV infection (Vitti et al., 2016). Elsharkawy et al. (2012) reported that the plant growth-promoting fungus *Fusarium equiseti* and the arbuscular mycorrhizal fungus *Glomus mosseae* induce systemic resistance against CMV in cucumber plants. The induction and accumulation of PR proteins activity took place during the early stages of root colonization by the mycoparasite *Trichoderma harzianum* strain T-203 (Yedidia et al., 2000). Sultana et al. (2008) reported elicitation of systemic resistance against the bacterial speck pathogen in *Arabidopsis thaliana* by culture filtrates of plant growth-promoting fungi. *Trichoderma* is also effective in triggering systemic resistance against rust (*Uromyces appendiculatus*) on *Phaseolus vulgaris* (Abeysinghe, 2009).

Arabidopsis thaliana treated with *Penicillium simplicissimum* or with its culture filtrate (CF) showed distinct resistance to *Pseudomonas syringae* pv. *tomato* DC3000 (Pst) (Hossain et al., 2007). The examination of the local and systemic gene expression revealed that *P. simplicissimum* and its CF modulate the expression of genes involved in both the SA and JA/ET signaling pathways. The subsequent challenge of fungal-colonized plants with Pst was accompanied by direct activation of SA-inducible PR-2 and PR-5 genes as well as potentiated expression of the JA-inducible Vsp gene. On the other hand, CF-treated plants infected with Pst exhibited elevated expression of most defense-related many genes (e.g., PR-1, PR-2, PR-5, etc.). Here Arabidopsis genotypes such as salicylic acid (SA) signaling expressing the NahG transgene or carrying disruption in NPR1 (npr1), jasmonic acid (JA) signaling (jar1), and ethylene (ET) signaling (ein2) were used. All genotypes screened were protected by *Penicillium simplicissimum* or with its culture filtrate. Workers concluded that multiple defense mechanisms were involved leading to an ISR of Arabidopsis by this fungus (Hossain et al., 2007). Although SA and JA/ET defense pathways are mutually antagonistic, evidence of synergistic interactions has also been reported mediated by the transcription factor NPR1 (nonexpression of PR genes 1). The NPR1 protein is an important transcriptional coactivator of SA-responsive PR genes; NPR1 is also a key regulator in SA-mediated suppression of JA/ET signaling. Furthermore, NPR1 has been implicated in several other JA/ET-dependent defense responses, including beneficial rhizobacteria-mediated ISR (van Loon et al., 2006) and JA/ET-dependent resistance against the soilborne fungus *Verticillium longisporum* (Pieterse et al., 2009); however, the pathway involved in response to beneficial microorganisms has not been elucidated. Colonization of *Arabidopsis* roots by *Trichoderma atroviride* enhances systemic disease resistance through jasmonic acid/ethylene and salicylic acid pathways (Salas-Marina et al., 2011).

26.4 Elicitors produced by FBCA

Use of elicitors is a valuable strategy for activation of plant defense; it may act as an alternative to the transgenic approach to protect plants against pathogens (Benítez et al., 2004; Jogaiah et al., 2019). It is quite obvious that there is an exchange of molecular signals between BCA and the plant. Now the question is what are these highly effective compounds that can dramatically change the plant proteome and trigger the resistance status. The answer must include some elicitors by which they exchange molecular talks/signals. The induction of plant resistance by colonization with some *Trichoderma* species is similar to that elicited by rhizobacteria (Van Loon et al., 1998; Stacey and Keen, 1996; Harman et al., 2004). During the interaction of *Trichoderma* with the plant, different classes of metabolites may act as elicitors or resistance inducers. These molecules include (i) proteins with enzymatic activity, such as xylanase (Lotan and Fluhr, 1990); (ii) avirulence-like gene products able to induce defence reactions in plants (Woo et al., 2004); (iii) low-molecular-weight compounds released from fungal or plant cell walls by the activity of *Trichoderma* enzymes (Harman et al., 2004; Woo et al., 2006; Woo and Lorito, 2007). Some of the low-molecular-weight degradation products released from fungal cell walls were also found, and they consist of short oligosaccharides, with and without an amino acid residue (Woo et al., 2006; Woo and Lorito, 2007). These compounds elicited a reaction in the plant, and further, they also stimulated the biocontrol ability of *Trichoderma* by activating the mycoparasitic gene expression cascade. Djonović et al. (2006) revealed a small protein (Sm1) elicitor secreted by *T. virens*. In a recent work, Alfano et al. (2007) investigated at a molecular level the plant genes involved in *Trichoderma hamatum* resistance induction by using a high-density oligonucleotide microarray approach. Interestingly, *Trichoderma*-induced genes were associated with biotic or abiotic stresses, as well as RNA, DNA, and protein metabolism. In particular, genes

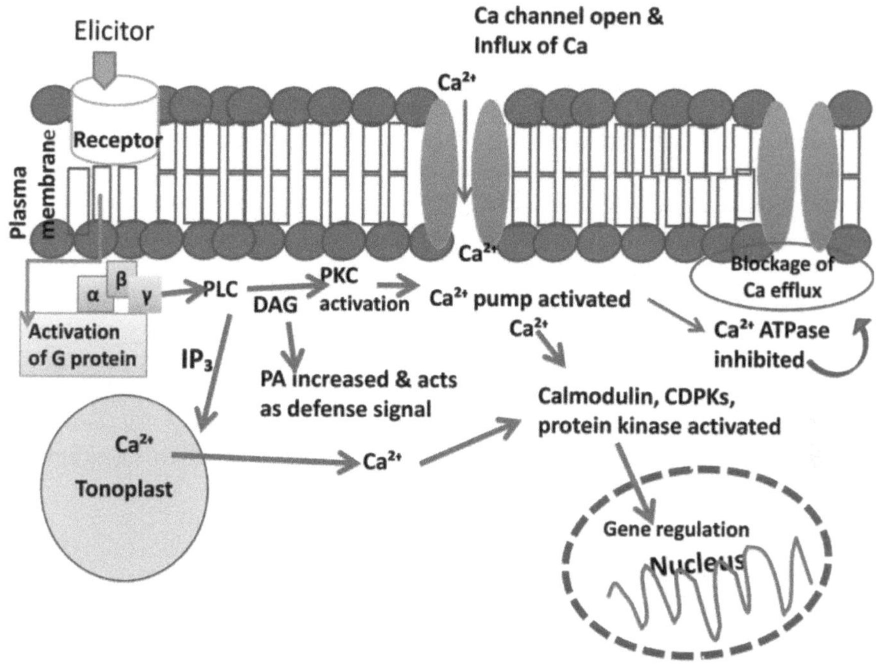

Fig. 26.10 Role of G protein in the calcium influx in response to pathogen elicitor. Elicitor activates G proteins, as a result Ca releases from tonoplast, Ca channel in plasma membrane opens but calcium ATPase becomes inhibited. Moreover, PLC activates PKC and Ca influx through DAG. DAG becomes converted to phosphatidic acid (PA) which itself acts as a signal for defense. IP3 being generated helps to release Ca from tonoplast (Dickinson, 2003). *PKC*, protein kinase C; *PLC*, phospholipase C; *DAG*, diacylglycerol; *IP3*, inositol 1,4,5 triphosphate; *CDPKs*, calcium-dependent protein kinases.

that codify for extensin and extensin-like proteins were found to be induced by the BCA, but not those codifying for proteins belonging to the PR-5 family (thaumatin-like proteins), which are considered the main molecular markers of SAR Elicitor activates G proteins for disease resistance (Dickinson, 2003) (Fig 26.10).

26.5 Transgenic approach for plant protection using BCA genes

The insertion of BCA genes which are responsible for the production of pathogen killing enzymes or phytoalexins or compound in the plant genome and their expression with plant promoter is a new edition of biocontrol of plant pathogens. Lorito et al. (1993) did pioneering work in this line and documented that *Trichoderma* genes can be expressed in plants for the control of plant pathogens. They observed that the endochitinase gene chit42 of *Trichoderma* was expressed in a high level in different plants. Transgenic tobacco and potato were induced to resistant to the leaf pathogens *Alternaria alternata, Alternaria solani,* and *B. cinerea*, and the soilborne pathogen *Rhizoctonia solani*. Bolar et al. (2001) observed that chit 42 expression increased resistance to *Venturia inaequalis*, and similarly broccoli significantly become resistant to attack by *Alternaria brassicicola* (Mora and Earle, 2001). The expression of chit42 in lemon enhanced resistance to *Phoma tracheiphila* and *B. cinerea* was documented by some workers (Gentile et al., 2007; Distefano et al., 2008). They also found a significant correlation between resistance and transgene expression, with upregulation of ROS and JA/ET-responsive genes (Gentile et al., 2007; Distefano et al., 2008). In the case of Rice, the homologous chit42 gene from *T. virens* was able to increase resistance against *R. solani* (Shah et al., 2009). Other *Trichoderma* chitinase genes when inserted in plants, showed same resistance generation as reported by like *T. harzianum* endochitinase chit40 gene in Tobacco against released the enzyme into the medium and were able to *Penicillium digitatum* (Brants and Earle, 2001); *T. virens* endochitinase gene Tv-ech1 in cotton against *A. alternata* and *R. solani* and the endochitinase chit36 gene of *T. harzianum* in carrot against *A. radicina* and *B. cinerea* (Baranski et al., 2008).

26.6 Siderophore in plant immune defense response

In response to the siderophore treatment plant immune system shows a different pattern and mechanism. In the case of maize plant the effect of "coprogen" helps to prime immune response while direct elicitation effect of coprogen shown in the *Arabidopsis* (Mauch and Staehelin, 1989; Oostendorp et al., 2001). In tomato cells, pyochelin and pyocyanin combination is associated with ROS production (Van Loon, 1999), but in *A. thaliana* siderophore differential ROS metabolism-linked gene also observed (Oostendorp et al., 2001). A modification of the iron distribution in cellular and organism level of the plant can rise ROS production. In tobacco plant, ROS only regulate siderophore effects on plant cell. Siderophore activates immunity dependent on different hormone of the plant body. In the case of rice, JA and ET required for immune system induction while salicylic acid required for tomato plant effect. That noted difference of hormonal signaling pathway is due to organ-specific siderophore perception (Albarouki et al., 2014). Pattern recognition receptor such as PRR, FLS2, CERK1, etc., recognizes MAMP or DAMP. Each recognition receptor recognizes specific MAMP or DAMP. Pathogen infection suppresses PRR activation. NB-LRR family triggers a hypersensitive response after recognition of pathogen effectors. Siderophore may target a metalloprotein if this protein is guarded by NB-LRR then the immune response may be activated.

26.7 ACCD [1-aminocyclopropane-1-carboxylate (ACC) deaminase] mediated plant defense

ACC deaminase enzymes are produced and secreted from some FBCAs and their activity has been observed in *Trichoderma asperellum* (Viterbo et al., 2010), *T. atroviride* (Gravel et al., 2007), and *Penicillium citrinum* (Jia et al., 2000). This enzyme cleaves the ACC into α ketoglutarate and ammonia (Todorovic and Glick, 2008; Van de Poel and Van Der Straeten, 2014). Ethylene is an important signaling molecule of the ET signaling pathway for induction of defense. So, ACCD influences other pathways either downregulating (ET/JA signaling, ET signaling, Auxin signaling, and ABA synthesis) or upregulating some pathways (SA signaling and GA signaling) and on the other side, *Trichoderma* secreted and supply to IAA to plants and IAA mediated the signaling pathways in reverse way as done by ACCD, as a result, a cross-talking involved (Gravel et al., 2007; Stepanova et al., 2007) and phytohormone homeostasis will control plant development and immune responses (Hermosa et al., 2012) (Fig. 26.11). Two putative acdS genes were isolated from the genome of *Arabidopsis thaliana* and data were placed validating the hypothesis that these genes can act as regulators of ACC levels in *A. thaliana* and also in tomato fruit development (McDonnell et al., 2009). Some *Trichoderma* spp. that have beneficial effects on plant growth and protection (Harman et al., 2004) also possess ACCD putative sequences (http://genome.jgi-psf.org/Trive1/Trive1.home.html; https://www.sciencedaily.com/releases/2011/04/110411194819.html).

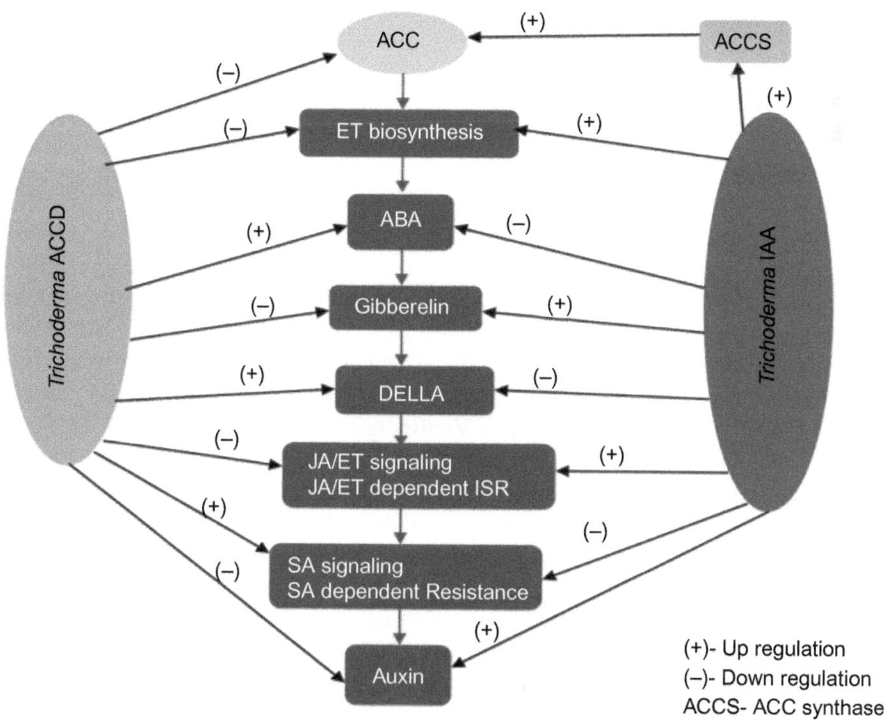

Fig. 26.11 Influence of ACCD and IAA on different cell signaling related to plant defense system and their cross-talks (Hermosa et al., 2012).

26.8 Induction of plant resistance/plant protection mechanisms by mycorrhizal fungi–plant interaction

Mycorrhiza *Glomus* sp. reduced pink root disease of Onion root (Safir, 1968; Inoue et al., 2009), the peanut plant pathogens (Bødker et al., 1998; Dehne, 1982; Abdalla and Abdel-Fattah, 2000;), the tomato plant pathogen (Cordier et al., 1998; Lingua et al., 2002; Hooker et al., 1994; Trotta et al., 1996; Abdalla and Abdel-Fattah, 2000) and later found that several other mycorrhizal fungi can enhance the tolerance of plants to many pathogens (Azcón-Aguilar and Barea, 1997; Caron et al., 1986; Caron, 1989; Cordier et al., 1996). Mycorrhizal protection against the pathogen in the case of a tomato plant, the accumulation of the phenolic compounds and mycorrhizal development were related to each other (Cordier et al., 1998). Cell defense responses associated with localized and systemic resistance to *Phytophthora parasitica* induced in tomato by an arbuscular mycorrhizal fungus (Cordier et al., 1998). Pozo et al. (2002) recorded localized vs the systemic effect of arbuscular mycorrhizal fungi on defense responses against *Phytophthora* infection in tomato crop. In the case of the potato plant, the postharvest disease has been reported to reduce as a result of the inoculation of the mycorrhizae (Niemira et al., 1996). Mycorrhizal fungi were also effective in annual grass, onion, and potato, citrus, common bean, cow pea to induce tolerance against pathogens as reported by some workers (Newsham et al., 1995; Torres-Barragán et al., 1996; Niemira et al., 1996; Davis and Menge, 1980; Ramraj et al., 1988).

26.9 Chemical interaction of the mycorrhizal fungi with the host

The underlying mechanism behind the fungi induced pathogen control has been well established after few decades of research (Dar et al., 1997; Davis and Menge, 1980; Ramraj et al., 1988). In case of the mycorrhizal association, some pathogen-resistant marker molecules like phytoalexins, callose, peroxidase, chitinase, β-1,3-glucanase, PR-1 protein are formed on the plant tissue and mycorrhiza has been reported to induce secondary metabolites (St-Arnaud et al., 1994; Mukerji, 1999; Gianinazzi-Pearson et al., 1996; Ramraj et al., 1988; Vierheilig et al., 1998). After an abiotic agent or biotic agent infection, the plant synthesizes and accumulates phytoalexins, which are low molecular weight compound (Paxton, 1981; Azcón-Aguilar and Barea, 1997; Morandi, 1996). There are some other molecules such as Rishitin; Solavetivone etc. also were produced by potato plant, which were reported as antimicrobial compounds (Yao et al., 2003; Vierheilig et al., 1998). Mycorrhizal colonization of *Glomus* sp. has been reported to the accumulation of the phytoalexins, Rishitins, Solavetivone like compounds in roots of the plants (Yao et al., 2003). During mycorrhiza formation, an increase of accumulation of phenolpropanoid compounds, like phytoalexins have been reported which is necessary for plant defense mechanism as stated earlier (Paxton, 1981; Engström et al., 1999; Yao et al., 2003). Isoflavonoids like glyceollin and coumestrol also have been reported to accumulate in the Soybean roots (Lambais, 2000; García-Garrido and Ocampo, 2002). Another chemical like medicarpin-3-*O*-glycoside and fibronectin which is a precursor of phytoalexin, have been found to accumulate in alfalfa root mycorrhizae (García-Garrido and Ocampo, 2002; Morandi and Le Quere, 1991; Morandi et al., 1984; Volpin et al., 1994). Within mycorrhizal plants, there is a highly complex mixture like glycosylated C^{13} cyclohexenone derivatives which become accumulated. While in the case of barley and wheat plants accumulation of blumenina terpenoid glycoside has been reported by Fyson and Oaks (1992). Colonization of the mycorrhizal fungi has been related to the accumulation of the level of a compound like apocarotenoids (Fyson and Oaks, 1992; Klingner et al., 1995; Peipp et al., 1997). Cyclohexanone derivatives accumulation has been found while *Glomus* sp. mycorrhiza has been colonized in wheat, barley, maize roots (Gianinazzi-Pearson et al., 1996). After mycorrhizal infection, in the plant body level of some important enzymes become significantly increased (Yao et al., 2003) e.g., tomato plants, chitinase, chitosanase, and beta-1,3-glucanase, superoxide dismutase, etc. (Walter et al., 2000). Transcription of methylerythritol phosphate pathway of isoprenoid biosynthesis enzyme like 1-deoxy-D-xylulose5-phosphate synthase (DXS) and 1-deoxy-D-xylulose 5-phosphate reductoisomerase (XDR) has been reported to increase by *Glomus* sp. In the case of the maize roots carotenoids and apocarotenoids accumulation related to the DXS2 transcription, which is stimulated by the mycorrhiza formation (Klingner et al., 1995). Along with the above reports, it has also found that mycorrhizae may cause a change in the plant defense system (Devi and Reddy, 2002; Dumas-Gaudot et al., 1992). Another chemical DIMBOA (2,4-dihydroxy-7-methoxy-1,4-benzoxazine-3-one) in the plant has been reported to have a role in the defense system (García-Garrido and Ocampo, 2002). In the maize plant, DIMBOA accumulation has increased with the mycorrhizal association which later increases the transcript level of gene BX9. When maize plant inoculated with the *Glomus* sp. or *Rhizoctonia* sp. DIMOBA level was found to be significantly high. Another important compound in plant defense is phenol derivative (Walter et al., 2000), more phenolic compounds have been accumulated in the mycorrhizal plant than in nonmycorrhizal plants (Shaul et al., 1999). In carrot root mycorrhizal Ri T-DNA transformation has been reported when it effected (Guenoune et al., 2001) while in pea plant phenolic compound accumulation is closely associated with the disease intensity (Mohr et al., 1998). Morandi (1996) observed the occurrence of phytoalexins and phenolic compounds in endomycorrhizal interactions and their potential role in biological control. Furthermore, the AMF is likely to have a role in the induction of hydrolytic enzymes (Pozo et al., 1999), enhanced levels of Pathogenesis-related (PR) proteins, accrual of phytoalexins (Dixon and Harrison, 1990; Morandi, 1996; Larose et al., 2002), callose deposition (Cordier et al., 1998), and ROS (reactive oxygen species) generation (Salzer et al., 1999). Hence, there are several reports exemplifying the potential of MF in reducing the severity and incidence of phytopathogens for a long time.

Phenylalanine ammonia lyase (PAL) stimulates the flow of carbon from primary to secondary metabolites in plants (Hahlbrock and Scheel, 1989). It is the first enzyme in the phenylpropanoid pathway, leading to the conversion of L-phenylalanine into trans-cinnamic acid (Da Cunha, 1987) which leads to the production of phytoalexins and phenolic substances. AMF after colonization in root triggers the phenylpropanoid pathway (Morandi, 1996). Yedidia et al. (1999) showed that inoculation of cucumber roots with *T. harzianum* (T-203) induced an array of pathogenesis-related proteins, including some hydrolytic enzymes.

26.10 Genes and signaling pathway involved in the induction of resistance of host by mycorrhizal fungi

During mycorrhizal establishment, some defense-related genes like phenylalanine ammonia-lyase (PAL), chalcone isomerase (CHI), chalcone synthase (CHS), and cell wall structural protein gene (HRGP) have been reported to be activated (Fester et al., 1999). The function of CHI and chitinase activity was increased in pre-colonization of mycorrhizae (Morandi and Le Quere, 1991; Dixon et al., 1983), PAL gene activity increases at the time of the colonization. During the inoculation with the mycorrhizal fungi, production of some new compounds including also some enzymes like peroxidase and polyphenol oxidase activity has been recorded (Dixon and Harrison, 1990; Tagu and Martin, 1996). *Glomus* sp. has been reported to induce new chitinase activity in the tomato plants (Walter et al., 2000). In mycorrhizal roots, expression of some genes responsible for phenolpropanoid compounds has observed by (Pozo et al., 2002). In Tobacco leaves mycorrhiza induced changes in the PR protein expression (Kapulnik et al., 1996). After many studies, it has found that mycorrhizal fungi initiate a host defense response which is primarily suppressed. Endochitinase (Niemeyer, 1988) and endoglucanase isoforms encoding mRNA level activity have been reported to decrease along with the infection of the pathogens, which were activated later by interaction with the mycorrhizal fungi (St-Arnaud et al., 1994).

AMF colonized rice was found to have induced lipid transfer protein (*Ltp* gene) which is responsible for plant defense response for its antimicrobial property. Phenylalanine ammonia- lyase (PAL) enzyme (*Pal* gene), which leads to the production of phytoalexins and phenolic compounds, was also upregulated in the rice colonized with AMF (Blilou et al., 2000). Moreover, *VCH3* gene, which encodes chitinase enzyme, was remarkably overexpressed by AMF in grapevine roots against *Meloidogyne incognita* (Li et al., 2006). Conrath et al. (2006) found that upon pathogen infection, mycorrhized plants accumulate a considerable amount of PR-1a, which leads to SAR and basic β-1,3-glucanase proteins than nonmycorrhized plants. Pozo and Azcón-Aguilar (2007) documented that mycorrhiza enhanced the attenuated plant defense responses by altering the host's signaling pathways and this is accomplished primarily through modulation in jasmonic acid (JA) and salicylic acid (SA) dependent pathways. Induction of *Medicago truncatula* genes TC104515 (6659-fold), TC101060 and TC98064 were observed in the roots colonized with *Glomus intraradices* (Liu et al., 2007). After receiving a molecular talk from the plant, AMF produces a small molecule known as Myc factor (Ludwig-Müller, 2000) and Ca-calmodulin-dependent protein kinase (ccaMK) (Lévy et al., 2004; Mitra et al., 2004; Kosuta et al., 2008). These promptly induce the cytosolic calcium ($[Ca^{2+}]$ cyt) enrichment (Navazio et al., 2007), which enhances the ROS generation (Strange, 2006) and Ca induces defense gene activation through MAPK (Fig. 26.10) and G protein (Fig. 26.11). Moreover, lipid peroxidation through ROS generation leads to membrane integrity, tissue necrosis and induction of phytoalexins, which is synthesized by the action of lipoxygenases (LOXs) (Shoresh et al. 2006). LOX metabolites through an octadecanoid pathway (Hause et al., 2007) lead to Jasmonate biosynthesis. Jasmonates play signal in mycorrhiza-induced systemic resistance (El-Khallal, 2007), which still requires more research to clarify it (Khan et al., 2010). *Glomus macrocarpum* or *Glomus fasciculatum* after colonization in tomato roots induce resistance against *Fusarium oxysporum* f. sp. *lycopersici* by increasing level of PAL and JA which indicates resistance is operated by JA signaling pathway (Kapoor, 2008).

The genes which were predicted to encode cysteine-rich proteins, display antifungal activity (Terras et al., 1995), their function is to trigger the hypersensitivity response with the matching resistance gene (de Wit, 1992) and this response is mediated by ROS produced early in the plant-pathogen interaction (Levine et al., 1994). However, TC104515 transcripts were found only in roots colonized with *G. intraradices* and not in roots colonized with *G. versiforme* or *Gigaspora gigantea.* This gene was also not expressed in *M. truncatula* and *G. mosseae* colonized roots (Hohnjec et al., 2005). So, there is a considerable variation in the genetic shifts of different plant-related defense genes colonized with diverse AMF species. It was also validated from experiments conducted by others. Liu et al. (2007) experimented taking different AM fungi, including *Gigaspora gigantea* and evaluated the transcriptional response of *Medicago truncatula* to AM fungi. A core set of genes were induced in response to the different AM fungi, advocating that gene induction specific to a single AM fungus might be rare. One gene defensin (encodes a cysteine-rich protein) was only upregulated in roots colonized by *Glomus intraradices*. When comparing between the capacity of *G. mosseae* and *G. intraradices* to induce local and systemic resistance to *Phytophthora parasitica,* it was observed that only the former was effective at inducing resistance (Pozo et al., 2002, 2009). After all, it is concluded that the recognition dialogue established between the host plant and AM fungi and colonization of AMF, leads to specific the host's gene regulation (Genre et al., 2009) with the serial production of specific multifunctional compounds responsible for transduction pathways and disease resistance (Pozo and Azcón-Aguilar, 2007; Liu et al., 2007; Van Wees et al., 2008).

26.11 Yeasts as BCA, induction of disease resistance signaling pathways in host plant

Yeasts are a heterogeneous group of unicellular fungi, multiply by budding and other asexual and sexual processes. They may be ascomycetous or basidiomycetes or deuteromycotous fungi—about 100 genera and 700 species (Kurtzman et al., 2011; Kurtzman and Fell, 1999). Their application as BCA is relatively new. *Pichia guilliermondii* parasitizes and inhibit the growth of many fungal pathogens like *Botrytis cenerea* and *Penicillium expansum* (Wisniewski et al., 1991). The yeast *Torulapsis candida* (*Candida famata*) effectively controls *Penicillium digitatum* infection on citrus fruits (Arras et al., 1997), and *Debaryomyces hansenii* reduces *Penicillium digitatum* decay on orange fruits. The mechanism of disease suppression by yeasts lies on the fact that the yeast-induced production of phytoalexins, scopoletin, and scoparone but did not produce toxic substances against the pathogens *Penicillium digitatum* and *Botrytis cinerea* (Arras and Arru, 1999); thus it may be implied that the yeast inhibits fungal infection by activating host's defense mechanisms. *Candida saitoana* and *C. oleophila* control postharvest diseases of apple and *Citrus* fruits (El-Ghaouth et al., 2000).

Nearly about 15 yeasts exhibited inhibitory potentiality against *P. expansum* based on nutrition and space competition (Coelho et al., 2007). The expression of an antifungal peptide (cecropin-A-based peptide) in *Saccharomyces cerevisiae*, arrested the germination of *Colletotrichum coccodes* (causal organism of tomato fruit decay) was recorded (Jones and Prusky, 2002). Capdeville et al. (2007) working on the interaction between *Candida magnus* and *C. gloeosporioides* on papaya fruits under an electron microscope and noted that the yeast colonized wound surface of fruits much faster than the pathogen, over competing with the pathogen for space and nutrition. Successful biocontrol of *Penicillium digitatum* in the grapefruit by yeast biocontrol agent *Candida oleophila* was done (Droby et al., 2002). A similar report is available from the results of experiments conducted by others and it indicated that *Candida* sp., *Rhodotorula* sp., *Cryptococcus* sp., and *Saccharomyces* sp. have the antagonistic potentiality on *P. expansum* (He et al., 2003), and *P. digitatum* (Ghosh et al., 2013, 2018) (Fig. 26.12). Pimenta et al. (2008) tested the efficacy of predaceous yeast *Saccharomycopsis schoeni* for biological control of *Penicillium italicum, P. digitatum and P. expansum.* About 26 different species of yeasts isolated exhibited antagonistic action against all molds (*Botrytis cinerea*, *Aspergillus carbonarius* and *Penicillium expansum* pathogen species) (Cordero-Bueso et al., 2017). In the latest experiment,

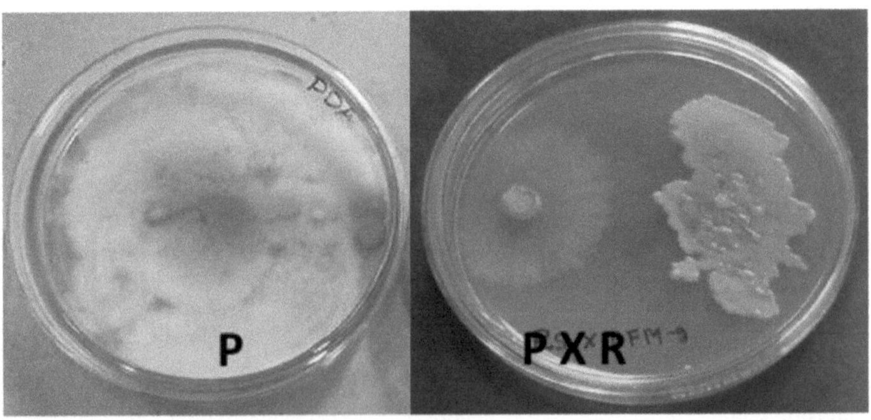

Fig. 26.12 Yeast as BCA in vitro. Left plate shows *P. digitatum* pathogen in full plate (control); right plate dual culture (*R. mucilaginosa* × *P. digitatum*) shows inhibition of growth of *P. digitatum* after 7 days.

three yeasts (*Candida famata*, *Rhodotorula mucilaginosa,* and *Debaryomyces hansenii*) were evaluated as potential antagonistic to *P. digitatum* and out of these three, *C. famata* was best to protect the *Citrus* (Orange) fruits both in vitro and in vivo and all yeasts applied here secreted volatile and nonvolatile chemicals for killing the *P. digitatum* (Ghosh et al., 2018). Although we (Ghosh and Santra, 2016) have recorded in our earlier experiments that *C. famata* secrets β-1,3 glucanase enzyme and siderophore in its culture filtrate and its high colonization capacity on citrus fruits, however, to acquire the key knowledge of the mechanism of action of *C. famata*, it would be useful in future studies to purify and characterize the substances released in the culture filtrates. Moreover, Calvente et al. (1999) observed that the *Rhodotorula glutenin* produced a high concentration of rhodotorulic acid.

Buxdorf et al. (2013) noted that the inoculation of *Pseudozyma aphidis* elicited induced resistance in *Arabidopsis*. Similarly, the leaf colonizing yeast *Pseudozyma churashimaensis* elicits systemic defense of pepper against pathogens (Lee et al., 2017) and induced resistance was confirmed by gene expression in *Arabidopsis*. The defense signaling pathways were reported to be on jasmonic acid (JA) and salicylic acid (SA) signaling but independent of ethylene signaling for the induced resistance by *P. aphids* (Barda et al., 2015; Gafni et al., 2015).

26.12 The three-way talk/interaction analysis: *Trichoderma*-plant-pathogen

The three-way interactions involving BCA (*Trichoderma*), plant, and fungal pathogen are going on in nature. But maximum works are concentrated on documentation of two-way interaction plant-pathogen, plant-antagonist or pathogen-antagonist. Although some workers are involved in three-way interactions in plant-pathogen–antagonist interactions by using novel methods such as proteomics (Marra et al., 2006) and gene reporter systems (Lu et al., 2004), their magnitudes are not spread to other workers as new works are very less in this line. To monitor the molecular cross-talk in three-way interactions needs some important experiments that investigate gene expression occurring in each partner involved, singly and later in all possible combinations. Marra et al. (2006) applied proteomic approach for the three-way interactions of *Trichoderma* with plant and different fungal pathogens and their results revealed that in the plant proteome-specific PR proteins and other disease-related factors (i.e., potential resistance genes) may regulate the three-way interaction and that the presence of the BCA modulates the plant response to a pathogen attack. Moreover, they noted in few cases, the FBCA reduced production of some defense proteins, but at the same time enhanced others in a higher concentration.

In order to study the mechanisms involved in this intricate three-way cross-talk among the plant, virus, and antagonist, thoroughly analyzed the plant growth-promoting effect of *Trichoderma*, the photosynthetic performance of the plant, as well as the role of ROS scavenging enzymes and hormones in the regulation of tomato-CMV-T22 interaction. The possible use of the BCA *Trichoderma* against CMV infection could be crucial not only for the development of an integrated understanding of the plant-pathogen-BCA complex but also for developing rational strategies to manage viral diseases (Vitti et al., 2016).

26.13 Conclusion

From the above discussion, it is very clear that there are many rhizospheric, mycorrhizal, and yeast fungi which have a definite role for plant protection from plant pathogen and pests and induction of plant immunity. FBCAs interact with plant pathogens (two-way talk) and eliminate them by various mechanisms such as competition, antibiosis, hyperparasitism, detoxifying pathogen toxic, etc. Hyperparasitisms of *Trichoderma* are being regulated by cAMP pathway and MAPkinase pathway FBCAs interact with host plant (another two-way talk) for induction of host immunity or resistance by different mechanisms or cell signaling pathways such as elicitor production, phytoalexin triggering, PR-protein formation, induction of different resistance-related genes, etc. Fungal Siderophore causes iron deficiency in plant pathogens and also induce disease resistance has both side positive role. The cell signaling pathways operated here are SA, JA/ET, MAPkinase pathways and their cross-talks for SAR and ISR immunity. Various genes have been isolated from FBCAs and artificially inserted in plant genomes (transgenic approaches), and they are operating for disease resistance. Several elicitors have detected and isolated from FBCAs, and they are active for triggering phytoalexin production and PR chemical production for plant resistance enhancement. Novel work on the three-way molecular interaction (FBCA-Pathogen-plant) is very limited. More two-way molecular interactions and three-way interaction will exhibit more clear scientific picture for the role of fungi as plant protection, mechanism of actions, and more sustainable application in agrifield.

Acknowledgment

Authors are grateful to the Department of Science and Biotechnology, GoWB for financial support and Principal, RKMVC College, Rahara for Lab facility.

References

Abdalla, M.E., Abdel-Fattah, G.M., 2000. Influence of the endomycorrhizal fungus *Glomus mosseae* on the development of peanut pod rot disease in Egypt. Mycorrhiza 10 (1), 29–35.

Abdalla, M., Gutierrez-Mohamed, M., Farah, I., 2003. Association of pesticide exposure and risk of breast cancer mortality in Mississippi. Biomed. Sci. Instrum. 39, 397–401.

Abdelrahman, M., Jogaiah, S., 2019. Defense mechanism and diverse actions of fungal biological control agents against plant biotic stresses. In: Khurana, S., Gaur, R. (Eds.), Plant Biotechnology: Progress in Genomic Era. Springer, Singapore, pp. 461–478.

Abeysinghe, S., 2009. Systemic resistance induced by *Trichoderma harzianum* RU01 against *Uromyces appendiculatus* on *Phaseolus vulgaris*. J. Natl. Sci. Found. 37(3).

Aktar, W., Sengupta, D., Chowdhury, A., 2009. Impact of pesticides use in agriculture: their benefits and hazards. Interdiscip. Toxicol. 2 (1), 1–12.

Albarouki, E., Schafferer, L., Ye, F., von Wirén, N., Haas, H., Deising, H.B., 2014. Biotrophy-specific downregulation of siderophore biosynthesis in *Colletotrichum graminicola* is required for modulation of immune responses of maize. Mol. Microbiol. 92 (2), 338–355.

Alfano, G., Ivey, M.L., Cakir, C., Bos, J., Miller, S., Madden, L., Kamoun, S., Hoitink, H., 2007. Systemic modulation of gene expression in tomato by *Trichoderma hamatum* 382. Phytopathology 97 (4), 429–437.

Arras, G., Arru, S., 1999. Integrated control of postharvest citrus decay and induction of phytoalexins by *Debaryomyces hansenii*. Adv. Horticult. Sci., 76–81.

Arras, G., Dessi, R., Sanna, P., Arru, S., 1997. Inhibitory activity of yeasts isolated from fig fruits against *Penicillium digitatum*. Acta Hortic. 485, 37–46.

Azcón-Aguilar, C., Barea, J., 1997. Applying mycorrhiza biotechnology to horticulture: significance and potentials. Sci. Hortic. 68 (1–4), 1–24.

Babu, N.A., Jogaiah, S., Ito, S.-I., Nagaraju, K.N., Tran, L.-S.P., 2015. Improvement of growth, fruit weight and early blight disease protection of tomato plants by rhizosphere bacteria is correlated with their beneficial traits and induced biosynthesis of antioxidant peroxidase and polyphenol oxidase. Plant Sci. 231, 62–73.

Baek, J.M., Howell, C.R., Kenerley, C.M., 1999. The role of an extracellular chitinase from *Trichoderma virens* Gv29-8 in the biocontrol of *Rhizoctonia solani*. Curr. Genet. 35 (1), 41–50.

Banerjee, S., Pal, S., Mukherjee, S., Podder, D., Mukherjee, A., Nandi, A., Debnath, P., Sur, P., Ghosh, S., 2016. Cellular abnormalities induced by *Trichoderma* spp. during in vitro interaction and control of white muscardine (*Beauveria bassiana*) and green muscardine (*Metarhizium anisopliae*) disease of silkworm *Bombyx mori*. J. Biopest. 9 (2), 104.

Baranski, R., Klocke, E., Nothnagel, T., 2008. Chitinase CHIT36 from *Trichoderma harzianum* enhances resistance of transgenic carrot to fungal pathogens. J. Phytopathol. 156 (9), 513–521.

Barda, O., Shalev, O., Alster, S., Buxdorf, K., Gafni, A., Levy, M., 2015. *Pseudozyma aphidis* induces salicylic-acid-independent resistance to *Clavibacter michiganensis* in tomato plants. Plant Dis. 99 (5), 621–626.

Benítez, T., 1998. Biofungicides: *Trichoderma* as a biocontrol agent against phytopathogenic fungi. Recent Res. Devel. Microbiol. 2, 129–150.

Benítez, T., Rincón, A.M., Limón, M.C., Codon, A.C., 2004. Biocontrol mechanisms of *Trichoderma* strains. Int. Microbiol. 7 (4), 249–260.

Bissegger, M., Rigling, D., Heiniger, U., 1997. Population structure and disease development of *Cryphonectria parasitica* in European chestnut forests in the presence of natural hypovirulence. Phytopathology 87 (1), 50–59.

Blilou, I., Ocampo, J.A., García-Garrido, J.M., 2000. Induction of LTP (lipid transfer protein) and PAL (phenylalanine ammonialyase) gene expression in rice roots colonized by the arbuscular mycorrhizal fungus *Glomus mosseae*. J. Exp. Bot. 51 (353), 1969–1977.

Bødker, L., Kjøller, R., Rosendahl, S., 1998. Effect of phosphate and the arbuscular mycorrhizal fungus *Glomus intraradices* on disease severity of root rot of peas (*Pisum sativum*) caused by *Aphanomyces euteiches*. Mycorrhiza 8 (3), 169–174.

Bolar, J.P., Norelli, J.L., Harman, G.E., Brown, S.K., Aldwinckle, H.S., 2001. Synergistic activity of endochitinase and exochitinase from *Trichoderma atroviride* (*T. harzianum*) against the pathogenic fungus (*Venturia inaequalis*) in transgenic apple plants. Transgenic Res. 10 (6), 533–543.

Brants, A., Earle, E., 2001. Transgenic tobacco cell cultures expressing a *Trichoderma harzianum* endochitinase gene release the enzyme into the medium. Plant Cell Rep. 20 (1), 73–78.

Buxdorf, K., Rahat, I., Levy, M., 2013. Pseudozyma aphidis induces ethylene-independent resistance in plants. Plant Signal. Behav. 8(11), e26273.

Calvente, V., Benuzzi, D., de Tosetti, M.S., 1999. Antagonistic action of siderophores from Rhodotorula glutinis upon the postharvest pathogen *Penicillium expansum*. Int. Biodeter. Biodegr. 43 (4), 167–172.

Capdeville, G., Souza Júnior, M.T., Santos, J.R.P., Miranda, S.P., Caetano, A.R., Falcão, R., Gomes, A.C.M.M., 2007. Scanning electron microscopy of the interaction between *Cryptococcus magnus* and *Colletotrichum gloeosporioides* on papaya fruit. Pesquisa Agropecuária Brasileira 42 (11), 1537–1544.

Caron, M., 1989. Potential use of mycorrhizae in control of soil-borne diseases. Can. J. Plant Pathol. 11 (2), 177–179.

Caron, M., Richard, C., Fortin, J., 1986. Effect of preinfestation of the soil by a vesicular-arbuscular mycorrhizal fungus, *Glomus intraradices*, on Fusarium crown and root rot of tomatoes. Phytoprotection 67 (1), 15–19.

Carsolio, C., Gutiérrez, A., Jiménez, B., Van Montagu, M., Herrera-Estrella, A., 1994. Characterization of ech-42, a *Trichoderma harzianum* endochitinase gene expressed during mycoparasitism. Proc. Natl. Acad. Sci. USA. 91 (23), 10903–10907.

Carsolio, C., Benhamou, N., Haran, S., Cortés, C., Gutiérrez, A., Chet, I., Herrera-Estrella, A., 1999. Role of the *Trichoderma harzianum* endochitinase gene, ech42, in mycoparasitism. Appl. Environ. Microbiol. 65 (3), 929–935.

Cherif, M., Benhamou, N., 1990. Cytochemical aspects of chitin breakdown during the parasitic action of a *Trichoderma* sp. on *Fusarium oxysporum* f. sp. radicis-lycopersici. Phytopathology 80 (12), 1406–1414.

Chet, I., Inbar, J., 1994. Biological control of fungal pathogens. Appl. Biochem. Biotechnol. 48, 37–43.

Chet, I., Inbar, J., Hadar, I., 1997. Fungal antagonists and mycoparasites. In: Wicklow, D.T., Soderstrom, B. (Eds.), The Mycota IV: Environmental and Microbial Relationships. Springer-Verlag, Berlin, pp. 165–184.

Coelho, A.R., Celli, M.G., Ono, E.Y.S., Wosiacki, G., Hoffmann, F.L., Pagnocca, F.C., Hirooka, E.Y., 2007. *Penicillium expansum* versus antagonist yeasts and patulin degradation in vitro. Braz. Arch. Biol. Technol. 50 (4), 725–733.

Conrath, U., Beckers, G., Flors, V., García-Agustín, P., Jakab, G., Mauch, F., Newman, M., Pieterse, C., Poinssot, B., Pozo, M., 2006. Priming: getting ready for battle. Mol. Pant-Microbe. Interact. 19 (10), 1062–1071.

Contreras-Cornejo, H.A., Macías-Rodríguez, L., Beltrán-Peña, E., Herrera-Estrella, A., López-Bucio, J., 2011. *Trichoderma*-induced plant immunity likely involves both hormonal-and camalexin-dependent mechanisms in *Arabidopsis thaliana* and confers resistance against necrotrophic fungi *Botrytis cinerea*. Plant Signal. Behav. 6 (10), 1554–1563.

Copping, L.G., 2004. The Manual of Biocontrol Agents. British Crop Protection Council.

Cordero-Bueso, G., Mangieri, N., Maghradze, D., Foschino, R., Valdetara, F., Cantoral, J.M., Vigentini, I., 2017. Wild grape-associated yeasts as promising biocontrol agents against *Vitis vinifera* fungal pathogens. Front. Microbiol. 8, 2025.

Cordier, C., Gianinazzi, S., Gianinazzi-Pearson, V., 1996. Colonisation patterns of root tissues by *Phytophthora nicotianae* var. parasitica related to reduced disease in mycorrhizal tomato. Plant and Soil 185 (2), 223–232.

Cordier, C., Pozo, M.J., Barea, J.-M., Gianinazzi, S., Gianinazzi-Pearson, V., 1998. Cell defense responses associated with localized and systemic resistance to *Phytophthora parasitica* induced in tomato by an arbuscular mycorrhizal fungus. Mol. Plant Microbe Interact. 11 (10), 1017–1028.

Da Cunha, A., 1987. The estimation of L-phenylalanine ammonialyase shows phenylpropanoid biosynthesis to be regulated by L-phenylalanine supply and availability. Phytochemistry 26 (10), 2723–2727.

Dar, G.H., Zargar, M., Beigh, G., 1997. Biocontrol of Fusarium root rot in the common bean (*Phaseolus vulgaris* L.) by using symbiotic *Glomus mosseae* and *Rhizobium leguminosarum*. Microb. Ecol. 34 (1), 74–80.

Davis, R., Menge, J., 1980. Influence of *Glomus fasciculatus* and soil phosphorus on *Phytophthora* root rot of citrus. Phytopathology 70 (5), 447–452.

De Meyer, G., Bigirimana, J., Elad, Y., Höfte, M., 1998. Induced systemic resistance in *Trichoderma harzianum* T39 biocontrol of *Botrytis cinerea*. Eur. J. Plant Pathol. 104 (3), 279–286.

Deacon, J., 2006. Fungal Biology, 4th edn Blackwell Publication, Oxford, UK.

Dehne, H., 1982. Interaction between vesicular-arbuscular mycorrhizal fungi and plant pathogens. Phytopathology 72, 1115–1119.

Devi, M.C., Reddy, M., 2002. Phenolic acid metabolism of groundnut (*Arachis hypogaea* L.) plants inoculated with VAM fungus and Rhizobium. Plant Growth Regul. 37 (2), 151–156.

Di Pietro, A., Lorito, M., Hayes, C., Broadway, R., Harman, G., 1993. Endochitinase from *Gliocladium virens*: isolation, characterization, and synergistic antifungal activity in combination with gliotoxin. Phytopathology 83, 308–313.

Dickinson, M., 2003. Molecular Plant Pathology. BIOS Scientific Publishers, London.

Distefano, G., La Malfa, S., Vitale, A., Lorito, M., Deng, Z., Gentile, A., 2008. Defence-related gene expression in transgenic lemon plants producing an antimicrobial *Trichoderma harzianum* endochitinase during fungal infection. Transgenic Res. 17 (5), 873–879.

Dixon, R.A., Harrison, M.J., 1990. Activation, structure, and organization of genes involved in microbial defense in plants. Adv. Genet. 28, 165–234.

Dixon, R.A., Gerrish, C., Lamb, C.J., Robbins, M.P., 1983. Elicitor-mediated induction of chalcone isomerase in *Phaseolus vulgaris* cell suspension cultures. Planta 159 (6), 561–569.

Djonović, S., Pozo, M.J., Dangott, L.J., Howell, C.R., Kenerley, C.M., 2006. Sm1, a proteinaceous elicitor secreted by the biocontrol fungus *Trichoderma virens* induces plant defense responses and systemic resistance. Mol. Plant Microbe Interact. 19 (8), 838–853.

Djonović, S., Vargas, W.A., Kolomiets, M.V., Horndeski, M., Wiest, A., Kenerley, C.M., 2007. A proteinaceous elicitor Sm1 from the beneficial fungus *Trichoderma virens* is required for induced systemic resistance in maize. Plant Physiol. 145 (3), 875–889.

Droby, S., Vinokur, V., Weiss, B., Cohen, L., Daus, A., Goldschmidt, E., Porat, R., 2002. Induction of resistance to *Penicillium digitatum* in grapefruit by the yeast biocontrol agent *Candida oleophila*. Phytopathology 92 (4), 393–399.

Dumas-Gaudot, E., Furlan, V., Grenier, J., Asselin, A., 1992. New acidic chitinase isoforms induced in tobacco roots by vesicular-arbuscular mycorrhizal fungi. Mycorrhiza 1 (3), 133–136.

Eisendle, M., Oberegger, H., Buttinger, R., Illmer, P., Haas, H., 2004. Biosynthesis and uptake of siderophores is controlled by the PacC-mediated ambient-pH regulatory system in *Aspergillus nidulans*. Eukaryot. Cell 3 (2), 561–563.

Elad, Y., Kapat, A., 1999. The role of *Trichoderma harzianum* protease in the biocontrol of *Botrytis cinerea*. Eur. J. Plant Pathol. 105 (2), 177–189.

Elad, Y., Chet, I., Katan, J., 1980. *Trichoderma harzianum*: a biocontrol agent effective against *Sclerotium rolfsii* and *Rhizoctonia solani*. Phytopathology 70 (2), 119–121.

Elad, Y., Barak, R., Chet, I., Henis, Y., 1983a. Ultrastructural studies of the interaction between *Trichoderma* spp. and plant pathogenic fungi. J. Phytopathol. 107 (2), 168–175.

Elad, Y., Chet, I., Boyle, P., Henis, Y., 1983b. Parasitism of *Trichoderma* spp. on *Rhizoctonia solani* and *Sclerotium rolfsii*-scanning electron microscopy and fluorescence microscopy. Phytopathology 73 (1), 85–88.

El-Ghaouth, A., Smilanick, J.L., Wilson, C.L., 2000. Enhancement of the performance of *Candida saitoana* by the addition of glycolchitosan for the control of postharvest decay of apple and citrus fruit. Postharvest. Biol. Technol. 19 (1), 103–110.

El-Khallal, S.M., 2007. Induction and modulation of resistance in tomato plants against Fusarium wilt disease by bioagent fungi (arbuscular mycorrhiza) and/or hormonal elicitors (jasmonic acid & salicylic acid): 2-changes in the antioxidant enzymes, phenolic compounds and pathogen related-proteins. Aust. J. Basic Appl. Sci. 1 (4), 717–732.

Elsharkawy, M.M., Shimizu, M., Takahashi, H., Hyakumachi, M., 2012. The plant growth-promoting fungus *Fusarium equiseti* and the arbuscular mycorrhizal fungus *Glomus mosseae* induce systemic resistance against cucumber mosaic virus in cucumber plants. Plant and Soil 361 (1-2), 397–409.

Engelberth, J., Koch, T., Schüler, G., Bachmann, N., Rechtenbach, J., Boland, W., 2001. Ion channel-forming alamethicin is a potent elicitor of volatile biosynthesis and tendril coiling: cross talk between jasmonate and salicylate signaling in lima bean. Plant Physiol. 125 (1), 369–377.

Engström, K., Widmark, A., Brishammar, S., Helmersson, S., 1999. Antifungal activity to *Phytophthora infestans* of sesquiterpenoids from infected potato tubers. Potato Res. 42 (1), 43–50.

Fester, T., Maier, W., Strack, D., 1999. Accumulation of secondary compounds in barley and wheat roots in response to inoculation with an arbuscular mycorrhizal fungus and co-inoculation with rhizosphere bacteria. Mycorrhiza 8 (5), 241–246.

Finley, R.D., 2007. The fungi in soil. In: van Elsas, J.D., Jansson, J.K., Travors, J.T. (Eds.), Modern Soil Microbiology. CRC Press, New York, pp. 107–146.

Fletcher, J., Bender, C., Budowle, B., Cobb, W., Gold, S., Ishimaru, C., Luster, D., Melcher, U., Murch, R., Scherm, H., 2006. Plant pathogen forensics: capabilities, needs, and recommendations. Microbiol. Mol. Biol. Rev. 70 (2), 450–471.

Fyson, A., Oaks, A., 1992. Rapid methods for quantifying VAM fungal infections in maize roots. Plant and Soil 147 (2), 317–319.

Gafni, A., Calderon, C.E., Harris, R., Buxdorf, K., Dafa-Berger, A., Zeilinger-Reichert, E., Levy, M., 2015. Biological control of the cucurbit powdery mildew pathogen podosphaera xanthii by means of the epiphytic fungus *Pseudozyma aphidis* and parasitism as a mode of action. Front. Plant Sci. 6, 132.

Gallou, A., Cranenbrouck, S., Declerck, S., 2009. *Trichoderma harzianum* elicits defence response genes in roots of potato plantlets challenged by *Rhizoctonia solani*. Eur. J. Plant Pathol. 124 (2), 219–230.

Gao, K., Liu, X., Guo, R., Huai, W., Zhang, M., 2001. Study on the antagonism of *Trichoderma* species on canker pathogen fungi of popular. Scientia Silvae Sinicae 37 (5), 82–86.

García, I., Lora, J.M., de la Cruz, J., Benítez, T., Llobell, A., Pintor-Toro, J.A., 1994. Cloning and characterization of a chitinase (CHIT42) cDNA from the mycoparasitic fungus *Trichoderma harzianum*. Curr. Genet. 27 (1), 83–89.

García-Garrido, J.M., Ocampo, J.A., 2002. Regulation of the plant defence response in arbuscular mycorrhizal symbiosis. J. Exp. Bot. 53 (373), 1377–1386.

Genre, A., Ortu, G., Bertoldo, C., Martino, E., Bonfante, P., 2009. Biotic and abiotic stimulation of root epidermal cells reveals common and specific responses to arbuscular mycorrhizal fungi. Plant Physiol. 149 (3), 1424–1434.

Gentile, A., Deng, Z., La Malfa, S., Distefano, G., Domina, F., Vitale, A., Polizzi, G., Lorito, M., Tribulato, E., 2007. Enhanced resistance to *Phoma tracheiphila* and *Botrytis cinerea* in transgenic lemon plants expressing a *Trichoderma harzianum* chitinase gene. Plant Breed. 126 (2), 146–151.

Ghisalberti, E., Sivasithamparam, K., 1991. Antifungal antibiotics produced by *Trichoderma* spp. Soil Biol. Biochem. 23 (11), 1011–1020.

Ghosh, S., Pal, S., 2016. Entomopathogenic potential of *Trichoderma longibrachiatum* and its comparative evaluation with malathion against the insect pest *Leucinodes orbonalis*. Environ. Monit. Assess. 188, 37.

Ghosh, S.K., Pal, S., 2017. Growth promotion and *Fusarium* wilt disease management eco-friendly in chickpea by *Trichoderma asperellum*. Int. J. Curr. Res. Aca. Rev. 5 (1), 14–26.

Ghosh, S.K., Santra, T., 2016. Scanning Electron Microscopy (SEM) of interaction between *Penicillium digitatum* and antagonistic yeast Candida famata on fruits of orange. JOBARI 4, 225–228.

Ghosh, S., Santra, T., Chakravarty, A., 2013. Study of antagonistic yeasts isolated from some natural sources of West Bengal. Agric. Biol. J. N. Am. 4, 33–40.

Ghosh, S.K., Banerjee, S., Pal, S., Chakraborty, N., 2018. Encountering epidemic effects of leaf spot disease (*Alternaria brassicae*) on *Aloe vera* by fungal biocontrol agents in agrifields—an ecofriendly approach. PLoS One. 13(3), e0193720.

Gianinazzi-Pearson, V., Dumas-Gaudot, E., Gollotte, A., Alaoui, A.T., Gianinazzi, S., 1996. Cellular and molecular defence-related root responses to invasion by arbuscular mycorrhizal fungi. New Phytol. 133 (1), 45–57.

Goetz, A.K., Ren, H., Schmid, J.E., Blystone, C.R., Thillainadarajah, I., Best, D.S., Nichols, H.P., Strader, L.F., Wolf, D.C., Narotsky, M.G., 2006. Disruption of testosterone homeostasis as a mode of action for the reproductive toxicity of triazole fungicides in the male rat. Toxicol. Sci. 95 (1), 227–239.

Gravel, V., Antoun, H., Tweddell, R.J., 2007. Growth stimulation and fruit yield improvement of greenhouse tomato plants by inoculation with *Pseudomonas putida* or *Trichoderma atroviride*: possible role of indole acetic acid (IAA). Soil Biol. Biochem. 39 (8), 1968–1977.

Guenoune, D., Galili, S., Phillips, D.A., Volpin, H., Chet, I., Okon, Y., Kapulnik, Y., 2001. The defense response elicited by the pathogen *Rhizoctonia solani* is suppressed by colonization of the AM-fungus *Glomus intraradices*. Plant Sci. 160 (5), 925–932.

Hahlbrock, K., Scheel, D., 1989. Physiology and molecular biology of phenylpropanoid metabolism. Annu. Rev. Plant Biol. 40 (1), 347–369.

Hanson, L., Howell, C., 2004. Elicitors of plant defense responses from biocontrol strains of *Trichoderma viren*. Phytopathology 94 (2), 171–176.

Harman, G.E., Howell, C.R., Viterbo, A., Chet, I., Lorito, M., 2004. *Trichoderma* species—opportunistic, avirulent plant symbionts. Nat. Rev. Microbiol. 2 (1), 43.

Hause, B., Mrosk, C., Isayenkov, S., Strack, D., 2007. Jasmonates in arbuscular mycorrhizal interactions. Phytochemistry 68 (1), 101–110.

He, D., Zheng, X., Yin, Y., Sun, P., Zhang, H., 2003. Yeast application for controlling apple postharvest diseases associated with *Penicillium expansum 1*. Bot. Bull. Acad. Sinica. 44.

Hermosa, R., Viterbo, A., Chet, I., Monte, E., 2012. Plant-beneficial effects of *Trichoderma* and of its genes. Microbiology 158 (1), 17–25.

Hohnjec, N., Vieweg, M.F., Pühler, A., Becker, A., Küster, H., 2005. Overlaps in the transcriptional profiles of *Medicago truncatula* roots inoculated with two different Glomus fungi provide insights into the genetic program activated during arbuscular mycorrhiza. Plant Physiol. 137 (4), 1283–1301.

Hooker, J., Jaizme-Vega, M., Atkinson, D., 1994. Biocontrol of plant pathogens using arbuscular mycorrhizal fungi. In: Impact of Arbuscular Mycorrhizas on Sustainable Agriculture and Natural Ecosystems. Springer, pp. 191–200.

Hossain, M.M., Sultana, F., Kubota, M., Koyama, H., Hyakumachi, M., 2007. The plant growth-promoting fungus *Penicillium simplicissimum* GP17-2 induces resistance in *Arabidopsis thaliana* by activation of multiple defense signals. Plant Cell Physiol. 48 (12), 1724–1736.

Hossain, M.M., Sultana, F., Islam, S., 2017. Plant growth-promoting fungi (PGPF): phytostimulation and induced systemic resistance. In: Plant-Microbe Interactions in Agro-Ecological Perspectives. Springer, pp. 135–191.

Howell, C., 1998. The role of antibiosis in biocontrol. In: Harman, G., Kubicek, C. (Eds.), *Trichoderma* and *Gliocladium*. In: vol. 2. Taylor & Francis, pp. 173–184.

Howell C (2003) Mechanisms employed by *Trichoderma* species in the biological control of plant diseases: the history and evolution of current concepts. Plant Dis. 87 (1): 4-10. Available from: http://genome.jgi-psf.org/Trive1/Trive1.home.html

Inoue, H., Kobayashi, T., Nozoye, T., Takahashi, M., Kakei, Y., Suzuki, K., Nakazono, M., Nakanishi, H., Mori, S., Nishizawa, N.K., 2009. Rice OsYSL15 is an iron-regulated iron (III)-deoxymugineic acid transporter expressed in the roots and is essential for iron uptake in early growth of the seedlings. J. Biol. Chem. 284 (6), 3470–3479.

Jia, Y.-J., Ito, H., Matsui, H., HONMA, M., 2000. 1-aminocyclopropane-1-carboxylate (ACC) deaminase induced by ACC synthesized and accumulated in *Penicillium citrinum* intracellular spaces. Biosci. Biotechnol. Biochem. 64 (2), 299–305.

Jogaiah, S., Mostafa, A., Tran, L.S.P., Shin-ichi, I., 2013. Characterization of rhizosphere fungi that mediate resistance in tomato against bacterial wilt disease. J. Exp. Bot. 64, 3829–3842.

Jogaiah, S., Abdelrahman, M., Tran, L.-S.P., Ito, S.-I., 2018. Different mechanisms of *Trichoderma virens*-mediated resistance in tomato against Fusarium wilt involve the jasmonic and salicylic acid pathways. Mol. Plant Pathol. 19, 870–882.

Jogaiah, S., Govind, S.R., Shetty, H.S., 2019. Role of oomycete elicitors in plant defense signaling. In: Jogaiah, S., Abdelrahman, M. (Eds.), Bioactive Molecules in Plant Defense. Springer, Cham, pp. 59–74.

Jones, R.W., Prusky, D., 2002. Expression of an antifungal peptide in Saccharomyces: a new approach for biological control of the postharvest disease caused by *Colletotrichum coccodes*. Phytopathology 92 (1), 33–37.

Joshi, S.M., De Britto, S., Jogaiah, S., Ito, S., 2019. Mycogenic selenium nanoparticles as potential new generation broad spectrum antifungal molecules. Biomolecules 9 (9), 419.

Kachroo, A., Kachroo, P., 2009. Fatty acid-derived signals in plant defense. Annu. Rev. Phytopathol. 47, 153–176.

Kapoor, R., 2008. Induced resistance in mycorrhizal tomato is correlated to concentration of jasmonic acid. Online J. Biol. Sci. 8 (3), 49–56.

Kapulnik, Y., Volpin, H., Itzhaki, H., Ganon, D., Galili, S., David, R., Shaul, O., Elad, Y., Chet, I., Okon, Y., 1996. Suppression of defence responses in mycorrhizal alfalfa and tobacco roots. New Phytol. 133 (1), 59–64.

Khan, M.H., Meghvansi, M., Panwar, V., Gogoi, H., Singh, L., 2010. Arbuscular mycorrhizal fungi-induced signalling in plant defence against phytopathogens. J. Phytology 2 (7), 53–69.

Klingner, A., Bothe, H., Wray, V., Marner, F.-J., 1995. Identification of a yellow pigment formed in maize roots upon mycorrhizal colonization. Phytochemistry 38 (1), 53–55.

Komatsu, M., 1968. *Trichoderma viride* as an Antagonist of Wood Inhabiting. Hymenomycetes, VIII. The Antibiotic Activity Against the Mycelial Growth of *Lentinus edodes* (Berk) Sig, of Three Genera *T. pachybasium*, Gliocladium and other Sterile Forms. Tottori Mycological Institute, Japan.

Korolev, N., David, D.R., Elad, Y., 2008. The role of phytohormones in basal resistance and Trichoderma-induced systemic resistance to Botrytis cinerea in *Arabidopsis thaliana*. BioControl 53 (4), 667–683.

Kosuta, S., Hazledine, S., Sun, J., Miwa, H., Morris, R.J., Downie, J.A., Oldroyd, G.E., 2008. Differential and chaotic calcium signatures in the symbiosis signaling pathway of legumes. Proc. Natl. Acad. Sci. 105 (28), 9823–9828.

Kurtzman, C., Fell, J., 1999. The Yeasts—A Taxonomic Study. 1055North Holland, Amsterdam.

Kurtzman, C., Fell, J.W., Boekhout, T., 2011. The Yeasts: A Taxonomic Study. Elsevier.

Laing, S., Deacon, J., 1991. Video microscopical comparison of mycoparasitism by *Pythium oligandrum*, *P. nunn* and an unnamed Pythium species. Mycol. Res. 95 (4), 469–479.

Lambais, M., 2000. Regulation of plant defense-related genes in arbuscular mycorrhizae. Curr. Adv. Mycorrhizae Res. 45–60.

Larose, G., Chênevert, R., Moutoglis, P., Gagné, S., Piché, Y., Vierheilig, H., 2002. Flavonoid levels in roots of *Medicago sativa* are modulated by the developmental stage of the symbiosis and the root colonizing arbuscular mycorrhizal fungus. J. Plant Physiol. 159 (12), 1329–1339.

Laville, J., Blumer, C., Von Schroetter, C., Gaia, V., Défago, G., Keel, C., Haas, D., 1998. Characterization of the hcnABC gene cluster encoding hydrogen cyanide synthase and anaerobic regulation by ANR in the strictly aerobic biocontrol agent *Pseudomonas fluorescens* CHA0. J. Bacteriol. 180 (12), 3187–3196.

Lee, W.J., Sandler, D.P., Blair, A., Samanic, C., Cross, A.J., Alavanja, M.C., 2007. Pesticide use and colorectal cancer risk in the agricultural health study. Int. J. Cancer 121 (2), 339–346.

Lee, G., Lee, S.-H., Kim, K.M., Ryu, C.-M., 2017. Foliar application of the leaf-colonizing yeast *Pseudozyma churashimaensis* elicits systemic defense of pepper against bacterial and viral pathogens. Sci. Rep. 7, 39432.

Levine, A., Tenhaken, R., Dixon, R., Lamb, C., 1994. H_2O_2 from the oxidative burst orchestrates the plant hypersensitive disease resistance response. Cell 79 (4), 583–593.

Lévy, J., Bres, C., Geurts, R., Chalhoub, B., Kulikova, O., Duc, G., Journet, E.-P., Ané, J.M., Lauber, E., Bisseling, T., 2004. A putative Ca^{2+} and calmodulin-dependent protein kinase required for bacterial and fungal symbioses. Science 303 (5662), 1361–1364.

Li, H.-Y., Yang, G.-D., Shu, H.-R., Yang, Y.-T., Ye, B.-X., Nishida, I., Zheng, C.-C., 2006. Colonization by the arbuscular mycorrhizal fungus *Glomus versiforme* induces a defense response against the root-knot nematode Meloidogyne incognita in the grapevine (*Vitis amurensis* Rupr.), which includes transcriptional activation of the class III chitinase gene VCH3. Plant Cell Physiol. 47 (1), 154–163.

Lim, T.K., Teh, B.K., 1990. Antagonism in vitro of *Trichoderma* sp. against several basidiomycetous, soil borne pathogens and *Sclerotium rolfsii*. Zeitschrift Fci Pflanzenkarkheiten Pur Pflanzenschutz 97, 33–34.

Lingua, G., D'Agostino, G., Massa, N., Antosiano, M., Berta, G., 2002. Mycorrhiza-induced differential response to a yellows disease in tomato. Mycorrhiza 12 (4), 191–198.

Liu, J., Maldonado-Mendoza, I., Lopez-Meyer, M., Cheung, F., Town, C.D., Harrison, M.J., 2007. Arbuscular mycorrhizal symbiosis is accompanied by local and systemic alterations in gene expression and an increase in disease resistance in the shoots. Plant J. 50 (3), 529–544.

van Loon, L.C., Rep, M., Pieterse, C.M., 2006. Significance of inducible defense-related proteins in infected plants. Annu. Rev. Phytopathol. 44, 135–162.

Lorenzo, O., Solano, R., 2005. Molecular players regulating the jasmonate signalling network. Curr. Opin. Plant Biol. 8 (5), 532–540.

Lorito, P., Emerson, O., Lomas, N., 1993. The isolation of toxic substances from the culture filtrate of *Trichoderma*. Phytopathology 26, 1068.

Lorito, M., Peterbauer, C., Hayes, C.K., Harman, G.E., 1994. Synergistic interaction between fungal cell wall degrading enzymes and different antifungal compounds enhances inhibition of spore germination. Microbiology 140 (3), 623–629.

Lorito, M., Woo, S.L., Harman, G.E., Monte, E., 2010. Translational research on *Trichoderma*: from omics to the field. Annu. Rev. Phytopathol. 48, 395–417.

Lotan, T., Fluhr, R., 1990. Xylanase, a novel elicitor of pathogenesis-related proteins in tobacco, uses a non-ethylene pathway for induction. Plant Physiol. 93 (2), 811–817.

Lu, Z., Tombolini, R., Woo, S., Zeilinger, S., Lorito, M., Jansson, J.K., 2004. In vivo study of *Trichoderma*-pathogen-plant interactions, using constitutive and inducible green fluorescent protein reporter systems. Appl. Environ. Microbiol. 70 (5), 3073–3081.

Ludwig-Müller, J., 2000. Hormonal balance in plants during colonization by mycorrhizal fungi. In: Arbuscular Mycorrhizas: Physiology and Function. Springer, pp. 263–285.

Luo, Y., Zhang, D.-D., Dong, X.-W., Zhao, P.-B., Chen, L.-L., Song, X.-Y., Wang, X.-J., Chen, X.-L., Shi, M., Zhang, Y.-Z., 2010. Antimicrobial peptaibols induce defense responses and systemic resistance in tobacco against tobacco mosaic virus. FEMS Microbiol. Lett. 313 (2), 120–126.

Mach, R., Zeilinger, S., 2003. Regulation of gene expression in industrial fungi: *Trichoderma*. Appl. Microbiol. Biotechnol. 60 (5), 515–522.

Marra, R., Ambrosino, P., Carbone, V., Vinale, F., Woo, S.L., Ruocco, M., Ciliento, R., Lanzuise, S., Ferraioli, S., Soriente, I., 2006. Study of the three-way interaction between *Trichoderma atroviride*, plant and fungal pathogens by using a proteomic approach. Curr. Genet. 50 (5), 307–321.

Martinez, C., Blanc, F., Le Claire, E., Besnard, O., Nicole, M., Baccou, J.-C., 2001. Salicylic acid and ethylene pathways are differentially activated in melon cotyledons by active or heat-denatured cellulase from *Trichoderma longibrachiatum*. Plant Physiol. 127 (1), 334–344.

Mauch, F., Staehelin, L.A., 1989. Functional implications of the subcellular localization of ethylene-induced chitinase and [beta]-1,3-glucanase in bean leaves. Plant Cell 1 (4), 447–457.

McDonnell, L., Plett, J.M., Andersson-Gunnerås, S., Kozela, C., Dugardeyn, J., Van Der Straeten, D., Glick, B.R., Sundberg, B., Regan, S., 2009. Ethylene levels are regulated by a plant encoded 1-aminocyclopropane-1-carboxylic acid deaminase. Physiol. Plant. 136 (1), 94–109.

McIntyre M, Nielsen J, Arnau J, Brink HVD, Hansen K, Madrid S (2004). Proceedings of the 7th European Conference on Fungal Genetics.

Meinert, R., Schüz, J., Kaletsch, U., Kaatsch, P., Michaelis, J., 2000. Leukemia and non-Hodgkin's lymphoma in childhood and exposure to pesticides: results of a register-based case-control study in Germany. Am. J. Epidemiol. 151 (7), 639–646.

Metcalf, D., Wilson, C., 2001. The process of antagonism of *Sclerotium cepivorum* in white rot affected onion roots by *Trichoderma koningii*. Plant Pathol. 50 (2), 249–257.

Mitra, R.M., Gleason, C.A., Edwards, A., Hadfield, J., Downie, J.A., Oldroyd, G.E., Long, S.R., 2004. A Ca^{2+}/calmodulin-dependent protein kinase required for symbiotic nodule development: gene identification by transcript-based cloning. Proc. Natl. Acad. Sci. 101 (13), 4701–4705.

Mohr, U., Lange, J., Boller, T., Wiemken, A., Vögeli-Lange, R., 1998. Plant defence genes are induced in the pathogenic interaction between bean roots and *Fusarium solani,* but not in the symbiotic interaction with the arbuscular mycorrhizal fungus *Glomus mosseae*. New Phytol. 138 (4), 589–598.
Monte, E., 2001. Understanding Trichoderma: between biotechnology and microbial ecology. Int. Microbiol. 4 (1), 1–4.
Mora, A., Earle, E., 2001. Resistance to *Alternaria brassicicola* in transgenic broccoli expressing a *Trichoderma harzianum* endochitinase gene. Mol. Breed. 8 (1), 1–9.
Morandi, D., 1996. Occurrence of phytoalexins and phenolic compounds in endomycorrhizal interactions, and their potential role in biological control. Plant and Soil 185 (2), 241–251.
Morandi, D., Le Quere, J., 1991. Influence of nitrogen on accumulation of isosojagol (a newly detected coumestan in soybean) and associated isoflavonoids in roots and nodules of mycorrhizal and non-mycorrhizal soybean. New Phytol. 117 (1), 75–79.
Morandi, D., Bailey, J., Gianinazzi-Pearson, V., 1984. Isoflavonoid accumulation in soybean roots infected with vesicular-arbuscular mycorrhizal fungi. Physiol. Plant Pathol. 24 (3), 357–364.
Morán-Diez, E., Hermosa, R., Ambrosino, P., Cardoza, R.E., Gutiérrez, S., Lorito, M., Monte, E., 2009. The ThPG1 endopolygalacturonase is required for the *Trichoderma harzianum*—plant beneficial interaction. Mol. Plant Microbe Interact. 22 (8), 1021–1031.
Morley, G.F., Gadd, G.M., 1995. Sorption of toxic metals by fungi and clay minerals. Mycol. Res. 99 (12), 1429–1438.
Mukerji, K., 1999. Mycorrhiza in control of plant pathogens: molecular approaches. In: Biotechnological Approaches in Biocontrol of Plant Pathogens. Springer, pp. 135–155.
Mukherjee, P.K., Horwitz, B.A., Herrera-Estrella, A., Schmoll, M., Kenerley, C.M., 2013. *Trichoderma* research in the genome era. Annu. Rev. Phytopathol. 51, 105–129.
Murali, M., Jogaiah, S., Amruthesh, K.N., Ito, S.-i., Shekar Shetty, H., 2013. Rhizosphere fungus *Penicillium chrysogenum* promotes growth and induces defense-related genes and downy mildew disease resistance in pearl millet. Plant Biol. 15, 111–118.
Nagaraju, A., Jogaiah, S., Mahadevamurthy, S., Ito, S.-I., 2012. Seed priming with *Trichoderma harzianum* isolates enhances plant growth and induces resistance against *Plasmopara halstedii*, an incitant of sunflower downy mildew disease. Aust. J. Plant Path. 41, 609–620.
Nandini, B., Puttaswamy, H., Prakash, H.S., Adhikari, S., Jogaiah, S., Geetha, N., 2020. Elicitation of novel trichogenic-lipid nanoemulsion signaling resistance against pearl millet downy mildew disease. Biomolecules 10 (1), 25.
Navazio, L., Moscatiello, R., Genre, A., Novero, M., Baldan, B., Bonfante, P., Mariani, P., 2007. A diffusible signal from arbuscular mycorrhizal fungi elicits a transient cytosolic calcium elevation in host plant cells. Plant Physiol. 144 (2), 673–681.
Newsham, K., Fitter, A., Watkinson, A., 1995. Arbuscular mycorrhiza protect an annual grass from root pathogenic fungi in the field. J. Ecol. 991–1000.
Niemeyer, H.M., 1988. Hydroxamic acids (4-hydroxy-1, 4-benzoxazin-3-ones), defence chemicals in the Gramineae. Phytochemistry 27 (11), 3349–3358.
Niemira, B.A., Hammerschmidt, R., Safir, G.R., 1996. Postharvest suppression of potato dry rot (*Fusarium sambucinum*) in prenuclear minitubers by arbuscular mycorrhizal fungal inoculum. Am. Potato J. 73 (11), 509–515.
Nigam, N., Kumar, R., Mukerji, K., Upadhyay, R., 1997. Fungi-a tool for biocontrol. In: - Upadhyay, R., Mukerji, K., Rajak, R. (Eds.), IPM System in Agriculture, Biocontrol in Emerging Biotechnology. Aditya Books Pvt Ltd, New Delhi II, pp. 503–526.
Nürnberger, T., Brunner, F., Kemmerling, B., Piater, L., 2004. Innate immunity in plants and animals: striking similarities and obvious differences. Immunol. Rev. 198 (1), 249–266.

Omero, C., Inbar, J., Rocha-Ramirez, V., Herrera-Estrella, A., Chet, I., Horwitz, B.A., 1999. G protein activators and cAMP promote mycoparasitic behaviour in *Trichoderma harzianum*. Mycol. Res. 103 (12), 1637–1642.

Omero, C., Dror, Y., Freeman, A., 2004. *Trichoderma* spp. antagonism to the dermatophyte *Trichophyton rubrum*: implications in treatment of onychomycosis. Mycopathologia 158 (2), 173–180.

Oostendorp, M., Kunz, W., Dietrich, B., Staub, T., 2001. Induced disease resistance in plants by chemicals. Eur. J. Plant Pathol. 107 (1), 19–28.

PAN. 2007. Available from: http://www.paninternational.org/panit/file/WG1/Eliminatingtheworstpesticide.pdf

Paxton, J., 1981. Phytoalexins—a working redefinition. J. Phytopathol. 101 (2), 106–109.

Peipp, H., Maier, W., Schmidt, J., Wray, V., Strack, D., 1997. Arbuscular mycorrhizal fungus-induced changes in the accumulation of secondary compounds in barley roots. Phytochemistry 44 (4), 581–587.

Pieterse, C.M., Leon-Reyes, A., Van der Ent, S., Van Wees, S.C., 2009. Networking by small-molecule hormones in plant immunity. Nat. Chem. Biol. 5 (5), 308.

Pimenta, R.S., Silva, F.L., Silva, J.F., Morais, P.B., Braga, D.T., Rosa, C.A., Corrêa Jr., A., 2008. Biological control of *Penicillium italicum, P. digitatum and P. expansum* by the predacious yeast *Saccharomycopsis schoenii* on oranges. Braz. J. Microbiol. 39 (1), 85–90.

Pozo, M.J., Azcón-Aguilar, C., 2007. Unraveling mycorrhiza-induced resistance. Curr. Opin. Plant Biol. 10 (4), 393–398.

Pozo, M.J., Azcón-Aguilar, C., Dumas-Gaudot, E., Barea, J.M., 1999. β-1,3-glucanase activities in tomato roots inoculated with arbuscular mycorrhizal fungi and/or *Phytophthora parasitica* and their possible involvement in bioprotection. Plant Sci. 141 (2), 149–157.

Pozo, M.J., Cordier, C., Dumas-Gaudot, E., Gianinazzi, S., Barea, J.M., Azcón-Aguilar, C., 2002. Localized versus systemic effect of arbuscular mycorrhizal fungi on defence responses to *Phytophthora* infection in tomato plants. J. Exp. Bot. 53 (368), 525–534.

Pozo, M.J., Verhage, A., García-Andrade, J., García, J.M., Azcón-Aguilar, C., 2009. Priming plant defence against pathogens by arbuscular mycorrhizal fungi. In: Mycorrhizas-Functional Processes and Ecological Impact. Springer, Berlin, pp. 123–135.

Pushpalatha, H.G., Sudisha, J., Geetha, N.P., Amruthesh, K.N., Shekar Shetty, H., 2011. Thiamine seed treatment enhances LOX expression, promotes growth and induces downy mildew disease resistance in pearl millet. Biol. Plantarum. 55, 522–527.

Ramette, A., Frapolli, M., Défago, G., Moënne-Loccoz, Y., 2003. Phylogeny of HCN synthase-encoding hcnBC genes in biocontrol fluorescent pseudomonads and its relationship with host plant species and HCN synthesis ability. Mol. Plant Microbe Interact. 16 (6), 525–535.

Ramraj, B., Shanmugam, N., Reddy, D., 1988. Biocontrol of Macrophomina root rot of cowpea and Fusarium wilt of tomato by VAM fungi. In: Mycorrhizae for Green Asia.Proceedings of First Asian Conference on Mycorrhiza, pp. 29–31.

Reithner, B., Brunner, K., Schuhmacher, R., Peissl, I., Seidl, V., Krska, R., Zeilinger, S., 2005. The G protein α subunit Tga1 of *Trichoderma atroviride* is involved in chitinase formation and differential production of antifungal metabolites. Fungal Genet. Biol. 42 (9), 749–760.

Reithner, B., Schuhmacher, R., Stoppacher, N., Pucher, M., Brunner, K., Zeilinger, S., 2007. Signaling via the *Trichoderma atroviride* mitogen-activated protein kinase Tmk1 differentially affects mycoparasitism and plant protection. Fungal Genet. Biol. 44 (11), 1123–1133.

Rishbeth, J., 1963. Stump protection against Fomes annosus: III—inoculation with *Peniophora gigantea*. Ann. Appl. Biol. 52 (1), 63–77.

Robin, C., Anziani, C., Cortesi, P., 2000. Relationship between biological control, incidence of hypovirulence, and diversity of vegetative compatibility types of *Cryphonectria parasitica* in France. Phytopathology 90 (7), 730–737.

Rocha-Ramírez, V., Omero, C., Chet, I., Horwitz, B.A., Herrera-Estrella, A., 2002. *Trichoderma atroviride* G-protein α-subunit gene tga1 is involved in mycoparasitic coiling and conidiation. Eukaryot. Cell 1 (4), 594–605.

Rockett, J.C., Narotsky, M.G., Thompson, K.E., Thillainadarajah, I., Blystone, C.R., Goetz, A.K., Ren, H., Best, D.S., Murrell, R.N., Nichols, H.P., 2006. Effect of conazole fungicides on reproductive development in the female rat. Reprod. Toxicol. 22 (4), 647–658.

Rotblat, B., Enshell-Seijffers, D., Gershoni, J.M., Schuster, S., Avni, A., 2002. Identification of an essential component of the elicitation active site of the EIX protein elicitor. Plant J. 32 (6), 1049–1055.

Safir, G.R., 1968. The Influence of Vesicular-Arbuscular Mycorrhiza on the Resistance of Onion to *Pyrenochaeta terrestris*. University of Illinois at Urbana-Champaign.

Salas-Marina, M., Silva-Flores, M., Uresti-Rivera, E., Castro-Longoria, E., Herrera-Estrella, A., Casas-Flores, S., 2011. Colonization of Arabidopsis roots by *Trichoderma atroviride* promotes growth and enhances systemic disease resistance through jasmonic acid/ethylene and salicylic acid pathways. Eur. J. Plant Pathol. 131 (1), 15–26.

Salzer, P., Corbière, H., Boller, T., 1999. Hydrogen peroxide accumulation in *Medicago truncatula* roots colonized by the arbuscular mycorrhiza-forming fungus *Glomus intraradices*. Planta 208 (3), 319–325.

Satapute, P., Milan, V.K., Shivakantkumar, S.A., Jogaiah, S., 2019a. Influence of triazole pesticides on tillage soil microbial populations and metabolic changes. Sci. Total Environ. 651, 2334–2344.

Satapute, P., Murali, K.P., Kurjogi, M., Jogaiah, S., 2019b. Physiological adaptation and spectral annotation of arsenic and cadmium heavy metal-resistant and susceptible strain *Pseudomonas taiwanensis*. Environ. Pollut. 251, 555–563.

Schaefer, K., Reeves, M., Spitzer, S., Kegley, S., 2004. Chemical trespass; pesticides in our bodies and corporate accountability. Global Pesticide Campaigner 14 (2), 4.

Schirmböck, M., Lorito, M., Wang, Y.-L., Hayes, C.K., Arisan-Atac, I., Scala, F., Harman, G.E., Kubicek, C.P., 1994. Parallel formation and synergism of hydrolytic enzymes and peptaibol antibiotics, molecular mechanisms involved in the antagonistic action of *Trichoderma harzianum* against phytopathogenic fungi. Appl. Environ. Microbiol. 60 (12), 4364–4370.

Segarra, G., Casanova, E., Bellido, D., Odena, M.A., Oliveira, E., Trillas, I., 2007. Proteome, salicylic acid, and jasmonic acid changes in cucumber plants inoculated with *Trichoderma asperellum* strain T34. Proteomics 7 (21), 3943–3952.

Segarra, G., Van der Ent, S., Trillas, I., Pieterse, C., 2009. MYB72, a node of convergence in induced systemic resistance triggered by a fungal and a bacterial beneficial microbe. Plant Biol. 11 (1), 90–96.

Shah, J.M., Raghupathy, V., Veluthambi, K., 2009. Enhanced sheath blight resistance in transgenic rice expressing an endochitinase gene from *Trichoderma virens*. Biotechnol. Lett. 31 (2), 239.

Shaul, O., Galili, S., Volpin, H., Ginzberg, I., Elad, Y., Chet, I., Kapulnik, Y., 1999. Mycorrhiza-induced changes in disease severity and PR protein expression in tobacco leaves. Mol. Plant Microbe Interact. 12 (11), 1000–1007.

Shoresh, M., Yedidia, I., Chet, I., 2005. Involvement of jasmonic acid/ethylene signaling pathway in the systemic resistance induced in cucumber by *Trichoderma asperellum* T203. Phytopathology 95 (1), 76–84.

Shoresh, M., Gal-On, A., Leibman, D., Chet, I., 2006. Characterization of a mitogen-activated protein kinase gene from cucumber required for *Trichoderma*-conferred plant resistance. Plant Physiol. 142 (3), 1169–1179.

Stacey, G., Keen, N.T., 1996. Plant-Microbe Interactions. Springer Science & Business Media.

St-Arnaud, M., Hamel, C., Fortin, J., 1994. Inhibition of *Pythium ultimum* in roots and growth substrate of mycorrhizal *Tagetes patula* colonized with *Glomus intraradices*. Can. J. Plant Pathol. 16 (3), 187–194.

Stepanova, A.N., Yun, J., Likhacheva, A.V., Alonso, J.M., 2007. Multilevel interactions between ethylene and auxin in *Arabidopsis* roots. Plant Cell 19 (7), 2169–2185.

Strange, R.N., 2006. Introduction to Plant Pathology. John Wiley & Sons.

Sudisha, J., Niranjana, S.R., Sukanya, S.L., Girijamba, R., Lakshmidevi, N., Shekar Shetty, H., 2010. Relative efficacy of strobilurin formulations in the control of downy mildew of sunflower. J. Pestic. Sci. 83, 461–470.

Sudisha, J., Sharathchandra, R., Amruthesh, K., Kumar, A., Shetty, H.S., 2012. Pathogenesis Related Proteins in Plant Defense Response. Plant Defence: Biological Control. Springer, pp. 379–403.

Sultana, F., Hossain, M.M., Kubota, M., Hyakumachi, M., 2008. Elicitation of systemic resistance against the bacterial speck pathogen in *Arabidopsis thaliana* by culture filtrates of plant growth-promoting fungi. Can. J. Plant Pathol. 30 (2), 196–205.

Tagu, D., Martin, F., 1996. Molecular analysis of cell wall proteins expressed during the early steps of ectomycorrhiza development. New Phytol. 133 (1), 73–85.

Terras, F.R., Eggermont, K., Kovaleva, V., Raikhel, N.V., Osborn, R.W., Kester, A., Rees, S.B., Torrekens, S., Van Leuven, F., Vanderleyden, J., 1995. Small cysteine-rich antifungal proteins from radish: their role in host defense. Plant Cell 7 (5), 573–588.

Tjamos, E., Papavizas, G.C., Cook, R.J., 2013. Biological Control of Plant Diseases: Progress and Challenges for The Future. vol. 230. Springer Science & Business Media.

Todorovic, B., Glick, B.R., 2008. The interconversion of ACC deaminase and D-cysteine desulfhydrase by directed mutagenesis. Planta 229 (1), 193–205.

Torres-Barragán, A., Zavaleta-Mejía, E., González-Chávez, C., Ferrera-Cerrato, R., 1996. The use of arbuscular mycorrhizae to control onion white rot (*Sclerotium cepivorum* Berk.) under field conditions. Mycorrhiza 6 (4), 253–257.

Trotta, A., Varese, G., Gnavi, E., Fusconi, A., Sampo, S., Berta, G., 1996. Interactions between the soilborne root pathogen *Phytophthora nicotianae* var. parasitica and the arbuscular mycorrhizal fungus *Glomus mosseae* in tomato plants. Plant and Soil 185 (2), 199–209.

Van de Poel, B., Van Der Straeten, D., 2014. 1-Aminocyclopropane-1-carboxylic acid (ACC) in plants: more than just the precursor of ethylene. Front. Plant Sci. 5, 640.

Van Loon, L.C., 1999. Occurrence and properties of plant pathogenesis-related proteins. In: Pathogenesis-related Proteins in Plants. CRC Press, Boco Raton, pp. 1–19.

Van Loon, L., Bakker, P., Pieterse, C., 1998. Systemic resistance induced by rhizosphere bacteria. Annu. Rev. Phytopathol. 36 (1), 453–483.

Van Wees, S.C., Van der Ent, S., Pieterse, C.M., 2008. Plant immune responses triggered by beneficial microbes. Curr. Opin. Plant Biol. 11 (4), 443–448.

Vey, A., Hoagland, R.E., Butt, T.M., 2001. 12 Toxic metabolites of fungal biocontrol agents. Fungi as Biocontrol Agents. 311.

Vierheilig, H., Bago, B., Albrecht, C., Poulin, M.-J., Piché, Y., 1998. Flavonoids and arbuscular-mycorrhizal fungi. In: Flavonoids in the Living System. Springer, pp. 9–33.

Vinale, F., Sivasithamparam, K., Ghisalberti, E., Marra, R., Barbetti, M., Li, H., Woo, S., Lorito, M., 2008. A novel role for *Trichoderma* secondary metabolites in the interactions with plants. Physiol. Mol. Plant Pathol. 72 (1-3), 80–86.

Viterbo, A., Horwitz, V., 2010. Mycoparasitism. In: Cellular and Molecular Biology of Filamentous Fungi. American Society of Microbiology, pp. 676–693.

Viterbo, A., Haran, S., Friesem, D., Ramot, O., Chet, I., 2001. Antifungal activity of a novel endochitinase gene (chit36) from *Trichoderma harzianum* Rifai TM. FEMS Microbiol. Lett. 200 (2), 169–174.

Viterbo, A., Wiest, A., Brotman, Y., Chet, I., Kenerley, C., 2007. The 18mer peptaibols from *Trichoderma* virens elicit plant defence responses. Mol. Plant Pathol. 8 (6), 737–746.

Viterbo, A., Landau, U., Kim, S., Chernin, L., Chet, I., 2010. Characterization of ACC deaminase from the biocontrol and plant growth-promoting agent *Trichoderma asperellum* T203. FEMS Microbiol. Lett. 305 (1), 42–48.

Vitti, A., Pellegrini, E., Nali, C., Lovelli, S., Sofo, A., Valerio, M., Scopa, A., Nuzzaci, M., 2016. *Trichoderma harzianum* T-22 induces systemic resistance in tomato infected by cucumber mosaic virus. Front. Plant Sci. 7, 1520.

Vlot, A.C., Dempsey, D.A., Klessig, D.F., 2009. Salicylic acid, a multifaceted hormone to combat disease. Annu. Rev. Phytopathol. 47, 177–206.

Voisard, C., Keel, C., Haas, D., Dèfago, G., 1989. Cyanide production by *Pseudomonas fluorescens* helps suppress black root rot of tobacco under gnotobiotic conditions. EMBO J. 8 (2), 351–358.

Volpin, H., Elkind, Y., Okon, Y., Kapulnik, Y., 1994. A vesicular arbuscular mycorrhizal fungus (*Glomus intraradix*) induces a defense response in alfalfa roots. Plant Physiol. 104 (2), 683–689.

Walter, M.H., Fester, T., Strack, D., 2000. Arbuscular mycorrhizal fungi induce the non-mevalonate methylerythritol phosphate pathway of isoprenoid biosynthesis correlated with accumulation of the 'yellow pigment'and other apocarotenoids. Plant J. 21 (6), 571–578.

Ward, M.H., DeKok, T.M., Levallois, P., Brender, J., Gulis, G., Nolan, B.T., VanDerslice, J., 2005. Workgroup report: drinking-water nitrate and health—recent findings and research needs. Environ. Health Perspect. 113 (11), 1607–1614.

Weindling, R., 1934. Studies on a lethal principle effective in the parasitic action of *Trichoderm lignorum on Rhizoctonia solani* and other soil fungi. Phytopathology 24, 1153–1179.

Wisniewski, M., Biles, C., Droby, S., McLaughlin, R., Wilson, C., Chalutz, E., 1991. Mode of action of the postharvest biocontrol yeast, *Pichia guilliermondii*. I. Characterization of attachment to *Botrytis cinerea*. Physiol. Mol. Plant Pathol. 39 (4), 245–258.

de Wit, P.J., 1992. Molecular characterization of gene-for-gene systems in plant-fungus interactions and the application of avirulence genes in control of plant pathogens. Annu. Rev. Phytopathol. 30 (1), 391–418.

de Wit, P.J., 2007. How plants recognize pathogens and defend themselves. Cell. Mol. Life Sci. 64 (21), 2726–2732.

Woo, S.L., Lorito, M., 2007. Exploiting the interactions between fungal antagonists, pathogens and the plant for biocontrol. In: Novel Biotechnologies for Biocontrol Agent Enhancement and Management. Springer, pp. 107–130.

Woo, S., Formisano, E., Fogliano, V., Cosenza, C., Mauro, A., Turra, D., Soriente, I., Ferraioli, S., Scala, F., Lorito, M., 2004. Factors that contribute to the mycoparasitism stimulus in *Trichoderma atroviride* strain P1. J. Plant Pathol. 30 (4), 421.

Woo, S., Scala, F., Ruocco, M., Lorito, M., 2006. The molecular biology of the interactions between *Trichoderma* spp., phytopathogenic fungi, and plants. Phytopathology 96 (2), 181–185.

Yao, M., Desilets, H., Charles, M., Boulanger, R., Tweddell, R., 2003. Effect of mycorrhization on the accumulation of rishitin and solavetivone in potato plantlets challenged with *Rhizoctonia solani*. Mycorrhiza 13 (6), 333–336.
Yedidia, I., Benhamou, N., Chet, I., 1999. Induction of defense responses in cucumber plants (*Cucumis sativus* L.) by the biocontrol agent *Trichoderma harzianum*. Appl. Environ. Microbiol. 65 (3), 1061–1070.
Yedidia, I., Benhamou, N., Kapulnik, Y., Chet, I., 2000. Induction and accumulation of PR proteins activity during early stages of root colonization by the mycoparasite *Trichoderma harzianum* strain T-203. Plant Physiol. Biochem. 38 (11), 863–873.
Zeilinger, S., Gruber, S., Bansal, R., Mukherjee, P.K., 2016. Secondary metabolism in *Trichoderma*–chemistry meets genomics. Fungal Biol. Rev. 30 (2), 74–90.

Further reading

Gaddie, G., 1993. Interaction of fungi with toxic metals. New Phytol. 124, 25–60.
Garon, D., Krivobok, S., Seigle-Murandi, F., 2000. Fungal degradation of fluorescence. Chemosphere 40 (1), 91–97.
Oldroyd, G.E., Harrison, M.J., Paszkowski, U., 2009. Reprogramming plant cells for endosymbiosis. Science 324 (5928), 753–754.
Pinedo-Rivilla, C., Aleu, J., Collado, I., 2009. Pollutants biodegradation by fungi. Curr. Org. Chem. 13 (12), 1194–1214.
Smith, P., Martino, D., Cai, Z., Gwary, D., Janzen, H., Kumar, P., McCarl, B., Ogle, S.O., Mara, F., Rice, C., Scholes, B., Sirotenko, O., 2007. Agriculture. In: Metz, B., Davidson, O.R., Bosch, P.R., Dave, R., Meyer, L.A. (Eds.), Climate Change 2007: Mitigation. Contribution of Working Group III to the Fourth Assessment Report of the Intergovernmental Panel on Climate Change. Cambridge University Press, Cambridge, UK, pp. 497–540.
Tigini, V., Prigione, V., Di Toro, S., Fava, F., Varese, G.C., 2009. Isolation and characterisation of polychlorinated biphenyl (PCB) degrading fungi from a historically contaminated soil. Microb. Cell Fact. 8 (1), 5.

Ecological studies of fungal biodiversity in freshwater and their broad-spectrum applications

27

K.S. Divya[a], *S. Mahadeva Murthy*[a], *and Sudisha Jogaiah*[b]
[a]Department of Microbiology, Yuvaraja's College, University of Mysore, Mysore, Karnataka, India, [b]Laboratory of Plant Healthcare and Diagnostics, P.G. Department of Biotechnology and Microbiology, Karnatak University, Dharwad, Karnataka, India

27.1 Introduction

Freshwater habitat occupies a relatively small portion of the earth's surface as compared to marine and terrestrial habitats, but their importance to man is far greater because freshwater habitat is the convenient and the cheapest source of freshwater for domestic and industrial needs. Freshwater habitat may consist of lentic (standing or stationary water bodies like lakes, ponds, and puddles) or lotic (running freshwater bodies like river and stream) ecosystems. Lotic systems are affected directly by using them as wastewater devices and indirectly by contaminating the soil (Neill, 1993).

The assumption that "Freshwater is a gift of God which would continue to be available in perpetuity and in abundance" stands seriously challenged. The UN water conference recorded that if all the world's waters are represented by half-gallon bottle, the quantity of freshwater would be about half a teaspoon and a single droplet would suffice to represent the surface flowing water (UNEP, 1978). It is apprehended that the day is not far when a drop of water will cost more than a drop of oil.

The term ecology is derived from the two Greek words, "oikos" meaning "household" or "dwelling" and "logos" meaning "study" or "law" or "discourse." Thus, ecology is the law of the household or the science that explores the interrelationships between organisms and their living (biotic) and nonliving (abiotic) environments. Aquatic ecosystems (consisting of ocean, lakes, ponds, pools, and lotic waters) play an important ecological role on a global scale, as the greater part of many natural microbial conversions occur in water. Ecology deals with the various principles which govern relationships between organisms and their environment. Further, the microorganisms occupy a key position in the orderly flow of materials and energy through the global ecosystem by virtue of their metabolic abilities to transform organic and inorganic substances. With population increase, the ecological cycles of water become complex: residues are dumped into the rivers without proper purification, polluting them to the point of making them dangerous to human health. Industrial effluents, chemical waste, and heavy metals are finding ways to water bodies in such quantities that often surpass the self-purification capacity of the bacteria in the river. Pollution of

Biocontrol Agents and Secondary Metabolites. https://doi.org/10.1016/B978-0-12-822919-4.00027-2

rivers first affects its physicochemical quality and then systemically destroys the microbial and plankton communities. Thus, unbalancing the delicate microbial food web affects the food chain of the ecosystem.

A change in water chemistry due to surface water contamination negatively affects all levels of an ecosystem. It poses a threat to the health of lower food chain organisms and, consequently, the availability of the food supply up through the food chain. It can also impact the health of wetlands and impair their ability to support healthy ecosystems, control flooding, and filter pollutants from stormwater runoff. Contaminated surface water can also affect the health of animals and humans when they drink or bathe in contaminated water or, for aquatic organisms, when they ingest contaminated sediments. One of the major concerns associated with surface water contamination is the ability of aquatic organisms, like fish. When other animals or humans ingest these organisms, they receive a higher dose of contamination than they would have if they had been directly exposed to the original source of the contamination. Several physicochemical factors (Homyra and Naz, 2006) and varying flux conditions can markedly influence the aquatic system and cause changes in abundance, diversity, and succession.

A fungus is a member of a large group of eukaryotic organisms that includes microorganisms such as yeasts and molds, as well as the more familiar mushrooms. One major difference is that fungal cells have cell walls that contain chitin, unlike the cell walls of plants and some protists, which contain cellulose, and unlike the cell walls of bacteria. These and other differences show that the fungi form a single group of related organisms, named the *Eumycota* (*true fungi* or *Eumycetes*), that share a common ancestor (is a *monophyletic group*). This fungal group is distinct from the structurally similar myxomycetes (slime molds) and oomycetes (water molds). The discipline of biology devoted to the study of fungi is known as mycology. Mycology has often been regarded as a branch of botany, even though it is a separate kingdom in biological taxonomy. Genetic studies have shown that fungi are more closely related to animals than to plants.

Abundant worldwide, most fungi are inconspicuous because of the small size of their structures, and their cryptic lifestyles in soil, on dead matter, and as symbionts of plants, animals, or other fungi. They may become noticeable when fruiting, either as mushrooms or molds. Fungi perform an essential role in the decomposition of organic matter and have fundamental roles in nutrient cycling and exchange. The fungus kingdom encompasses an enormous diversity of taxa with varied ecologies, life cycle strategies and morphologies ranging from single-celled aquatic chytrids to large mushrooms. Ever since the pioneering 18th and 19th century taxonomical works of Carl Linnaeus, Christian Hendrik Persoon, and Elias Magnus Fries, fungi have been classified according to their morphology (e.g., characteristics such as spore color or microscopic features) or physiology. Advances in molecular genetics have opened the way for DNA analysis to be incorporated into taxonomy, which has sometimes challenged the historical groupings based on morphology and other traits. Phylogenetic studies published in the last decade have helped reshape the classification of Kingdom Fungi, which is divided into one subkingdom, seven phyla, and 10 subphyla.

27.2 Diversity of fungi

Fungi are eukaryotic, heterotrophic organisms, including both single-celled yeasts and multicellular filamentous fungi. They primarily function as recyclers of organic material. Many fungal species can survive in an oligotrophic environment, through scavenging nutrients from the substrate which they colonize, or the air or water in which they live. To maximize nutrient uptake, filamentous fungi form mats of fine hyphae. Dispersion is via spores. Fungi also produce secondary metabolites, some of which are toxins. Some of the fungal species and the metabolites they produce are human pathogens or allergens (Paterson and Lima, 2005).

The presence of fungi in water distribution systems may cause other indirect challenges for water companies. For instance, the secondary metabolites produced by some species can alter the taste and smell of water, generating complaints from end users. Organic acids produced by fungal metabolic processes can increase the rate of corrosion of water pipes, especially when it is difficult to maintain sufficient concentrations of water disinfectants, such as chlorine, throughout the distribution system (Grabinska-Loniewska et al., 2007; Thamizhmani et al., 2013). Due to their tolerance of oligotrophic environment, some species of fungi are able to colonize drinking water distribution systems, which are typically low in nutrients. The significance of drinking water as an exposure pathway to pathogenic, allergenic, or toxic fungal species or their metabolites is not well known.

The variety and galaxy of fungi and their natural beauty occupy a prime place in the biological world and India has been the cradle for such fungi. Only a fraction of total fungal wealth has been subjected to scientific scrutiny and mycologists have to unravel the unexplored and hidden wealth. One third of fungal diversity of the globe exists in India. Out of 1.5 million fungi, only 50% are characterized until now. Unfortunately, only around 5%–10% of fungi can be cultured artificially. Fungi are not only beautiful but play a significant role in the daily life of human beings besides their utilization in industry, agriculture, medicine, food industry, textiles, bioremediation, natural cycling, as biofertilizers and many other ways. Fungal biotechnology has become an integral part of human welfare (Manoharachary et al., 2005).

Most studies take samples of water from the tap or various places in the distribution system, often as part of routine bacteriological monitoring. It is difficult to obtain a representative sample; fungi are often unevenly distributed through water. Many are held in biofilms, fragments of which occasionally break off. Therefore, quantities of fungi are likely to be highly variable with time with occurrences in mobile phases often for short durations and small volumes (Paterson and Lima, 2005; Hageskal et al., 2009).

Fungal infections are becoming of increasing concern due to the increasing numbers of immunocompromised patients and those with other risk factors (Annaisie et al., 2002). Therefore, there is a need to ascertain what the exposure pathways are and whether treated drinking water has a role as a source of exposure to pathogenic fungi. There is a need to determine the extent of current knowledge regarding which fungal species have been reliably identified as present in treated drinking water and its distribution systems, their ecology and the extent to which they are a hazard to human health. Abundance and diversity of fungi in a salt pan and marine of the East coast of Tamil Nadu, India was studied in terms of species diversity, frequency of occurrence in Marakkanam, Pillaichavadi, T.R. Pattanam, Muthukuda, and Kanyakumari of the marine and salt pane ecosystem. Among the 46 species of fungi isolated, totally 35 species were isolated from sediment samples followed by water with 31 species. Among the fungal members, *Aspergillus* was common genus represented with 10 species followed by *Penicillium* with five species (Thamizhmani et al., 2013).

Fungi have a worldwide distribution and grow in a wide range of habitats, including extreme environments such as deserts or areas with high salt concentrations (Vaupotic et al., 2008) or ionizing radiation (Dadachova et al., 2007), as well as in deep-sea sediments (Raghukumar and Raghukumar, 1998). Some can survive the intense UV and cosmic radiation encountered during space travel (Sancho et al., 2007). Most grow in terrestrial environments though several species live partly or solely in aquatic habitats, such as the chytrid fungus *Batrachochytrium dendrobatidis*, a parasite that has been responsible for a worldwide decline in amphibian populations. This organism spends part of its life cycle as a motile zoospore, enabling it to propel itself through water and enter its amphibian host (Brem and Lips, 2008). Other examples of aquatic fungi include those living in hydrothermal areas of the ocean (Le Calvez et al., 2009).

Nearly 100,000 species of fungi have been formally described by taxonomists, but the global biodiversity of the fungus kingdom is not fully understood (Mueller and Schmit, 2006). On the basis of observations of the ratio of the number of fungal species to the number of plant species in selected environments, the fungal kingdom has been estimated to contain about 1.5 million species (Hawksworth, 2006); a recent (2011) estimate suggests there may be over 5 million species (Blackwell, 2011). In mycology, species have historically been distinguished by a variety of methods and concepts. Classification based on morphological characteristics, such as the size and shape of spores or fruiting structures, traditionally dominated fungal taxonomy. Species may also be distinguished by their biochemical and physiological characteristics, such as their ability to metabolize certain biochemicals or their reaction to chemical tests. The biological species concept discriminates species based on their ability to mating in fungi. The application of molecular tools, such as DNA sequencing and phylogenetic analysis, to study diversity has greatly enhanced the resolution and added robustness to estimates of genetic diversity within various taxonomic groups (Jogaiah et al., 2009).

Aquatic fungi, in the broadest sense, include fungi present transiently in water, terrestrial fungi that release spores which are dispersed in water, and species that function entirely within the water. Water is found in transient pools after rain, swamps, estuaries, permanent rivers, and oceans. Within a single environment, the water can have predictable characteristics. At the edge, however, water bodies may contain soil and plant materials and will be a changing environment.

Fungi living in these environments require a wide array of characteristics to cope with the constraints. Conditions in freshwater are markedly different from those in the ocean. The open sea has quite different salinity to salterns where the water is slowly evaporating. Sometimes the water is extremely acidic or toxic. Fungi are found in all these environments. The mechanisms that enable the successful occupation of these environments are largely unexplored. Indeed, fungi have only recently been acknowledged as important inhabitants of extreme aquatic habitats. This section provides some details of the simple responses widely studied in the fungi.

27.3 Ecological impact on fungal biodiversity

Fungi, like all living organisms, require three essential factors for life viz., air, food, and water. Fungal growth in the indoor environment is dictated by the availability of these three factors. Just like humans, what most fungi "breathe" is oxygen, which is generally abundant in indoor environments. Fungi can digest human food, but also many materials that are commonly used in buildings. Water availability, however, is a factor that can be controlled to prevent or reduce fungal growth.

Fungi do not ingest their food like animals. Instead, they release enzymes into their surrounding environment to break down complex materials into simpler ones that can be absorbed by the fungus in a process known as absorptive nutrition. For these enzymes to leave the fungus, remain functional, and to break down complex substances, water is necessary. Once the complex substances have been broken down and dissolved in water they are absorbed by the fungus.

Many physical factors in the environment can affect the water requirements of fungi. Two of the most important factors in the indoor environment are temperature and the type of food that is present. For each species of fungi, there is an optimal temperature range for growth to occur. Outside of that optimal temperature range, water requirement for growth increases. The same is true if the food source is not something the fungi normally would grow on in nature. Many fungi have evolved to break down plant compounds; thus fungi tend to grow more frequently and quickly on plant-derived materials such as wood and paper. When building materials do not come from plant origin or have been altered in some way, fungi require a larger amount of water to grow.

Studies have indicated that some fungi can survive periodic drying and rewetting, continuing their growth once free water is available again. Fungi are also susceptible to the rate of drying for a given material or substrate that they grow on. This process is more detrimental to the fungus when the drying occurs rapidly. The slower the drying process occurs, the more likely the fungus will be able to continue its growth once water is reintroduced to the environment. A series of research work has been carried to study the ecological factors influencing a diverse group of fungi to flourish.

Natural surface water bodies like rivers and streams are subjected to pollution comprising organic and inorganic constituent. The water quality of Narmada River was monitored to evaluate the pollution of the riverine ecosystem using physicochemical parameters such as pH, EC, and turbidity to study the influence of the water runoff from various fields directly or indirectly (Sharma et al., 2011).

Different aspects of water and wastewater have been studied by many workers such as physicochemical characteristics of sewage entering into the river (Matkar and Gangotri, 2002). The physical factors contributing to water quality include temperature, pH and turbidity, water level, and intensity of illumination (Sivakumar et al., 2003). Increase in the pollution input downstream gives a sufficient indication of the deteriorating quality of water due to entry of wastewater in the river. Extensive investigations on physicochemical parameters of river Ganga concluded that Ganga water is most probably not fit for drinking and its need to be treated to reduce the contaminations especially the alkalinity and hardness (Khare et al., 2011; Yadav and Srivastava, 2011).

Physicochemical characteristics of aquatic fungi were studied in relation to physicochemical factors of Kosi River (Paliwal and Sati, 2009) and the water quality of lake in Kalyani, West Bengal (Sinha and Biswas, 2011). Low DO and higher concentration of nutrients like phosphate, nitrate, sulfate and chloride combined with high value of water quality index (WQI) indicated a high degree of pollution. The study was carried out to determine the physicochemical properties of Yamuna River water from nine different sampling sites in Agra City. The analysis was done for the parameters like turbidity, pH, total dissolved solids, electrical conductivity, total hardness, total alkalinity, chloride, calcium, magnesium, and pH. The results showed that Yamuna river water is alkaline in nature; turbidity and total dissolved solids were found above the WHO permissible limits (Gupta et al., 2013).

Aquatic fungi and fungus-like organisms isolated from surface water situated near Bialystok in Podlasie Province of Poland using the insect *Notonecta glauca* as bait (Kiziewicz and Kurzatkowska, 2004). The review of literature on the occurrence of fungi in drinking water which has received increased attention in the last decades and fungi are now generally accepted as drinking water contaminants (Hageskal et al., 2009).

In all, 34 identified in addition to five unidentified species appertaining to 10 genera of zoosporic fungi were identified and isolated from 84 polluted water samples, and showed that pH values of the polluted water samples had no influence on the occurrence of zoosporic fungi. Water samples characterized by high organic matter content and low total soluble salts were the richest in zoosporic fungal species (Ali, 2007). Comprehensive phylogeny derived from 5 genes for 356 isolates and 41 families in Dothideomycetes (Schoch et al., 2009).

Ecological studies of aquatic hypomycetes were carried out in the canal and its associated irrigation water channels and compared physicochemical properties and the spora of the aquatic hypomycetes (Arshad and Bareen, 2009). Aquatic filamentous fungi were surveyed and this study showed for the first time the existence of aquatic fungi in four lakes. Amorphic fungi dominated the mycobiota (El-Sharouny et al., 2009). A review was made on the aquatic fungi and documented that aquatic fungi are common only in various lentic habitats and play a crucial role in nutrient and carbon cycling and interact with the organisms (Wurzbacher et al., 2010). The potentiality of the marine fungi was reviewed and inferred that the physicochemical properties of the marine environment are likely to confer the marine fungi with special physiological adaptations (Raghukumar, 2008). The vital role of fungi in detritus decomposition, nutrient cycling, and energy flow was also studied.

The qualitative and quantitative fungal composition of dam water showed the domination of *Aspergillus niger*. Fungus and fungus-like organisms as a biological factor of ecological water systems have a significant influence on the environment and its modifications. Filamentous fungi present in hypersaline environment (Dead Sea) suggest that the sea could be considered a possible transmission route for filamentous fungi and many potential health hazards. The qualitative and quantitative composition of river water and filamentous fungi was surveyed to understand the overall diversity among different taxonomic groups of fungi (Parveen et al., 2011).

A major problem one encounters when reviewing the literature on aquatic fungi is that often only one group is addressed and all others are neglected. For example, after several decades of intensive research on leaves decaying in streams, much is known about aquatic *Hyphomycetes*, and they are widely regarded as dominating this process. Recently, however, molecular methods have revealed an astonishing diversity of fungi belonging to other groups suggesting a hitherto unknown ecological role of these taxa (Nikolcheva and Bärlocher, 2004). The Kingdom Fungi was only recognized some 40 years ago, and its diversity is vastly understudied (only approximately 7% of total estimated species have been described). Accordingly, there is an urgent need for better documentation of the numerous undescribed species, especially in aquatic habitats (Shearer et al., 2007). Only recently has a consistent fungal tree of life been generated and it excludes several phylogenetic groups formerly considered fungi, for example, Oomycetes and various slime molds (Lutzoni et al., 2004; James et al., 2006).

Based on their evolutionary history, fungi in aquatic habitats are either primarily (Chytridiomycetes) or secondarily adapted to life in water (aquatic hyphomycetes and yeasts). In addition, primarily terrestrial forms (various fungi imperfecti and endophytes) are found in lakes, suggesting that their occurrence is not necessarily restricted to a single habitat by stringent morphological or physiological adaptations. Like Oomycetes (fungus-like organisms belonging to the Kingdom Chromista), Chytridiomycetes (Kingdom Fungi) possess chemotactic, flagellated zoospores, which can disperse in the water column. They are located at the base of the fungal phylogenetic tree, which means they had a window of 900–480 million years to potentially colonize and adapt to all aquatic habitats and niches. They may well be ubiquitous within freshwater ecosystems but have often been overlooked or misidentified, for example, as flagellates (Lefèvre et al., 2007).

Analysis of physical and chemical parameters in the water samples helps to determine the type of pollution that may be affecting a waterway and can provide some clues as to the sources. Water quality is good if naturally occurring substances are present at levels that support aquatic life. Problems occur when activities alter natural levels or introduce substances that are toxic to aquatic life. The important physicochemical parameters used to analyze water quality are pH, temperature, turbidity, electrical conductivity (EC), total hardness (TH), Calcium (Ca), Magnesium (Mg), alkalinity, Chloride (Cl), Fluoride (Fl), Nitrate (NO_3), Sulfate (SO_4), Phosphate (PO_4), Ferrum (Fe), Dissolved oxygen (DO), Biochemical oxygen demand (BOD), and Chemical oxygen demand (COD).

The pH and temperature can be recorded on the spot. For the estimation of DO, water samples should be collected in separate sterilized 300 mL BOD bottles and fixed immediately using Winkler's reagent and later analyzed in the laboratory. In addition to this raw sample, samples acidified with H_2SO_4 and nitric acid can be collected in three separate sterile polythene bottles of 50 mL. The acidified samples are used for the analysis of BOD and the raw samples for immediate analysis in the laboratory for acidity, alkalinity, and inorganic ions.

The fungal assay is done to understand the overall diversity among different taxonomic groups of fungi present, as well as their occurrence, frequency, and composition of the fungal species in the freshwater. The qualitative fungal composition can be surveyed by spreading the samples over presterilized and solidified potato dextrose agar medium and incubated for 5–7 days. Fungal colonies are identified using stereomicroscope and by studying the colony characteristics using fungal manuals (Gilman, 1956; Singh et al., 1991).

The variety and galaxy of fungi and their natural beauty occupy a prime place in the biological world and India has been the cradle for such fungi. Only a fraction of total fungal wealth has been subjected to scientific scrutiny and mycologists have to unravel the unexplored and hidden wealth. One third of fungal diversity of the globe exists in India. Out of 1.5 million fungi, only 50% are characterized until now. Unfortunately, only around 5%–10% of fungi can be cultured artificially. Fungi are not only beautiful but play a significant role in the daily life of human beings besides their utilization in industry, agriculture, medicine, food industry, textiles, bioremediation, natural cycling, biofertilizers, and many other ways. Fungal biotechnology has become an integral part of human welfare (Manoharachary et al., 2005).

Most studies take samples of water from the tap or from various places in the distribution system, often as part of routine bacteriological monitoring. It is difficult to obtain a representative sample. Fungi are often unevenly distributed through the water. Many are held in biofilms, fragments of which occasionally break off. Therefore, quantities of fungi are likely to be highly variable with time with occurrences in mobile phases often for short durations and small volumes (Hageskal et al., 2009).

Monitoring of freshwaters is primarily done to detect the status and trends on water quality and to identify whether observed trends arise from natural or anthropogenic causes. Most important environmental problems in river water quality are eutrophication, acidification, and emission dispersion where nonpoint source pollution has become increasingly important within the last decades. Monitoring of water quality is important to detect and control contamination in surface and underground water (Gray, 1999). In India ponds, rivers, and groundwater are used for domestic and agricultural purposes. The quality of water may be described according to their physicochemical and microbiological characteristics. For effective maintenance of water quality through appropriate control measures, continuous monitoring of large number of quality parameters is essential.

Water analysis is a statement of the kind and amount of impurities contained in the water. The form of the statement will depend on the point of view from which the water is being tested, that is to say, whether it is being examined for its sustainability for drinking, domestic, or its suitability for industrial processes.

27.4 Occurrence

This field of research highlights the presence and ecological roles of fungi in lakes and aims to stimulate research in aquatic mycology. In the study of lentic systems, this field is an almost completely neglected topic and, if considered at all, has often been restricted to specific groups of fungi, such as yeasts, filamentous fungi (e.g., aquatic hyphomycetes), or phycomycetes (an obsolete taxonomic category that included various fungal and fungal-like organisms). Aquatic fungi are common in various lentic habitats (Wurzbacher et al., 2010). They play a potentially crucial role in nutrient and carbon cycling and interact with other organisms, thereby influencing food web dynamics.

Man's concern with environmental aspects may also account for the increased interest in aquatic organisms. Pollution of streams, rivers, and local shores is common. Therefore, the role of microfungi in these situations is important. The occurrence and distribution of zoosporic fungi in relation to water characteristics as well as to the various geographical regions of the world have been intensively studied. The aquatic hyphomycetes have a worldwide distribution and have been repeatedly observed on decaying leaf litter in temperate and tropical streams.

The richest samples of fungi are seen in water areas with low or moderate temperature and comparatively high total organic matter and DO. The important role of higher fungi in freshwater ecosystems is in the utilization of dead plant material as sources of nutrients and also responsible for the biodegradation of organic materials in aquatic habitats, particularly plant materials in the form of leaf litter and other plant debris. Several studies on the diversity of freshwater fungi had previously been carried out in temperate regions. Recently, studies in tropical regions and subtropical regions have witnessed an increase. A number of novel taxa have been discovered in the last decade. Water fungi play a crucial role in the freshwater ecosystem in nutrient cycling by breaking down of leaves and woody substrates and also as symbionts. *Aspergillus* is one of the more commonly isolated genera in water and also found that the genera *Aspergillus* (9 species out of 45 species, 166 colonies out of 400 fungal total counts) and *Penicillium* (7 species, 68 colonies) had the greatest diversity of the isolated species as well as the highest fungal total count (Brandi et al., 2007; Sharma and Parveen, 2011). The complete surveys of fungal diversity at the species level, even for a small geographic region, can be an exhaustive task. It is a well-known fact that due to the annihilation of our natural areas is leading to an ever-increasing decline in biodiversity worldwide.

27.5 Reproduction

Lakes, rivers, and streams have important multiusage components, such as sources of drinking water, irrigation, fishery, and energy production. Water is a scarce and fading resource, and its management can have an impact on the flow and the biological quality of rivers and streams. In the recent past, expanding human population, industrialization, intensive agricultural practices, and discharges of a massive amount of wastewater into the river and stream have resulted in deterioration of water quality. The impact of these anthropogenic activities has been so extensive that the water bodies have lost their self-purification capacity to a large extent (Sood et al., 2008).

These parameters could be affected by external and internal factors. There is an intricate relationship between the external and internal factors in an aquatic environment. Meteorological events and pollution are a few of the external factors which affect physicochemical parameters such as temperature, pH, and DO of the water. These parameters have major influences on biochemical reactions that occur within the water. Sudden changes in these parameters may be indicative of changing conditions in water. Internal factors, on the other hand, include events which occur between and within bacterial and plankton populations in the water body.

Ecological status of any water body depends on the healthy interaction of abiotic and biotic components. Wide fluctuation in abiotic variables (seasonal) has a significant impact on biotic life. Lentic water bodies, which are characterized by a continual movement of dissolved substances and suspended materials result in dynamic interaction between components of the ecosystem. The chemical composition of natural water bodies depends on weathering from the catchment, atmospheric deposition, and exchange with sediments with the water body. Physicochemical analysis is of immense importance to assess the quality of water for use in drinking, bathing, fishing, industrial process, and others. The physicochemical characteristics of the six lentic water bodies are analyzed and discussed as per their seasonal variations and their interrelationship.

Fungal reproduction is complex, reflecting the differences in lifestyles and genetic makeup within this diverse kingdom of organisms. It is estimated that a third of all fungi reproduce using more than one method of propagation; for example, reproduction may occur in two well-differentiated stages within the life cycle of a species, the teleomorph and the anamorph. Environmental conditions trigger genetically determined developmental states that lead to the creation of specialized structures for sexual or asexual reproduction. These structures aid reproduction by efficiently dispersing spores or spore-containing propagules.

27.6 Uses of fungi

The freshwater is a vital concern for mankind since it is directly linked with human welfare. The surface lentic water bodies are important sources of water for human activities. Unfortunately, these surface water bodies have been put under severe activities. There is an increasing concern for the fast depleting quantity of utilizable water. The cause of such a situation is aplenty, but the gross population of water has its origin mainly in urbanization, industrialization, and increasing human population. Water being the internal medium of all organisms holds a unique place.

The ecological conditions existing in a water body are disturbed under anthropogenic activities influences, which not only modify the physicochemical qualities but also alter the biological composition. In general, chemical and biological analyses have been largely used to evaluate the quality of the water body and to assess its trophic status. Therefore, prior knowledge of the physicochemical and biological characteristics of the water bodies is a prerequisite before it is exploited for the benefit of human welfare.

The physicochemical characteristics of water play an important role in the distribution and diversity of aquatic organisms. They also maintain the quality of the water for a sustainable life. These characteristics have been derived mainly from the atmospheric precipitation, mineral matter, and the atmospheric gases dissolved in it.

Fungal habitats in lake ecosystems are numerous, diverse, and often hidden. Not only fungal diversity but also important energy transformation processes related to fungal activities need to be scientifically assessed. Although emergent macrophytes have been well studied with respect to fungi, many potential fungal habitats and mechanisms remain understudied. Study reveals that the richest samples of fungi can be seen in water areas with low-or-moderate temperature and comparatively high total organic matter and DO.

The human use of fungi for food preparation or preservation and other purposes is extensive and has a long history. Mushroom farming and mushroom gathering are large industries in many countries. The study of the historical uses and sociological impact of fungi is known as ethno-mycology. Because of the capacity of this group to produce an enormous range of natural products with antimicrobial or other biological activities, many species have long been used or are being developed for industrial production of antibiotics, vitamins, and anticancer and cholesterol-lowering drugs. More recently, methods have been developed for genetic engineering of fungi (Fincham, 1989), enabling metabolic engineering of fungal species. For example, genetic modification of yeast species (Hawkins and Smolke, 2008), which are easy to grow at fast rates in large fermentation vessels, has opened up ways of pharmaceutical production that are potentially more efficient than production by the original source organisms (Huang et al., 2008).

27.6.1 Drugs

Many species produce metabolites that are major sources of pharmacologically active drugs. Particularly important are the antibiotics, including the penicillins, a structurally related group of β-lactam antibiotics that are synthesized from small peptides. Although naturally occurring penicillins such as penicillin G (produced by *Penicillium chrysogenum*) have a relatively narrow spectrum of biological activity, a wide range of other penicillins can be produced by chemical modification of the natural penicillins. Modern penicillins are semisynthetic compounds, obtained initially from fermentation cultures, but then structurally altered for specific desirable properties (Brakhage et al., 2004). Other antibiotics produced by fungi include cyclosporin, commonly used as an immunosuppressant during transplant surgery; and fusidic acid, used to help control infection from methicillin-resistant *Staphylococcus aureus* bacteria (Pan et al., 2008). Widespread use of these antibiotics for the treatment of bacterial diseases, such as tuberculosis, syphilis, leprosy, and many others began in the early 20th century and continues to play a major part in antibacterial chemotherapy. In nature, antibiotics of fungal or bacterial origin appear to play a dual role: at high concentrations they act as a chemical defense against the competition with other microorganisms in species-rich environments, such as the rhizosphere, and at low concentrations as quorum-sensing molecules for intra- or interspecies signaling (Fajardo and Martínez, 2008).

Other drugs produced by fungi include griseofulvin isolated from *Penicillium griseofulvum*, used to treat fungal infections, and statins (HMG-CoA reductase inhibitors), used to inhibit cholesterol synthesis. Examples of statins found in fungi include mevastatin from *Penicillium citrinum* and lovastatin from *Aspergillus terreus* and the oyster mushroom (*Pleurotus ostraeatus*).

27.6.2 Cultured foods

Shoyu koji mold (*Aspergillus oryzae*) is an essential ingredient in brewing Shoyu (soy sauce) and sake, and the preparation of miso, while *Rhizopus* sp. are used for making tempeh. Several of these fungi are domesticated species that were bred or selected according to their capacity to ferment food without producing harmful mycotoxins (see below), which are produced by very closely related *Aspergilli*. Quorn, a meat substitute, is made from *Fusarium venenatum*.

27.6.3 Pest control

Biological control of plant pathogens have received attention and is a promising supplement to chemical control because of the lack of deleterious effects on human health and the environment. Biotic inducers and their metabolites/molecules primed to plants can be used in commercial formulations to induce systemic resistance. Plants priming leads to high and rapid induction of defense responses and ultimately to disease resistance during subsequent phytopathogen attack (McFarland, 2010).

In agriculture, fungi may be useful if they actively compete for nutrients and space with pathogenic microorganisms such as bacteria or other fungi via the competitive exclusion principle, or if they are parasites of these pathogens. For example, certain species may be used to eliminate or suppress the growth of harmful plant pathogens, such as insects, mites, weeds, nematodes, and other fungi that cause diseases of important crop plants. This has generated strong interest in practical applications that use these fungi in the biological control of these agricultural pests. Entomopathogenic fungi can be used as biopesticides, as they actively kill insects. Examples that have been used as biological insecticides are *Beauveria bassiana*, *Metarhizium* spp., *Hirsutella* spp., *Paecilomyces* (*Isaria*) spp., and *Lecanicillium lecanii*. Endophytic fungi of grasses of the genus *Neotyphodium*, such as *N. coenophialum*, produce alkaloids that are toxic to a range of invertebrate and vertebrate herbivores. These alkaloids protect grass plants from herbivory, but several endophyte alkaloids can poison grazing animals, such as cattle and sheep. Infecting cultivars of pasture or forage grasses with *Neotyphodium* endophytes is one approach being used in grass breeding programs; the fungal strains are selected for producing only alkaloids that increase resistance to herbivores such as insects while being nontoxic to livestock.

Many plant growth-promoting fungi, viz., *Trichoderma*, *Penicillium*, *Aspergillus*, and *Fusarium* spp. have the ability to stimulate the plant immune response upon enemy attack and are considered as one of the safest modes for ISR and growth promotion in crop plants (Conrath et al., 2006; Jogaiah et al., 2018; Van Wees et al., 2008). A growing body of evidence indicates that, in addition to the promotion of plant growth (Jogaiah et al., 2013; Fontenelle et al., 2011), PGPFs are beneficial to plants in reducing the impacts of various fungi (Nagaraju et al., 2012), bacteria (Murali et al., 2013; Hossain et al., 2008; Sultana et al., 2009), viruses (Yoshioka et al., 2012), and nematodes (Elsharkawya et al., 2012) by eliciting ISR.

It is said that the phosphate solubilization, production of IAA, and other related compounds by the fungus will interact with plants as part of its colonization, leading to growth promotion, induced resistance, and modification of basal plant defense mechanisms (Gotlieb et al., 2003; Altomare et al., 1999; Contreras et al., 2009). Of all, *Trichoderma* spp. is known for facilitating increased availability and efficient uptake of soil nutrients, thereby improving the yield of a ratoon crop. Recently, treatment of tomato plants with *Penicillium* sp. EU0013, which is known to induce resistance against Fusarium wilt disease in tomato and cabbage, was also shown to significantly mobilize nitrogen from the soil to plants and enhance the growth-yield parameters (Yadav et al., 2011; Alam et al., 2011). There are many reports on applications of PGPF isolates in agricultural practices as biocontrol agents, biofertilizers, and soil amendments for the management of phytopathogens and crop improvement in many crop plants have been documented (Saldajeno and Hyakumachi, 2011). Recently, *T. harzianum* and *Penicillium chrysogenum* was shown to induced systemic resistance against downy mildews of sunflower and pearl millet (Fontenelle et al., 2011; Nagaraju et al., 2012).

27.6.4 Bioremediation

Certain fungi, in particular "white rot" fungi, can degrade insecticides, herbicides, pentachlorophenol, creosote, coal tars, and heavy fuels and turn them into carbon dioxide, water, and basic elements. Fungi have been shown to biomineralize uranium oxides, suggesting they may have application in the bioremediation of radioactively polluted sites.

27.7 Significance

The problem of pollution has been explored in freshwater worldwide. Possible relationship to human activities and other management issues covering the pattern of use of surface water has an impact on the quality of water used for drinking as well as on the foods we consume. Deteriorating natural environment and increasing anthropogenic activities needs elucidation for management of water resources. The impact of increasing human population and abuse of this vital resource gives an alarm for sustaining the quality of life. Freshwater resources harbor a unique biota; however, the integrity of this community is being lost at a rapid rate. Moreover, the demand for water for human consumption is increasing in urban areas. Hence, the freshwater resources are being degraded and face enormous peril of being lost.

The brief study with the fungal occurrence, impact of physical and chemical factors that change the quality of water, their uses determines survival, and health of future generations. The overall picture warrants strict vigilance and continuous monitoring of freshwater bodies for conservation and sustainable management.

Water used in drinking and processing of foods needs to conform to WHO guidelines. When contaminated, water acts as a conduit and as an additive of various types of hazards-physical (like extreme ranges of pH), chemicals like pesticides, weedicides, agricultural chemicals, various minerals (including toxic chemicals and biochemicals comprising toxins) and biologicals such as pathogens, spoilers, allergens, and so on. With groundwater exposed to sewage and industrial wastes, humans have created an alarmingly high dose of avoidable risk constituents in drinking water. While pathogens have been easily diagnosed as hazardous entities, the role of molds is generally ignored.

Microbial investigations in foods, beverages, and drinking water are largely focused on bacteriological constituents. However, the role of microscopic propagules of helminths and protozoans has been considered. Filamentous fungi have been isolated from drinking water. Using water agar medium, visible growth of *A. flavus*, *Penicillium patulum,* and *Fusarium graminearum* from drinking water was seen in the laboratory at the Biological Engineering Centre, University of Minho, Braga, Portugal; they recommend the use of ergosterol (fungal cell wall constituent) for quantifying mold contamination in drinking water. By the use of immuno-affinity column and HPLC, evidence of various mycotoxins like aflatoxin, patulin, and zearalenone have been reported in stored water. The mycotoxins affect biofilms in water apart from indoor toxicity and oestrogenicity in water.

Apart from water content and water activity, food parameters like texture, composition, pH; implicit factors like intrinsic as well as exogenous microflora together with process parameters affect quality, shelf life, and safety of various commodities. Process water used for cleaning, washing, filling, and in several other process steps greatly impact the status of food materials and processed food products. The heterogeneous growth requirements, shorter generation time together with neutral pH seems to favor the growth of bacteria over molds. Interaction of molds with bacteria also needs elucidation in overall biological status of foods. Holistic assessment is essential to relate the total impact of the biological contaminants in water. The interaction between various entities impacts their survival, growth, and metabolism. The dominance of a few species under a given set of process parameters usually follows a series of succession dynamics. Predictable models now greatly help to increase control in the management of quality and safety issues.

References

Alam, S.S., Sakamoto, K., Inubushi, K., 2011. Effect of *Penicillium* sp. EU0013 inoculation on tomato growth and *Fusarium* wilt. Hort. Res. 65, 69–73.

Ali, E.H., 2007. Biodiversity of zoosporic fungi in polluted water drainages across Niles'Delta region, Lower Egypt. Acta Mycol. 42, 99–111.

Altomare, C., Norvell, W.A., Bjorkman, T., Harman, G.E., 1999. Solubilization of phosphates and micronutrients by the plant growth promoting and biocontrol fungus *Trichoderma harzianum* Rifai 1295–22. Appl. Environ. Microbiol. 65, 2926–2933.

Annaisie, E.J., Stratton, S.L., Dignani, M.C., Summerbell, R.C., Rex, J.H., Monson, T.P., Spencer, T., Kasai, M., Francesconi, A., Walsh, T.J., 2002. Pathogenic *Aspergillus* species recovered from a hospital water system: a 3-year prospective study. Clin. Infect. Dis. 34, 780–789.

Arshad, M., Bareen, F., 2009. Ecological studies of aquatic hyphomycetes in a canal and its connecting irrigation channels. Biol. Diver. Conserv. 2 (3), 99–106.

Blackwell, M., 2011. The Fungi: 1, 2, 3 … 5.1 million species? Am. J. Bot. 98, 426–438.

Brakhage, A.A., Spröte, P., Al-Abdallah, Q., Gehrke, A., Plattner, H., Tüncher, A., 2004. Regulation of penicillin biosynthesis in filamentous fungi. Adv. Biochem. Eng. Biotechnol. 88, 45–90.

Brandi, B., Sisti, M., Paparini, A., Gianfranceschi, G., Schiavano, F.G., De Santi, M., Santoni, D., Magini, V., Vincenzo, R.S., 2007. Swimming pools and fungi: an environmental epidemiology survey in Italian indoor swimming facilities. Int. J. Environ. Health Res. 17, 197–206.

Brem, F.M., Lips, K.R., 2008. *Batrachochytrium dendrobatidis* infection patterns among Panamanian amphibian species, habitats and elevations during epizootic and enzootic stages. Dis. Aquat. Org. 81, 189–202.

Conrath, U., Beckers, G.J.M., Flors, V., Garcia-Agustin, P., Jakab, G., Mauch, F., Newman, M.A., Pieterse, C.M.J., Poinssot, B., Pozo, M.J., Pugin, A., Schaffrath, U., Ton, J., Wendehenne, D., Zimmerli, L., Mauch-Mani, B., 2006. Priming: getting ready for battle. Mol. Plant-Microbe Inter. 19, 1062–1071.

Contreras, C.H., Macias, R.L., Cortes, P.C., Lopez, B.J., 2009. *Trichoderma virens*, a plant beneficial fungus, enhances biomass production and promotes lateral root growth through an auxin–dependent mechanism in *Arabidopsis*. Plant Physiol. 149, 1579–1592.

Dadachova, E., Bryan, R.A., Huang, X., Moadel, T., Schweitzer, A.D., Aisen, P., Nosanchuk, J.D., Casadevall, A., 2007. Ionizing radiation changes the electronic properties of melanin and enhances the growth of melanized fungi. PLoS One 2 (5), 457.

Elsharkawya, M.M., Shimizu, M., Takahashi, H., Hyakumachi, M., 2012. Induction of systemic resistance against cucumber mosaic virus by *Penicillium simplicissimum* GP17–2 in *Arabidopsis* and tobacco. Plant Pathol. 61, 964–976.

El-Sharouny, H.M., Gherbawy, Y.A.M.H., Abdel-Aziz, F.A., 2009. Fungal diversity in brackish and saline lakes in Egypt. Nova Hedwigia 89, 437–450.

Fajardo, A., Martínez, J.L., 2008. Antibiotics as signals that trigger specific bacterial responses. Curr. Opin. Microbiol. 11, 161–167.

Fincham, J.R.S., 1989. Transformation in fungi. Microbiol. Rev. 53, 148–170.

Fontenelle, A.D.B., Guzzo, S.D., Lucon, C.M.M., Harakaya, R., 2011. Growth promotion and induction of resistance in tomato plant against *Xanthomonas euvesicatoria* and *Alternaria solani* by *Trichoderma* spp. Crop Protect. 30, 1492–1500.

Gilman, J.C., 1956. A Manual of Soil Fungi, second ed. Iowa State College Press, Ames, pp. 1–402.

Gotlieb, D., Oka, Y., Ben-Daniel, B., Cohen, Y., 2003. Dry mycelium of *Penicillium chrysogenum* protects cucumber and tomato plants against the root–knot nematode *Meloidogyne javanica*. Phytoparasitica 31, 217–225.

Grabinska-Loniewska, A., Konillowicz-Kowalska, T., Wardzynska, G., Boryn, K., 2007. Occurrence of fungi in water distribution system. Polish J. Environ. Stud. 16, 539–547.

Gray, N.F., 1999. Water Technology: An Introduction for Environmental Students. Elsevier Science and Technology Books, New York, pp. 133–189.

Gupta, N., Yadav, K.K., Kumar, V., Singh, D., 2013. Assessment of physico-chemical properties of Yamuna River in Agra city. Int. J. Chem. Technol. Res. 5, 528–531.

Hageskal, G., Lima, N., Skaar, I., 2009. The study of fungi in drinking water. Mycobiol. Res. 113, 165–172.

Hawkins, K.M., Smolke, C.D., 2008. Production of benzylisoquinoline alkaloids in *Saccharomyces cerevisiae*. Nat. Chem. Biol. 4, 564–573.

Hawksworth, D.L., 2006. The fungal dimension of biodiversity: magnitude, significance, and conservation. Mycol. Res. 95, 641–655.

Homyra, N.E.S., Naz, S., 2006. Diversity of phytoplankton in central Park lake, Rajshahi, Bangladesh. Pollut. Res. 25, 223–226.

Hossain, M.M., Sultana, F., Kubota, M., Koyama, H., Hyakumachi, M., 2008. Differential inducible defense mechanisms against bacterial speck pathogen in *Arabidopsis thaliana* by plant growth-promoting fungus *Penicillium* sp GP16–2 and its cell–free filtrate. Plant Soil 304, 227–239.

Huang, B., Guo, J., Yi, B., Yu, X., Sun, L., Chen, W., 2008. Heterologous production of secondary metabolites as pharmaceuticals in *Saccharomyces cerevisiae*. Biotechnol. Lett. 30, 1121–1137.

James, T.Y., Kauff, F., Schoch, C.L., Matheny, P.B., 2006. Reconstructing the early evolution of fungi using a sixgene phylogeny. Nature 443, 818–822.

Jogaiah, S., Anand Kumar, S., Thakur, R.P., Rao, V.P., Shekar Shetty, H., 2009. Molecular characterization of *Sclerospora graminicola*, the incitant of pearl millet downy mildew revealed by ISSR markers. J. Phytopathol. 157, 748–755.

Jogaiah, S., Mostafa, A., Tran, L.S.P., Shin-ichi, I., 2013. Characterization of rhizosphere fungi that mediate resistance in tomato against bacterial wilt disease. J. Exp. Bot. 64, 3829–3842.

Jogaiah, S., Abdelrahman, M., Tran, L.-S.P., Ito, S.-I., 2018. Different mechanisms of *Trichoderma virens*-mediated resistance in tomato against Fusarium wilt involve the jasmonic and salicylic acid pathways. Mol. Plant Pathol. 19, 870–882.

Khare, R., Khare, S., Kamboj, M., Pandey, J., 2011. Physico-chemical analysis of Ganga river water. Asian J. Biochem. Pharm. Res. 1, 232–239.

Kiziewicz, B., Kurzatkowska, A., 2004. Aquatic fungi and fungus like organisms isolated from surfacewater situated near Bialystok in Podlasie. Mycol. Bal. 1, 117–123.

Le Calvez, T., Burgaud, G., Mahé, S., Barbier, G., Vandenkoornhuyse, P., 2009. Fungal diversity in deep sea hydrothermal ecosystems. Appl. Environ. Microbiol. 75, 6415–6421.

Lefèvre, C., Domaizon, I., Debroas, D., 2007. Community composition of lacustrine small eukaryotes in hyper-eutrophic conditions in relation to top–down and bottom–up factors. FEMS Microbiol. Ecol. 61, 483–495.

Lutzoni, F., Kauff, F., Cox, C.J., McLaughlin, D., 2004. Assembling the fungal tree of life: progress, classification, and evolution of subcellular traits. Am. J. Bot. 91, 1446–1480.

Manoharachary, C., Sridhar, K., Singh, R., Adholeya, A., Suryanarayanan, T.S., Rawat, S., Johri, B.N., 2005. Fungal biodiversity: distribution, conservation and prospecting of fungi from India. Curr. Sci. 89, 58–71.

Matkar, L.S., Gangotri, M.S., 2002. Physico-chemical analysis of sugar industrial effluents. J. Indl. Pollut. Cont. 18, 139–144.

McFarland, V.C., 2010. Systematic review and meta-analysis of *Saccharomyces boulardii* in adult patients. World J. Gastroenterol. 16, 2202–2222.

Mueller, G.M., Schmit, J.P., 2006. Fungal biodiversity: what do we know? What can we predict? Biodivers. Conserv. 16, 1–5.

Murali, M., Sudisha, J., Amruthesh, K.N., Ito, S.-i., Shekar Shetty, H., 2013. Rhizosphere fungus *Penicillium chrysogenum* promotes growth and induces defence-related genes and downy mildew disease resistance in pearl millet. Plant Biol. 15, 111–118.

Nagaraju, A., Sudisha, J., Mahadevamurthy, S., Ito, S.-i., 2012. Seed priming with *Trichoderma harzianum* isolates enhances plant growth and induces resistance against *Plasmopara halstedii*, an incitant of sunflower downy mildew disease. Aust. J. Plant Pathol. 41, 609–620.

Neill, P.O., 1993. Environmental Chemistry. Chapman and Hall, London, UK, pp. 48–49.

Nikolcheva, L., Bärlocher, F., 2004. Taxon-specific fungal primers reveal unexpectedly high diversity during leaf decomposition in a stream. Mycol. Prog. 3, 41–49.

Paliwal, P.C., Sati, S.C., 2009. Physico-chemical characteristics distribution of aquatic fungi in relation two physico-chemical factors of Kosi river in Kumaun, Himalaya. Nat. Sci. 7, 1545–1548.

Pan, A., Lorenzotti, S., Zoncada, A., 2008. Registered and investigational drugs for the treatment of methicillin-resistant *Staphylococcus aureus* infection. Recent Patents Anti-Infect. Drug Dis. 3, 10–33.

Parveen, S., Lanjewar, S., Sharma, K., Kutti, U., 2011. Isolation of fungi from the surface water of river. J. Exp. Sci. 2, 58–59.

Paterson, R.R.M., Lima, N., 2005. Fungal contamination of drinking water. In: Lehr, J., Keeley, J., Kingery III, T.B. (Eds.), Water Encyclopedia. John Wiley and Sons.

Raghukumar, C., 2008. Marine fungal biotechnology: an ecological perspective. Fungal Diver. 3, 19–35.

Raghukumar, C., Raghukumar, S., 1998. Barotolerance of fungi isolated from deep-sea sediments of the Indian Ocean. Aquat. Microb. Ecol. 15, 153–163.

Saldajeno, M.G.B., Hyakumachi, M., 2011. The plant growth promoting fungus *Fusarium equiseti* and the arbuscular mycorrhizal fungus *Glomus mosseae* stimulate plant growth and reduce severity of anthracnose and damping–off diseases in cucumber (*Cucumis sativus*) seedlings. Ann. Appl. Biol. 159, 28–40.

Sancho, L.G., de la Torre, R., Horneck, G., Ascaso, C., de Los, R.A., Pintado, A., Wierzchos, J., Schuster, M., 2007. Lichens survive in space: results from the 2005 LICHENS experiment. Astrobiology 7, 443–454.

Schoch, C.L., Crous, P.W., Groenewald, J.Z., Boehm, E.W.A., Burgess, T.I., de Gruyter, J., de Hoog, G.S., Dixon, L.J., Grube, M., Gueidan, C., Harada, Y., Hatakeyama, S., Hirayama, K., Hosoya, T., Huhndorf, S.M., Hyde, K.D., EBG, J., Kohlmeyer, J., Kruys, Å., Li, Y.M., Lücking, R., Lumbsch, H.T., Marvanová, L., Mbatchou, J.S., AH, M.V., Miller, A.N., Mugambi, G.K., Muggia, L., Nelsen, M.P., Nelson, P., Owensby, C.A., AJL, P., Phongpaichit, S., 2009. A class-wide phylogenetic assessment of *Dothideomycetes*. Stud. Mycol. 64, 1–15.

Sharma, K., Parveen, S., 2011. Ecological study of fungi isolated from the surface water of Dudhawa Dam Dhamtari, Chhattisgarh, India. J. Phytol. 3, 6–8.

Sharma, S., Vishwakarma, R., Dixit, S., Jain, P., 2011. Evaluation of water quality of Narmada River with reference to physico-chemical parameters at Hoshangabad city, MP, India. Res. J. Chem. Sci. 1, 40–48.

Shearer, C.A., Descals, E., Kohlmeyer, B., Kohlmeyer, J., Marvanova, L., Padgett, D., Porter, D., Raja, H.A., Schmit, J.P., Thorton, H.A., Voglymayr, H., 2007. Fungal biodiversity in aquatic habitats. Biodivers. Conserv. 16, 49–67.

Singh, K., Frisvad, J.C., Thrane, U., Mathur, S.B., 1991. An Illustrated Manual on Identification of some Seed–borne Aspergilli, Fusaria, Penicillia and their Mycotoxins, Danish

Government Institute of Seed Pathology for Developing Countries, first ed. Department of Bio-technology the Technical University of Denmark, Hellerup, Denmark, pp. 31–122.

Sinha, S.N., Biswas, M., 2011. Physico-chemical characteristics and water quality of Lake Kalyani, West Bengal. Asian J. Exp. Biol. Sci. 2, 12–18.

Sivakumar, A.A., Arunadevi, P., Aruchami, M., 2003. Studies on water quality of the river Ambarapalaym, Coimbatore district, Tamil Nadu. Nat. Environ. Pollut. Technol. 2, 305–308.

Sood, A., Singh, K.D., Pandey, P., Sharma, S., 2008. Assessment of bacterial indicators and physico–chemical parameters to investigate pollution status of Gangetic river system of Uttarakhand (India). Ecol. Indic. 8, 709–717.

Sultana, F., Hossain, M.M., Kubota, M., Hyakumachi, M., 2009. Induction of systemic resistance in *Arabidopsis thaliana* in response to a culture filtrate from a plant growth–promoting fungus, *Phoma* sp GS8–3. Plant Biol. 11, 97–104.

Thamizhmani, R., Raj, R.V., Sivakumar, T., 2013. Abundance and diversity of fungi in salt pan and marine ecosystem of the east coast of Tamil Nadu, India. Int. J. Curr. Microbiol. Appl. Sci. 2, 67–75.

UNEP, 1978. Environmental issues in river basin development. In: Proceedings of the UN Water Conference on Water Management and Development. 1, Pergamon Press, New York, pp. 1163–1172.

Van Wees, S.C.M., Van der Ent, S., Pieterse, C.M.J., 2008. Plant immune responses triggered by beneficial microbes. Curr. Opin. Plant Biol. 11, 443–448.

Vaupotic, T., Veranic, P., Jenoe, P., Plemenitas, A., 2008. Mitochondrial mediation of environmental osmolytes discrimination during osmoadaptation in the extremely halotolerant black yeast *Hortaea werneckii*. Fungal Genet. Biol. 45, 994–1007.

Wurzbacher, C.M., Bärlocher, F., Grossart, H.P., 2010. Fungi in lake ecosystems. Aquat. Microb. Ecol. 59, 125–149.

Yadav, R.C., Srivastava, V.C., 2011. Physico-chemical properties of the water of river Ganga at Ghazipur. Indian J. Sci. Res. 2, 41–44.

Yadav, J., Verma, J.P., Yadav, S.K., Tiwari, K.N., 2011. Plant growth promoting activities of fungi and their effect on chickpea plant growth. Asian J. Biol. Sci. 4, 291–299.

Yoshioka, Y., Ichikawa, H., Naznin, H.A., Kogure, A., Hyakumachi, M., 2012. Systemic resistance induced in *Arabidopsis thaliana* by *Trichoderma asperellum* SKT–1, a microbial pesticide of seed borne diseases of rice. Pest Manag. Sci. 68, 60–66.

CRISPR/Cas system: A powerful approach for enhanced resistance against rice blast

28

Muntazir Mushtaq[a], *Hilal Ahmad Pir*[b], *Abbu Zaid*[c], *and Shabir Hussain Wani*[d]
[a]School of Biotechnology, Sher-e-Kashmir University of Agricultural Sciences and Technology of Jammu, Jammu, Jammu and Kashmir, India, [b]Division of Plant Breeding and Genetics, Sher-e-Kashmir University of Agricultural Sciences and Technology of Jammu, Jammu, Jammu and Kashmir, India, [c]Plant Physiology and Biochemistry Section, Department of Botany, Aligarh Muslim University, Aligarh, Uttar Pradesh, India, [d]Mountain Research Centre for Field Crops, Sher-e-Kashmir University of Agricultural Sciences and Technology of Kashmir, Khudwani Anantnag, Jammu and Kashmir, India

28.1 Introduction

The rice (*Oryza sativa* L.), a crop that provides 20% of the global dietary energy supply and more than half of the world's population consumes rice daily (Birla et al., 2017; Mishra et al. 2018). This is attributed to its abundance of genetic resources, wider adaptability, genomic synteny with other cereals, amenable to genetic transformation, and rich history of comprehensive research efforts. Rice has been considered as a strategic commodity for global food security by Food and Agriculture Organization (FAO) (Mishra et al., 2018). Rice has been used as a model plant for establishing hi-tech platforms and testing strategies for improvement to itself and other crops as well. It has been estimated that 40% more rice needs to be cultivated by 2030 to meet the growing demand (Khush, 2005; Li et al., 2017). The last few decades have witnessed enhanced growth in rice productivity owing to the enormous contribution of conventional, mutational, and molecular breeding strategies. Moreover, complexity in the genetic makeup of rice, as with other crop species have witnessed improvement in increasing yields. However, rice yields have been progressively declined in recent decades due to rapid population explosion, climate change, the emergence of pathogens and insect pests, and other environmental menaces. Thus, there is an urgent need to have more advanced biotechnologies that can generate new rice varieties with higher yield potential, enhanced abiotic stress tolerance, and improved resistance to biotic stresses. From the past two decades, sequence-specific nucleases (SSNs) have been demonstrated to be powerful tools for crop improvement and CRISPR/Cas9 is thought to be the most effective SSN.

Rice blast caused by *Magnaporthe oryzae*, is one of the most devastating diseases that affect rice, right from the vegetative stage (leaf blast) to the reproductive stage (neck blast) resulting in 30% of yield loss annually, therefore intimidating food

Biocontrol Agents and Secondary Metabolites. https://doi.org/10.1016/B978-0-12-822919-4.00028-4

security worldwide (Sakulkoo et al., 2018; Devi et al., 2020). The outbreak of this disease is a serious concern in many Asian and African countries, where the yield loss under endemic conditions ranged from 60% to 100% (Kihoro et al., 2013). In recent years, various technologies have been used to control blast disease in rice. In order to control the spread of this destructive fungal disease, chemical and biological methods, disease forecasting, and improved cultivation practices have been adapted widely (Miah et al., 2013). Unfortunately, these measures were not very effective. On the other hand, pesticide use proved to be expensive, theoretical, and unsafe for the environment. Some strategies such as multilines, mixtures, and pyramiding use complete and specific resistance genes while others are based on the accumulation of partial resistance. Exploiting the host resistance is the most economical and eco-friendly way to manage this havoc as the diversity in germplasm with variation is available for this crop. Further, until now, 100 blast resistance genes (Zhao et al., 2018) and 350 QTL were mapped on rice chromosomes (Singh et al., 2015; Devi et al., 2020). Dominant R genes offer vertical resistance which is race specific and prone to rapid breakdown in nature leading to "boom and bust" scenario. On the other hand, field resistance contributed by QTL offer horizontal or partial resistance, which is known for its durability in combating the disease outbreaks in a race nonspecific manner (Fukuoka et al., 2015). Various rice research groups and breeders have developed new improved blast-resistant rice varieties employing molecular breeding approach involving DNA markers, QTLs mapping, marker-assisted selection (MAS), and genetic transformation. Conventional breeding takes almost 10–15 years from initiation to varietal release and is dependent on environmental conditions (Miah et al., 2013). In contrast, MAS is efficient, effective, and reliable than phenotypic selection. MAS also allows the breeding of complex traits which is not possible through conventional approaches. Furthermore, MAS can reduce time in developing new varieties significantly. However, the high cost of MAS has been a major obstacle for plant breeding in developing countries (Collard and Mackill, 2008).

Genome editing is becoming the most important biotechnological technique that has stimulated the progress of our biological information and the field of biotechnology with tremendous contributions to rapid advancement to industry, medicine, and agriculture. The terms "genome engineering," "genome editing," and "gene editing" refer to allowing precise genetic modifications, such as site-specific insertion, deletion, substitution, and epiallelic changes of the DNA in a cell or organism (El-Mounadi et al., 2020). In the past decade, the emergence of genome editing technologies and their potential applications in crop plants have revolutionized almost every aspect of plant science (Mushtaq et al., 2019). Plant genome editing includes the engineering of sequence-specific nucleases, delivery of editing reagents, generation and selection of edit events, and characterization and utilization of intact plants, which further have implications for the public acceptance and regulatory approval of genome-edited plants. Over the last decade, genome editing platforms have been established and applied in more than 45 genera of plants (Shan et al., 2020; Yang, 2020). Genome editing is principally based on in vivo DNA double-strand breaks (DSBs) induced by engineered endonucleases that are programmed to recognize the preselected genomic sites and exploitation of cellular DSB repair mechanisms

(El-Mounadi et al., 2020; Yang, 2020). These programmable nucleases include meganucleases (Gong and Golic, 2003), zinc-finger nucleases (Urnov et al., 2005), transcription activator-like effector nucleases (TALENs) (Liu et al., 2012), and clustered regularly interspaced short palindromic repeats (CRISPR) associated nucleases (Cas) (Jinek et al., 2012; Cong et al., 2003; Mali et al., 2013; Choudhary et al., 2017; Mushtaq et al., 2019; El-Mounadi et al., 2020; Yang, 2020). In addition, engineered nuclease variants can target single-stranded (SS) DNA (Rees and Liu, 2018) or can perform epigenome editing without causing DNA breakage and no DSB repair (Holtzman and Gersbach, 2018; Yang, 2020). In this chapter, we focus on CRISPR-Cas systems for improving blast resistance in rice.

CRISPR/Cas9 system in contrast to ZFNs, TALENs, and meganucleases is independent of any protein engineering steps and can be retargeted to new DNA sequences by simply altering single-guide RNA (sgRNA) sequence. CRISPR/Cas9 is composed of duplex-RNA structure [CRISPR RNA (crRNA): trans-activating crRNA (tracrRNA)] that directs Cas9 nucleases to the target DNA (Mushtaq et al., 2019). In order to increase efficiency, the dual crRNA: tracrRNA structure is engineered into a single-guide RNA (sgRNA) chimera targeted to specific genomic loci. Generally, CRISPR-Cas9-based editing in plant genomes comprises mainly of four steps (Fig. 28.1). In the first step, gene-specific sgRNA is designed and can be performed using many online tools specifically developed for the computer-based design of sgRNAs. In order to enhance the accuracy of computational sgRNA selection, organized study of sgRNA efficiencies in plant cells and large-scale data collection is needed. The second step involves transient transformation systems like hairy-root or protoplast transformation with CRISPR before being used in genome editing, the best way to validate the activity of sgRNA. Subsequently, the third step involves the delivery of genome-editing construct into plant cells usually via *Agrobacterium*-mediated transformations or particle bombardment and the stable integration of sgRNA and Cas9 into the plant genome. In the fourth step, the

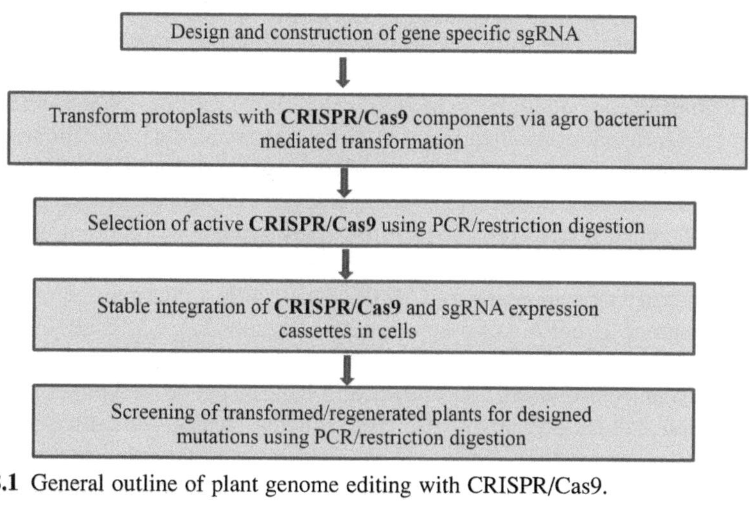

Fig. 28.1 General outline of plant genome editing with CRISPR/Cas9.

screening of transformed or regenerated plants for mutation via PCR/RE genotyping and confirmed by sequencing.

28.2 Concept-proof demonstration of CRISPR/Cas system in rice

In recent years, the use of CRISPR/Cas9 has facilitated genome editing technologies and has become the leading tool used to generate specific changes to DNA sequences in a wide range of plant species including rice crops. In the year 2013 only, five major reports were published demonstrating targeted mutagenesis by CRISPR/Cas9 system in rice (Jiang et al., 2013; Miao et al., 2013; Shan et al., 2013; Xie and Yang, 2013). Shan et al. (2013) used a single customized sgRNA to introduce targeted mutations in three rice genes, *OsBADH2*, *Os02g23823*, and *OsMPK2* producing higher mutation frequency of CRISPR/Cas9 system in contrast to TALENs. In another study, Jiang et al. (2013) targeted *OsSWEET14* and *OoSWEET11* in rice using CRISPR/Cas9 resulting in the deletion of nine and seven nucleotides from the promoter regions of these targeted genes. In the same year, Feng et al. (2013) designed CRISPR/Cas9 construct to target three rice genes [rice outermost cell-specific gene 5 (*ROC5*), stromal processing peptidase (SPP), and young seedling albino (YSA)] simultaneously, resulting in the generation of homozygous or bi-allelic mutants with the mutation frequency as high as 84% in the T_0 and T_1 rice lines. Further, Zhou et al. (2014) targeted four efflux transporter genes (*OsSWEET11*, *OsSWEET12*, *OsSWEET13*, and *OsSWEET14*) in rice using CRISPR/Cas9 vectors resulting in large chromosomal deletions. These reports clearly suggest that CRISPR/Cas9 system can be utilized effectively for genome editing generating insertions, deletion, substitution, and translocation lines demonstrating greater efficiency for the generation of new improved varieties with novel traits.

Endo et al. (2016) successfully achieved multi-paralogous gene knockout in rice by targeting phytoene desaturase (*OsPDS*), *Os02g23823*, and *OsMPK2* genes using the CRISPR/Cas9 system. This study revealed a high co-mutation rate with the mutation frequency varying between 66.4% and 81%. In another study, Ma et al. (2015) demonstrated a high-efficiency multiplex gene editing in rice by targeting 46 target sites using multiple sgRNA cassettes resulting in the mutation frequency of 85.4%. The study also revealed the editing of three sites within *OsWaxy* gene simultaneously, causing up to 14% reduction in amylose content. In a similar study, the endogenous tRNA processing system was targeted in rice crops using a multiplex genome editing tool, generating large genomic deletions in T_0 generation (Xie et al., 2015). Similarly, Liang et al. (2016) established a new approach for CRISPR/Cas9-sgRNA multiplex editing system in rice by designing 21 sgRNAs for constructing corresponding Cas9/sgRNAs expression vectors. It was shown that transformed plants were substantially edited representing deletion, insertion, substitution, and inversion in 82% of the desired target sites, thus exhibiting high editing efficiency. All these studies clearly demonstrate that the CRISPR/Cas9 system is highly efficient and applicable in targeting multiple genes simultaneously generating multiple gene mutations using the conventional strategy that could be subsequently used for future rice improvement programs.

Lee et al. (2019) demonstrated that *Agrobacterium*-mediated T-DNA integration using CRISPR-Cas9 system can lead to targeted T-DNA integration in the rice genome. They constructed a T-DNA containing a CRISPR/Cas9 system using SpCas9 and a gRNA targeting the coding region of rice AP2 domain-containing protein gene *Os01g04020* using a standard *Agrobacterium* binary vector. In addition, T-DNA also carried hygromycin resistance (*hptII*) gene and a fluorescent protein. In order to detect targeted T-DNA insertion events, another T-DNA with a promoterless *hptIII* gene was constructed. They reported that these constructs could generate targeted T-DNA insertions with mutation frequencies ranging between 4% and 5.3% of transgenic callus events, besides producing high efficiency targeted mutations of 50%–80%. Thus, they showed that *Agrobacterium*-mediated transformation combined with a CRISPR/Cas9 system can be used effectively to generate targeted T-DNA insertions.

Clustered regularly interspaced short palindromic repeats from *Prevotella* and *Francisellal* system is the advanced genome editing system and has been used for gene editing in plants since 2016 (Endo et al., 2016). Xu et al. (2017) used CRISPR/Cpf1 for generating stable and heritable mutations in two endogenous rice genes *OsPDS* and *OsBEL*. This study suggested that CRISPR/Cpf1 is an efficient system in generating targeted mutations in rice with mutation efficiencies ranging from 13.6% to 43.1%. Tang et al. (2017) used a dual RNA polymerase II promoter expression system in CRISPR/Cpf1 genome editing strategy to evaluate the activity of AsCpf1 and LbCpf1 from *Acidaminococcus* sp. BV3L6 and *Lachnospiraceae* bacterium *ND2006*, respectively, and study their role in targeted gene repression in vivo. They reported that the *LbCpf1* coupled with Pol II-promoter and a double ribozyme system can be efficiently used in the maturation and expression of crRNAs generating mutations in rice effectively. Moreover, CRISPR/Cpf1 system has been used for multiplex genome editing in rice by Wang et al. (2017) and Mishra et al. (2018). They compared the activity of *Francisella novicida* Cpf1 (FnCpf1) and *Lachnospiraceae* bacterium *ND2006* Cpf1 (LbCpf1) by designing two T-DNA constructs on the pCambia binary vector in order to express FnCpf1 and LbCpf1 systems and targeting 22–24 nt sequences to induce mutations at six sites of *5-enolpyruvylshikimate 3-phosphate synthase* (*OsEPSPS*, *LOC_Os06g04280*), *bentazon-sensitive lethal* (*OsBEL*, *LOC_Os03g55240*), and *phytoene desaturase* (*OsPDS*, *LOC_Os03g08570*). Reports showed that both FnCpf1 and LbCpf1 with a simple direct repeat (DR) guide array introduced targeted mutations in transgenic plants. Further, this investigation indicated that the LbCpf1 system exhibited higher editing efficiency with a mutation frequency of 40%–60% compared to FnCpf1 system with mutation frequency ranging from 40% to 60%. These results illustrated the faithfulness of CRISPR/Cpf1 system in multiplex gene editing in rice. Recently, Li et al. (2017) reported the multiplex editing of target genes containing TYCV PAMs in rice. They generating two modified LbCpf1 variants to study their gene editing efficiency and broaden the range of CRISPR/Cpf1 system in rice. Results showed that this system is highly significant as it enlarges the genome editing range by targeting and editing the genomic sequences containing noncanonical PAMs, hence CRISPR/LbCpf1 (RR) variant of gene editing system may be used in basic plant science research and crop breeding.

28.3 Engineering rice blast resistance through CRISPR tool-kit

There is evidence that plant ethylene-responsive factors (ERFs) are involved in the modulation of multiple stress tolerance (Müller and Munné-Bosch, 2015). Furthermore, the knockdown expression of rice *ERF922* by RNAi enhances resistance to *M. oryzae*, indicating that this gene may act as a negative regulator of resistance (Liu et al., 2012). Wang et al. (2017) have reported the improvement of rice blast resistance by engineering a CRISPR/Cas9 SSN (C-ERF992) targeting the *OsERF992* gene in rice. Further, the knockdown of the rice *ERF922* gene by RNAi enhanced resistance to fungus signifies that ERF22 acts as the negative regulator of resistance (Liu et al., 2012). Targeted mutagenesis showed insertion or deletion at the target site and the mutation frequency was observed up to 42% in T_0 plants. Phenotypic analysis of six T_2 homozygous mutant lines revealed that the number of blast lesions was significantly reduced as compared with wild-type plants suggesting that this system can allow highly effective multi-locus mutations in rice.

In another study, Ma et al. (2018) revealed the defensive role of *OsSEC3A* gene to blast fungus by designing two sgRNAs targeting the third and tenth exons of this gene. Results showed deletions, insertions, or both in the targeted sites of five rice mutant lines. Furthermore, one line out of the five was selected for further analysis depicting the enhanced activation of defense response to *M. oryzae* due to the upregulation of pathogenesis-related proteins and salicylic acid synthesis-related genes and enhanced levels of salicylic acid. Therefore, *ossec3a* mutants were found more resistant to *M. oryzae* infection owing to this augmented defense mechanism.

In another line of defense, Foster et al. (2018) have designed a rapid plasmid-free CRISPR/Cas9-based gene-editing system. They demonstrated that the stable expression of Cas9 is extremely toxic to *M. oryzae*. So, in order to achieve efficient targeted gene editing in *M. oryzae*, they introduced a purified Cas9 precomplexed with guide RNAs to form ribonucleoproteins (RNPs) Further, when this system was used in combination with oligonucleotide or PCR-generated donor DNAs, the strains generated with specific base-pair edits, in-locus gene replacements, or multiple gene edits, or multiple gene edits was very fast and straightforward. They showed a coediting approach for the generation of single-base changes at specific loci. In addition, they reported a novel selection scheme that exploits negative cross-resistance to two fungicides that allow CRISPR-mediated counter selection. This counter selection strategy allows the generation of mutants that contain no foreign DNA and are isogenic to an original wild-type strain. Hence, this investigation paves a way for precise and rapid gene manipulation in *M. oryzae* and other fungal species, thereby accelerating the speed of research in this economically important disease.

Digu is a rice variety with durable, elevated resistance to a broad spectrum of *M. oryzae* races (Li et al., 2017), rendering it an excellent resource for discovery of novel genes conferring broad-spectrum resistance. Li et al. (2017) reported that the novel allele, *bsr-dl* is involved in the broad spectrum, durable resistance to *M. oryzae* in Digu. *Bsr-dl* encodes a C2H2-type transcription factor, which is directly regulated by a MYB family transcription factor. These two transcription factors regulate the expression of H_2O_2-degradation enzymes to accomplish resistance to *M. oryzae*, constituting a novel mechanism employed by rice blast resistance. Furthermore, they used CRISPR/Cas9 technology to validate the role of *Bsr-d1* in blast resistance. CRISPR/Cas9 technology was used to knock out the endogenous *Bsr-d1* gene (Bsr-d1KO) in TP309 which revealed that the natural allele, bsr-d1 of a transcription factor is involved in broad-spectrum resistance in rice.

Yamato et al. (2019) established a novel genome editing approach by the use of single crossover-mediated HR in the fungus *Pyricularia* (*Magnaporthe*) *oryzae*. This strategy includes the CRISPR/Cas9 system and a donor vector harboring a single homology arm with point mutations at the CRISPR/Cas9 cleavage site. Using this method, they validated highly capable and freely programmable base substitutions within the desired genomic locus, and target gene disrupted mutants were also obtained via a shortened (100–1000 bp) single homology arm. Further, they demonstrated that this method allowed a one-step GFP gene knock-in at the C-terminus of the targeted gene. Since the genomic recombination does not require an intact protospacer-adjacent motif within the donor construct and any additional modifications of host components, this method can be used in various filamentous fungi for CRISPR/Cas9-based basic and applied biological analyses.

28.4 Perspectives for genome-edited blast-resistant rice

The CRISPR-Cas system became the benchmark for genome editing owing to its simplicity, efficiency, and versatility. It is one of the genetic switches that can be implemented, and disease resistance is repeatedly mentioned as the most promising use of CRISPR/Cas9 technology in agriculture. There are three principal ways for this: firstly, understanding of the molecular mechanisms underlying various pathosystems is adequately advanced to facilitate the proposal of genes to be edited to accomplish disease resistance. Secondly, in order to achieve plant disease resistance, single genes can be modified which is in principle less challenging. This is analogous to the modification of metabolic pathways which involves the editing of a single gene resulting in an all-or-nothing effect, however, unlike abiotic stress tolerance, where by and large many genes have to be modified in a synchronized manner to achieve incremental improvements. Thirdly, targeted mutagenesis, the only use of CRISPR/Cas9 technology at present mastered with respect to crops, is readily applicable to disease resistance, as the inactivation of susceptibility genes leads to protection. Till now, various findings have demonstrated the efficient generation of CRISPR-edited blast-resistant rice crops, thus accelerating the pace of research in this economically important pathogen. Currently, CRISPR/Cas-mediated base editing, a novel genome editing strategy that enables irreversible nucleotide changes at target loci without double-stranded DNA cleavage or any donor template, can be adopted for generating gain-of-function germplasms in functional genomics research and crop genetic improvement. The recently reported "prime editing" improves the researcher's chances of getting only the edits they want by reducing the off-target effects, thus offering more control than CRISPR/Cas9 gene-editing tool. Prime editing utilizes Cas9 and guides RNA variants to precisely alter a specific DNA sequence by allowing the introduction of indels and all 12 base-to-base conversions (both transitions and transversions) without introducing a double-strand break (Anzalone et al., 2019; Marzec et al., 2020). Thus, prime editing is superior to base editing which generates mutations at a single-base resolution within a window of a guide RNA-specified DNA site (Holtzman and Gersbach, 2018; Rees and Liu, 2018; Molla and Yang, 2019).

References

Anzalone, A.V., Randolph, P.B., Davis, J.R., Sousa, A.A., Koblan, L.W., Levy, J.M., Chen, P.J., Wilson, C., Newby, G.A., Raguram, A., Liu, D.R., 2019. Search-and-replace genome editing without double-strand breaks or donor DNA. Nature 576 (7785), 149–157.

Birla, D.S., Malik, K., Sainger, M., Chaudhary, D., Jaiwal, R., Jaiwal, P.K., 2017. Progress and challenges in improving the nutritional quality of rice (*Oryza sativa* L.). Crit. Rev. Food Sci. Nutr. 57 (11), 2455–2481.

Choudhary, I.C., Mushtaq, M., Singh, A.K., Mukhtar, S., Shah, A.A., 2017. Genome editing using CRISPR/Cas system: new era genetic technology in agriculture to boost crop output. Eur. Exp. Biol. 7 (3), 20.

Collard, B.C., Mackill, D.J., 2008. Marker-assisted selection: an approach for precision plant breeding in the twenty-first century. Philos. Trans. R. Soc. B Biol. Sci. 363 (1491), 557–572.

Cong, L., Ran, F.A., Cox, D., Lin, S., Barretto, R., Habib, N., Hsu, P.D., Wu, X., Jiang, W., Marraffini, L.A., Zhang, F., 2003. Multiplex genome engineering using CRISPR/Cas systems. Science 339 (6121), 819–823.

Devi, S.R., Singh, K., Umakanth, B., Vishalakshi, B., Rao, K.V.S., Suneel, B., Sharma, S.K., Kadambari, G.K.M., Prasad, M.S., Senguttvel, P., Syamaladevi, D.P., Madhav, M.S., 2020. Identification and characterization of a large effect QTL from oryzaglumaepatula revealed Pi68 (t) as putative candidate gene for rice blast resistance. Rice 13 (1), 1–13.

El-Mounadi, K., Morales-Floriano, M.L., Garcia-Ruiz, H., 2020. Principles, applications, and biosafety of plant genome editing using CRISPR-Cas9. Front. Plant Sci. 11.

Endo, A., Masafumi, M., Kaya, H., Toki, S., 2016. Efficient targeted mutagenesis of rice and tobacco genomes using Cpf 1 from *Francisellanovicida*. Sci. Rep. 6, 38169.

Feng, Z., Zhang, B., Ding, W., Liu, X., Yang, D., Wei, P., Cao, F., Zhu, S., Zhang, F., Mao, Y., Zhu, J., 2013. Efficient genome editing in plants using a CRISPR/Cas system. Cell Res. 23, 1229–1232.

Foster, A.J., Martin-Urdiroz, M., Yan, X., Wright, H.S., Soanes, D.M., Talbot, N.J., 2018. CRISPR-Cas 9 ribonucleoprotein-mediated co-editing and counterselection in the rice blast fungus. Sci. Rep. 8 (1), 1–12.

Fukuoka, S., Saka, N., Mizukami, Y., Koga, H., Yamanouchi, U., Yoshioka, Y., Hayashi, N., Ebana, K., Mizobuchi, R., Yano, M., 2015. Gene pyramiding enhances durable blast disease resistance in rice. Sci. Rep. 5 (1), 1–7.

Gong, W.J., Golic, K.G., 2003. Ends-out, or replacement, gene targeting in *Drosophila*. Proc. Natl. Acad. Sci. U. S. A. 100 (5), 2556–2561.

Holtzman, L., Gersbach, C.A., 2018. Editing the epigenome: reshaping the genomic landscape. Annu. Rev. Genomics Hum. Genet. 19, 43–71.

Jiang, W., Zhou, H., Bi, H., Fromm, M., Yang, B., Weeks, D.P., 2013. Demonstration of CRISPR/Cas9/sgRNA-mediated targeted gene modification in *Arabidopsis*, tobacco, sorghum and rice. Nucleic Acids Res. 41 (20), e188.

Jinek, M., Chylinski, K., Fonfara, I., Hauer, M., Doudna, J.A., Charpentier, E., 2012. A programmable dual-RNA-guided DNA endonuclease in adaptive bacterial immunity. Science 337 (6096), 816–821.

Khush, G.S., 2005. What it will take to feed 5.0 billion rice consumers in 2030. Plant Mol. Biol. 59 (1), 1–6.

Kihoro, J., Bosco, N.J., Murage, H., Ateka, E., Makihara, D., 2013. Investigating the impact of rice blast disease on the livelihood of the local farmers in greater Mwea region of Kenya. SpringerPlus 2(1).

Lee, K., Eggenberger, A.L., Banakar, R., McCaw, M.E., Zhu, H., Main, M., et al., 2019. CRISPR/Cas9-mediated targeted T-DNA integration in rice. Plant Mol. Biol. 99 (4–5), 317–328.

Li, W., Zhu, Z., Chern, M., Yin, J., Yang, C., Ran, L., Cheng, M., He, M., et al., 2017. A natural allele of a transcription factor in rice confers broad-spectrum blast resistance. Cell 170, 114–126.

Liang, G., Zhang, H., Lou, D., Yu, D., 2016. Selection of highly efficient sgRNAs for CRISPR/Cas9-based plant genome editing. Sci. Rep. 6 (1), 1–8.

Liu, D., Chen, X., Liu, J., Ye, J., Guo, Z., 2012. The rice ERF transcription factor *OsERF922* negatively regulates resistance to *Magnaportheoryzae* and salt tolerance. J. Exp. Bot. 63 (10), 3899–3911.

Ma, J., Lei, C., Xu, X., Hao, K., Wang, J., Cheng, Z., Ma, X., Ma, J., Zhou, K., Zhang, X., Guo, X., Wu, F., Lin, Q., Wang, C., Zhai, H., Wang, H., Wan, J., 2015. *Pi64*, encoding a novel CC-NBS-LRR protein, confers resistance to leaf and neck blast in rice. Mol. Plant-Microbe. Interact. 28, 558–568.

Ma, J., Chen, J., Wang, M., Ren, Y.L., Wang, S., Lei, C.L., Cheng, Z.J., Sodmergen, 2018. Disruption of OsSEC3A increases the content of salicylic acid and induces plant defense responses in rice. J. Exp. Bot. 69 (5), 1051–1064.

Mali, P., Esvelt, K.M., Church, G.M., 2013. Cas9 as a versatile tool for engineering biology. Nat. Methods 10 (10), 957–963.

Marzec, M., Brąszewska-Zalewska, A., Hensel, G., 2020. Prime editing: a new way for genome editing. Trends Cell Biol. 30 (4), 257–259.

Miah, G., Rafii, M.Y., Ismail, M.R., Puteh, A.B., Rahim, H.A., Asfaliza, R., Latif, M.A., 2013. Blast resistance in rice: a review of conventional breeding to molecular approaches. Mol. Biol. Rep. 40 (3), 2369–2388.

Miao, J., Guo, D., Zhang, J., Huang, Q., Qin, G., Zhang, X., Wan, J., Gu, H., Qu, L., 2013. Targeted mutagenesis in rice using CRISPR-Cas system. Cell Res. 23 (10), 1233–1236.

Mishra, R., Joshi, R.K., Zhao, K., 2018. Genome editing in rice: recent advances, challenges, and future implications. Front. Plant Sci. 9, 1361.

Molla, K.A., Yang, Y., 2019. CRISPR/Cas-mediated base editing: technical considerations and practical applications. Trends Biotechnol.

Müller, M., Munné-Bosch, S., 2015. Ethylene response factors: a key regulatory hub in hormone and stress signaling. Plant Physiol. 169 (1), 32–41.

Mushtaq, M., Sakina, A., Wani, S.H., Shikari, A.B., Tripathi, P., Zaid, A., Galla, G., Abdelrahman, M., Sharma, M., Singh, A.K., Salgotra, R.K., 2019. Harnessing genome editing techniques to engineer disease resistance in plants. Front. Plant Sci. 10.

Rees, H.A., Liu, D.R., 2018. Base editing: precision chemistry on the genome and transcriptome of living cells. Nat. Rev. Genet. 19 (12), 770–788.

Sakulkoo, W., Osés-Ruiz, M., Garcia, E.O., Soanes, D.M., Littlejohn, G.R., Hacker, C., Correia, A., Valent, B., Talbot, N.J., 2018. A single fungal MAP kinase controls plant cell-to-cell invasion by the rice blast fungus. Science 359, 1399–1403.

Shan, Q., Wang, Y., Li, J., Zhang, Y., Chen, K., Liang, Z., Zhang, K., Liu, J., Xi, J.J., Qiu, J., Gao, C., 2013. Targeted genome modification of crop plants using a CRISPR-Cas system. Nat. Biotechnol. 31 (8), 686–688.

Shan, S., Soltis, P.S., Soltis, D.E., Yang, B., 2020. Considerations in adapting CRISPR/Cas9 in nongenetic model plant systems. Appl. Plant Sci., e11314.

Singh, W.H., Kapila, R.K., Sharma, T.R., Rathour, R., 2015. Genetic and physical mapping of anewallele of *Pik* locus from *japonica* rice 'Liziangxintuanheigu'. Euphytica 205, 889–901.

Tang, X., Lowder, L.G., Zhang, T., Malzahn, A.A., Zheng, X., Voytas, D.F., et al., 2017. A CRISPR–Cpf1 system for efficient genome editing and transcriptional repression in plants. Nat. Plants 3 (3), 1–5.

Urnov, F.D., Miller, J.C., Lee, Y., Beausejour, C.M., Rock, J.M., Augustus, S., Jamieson, A.C., Porteus, M.H., Gregory, P.D., Holmes, M.C., 2005. Highly efficient endogenous human gene correction using designed zinc-finger nucleases. Nature 435 (7042), 646–651.

Wang, M., Mao, Y., Lu, Y., Tao, X., Zhu, J., 2017. Multiplex gene editing in rice using the CRISPR-Cpf1 system. Mol. Plant 10 (7), 1011–1013.

Xie, K., Yang, Y., 2013. RNA-guided genome editing in plants using a CRISPR–Cas system. Mol. Plant 6 (6), 1975–1983.

Xie, K., Minkenberg, B., Yang, Y., 2015. Boosting CRISPR/Cas9 multiplex editing capability with the endogenous tRNA-processing system. Proc. Natl. Acad. Sci. 112 (11), 3570–3575.

Xu, R., Qin, R., Li, H., Li, D., Li, L., Wei, P., Yang, J., 2017. Generation of targeted mutant rice using a CRISPR-Cpf1 system. Plant Biotechnol. J. 15 (6), 713–717.

Yamato, T., Handa, A., Arazoe, T., Kuroki, M., Nozaka, A., Kamakura, T., et al., 2019. Single crossover-mediated targeted nucleotide substitution and knock-in strategies with CRISPR/Cas9 system in the rice blast fungus. Sci. Rep. 9, 7427.

Yang, B., 2020. Grand challenges in genome editing in plants. Front. Genome Edit. 2, 2.

Zhou, H., Liu, B., Weeks, D.P., Spalding, M.H., Yang, B., 2014. Large chromosomal deletions and heritable small genetic changes induced by CRISPR/Cas9 in rice. Nucl. Acids Res. 42 (17), 10903–10914.

Zhao, H., Wang, X., Jia, Y., Minkenberg, B., Wheatley, M., Fan, J., Jia, M.H., Famoso, A., Edwards, J.D., Wamishe, Y., Valent, B., Wang, G., Yang, Y., 2018. The rice blast resistance gene Ptr encodes an atypical protein required for broad-spectrum disease resistance. Nat. Commun. 9 (1), 1–12.

Regulatory requirement for commercialization of biocontrol agents

29

A.B. Vedamurthy[a], S.L. Varsha[a], and S.D. Shruthi[b]
[a]P.G. Department of Studies in Microbiology and Biotechnology, Karnatak University, Dharwad, Karnataka, India, [b]Microbiology and Molecular Biology Lab, BioEdge Solutions, Bangalore, Karnataka, India

29.1 Introduction

With the increasing requirement for increased and quality agricultural produce, the need of the hour is a greener approach in plant/crop protection. Biocontrol agents are being used in place of chemical pesticides. These are environmentally friendly and could definitely replace the use of chemicals. One major attribute of a successful biocontrol agent (BCA) is that, upon its release into the soil/environment, it should be in a position to withstand competition and/or antagonism, if any, offered by native microflora (Nagaraju et al., 2012a).

The use of the term "Biocontrol agents" came into existence as there was inconsistency in using the term biopesticide. This was promoted by the International Biocontrol Manufacturers' Association (IBMA) and the International Organization for Biological Control. IBMA categorizes biocontrol agents into four groups, viz., (i) Macrobials; (ii) Microbials; (iii) Natural products; and (iv) Semiochemicals (insect behavior-modifying agents). Microbials are the most important products among all the biocontrol agents, followed by Macrobials and lastly by Natural products. Microbials are composed mainly of microorganisms such as Bacteria, Fungi, and Viruses as Biocontrol agents (Jogaiah et al., 2013). These are the living systems that suppress the target pests. Hence, the term biocontrol agents. These include different microorganisms like *Trichoderma*, *Phytophthora*, *Beauveria*, *Bacillus*, *Pseudomonas*, etc. (Murali et al., 2013; Babu et al., 2015), which act as nano-based biofungicide against plant diseases (Joshi et al., 2019; Nandini et al., 2020), *Phytophthora* acting as a bioherbicide, *Bacillus* and *Pseudomonas* acting as bioinsecticides. They synthesize toxic metabolites which help in preventing the attack of microbes, resulting in poor establishment of pathogens on healthy plants. This property of higher selectivity and much lower toxicity makes the biocontrol agents more advantageous over chemical pesticides (Abdelrahman et al., 2016). This chapter attempts to throw light on various BCAs which have been explored and exploited for commercialization with a single aim of suppression of pests. The term pests is sometimes used as an umbrella term to include fungi, bacteria, viruses, nematodes, protozoans, insects, mites, and last

Biocontrol Agents and Secondary Metabolites. https://doi.org/10.1016/B978-0-12-822919-4.00029-6

but not least, weeds. Following their commercialization, the regulatory measures in place aimed at evolving policies in different regions of the world have been looked into.

29.2 Biocontrol agents

29.2.1 Bacteria

Naturally occurring living organisms which act on pathogens of plants or animals can be considered as biocontrol agents. The toxins from microorganisms and the organisms themselves can be used to manage the diseases of plants caused by different living forms. Biopesticides are the biocontrol microbials: their toxic metabolic products and other chemicals derived from living organisms are known as Biopesticides by the Environmental Protection Agency (EPA). These offer an alternative to the chemical pesticides. Keeping in view the ill effects of chemical pesticides on the environment, numerous trials were made to establish microbes such as *Bacillus thuringiensis* (Bt), which is in use commercially, as a microbial insecticide since 40 years (Gelernter and Schwab, 1993).

The most common form of biocontrol agents which are in wider use is bacteria. Though most of the targets are insect pests, the growth of many plant pathogens such as bacteria, fungi is also inhibited (Abdelrahman and Jogaiah, 2019). Bacterial biocontrol agents, when acting against pathogenic fungi, perch on the mycelia and prevent the colonization by fungal pathogens (O'Brien et al., 2009; Jogaiah et al., 2018). In some cases, bacteria are able to synthesize harmful toxins which when ingested by insects lead to the paralysis of local tissue, leading to the death of the insect pests (Schallmey et al., 2004). The insecticides based on many species of *Bacillus thuringiensis* are considered most predominant. These include *B. thuringiensis* ssp. *kurstaki* and *aizawi* which confer the highest activity against lepidopteron larvae. Whereas the activity on mosquito larva, black fly (a simulid), and fungus gnats is exhibited by *B. thuringiensis* ssp. *israelensis*, *B. thuringiensis tenebrionis* shows its potency against coleopteran larvae and adults, especially on the Colorado potato beetle (*Leptinotarsa decemlineata*); *Bacillus thuringiensis* is powerful against *Aedes aegypti*; the species *B. sphaericus* is found to be effective on different types of mosquitoes like *Culex quinquefasciatus* (Lacey et al., 2001); *B. thuringiensis japonensis* strain buibui is active against soil-inhabiting beetles (Carlton, 1993; Copping and Menn, 2000). Several commercial *Bacillus* spp. like *Bacillus thuringiensis israelensis* (Bti) and *Bacillus sphaericus* 2362 are found effective against dipteran larvae (Goldberg and Margalit, 1977) and mosquitoes (Revathi et al., 2013).

The basic understanding of all commercially available biocontrol agents is based on the information available on *Bacillus thuringiensis*, although the matter pertaining to individual biocontrol agents may vary. *Bacillus thuringiensis*, during its sporulation, mainly synthesizes an inclusion body containing a crystalline protein called Cry protein or δ endotoxin which has insecticidal property. The endotoxin is mainly active against the larval stage of different moths. *B. thuringiensis* ssp. *kurstaki* is potent against the larval stage of a wide variety of moths and butterflies. This attribute is exploited in developing numerous biopesticides, bioinsecticides, and also biofungicides. In addition to the use as a biocontrol agent, the genes coding for the Cry protein possessing the insecticidal property are successfully incorporated into different crops such as cotton, brinjal, tomato, etc. yielding economic benefits. Owing to their eco-friendly and high specificity, this is used as a biocontrol agent against insect pests (Roy et al., 2007; Kumar, 2012). Reports suggest the beneficial aspects of many Bacillus species being effective against the diseases caused by Fungi at large. *Bacillus* spp. exhibit its action on Root diseases like Avocado root rot, Tomato damping off, Wheat take-all, etc. It is also efficacious against many foliar diseases such as Cucurbit and Strawberry Powdery Mildews and also on postharvest diseases, namely Green, Gray, and Blue molds (Cazorla et al., 2007; Perazzolli et al., 2008; Arrebola et al., 2010). Few strains of *B. subtilis* are efficacious against fungal diseases like Damping off and Soft rots.

Another group of bacteria, Pseudomonads, is considered as an important class of plant growth-promoting rhizobacteria. Because of their root colonizing ability and metabolic activity, they are also studied for their role as biocontrol agents. In addition, they are also known to increase plant growth, yield, and reduce the intensity of many diseases (Sudisha et al., 2006). Many species of *Pseudomonas* such as *P. fluorescens* antagonize pathogenic bacteria. They are effective against Late blight of Potato fungus as well. Some of them have shown their potency against the dreaded bacterial plant pathogen *Ralstonia solanacearum* (Bora and Deka, 2007). Many strains of *P. syringae, P. aeruginosa,* and also *P. fluorescens* are effective against pathogens causing damping off and soft rots (Kloepper et al., 2004; Haas and Défago, 2005; Berg, 2009). Fluorescent pseudomonads also show their potency against *Aspergillus, Alternaria, Fusarium, Pythium, Sclerotinia,* and *Rhizoctonia* (Dunne et al., 1998). Under controlled conditions, *Pseudomonas maltophilia* is potent against soilborne fungi *Rhizoctonia bataticola, R. solani, Fusarium oxysporum,* and *Sclerotinia sclerotiorum* causing root rot diseases of many economically important crops. Some strains of a well-known biocontrol agent, *Agrobacterium radiobacter*, are effective against *Agrobacterium tumefaciens* (causing crown gall symptoms on many temperate fruit crops).

Among the different types of microorganisms utilized as a biocontrol agent, bacterial communities inhibit phytopathogens by their antagonistic effects. The selection of bacteria as a potential biocontrol agent starts from their isolation from soil samples or from any part of the plant (Jogaiah et al., 2010). Followed by isolation, they are identified by chemical and molecular biology methods. Several studies should be conducted to understand their antagonistic mechanisms against phytopathogens and to determine whether they are nontoxic for the environment, plants, humans, and animal health (Jogaiah et al., 2018). In general, these kinds of biocontrol agents should harm phytopathogens, and at the same time should be beneficial to host plants, which is depicted in Fig. 29.1.

29.2.2 Fungi

Another major group of microorganisms that is used as biocontrol agents are the Fungi. These are ubiquitous in nature. Their use as a biocontrol agent lies in the fact that they show a high degree of host specificity, persistence, and dispersal efficiency

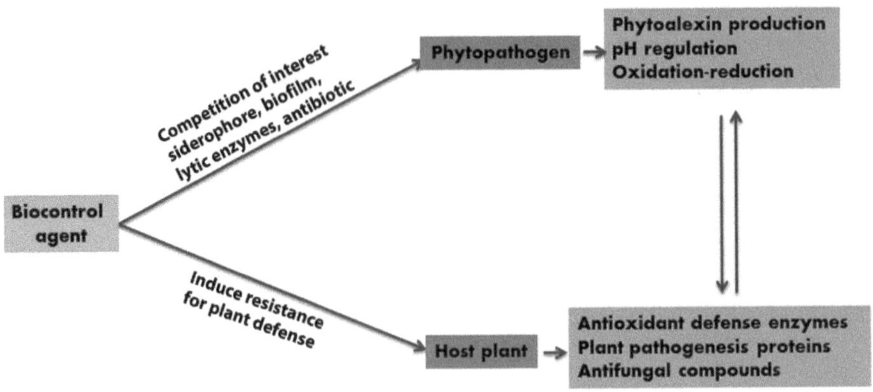

Fig. 29.1 Antagonistic mechanisms of bacteria as biocontrol agents.

(Murali et al., 2013). They need moist soil and low temperature for their growth and most can be easily cultured and maintained in the laboratory.

Fungi exhibit their pathogenicity on other organisms mainly in three different modes, namely,

(i) Direct antagonism: Here, the fungi parasitize on other microorganisms, which results in death of the organism (Baker and Cook, 1974). This method of elimination of pathogen by fungi is called mycoparasitism. Fungal hyphae of *Trichoderma* coil around the hypha of the target fungi. At certain points of contacts, the cell wall degrading enzymes synthesized by *Trichoderma* puncture the target fungal hypha, eventually leading to the suppression of the target fungal mycelia. This process is observed in case of *Trichoderma viride* parasitizing the hyphae of *Rhizoctonia solani,* a common oil-borne pathogen (Harman et al., 2004; Sivan and Chet, 1989; Vinale et al., 2008; Zeilinger et al., 1999). *Piriformospora indica* is known to produce a certain chemical compound which is known to hinder the growth of fungi belonging to order Sebacinales by a process called pathogen-derived cell death (Deshmukh et al., 2006; Schäfer et al., 2007, 2009; Serfling et al., 2007; Waller et al., 2005).

(ii) Antibiosis: Antibiosis is a process in which an antibiotic substance is synthesized in close proximity to suppress the growth of other microorganisms. Fungal antimicrobial compounds can affect the sporulation and hyphal elongation of target fungi in their life cycle (Keller et al., 2005). Enzymes such as xylanases, cellulose, and swollenin are secreted by different *Trichoderma* spp. which obstruct the growth of many fungal pathogens. *Trichoderma* spp. are effective in suppressing the viability of sclerotia produced by pathogenic fungi.

(iii) Competition: The microorganisms present in an ecological habitat try to compete for the available nutrients and space with each other. In doing so, the antagonistic fungi outgrow the pathogenic fungal forms and establish themselves (Chet et al., 1997). The short supply of easy availability of nutrients for the microbes living in the soil or on plant surfaces results in competition among microorganisms. Many *Trichoderma* BCAs produce efficient siderophores that chelate iron and stop the growth of other fungi (Chet and Inbar, 1994; Jogaiah et al., 2013).

29.2.3 Viruses

Like Bacteria and Fungi, Viruses also play an important role as biocontrol agents. Viruses are obligate parasites and are often host-specific. Viruses infecting bacteria are called Bacteriophages. Recent studies have demonstrated a significant reduction in disease severity in respect of a range of pathogens including *Agrobacterium, Xanthomonas, Ralstonia solanacearum, Erwinia amylovora*, and *Streptomyces* pathogenic on a variety of crops Likewise, viruses acting on fungi are known as Mycoviruses. This property of the viruses is put to use as a biocontrol agent. The causal organism of bacterial spot disease of tomato and pepper, *Xanthomonas campestris* pv. *Vesicatoria*, is controlled by the bacteriophage specific to the pathogen. Likewise, the virus attacking the pathogen *Pseudomonas syringae* is being used as a biocontrol agent (Frampton et al., 2012). Certain mycoviruses, through the phenomenon of hyphal anastosmosis, are successful in controlling the Chestnut Blight pathogen *Cryphonectriaparasitica* in certain geographical locations. Hence, these are likely to be successfully used as biocontrol agents (Ghabrial et al., 2015).

Another group of BCA viruses is the Baculovirus active against insect pests. Baculovirus comprises two groups, namely Granulosis virus and Nuclear Polyherdosis Virus (NPV). These mainly control the caterpillar stage of some insect pests (Arthurs et al., 2005). They are particularly effective against the lepidopteran pests of cotton, rice, and vegetables. Viral products for codling moth, *Heliothis* NPV, and beet armyworm NPV are effective. NPV is active in controlling the dreaded lepidopteran pests such as the cotton bollworm and cotton budworm (Arthurs and Lacey, 2004; Arthurs et al., 2005).

29.2.4 Screening for commercialization of biocontrol agents

With this background of the concepts spanning biocontrol, biocontrol agents (BCAs), exploitable BCAs from among bacteria, fungi, and viruses based on their efficacies against the plant pathogens (fungi, bacteria, viruses, etc.) which bring down the yields of our cultivated crops, it is only logical and interesting to know about the commercialization aspects pertaining to those promising BCAs. It will then become pertinent to learn about the prevailing regulatory practices evolved nationally and/or internationally (Fig. 29.2).

With the advancement in the field of science, in order to feed the ever-increasing human and animal population, and with the increased awareness to have organically grown foods, it becomes a necessity to use biocontrol agents to cope with this situation (Nagaraju et al., 2012b). This situation has led to the inundatory use of BCAs to control a wide variety of plant pathogens as also insect pests. But in doing so, we are replacing the native microbial load with the increasing load of BCAs. Hence, these call for measures to be followed for the efficient use of BCAs. Regularization and commercialization have to be brought about in each of the forthcoming steps, right from isolation of a potent strain of the BCA organism till the end user.

Isolation of microorganisms is the basic step. Organisms are isolated from their natural habitat in the pure form using different culturing methods. Culturing of the

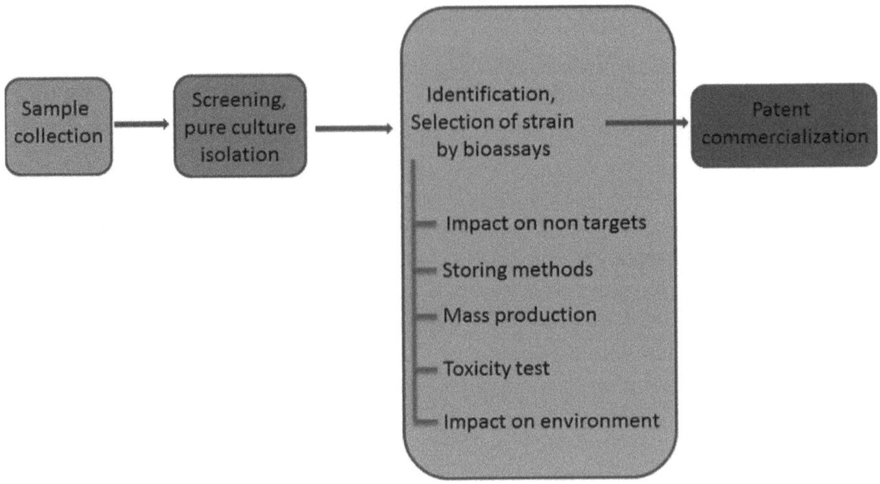

Fig. 29.2 Showing how a novel biocontrol agent can be brought into market.

organisms is usually carried out in laboratory conditions using suitable growth media comprising a proper carbon source, nitrogen source, and other special ingredients, if any, required by the organism. For example, culturing of the entamopathogenic fungus *Paecilomyces chlororaphis* is much easier than culturing of a nematode-pathogenic bacterium *Pasteuria penetrans* (Agrios, 1997), and it is still more difficult to culture viruses as they are obligate parasites. The primary isolation is done repeatedly. Once the primary culture is obtained in pure form, it is subjected to mass multiplication for enrichment of the propagules of BCAs so as to obtain the organism of interest in a sizable amount. Colony forming units (cfu) are a case in point: Irrespective of the type of formulation (carrier-based talc, soil, or liquid), a minimum number of cfu of any BCA must be present in a gram or milliliter of the formulation product.

The crucial determining step is the screening process. The success of this step mainly depends on the target pathogen and the crop intended. Screening for the antagonistic activity is mainly based on BCA's mode of action, viz., Antibiosis, parasitism, or competition, against the target organism. For example, for screening microbes to address the postharvest diseases, it is required to screen for the microbes which are present on the surface of the fruit/vegetable, rather than screening for microbes which compete for soil colonization. Isolates are subjected to bioassay tests, done to find out its efficacy in a laboratory scale in a controlled environment. This is a time-consuming process. A three-partner model system, comprising the said BCA, target organism, and host plant in the laboratory under controlled conditions, is recommended. Nowadays, even though the success rate is low, molecular markers are used to screen for BCAs (Mathre et al., 1999). The baiting method is the most followed method to screen for potent BCAs. Annexure provides a list of potent BCAs.

Pilot tests are undertaken in order to test the isolates of organism of interest against different pathogens under different conditions such as light, temperature, pH, humidity, soil type, cultivar type, type of crop, and also the cropping system and also to find out its spectrum of activity. For example, the efficacy of nonpathogenic *Fusarium oxysporum is* affected markedly with the difference in temperature and light (Larkin and Fravel, 2002). Only a few are able to sustain their broad spectrum of activity and yield consistent results from numerous trials (Harman, 2000).

In addition to the abovementioned steps, it is advisable to include the toxicology tests for the biosafety on nontarget organisms (plants and animals). For example, absence of hypersensitivity reactions in solanaceous plants by bacteria indicates its nonpathogenicity. Mammalian toxicology studies are to be carried out to assure safety of consumers' health. Median Lethal Dose of LD_{50} is used in oral acute toxicity tests in rats. For bacteria, LD50 denotes 10^{11} cfu/kg of animal weight. In some cases, keeping in view the delayed hypersensitivity reactions, the toxicity tests should be carried out by carrying out Magnusson & Klingman Tests. In some cases, the Ames Mutagenicity Test is also carried out mainly to detect the toxic secondary metabolites.

Large-scale production is carried out in fermentors based on the type of organism (bacteria, fungi, or viruses) to be used as BCAs. For bacteria, continuous batch fermentation is carried out preferably using a liquid medium. For fungi, both solid and liquid state fermentation is preferred based on the site of product synthesis (intracellular or extracellular), whereas for obligate parasites like viruses, large-scale production is done using cell cultures or tissue cultures.

The next step is the preservation/storage of microbes acting as BCAs. This is done to increase safety and shelf life. Different methods such as cold storage, refrigeration, drying, etc. are used. Two broad methods, namely Cold storage and Dehydration, are followed in order to maintain the cells in viable condition for a longer period. The best method to do so is Lyophilization. Lyophilized and air-dried preparations of BCAs are suitable (Nakkeeran et al., 2005). Many formulations could have extended the shelf life of bioformulations if they are granules, pellets, and dry powder based. But it is not economical, so industries do not prefer it. Industries mostly prefer the Encapsulation method followed by drying. Encapsulation is done using a matrix-forming material such as a polysaccharide or lipid emulsion.

In order to be used in the fields, for commercial purposes, the final product needs to be formulated by adding biocompatible supplements to improve its stability and application (Boyetchko et al., 1998). The method of application of the biocontrol agent is similar to the methods used for chemical pesticides. Dry powder based is widely preferred, as it can be suspended in water and applied to a larger area. The granules protect the active ingredient present. Currently, efforts are on to optimize their controlled release by using nanoemulsion, nanosuspension, and nanocapsule suspension with the help of nanotechnology (Ranga Rao et al., 2007; Ghormade et al., 2011; Glare et al., 2012). In addition to the abovesaid measures, trials are ongoing to find out Residue analysis and traceability using molecular approaches to distinguish the BCA strain with the indigenous variety present in the natural habitat (Louws et al., 1999; Schena et al., 2000). The features of an agricultural economy make it difficult for farmers to adapt to new technology. It also makes it difficult for the industries to provide funds for developing new BCAs. Overall, there are various factors which contribute to the success of bringing efficient bioagents into the market (Fig. 29.3).

Flow diagram:

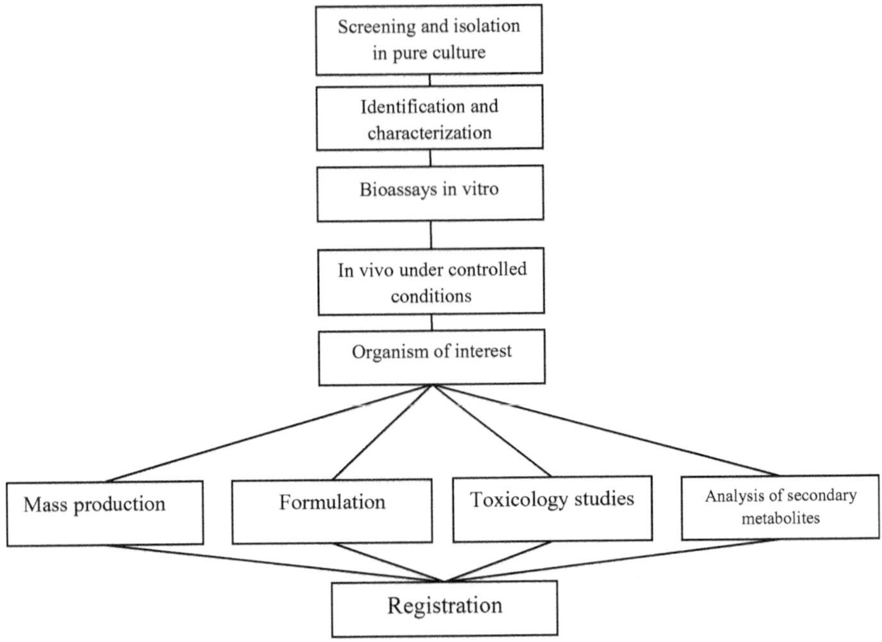

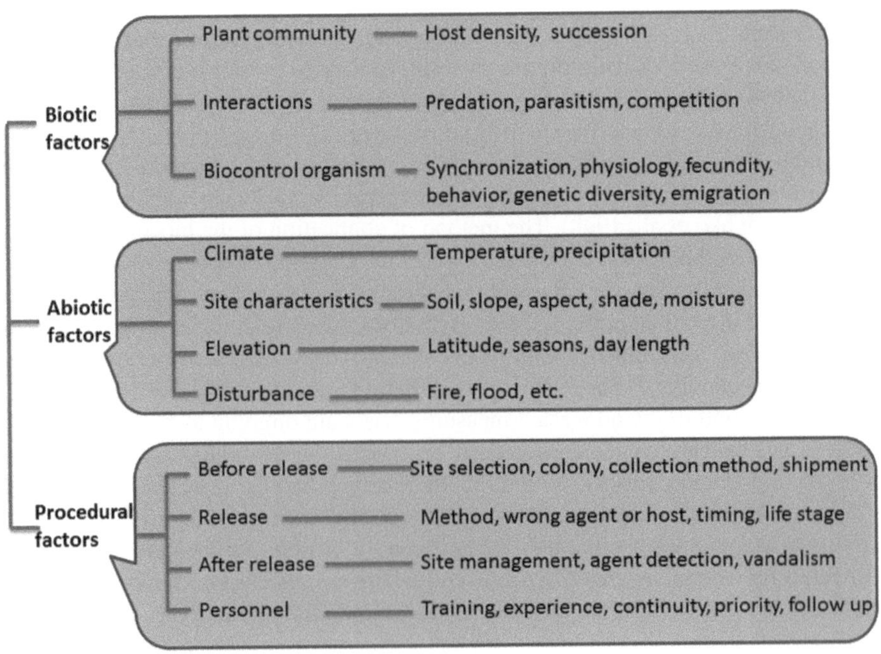

Fig. 29.3 Factors affecting the success of bioagents.

29.3 Regulatory requirements: Indian and global perspective

The abovementioned regulatory practices call for the legal and standard use of BCAs for biosafety purposes. Hence, it is mandatory to obtain permission for registration or licensing of BCAs. Risk assessment is important, as it provides the basis for approval/registration by the government authority known as the National Regulatory Authority. The registration involves the compilation and submission of data which are collected for regulation and commercialization of BCAs. Based on the dossier, the government then sets up a regulatory framework to be accepted globally. The preparation of data is a tedious process. Reasons that make data submission difficult are (Ravensberg, 2011):

(a) Data requirements are not well-defined and they are the same for chemical pesticides too.
(b) Data relating to toxicity test and validation are the same for both chemical pesticides and biocontrol agents.
(c) Risk assessments are not well-defined.
(d) Regulators and Assessors lack the technical know-how.

Regulatory bodies like the International Organization for Biological Control (IOBC), the Organization for Economic Cooperation and Development (OECD), and the European and Mediterranean Plant Protection Organization (EPPO) have time and again tried to set right the registration process, but not much success ensued. Policymakers, industries, scientists must come forward to address the situation emphasizing mainly on biosafety protection of humans and environment and, secondly, properties of products, their quality and consistency (Chandler et al., 2011).

In India, the use of biocontrol agents is encouraged by that part of the Government program called IPM—Integrated Pest Management and also through the Department of Biotechnology. India has been a forerunner in vouching for biosafety legislations with special importance to Biocontrol agents. Section 3 of the Insecticides Act, 1968 has a comprehensive list of BCAs which is amended from time to time. The microbes listed under the Act are compulsorily required to undergo Toxicology tests and are to be registered with the Central Insecticide Board & Registration Committee (CIBRC). The guidelines are based on OECD guidelines. The Ministry of Science and Technology, Department of Biotechnology has proposed the Recombinant DNA Safety Guidelines 1990. The guidelines are based on the host range of BCAs, which have the capacity to cause diseases to humans/animals/plants, availability of curative measures, etc. In addition to these, The Rules for Manufacture, Use, Import, Export, and Storage of Hazardous Microorganisms/Genetically Engineered Organisms or Cells 1989 under the Environmental Protection Act 1986 also look into the use of BCAs. It has a special section called "Plant Pests" which emphasizes that no person shall import, export, manufacture, process, use, or sell any hazardous microbes or genetically engineered organisms/substances or cells without the approval of the Genetic Engineering Approval Committee (GEAC), functioning under the Ministry of Environment and Forests, Government of India. The National Agricultural Research

System, comprising the Indian Council of Agricultural Research and State Agricultural Universities (SAUs), also plays a role in supporting the use of these biocontrol agents.

Asian countries like Japan, South Korea, China, etc. are emerging as key players in the regulation and use of biocontrol agents. Japanese Agricultural Standard (JAS) Law 2000 regulates the use of chemicals, fertilizers, soil enhancing materials along with BCAs. In South Korea, the registration is done in a phase-wise manner. Initially controlled by Agromaterials Management Division (AMD) and Rural Development Administration (RDA), it is later processed by Pesticide Safety and Evaluation Division (PSED) and National Academy of Agricultural Sciences (NAAS). The Microbial Pesticide Regulation 2000 was shifted to Management of Agricultural Chemicals (Arora et al., 2016). The Regulation on Pesticide Administration was introduced in China in 1997. It regulates the registration and use of BCAs before they enter the market. The National Regulatory Authority, Institute for the Control of Agrochemicals, Ministry of Agriculture (ICAMA) 2008 governs the registration of Pesticides (Arora et al., 2016). The Ministry of Forestry supports the use of BCAs by providing funds. It is only the General Administration of Quality Supervision Inspection and Quarantine of the People's Republic of China that has the right to apply for registration (Kabaluk et al., 2010).

The European Union comprises 27 countries. The use of biological control agents is supported under the IPM system. Their use is controlled by Council of Directive 91/414/EEC and its succeeding amendments applying to all plant protection products. In November 2009, a regulation was passed to replace Council of Directive 91/414/EEC to provide direction for registration of microbial pesticides. Although Europe does not have microbial pesticide-based directives, the guidelines for the same could be found in the Reference Section. Some member countries like Belgium have a web page dedicated to Plant protection product registration service named the—Program for Reduction of Pesticides and Biopesticides. The United Kingdom has also established a Biopesticide Scheme to increase the production of Biocontrol agents. On lines similar to the United Kingdom, the Netherlands also has the Gewasbeschermingsmiddelen van Natuurlijke Oorsprong Effectief Gebruiken GENOEG project which assists the production and use of natural pesticides (Kabaluk et al., 2010).

On lines similar to the European Union, the newly Independent states—Russia, Ukraine, and Moldova— also favor the use of biocontrol agents. With the reemergence of IPM, during 1957, the biocontrol was started through the setting up of the Institute of Plant Protection which mainly focus on the use of *Beauveria bassiana*. The regulatory authority in Ukraine, Main State Inspection of Plant Protection (MSIPP) is in charge of registration of biopesticides and other products of plant protection. The National Academy of Sciences (NANU) and the Ukrainian Agricultural Academy of Sciences (UAAN) are currently helping in the use of biocontrol agents. In Russia, the registration is managed by the Russian Agricultural Control (RAC). These days, the All-Russian Scientific Institute for Plant Protection (VIZR) is carrying out research and helping in the use of biocontrol agents. The Institute for Plant Protection and Ecologic Agriculture, Academy of Sciences of Moldova is carrying out research and registration of biocontrol agents.

In North America, Canada and the United States together have a regulatory authority named the North American Free Trade Association (NAFTA) for assessment of biocontrol agents. USEPA is the regulatory authority which comprehensively evaluates the process of regulation and registration of biocontrol agents. The regulation falls under the preview of the Office of Chemical Safety and Pollution Prevention (OCSPP) and the Office of Pesticide Programs (OPP). Biopesticide Industry Alliance (BPIA) fosters the use of biocontrol agents and also collaborates with Biopesticide Regulation Authorities to verify correct regulatory requirements. Canada regulates the use of BCAs by the Pest Management Regulatory Agency, PMRA under the Pest Control Product Act 2002. The use of nonnative microbials is regulated by the Canadian Food Inspection Agency (Arora et al., 2016).

South America too is not lagging behind in implementing measures for the use of biocontrol agents. In Argentina, the Coordination of Agrochemical and Biological Products' wing of SENASA regulates the registration, commercialization, and use of bioproducts. The National Agricultural Department (NAD) and Environment Policy Secretary (EPS) are involved in the regulation of biocontrol agents. The Brazilian Institute of Environment and Renewable Natural Resources (IBAMA) and the National Health Surveillance Agency (ANVISA) monitor the use of pesticide registered by the Brazilian Ministry Of Agriculture, Livestock, and Food Supply (MAPA). In 2009, the law was enforced to use macro- and microorganisms for plant protection. To organize this, the Brazilian Biocontrol Manufacturers' Association was created.

In the Australian continent, the regulation is controlled by the National Registration Scheme (NRS), now renamed the Australian Pesticide and Veterinary Medicine Authority, which supervises the guidelines issued on evaluation of toxicity studies, host range, effect on native flora and fauna. New Zealand has the Hazardous Substances and New Organisms (HSNO) Act 1996 which supervises the regulation of microbial pesticides. Prior to commercialization of microbials, consent has to be taken from the Environmental Risk Management Authority (ERMA). Upon acceptance, registration has to be done by the Agricultural Compounds and Veterinary Medicines (ACVM) Act 1997. Their identity and purity are checked by the Ministry of Agriculture in New Zealand.

Africa has robust regulatory mechanisms in place in different countries of the continent. Regarding the use of biocontrol agents, Kenya, being a pioneer there, developed specific guidelines for registration, representing a rational system that evaluates the safety and dangers associated with biocontrol agents. The regulatory authority, the Pest Control Products Board (PCPB), regulates the use of pesticides. Along with the Agriculture Act Chapter (Cap318), Kenya Plant Health Inspectorate Service (KEPHIS) Act, Cap 512/2013, and Plant Protection Act are responsible for the import and export of live organisms. The Pest Control Products Act is also involved in registration and regulations of biocontrol agents' use. Other countries like Uganda, Ethiopia, Tanzania, Ghana, Ivory Coast, South Africa, etc. have also followed suit on the lines of Kenya, for the regulation and commercialization of biocontrol agents (Fig. 29.4).

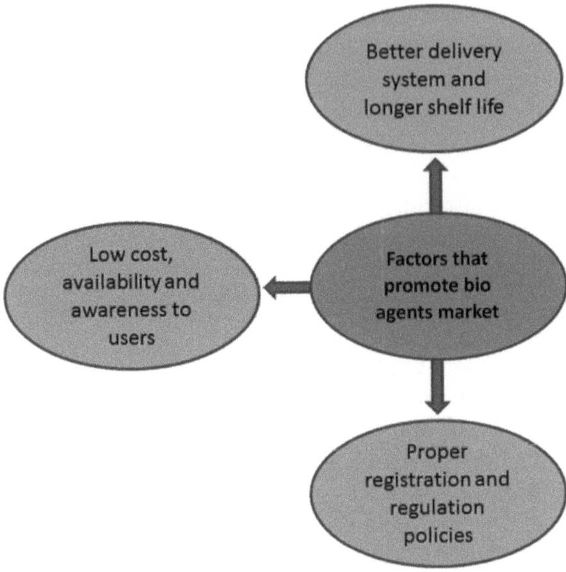

Fig. 29.4 Various factors which promote the market of bioagents.

29.4 Summary and conclusion

The practice of biocontrol of pathogens and insect pests has emerged as a viable alternative to chemical control in plant protection. This has led to the exploitation of native BCAs with the concomitant practice of their commercialization. Inevitably, this calls for evolving suitable regulatory measures in areas/countries where appreciable commercialization activity has taken place. The overriding feature observable in all regulatory mechanisms/policies is that the BCAs be necessarily registered with a statutory body in different countries. The relevant toxicology and biosafety data pertaining to those BCAs are be provided while registering.

Annexure. List of efficacious biocontrol agents

Type	Species	Target
Bacterial BCAs		
Insecticide	*Bacillus papillae*	*Papilla japonica*
	Bacillus thuringiensis ssp. *aizawi*	*Galleria mellonella*
	Bacillus thuringiensis ssp. *israelensis*	Dipteran larvae
	Bacillus thuringiensis ssp. *kurstaki*	Lepidopteran larvae

	Bacillus thuringiensis ssp. *xentari*	Lepidopteran larvae
	Bacillus thuringiensis ssp. San Diego	Coleopteran larvae
	Bacillus thuringiensis ssp. *tenebrionis*	Coleopteran larvae
Fungicide	*Burkholderia cepacia*	Soilborne fungi, nematodes
	Pseudomonas fluroscens	Soilborne fungi
	P. syringae	Postharvest fungi
	P. chlororaphis	Soilborne fungi
	P. aureofaciens	Soilborne fungi, Antracnose
	Bacillus subtilis	Soilborne fungi
Bactericide	*Agrobacterium radiobacter*	*Agrobacterium tumefaciens*
	Ralstonia solanacearum	*Ralstonia solanacearum (pathogenic)*
	Pseudomonas fluroscens	*Erwinia amylovora*- fire blight
Nematicide	*Bacillus firmus*	Nematodes
Fungal BCAs		
Insecticide	*Beauveria bassiana*	White fly
	Vertricillum lecanii	White fly
	Paecilomyces fumosoroseus	White fly
	Metarhizium anisopliae	Black beetle
Fungicide	*Trichoderma polysporum, T. harzianum*	Soilborne fungi
	Trichoderma harzianum	Foliar fungi
	Trichoderma harzianum, T. viride	Various fungi
	Trichoderma lignorum	Vascular wilt
	Trichoderma spp.	Soilborne fungi
	Ampelomyces quisqualis	Powdered mildew
	Gliocladium catenulatum	Soilborne fungi
	Fusarium oxysporum	*Fusarium oxysporum(pathogenic)*
	Pythium oligandrum	*Pythium ultimum*
	Candida oleophila	Postharvest rot
	Coniothyrium minitans	*Sclerotinia sclerotiorum*
Nematicide	*Paecilomyces ilacinus*	Nematode
Viral BCAs		
Insecticide	Granulosis virus	Codling moth
	Heliothis NPV	*Helicoverpa zeae*
	Mamestria brassicae	*Heliothis*
	Gypsy moth NPV	*Lymantria dispar*
Fungicide	Bacteriophage of *P. tolaasii*	*Bacterial rot of mushroom*

Acknowledgment

All the authors are grateful to the DST, New Delhi for providing Financial support under DST-PURSE-PHASE-II Program (grant no. SR/PURSE-PHASE-2/13/G).

References

Abdelrahman, M., Jogaiah, S., 2019. Defense mechanism and diverse actions of fungal biological control agents against plant biotic stresses. In: Khurana, S., Gaur, R. (Eds.), Plant Biotechnology: Progress in Genomic Era. Springer, Singapore, pp. 461–478.

Abdelrahman, M., Abdel-Motaal, F., El-Sayed, M., Jogaiah, S., Shigyo, M., Ito, S.-i., Tran, L.S.P., 2016. Dissection of *Trichoderma longibrachiatum*-induced defense in onion (*Allium cepa* L.) against *Fusarium oxysporum* f. sp. *cepa* by target metabolite profiling. Plant Sci. 246, 128–138.

Agrios, G.N., 1997. Plant Pathology, 4th ed. Academic Press, San Diego, CA.

Arora, N.K., Verma, M., Prakash, J., Mishra, J., 2016. Regulation of biopesticides: global concerns and policies. In: Bioformulations: For Sustainable Agriculture. Springer, New Delhi, pp. 283–299.

Arrebola, E., Jacobs, R., Korsten, L., 2010. Iturin A is the principal inhibitor in the biocontrol activity of *Bacillus amyloliquefaciens* PPCB004 against postharvest fungal pathogens. J. Appl. Microbiol. 108 (2), 386–395.

Arthurs, S.P., Lacey, L.A., 2004. Field evaluation of commercial formulations of the codling moth granulovirus: persistence of activity and success of seasonal applications against natural infestations of codling moth in Pacific Northwest apple orchards. Biol. Control 31 (3), 388–397.

Arthurs, S.P., Lacey, L.A., Fritts Jr., R., 2005. Optimizing use of codling moth granulovirus: effects of application rate and spraying frequency on control of codling moth larvae in Pacific Northwest apple orchards. J. Econ. Entomol. 98 (5), 1459–1468.

Babu, N.A., Jogaiah, S., Ito, S.I., Nagraj, K.A., Tran, L.S.P., 2015. Improvement of growth, fruit weight and early blight disease protection of tomato plants by rhizosphere bacteria is correlated with their beneficial traits and induced biosynthesis of antioxidant peroxidase and polyphenol oxidase. Plant Sci. 231, 62–73.

Baker, K.F., Cook, R.J., 1974. Biological Control of Plant Pathogens. WH Freeman and Company.

Berg, G., 2009. Plant–microbe interactions promoting plant growth and health: perspectives for controlled use of microorganisms in agriculture. Appl. Microbiol. Biotechnol. 84 (1), 11–18.

Bora, L.C., Deka, S.N., 2007. Wilt disease suppression and yield enhancement in tomato (*Lycopersicon esculentum*) by application of *Pseudomonas fluorescens*-based biopesticide (Biofor-Pf) in Assam. Indian J. Agric. Sci. 77 (8), 490–494.

Boyetchko, S., Pedersen, E., Punja, Z., Reddy, M., 1998. Formulations of biopesticides. In: - Hall, F.R., Barry, J.W. (Eds.), Methods in Biotechnology. In: vol. 5. Humana Press, Totowa, NJ, pp. 487–508.

Carlton, B.C., 1993. Genetics of BT insecticidal crystal proteins and strategies for the construction of improved strains. In: Pest Control With Enhanced Environmental Safety.ACS Symposium Series 524, pp. 326–337.

Cazorla, F.M., Romero, D., Pérez-García, A., Lugtenberg, B.J.J., Vicente, A.D., Bloemberg, G., 2007. Isolation and characterization of antagonistic *Bacillus subtilis* strains from the avocado rhizoplane displaying biocontrol activity. J. Appl. Microbiol. 103 (5), 1950–1959.

Chandler, D., Bailey, A.S., Tatchell, G.M., Davidson, G., Greaves, J., Grant, W.P., 2011. The development, regulation and use of biopesticides for integrated pest management. Philos. T. R. Soc. B. 366 (1573), 1987–1998.

Chet, I., Inbar, J., 1994. Biological control of fungal pathogens. Appl. Biochem. Biotechnol. 48 (1), 37–43.

Chet, I., Inbar, J., Hadar, I., 1997. Fungal Antagonists and Mycoparasites. The Mycota IV: Environmental and Microbial Relationships. Springer-Verlag, Berlin, pp. 165–184.

Copping, L.G., Menn, J.J., 2000. Biopesticides: a review of their action, applications and efficacy. Pest Manag. Sci. 56 (8), 651–676.

Deshmukh, S., Hückelhoven, R., Schäfer, P., Imani, J., Sharma, M., Weiss, M., et al., 2006. The root endophytic fungus *Piriformospora indica* requires host cell death for proliferation during mutualistic symbiosis with barley. Proc. Natl. Acad. Sci. U.S.A. 103 (49), 18450–18457.

Dunne, C., Moënne-Loccoz, Y., McCarthy, J., Higgins, P., Powell, J., Dowling, D.N., O'gara, F., 1998. Combining proteolytic and phloroglucinol-producing bacteria for improved biocontrol of Pythium-mediated damping-off of sugar beet. Plant Pathol. 47 (3), 299–307.

Frampton, R.A., Pitman, A.R., Fineran, P.C., 2012. Advances in bacteriophage-mediated control of plant pathogens. Int. J. Microbiol.. 2012.

Gelernter, W., Schwab, G.E., 1993. Transgenic bacteria, viruses, algae and other microorganisms as *Bacillus thuringiensis* toxin delivery systems. In: *Bacillus thuringiensis*, an Environmental Biopesticide: Theory and Practice. John Wiley & Sons, New York, pp. 89–104.

Ghabrial, S.A., Castón, J.R., Jiang, D., Nibert, M.L., Suzuki, N., 2015. 50-plus years of fungal viruses. Virology 479, 356–368.

Ghormade, V., Deshpande, M.V., Paknikar, K.M., 2011. Perspectives for nano-biotechnology enabled protection and nutrition of plants. Biotechnol. Adv. 29 (6), 792–803.

Glare, T., Caradus, J., Gelernter, W., Jackson, T., Keyhani, N., Köhl, J., Stewart, A., 2012. Have biopesticides come of age? Trends Biotechnol. 30 (5), 250–258.

Goldberg, L.J., Margalit, J., 1977. A bacterial spore demonstrating rapid larvicidal activity against *Anopheles sergentii, Uranotaenia unguiculata, Culex univitattus, Aedes aegypti* and *Culex pipiens*. Mosq. News 37 (3), 355–358.

Haas, D., Défago, G., 2005. Biological control of soil-borne pathogens by fluorescent pseudomonads. Nat. Rev. Microbiol. 3 (4), 307–319.

Harman, G.E., 2000. Myths and dogmas of biocontrol changes in perceptions derived from research on *Trichoderma harzinum* T-22. Plant Dis. 84 (4), 377–393.

Harman, G.E., Howell, C.R., Viterbo, A., Chet, I., Lorito, M., 2004. Trichoderma species—opportunistic, avirulent plant symbionts. Nat. Rev. Microbiol. 2 (1), 43–56.

Jogaiah, S., Roopa, K.S., Pushpalatha, H.G., Shekar Shetty, H., 2010. Evaluation of plant growth-promoting rhizobacteria for their efficiency to promote growth and induce systemic resistance in pearl millet against downy mildew disease. Arch. Phytopathol. Plant Prot. 43, 368–378.

Jogaiah, S., Mostafa, A., Tran, L.S.P., Shin-ichi, I., 2013. Characterization of rhizosphere fungi that mediate resistance in tomato against bacterial wilt disease. J. Exp. Bot. 64, 3829–3842.

Jogaiah, S., Abdelrahman, M., Tran, L.-S.P., Ito, S.-I., 2018. Different mechanisms of *Trichoderma virens*-mediated resistance in tomato against Fusarium wilt involve the jasmonic and salicylic acid pathways. Mol. Plant Pathol. 19, 870–882.

Joshi, S.M., De Britto, S., Jogaiah, S., Ito, S., 2019. Mycogenic selenium nanoparticles as potential new generation broad spectrum antifungal molecules. Biomol. Ther. 9 (9), 419.

Kabaluk, J.T., Antonet, M.S., Mark, S.G., Stephanie, G.W. 2010. *The Use and Regulation of Microbial Pesticides in Representative Jurisdictions Worldwide*. IOBC Global. Available from: www.IOBC-Global.org

Keller, N.P., Turner, G., Bennett, J.W., 2005. Fungal secondary metabolism—from biochemistry to genomics. Nat. Rev. Microbiol. 3 (12), 937–947.

Kloepper, J.W., Ryu, C.M., Zhang, S., 2004. Induced systemic resistance and promotion of plant growth by *Bacillus* spp. Phytopathology 94 (11), 1259–1266.

Kumar, S., 2012. Biopesticides: a need for food and environmental safety. J. Biofertil. Biopestic. 3 (4), 1–3.

Lacey, L.A., Frutos, R., Kaya, H.K., Vail, P., 2001. Insect pathogens as biological control agents: do they have a future? Biol. Control 21 (3), 230–248.

Larkin, R.P., Fravel, D.R., 2002. Effects of varying environmental conditions on biological control of Fusarium wilt of tomato by nonpathogenic *Fusarium* spp. Phytopathology 92 (11), 1160–1166.

Louws, F.J., Rademaker, J.L.W., De Bruijn, F.J., 1999. The three Ds of PCR-based genomic analysis of phytobacteria: diversity, detection, and disease diagnosis. Annu. Rev. Phytopathol. 37 (1), 81–125.

Mathre, D.E., Cook, R.J., Callan, N.W., 1999. From discovery to use: traversing the world of commercializing biocontrol agents for plant disease control. Plant Dis. 83 (11), 972–983.

Murali, M., Jogaiah, S., Amruthesh, K.N., Ito, S.-I., Shekar Shetty, H., 2013. Rhizosphere fungus *Penicillium chrysogenum* promotes growth and induces defense-related genes and downy mildew disease resistance in pearl millet. Plant Biol. 15, 111–118.

Nagaraju, A., Jogaiah, S., Mahadevamurthy, S., Ito, S.-I., 2012a. Seed priming with *Trichoderma harzianum* isolates enhances plant growth and induces resistance against *Plasmopara halstedii*, an incitant of sunflower downy mildew disease. Aust. J. Plant Path. 41, 609–620.

Nagaraju, A., Murali, M., Jogaiah, S., Amruthesh, K.N., Mahadeva Murthy, S., 2012b. Beneficial microbes promote plant growth and induce systemic resistance in sunflower against downy mildew disease caused by *Plasmopara halstedii*. Curr. Bot. 3, 12–18.

Nakkeeran, S., Fernando, W.D., Siddiqui, Z.A., 2005. Plant growth promoting rhizobacteria formulations and its scope in commercialization for the management of pests and diseases. In: PGPR: Biocontrol and Biofertilization. Springer, Dordrecht, pp. 257–296.

Nandini, B., Puttaswamy, H., Prakash, H.S., Adhikari, S., Jogaiah, S., Geetha, N., 2020. Elicitation of novel trichogenic-lipid nanoemulsion signaling resistance against pearl millet downy mildew disease. Biomol. Ther. 10 (1), 25.

O'Brien, K.P., Franjevic, S., and Jones, J. 2009. *Green Chemistry and Sustainable Agriculture: The Role of Biopesticides*. Advancing Green Chemistry. Available from: http://advancinggreenchemistry.org/wp-content/uploads/Green-Chem-and-Sus.-Ag.-the-Role-of-Biopesticides.pdf.

Perazzolli, M., Dagostin, S., Ferrari, A., Elad, Y., Pertot, I., 2008. Induction of systemic resistance against *Plasmopara viticola* in grapevine by *Trichoderma harzianum* T39 and benzothiadiazole. Biol. Control 47 (2), 228–234.

Ranga Rao, G.V., Rupela, O.P., Rao, V.R., Reddy, Y.V.R., 2007. Role of biopesticides in crop protection: present status and future prospects. Indian J. Plant Protect. 35 (1), 1–9.

Ravensberg, W.J., 2011. A Roadmap to the Successful Development and Commercialization of Microbial Pest Control Products for Control of Arthropods. Vol. 10 Springer Science & Business Media.

Revathi, K., Chandrasekaran, R., Thanigaivel, A., Kirubakaran, S.A., Sathish-Narayanan, S., Senthil-Nathan, S., 2013. Effects of *Bacillus subtilis* metabolites on larval *Aedes aegypti* L. Pestic. Biochem. Physiol. 107, 369–376.

Roy, A., Moktan, B., Sarkar, P.K., 2007. Characteristics of *Bacillus cereus* isolates from legume-based Indian fermented foods. Food Control 18 (12), 1555–1564.

Schäfer, P., Khatabi, B., Kogel, K.H., 2007. Root cell death and systemic effects of *Piriformospora indica*: a study on mutualism. FEMS Microbiol. Lett. 275 (1), 1–7.

Schäfer, P., Pfiffi, S., Voll, L.M., Zajic, D., Chandler, P.M., Waller, F., et al., 2009. Manipulation of plant innate immunity and gibberellin as factor of compatibility in the mutualistic association of barley roots with *Piriformospora indica*. Plant J. 59 (3), 461–474.

Schallmey, M., Singh, A., Ward, O.P., 2004. Developments in the use of *Bacillus* species for industrial production. Can. J. Microbiol. 50 (1), 1–17.

Schena, L., Ippolito, A., Zahavi, T., Cohen, L., Droby, S., 2000. Molecular approaches to assist the screening and monitoring of postharvest biocontrol yeasts. Eur. J. Plant Pathol. 106 (7), 681–691.

Serfling, A., Wirsel, S.G., Lind, V., Deising, H.B., 2007. Performance of the biocontrol fungus *Piriformospora indica* on wheat under greenhouse and field conditions. Phytopathology 97 (4), 523–531.

Sivan, A., Chet, I., 1989. Degradation of fungal cell walls by lytic enzymes of *Trichoderma harzianum*. Microbiology 135 (3), 675–682.

Sudisha, J., Niranjana, S.R., Umesha, S., Prakash, H.S., Shekar Shetty, H., 2006. Transmission of seed-borne infection of muskmelon by *Didymella bryoniae* and effect of seed treatments on disease incidence and fruit yield. Biol. Control 37, 196–205.

Vinale, F., Sivasithamparam, K., Ghisalberti, E.L., Marra, R., Woo, S.L., Lorito, M., 2008. Trichoderma–plant–pathogen interactions. Soil Biol. Biochem. 40 (1), 1–10.

Waller, F., Achatz, B., Baltruschat, H., Fodor, J., Becker, K., Fischer, M., et al., 2005. The endophytic fungus *Piriformospora indica* reprograms barley to salt-stress tolerance, disease resistance, and higher yield. Proc. Natl. Acad. Sci. USA 102 (38), 13386–13391.

Zeilinger, S., Galhaup, C., Payer, K., Woo, S.L., Mach, R.L., Fekete, C., Kubicek, C.P., 1999. Chitinase gene expression during mycoparasitic interaction of *Trichoderma harzianum* with its host. Fungal Genet. Biol. 26 (2), 131–140.

Index

Note: Page numbers followed by *f* indicate figures and *t* indicate tables.

G

Q

R

T

FSC
www.fsc.org
MIX
Papier | Fördert gute Waldnutzung
FSC® C083411

Kolibri 360 GmbH
Ferdinand-Jühlke-Straße 7
99095 Erfurt, Deutschland
produktsicherheit@kolibri360.de